See also Appendix I

Name of quantity	Common symbol	Value
"Mechanical equivalent of heat"	J	4.18400 J/cal
Stefan-Boltzmann constant	σ	5.6696×10^{-8} J/m²-s-K⁴
Coulomb force constant	$k = 1/4\pi\varepsilon_0$	8.987554×10^9 N-m²/C²
Permittivity of free space	ε_0	8.854185×10^{-12} C²/N-m²
Permeability of free space	μ_0	$4\pi \times 10^{-7}$ Wb/A-m
Speed of light (in vacuum)	c	2.99792456×10^8 m/s
Electron charge	e	1.602192×10^{-19} C
Electron volt	1 eV	1.602192×10^{-19} J
Mass-energy conversion factor (atomic mass unit = u)	1 u = $(\frac{1}{12})^{12}_6 C$	931.481 MeV/c^2 1.6604×10^{-27} kg
Electron mass	m_e	9.10956×10^{-31} kg 0.000548593 u 0.511004 MeV/c^2
Proton mass	m_p	1.67261×10^{-27} kg 1.00727661 u 938.259 MeV/c^2 = 1836.11 m_e
Neutron mass	m_n	1.67492×10^{-27} kg 1.0086652 u 939.553 MeV/c^2
Planck's constant	h $\hbar = h/2\pi$	6.626196×10^{-34} J-s 1.054591×10^{-34} J-s

Elementary Physics: Classical and Modern

ELEMENTARY PHYSICS: Classical and Modern

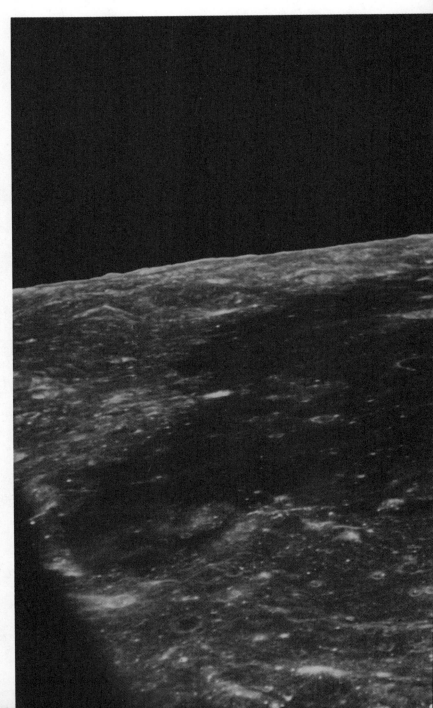

The Apollo 11 lunar module ascent stage photographed from the command service module during rendezvous in lunar orbit. The Earth rises above the lunar horizon. (Courtesy National Aeronautics and Space Administration.)

Richard T. Weidner
Professor of Physics
Rutgers University
New Brunswick, New Jersey

Robert L. Sells
Professor of Physics
State University College of Arts and Sciences
Geneseo, New York

with the assistance of Arthur E. Walters

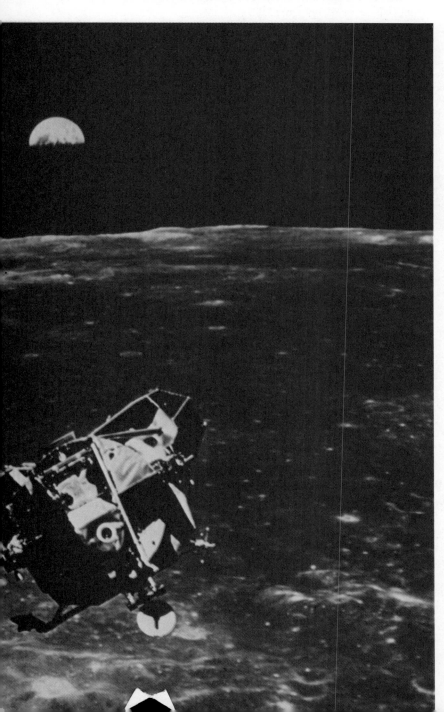

Allyn and Bacon, Inc.
Boston
London
Sydney
Toronto

© COPYRIGHT 1975 BY ALLYN AND BACON, INC.
470 Atlantic Avenue, Boston, Massachusetts 02210

Portions of this book appeared previously in *Elementary Classical Physics* (Volumes 1 and 2), Second Edition, by **Weidner and Sells** © copyright 1973, 1965 by Allyn and Bacon, Inc., and *Elementary Modern Physics,* Alternate Second Edition, by Weidner and Sells, © copyright 1973, 1968, and 1960 by Allyn and Bacon, Inc.

All rights reserved.

No part of the material protected by this copyright notice may be reproduced or utilized in any form or by any means, electronic or mechanical, including photocopying, recording, or by any informational storage and retrieval system, without written permission from the copyright owner.

Printed in the United States of America

LIBRARY OF CONGRESS CATALOGING IN PUBLICATION DATA

Weidner, Richard T
   Elementary physics, classical and modern.

  Includes index.
  1. Physics. I. Sells, Robert L., joint author.
II. Walters, Arthur E., joint author. III. Title.
QC23.W414   530   74-28121

ISBN 0-205-04647-9   Fifth printing...August, 1979
      0-205-04778-5 *(International)*   Second printing...August, 1976

Contents

Preface

CHAPTER 1 INTRODUCTION   1

1-1 Events 2 / 1-2 Length 3 / 1-3 Time 4 / 1-4 Systems of Units 4

CHAPTER 2 VECTORS   6

2-1 Vectors and Scalars 6 / 2-2 Displacement as a Vector 7 / 2-3 Component Representation of Vectors 9 / 2-4 Vector Addition by the Component Method 10 / 2-5 Vectors in Three Dimensions 12 / 2-6 The Scalar and Vector Products 13

CHAPTER 3 STRAIGHT-LINE KINEMATICS   17

3-1 The Meaning of a Particle 17 / 3-2 Average Velocity 17 / 3-3 Instantaneous Velocity 19 / 3-4 Acceleration 22 / 3-5 Rectilinear Motion at Constant Acceleration 24 / 3-6 Freely Falling Bodies 28 / **Summary** 30

CHAPTER 4 KINEMATICS IN TWO DIMENSIONS   33

4-1 Velocity and Acceleration as Vector Quantities 33 / 4-2 Motion at Constant Acceleration 37 / 4-3 Pro-

jectile Motion 39 / *4-4 Uniform Circular Motion* 42 / *4-5 Angular Speed* 45 / **Summary** 46

CHAPTER 5

THE LAWS OF INERTIA, MASS CONSERVATION, AND MOMENTUM CONSERVATION 50

5-1 The Law of Inertia 51 / *5-2 Inertial Mass* 52 / *5-3 The Standard of Mass* 54 / *5-4 The Law of Conservation of Mass* 54 / *5-5 Momentum Conservation in Head-on Collisions* 55 / *5-6 Momentum Conservation in Two and Three Dimensions* 60 / *5-7 Momentum Conservation in Atomic Collisions* 63 / **Summary** 64

CHAPTER 6

CENTER OF MASS AND REFERENCE FRAMES 66

6-1 Velocity of the Center of Mass 66 / *6-2 Location of the Center of Mass* 68 / *6-3 Reference Frames and Relative Velocities* 71 / *6-4 Momentum Conservation in the Center of Mass Reference Frame* 75 / **Summary** 76

CHAPTER 7

FORCE 80

7-1 The Fundamental Origins of Force 80 / *7-2 Force Defined* 81 / *7-3 The Superposition Principle for Forces* 83 / *7-4 Force Pairs* 84 / *7-5 Weight* 86 / *7-6 Impulse and Momentum* 88 / **Summary** 91

CHAPTER 8

NEWTONIAN MECHANICS AND ITS APPLICATIONS 94

8-1 Newton's Laws of Motion 94 / *8-2 Applications of Newton's Laws* 96 / *8-3 Friction* 101 / *8-4 Systems of Interacting Particles* 104 / *8-5 Uniform Circular Motion* 106 / **Summary** 109

CHAPTER 9

WORK AND KINETIC ENERGY 112

9-1 Kinetic Energy Defined 112 / *9-2 Work Defined* 113 / *9-3 The General Work-Energy Theorem* 115 / *9-4 Work Done by a Spring* 121 / *9-5 Power* 124 / **Summary** 126

CHAPTER 10

POTENTIAL ENERGY AND ENERGY CONSERVATION 129

10-1 Potential Energy of a Spring 129 / *10-2 Gravitational Potential Energy* 133 / *10-3 General Definition of Potential Energy* 139 / *10-4 Properties of Potential Energy* 141 / *10-5 The Conservation of Energy Law* 144 / **Summary** 150

CHAPTER 11 — ANGULAR MOMENTUM 154

11-1 Angular Velocity as a Vector 154 / 11-2 Some Qualitative Features of Angular Momentum 155 / 11-3 Torque as Time Rate of Change of Angular Momentum 158 / 11-4 A Particle with Constant Angular Momentum 163 / 11-5 Angular Momentum of a Rigid Body 166 / 11-6 Kinetic Energy of a Spinning Rigid Body 167 / 11-7 The Law of Angular-momentum Conservation 168 / **Summary 172**

CHAPTER 12 — ROTATIONAL DYNAMICS 177

12-1 Rotational Kinematics 177 / 12-2 Newton's Second Law for Rotation 179 / 12-3 Work and Kinetic Energy in Rotational Motion 185 / 12-4 Center of Gravity 188 / 12-5 Equilibrium of a Rigid Body 190 / **Summary 192**

CHAPTER 13 — GRAVITATION 196

13-1 The Law of Universal Gravitation 196 / 13-2 The Cavendish Experiment 201 / 13-3 The Gravitational Field 203 / 13-4 Kepler's Laws of Planetary Motion 204 / 13-5 Gravitational Potential Energy 209 / **Summary 215**

CHAPTER 14 — SIMPLE HARMONIC MOTION 218

14-1 Dynamics and Kinematics of Simple Harmonic Motion 218 / 14-2 Energetics of Simple Harmonic Motion 222 / 14-3 The Simple Pendulum 223 / 14-4 Small Oscillations and Simple Harmonic Motion 225 / 14-5 Damped Oscillations and Resonance 229 / **Summary 230**

CHAPTER 15 — WAVES ON A STRING 233

15-1 Basic Wave Behavior 233 / 15-2 The Superposition Principle and Interference 236 / 15-3 Reflection of Waves 238 / 15-4 Sinusoidal Waves 239 / 15-5 Standing Waves 244 / 15-6 Resonance 248 / **Summary 250**

CHAPTER 16 — COMPRESSIONAL WAVES 253

16-1 Longitudinal Waves 253 / 16-2 Compressional Wavespeeds 255 / 16-3 Superposition, Reflection, and Standing

Waves 257 / 16-4 Sound and Acoustics 259 / 16-5 Wavefronts and Rays 260 / 16-6 Intensity Variation with Distance from Source 260 / 16-7 Huygen's Principle 262 / **Summary 262**

CHAPTER 17 — TEMPERATURE AND THE MACROSCOPIC PROPERTIES OF AN IDEAL GAS 265

17-1 Temperature and the Zeroth Law of Thermodynamics 266 / 17-2 Thermometry 267 / 17-3 Pressure in a Gas 269 / 17-4 The Constant-volume Gas Thermometer 271 / 17-5 The General-gas Law 273 / 17-6 Changes in the State of a Gas 276 / **Summary 277**

CHAPTER 18 — THE KINETIC THEORY AND THE FUNDAMENTAL CONCEPTS OF THERMODYNAMICS 280

18-1 Molecular Size 280 / 18-2 The Kinetic Theory Applied to an Ideal Gas 281 / 18-3 The Meaning of Temperature and Internal Energy 285 / 18-4 The Meaning of Heat and the First Law of Thermodynamics 287 / 18-5 Disorder and the Second Law of Thermodynamics 291 / **Summary 295**

CHAPTER 19 — THERMAL PROPERTIES OF SOLIDS AND LIQUIDS 298

19-1 Solids and Liquids as Thermal Systems 298 / 19-2 The First Law of Thermodynamics 300 / 19-3 Specific Heats and Heat Units 302 / 19-4 Thermal Conduction 305 / 19-5 Thermal-energy Transfer through Radiation 308 / **Summary 310**

CHAPTER 20 — THE ELECTRIC FORCE BETWEEN CHARGES 313

20-1 Electric Charge 313 / 20-2 Coulomb's Law 316 / 20-3 Electric Units 319 / 20-4 Charge Conservation 321 / 20-5 Charge Quantization 322 / **Summary 323**

CHAPTER 21 — THE ELECTRIC FIELD 326

21-1 Electric Field Defined 326 / 21-2 Electric Field Lines 329 / 21-3 Electric Fields for Three Simple Geometries 330 / 21-4 Electric Fields and Conductors 333 / 21-5 The Millikan Experiment 334 / **Summary 335**

CHAPTER 22 — GAUSS' LAW   339

22-1 Electric Flux 339 / 22-2 Gauss' Law 341 / 22-3 Electric Field Lines 345 / 22-4 Spherically Symmetric Charge Distributions 346 / 22-5 Gauss' Law and Conductors 347 / **Summary** 351

CHAPTER 23 — ELECTRIC POTENTIAL, CAPACITANCE, AND DIELECTRICS   354

23-1 Electric Potential Energy of Two Point-Charges 354 / 23-2 Electric Potential Defined 359 / 23-3 Electric Potential of Point-Charges 361 / 23-4 Relations Between V and **E** 362 / 23-5 Equipotential Surfaces 364 / 23-6 Electric Potential and Conductors 366 / 23-7 Capacitance Defined 368 / 23-8 Capacitance for Some Simple Configurations 369 / 23-9 Capacitor Circuits 372 / 23-10 Energy of a Charged Capacitor and of the Electric Field 375 / **Summary** 377

CHAPTER 24 — ELECTRIC CURRENT AND RESISTANCE   382

24-1 Electric Current 382 / 24-2 Current and Energy Conservation 384 / 24-3 Resistance and Ohm's Law 386 / **Summary** 389

CHAPTER 25 — DC CIRCUITS   393

25-1 The EMF of a Battery 393 / 25-2 Single-loop DC Circuits 396 / 25-3 Resistors in Series and in Parallel 399 / 25-4 Multiloop Circuits 402 / 25-5 DC Circuit Instruments 404 / **Summary** 408

CHAPTER 26 — THE MAGNETIC FORCE   412

26-1 Magnetic Induction Field Defined 412 / 26-2 Magnetic Flux 415 / 26-3 Motion of Charged Particles in a Uniform Magnetic Field 416 / 26-4 Charged Particles in Uniform **B** and **E** Fields 419 / 26-5 Magnetic Force on a Current-carrying Conductor 423 / 26-6 Magnetic Torque on a Current Loop 425 / **Summary** 427

CHAPTER 27 — THE SOURCES OF THE MAGNETIC FIELD   431

27-1 The Oersted Experiment 431 / 27-2 The Biot-Savart Law 432 / 27-3 The Magnetic Field for Some Simple Geometries 434 / 27-4 The Magnetic Force Between Current-carrying Conductors 438 / 27-5 Gauss' Law for Magnetism 439 /

27-6 Ampère's Law 440 / 27-7 The Solenoid 444 / **Summary 445**

CHAPTER 28

ELECTROMAGNETIC INDUCTION AND INDUCTANCE   450

28-1 Motional EMF 450 / 28-2 Examples of Electromagnetic Induction 455 / 28-3 Lenz' Law 456 / 28-4 Faraday's Law and the Electric Field 458 / 28-5 Self-inductance Defined 461 / 28-6 Energy of an Inductor and of the Magnetic Field 464 / **Summary 465**

CHAPTER 29

ELECTRIC OSCILLATIONS, RC AND RL CIRCUITS   468

29-1 Electric Free Oscillations 468 / 29-2 Electrical-Mechanical Analogs 473 / 29-3 The RC Circuit 474 / 29-4 The LR Circuit 475 / **Summary 477**

CHAPTER 30

MAXWELL'S EQUATIONS AND ELECTROMAGNETIC WAVES   480

30-1 Ampère's Law and Maxwell's Displacement Current 480 / 30-2 Maxwell's Equations 483 / 30-3 Electromagnetic Waves from Maxwell's Equations 484 / 30-4 Electromagnetic Energy Density, Intensity, and the Poynting Vector 489 / 30-5 Radiation Force and Pressure: The Linear Momentum of an Electromagnetic Wave 491 / 30-6 Accelerating Charges and Electromagnetic Waves 495 / 30-7 Sinusoidal Electromagnetic Waves 498 / 30-8 The Electromagnetic Spectrum 500 / 30-9 Measurements of the Speed of Light 502 / **Summary 505**

CHAPTER 31

RAY OPTICS: REFLECTION, REFRACTION, AND LENSES   508

31-1 Ray Optics and Wave Optics 508 / 31-2 The Reciprocity Principle 510 / 31-3 Rules of Reflection and Refraction 511 / 31-4 Reflection 513 / 31-5 Index of Refraction 514 / 31-6 Refraction 516 / 31-7 Total Internal Reflection 518 / 31-8 Thin Lenses 519 / 31-9 Optical Instruments 527 / **Summary 529**

CHAPTER 32

WAVE OPTICS: INTERFERENCE AND DIFFRACTION 533

32-1 Superposition and Interference of Waves 533 / 32-2 Interference from Two Point Sources 535 / 32-3 Reflection and Change of Phase 538 / 32-4 Coherent and Incoherent Sources 540 / 32-5 Young's Double-slit Experiment 542 / 32-6 Interference in Thin Films 544 / 32-7 The Michelson Interferometer 546 / 32-8 Radiation from a Row of Point Sources 548 / 32-9 Diffraction by a Single Slit 550 / 32-10 The Diffraction Grating 552 / 32-11 Other Examples of Diffraction 554 / 32-12 Diffraction and Resolution 556 / **Summary 558**

CHAPTER 33

POLARIZATION 561

33-1 Superposition of Simple Harmonic Motions 561 / 33-2 Polarization Properties of Waves from an Electric Dipole Oscillator 562 / 33-3 Polarization of Visible Light 563 / 33-4 Polarization in Scattering 568

CHAPTER 34

RELATIVISTIC KINEMATICS: SPACE AND TIME 571

34-1 The Constancy of the Speed of Light 571 / 34-2 Time Dilation 574 / 34-3 Space Contraction 579 / 34-4 The Lorentz Coordinate Transformations 581 / 34-5 Relativistic Velocity Relations 583 / **Summary 585**

CHAPTER 35

RELATIVISTIC DYNAMICS: MOMENTUM AND ENERGY 590

35-1 The Invariance of the Laws of Physics 590 / 35-2 Relativistic Mass and Momentum 593 / 35-3 Relativistic Energy 600 / 35-4 Mass-Energy Equivalence and Bound Systems 604 / 35-5 Computations and Units in Relativistic Mechanics 607 / **Summary 610**

CHAPTER 36

QUANTUM EFFECTS: THE PARTICLE ASPECTS OF ELECTROMAGNETIC RADIATION 613

36-1 Quantization in Classical Physics 613 / 36-2 The Photoelectric Effect 615 / 36-3 X-ray Production and Bremsstrahlung 622 / 36-4 The Compton Effect 625 / 36-5 Pair Production and Annihilation 631 / 36-6 Photon-Electron Interactions 634 / **Summary 637**

CHAPTER 37

QUANTUM EFFECTS: THE WAVE ASPECTS OF MATERIAL PARTICLES 642

37-1 De Broglie Waves 642 / 37-2 The Bragg Law 644 / 37-3 X-ray and Electron Diffraction 647 / 37-4 The Principle of Complementarity 649 / 37-5 The Probability Interpretation of De Broglie Waves 652 / 37-6 The Uncertainty Principle 656 / 37-7 The Quantum Description of a Confined Particle 664 / **Summary 671**

CHAPTER 38

ATOMIC STRUCTURE 675

38-1 α-Particle Scattering 675 / 38-2 The Hydrogen Spectrum 678 / 38-3 The Bohr Theory of Hydrogen 681 / 38-4 The Four Quantum Numbers for Atomic Structure 688 / 38-5 The Pauli Exclusion Principle and the Periodic Table 693 / **Summary 698**

CHAPTER 39

NUCLEAR STRUCTURE 703

39-1 The Nuclear Constitutents 703 / 39-2 The Forces between Nucleons 704 / 39-3 The Deuteron 706 / 39-4 Stable Nuclei 708 / 39-5 The Radioactive Decay Law 712 / 39-6 α, β, and γ Decay 714 / 39-7 High-energy Particle Accelerators 723 / 39-8 Nuclear Reactions 730 / **Summary 734**

CHAPTER 40

THE ELEMENTARY PARTICLES 739

40-1 The Electromagnetic Interaction 740 / 40-2 Other Fundamental Interactions 745 / 40-3 Properties of Fundamental Particles 747 / 40-4 The Universally Valid Conservation Laws 751 / 40-5 Isospin, Strangeness, and Parity 755 / 40-6 Resonance Particles 760

APPENDIXES

I Fundamental Constants 767 / II Conversion Factors 769 / III Electric and Magnetic Units and Conversion Factors 772 / IV Atomic Masses 775 / V Mathematical Relations 780 / VI Natural Trigonometric Functions 781 / VII Greek Alphabet 783 / VIII Moment-of-Inertia Calculations 784

Answers to Odd-numbered Problems   787

Index   795

Preface

Elementary Physics: Classical and Modern is intended primarily as a basis for an introductory course in college physics for students of science and engineering with a strong emphasis, not only on the basic principles of classical physics, but also on fundamental topics in modern physics. The text aims at giving a thorough introduction to the whole range of essential topics in physics, but at a realistic level of conceptual and mathematical sophistication for beginning students in college physics. Concurrent study of elementary calculus would more than suffice for any applications involved herein; indeed, elementary calculus is first required in Chapter 6.

Based upon the authors' three-volume set, *Elementary Classical Physics,* volumes 1 and 2, and *Elementary Modern Physics,* this text gives primary attention to the absolutely central topics. This was accomplished, not by scaling down the treatment of all topics appearing in the parent volumes, but rather by very intentionally retaining the fundamental topics in classical and modern physics and treating them in such depth that students need not unlearn them in subsequent courses. Less central topics have been eliminated. In short, we have not chosen to treat topics more briefly in order to "cover" more material; we have revised the previous volumes to function as a complete and new unit for a shorter course.

Chapters devoted to modern physics constitute approximately one-fourth of the entire book. Recognizing the need of future scientists and technologists for an early introduction to contemporary physics, here too we have chosen to give a reasonably full treatment of special relativity, the quantum theory, atomic and nuclear structure, and elementary particles, rather than superficial coverage to the whole range of topics that may properly be considered to comprise contemporary physics. Although topics in so-called modern physics appear in the last chapters, they may, at the option of the instructor, be introduced much earlier. For example, Chapter 34 on relativistic kinematics can be taken up after Chapter 4, and Chapter 35 on relativistic dynamics after Chapter 10. Indeed, our intent throughout has been to make the treatment of each principal topic essentially complete and independent of others, so that the sequence may be altered, thereby allowing for flexibility in meeting the needs of particular instructors or students.

The general strategy in treating classical mechanics has been to give primary emphasis to the conservation laws. After some preliminaries on vectors and kinematics, the conservation law of linear momentum is introduced before Newton's laws, a sequence which is, in our view and that of many past users, logically and pedagogically preferable to the reverse order. Similarly, angular momentum and its conservation precedes the more specialized (and optional) topic of rotational dynamics. Another general emphasis is on the atomic point of view and the correlation of classical physics with its contemporary counterparts. Such specialized topics as hydrodynamics, formal thermodynamics and heat engines, and ac circuits have been excluded.

Special features of the text include: a large number of straightforward but nontrivial completely worked-out examples; succinct chapter summaries, set off in distinctive color; (almost) exclusive use of the International System (mksa) of units; and numerous problems at the chapter ends, one-third of which are new and place a special emphasis on applications of physics to related areas.

In arriving at the relatively limited number of central topics which were to be included in this book, we have had the benefit of the views of many college physics professors, as well as those of their students, transmitted through surveys by the publisher's representatives and editorial staff. The final selection corresponds closely to this preponderant opinion. The detailed reviews of the manuscript by Professor Bernard Chasan of Boston University, Professor Hans Courant of the University of Minnesota, Professor F. Alan McDonald of Southern Methodist University, and Professor Dan W. Schlitt of the University of Nebraska have been particularly helpful. The authors, however, assume responsibility for all errors of omission or commission. The authors gratefully acknowledge the assistance and collaboration of Dr. Arthur E. Walters in preparing a large fraction of the revised manuscript, including construction of new problems, for the chapters in classical physics. The rewriting of chapters in modern physics has been the work of the senior co-author.

The publishers and authors solicit the opinions of this text from its users, both students and professors, and a simple questionnaire has been provided in the back of the book for your convenience.

Richard T. Weidner
Robert L. Sells

CHAPTER 1

Introduction

Physics is *the* fundamental experimental science. Its purpose is to make sense out of the behavior of the physical universe. Physics begins with controlled observation, or experiment, in which some one phenomenon is examined quantitatively through measurements. The relations among the physical quantities observed in experimentation are expressed with precision and economy in the language of mathematics. When a relation summarizes many experiments with a realiability so great that it can be said to reflect universal behavior in nature, then it is said to be a "law" of physics. Happily, the laws of physics are few, and the whole variety of physical phenomena is comprehended in a remarkably small number of fundamental laws.

Theory and experiment both play essential roles in the development of physics. Experiment discloses the facts of nature; theory makes sense out of them. Theory, moreover, suggests still further experiments as tests of the laws of physics, and experiment reveals points at which a theory may be defective.

Physics is not complete—and probably will never be. As contemporary physicists probe the nucleus of the atom and its constituent particles, they find phenomena unaccounted for in present-day laws of physics. Presumably, if one knew completely the "elementary" particles of physics and the ways in which they interact with one another, all other physical phenomena—atomic and nuclear structure, the behavior and properties of ordinary materials, and even the collisions of galaxies—would be explainable. But that day has not arrived.

Although all laws in physics must be regarded as incomplete and tentative to a greater or lesser degree, there is one body of knowledge in physics that can be regarded as essentially complete and correct: This is *classical physics*. So-called classical physics deals with the behavior of bodies of ordinary size (greater than that of the atom, which is 10^{-10} m) moving at speeds much less than that of light (3×10^8 m/s). It had its origins in mechanics, in the work of Galileo and Newton. Its last chapters were written at the end of the nineteenth century, when the theories of thermodynamics and electromagnetism reached their full classical development. The first 33 chapters of this text are devoted primarily to classical physics.

Classical mechanics and electromagnetism can be said to be "right" because their theories adequately describe the behavior of all bodies of ordinary size. But classical physics fails when applied to the motions of objects of atomic or nuclear size (for which the quantum theory must be invoked) or to high-speed particles (for which the theory of relativity must be used). We study classical physics, not only because it is correct over a broad domain, but also because the same concepts and language of classical physics appear in the modern physics of high-speed atomic particles.

The term *modern physics* is still used to describe the new physics originating at the beginning of the twentieth century, especially the theory of relativity and the quantum theory and their applications to the structure of the atomic nucleus and to the characteristics of the fundamental elementary particles. Modern physics is the subject of Chaps. 34 through 40.

Since the primary aim of this text is the exposition of fundamental principles, relatively less attention is given to such related matters as the historical development of classical and modern physics; the rejection of the one-time plausible, but inadequate, theories; the many careful experiments performed to verify fundamental physical principles, including the analysis of errors and treatment of experimental data; and the influence of society on the enterprise of physics (and the converse). These are all important parts of the story of physics. The student should, of course, experience through the laboratory the fact that physics is an *experimental* science. As in all experimental sciences, there is no single "scientific method" through which advances in physics are made. Imaginative, as well as pedestrian experiments, intuitive conjecture and even bold flights of fantasy, tedious mathematical analysis, and the radical reexamination of untested but apparently obvious presuppositions all play a role in scientific work. But these criteria must always be met: The experimental data must truly reflect the behavior of natural phenomena, and the theoretical framework must summarize the observations not only with the greatest possible generality, but also with the greatest simplicity.

1-1 EVENTS

The birth of Galileo took place at latitude 43.7° north, longitude 10.4° east of Greenwich, altitude 1,500 ft above sea level, in the year A.D. 1564. Four numbers are required in the specification of this historical event—three for its location in space and one for its location in time. Classical physics also deals with events in space and time. An event in physics may be nothing more than the appearance of a particle, but it is distinct from an historical event in that it represents a universal behavior illustrating the simple and general laws of physics.

An event in physics is also completely described by three quantities giving the location relative to some arbitrarily chosen origin (not necessarily the earth's center or the town of Greenwich) and a single quantity giving the time relative to some arbitrarily chosen zero of time. That is to say, the events in physics, like ordinary events, are described in terms of position and time.

1–2 LENGTH

Measurement of position requires a measurement of length, and a length measurement requires the choice of a universally agreed upon standard of length, relative to which all distance measurements are ultimately related. In the International System of Units (abbreviated SI) the standard of length is the *meter*. The meter was first chosen so that the distance from the earth's pole to its equator along the meridian line through Paris would be 10,000,000 m. Later the meter was defined as the distance between two fine scratches on a carefully preserved bar of platinum-iridium. At present the meter is defined in terms of the wavelength in vacuum of the orange-red radiation corresponding to the unperturbed transition between the levels $2p_{10}$ and $5d_5$ of the atom krypton-86:

1 m = 1,650,763.73 wavelengths of krypton-86 light

A standard of length must meet the requirements of easy accessibility and high accuracy, inappreciably influenced by such disturbances as changes in temperature or pressure. The platinum-iridium bar, which was the standard of length before 1960, can expand with a temperature rise or shrink with a pressure rise. This leads to ambiguities compounded by the width of the scratches. And, worst of all, the bar can be lost or destroyed!

The krypton-86 spectral line overcomes these difficulties in large measure. The wavelength of this electromagnetic radiation is (nearly) independent of external changes and depends only on the structure of the atoms of krypton-86, which is, of course, identical for all such atoms.

The standard of length in the U.S. Customary (or English) System of Units is the *foot*. It is, by definition, exactly one-third of a yard where, by legal agreement, 1 yd = 0.9144 m. For the conversion factors between other length units see Appendix II.

In its simplest form a measurement of the length of an object consists simply of counting the number of times the standard of length — for example, an ordinary meterstick — is contained in the object to be measured. *All* measurements in physics are characterized by the same steps: the choice of a standard and a counting of the multiples (or submultiples) of the standard in the measured quantity.

1-3 TIME

Just as one can describe the length of an object as that which one measures with a meterstick, time can best be defined as the physical quantity which one reads from a clock. But what is a clock, and especially, what is a reliable clock? Any object or collection of objects showing regular repetitive motion may be used as a clock. The heartbeat, the pulse, may be used as a very crude clock. A swinging pendulum or an oscillating spring are better clocks. The daily rotation of the earth about its axis provides still a better basis for measurements of time intervals. But the most precise clock is an *atomic clock*. Indeed, the *second,* the standard unit for time, is now defined in terms of the time for one oscillation in the transition between two hyperfine levels of the fundamental state of cesium-133 atoms:

$$1 \text{ s} = 9,192,631,770 \text{ periods of cesium-133 radiation}$$

The advantages of the atomic clock are high precision (1 part in 10^{11} or better) and accessibility to anyone who can build such a clock.

As defined above the second is such that there are 86,400 ($60 \times 60 \times 24$) in one day, the time for the earth to make one rotation.

The ranges of length and time measurements encountered in human observation are: *Length:* from 10^{-15} m, the size of an atom's nucleus, to 10^{25} m, the size of the universe. *Time:* from 10^{-23} s, nuclear time, to 10^{17} s, the age of the universe. Man, in space and in time, seems to lie between the microscopic and macroscopic limits.

1-4 SYSTEMS OF UNITS

The International (or metric) System is used almost universally in scientific work because it is simple. Unlike the U.S. Customary System, in which the several units of length are related to one another by nonsimple multiples (the mile is 5,280 ft, the foot is 12 in), the metric units are always related by multiples of 10. For example, the kilometer is 1,000 m.

Listed in Table 1-1 are the prefixes which are commonly used to change the size of units by multiples of 10.

Table 1-1

Prefix	Abbreviation	Meaning
pico	p	10^{-12}
nano	n	10^{-9}
micro	μ	10^{-6}
milli	m	10^{-3}
centi	c	10^{-2}
deci	d	10^{-1}
kilo	k	10^{3}
mega	M	10^{6}
giga	G	10^{9}
tera	T	10^{12}

Thus, 1 km is a thousand meters, and 1 ns is a billionth of a second.

Length and time, represented by the meter and second, are two of the three fundamental dimensions of measurement. The third fundamental dimension is mass, for which the metric standard is the kilogram. The mass standard will be discussed in Sec. 5-3; we merely note here that *all* physical quantities in mechanics can be expressed in terms of these fundamental and independent quantities: length L, time T, and mass M. That is, a given physical quantity is described by its *dimensions,* by a certain combination of $M, L,$ and T only. Clearly, a distance has the dimension length L, and the dimension of speed is length divided by time L/T. As we shall see, the dimensions of force are ML/T^2 and the dimensions of energy are ML^2/T^2.

CHAPTER 2

Vectors

This chapter treats vector algebra, a particularly useful mathematical language for expressing relationships in physics. In addition to defining the basic relations for adding vector quantities and for decomposing vectors into their equivalent components, we give the definitions of the scalar (or dot) and the vector (or cross) product of two vectors.†

2-1 VECTORS AND SCALARS

In elementary physics we encounter two types of physical quantities, *vector quantities* and *scalar quantities* (or simply *vectors* and *scalars*). A scalar quantity has magnitude only, and it obeys the usual laws of algebra. Examples of scalars are the pure numbers of arithmetic and such physical quantities as time, volume, and mass. A scalar physical quantity is fully specified by giving a number together with an appropriate unit, as in 10.4 s, 10^4 m³, or 9.11×10^{-31} kg.

A vector quantity, on the other hand, has (1) direction, as well as (2) magnitude, and (3) it obeys the rules of *vector algebra* which are found to hold for displacement vectors. We shall be concerned only with the vector properties of displacement in this chapter, because displacement serves as a prototype for other vector quantities.

---

†The dot and cross products will first be required in Secs. 9-2 and 11-3, respectively, and discussion of them may be postponed until these topics are treated.

2-2 DISPLACEMENT AS A VECTOR

Consider Fig. 2-1, which shows the position in a plane of a small object at three different times. Initially the object was at point a, later at b, and finally at c. *Displacement* is defined as the *change* in the *position* of the object. We can represent the displacement from a to b by a directed line segment between the points, locating the arrowhead at the terminus b and indicating thereby that the displacement was from a to b. The length of the line segment represents, in any appropriate scale, the distance along the straight line from a to b. Note particularly that we do *not* say (or know) whether the object has traveled along a *straight* line from a to b—its path between the two end points may, in fact, have been quite tortuous; all we know is that b is the final *position* and a the initial one.

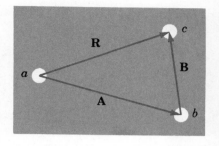

FIG. 2-1. An object at positions a, b, and c.

We represent the vector a to b by the boldface symbol **A**. Because it is difficult to write boldface symbols on paper or on the blackboard, a vector quantity such as **A** is commonly distinguished from a scalar quantity by an arrow above the symbol, $\vec{A}$, or by using a wavy underscore, $\underset{\sim}{A}$, the printer's symbol for boldface type. The magnitude of the vector **A** is a scalar quantity and is symbolized by lightface type, as A, or by $|\mathbf{A}|$. Thus, if b is 10 m distant from a, $A = 10$ m.

In Fig. 2-1 the displacement from point b to point c is written **B,** and **R** represents the vector displacement from a to c. It is clear from the geometry of the figure that if an object is first displaced from a to b, and then from b to c, the overall, or *resultant,* displacement is equivalent to a single displacement directly from a to c. In the symbolism of vector algebra we write the *vector equation*

$$\mathbf{A} + \mathbf{B} = \mathbf{R} \qquad (2\text{-}1)$$

This implies that the single displacement **R** gives the same change in position as the two successive displacements **A** and **B**. See Fig. 2-2.

Equation (2-1) represents a very special sort of algebra, vector algebra, and the plus sign has a special meaning. The resultant **R** is the *vector sum* of **A** and **B**. Unless **A** and **B** point in the same direction, the scalar equation $A + B = R$ is *not* true. Indeed, if **A** is fixed in both magnitude and direction whereas **B** is fixed in magnitude only, the *magnitude* of **R** can assume any value ranging from $A + B$, when **A** and **B** are aligned, to the magnitude of $A - B$ when they are in opposite directions.

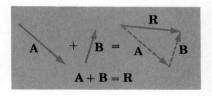

FIG. 2-2. Graphical representation of the vector sum $A + B = R$.

We restate the definition of a vector. *It is a quantity,* such as displacement, *that has magnitude and direction and which follows the law of vector addition illustrated in Fig. 2-2 and symbolized by Eq. (2-1).* Conversely, a quantity having magnitude and direction is *not* a vector *unless* it obeys the law of vector addition.

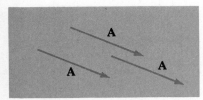

FIG. 2-3. All three vectors are the same vector A.

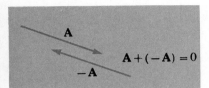

FIG. 2-4. To find the negative of a vector, one merely reverses the direction of the arrow.

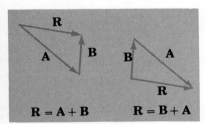

FIG. 2-5. The commutative law of vector addition illustrated: $A + B = B + A$.

Equality of Two Vectors. Two vectors are equal when they are of the same magnitude and point in the same direction, even though located at different points in space. Remembering that displacement is the *change* in position, it follows that any two displacements of, say, 10 m north are identical although they may have different starting points. Thus, all the vectors shown in Fig. 2-3 are, in fact, the same vector. This property is important because it allows us to shift a vector from one location to another in a diagram without changing the vector in any way, so long as the magnitude and direction are unchanged.

Negative of a Vector. If **A** represents the displacement from a to b, then $-\mathbf{A}$ is the displacement from b to a. Clearly, $\mathbf{A} + (-\mathbf{A}) = 0$, as shown in Fig. 2-4. Vectors **A** and $-\mathbf{A}$ have the same magnitude but are oppositely directed. To find the negative of **A**, reverse the direction of its arrow.

Commutative Law in Vector Addition. The resultant vector **R** can be written either as $\mathbf{A} + \mathbf{B}$ or as $\mathbf{B} + \mathbf{A}$. See Fig. 2-5. Because the order is of no consequence, we may add **B** to **A**, or **A** to **B**, to yield **R**:

$$\mathbf{R} = \mathbf{A} + \mathbf{B} = \mathbf{B} + \mathbf{A}$$

Associative Law in Vector Addition. The vector sum of three or more successive displacements does not depend on the order in which we add the displacements. Therefore $(\mathbf{A} + \mathbf{B}) + \mathbf{C} = \mathbf{A} + (\mathbf{B} + \mathbf{C})$, as shown in Fig. 2-6.

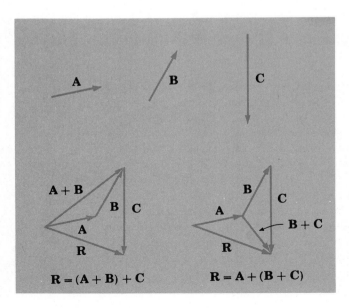

FIG. 2-6. The associative law of vector addition illustrated: $(A + B) + C = A + (B + C)$.

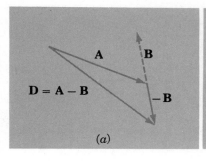
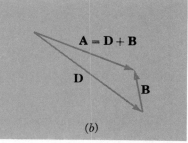

FIG. 2-7. The vector difference $D = A - B$: (a) Adding $-B$ to A and (b) adding D to B to yield A.

Subtraction of Vectors. To subtract **B** from **A** we add **A** and $-$**B**. The vector difference **D** is $D = A - B = A + (-B)$. See Fig. 2-7.

We can look at this differently. When we add **B** to both sides of $D = A + (-B)$, we obtain $D + B = A$. Thus, the vector difference **D** between **A** and **B** is that vector which must be added to **B** to yield **A**.

Multiplication of a Vector by a Scalar. The vector 2**A** is defined as the sum of **A** and **A**, inasmuch as two successive and identical displacements, **A** and **A**, yield a resultant vector of the same direction as **A** but with twice the magnitude of **A**. Thus, $2A = A + A$.

In general, the multiplication of vector **A** by the scalar s yields another vector having the direction of **A** but a magnitude sA. If s is negative, the direction of sA is opposite to that of **A**. See Fig. 2-8.

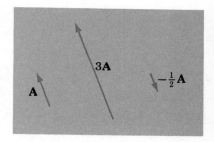

FIG. 2-8. Multiplication of a vector by a scalar changes the length and/or sense of the vector, but not its direction.

2–3 COMPONENT REPRESENTATION OF VECTORS

Since vectors have a simple graphical representation, it is possible to find the magnitude and direction of the sum of several vectors simply by drawing the vectors to scale, head-to-tail fashion, with a ruler and protractor. See Fig. 2-9. The graphical method for solving problems in vector algebra is, however, often inconvenient and limited in accuracy, and the computation of vector sums is best made through the analytical method based on the components of a vector.

The component of a vector **A** along any line is the orthogonal projection of **A** along that line. For **A** shown in Fig. 2-10 the component of **A** along line p is the vector A_p (that is, A_p is the orthogonal projection of **A** along line p). The component along line q is A_q, and the component along a line perpendicular to **A** is zero. The magnitude of the component of **A** along any line is $A \cos \theta$, where θ is the angle between **A** and the chosen line.

If we find the components of **A** along any two mutually perpendicular directions, then the vector sum of these two component

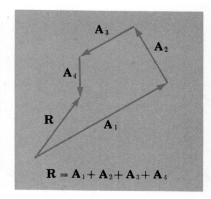

FIG. 2-9. One can find graphically the resultant R of the several vectors A_1, A_2, A_3, and A_4 by arranging the vectors in head-to-tail fashion.

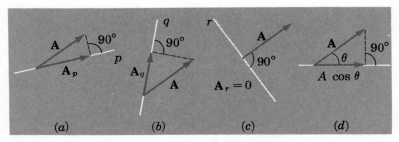

FIG. 2-10. Components of the vector A along different directions: *(a)* Along the line *p*; *(b)* along the line *q*; *(c)* along the line *r*, at right angles to A; and *(d)* in general, the component as $A \cos \theta$.

vectors is equal to the original vector **A** (Fig. 2-11). Any vector (in the plane) can then be represented by, or resolved into, two mutually perpendicular, or rectangular, vector components. The two perpendicular coordinate axes are usually identified as the X and Y axes. The location of the origin and the orientation of the axes are, of course, arbitrary. By assigning positive and negative directions to the coordinate axes, it is possible to specify any vector in a plane completely by giving its X and Y components, the signs $+$ or $-$ signifying the direction of the component (left or right, up or down). In Fig. 2-11 vector **A** has the rectangular components

$$A_x = A \cos \theta \\ A_y = A \sin \theta \quad (2\text{-}2)$$

where, following the usual convention, we measure the angle θ counterclockwise from the positive X axis. Conversely, the two rectangular components of a vector (in a plane) uniquely determine the vector. If we know A_x and A_y, then we know that vector **A** has

Magnitude given by: $\quad A = \sqrt{A_x^2 + A_y^2}$

Direction given by: $\quad \tan \theta = \dfrac{A_y}{A_x} \quad (2\text{-}3)$

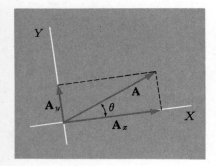

FIG. 2-11. Vector sum of the rectangular components of a vector A is equal to A.

where θ is measured relative to the X axis. In short, two numbers are required to specify vector **A** in a plane: either A and θ, or A_x and A_y.

Example 2-1

Consider the displacement whose magnitude is 10.0 m and whose direction is 30° north of east. Find the components of this vector along the north-south and east-west lines.

We identify $+X$ as east, $-X$ as west, $+Y$ as north, and $-Y$ as south. From Eq. (2-2):

$$A_x = A \cos 30° = (10.0 \text{ m})(0.866) = 8.66 \text{ m east} \\ A_y = A \sin 30° = (10.0 \text{ m})(0.500) = 5.00 \text{ m north}$$

2–4 VECTOR ADDITION BY THE COMPONENT METHOD

Rectangular components are useful for finding the sum of several vectors. For example, consider the two vectors **A** and **B** in Fig. 2-12 and their sum **R**.

From the geometry of Fig. 2-12 we see that the component of the vector **R** along any line is equal to the sum of the components of the vectors **A** and **B** along that line. Along the X axis, $R_x = A_x + B_x$, and along the Y axis, $R_y = A_y + B_y$. The magnitude and direction of the resultant vector are then:

$$R = \sqrt{R_x^2 + R_y^2} \quad \text{and} \quad \tan\theta = \frac{R_y}{R_x}$$

The general procedure for finding vector sums by first resolving the individual vectors into rectangular components is this:

1. Choose conveniently oriented rectangular coordinate axes (such that as many vectors as possible lie along X or Y).
2. Resolve each vector into its X and Y components.
3. Sum (algebraically) the X components of the individual vectors to find the X component of the resultant; similarly for the Y components.
4. Find the magnitude and direction of the resultant from its X and Y components.
5. To detect gross errors in computation, sketch the vectors roughly in head-to-tail fashion to find the approximate resultant by geometrical construction, and check this result against that obtained analytically.

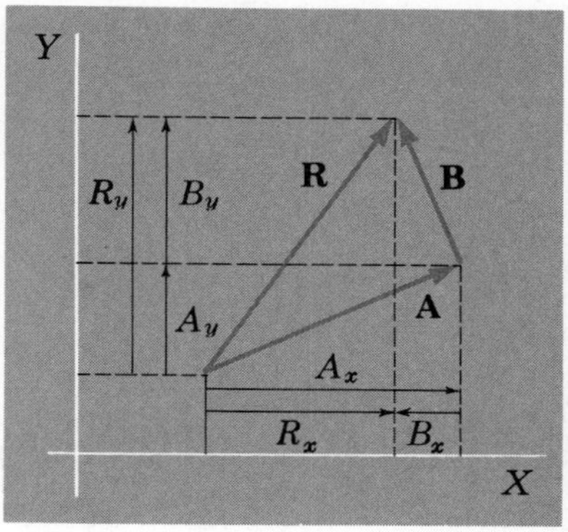

FIG. 2-12. The rectangular component of the resultant equals the algebraic sum of the components of the vectors along the same axis: $R_y = A_y + B_y$; $R_x = A_x - B_x$.

Example 2-2

A man walks 7.00 km 30° north of east, then 3.46 km 60° north of west, then 3.00 km 30° south of west, and finally 2.00 km south. Find the man's final distance from and orientation relative to his starting point.

These four displacements and their sum are shown in head-to-tail fashion in Fig. 2-9. For convenience in finding components, we locate the tails of all four displacements at the origin; see Fig. 2-13a. (Recall that any vector may be relocated without its value being changed, so long as the magnitude and direction are unchanged.) The four displacements are labeled $\mathbf{A}_1$, $\mathbf{A}_2$, $\mathbf{A}_3$, and $\mathbf{A}_4$, respectively. Because of the orientation of the X and Y axes shown in Fig. 2-13a, all but one of the vectors lie along the coordinate axes.

Next, we compute the rectangular components of each vector. For example, the components of $\mathbf{A}_4$ (2.00 km south) are:

$$A_{4x} = (2.00 \text{ km}) \cos 240° = -(2.00 \text{ km}) \cos 60° = -1.00 \text{ km}$$
$$A_{4y} = (2.00 \text{ km}) \sin 240° = -(2.00 \text{ km}) \sin 60° = -1.73 \text{ km}$$

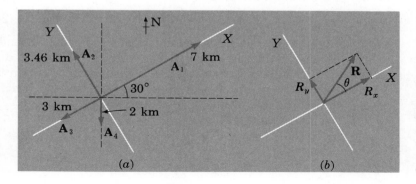

FIG. 2-13. (a) Four displacement vectors whose resultant is to be computed. Note that the rectangular components are easily computed by orienting the X and Y axes as shown. (b) The resultant and its components along the chosen X and Y axes.

The components of all vectors are shown in the table below (such a table is a useful device for keeping one's information organized).

Vector	X component (km)	Y component (km)
$A_1 = 7.00$ km 30°N of E	+7.00	0
$A_2 = 3.46$ km 60°N of W	0	+3.46
$A_3 = 3.00$ km 30°S of W	−3.00	0
$A_4 = 2.00$ km S	−1.00	−1.73
	$R_x = +3.00$ km	$R_y = +1.73$ km

The resultant **R** has components $R_x = 3.00$ km and $R_y = 1.73$ km. The magnitude R and direction θ of the resultant then is

$$R = \sqrt{R_x^2 + R_y^2} = \sqrt{(3.00)^2 + (1.73)^2}\text{ km} = 3.46\text{ km}$$

$$\tan\theta = \frac{R_y}{R_x} = \frac{1.73\text{ km}}{3.00\text{ km}} = 0.577$$

$$\theta = 30°$$

The resultant is 3.46 km 60° north of east, or 30° east of north, as shown in Fig. 2-13b.

2–5 VECTORS IN THREE DIMENSIONS

The three mutually perpendicular X, Y, and Z coordinate axes shown in Fig. 2-14 form a *right*-handed set. If one imagines the $+X$ axis to be turned into the $+Y$ axis through the smaller (90°) angle with the fingers of the *right* hand, the direction of the $+Z$ axis, which is perpendicular to the XY plane, is given by the direction of the right-hand thumb. Thus, in a right-hand set of axes, if $+X$ is to the right and $+Y$ upward, then $+Z$ is out of the paper.

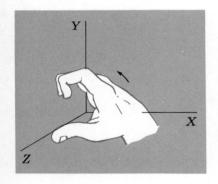

FIG. 2-14. The X, Y, and Z axes shown form a right-handed set.

The rectangular components of a vector in three dimensions are the orthogonal projections of that vector along the axes X, Y, and Z. The components A_x, A_y, and A_z of vector **A** are shown in Fig. 2-15. From the geometry it follows that

$$A^2 = A_x^2 + A_y^2 + A_z^2$$

Therefore, the magnitude of **A** is

$$A = \sqrt{A_x^2 + A_y^2 + A_z^2} \tag{2-4}$$

The analytical method of adding vectors is easily extended to three dimensions. Resolve each vector into its X, Y, and Z components. Then add algebraically all X components to obtain the X component of the resultant, and likewise for Y and Z. Finally, use (2-4) to find the magnitude of the resultant.

To such a *vector* equation as

$$\mathbf{R} = \mathbf{A} + \mathbf{B} + \mathbf{C}$$

there correspond *three* component *scalar* equations:

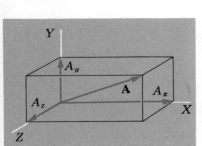

FIG. 2-15. The rectangular components of a vector in three dimensions.

$$R_x = A_x + B_x + C_x$$
$$R_y = A_y + B_y + C_y$$
$$R_z = A_z + B_z + C_z$$

Herein lie the economy, generality, and elegance of the vector algebra. A single vector equation replaces three scalar equations. Moreover, the vector equation expresses a relationship which is independent of the particular choice of coordinate axes.

2–6 THE SCALAR AND VECTOR PRODUCTS

In addition to the multiplication of a vector quantity by a scalar quantity, there are two important products between a pair of vector quantities: the scalar (or dot) product and the vector (or cross) product. As we shall see, the scalar product simplifies the definition of work (Sec. 9-2), and the vector product simplifies the treatment of angular momentum and torque (Sec. 11-3).

The Scalar Product. Consider two vectors **A** and **B** with angle θ between them. The scalar product of **A** and **B** is symbolized by **A · B**. By definition, **A · B** is

$$\mathbf{A} \cdot \mathbf{B} = AB \cos \theta \qquad (2\text{-}5)$$

That is, the scalar product of **A** and **B** is the product of the magnitudes of **A** and **B** multiplied by the cosine of θ. Alternatively, the magnitude of **A · B** is the magnitude of **A** multiplied by the component of **B** along the direction of **A**, as in Fig. 2-16 (or |**B**| multiplied by the component of **A** along **B**). The dimensions of **A · B** are the product of the dimensions of **A** and **B**. When **A** and **B** are

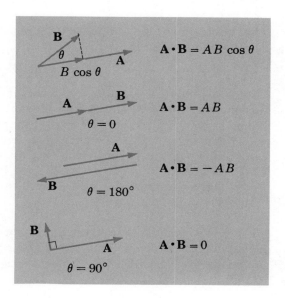

FIG. 2-16. The scalar product of vectors A and B for several values of the angle θ between them.

aligned ($\theta = 0°$), then $\mathbf{A} \cdot \mathbf{B} = AB$. When **A** and **B** are antialigned ($\theta = 180°$), then $\mathbf{A} \cdot \mathbf{B} = -AB$. When **A** and **B** are oriented at right angles with respect to one another ($\theta = 90°$), then $\mathbf{A} \cdot \mathbf{B} = 0$. See Fig. 2-16.

The *commutative* and *distributive laws* of the scalar product follow directly from the definition of the scalar product. Thus,

$$\mathbf{A} \cdot \mathbf{B} = AB \cos \theta = BA \cos \theta = \mathbf{B} \cdot \mathbf{A}$$

By referring to Fig. 2-17, we can easily prove the distributive relationship: If $\mathbf{C} = \mathbf{A} + \mathbf{B}$, then

$$\mathbf{D} \cdot \mathbf{C} = \mathbf{D} \cdot (\mathbf{A} + \mathbf{B}) = \mathbf{D} \cdot \mathbf{A} + \mathbf{D} \cdot \mathbf{B} \qquad (2\text{-}6)$$

because the component of vector **C** along the direction of **D** is the sum of the components of vectors **A** and **B** along **D**. Thus the distributive law holds.

Consider vectors **A** and **B** written as a sum of their rectangular vector components:

and
$$\mathbf{A} = \mathbf{A}_x + \mathbf{A}_y + \mathbf{A}_z$$
$$\mathbf{B} = \mathbf{B}_x + \mathbf{B}_y + \mathbf{B}_z$$

Because of the mutual perpendicularity of the coordinate axes, $\mathbf{A}_x \cdot \mathbf{B}_x = A_x B_x$ but $\mathbf{A}_x \cdot \mathbf{B}_y = 0$, and similarly for other combinations. It follows then that

$$\mathbf{A} \cdot \mathbf{B} = \mathbf{A}_x \cdot \mathbf{B}_x + \mathbf{A}_y \cdot \mathbf{B}_y + \mathbf{A}_z \cdot \mathbf{B}_z$$
or
$$\mathbf{A} \cdot \mathbf{B} = A_x B_x + A_y B_y + A_z B_z \qquad (2\text{-}7)$$

As a special case, when $\mathbf{B} = \mathbf{A}$,

$$\mathbf{A} \cdot \mathbf{A} = A_x^2 + A_y^2 + A_z^2$$
or
$$\mathbf{A} \cdot \mathbf{A} = A^2$$

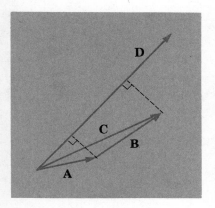

FIG. 2-17. Graphical proof of the distributive law for scalar products: $\mathbf{D} \cdot \mathbf{C} = \mathbf{D} \cdot (\mathbf{A} + \mathbf{B}) = \mathbf{D} \cdot \mathbf{A} + \mathbf{D} \cdot \mathbf{B}$.

The Vector Product. The vector product of **A** and **B** is another vector **C** symbolized by

$$\mathbf{C} = \mathbf{A} \times \mathbf{B} \qquad (2\text{-}8)$$

The cross product **C** is along a line perpendicular to the plane containing the vectors **A** and **B**. By convention, the direction of **C** is taken to be the direction of the right-hand thumb when the first vector **A** is imagined to be turned by the right-hand fingers through the smaller angle between **A** and **B** to align with the second vector **B**. See Fig. 2-18.

The magnitude of $\mathbf{A} \times \mathbf{B}$ is given by

$$C = AB \sin \theta \qquad (2\text{-}9)$$

where θ is the angle between **A** and **B**. If **A** and **B** are the sides of a parallelogram, $\mathbf{A} \times \mathbf{B}$ is a vector along the normal to the plane of the parallelogram having a magnitude equal to the parallelogram's area.

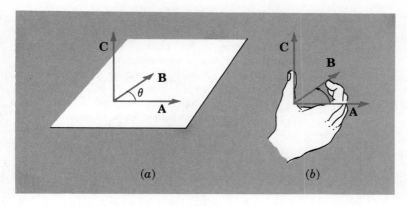

FIG. 2-18. *(a)* The cross product $C = A \times B$ and *(b)* the right-hand rule for cross products.

The order in which the two vectors in a cross product are written is important, since $\mathbf{A} \times \mathbf{B} \neq \mathbf{B} \times \mathbf{A}$. The cross product does *not* obey the commutative law. Indeed, if the order of the vectors is reversed, so is the direction of their cross product. Therefore,

$$\mathbf{A} \times \mathbf{B} = -\mathbf{B} \times \mathbf{A}$$

The distributive law does, however, hold for cross products:

$$\mathbf{A} \times (\mathbf{D} + \mathbf{E}) = (\mathbf{A} \times \mathbf{D}) + (\mathbf{A} \times \mathbf{B})$$

If the vectors $\mathbf{A}$ and $\mathbf{B}$ are parallel or antiparallel, the angle between them being either 0 or 180°, then $\mathbf{A} \times \mathbf{B} = 0$, inasmuch as the factor $\sin \theta$ is then zero. Thus, $\mathbf{A} \times \mathbf{A} = 0$.

The cross product and dot product must be carefully distinguished. The cross product is a vector; the dot product, a scalar. The cross product does not obey the commutative law; the scalar product does. The cross product is zero when the vectors are parallel; the dot product is zero when the vectors are perpendicular.

PROBLEMS

2-1. Displacement **D** is 12 m north. What are *(a)* −**D**, *(b)* **D**/2, and *(c)* −**D**/3?

2-2. Starting from point *A* a man walks 5.0 blocks east and 15.0 blocks south to reach point *B*. What is the displacement (in blocks) of point *B* relative to point *A*, and of point *A* relative to *B*?

2-3. Vector **A** is a displacement of 60.0 m at 210°. What are the *X* and *Y* components of *(a)* **A**, *(b)* −**A**/3, and *(c)* −3**A**?

2-4. Vectors **A** and **B** have magnitudes 12 and 5 m, respectively. How must **A** and **B** be oriented to give a resultant of *(a)* 7 m, *(b)* 12 m, *(c)* 13 m, *(d)* 17 m?

2-5. An object is displaced 12 m down from the top of an incline that makes an angle of 60° with respect to the horizontal. What are the components of this displacement *(a)* along and *(b)* perpendicular to the incline?

2-6. Find *(a)* the vector sum **A** + **B** and *(b)* the vector difference **A** − **B**, of the vectors **A** = 15.0 mi southeast and **B** = 25.0 mi at 60° north of west.

2-7. Find the resultant of the following displacements: 4.0 m 30° north of east, 5.0 m 60° west of north, and 6.0 m southeast.

2-8. Find the vector sum of the following displacements: 50 mi 37.1° east of north, 60 mi 30° north of west, and 30 mi south.

2-9. Vector **A** is 5.0 m at 45°, **B** is 10.0 m at 60°, **C** is 15.0 m at 270°, and **A** + **B** + **C** + **D** = 0. What is **D**?

2-10. Vector **A** makes an angle θ with vector **B**. Show that the magnitude of **A** + **B** is $(A^2 + B^2 + 2AB \cos \theta)^{1/2}$.

2-11. Suppose that **A** + **B** = **C**, where **C** is 5.0 mi northeast and **B** is 8.0 mi west. Find **A**.

2-12. Find *A* and *B* given the following data: **A** + **B** = **C**; **A** points 30° east of north, **B** is directed to the northwest, and **C** is 12.0 m north.

2-13. What must be the relative orientations of **A** and **B** if the magnitude of **A** + **B** equals the magnitude of **A** − **B**?

2-14. The rectangular components of a vector along the *X* and *Y* axes are *x* and *y*, respectively. A second set of mutually perpendicular axes, *X'* and *Y'*, have their origin at the origin of *X* and *Y*, but the *X'* axis is at an angle of θ with respect to *X*. *(a)* Show that the components *x'* and *y'* along *X'* and *Y'* are given by $x' = x \cos \theta + y \sin \theta$ and $y' = y \cos \theta - x \sin \theta$, respectively. (These equations are known as the coordinate transformation equations for pure rotation.) *(b)* Prove that the magnitude of a vector is independent of the choice of coordinate axes by showing that $x^2 + y^2 = x'^2 + y'^2$.

2-15. Vector **A** = 5 units east and **B** = 10 units 37.1° north of east. *(a)* Find **A** · **B**. *(b)* What is the magnitude of a vector having the direction 30° north of east which also gives the same scalar product with vector **A**?

2-16. Consider the three vectors **A** = 15 units north, **B** = 10 units 30° east of north, and **C** = 5 units east. *(a)* Find **A**(**B** · **C**). *(b)* Find **C**(**B** · **A**).

2-17. Derive the law of cosines for oblique triangles from the rules governing the scalar product. (*Hint:* Take vector **C** = **A** + **B** and use the results of Problem 2-10.)

2-18. Vector **A** is 20 in the direction west. Vector **B** is 30 in the direction southeast. What are the magnitude and direction of *(a)* **A** × **B** and *(b)* **B** × **A**?

2-19. Vectors **A** and **B** represent the sides of a parallelogram. Show that the magnitude of **A** × **B** is the area of the parallelogram and that the direction of **A** × **B** is along the normal to the plane of the parallelogram.

2-20. Vector **A** is 3 in the upward direction, **B** has a magnitude 5, and the scalar product **A** · **B** is zero. What are the possible directions (!) and the magnitude of **A** × **B**?

2-21. Vector **A** is 5 units south and **B** is 10 units 30° west of north. What are *(a)* **A** · **B** and *(b)* **A** × **B**?

2-22. Vector **A** is 7 in the +*X* direction; **B** is 3 in the −*Y* direction. What are *(a)* **A** · **B** and *(b)* **A** × **B**?

2-23. Given $C = \mathbf{A} \cdot \mathbf{B} = 7.5$, $D = |\mathbf{A} \times \mathbf{B}| = 13.0$, and $A = 1.5$, find *B*.

2-24. Show that **A** × **B** · **C** represents the volume of a parallelopiped with edges **A**, **B**, and **C** and that, therefore, **A** × **B** · **C** = **A** · **B** × **C**.

CHAPTER 3

Straight-line Kinematics

Dynamics is concerned with the motions of objects as they are related to the physical concepts of mass, force, momentum, and energy, and with the general laws unifying these concepts. Kinematics, on the other hand, has a more modest program. Combining the ideas of geometry and time, it describes motion without giving attention to its causes. Thus, kinematics is a necessary preliminary to the study of dynamics and the search for physical laws. In this chapter we explore the kinematics of a particle moving along a straight line (in one dimension), the meaning of the terms displacement, velocity, and acceleration, and the mathematical relations that apply to the special case of motion with constant acceleration.

3–1 THE MEANING OF A PARTICLE

What is a particle? It is not, necessarily, a tiny, hard sphere as commonly imagined; rather, we may take a particle to be simply an object that is small enough for its size to be unimportant for the scale of our observations. It has no internal structure with which we need be concerned. Thus, a star may properly be regarded as a particle when viewed as one part of a galaxy, but an atom is too large to be considered a particle when we examine its component parts. It is all a matter of scale. The particle is the physical counterpart of the point in mathematics; it has not only the property of precise localizability but also such physical attributes as mass and electric charge.

3–2 AVERAGE VELOCITY

We restrict our discussion in this chapter to a particle in *rectilinear motion*, motion along a straight line. A straight line may be defined as the path of a beam of light through a vacuum. The particle's location along the straight line, here called the X axis, is designated by the coordinate x, with some appropriate unit of length. Locations to the right or left of the origin are distinguished by the algebraic sign of x.

The *displacement* of the particle is defined as the *change in its position* (for one dimension, the change in its coordinate). We represent the displacement by the symbol Δx (where the symbol delta means "the change in"). If x_i is the initial coordinate and x_f the final coordinate at some later time, then

$$\Delta x = x_f - x_i \tag{3-1}$$

Clearly, the sign of Δx denotes the sense in which the particle shifts position along the axis; when $x_f > x_i$, Δx is positive for a displacement to the right, and conversely. Displacement must be distinguished from the distance traveled, or the total length of path traversed by the particle, which is always positive. A particle making a round trip to its starting point has zero displacement, although it travels a finite distance.

Consider Fig. 3-1 in which the coordinate x of some particle in motion is plotted as a function of the time t at which the particle was at each position. (Data of this sort might, for example, be derived from a motion picture in which each frame shows one event in the history of the particle, or from a stroboscopic photograph with multiple images giving the position of a particle at equal time intervals.)

The *average velocity* $\bar{v}$ is defined as the displacement a particle undergoes *divided by the elapsed time* for this displacement. In symbols,

$$\bar{v} = \frac{\Delta x}{\Delta t} \tag{3-2}$$

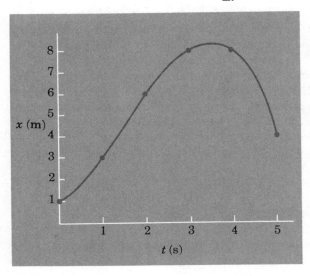

FIG. 3-1. Displacement x of a particle plotted as a function of time t.

where $\Delta t = t_f - t_i$ and $\Delta x = x_f - x_i$. From its definition, velocity has the dimensions of length divided by time (L/T), for example, m/s, mi/h, or light-yr/yr.

Applying this definition to the data in Fig. 3-1, we compute the average velocity for a number of time intervals:

For $t_i = 1$ s and $t_f = 2$ s:

$$\bar{v}_{12} = \frac{\Delta x}{\Delta t} = \frac{x_f - x_i}{t_f - t_i} = \frac{(6-3) \text{ m}}{(2-1) \text{ s}} = 3 \text{ m/s}$$

For $t_i = 2$ s and $t_f = 4$ s:

$$\bar{v}_{24} = \frac{\Delta x}{\Delta t} = \frac{x_f - x_i}{t_f - t_i} = \frac{(8-6) \text{ m}}{(4-2) \text{ s}} = 1 \text{ m/s}$$

For $t_i = 3$ s and $t_f = 4$ s:

$$\bar{v}_{34} = \frac{\Delta x}{\Delta t} = \frac{x_f - x_i}{t_f - t_i} = \frac{(8-8) \text{ m}}{(4-3) \text{ s}} = 0 \text{ m/s}$$

And for $t_i = 4$ s and $t_f = 5$ s:

$$\bar{v}_{45} = \frac{\Delta x}{\Delta t} = \frac{x_f - x_i}{t_f - t_i} = \frac{(4-8) \text{ m}}{(5-4) \text{ s}} = -4 \text{ m/s}$$

For the data in Fig. 3-1 the value of the average velocity clearly depends upon the time interval chosen: $\bar{v}_{12}$ is positive, $\bar{v}_{24}$ has a smaller positive magnitude (a smaller average velocity over this interval), $\bar{v}_{34}$ is zero (no displacement over this interval), and $\bar{v}_{45}$ is negative (motion toward the origin). Only if the displacement-time graph consists of a single straight line will the average velocity have the same magnitude for all possible pairs of points on the line; then we have constant (uniform) velocity.

3-3 INSTANTANEOUS VELOCITY

The slope of any line is the change in the ordinate (here Δx) over the corresponding change in abscissa (here Δt). For a displacement-time line the *slope* $\Delta x/\Delta t$ is the *average velocity* over the chosen interval Δt. A horizontal line, with zero slope, indicates rest; positive and negative slopes imply motion away from and toward the coordinate origin, respectively (for positive values of x).

See Fig. 3-2, where the motion portrayed is that of an object which moves slowly at first, picks up speed, and then slows down. Figure 3-2 may not yield the same value for $\bar{v}$ for different pairs of points. To characterize the velocity more precisely, we need the *instantaneous velocity*, the velocity at an instant of time, which we will take to be *the average velocity computed over an infinitesimally small time interval*.

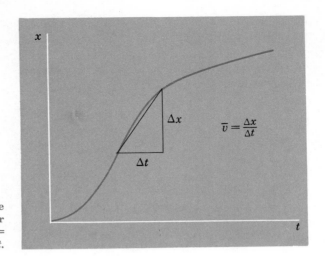

FIG. 3-2. Displacement-time graph. The average velocity over the time interval shown is $\bar{v} = \Delta x/\Delta t$.

The computation of the instantaneous velocity is illustrated in Fig. 3-3. The interval chosen for Δt is so small that the actual curve cannot be distinguished from a straight line drawn between the two end points. Since a line drawn tangent to the displacement-time curve will be identical with the curve over a small region at the point of contact, the instantaneous velocity is the slope of the curve. In formal terms: The instantaneous velocity v at time t is the limit, as the time interval Δt approaches zero duration, of the displacement Δx divided by the corresponding time interval Δt, the interval Δt being centered at time t. In symbols,

$$\text{Instantaneous velocity} = v = \lim_{\Delta t \to 0} \frac{\Delta x}{\Delta t} = \frac{dx}{dt} \qquad (3\text{-}3)$$

In the notation of the calculus† $v = dx/dt$, the derivative of x with respect to t, or the time rate of change of displacement.

Hereafter, the symbol v will represent the instantaneous velocity and so will the term velocity, unless otherwise specified.

The term *instantaneous speed* is used to designate the magnitude of the instantaneous velocity; speed is always positive. Velocity indicates, in addition, the direction of motion (plus if to the right, minus if to the left). Figure 3-4a gives the instantaneous velocity as a function of time. It was derived from the displacement-time curve of Fig. 3-3 by the procedure of taking slopes.

A velocity-time curve can be derived from the slope of a displacement-time curve; conversely, one can arrive at a displacement-time curve given first the velocity-time curve. Consider Fig. 3-4b. A number of narrow rectangles, each of width Δt, are fitted under the curve. Over a small time interval Δt the velocity may be re-

---

†Although calculus notation is used in defining instantaneous velocity and instantaneous acceleration, in this text the calculus is not required in any essential way until Chap. 6.

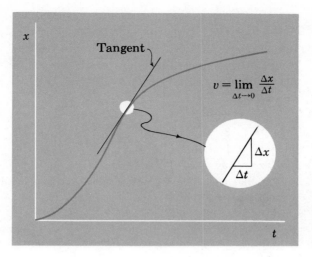

FIG. 3-3. The instantaneous velocity is the tangent of the displacement-time graph, which is the limit of the average velocity for a vanishingly small time interval.

garded as essentially constant. The corresponding displacement Δx is $v \, \Delta t$ which is, as we see from the figure, the area of the shaded rectangle. Then, if the entire displacement x (from the start to

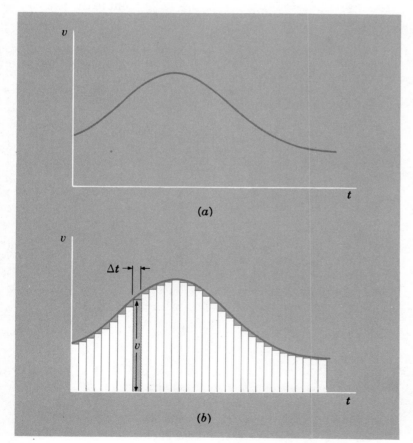

FIG. 3-4. (a) Instantaneous velocity of the function of time. (b) The displacement is equal to the area under the velocity-time graph.

Instantaneous Velocity SEC. 3-3

the time t) is to be found, we add all the elementary rectangles, which is to say that the displacement is equal to the entire area under the velocity-time curve. If the velocity is negative and the velocity-time curve lies below the time axis, the contribution of the area "under" the curve is negative. In the language of integral calculus,

$$x_f - x_i = \lim_{\Delta t \to 0} \Sigma v \, \Delta t = \int_{t_i}^{t_f} v \, dt$$

Example 3-1

The displacement of an object is given as a function of time by the relation $x = 4t^3$. Compute the average velocity for the following conditions: (a) $\Delta t = 2$ and centered at $t = 3$, and (b) $\Delta t = 0.2$ and again centered at $t = 3$.

(a) $x_i = 4(2)^3 = 32$ and $x_f = 4(4)^3 = 256$

$$\bar{v} = \frac{\Delta x}{\Delta t} = \frac{256 - 32}{2} = 112$$

(b) $x_i = 4(2.9)^3 = 97.556$ and $x_f = 4(3.1)^3 = 119.164$

$$\bar{v} = \frac{\Delta x}{\Delta t} = \frac{119.164 - 97.556}{0.2} = 108.04$$

Taking a still smaller Δt, we would find that the average velocity approaches closely 108.0000 . . . ; that is, the instantaneous velocity at $t = 3$ is $v = 108$ exactly.

The same result follows from taking the time derivative of $x = 4t^3$ to find the relation for the instantaneous velocity:

$$v = \frac{dx}{dt} = 12t^2 = 12(3)^2 = 108$$

3-4 ACCELERATION

The instantaneous velocity is the time rate of change of displacement. It tells us how displacement changes with time. Another useful quantity tells us how the velocity changes with time. This quantity, the *acceleration*, is the time rate of change of velocity. The average acceleration $\bar{a}$ is defined as the velocity change Δv divided by the corresponding elapsed time interval Δt:

$$\bar{a} = \frac{\Delta v}{\Delta t} \tag{3-4}$$

Thus, average acceleration is the slope of the straight line drawn between two points on a velocity-time curve. The *instantaneous acceleration* a is the average acceleration over an infinitesimally small time interval (see Fig. 3-5). Equivalently, a is the slope of the velocity-time curve. In symbols:

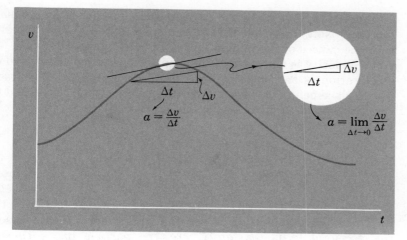

FIG. 3-5. The instantaneous acceleration is the tangent of the velocity-time graph, which is the limit of the average acceleration for a vanishingly small time interval.

$$\text{Instantaneous acceleration} = a = \lim_{\Delta t \to 0} \frac{\Delta v}{\Delta t} = \frac{dv}{dt} \qquad (3\text{-}5)$$

Hereafter "acceleration" shall mean instantaneous acceleration.

Acceleration has the units of velocity divided by time. Its dimensions are length/time² (L/T^2), for example mi/(h)(s), m/(s)(s), kn/s. It is customary to write m/(s)(s) as m/s². Thus, a body having a constant acceleration of +10 m/s² has its velocity increasing by 10 m/s during each 1-s interval.

Figure 3-6 is a graph of the instantaneous acceleration as a function of time, derived from the velocity-time curve of Fig. 3-4 or 3-5. The acceleration is at first positive, later zero, then negative, and finally zero again. Thus, Fig. 3-6 indicates that, respectively, first the velocity of the body was increasing to the right, later the velocity was momentarily unchanged, then the velocity to the right decreased, and finally the velocity was constant.

The procedure we have followed, in defining the velocity as the time rate of change of the displacement and the acceleration as the time rate of change of velocity, could be extended to yield a quantity giving the time rate of change of acceleration (which would give a measure of the "jerkiness" of the motion); but this is not usually necessary. The concepts of velocity and acceleration are sufficient to describe motion adequately. The reasons for this lie in the physics, or dynamics, of the motion, not in the mathematics or kinematics of the motion—as we shall see later.

As another example, consider the displacement-time, velocity-time, and acceleration-time curves shown in Fig. 3-7. The particle was first at rest in region A, then (positively) accelerated in region B. It coasted with constant velocity during C. In region D the body slowed down, came to rest momentarily, and reversed the direction of its motion. In region E the body moved with a constant speed to the left, approached and finally passed its starting point. This is a familiar motion exemplified by an object set

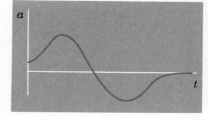

FIG. 3-6. Instantaneous acceleration as a function of time.

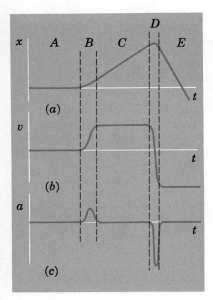

FIG. 3-7. (a) Displacement-time graph. (b) Velocity-time graph derived from (a). (c) Acceleration-time graph derived from (b). (Note that the acceleration is not constant in time.)

into motion by a push to the right and later returned by a push to the left. Note, however, that in kinematics we are not concerned with the causes of the motion, but only with its description.

Velocity is the slope of the displacement-time curve, and the acceleration is the slope of the velocity-time curve. Conversely, the displacement is the area under the velocity-time curve, and the velocity is the area under the acceleration-time curve.

3-5 RECTILINEAR MOTION AT CONSTANT ACCELERATION

A simple, special, but important, type of rectilinear motion is that in which the acceleration is constant, or uniform. Of course, if the acceleration is *constantly zero,* the particle has a *constant velocity:* It is either moving with unchanging speed in a straight line or remaining at rest.

Consider Fig. 3-8, a velocity-time graph for constant acceleration a. Here the slope is unchanged. The velocity increases by equal amounts in equal time intervals, the change in the velocity being proportional to the time interval. We label the velocity at $t = 0$ the *initial velocity* v_0. The velocity at any other time t is simply v.

Because the acceleration is assumed constant, the average acceleration for any time interval and the instantaneous acceleration are equal. We may choose any two points on the velocity-time curve to find the slope, or acceleration. For convenience, we choose the time interval running from time 0 to time t, during which the velocity has changed from v_0 to v. Thus,

$$a = \frac{\Delta v}{\Delta t} = \frac{v - v_0}{t - 0}$$

or
$$v = v_0 + at \qquad (3\text{-}6)$$

Equation (3-6) gives the velocity at time t in terms of the initial velocity, the acceleration, and the elapsed time. The final velocity v is comprised of two parts: the initial velocity v_0 and the change in velocity at, as shown in Fig. 3-8.

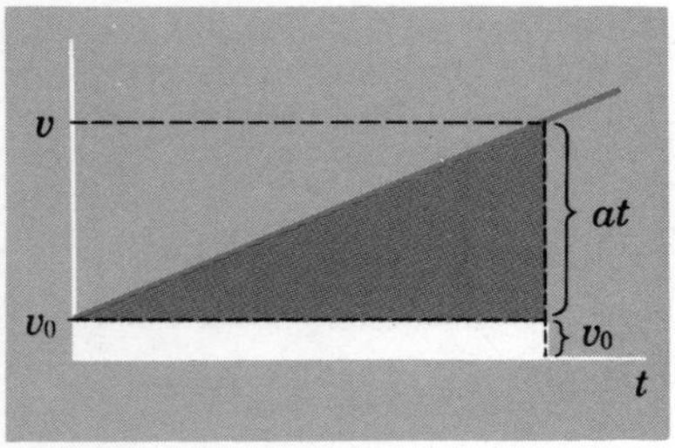

Let us express the displacement x in terms of the time t and the constants v_0 and a. We find the displacement from the velocity-time curve by computing the area under the curve. It is clear that the total area under the straight line in Fig. 3-8 is a rectangle of width t and height v_0 plus a right triangle of sides t and at. Thus, the displacement x is the sum of $v_0 t$ and $\tfrac{1}{2}(at)(t)$, or

$$x = x_0 + v_0 t + \tfrac{1}{2} a t^2 \qquad (3\text{-}7)$$

where the additional term x_0 is the displacement at $t = 0$, that is, $x = x_0$ when $t = 0$. Taking the initial displacement to be x_0 allows

FIG. 3-8. Velocity-time graph for constant acceleration.

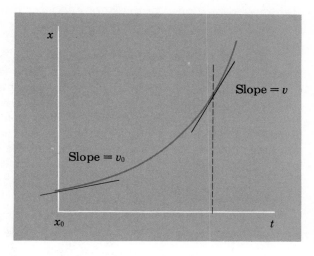

FIG. 3-9. Displacement-time graph for constant acceleration.

for situations in which the particle does not start at the origin. The terms in (3-7) have a simple interpretation: The first gives the initial displacement, the next the displacement that would occur if the velocity were to remain v_0, and the third is the additional displacement resulting from the acceleration.

Figure 3-9 is a displacement-time curve for constant acceleration. Equation (3-7) shows that it is a parabola. The slope increases uniformly, corresponding to the constant acceleration.

The kinematic formulas (3-6) and (3-7) permit the future (and past) history of a particle to be projected; that is, they give the displacement and velocity for any future time when the initial displacement x_0, the initial velocity v_0, and the constant acceleration a are known.

We can arrive at another useful kinematic formula by noting that, *for constant acceleration,* the average velocity $\bar{v}$, defined as the total displacement divided by the elapsed time, is simply the (algebraic) average of the initial and final velocities:

$$\bar{v} = \frac{v_0 + v}{2} \tag{3-8}$$

This follows from the fact that the area of the trapezoid under the velocity-time curve (the displacement) is identical with the area of the rectangle under a velocity-time curve for a constant velocity equal to $\bar{v}$, as given by (3-8). See Fig. 3-10. Since the net displacement $x - x_0 = \bar{v}t$,

$$x = x_0 + \frac{v_0 + v}{2} t \tag{3-9}$$

A fourth kinematic formula is easily obtained by solving for t from (3-6) and substituting in (3-9):

$$x - x_0 = \frac{(v + v_0)(v - v_0)}{2a}$$

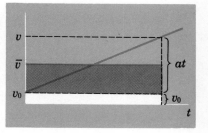

FIG. 3-10. Graph for constant acceleration: The area under the average velocity $\bar{v}$ versus time is equal to the area under the trapezoid in Fig. 3-8.

Rearranging yields
$$v^2 = v_0^2 + 2a(x - x_0) \qquad (3\text{-}10)$$

The four kinematic formulas (3-6), (3-7), (3-9), and (3-10) relate six quantities: the constants x_0, v_0, and a, and the variables x, v, and t. Equation (3-6) contains v_0, v, a, and t, but not x. Similarly, (3-7), (3-9), and (3-10) do not contain v, a, and t, respectively. These relations may be used to solve all problems involving motion at constant acceleration along a line. They express in mathematical language nothing more than the logical and algebraic consequences of the definitions of velocity and constant acceleration.

The kinematic relations for constant acceleration are easily verified by the differential calculus†:

$$x = x_0 + v_0 t + \tfrac{1}{2}at^2$$

$$v = \frac{dx}{dt} = v_0 + at$$

$$a = \frac{dv}{dt} = a$$

Conversely, by integrating we can obtain the velocity-time and displacement-time formulas directly from the definitions:

$$a = \frac{dv}{dt}$$

or
$$dv = a\,dt$$

We integrate velocity from the initial value v_0 to the final value v, the corresponding limits on time being zero and t:

$$\int_{v_0}^{v} dv = a \int_0^t dt$$

$$v - v_0 = at$$

$$v = v_0 + at \qquad (3\text{-}6)$$

We integrate once more to find x:

$$\frac{dx}{dt} = v_0 + at$$

or
$$dx = v_0\,dt + at\,dt$$

The displacement goes from x_0 at $t = 0$ to x at time t. Thus, the definite integrals are

$$\int_{x_0}^{x} dx = v_0 \int_0^t dt + a \int_0^t t\,dt$$

$$x - x_0 = v_0 t + \tfrac{1}{2}at^2 \qquad (3\text{-}7)$$

Note that the assumed constancies of v_0 and a were an essential part of the analysis.

---

†Calculus is not really required at this stage.

Example 3-2

A car starting from rest travels 100 m in 10 s. *(a)* What is the acceleration, assuming it to be constant throughout? *(b)* What is the car's final speed?

(a) We know x_0 (taken as zero), v_0, x, and t, and are to find a. Equation (3-7) is useful here:

$$x = x_0 + v_0 t + \tfrac{1}{2}at^2 = \tfrac{1}{2}at^2 \qquad \text{since } x_0 = 0 \text{ and } v_0 = 0$$

Solving for a we have

$$a = \frac{2x}{t^2} = \frac{2 \times 100 \text{ m}}{(10 \text{ s})^2} = 2.0 \text{ m/s}^2$$

(b) We are to compute v, knowing v_0, a, and t. Equation (3-6), $v = v_0 + at$, is applicable:

$$v = 0 + (2.0 \text{ m/s}^2)(10 \text{ s}) = 20 \text{ m/s}$$

or, if we wish to convert units to miles per hour,

$$v = \frac{20 \text{ m}}{\text{s}} \times \frac{3.28 \text{ ft}}{\text{m}} \times \frac{1 \text{ mi}}{5{,}280 \text{ ft}} \times \frac{3{,}600 \text{ s}}{\text{h}} = 45 \text{ mi/h}$$

Note two points, illustrated in this example and applicable to all problems in kinematics—indeed, applicable to all problems in physics:

1. We solve algebraically for the unknown *before* substituting numerical values for symbols.
2. The units are carried along with the numbers and treated algebraically. One necessary (but not sufficient) test of the correctness of the solution is whether the dimensions are appropriate. Any formula *in physics* must be dimensionally consistent; that is, all of its terms must have the same dimensions. For example, the terms x, x_0, $v_0 t$, and $\tfrac{1}{2}at^2$ in the equation $x = x_0 + v_0 t + \tfrac{1}{2}at^2$ must have the same *units*; for example, the lengths must be all meters. Of course, no answer is complete if it does not specify the units.

Example 3-3

An electron traveling at 4.0×10^6 m/s to the right enters the region between vertical, parallel, electrically charged, metal plates separated by 2.0 cm; see Fig. 3-11. Within this region the electron undergoes a constant acceleration of 7.9×10^{14} m/s² to the right. (The electron has a negative electric charge; the left and right plates carry minus and plus electric charges, respectively, by virtue of their connection to the terminals of a battery. The constant acceleration here has its origin in electric charges.) *(a)* With what velocity does the electron strike the right plate? *(b)* How long does it take to reach this plate?

(a) We know that $v_0 = 4.0 \times 10^6$ m/s and that $a = 7.9 \times 10^{14}$ m/s²; we ask for v when $x - x_0 = 2.0 \times 10^{-2}$ m. Equation (3-10), $v^2 = v_0^2 + 2a(x - x_0)$, is useful:

$$v^2 = (4.0 \times 10^6 \text{ m/s})^2 + 2(7.9 \times 10^{14} \text{ m/s}^2)(2.0 \times 10^{-2} \text{ m})$$
$$v = 6.9 \times 10^6 \text{ m/s}$$

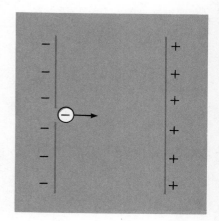

FIG. 3-11. An electron undergoing motion at constant acceleration between two oppositely charged parallel plates.

(b) We compute t from Eq. (3-9), $x - x_0 = (v_0 + v)t/2$, knowing v_0, v, and $x - x_0$:

$$t = \frac{2(x - x_0)}{v_0 + v} = \frac{2(2.0 \times 10^{-2} \text{ m})}{(4.0 + 6.9) \times 10^6 \text{ m/s}} = 3.7 \times 10^{-9} \text{ s}$$

The electron undergoes an extraordinarily large acceleration, but for only a very short time.

3–6 FREELY FALLING BODIES

Close to the earth's surface all freely falling bodies have a nearly constant magnitude of acceleration toward the center of the earth. By a "falling" body is meant not only one dropped from rest, but also a thrown object (in the vicinity of the earth's surface). Such a body is truly *freely* falling only when it moves through a vacuum, but for relatively low speeds and for smooth, compact, dense objects, the resistance of the air is negligible and their motion closely approximates the motion of free fall. Thus, descending parachutes, flying birds, and falling leaves do not qualify as freely falling objects. Unless it is stated otherwise, we hereafter ignore the effects of air resistance.

The acceleration of a freely falling body does not depend on the mass or weight of the object. Experiments show that an apple, a neutron (a small, electrically neutral particle with a mass of only 1.7×10^{-27} kg), or any other object near the earth's surface falls with a constant acceleration of 9.80 m/s² or 32.2 ft/s². This acceleration plays so important a role in physics that it is designated by a special symbol g and is referred to as the *acceleration due to gravity* (*not* as "gravity" and especially not as the "force of gravity"). How g is related to the general phenomenon of gravitation will be dealt with later; here we treat only falling-body kinematics.†

We must qualify slightly some of our assertions. Strictly, g is not constant. The value of g differs (no more than 0.4 percent) at various places on the earth's surface. Moreover, g decreases with increasing distance from the earth's center (the moon "falls" around the earth, but with an acceleration of only 2.7×10^{-3} m/s²). For altitudes of up to 40 mi above sea level, g is constant to within 2 percent.

Falling-body motion is one of the very few situations in physics illustrating motion at constant acceleration (another example is a charged particle between parallel, charged plates).

---

†The fundamental theoretical and experimental discoveries in the motion of falling bodies and the kinematics of accelerated motion were made by Galileo Galilei (1564–1642), the first physicist of the modern era and one of the greatest physicists of all time. Galileo's contributions to science are sketched in Chap. 5.

Example 3-4

An object is thrown vertically upward at 30 m/s. *(a)* What is its displacement after 4.0 s? *(b)* What is its velocity after 4.0 s? *(c)* What is the maximum height it attains? *(d)* How long does it take for the object to return to the ground? *(e)* What is its velocity upon striking the ground?

The kinematic relations for constant acceleration apply here. Since the direction of motion is along the vertical, it is appropriate to label the displacement y, rather than x, with the positive direction of the Y axis vertically upward. Then $a = -g$. To simplify the computation, we take $g = 10$ m/s^2 and put $y_0 = 0$.

(a) We know that $v_0 = +30$ m/s and $a = -10$ m/s^2; we are to find y for $t = 4.0$ s. Equation (3-7), $y - y_0 = v_0 t + \frac{1}{2} a t^2$, becomes

$$y = (30 \text{ m/s})(4.0 \text{ s}) + \tfrac{1}{2}(-10 \text{ m/s}^2)(4.0 \text{ s})^2 = +40 \text{ m}$$

(b) What is v after 4 s? Equation (3-6), $v = v_0 + at$, is useful here:

$$v = (30 \text{ m/s}) + (-10 \text{ m/s}^2)(4.0 \text{ s}) = -10 \text{ m/s}$$

Four seconds after the object is thrown, it is 40 m above the ground and traveling downward with an instantaneous speed of 10 m/s. The total *distance* traveled is, however, *not* 40 m.

(c) At its maximum height, where the object is changing direction of motion, its velocity is zero (its acceleration is still g!). Therefore, we want y when $v = 0$, using Eq. (3-10), $v^2 = v_0^2 + 2a(y - y_0)$,

$$y = \frac{v^2 - v_0^2}{2a} = \frac{0 - (30 \text{ m/s})^2}{2(-10 \text{ m/s}^2)} = +45 \text{ m}$$

(d) The displacement y is zero when the object strikes the ground (the magnitude of the total *distance* over the trip is twice the maximum height). We wish to find the corresponding time of flight. Using Eq. (3-7), $y - y_0 = v_0 t + \frac{1}{2} a t^2$, to find t for $y = 0$, we have:

$$0 = (v_0 + \tfrac{1}{2} at) t$$

$t = 0$ or $t = -\dfrac{2 v_0}{a} = -\dfrac{2(30 \text{ m/s})}{(-10 \text{ m/s})^2}$

$t = 0$ or $t = 6.0$ s

The time of flight is 6.0 s.

(e) The average velocity $\bar{v}$ of the entire motion is zero, inasmuch as the overall displacement y is zero. Therefore, Eq. (3-8), $\bar{v} = (v_0 + v)/2$, gives

$$v = 2\bar{v} - v_0 = 0 - 30 \text{ m/s} = -30 \text{ m/s}$$

[Alternatively, we can find v for $t = 6.0$ s by using (3-6).] The body strikes the ground with the same speed as that with which it was thrown. The entire motion is, in fact, symmetrical in time about the midtime. Stated differently, it is the same motion with time running backward; one cannot distinguish between motion pictures of a thrown object, one with the film run forward and the other backward.

The displacement-time, velocity-time, and acceleration-time curves in Fig. 3-12 display all the information we have computed

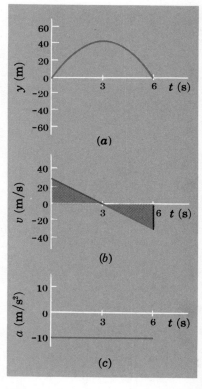

FIG. 3-12. *(a)* Displacement-time, *(b)* velocity-time, and *(c)* acceleration-time graphs for a projectile thrown vertically upward at 30 m/s.

here (and more). Note that the displacement is zero at 6.0 s, as is indicated not only on the displacement-time curve but also by the net area, shaded, on the velocity-time graph.

SUMMARY

For rectilinear motion along x, the instantaneous velocity v is the time rate of change of displacement:

$$v = \lim_{\Delta t \to 0} \frac{\Delta x}{\Delta t} = \frac{dx}{dt} \tag{3-3}$$

The instantaneous acceleration a is the time rate of change of velocity:

$$a = \lim_{\Delta t \to 0} \frac{\Delta v}{\Delta t} = \frac{dv}{dt} \tag{3-5}$$

The kinematic relations describing rectilinear motion at the *constant* acceleration a are:

$$v = v_0 + at \tag{3-6}$$

$$x = x_0 + v_0 t + \tfrac{1}{2} a t^2 \tag{3-7}$$

$$x - x_0 = \tfrac{1}{2}(v_0 + v)t \tag{3-9}$$

$$v^2 = v_0^2 + 2a(x - x_0) \tag{3-10}$$

where x is the displacement and v is the velocity at time t, x being x_0 and v being v_0 at $t = 0$.

All freely falling bodies near the earth's surface have a constant downward acceleration $g = 9.80$ m/s².

PROBLEMS

3-1. In the 1972 Olympics the winning time for the 1500-m run was 3 min 36.3 s. What was the average speed *(a)* in m/s and *(b)* in mi/h? *(c)* Assuming the same average speed, what would be the elapsed time for one mile?

3-2. Using the displacement-time graph of Fig. P3-2,

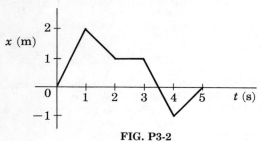

FIG. P3-2

compute the following quantities: *(a)* the average velocity $\bar{v}_{01}$ (from $t = 0$ to $t = 1$s), *(b)* $\bar{v}_{12}$, *(c)* $\bar{v}_{23}$, *(d)* $\bar{v}_{34}$, *(e)* $\bar{v}_{45}$, *(f)* $\bar{v}_{03}$, *(g)* $\bar{v}_{35}$, *(h)* $\bar{v}_{05}$. Sketch the velocity as a function of the time.

3-3. An automobile travels east at 80 km/h for 1 h, stops for 15 min, and then travels west at 120 km/h to the starting point. For the entire trip, what are *(a)* the average velocity, *(b)* the average speed, and *(c)* the distance traveled? *(d)* If the stopping time were extended to 30 min, with what speed would the motorist have to return to achieve the same average speed over the entire trip?

3-4. An automobile travels east at a constant speed of 20 m/s, and immediately upon reaching a point a distance d away, returns at the constant speed of 50 m/s. What are *(a)* the average velocity and *(b)* the average speed, both over the entire trip? (Neither is 35 m/s!)

3-5. A jet-powered racing car reaches a speed of 900 km/h after it has moved 1200 m from a standing start. Assuming constant acceleration find (a) the acceleration and (b) the elapsed time.

3-6. A particle has an initial velocity of -60 m/s. Its acceleration is a constant -4.0 m/s². (a) What are the body's displacement and velocity at $t = 2.0$ s? (b) What are the body's displacement and velocity at $t = 5.0$ s? (c) At what time is the body's velocity zero? (d) How long does it take the body to attain zero displacement? (e) At what time is the velocity -20 m/s?

3-7. In a typical picture tube for color television, electrons travel 40 cm from the electron gun to the screen, striking it at a speed of 8.0×10^7 m/s. (a) What is the acceleration of the electrons, assuming it to be constant? (b) How much time does it take for the electrons to move through this 40-cm distance if they start from rest?

3-8. Starting from rest an object increases in speed, from 40 m/s to 50 m/s, over a distance of 90 m. If the acceleration is constant throughout, what is the distance traveled before reaching 20 m/s?

3-9. An automobile traveling at 60 m/s can decelerate at 5 m/s². (a) How long does it take to come to rest? (b) How far has it traveled, assuming the acceleration to remain constant?

3-10. The following relations give the displacement x of a body as a function of the time t; the other symbols represent constants. Find the velocity and acceleration in each instance as a function of time by taking derivatives, and give the appropriate dimensions of the constants: (a) $x = At^4$, (b) $x = B \sin Ct^2$, (c) $x = De^{-kt}$, (d) $x = Et + Ft^2 + Gt^3$.

3-11. The following relations give the acceleration a of a body as a function of the time t; the other symbols represent constants. Assume $x_0 = 0$ and $v = v_0$ when $t = 0$. Find the velocity and displacement in each instance as a function of time by integrating and give the appropriate dimensions of the constants: (a) $a = -At^2$, (b) $a = B \cos Ct$, (c) $a = D + Et^3$.

3-12. Draw approximate velocity-time and acceleration-time graphs corresponding to the motions portrayed in the displacement-time graphs of Fig. P3-12.

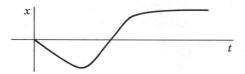

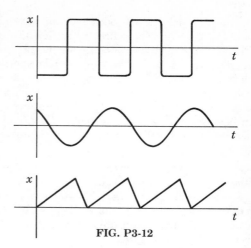

FIG. P3-12

3-13. Draw approximate displacement-time and acceleration-time graphs corresponding to the motions portrayed in the velocity-time graphs of Fig. P3-13.

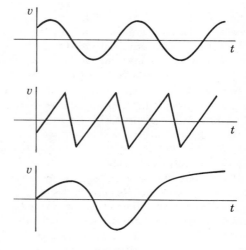

FIG. P3-13

3-14. Figure P3-14 shows the acceleration plotted as a function of time. Sketch the (a) velocity and (b) displacement as functions of time, assuming $x_0 = 0$ and $v_0 = 0$.

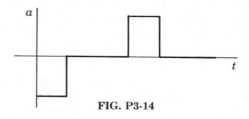

FIG. P3-14

3-15. Two automobiles are both moving at 90 km/h in the same direction, one directly behind the other. The driver of the lead car suddenly applies his brakes, decelerating at 7.5 m/s². The other driver applies his own brakes after a delay of 0.40 s, and his car slows down at a rate of 6.0 m/s². (His tires are worn.) If there is to be no collision, what is the minimum separation between the cars at the instant the lead car's brakes are applied? Show a plot of the square of the speed vs. distance for each vehicle.

3-16. A ball moving vertically strikes the floor at a speed of 15 m/s and rebounds at 9 m/s. The duration of its contact with the floor is 1.0×10^{-2} s. Assuming its acceleration to be constant while it is in contact with the floor, what is the direction and magnitude of this acceleration of the ball?

3-17. A truck traveling at 120 km/h is approaching an automobile moving in the same direction at 90 km/h. (a) Find the minimum separation between the vehicles at the instant the truck's brakes are applied if there is no collision, assuming that the truck can slow down at the rate of 6.0 m/s² and that the automobile's velocity does not change. (b) Find the minimum separation at the instant that the truck driver perceives the situation if his reaction time is 0.30 s. (c) Plot a graph showing the facts in this situation.

3-18. Suppose that an antiaircraft gun fires a shell vertically upward with a muzzle velocity of 1000 m/s. (a) By what distance must the gunner "lead" a jet plane moving horizontally at 2400 km/h at an altitude of 500 m? (b) What is the angle between the line of sight and the vertical at the instant of firing?

3-19. Show that the following assertion of Galileo, taken from *Two New Sciences* (1683), is correct: ". . . so far as I know no one has yet pointed out that the distances traversed during equal intervals of time, by a body falling from rest, stand to one another in the same ratio as the odd numbers beginning with unity."

3-20. It is found by experiment that the time taken for an electron starting from rest at the left plate, Fig. 3-11, to reach the right plate varies inversely as the square root of the number of batteries (connected in series) attached to the parallel metal plates. How does the acceleration of an electron in the region between the plates depend upon the number of batteries?

3-21. A girl throws a ball upward. It strikes the ground 3.0 s later. (a) With what speed was the ball thrown? (b) What height did it achieve?

3-22. Ball A is thrown vertically upward with a speed of 15 m/s. Two seconds later ball B is projected upward with a velocity of 30 m/s. When and where do the two balls collide? Assume $g = 10$ m/s².

3-23. A man drops a baseball from the window of a tall building and then 1.00 s later throws a second baseball vertically downward with an initial speed of 13.0 m/s. (a) How long after the second ball is thrown does it pass the first? (b) How far down? (Assume that the balls do not collide.)

3-24. A car starts from rest with a constant acceleration of 1.20 m/s² at the moment a traffic light turns green. At this same moment a truck is traveling at the constant speed of 30 m/s but is at a distance of 45 m from the traffic light. (a) Will the car and truck pass? (b) If so, where and when? (c) Draw a displacement-time graph for both vehicles on the same diagram.

3-25. An elevator 3.0 m from floor to ceiling ascends at the constant speed of 1.5 m/s. An object becomes detached from the ceiling and falls to the elevator floor. How long does it take for the object to strike the floor? (Work this problem both from the point of view of an observer at rest in the building and from the point of view of an observer traveling in the elevator.)

3-26. A balloon ascends at 8 m/s. A sandbag is dropped when the balloon is 40 m above the ground. (a) How long does it take for the sandbag to strike the ground? (b) What is its velocity then?

3-27. A number of stones are attached to a string in such a way that the distance between adjacent stones increases by a constant amount from one pair of stones to the next. The string is held at the large-spacing end, the lowest stone touching the ground. Show that if the string is dropped from rest, the stones strike the ground at equally spaced time intervals.

3-28. An apartment dweller sees a flowerpot (originally on a windowsill above) pass the 2.0-m-high window of his fifth-floor apartment in 0.12 s. The distance between floors is 4.0 m. From what floor did the flowerpot fall?

CHAPTER 4

Kinematics in Two Dimensions

In this chapter we consider the kinematics of a particle moving in a plane. We shall find that velocity and acceleration are vector quantities. The special cases of motion with constant acceleration and uniform circular motion are treated analytically.

4–1 VELOCITY AND ACCELERATION AS VECTOR QUANTITIES

Consider the simulated multiflash photograph, Fig. 4-1. The position of the small object relative to some arbitrary origin, here chosen as the starting point a, is specified by the displacement vector $\mathbf{r}$. If we designate the displacement from point g to point k by $\Delta \mathbf{r}$, then as seen from Fig. 4-1,

$$\mathbf{r}_g + \Delta \mathbf{r} = \mathbf{r}_k$$

(Note that $\Delta \mathbf{r}$ does *not* depend on the origin chosen for the vectors $\mathbf{r}_g$ and $\mathbf{r}_k$.) The time interval elapsing between the appearance of the body at g, and later at k, is Δt. Then the average velocity $\bar{\mathbf{v}}_{gk}$ over the interval Δt is $\bar{\mathbf{v}}_{gk} = (\mathbf{r}_k - \mathbf{r}_g)/\Delta t$, the average velocity in general being defined as the displacement $\Delta \mathbf{r}$ (a vector) divided by the elapsed time Δt (a scalar):

$$\bar{\mathbf{v}} = \frac{\Delta \mathbf{r}}{\Delta t} \qquad (4\text{-}1)$$

The average velocity is, of course, a vector quantity, inasmuch as it represents the vector quantity $\Delta \mathbf{r}$ multiplied by the scalar quantity $1/\Delta t$. Because Δt is always positive, the direction

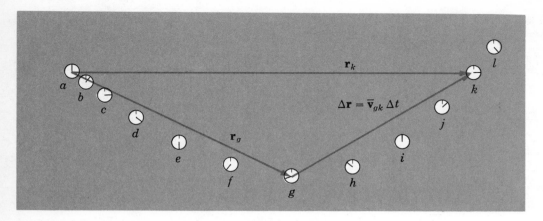

FIG. 4-1. A simulated multiflash photograph. The average velocity from g to k is $\bar{\mathbf{v}}_{gk}$. The successive disks represent the face of a clock in which the hand is seen rotating clockwise at a constant rate.

of the vector $\bar{\mathbf{v}}$ is the same as the direction of the vector $\Delta \mathbf{r}$. The motion from g to k in Fig. 4-1 is clearly not one of constant velocity, but the average velocity from g to k gives that *constant velocity* which the particle would have if it were to undergo the same displacement in the same time interval as it underwent during its actual displacement from g to k.

The instantaneous velocity $\mathbf{v}$, the time rate of change of displacement, is the limit of the average velocity as the time interval becomes indefinitely small:

$$\mathbf{v} = \lim_{\Delta t \to 0} \frac{\Delta \mathbf{r}}{\Delta t} = \frac{d\mathbf{r}}{dt} \qquad (4\text{-}2)$$

The time interval is sufficiently small to give the instantaneous velocity when the straight-line displacement $\Delta \mathbf{r}$ cannot be distinguished from the actual path of the particle over the interval Δt. Therefore, the instantaneous velocity of a particle is always tangent to the particle's path, pointing in the direction of the motion. It follows from (4-2) that, if Δt is small enough, the velocity $\mathbf{v}$ can be regarded as constant in direction and magnitude over this small time interval, and the incremental displacement $\Delta \mathbf{r}$ will be $\mathbf{v}\, \Delta t$. See Fig. 4-2. The dimensions of velocity are, of course, length divided by time.

Vectors showing the instantaneous velocity at several points along the path are shown in Fig. 4-3a. Note that the velocity vectors are all tangent to the path; their lengths represent (on an arbitrarily chosen scale for velocity) the distance covered per unit time over infinitesimal time intervals. The magnitude of the instantaneous velocity is called the instantaneous speed.

Note further that a diagram such as that of Fig. 4-3a, in which velocity vectors are shown superimposed on the path of the particle, is a combination, as it were, of two diagrams: one giving the spatial location of the particle (according to an arbitrarily

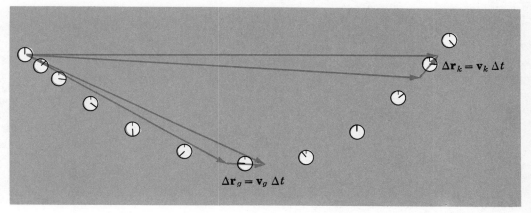

FIG. 4-2. The instantaneous velocities at g and k are v_g and v_k, respectively.

chosen scale of distances) and the second showing the instantaneous velocity vectors (with a different scale). For convenience each velocity vector is located with its tail at the position of the particle at that instant. Since we may relocate any vector without changing its value, the velocity vectors may be equally well displayed as in Fig. 4-3b, with all tails at the same origin.

FIG. 4-3. (a) Instantaneous velocity vectors at several points. The tails of the velocity vectors are here placed at the corresponding positions of the moving particle. (b) The instantaneous velocity vectors of (a) drawn from a common origin (with twice the scale).

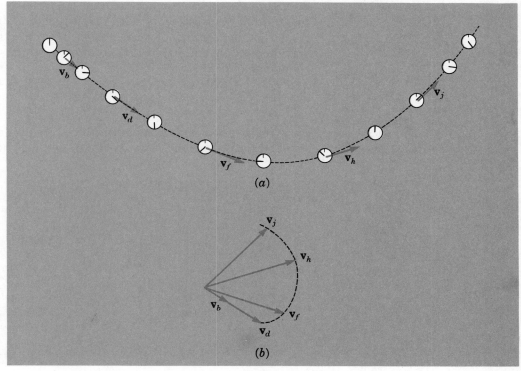

Velocity and Acceleration as Vector Quantities   SEC. 4-1

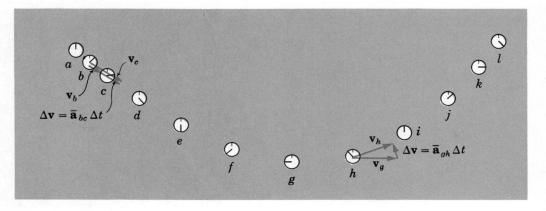

FIG. 4-4. Average acceleration vectors $\bar{a}_{bc}$ and $\bar{a}_{gh}$.

The average acceleration $\bar{\mathbf{a}}$ is defined as the change in (instantaneous) velocity per unit time:

$$\bar{\mathbf{a}} = \frac{\Delta \mathbf{v}}{\Delta t} \tag{4-3}$$

The average acceleration is a vector by virtue of being the product of the vector $\Delta \mathbf{v}$ and a scalar $1/\Delta t$. Its dimensions are those of velocity divided by time. Equation (4-3) shows that the change in the velocity $\Delta \mathbf{v}$ over the interval Δt is $\bar{\mathbf{a}} \, \Delta t$. See Fig. 4-4. Note that the direction of $\Delta \mathbf{v}$, and hence of $\bar{\mathbf{a}}$, is *not* necessarily along $\mathbf{v}$, the direction of motion. An acceleration occurs if the velocity changes magnitude, or direction, or both magnitude and direction.

The instantaneous acceleration is simply the average acceleration over an infinitesimal time interval, that is, the time rate of change of velocity:

$$\mathbf{a} = \lim_{t \to 0} \frac{\Delta \mathbf{v}}{\Delta t} = \frac{d\mathbf{v}}{dt} \tag{4-4}$$

FIG. 4-5. Instantaneous acceleration vectors at several points. The tails of the acceleration vectors are here placed at the corresponding positions of the moving particle.

Figure 4-5 shows instantaneous acceleration vectors superimposed on a diagram of the particle's path. The scale of the ac-

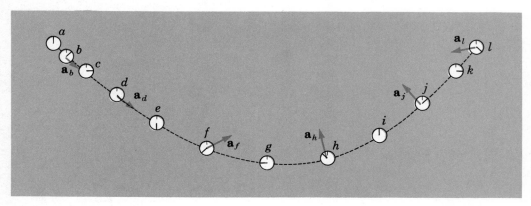

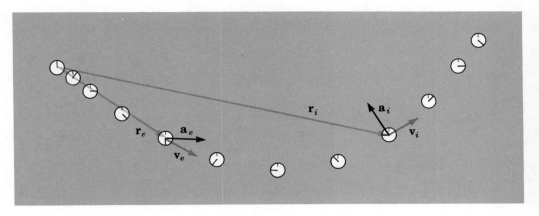

FIG. 4-6. Displacement, instantaneous velocity, and instantaneous acceleration vectors for different positions of a moving particle.

celeration vectors is arbitrary (but different from the scales for displacement and velocity). At the start of the motion (*a, b, c*) the acceleration is constant in direction and magnitude and lies along the direction of the motion. In the vicinity of points *g, h, i,* and *j,* where the body moves with constant speed but changes direction of motion, the acceleration vectors are at right angles to the path. At such points as *f* and *l* the acceleration is neither along nor perpendicular to the path. In every case the instantaneous acceleration vector gives the magnitude and direction of the *change* in the *velocity vector* per unit time.

It is often convenient to show different types of vectors on the same diagram. Figure 4-6 shows displacement, instantaneous velocity, and instantaneous acceleration vectors for two points along the path. It must be emphasized that displacements may properly be added to displacements, velocities to velocities, and accelerations to accelerations. We cannot, however, add together unlike vector quantities, such as displacement and velocity.

4–2 MOTION AT CONSTANT ACCELERATION

An important type of motion is that in which the acceleration is constant both in magnitude and in direction. Although special, this situation is realized in two important physical situations: a small object thrown at a moderate speed near the earth's surface and an electrically charged particle moving between oppositely charged parallel plates (a particle with gravitational mass in a uniform gravitational field and a particle with electric charge in a uniform electric field, respectively). We recall that in one-dimensional kinematics the displacement x and velocity v of a particle having an initial displacement x_0 and velocity v_0 and a constant acceleration a are given in terms of the time t by

$$v = v_0 + at$$
$$x = x_0 + v_0 t + \tfrac{1}{2}at^2$$

The corresponding relations in two (or three) dimensions are easily shown to be

For constant **a**:
$$\mathbf{v} = \mathbf{v}_0 + \mathbf{a}t \qquad (4\text{-}5)$$
$$\mathbf{r} = \mathbf{r}_0 + \mathbf{v}_0 t + \tfrac{1}{2}\mathbf{a}t^2 \qquad (4\text{-}6)$$

Here $\mathbf{r}_0$ and $\mathbf{r}$ are the *displacements* relative to the origin, $\mathbf{v}_0$ and $\mathbf{v}$ are the instantaneous vector velocities at the times 0 and t, respectively, and **a** is the constant vector acceleration. It is important to recognize that **r** does *not* represent the distance traveled by the particle.

Equation (4-5) follows directly from the definition of average acceleration; for if the acceleration is constant, the instantaneous acceleration **a** is identical with the average acceleration **ā**. Then the (vector) difference in velocity $\Delta \mathbf{v}$ is $\mathbf{v} - \mathbf{v}_0$, and the time interval Δt is the time elapsed after $t = 0$. Therefore, (4-3) becomes

$$\bar{\mathbf{a}} = \mathbf{a} = \frac{\Delta \mathbf{v}}{\Delta t} = \frac{\mathbf{v} - \mathbf{v}_0}{t}$$

$$\mathbf{v} = \mathbf{v}_0 + \mathbf{a}t \qquad (4\text{-}5)$$

Equation (4-6) is also easily arrived at. For simplicity we take $\mathbf{r}_0 = 0$. Suppose that the acceleration is zero; then the displacement **r** is along the direction of the constant velocity $\mathbf{v}_0$ and $\mathbf{r} = \mathbf{v}_0 t$. Now suppose that the initial velocity is zero, but not the acceleration; the displacement along some line is given by $\tfrac{1}{2}at^2$, where the acceleration a is along the line. This relation may be written in vector form as $\mathbf{r} = \tfrac{1}{2}\mathbf{a}t^2$, where the vectors **r** and **a** are now along the same direction. Inasmuch as displacements follow the rules of the vector algebra, the displacement **r** will, in general, be equal to the vector sum of the displacement $\mathbf{v}_0 t$ at constant velocity and the displacement $\tfrac{1}{2}\mathbf{a}t^2$ arising from a uniformly changing velocity:

$$\mathbf{r} = \mathbf{v}_0 t + \tfrac{1}{2}\mathbf{a}t^2 \qquad \text{with } \mathbf{r}_0 = 0 \qquad (4\text{-}6)$$

The vector equations (4-5) and (4-6) are illustrated in Fig. 4-7. Note that the displacement **r** is not necessarily along the direction of $\mathbf{v}_0$, **v**, or **a**.

If the origin of the vector **r** is chosen as the origin of the X and Y axes, we may write the vector equations (4-5) and (4-6) equivalently in scalar form by using the components of the displacement, velocity, and acceleration along the X and Y directions, as shown in Fig. 4-8:

$$\begin{aligned} v_x &= v_{0x} + a_x t \\ v_y &= v_{0y} + a_y t \end{aligned} \qquad (4\text{-}7)$$

$$\begin{aligned} x &= v_{0x} t + \tfrac{1}{2}a_x t^2 \\ y &= v_{0y} t + \tfrac{1}{2}a_y t^2 \end{aligned} \qquad (4\text{-}8)$$

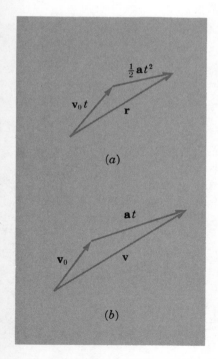

FIG. 4-7. Vector relations giving *(a)* the displacement and *(b)* the velocity for motion at constant acceleration.

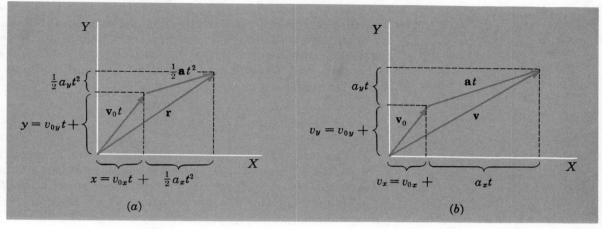

FIG. 4-8. Scalar kinematic relations giving the rectangular components of (a) the displacement (b) the velocity, as derived from the vector kinematic relations.

The motion of a particle at constant acceleration in a plane is described completely by two independent and simultaneous motions along the X and Y axes at the constant accelerations a_x and a_y, respectively.

The strategy for solving kinematic problems in two dimensions with constant acceleration is this: We resolve the vector displacements, velocities, and accelerations into components along the X and Y axes and use Eqs. (4-7) and (4-8) for the components of the velocities and displacements along X and Y. The link between these equations is the common time t.

4-3 PROJECTILE MOTION

Suppose that a projectile is thrown horizontally and falls under the influence of gravity. Figure 4-9 shows the displacement and velocity vector diagrams (superimposed on the X and Y axes and

FIG. 4-9. A projectile thrown horizontally: (a) the displacement vectors and (b) the velocity vectors.

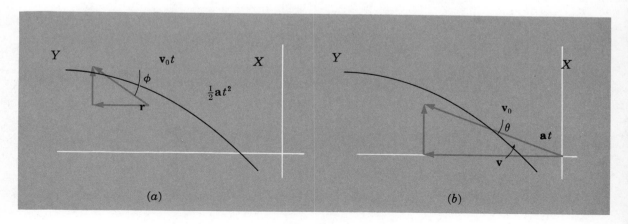

Projectile Motion   SEC. 4-3   39

the trajectory of the particle). For this simple case $\mathbf{v}_0$ is horizontal and $\mathbf{a}$ is vertically downward. The velocity is initially along the X direction; thus, $v_{0y} = 0$. In addition, the acceleration is in the vertical direction, and $a_x = 0$. Choosing the positive Y axis as upward, we write a_y as $-g$, where g is 9.8 m/s². Equations (4-7) and (4-8) become

$$v_x = v_{0x} \qquad v_y = -gt$$
$$x = v_{0x}t \qquad y = -\tfrac{1}{2}gt^2$$

The body coasts along the horizontal while it accelerates from rest at a constant rate along the vertical. By eliminating t between the equations for x and y, we can find the equation of the path, giving y in terms of x and the constants v_{0x} and g:

$$y = -\frac{g}{2v_{0x}^2} x^2 \qquad (4\text{-}9)$$

The path is a parabola whose axis of symmetry is parallel to the direction of the constant acceleration; the instantaneous velocity vector $\mathbf{v}$ is tangent to the path at all points, as shown in Fig. 4-9b.

Figure 4-10 shows the motion together with the projections of the motion along the X and Y axes for equal time intervals. Because the vertical and horizontal components of the motion are independent, the *vertical* motion of an object thrown horizontally is identical with that of an object dropped from rest. If one throws an object horizontally and simultaneously drops a second body from rest, the two objects strike a horizontal plane at the same instant. The thrown object strikes the surface with a larger velocity (it has both horizontal and vertical velocity components), but the *vertical components* of the horizontally thrown and the dropped objects are identical at any given time.

The more general case, in which the initial velocity $\mathbf{v}_0$ is not at right angles to the constant acceleration $\mathbf{a}$, is shown in Fig. 4-11. The horizontal velocity component, which is at right angles to the

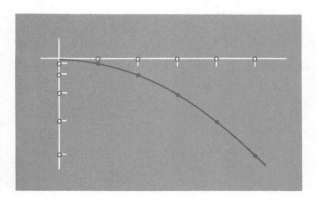

FIG. 4-10. Simulated multiflash photograph corresponding to Fig. 4-9.

acceleration, remains unchanged, while the vertical component of **v** changes steadily with time.

Figure 4-11 illustrates projectile motion over a range small compared with the earth's radius; the acceleration is constant and has a magnitude g. When a body is thrown obliquely into the air, it changes from its otherwise straight-line motion because of the acceleration due to gravity. Consider a bullet shot from a gun aimed directly at a distant object; this object is held above the ground and released from rest at the instant the bullet is fired. If one could, so to speak, "turn off gravity," the bullet would travel a straight-line path, the object would not fall when released, and the bullet would hit the object. Now suppose that bullet and object fall, as do all objects close to the earth, with the same constant vertical acceleration. The object falls a distance of $\frac{1}{2}gt^2$ from rest, and during this same time the bullet "falls" the same vertical distance $\frac{1}{2}gt^2$ from its straight-line motion. Consequently, the bullet always strikes the target, quite apart from the magnitude of the initial velocity, or even of the magnitude of g.

For an object having a constant acceleration, equations giving the X and Y components of displacement and velocity as a function of time are easily written from Eqs. (4-7) and (4-8) as follows:

$$v_x = v_{0x} \qquad v_y = v_{0y} - gt$$
$$x = v_{0x}t \qquad y = v_{0y}t - \tfrac{1}{2}gt^2 \tag{4-10}$$

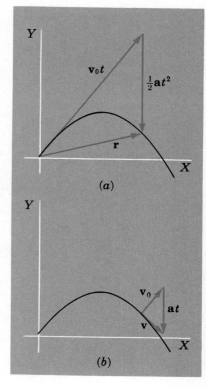

FIG. 4-11. A projectile thrown obliquely: *(a)* the displacement vectors and *(b)* the velocity vectors.

Example 4-1

A small object is thrown with a speed of 25 m/s at an angle of 53° above the horizontal. Ignoring air resistance, find *(a)* the maximum altitude achieved by the projectile, *(b)* the horizontal range (the horizontal distance traversed when the projectile is again at its initial elevation), and *(c)* the displacement and velocity of the projectile 3 s after firing.

The components of **v**₀ are:

$$v_{0x} = v_0 \cos \theta_0 = (25 \text{ m/s}) \cos 53° = 15 \text{ m/s}$$
$$v_{0y} = v_0 \sin \theta_0 = (25 \text{ m/s}) \sin 53° = 20 \text{ m/s}$$

The equations of motion are then (where we have taken $g = 10$ m/s²):

$$v_x = 15 \text{ m/s}$$
$$v_y = (20 \text{ m/s}) - (10 \text{ m/s}^2)t$$
$$x = (15 \text{ m/s})t$$
$$y = (20 \text{ m/s})t - (5 \text{ m/s}^2)t^2$$

These give the object's position and velocity as functions of the time t. All questions concerning the motion can be answered by applying these equations.

(a) We find the maximum height y_m by first finding the time

t_m at which the projectile reaches its highest point. At maximum altitude the projectile is momentarily traveling horizontally, and $v_y = 0$. Thus, with $v_y = 0$, we have

or
$$v_y = 0 = (20 \text{ m/s}) - (10 \text{ m/s}^2)t_m$$
$$t_m = 2.0 \text{ s}$$
Then
$$y_m = (20 \text{ m/s})(2.0 \text{ s}) - (5 \text{ m/s}^2)(2.0 \text{ s})^2$$
$$y_m = 20 \text{ m}$$

(b) To compute the horizontal range, we first find the time at which the projectile is again on the X axis; that is, we compute t for $y = 0$ and substitute this time of flight into the equation for x:

$$y = 0 = (20 \text{ m/s})t - (5 \text{ m/s}^2)t^2$$
$$t = 0 \quad \text{and} \quad 4 \text{ s}$$

The object was at $y = 0$ both at $t = 0$ and at $t = 4$ s, the time of flight. The horizontal range is then

Range: $\quad x = (15 \text{ m/s})t = (15 \text{ m/s})(4.0 \text{ s}) = 60 \text{ m}$

(c) We find the displacement and velocity components of the object after it has been in flight for 3 s.

$$v_x = 15 \text{ m/s}$$
$$v_y = (20 \text{ m/s}) - (10 \text{ m/s}^2)(3 \text{ s}) = -10 \text{ m/s}$$

The component displacements at $t = 3$ s are

$$x = (15 \text{ m/s})(3 \text{ s}) = 45 \text{ m}$$
$$y = (20 \text{ m/s})(3 \text{ s}) - (5 \text{ m/s}^2)(3 \text{ s})^2 = 15 \text{ m}$$

4–4 UNIFORM CIRCULAR MOTION

A particle moving in a circular arc at constant speed is said to be in uniform circular motion. It covers equal distances along the circumference in equal times, while continuously changing its direction of motion. The velocity, although constant in magnitude, continuously changes in direction, and there is, consequently, an acceleration. We wish to find the magnitude and direction of this acceleration.

It is convenient (but not necessary) to indicate the location of a particle traveling in a circle by means of a displacement vector, or *radius vector,* **r,** whose tail is at the center of the circle. Then, if the speed of the body is constant, the radius vector sweeps through equal angles in equal times. Consider the small displacement $\Delta \mathbf{r}$ occurring in a very small time interval Δt. The vector displacement $\Delta \mathbf{r}$ cannot be distinguished from the true path of the particle along a short segment of circular arc; see Fig. 4-12. The radius vectors $\mathbf{r}_i$ and $\mathbf{r}_f$, both of magnitude r, give the initial and final displacements over the small interval Δt. As the figure shows,

$$\mathbf{r}_f - \mathbf{r}_i = \Delta \mathbf{r} = \mathbf{v} \, \Delta t$$

where **v** is the velocity over the interval Δt.

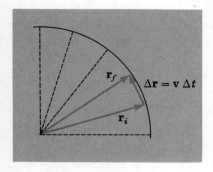

FIG. 4-12. The displacement $\Delta \mathbf{r}$ over the time interval Δt of a particle moving in uniform circular motion.

The velocity vector is always at right angles to the radius vector. As **r** rotates, so does **v**. See Fig. 4-13a, where velocity vectors, all of the same magnitude, are shown with their tails on the circumference of the circle, and Fig. 4-13b, where the same velocity vectors are displayed with their tails at a common point.

The initial and final velocities, $\mathbf{v}_i$ and $\mathbf{v}_f$, both of magnitude v, differ only in direction. Their vector difference $\Delta \mathbf{v}$ is given by

$$\mathbf{v}_f - \mathbf{v}_i = \Delta \mathbf{v} = \mathbf{a}\, \Delta t$$

where **a** is the acceleration in the time interval Δt during which the velocity has changed from $\mathbf{v}_i$ to $\mathbf{v}_f$. Now, the radius vectors $\mathbf{r}_i$, $\mathbf{r}_f$, and $\Delta \mathbf{r}$ form an isosceles triangle which is similar to the triangle formed by the corresponding velocity vectors $\mathbf{v}_i$, $\mathbf{v}_f$, and $\Delta \mathbf{v}$. This is so because the displacement vector sweeps through the same angle as does the velocity vector during the time Δt. Consequently, the magnitudes of corresponding sides are in the same ratio:

$$\frac{\Delta r}{r} = \frac{\Delta v}{v}$$

or

$$\frac{v\, \Delta t}{r} = \frac{a\, \Delta t}{v}$$

Thus,

$$a = \frac{v^2}{r} \tag{4-11}$$

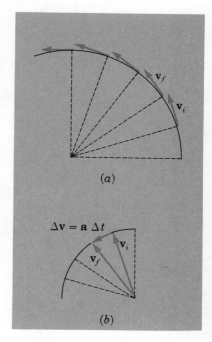

FIG. 4-13. (a) Instantaneous velocity vectors arranged with their tails at the corresponding positions of the particle. (b) The same velocity vectors (scale doubled) arranged with their tails at a common origin.

The acceleration of a particle moving at constant speed v in a circle of radius r has a constant magnitude of v^2/r. Now let us find the direction of the acceleration.

Vector $\Delta \mathbf{r} = \mathbf{v}\, \Delta t$ is perpendicular to **r**, and vector $\Delta \mathbf{v} = \mathbf{a}\, \Delta t$ is perpendicular to **v**, as seen from Figs. 4-12 and 4-13. Therefore, the acceleration has a direction always opposite to that of the displacement **r**; that is, it is directed toward the center of the circle. Uniform circular motion is, then, characterized by an acceleration that is constant in magnitude, changing in direction, and always radially inward. See Figs. 4-14 and 4-15. The term *radial*, or sometimes *centripetal*, *acceleration* is used to designate such an acceleration.

The converse is true. When a body moves with an acceleration which is constant in magnitude and always perpendicular to the direction of motion, the body executes circular motion at constant speed. We may write a vector equation for the centripetal acceleration as follows:

$$\mathbf{a} = -\left(\frac{v}{r}\right)^2 \mathbf{r} = -\frac{v^2}{r}\left(\frac{\mathbf{r}}{r}\right) \tag{4-12}$$

The minus sign indicates that the direction of **a** is opposite to that of **r**. The magnitude of **a** is $(v/r)^2$ multiplied by the magnitude of **r**; thus, $a = (v/r)^2 r = v^2/r$, in agreement with Eq. (4-11).

When a particle moves in a straight line, its velocity direction is constantly along the path, and changes in speed occur by

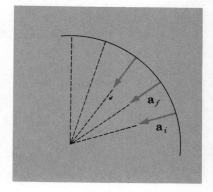

FIG. 4-14. Instantaneous acceleration vectors arranged with their tails at the corresponding positions of the particle.

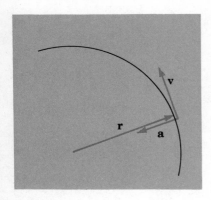

FIG. 4-15. Instantaneous velocity and acceleration vectors for a particle in uniform circular motion.

virtue of an acceleration which also must always lie along the path. On the other hand, when a particle moves in a circular arc at constant speed, its velocity direction changes continuously, and the acceleration now points at right angles to the path, radially inward toward the center of curvature. Thus, an acceleration *along* the path is identified with a change in the *magnitude* of the velocity, while an acceleration *perpendicular* to the path is identified with a change in the *direction* of the velocity.

Example 4-2

Suppose that an object is projected horizontally from the peak of an imaginary high mountain extending well above most of the earth's atmosphere, but still within an approximate range so that the acceleration due to gravity can be taken as 9.8 m/s² (Fig. 4-16). Its path is a parabola, and it strikes the earth after traveling for much less than several hundred miles.

If the object is thrown with a higher speed, so that it travels a distance comparable to the earth's radius before striking the earth, the acceleration can *not* be regarded as constant in direction and the path is not a parabola. On the other hand, if the initial speed is enormous, the object flies out into space. But there is one particular speed at which an object will always fall the same distance toward the earth that it curves away from its straight-line motion; which is to say, there is one particular speed for which the object will travel in a circle. The object becomes an earth satellite. The argument developed here was first put forth by Sir Isaac Newton (1642–1727).

Let us compute the speed and the period (the time for one complete revolution) of a satellite in circular orbit close to the earth's surface (but outside of the atmosphere). The earth's mean radius is 6.37×10^6 m.

Since the radial acceleration has the magnitude 9.8 m/s², we have from Eq. (4-11)

$$a = \frac{v^2}{r}$$

$$v = \sqrt{ar} = \sqrt{(9.8 \text{ m/s}^2)(6.37 \times 10^6 \text{ m})}$$
$$= 7.9 \times 10^3 \text{ m/s}$$

The time T for one revolution is the circumference divided by the speed:

$$T = \frac{2\pi(6.37 \times 10^6 \text{ m})}{7.9 \times 10^3 \text{ m/s}} = 5.1 \times 10^3 \text{ s} = 85 \text{ min}$$

We are now in a position to understand the simpler kinematical aspects of satellite launching and recovery. After firing, the rocket first follows a nearly parabolic path. At its highest point, when the velocity is parallel to the earth's surface, a booster rocket is fired to increase the speed of the released satellite to the value required for orbital motion. If the satellite is well above the earth's atmosphere, so that atmospheric drag is indeed negligible, the speed

FIG. 4-16. An object is projected horizontally from the peak of a high mountain at several different speeds.

remains constant and the satellite orbits indefinitely; on the other hand, if the drag of the atmosphere is appreciable, it spirals inward.

4-5 ANGULAR SPEED

In Fig. 4-17 the angle $\Delta\theta$ subtends an arc whose length is Δs. By definition, the angle in radians (2π rad = $360°$) is

$$\Delta\theta = \frac{\Delta s}{r}$$

or
$$\Delta s = r\,\Delta\theta \qquad (4\text{-}13)$$

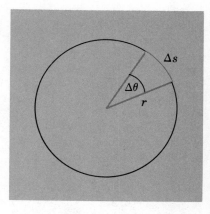

FIG. 4-17. Angle $\Delta\theta$ in radians corresponds to a distance Δs along the circumference for a circle of radius r.

Suppose that a particle travels the distance Δs along the arc in a time Δt while the radius vector of length r sweeps through the angle $\Delta\theta$. The particle's linear speed is

$$v = \frac{\Delta s}{\Delta t}$$

and the *angular speed* ω of the rotating radius vector is, by definition,

$$\omega = \frac{\Delta\theta}{\Delta t} \qquad (4\text{-}14)$$

Angular speed is the time rate of change of angular displacement. Appropriate units for angular speed are radians per second. Thus, one rotation per second is $\omega = 2\pi$ rad/s, or simply 2π s^{-1}.

From (4-13) and (4-14) it follows that

$$\frac{\Delta s}{\Delta t} = r\frac{\Delta\theta}{\Delta t}$$

or
$$v = r\omega \qquad (4\text{-}15)$$

Equation (4-15) shows that the *tangential* speed v of any particle moving in a circular arc a distance r from a center of rotation is given by $r\omega$, where ω is the rate at which the radius vector from the center to the particle turns through angles. See Fig. 4-18.

Using (4-15) we may rewrite the relation for the radial acceleration as follows:

$$a = \frac{v^2}{r} = \frac{(r\omega)^2}{r} = \omega^2 r \qquad (4\text{-}16)$$

A point at a fixed distance r from the center of a circle has a radial acceleration that is proportional to the square of the angular speed. We may write (4-16) in vector form also, recalling that the direction of **a** is opposite to that of the radius vector **r**:

$$\mathbf{a} = -\omega^2 \mathbf{r} \qquad (4\text{-}17)$$

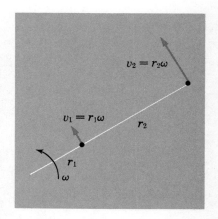

FIG. 4-18. The speeds of points on a rotating object are proportional to their respective distances from the axis of rotation.

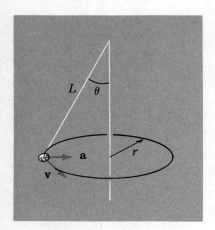

FIG. 4-19. A stone on a string which sweeps out a cone about the vertical at constant speed. Note that the acceleration lies in the plane of the horizontal circle.

Example 4-3
A stone is attached to the lower end of a string 1.0 m long and swung in a horizontal circle, as shown in Fig. 4-19. When the string makes a constant angle of 30° with the vertical, it is found that the stone executes 32 revolutions in 1 min. What is the direction and magnitude of the radial acceleration of the stone?

The angular speed is

$$\omega = (32 \text{ rev/min})\left(\frac{1 \text{ min}}{60 \text{ s}}\right)\left(\frac{2\pi \text{ rad}}{1 \text{ rev}}\right) = 3.35 \text{ rad/s}$$

The radius of the stone's circular path is (1.0 m) sin 30°, and the centripetal acceleration has the magnitude

$$a = \omega^2 r = (3.35 \text{ s}^{-1})^2 (1.0 \text{ m}) \sin 30° = 5.6 \text{ m/s}^2$$

The radial acceleration is in a horizontal plane and toward the center of the horizontal circle.

SUMMARY

The instantaneous velocity **v** is the time rate of change of the vector displacement:

$$\mathbf{v} = \lim_{\Delta t \to 0} \frac{\Delta \mathbf{r}}{\Delta t} = \frac{d\mathbf{r}}{dt} \quad (4\text{-}2)$$

The instantaneous acceleration **a** is the time rate of change of the vector velocity:

$$\mathbf{a} = \lim_{\Delta t \to 0} \frac{\Delta \mathbf{v}}{\Delta t} = \frac{d\mathbf{v}}{dt} \quad (4\text{-}4)$$

For the constant acceleration **a**, the velocity and displacement are given by

$$\mathbf{v} = \mathbf{v}_0 + \mathbf{a}t \quad (4\text{-}5)$$

$$\mathbf{r} = \mathbf{r}_0 + \mathbf{v}_0 t + \tfrac{1}{2}\mathbf{a}t^2 \quad (4\text{-}6)$$

where **r** is the displacement and **v** is the velocity at time t, **r** being $\mathbf{r}_0$ and **v** being $\mathbf{v}_0$ at $t = 0$.

A particle moving in a circular arc of radius of curvature r at speed v has a radially inward acceleration

$$a = \frac{v^2}{r} = \omega^2 r \quad (4\text{-}11), (4\text{-}16)$$

where ω is the angular speed of the radius vector in radians per unit time.

PROBLEMS

4-1. The following are the space and time coordinates (x, y, t) of a particle in meters and seconds, respectively: (3, 0, 0), (4, −2, 1), (6, 3, 2), and (4, 6, 3). What are the X and Y components of the particle's average velocity over the following time intervals: (a) Δt_{01} (from $t = 0$ to $t = 1$ s), (b) Δt_{12}, (c) Δt_{23}, (d) Δt_{03}, and (e) Δt_{13}?

4-2. The equations of motion of a particle are $x = +10t - 20 t^2$ and $y = -15t + 30t^2$, where x and y are in meters and t is in seconds. What are the magnitudes and directions of (a) the initial velocity and (b) the acceleration? (c) Sketch roughly the path of the particle.

4-3. In Fig. 4-1, the body is imagined to have started at point a and moved to b, c, etc. Suppose, now, that the motion is reversed (or that time is reversed) and the body starts at l, moves to k, j, etc., and comes finally to point a. How do (a) the velocity and (b) the acceleration vectors compare with those shown in Figs. 4-3 and 4-5? The reversal of the velocity direction and the nonreversal of the acceleration direction arise from the fact that the velocity is the *first* time derivative of the displacement and the acceleration is the *second* time derivative of the displacement. (c) How does the time derivative of the acceleration behave under time reversal?

4-4. A ball rolls off the edge of a 2.0-m-high table with a speed of 8.0 m/s. How far horizontally from the edge of the table does the ball strike the floor?

4-5. As neutrons go, one with a speed of 2,200 m/s (a thermal neutron) is a slow neutron. How far must such a neutron travel horizontally to fall vertically 1.0 mm?

4-6. An object is projected with an initial speed v_0 and at an elevation of θ above the horizontal from a point having an elevation h above a horizontal plane. Show that the speed with which the object strikes the horizontal plane is given by $(v_0^2 + 2gh)^{1/2}$, *independent of the angle θ*, whether positive or negative (above or below the horizontal).

4-7. A projectile is thrown with an initial speed v_0 at an angle of θ with respect to a horizontal surface. (a) What is the maximum height the projectile achieves (in terms of v_0, θ, and g)? (b) What is the time of flight for the entire motion? (c) Show that the horizontal range is given by $(v_0^2/g) \sin 2\theta$. (d) Show that any two complementary angles θ (such as 20 and 70°) give the same range (first proved by Galileo). (e) Show that the range is a maximum for $\theta = 45°$.

4-8. In the 1972 Olympics the winning throw in the shotput was for a distance of 21.18 m. What were (a) the initial speed, (b) the maximum height, and (c) the time of flight if the shot was launched at an angle of 45° relative to the vertical? Assume that $g = 9.80$ m/s². Give your results to three significant digits.

4-9. A plane releases a bomb from an altitude of 1,000 m; at the instant of release the plane has a speed of 900 km/h and is in a dive at an angle of 60° with respect to the vertical. (a) How long does it take for the bomb to reach the ground? (b) What are the speed and direction of the bomb at impact? Assume that $g = 10$ m/s².

4-10. (a) At what *two* angles relative to the horizontal must a gun be aimed so that the horizontal range of the projectile is 1,000 m, the muzzle velocity being 140 m/s? (b) Calculate the time of flight for each of the two trajectories in part (a).

4-11. Freely falling bodies on the moon have an acceleration due to gravity of 1.6 m/s². How would (a) the maximum height, (b) the time of flight, and (c) the range of a projectile thrown from the moon's surface compare with the corresponding quantities on earth?

4-12. A baseball hit at 45° above the horizontal just clears a 21.0-ft high wall 300 ft from home plate. Find (a) the initial speed of the ball, (b) the time it needs to reach the wall, and (c) its speed and direction as it clears the wall. Assume that the bat-ball impact takes place 5.00 ft above the ground and $g = 32.0$ ft/s².

4-13. An elevator has a light source at one side that shines a beam of light horizontally across to the opposite wall 3.0 m away. The light source may be thought to emit particles of light which travel at a speed of 3.0×10^8 m/s. At what constant acceleration upward must the elevator move so that the spot of light on the distant wall is displaced downward by 1.0 mm, as viewed by an observer in the elevator?

Actually, an observer in a closed elevator would not be able to distinguish between the upward accelerated motion just described and the effect of some gravitational attraction. He could conclude that the light beam is bent downward because it has weight. The equivalence of accelerated motion and gravitational attraction is the basic assumption of the general theory of relativity; the bending of light when it passes close to the sun, predicted by the general theory, has been observed.

4-14. Consider Fig. P4-14. Electrically charged particles are accelerated horizontally from rest between the two vertical plates. The particles then enter the region between horizontal plates, where they are accelerated vertically. When the particles finally strike the screen, their vertical displacement is y. Show that y is unchanged if the *ratio* of the accelerations between the vertical and horizontal plates, respectively, is not changed. Because the acceleration of a charged particle between a pair of parallel charged plates is directly proportional to the electric charge of the particle, *any* charged particle will strike at the same point y (neglecting free fall due to gravity, of course).

FIG. P4-14

4-15. An electron moving horizontally to the right at a speed v_0 enters the region between two horizontal, electrically charged, metal plates, as shown at the center of Fig. P4-14. The acceleration is constant and vertical in the region between the parallel plates and zero outside this region. Show that the vertical displacement y of electrons on the distant screen (relative to their position on the screen when the acceleration is zero) is directly proportional to the acceleration between the plates. The displacement y is proportional to the acceleration between the plates, which is, in turn, proportional to the potential difference, or voltage, between the plates. Thus, the displacement y is a measure of the applied voltage, and the device may be used as a voltmeter.

4-16. Consider the arrangement shown in Fig. P4-14. Electrons having an initial horizontal velocity of 1.0×10^6 m/s enter the region between the horizontal plates, where the acceleration is 80×10^{13} m/s^2 upward. The plates have a horizontal dimension of 3.0 cm. The screen is 20 cm to the right of the right ends of the plates. (a) Compute the magnitude and direction of the electrons' velocity as they leave the parallel plates. (b) Compute the vertical displacement y of the electron beam at the screen. (c) What is the time of flight from the moment an electron enters the horizontal plates until it strikes the screen?

4-17. The moon's radius is 1.74×10^6 m; the acceleration due to gravity as its surface is 1.6 m/s^2. What is the period of a satellite orbiting close to the moon's surface?

4-18. Suppose that the pilot of a jet plane wished to limit the radial acceleration to $5\,g = 50$ m/s^2. What is the minimum allowable radius of curvature if the pilot wishes to change direction while flying at 600 m/s? (This speed is representative of modern supersonic aircraft.)

4-19. A projectile thrown horizontally will fall a vertical distance $\tfrac{1}{2}gt^2$ in the time t. If the projectile has a sufficiently high speed, it becomes a satellite, and the distance it falls, $\tfrac{1}{2}gt^2$, will then be exactly the same as the distance that the earth curves away from a straight line by virtue of its spherical shape. Show, from the geometry of Fig. P4-19, that the magnitude of the acceleration g is v^2/r, where v is the speed of the object and r is its radius of curvature. This alternative method of deriving the radial acceleration relation was used by Newton.

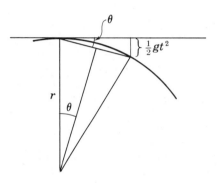

FIG. P4-19

4-20. (a) At what speed must an automobile round a turn having a radius of curvature of 40 m in order that its radial acceleration be equal to g? (b) Suppose that the automobile is traveling at this speed along a straight roadway but over a hill having a radius of curvature of 40 m. What is the behavior of unattached objects within the car?

4-21. A particle moves in a circle of radius r completing one cycle in time T. What is the magnitude of the particle's average acceleration over the time (a) T and (b) $\tfrac{1}{2}T$?

4-22. A wheel rotates at 1.0×10^3 rev/min (revolutions per minute). What are the speed and acceleration of a point (a) 1.0 cm and (b) 2.0 cm from the axis of rotation?

4-23. In the Bohr model of the hydrogen atom an electron is *imagined* to move at a speed of 2.2×10^6 m/s in a circular orbit 5.3×10^{-11} m in radius. (a) What is the radial acceleration of the electron? (b) How many revolutions does the electron complete in 1 s?

4-24. An LP phonograph record turns at an angular speed of $33\frac{1}{3}$ rev/min. The radial distance r from the center of rotation to the stylus ranges from a minimum of 5.0 cm to a maximum of 15.0 cm. *(a)* What is the linear speed of the record surface relative to the stylus for $r = 10.0$ cm, and *(b)* what is the corresponding distance along the groove which represents one vibration of a sound with a frequency of 15 kHz (15,000 vibrations per second)?

4-25. *(a)* What is the angular speed of the earth (in rad/s) due to its daily rotation? What is the instantaneous velocity of a point on the earth's surface *(b)* at the equator and *(c)* at a latitude of $\pm 60°$? The mean radius of the earth is 6.37×10^6 m.

4-26. Because the earth rotates about its axis, objects at rest on the earth's surface have a radial acceleration toward the axis of rotation. Compute the radial acceleration of an object at the latitudes *(a)* 0°, *(b)* 45°, and *(c)* 90° in units of g. The earth's mean radius is 6.37×10^6 m.

4-27. The planets orbit around the sun in nearly circular paths. The mean orbit radii and periods (to two significant figures) of some of the planets are:

	Orbit radius (m)	Period (days)
Mercury	5.8×10^{10}	88
Mars	2.3×10^{11}	690
Jupiter	7.8×10^{11}	4300
Saturn	1.4×10^{12}	11,000

(a) Show that if the acceleration of any planet toward the sun is inversely proportional to the square of its distance R from the sun, the quantity R^3/T^2 has the same value for all planets, where T is the period. *(b)* Show that the data above correspond to an inverse square radial acceleration. *(c)* Using the fact that the earth's mean orbit radius about the sun is 1.5×10^{11} m, compute the period of the earth's rotation and compare it with the known period (1 yr).

4-28. The moon circles the earth once every 27.3 days. Suppose that its average distance from earth is 240,000 mi. *(a)* What is the moon's acceleration relative to the earth? *(b)* Show that the moon's acceleration is smaller than g at the earth's surface by the factor $(60)^2$. (The moon's orbital radius is 60 times the radius of the earth.)

4-29. Show that the acceleration of a particle moving at constant speed must be perpendicular to its instantaneous velocity by using the fact that $d(v^2)/dt = d(\mathbf{v} \cdot \mathbf{v})/dt = 0$.

4-30. A centrifuge is a device for separating materials according to their relative densities. Particles rotated in a high-speed centrifuge achieve extraordinarily high accelerations. What is the centripetal acceleration (in units of g) of a particle in a centrifuge at a distance of 5.0 mm from the axis of rotation and making 10,000 turns per second?

4-31. What is the angular speed (in rev/min) of the 30-in diameter tires on a car traveling 60 mi/h, assuming that the car moves without slipping? (*Hint:* View the spinning wheel as an observer fixed to the car. Then the wheel axle is fixed but the roadway travels backward at the speed of 60 mi/h. If the wheel does not slip on the roadway, the speed of the point of the tire touching the roadway must also be 60 mi/h.)

CHAPTER 5

The Laws of Inertia, Mass Conservation, and Momentum Conservation

In this chapter we begin our study of mechanics, the oldest and most basic branch of physics. Mechanics is the foundation of all physics and engineering, and through the refinements of quantum mechanics and the theory of relativity during this century, mechanics is the basis of our present understanding of the atom, its nucleus, and the elementary constituents thereof.

We shall be concerned with *classical mechanics*. It treats the motions and interactions of ordinary objects—objects which are large compared with the size of the atom (10^{-10} m) and which move at speeds appreciably less than the speed of light (3×10^8 m/s). We go beyond kinematics, which merely describes the motion of objects but does not relate the motion of one to that of another nor does it give the causes of the motion: This is the function of mechanics, or dynamics.

Classical mechanics had its origins in the studies of Galileo Galilei (1564–1642) and, later, of Isaac Newton (1642–1727). Galileo was the first "physicist" in the modern sense of the word. His specific contributions to physics and astronomy were enormous.[†] But perhaps of even greater significance was Galileo's use

---

[†] Some of Galileo's contributions in physics: the recognition of the constant acceleration of all freely falling bodies, the kinematics of projectile motion, including the resolution into independent components, the galilean coordinate and velocity transformations, the independence of a pendulum's period from the mass and amplitude, and the law of inertia. In astronomy: the galilean telescope; observation of the phases of Venus, rings of Saturn, moons of Jupiter, sunspots, earth-shine (on the moon), height of the moon's mountains; the observation of the Milky Way as a collection of discrete stars; and the confirmation of Copernicus' heliocentric model of the universe.

of a strategy and program in attacking problems in physics that remains essentially unchanged and equally successful in present-day physics.

Galileo insisted that we first study simple things by controlled observation (experiment), removing or minimizing nonessential disturbances, and that the results be expressed in the compact and precise language of mathematics. These elements in the strategy of physics are illustrated in the motion of falling bodies. One first studies a falling body in an experiment in which the disturbing and complicating effect of air resistance and friction is minimized. Then the fact that a freely falling body has a constant acceleration is reflected in the experimental result that the displacement of a body dropped from rest is proportional to the square of the time elapsed ($y = \tfrac{1}{2}gt^2$). The same procedure applies when one studies an even simpler situation, the motion of a body on a smooth horizontal plane. Friction, which otherwise obscures and interferes with the motion, must be eliminated or minimized. Considerations of this kind led Galileo to give simple and profound answers to questions that had troubled thinkers for centuries.

What is required to keep a body at rest? *Nothing.* What is required to keep a body in motion? Said Galileo: *Nothing.* This is the essence of the *law of inertia*.

5-1 THE LAW OF INERTIA

A hockey puck slides on a smooth horizontal surface of ice. Friction between the puck and the surface is small, but not entirely negligible. If initially at rest, the puck remains at rest; if set into motion, but left undisturbed, the puck coasts in a straight line for considerable distances before coming finally to rest.

Suppose, now, that a puck is so constructed that it floats on air above a smooth horizontal surface and the effect of friction is thus reduced drastically. A simple type of air-suspended puck, or disk, is shown in Fig. 5-1. When such a disk is set in motion, it coasts for amazingly large distances before coming to rest. We can well imagine that if the disk were completely isolated from external disturbances, it would continue in its motion indefinitely in a straight line at constant speed. Observations of this sort lead us to the generalization: *If an object is isolated from external influence, its velocity is constant, either zero or nonzero. An isolated object continues in uniform motion—along a straight line covering equal*

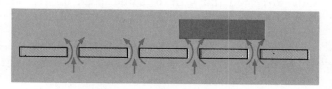

FIG. 5-1. A simple type of air-suspended disk.

FIG. 5-2. The velocity-time graph observed for the motion of a disk floating on a layer of air on a smooth horizontal surface.

distances in equal time intervals (see Fig. 5-2). This is the *law of inertia,* sometimes referred to as the first of Newton's three laws of motion.

We have referred above to an external influence, or force, acting on an object. In Chap. 7 we shall give a precise definition of force. For our present purposes an external force is an influence that can cause a body coasting on a smooth frictionless surface to depart from its motion at constant velocity.

The frame of reference in which an undisturbed object is found to have a constant velocity is called an *inertial frame.* The law of inertia holds when an observer is in an inertial reference frame, but not in a reference frame which accelerates relative to an inertial frame. The earth's surface is an approximate inertial frame—approximate because the earth rotates on its axis and revolves about the sun. The most nearly perfect inertial frame is one at rest with respect to the fixed stars.

In Chap. 6 we shall discuss how velocities are related when observed in two different reference frames in relative motion at constant velocity. Suffice it to say here that, if a particle has a constant velocity when observed in one reference frame, it has a different but still constant velocity when observed in a second reference frame moving at constant velocity relative to the first. Thus, if an observer moves relative to the earth at the same speed and in the same direction as a freely coasting disk, he finds the disk to be at rest. Conversely, in his reference frame such a moving observer finds a disk at rest on earth to be moving with a constant nonzero velocity. The states of uniform motion and of rest are equivalent. From the point of view of the law of inertia, motion is just as "natural" as rest.

Unless indicated specifically otherwise, we shall hereafter assume that all observations will be made from the point of view of an inertial frame. Indeed, the laws of physics ordinarily apply only to inertial frames.

5-2 INERTIAL MASS

Experiments with air-suspended disks lead to the conclusion that any isolated object will, when viewed from an inertial frame, maintain a state of constant velocity. This property of an object is called its inertia. Our task now is to develop a quantitative measure of inertia.

Consider two different frictionless disks set in motion in the same direction and with the same speed, on a smooth horizontal surface. As long as the disks remain isolated, they travel side by side, maintaining their same relative separation. See Fig. 5-3. If we attach a thin rigid rod to the two disks, nothing changes. Given the same velocity, they again move together, and the rod does not turn.

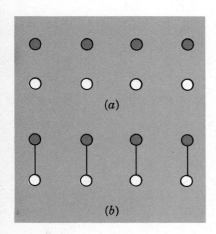

FIG. 5-3. Two disks moving to the right at the same constant velocity, *(a)* with no connection between them and *(b)* with a rigid rod connecting them.

What happens if we pull on the rod at its center? In general, the rod rotates, as shown in Fig. 5-4. The shaded disk in the figure resists a change in its state of motion to a greater extent than does the unshaded disk. If both disks are originally at rest, the shaded disk lags behind the unshaded disk when the rod is pulled at its center, Fig. 5-4a. If the disks are originally in motion and the rod is then pulled at its center to retard the motion, the shaded disk leads the unshaded disk, Fig. 5-4b. In short, the shaded disk has a greater resistance to a change in velocity. It is said to have a larger *inertia,* or *inertial mass,* than the unshaded disk. Hereafter we shall call the inertial mass simply "the mass." Note that it is *not* necessary that the pull at the rod's center give this point a *constant* acceleration; all that is required is that the velocity of this point change.

When then do two objects have the same mass? When two objects move together, neither body lagging or leading the other, *by definition,* they are identical in mass. See Fig. 5-5. The objects have equal masses because they exhibit the same inertial response to a change in their state of motion.

The simple device used here to compare masses is an *inertial balance.* If two masses balance when compared in one inertial frame, they balance in any other inertial frame, whether in the (approximate) inertial frame of the earth's surface, or in a reference frame moving with constant velocity relative to the earth, or even in an inertial frame in interstellar space (where the smooth horizontal surface is, of course, not required). Mass is an intrinsic property of an object. It is, unlike the *weight* of an object (about which much will be said later), the same at all points in space. An object's mass is independent of such properties as its temperature, shape, and color. Mass is a quantitative measure of inertia alone.

An inertial balance like that in Fig. 5-5 is not a convenient instrument. When two objects have equal inertial mass, as established by experiment with an inertial balance, we are not surprised (but should be) to find that the two objects are also in balance when placed on the pans of a sensitive beam balance of the familiar variety, shown in Fig. 5-6. That is, if two objects with identical inertial masses are simultaneously placed on the pans of a beam balance, the beam, like the rod in the inertial balance, does not turn. A beam balance is a *gravitational balance.* Actually, as we shall see later, the equivalence of the inertial balance, which compares the inertia of bodies, and the gravitational balance, which compares the gravitational force on bodies, is not obvious. The equivalence of the inertial mass as measured by an inertial balance and the gravitational mass as measured by a gravitational balance is a most remarkable and profound experimental result.

Masses add as scalar quantities: The mass of a body is the sum of the masses of its parts. This assertion also is not obvious a priori; it must be, and is, confirmed by experiment. (After all,

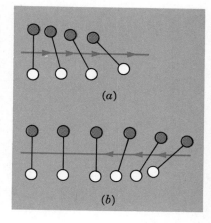

FIG. 5-4. The connecting rod is pulled suddenly at its center. *(a)* Both disks are initially at rest. *(b)* Both disks are initially in motion to the right at constant speed, and the pull of the connecting rod is to the left. The shaded disk shows a greater resistance to a change in its state of rest or motion. The arrows indicate the direction of the pull.

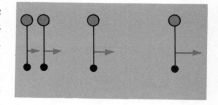

FIG. 5-5. An inertial balance. Both objects have the same inertial mass if the bodies move together when the rod is pulled at its midpoint.

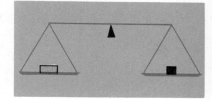

FIG. 5-6. A gravitational balance. Both objects have the same gravitational mass if the balance remains horizontal.

physical volumes do *not* always add as scalars: 1.0 m³ of water mixed with 1.0 m³ of alcohol is *not* 2.0 m³.) The scalar additivity of mass is established by noting that the mass of a composite object does not depend upon how its constituent parts are attached. The mass of the parts, added together, is equal to the mass of the whole. For example, the mass of a water molecule is, within experimental error, the sum of the masses of two hydrogen atoms and one oxygen atom. Thus, scalar additivity of masses holds extremely well even in the atomic domain. There are departures in the nuclear domain which are accounted for, however, by the relativistic mass-energy equivalence.

5-3 THE STANDARD OF MASS

As in the case of the fundamental units of length and of time, we must arbitrarily choose some object as a standard of mass. The standard of mass, to which ultimately all mass measurements are related, is a platinum-iridium cylinder carefully stored at the International Bureau of Weights and Measures near Paris. The mass of this cylinder is *by definition* exactly *one kilogram* (1 kg). Replicas of the standard kilogram have been made, and the one residing under double bell jars in the U.S. Bureau of Standards is the standard of mass for the United States.

The kilogram is the unit of mass in the mks (*meter-kilogram-second*) system, or International System of Units (SI).

The familiar unit the pound appears as a unit of weight, *not* mass, in the U.S. Customary (or English) System of Units. Roughly, 1 lb corresponds to 454 g; strictly, the pound is defined as the weight of a 0.45359237-kg mass at a location where $g = 32.17398$ ft/s².

Because the masses of atoms are small, a special system of units is employed for measuring them. The basic unit is called the *atomic mass unit,* u. By definition, the mass of an electrically neutral atom of carbon isotope 12 is exactly 12 u. Measurements show that $1\text{ u} = 1.660 \times 10^{-27}$ kg.

5-4 THE LAW OF CONSERVATION OF MASS

Suppose that we build a completely leakproof container through which no particles can pass, in or out. Whatever the nature of the contents of the container and despite any chemical or other changes that take place in its interior, the total mass of this isolated system is constant, as measured, for example, on a beam balance. This is the classical *law of conservation of mass,* the first of several fundamental conservation laws of physics: The total mass of an isolated system is constant.

Any object at a finite temperature radiates energy in the form of heat, light, or other types of electromagnetic radiation. Such radiation may also be absorbed by a body. Radiation has mass associated with it, as the special theory of relativity shows. In its modern, more general, form, the law of mass conservation can be stated as follows: The mass of an isolated system from which, or into which, no material particles *or radiation* can leak is constant. But the mass-conservation law includes the mass associated with radiation so long as the container is leakproof or the system isolated from radiation.

The mass equivalent of radiation is ordinarily so very small that it can be neglected. For example, a flashlight loses about 10^{-16} kg/s in the form of radiation; whereas the sun loses 4.2×10^9 kg/s (out of 10^{30}-kg total mass) as heat and light. See Chap. 35.

5-5 MOMENTUM CONSERVATION IN HEAD-ON COLLISIONS

We have already discussed the simplest conceivable physical situation, that of a single isolated particle. Its velocity is constant. Now we turn to the next most complicated situation, that of two particles isolated from the rest of the universe but interacting with each other. In studying the simplest type of two-particle interaction — a head-on collision — we shall be brought to the fundamental law of momentum conservation.

We consider a simple experiment in which one object collides head on with a second. We have found in advance that when set in motion on a frictionless horizontal surface, each moves with unchanged velocity. Each of the two objects will depart from uniform velocity only to the extent that it interacts with — which is to say, collides with — the other.

A 1.00-kg object, initially moving at 2.00 m/s to the right, collides with a 3.00-kg object originally at rest. The observed results are as shown in Fig. 5-7a; the lighter object rebounds to the left at 0.70 m/s, and the more massive object is set in motion and moves to the right at 0.90 m/s. The velocities are plotted as a function of time in Fig. 5-7b and c. Each object has constant velocity before the collision and a constant (but different) velocity after. The velocity changes only while the objects interact. The velocity of the 1.00-kg object changes by the amount $\Delta v_1 = -2.70$ m/s; the corresponding velocity change for the 3.00-kg object is $\Delta v_2 = 0.90$ m/s (subscripts 1 and 2 designate the 1- and 3-kg objects, respectively). The velocity changes Δv_1 and Δv_2 and the masses m_1 and m_2 are simply related:

$$\frac{\Delta v_1}{\Delta v_2} = \frac{-2.70 \text{ m/s}}{0.90 \text{ m/s}} = -3.00$$

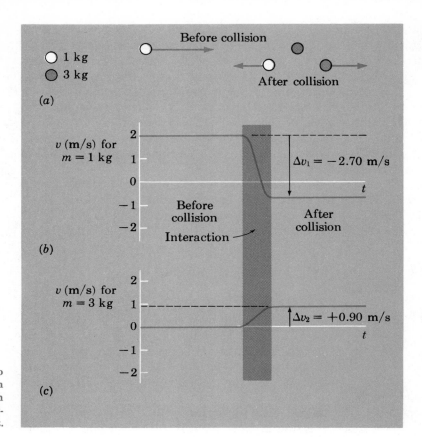

FIG. 5-7. (a) Velocities of two objects before and after a head-on collision. (b) Velocity-time graph for the 1.0-kg object. (c) Velocity-time graph with the 3.0-kg object.

and
$$\frac{m_2}{m_1} = \frac{3.00 \text{ kg}}{1.00 \text{ kg}} = 3.00$$

so that
$$m_1 \, \Delta v_1 = -m_2 \, \Delta v_2$$

or
$$m_1 \, \Delta v_1 + m_2 \, \Delta v_2 = 0 \tag{5-1}$$

Further experiments establish that the simple result expressed in (5-1) holds for other head-on collisions. Indeed, it holds for *all* collisions, quite apart from the nature of the interacting objects, so long as the system as a whole is isolated.

Equation (5-1) expresses a remarkable regularity in nature deserving the designation *law of physics*. This law can be written more simply when we define a quantity **p**, known as the *linear momentum* of a body, as the *product of mass and velocity*:

$$\text{Linear momentum} = \mathbf{p} = m\mathbf{v} \tag{5-2}$$

Linear momentum, or simply momentum (for short), is a vector quantity (the product of a scalar m and a vector $\mathbf{v}$). Its direction is that of the particle's velocity, and its magnitude is the product of the mass and speed. See Fig. 5-8. Linear momentum has the dimensions ML/T. Appropriate units for linear momentum are kilogram-meters per second (kg-m/s).

We can write (5-1) in terms of momentum. By definition,

$$\mathbf{p} = m\mathbf{v} \qquad \Delta\mathbf{p} = \Delta(m\mathbf{v}) = m\,\Delta\mathbf{v}$$

where it is assumed, in the last step, that the mass m remains constant. Equation (5-1) becomes

$$\Delta\mathbf{p}_1 = -\Delta\mathbf{p}_2$$

or
$$\Delta\mathbf{p}_1 + \Delta\mathbf{p}_2 = 0 \tag{5-3}$$

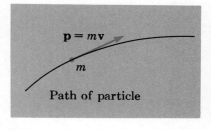

FIG. 5-8. Linear momentum p of a particle of mass m and velocity v.

Equation (5-3) says: When two isolated objects interact with one another, the loss in the momentum of one object equals the gain in momentum of the second object. Or, the total change in momentum for the system is zero. A collision is a process in which momentum is transferred from one body to another, with no change in the total momentum of the system. Thus, if $\mathbf{p}_1$ is the momentum of object 1 and $\mathbf{p}_2$ that of object 2, *the total momentum of the isolated system is constant*:

$$\mathbf{p}_1 + \mathbf{p}_2 = \text{constant (in direction and magnitude)} \tag{5-4}$$

This is the law of the *conservation of momentum*.

Returning to our original example (Fig. 5-7), we now plot graphs of momentum vs. time. See Fig. 5-9. The 1.00-kg object loses a momentum of 2.70 kg-m/s to the right, while the 3.00-kg object gains the same momentum to the right. There is no change in the system's total momentum.

$\mathbf{p}_1 + \mathbf{p}_2 = 2.00$ kg-m/s to the right *before* and *after* collision

Figure 5-9 shows, furthermore, that the total momentum is the same, not only before and after the collision but also at each instant *during* the collision.

The power and generality of the momentum-conservation law can hardly be overemphasized. So long as the particles are isolated from external influence, the total momentum is unchanged, quite apart from the details of, or even a knowledge of, the interaction. These details will, of course, concern us later. But whether the collision is "soft" (the momentum changing slowly) or "hard" (momentum changing rapidly), whether the particles separate after the collision (as with billard balls) or stick together as a single composite object (as with pieces of putty), whether they "touch" or merely come close together (as in the case of the repulsion through magnets)—in short, whatever the circumstances—for an

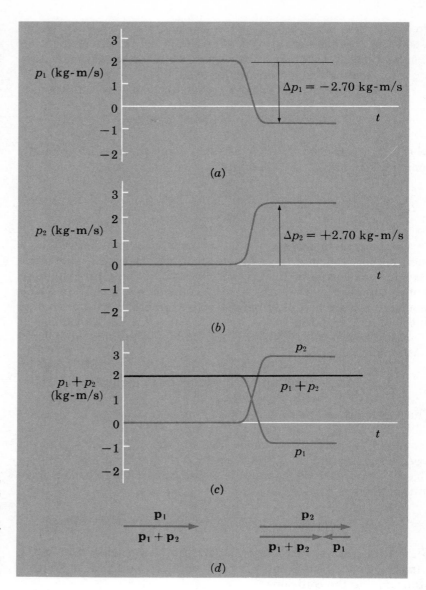

FIG. 5-9. Linear-momentum–time graphs corresponding to the collision of Fig. 5-7 for (a) the 1.0-kg object, (b) the 3.0-kg object, (c) the total linear momentum of the system of two objects. (d) Linear momentum vectors before and after the collision.

isolated system the total momentum **P** into the collision equals the total momentum out.†

$$\mathbf{P} = \Sigma \mathbf{p}_i = \text{constant (magnitude and direction)} \qquad (5\text{-}5)$$

---

†The relation $\mathbf{p} = m\mathbf{v}$ for linear momentum is restricted, however, to velocities much less than that of light, 3×10^8 m/s. For high speeds the correct relation for the vector quantity $\mathbf{p}$ that is conserved in all collisions for an isolated system is $\mathbf{p} = m_0\mathbf{v}/\sqrt{1-(v/c)^2}$, where c is the speed of light and m_0 denotes the so-called rest mass, the mass of an object at zero speed. This expression for the *relativistic momentum* reduces to the classical momentum relation, $\mathbf{p} = m_0\mathbf{v}$, for speeds much less than that of light, that is, $v/c \ll 1$. For example, with $v = 30{,}000$ km/s, $v/c = 1/10$, and the denominator differs from 1 by a mere 0.5 percent. The law of momentum conservation, Eq. (5-5), remains valid at all speeds. See Sec. 35-2 for a more detailed discussion of relativistic momentum.

Example 5-1

An object initially at rest explodes into two pieces with masses m_1 and m_2, respectively. How are the velocities of the two pieces related?

An explosion is simply a collision in which the two interacting objects are initially at rest. Here the total momentum is initially zero, and it must remain zero.

$$\mathbf{p}_1 + \mathbf{p}_2 = 0$$

$$m_1 v_1 + m_2 v_2 = 0$$

$$\frac{m_1}{m_2} = -\frac{v_2}{v_1}$$

The speeds are in the inverse ratio of the masses. The minus sign implies that the two pieces must fly off in opposite directions. For a fired rifle the bullet's speed exceeds the rifle recoil speed by the same factor that the rifle's mass exceeds that of the bullet. If the two masses are equal, $m_1 = m_2$, then the velocities are equal in magnitude as well as opposite, $v_1 = -v_2$. This may be used as a test for equality of masses: If two pieces fly off with equal speeds, their masses must be equal. Similarly, if two pieces fly together at equal speeds and stick together to form a composite object, which is at rest, we know that the two pieces have equal masses.

Example 5-2

A 5.00-g bullet is shot horizontally into a 10.0-kg block resting on a smooth horizontal surface. The bullet comes to rest within the block, and the block slides at 25.0 cm/s. What is the original speed of the bullet?

The conservation of momentum law requires that the momentum of the bullet before collision equal the momentum of the block (and bullet) after the collision. If m and v are the bullet's mass and initial speed, respectively, and M and V are the block's mass and final speed, respectively, then

$$mv = (M + m)V$$

$$v = \frac{(M+m)V}{m} = \frac{(10.0 \text{ kg} + 5.00 \times 10^{-3} \text{ kg})(0.25 \text{ m/s})}{5.00 \times 10^{-3} \text{ kg}} = 500 \text{ m/s}$$

Example 5-3

(a) A light ball of mass m moving at a speed v strikes a massive ball of mass M at rest. The light ball bounces back with nearly the same speed. What is the speed V of the massive ball after the collision?

Momentum conservation requires:

Momentum before collision = momentum after collision

$$m\mathbf{v} = M\mathbf{V} - m\mathbf{v}$$

$$V = \frac{2mv}{M}$$

where the symbols represent the same quantities as in Example 5-2.

(b) Suppose, now, that the light ball sticks to the massive ball. What is the speed V' of the massive ball?

$$m\mathbf{v} = (M + m)\mathbf{V}' \approx M\mathbf{V}' \qquad \text{since } m \ll M$$

$$V' = \frac{mv}{M}$$

Comparing V' in part *(b)* with V in *(a)*, we see that the massive ball moves only half as fast when the light ball sticks to it as it does when the light ball rebounds. If the light ball were to miss, it would transfer zero momentum. When it sticks, it transfers momentum mv; and when it rebounds, it transfers momentum $2mv$.

5-6 MOMENTUM CONSERVATION IN TWO AND THREE DIMENSIONS

Momentum conservation applies not only to head-on collisions in which the interacting particles travel along a single straight line but also to the most general type of collision in which the particle velocities may differ in direction. Experiment shows that the *vector sum* of the momenta of the individual particles of any isolated system is constant. Indeed, only this observation allows us to speak properly of momentum as a vector quantity and to write in general that $\mathbf{P} = \Sigma \mathbf{p}_1 = $ constant.

Consider the collision shown in Fig. 5-10. Here two particles approach each other, interact through a repulsive force, and then depart in different directions. At distances sufficiently far from the interaction site, each particle is isolated from the other and travels in a straight line with constant momentum. The momentum of both particles changes, both in magnitude and direction, as they

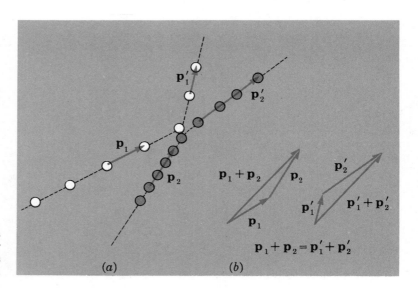

FIG. 5-10. Two objects in collision. The total vector momentum of the system is constant.

interact. Finally, they depart in straight lines. At each instant, the total vector momentum of the system is constant.

As the momentum vectors indicate, the total momentum before the collision, $\mathbf{p}_1 + \mathbf{p}_2$, is equal in direction and magnitude to the total momentum after the collision, $\mathbf{p}_1' + \mathbf{p}_2'$:

$$\mathbf{p}_1 + \mathbf{p}_2 = \mathbf{p}_1' + \mathbf{p}_2'$$

Since momentum is a vector, and since any vector may be replaced by its components, we know not only that the total vector momentum of an isolated system is constant in direction and magnitude but also that the algebraic sum of the momentum components of all particles along *any direction* in space is also a constant. Specifically, if

$$\Sigma \mathbf{p}_i = \text{constant vector}$$

then
$$\Sigma p_{ix} = \text{a constant}$$

and
$$\Sigma p_{iy} = \text{a constant}$$

where p_{ix} and p_{iy} are the X and Y components of $\mathbf{p}_i$. (The Z components are ignored, since motion is confined to the XY plane.)

Example 5-4

A package originally at rest explodes into three parts all of equal mass. One part A flies west at 80 m/s; another, B, flies south at 60 m/s. What are *(a)* the speed v_C and *(b)* the direction θ, with respect to east, of the third part C?

The total momentum of the system is initially zero, and it remains zero during and after the explosion. Therefore,

$$0 = \mathbf{p}_A + \mathbf{p}_B + \mathbf{p}_C$$

and the momentum vectors are arranged as shown in Fig. 5-11.

After the collision the components of momentum must add to zero algebraically along the east-west direction, along the north-south

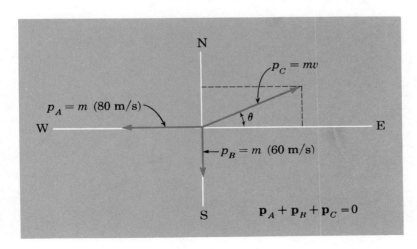

FIG. 5-11. A package explodes into three parts with momenta $\mathbf{p}_A$, $\mathbf{p}_B$, and $\mathbf{p}_C$, respectively.

direction, and along the direction perpendicular to the plane of the paper. C's momentum component perpendicular to the paper must be zero; thus C travels in the plane of A and B. Taking east as positive, and west as negative, we have for the east-west momentum components:

$$0 = mv \cos \theta - m(80 \text{ m/s})$$

Similarly, for the north-south components,

$$0 = mv \sin \theta - m(60 \text{ m/s})$$

Dividing the second equation by the first yields

$$\tan \theta = \tfrac{60}{80} = 0.75$$

$$\theta = 37° \text{ north of east}$$

The first equation yields

$$v = \frac{80 \text{ m/s}}{\cos 37°} = 100 \text{ m/s}$$

Example 5-5

A 1.0-kg object A with a velocity of 4.0 m/s to the right strikes a second object B of 3.0 kg originally at rest. In the collision A is deflected from its original direction through an angle of 50°; its speed after the collision is 2.0 m/s. What is (a) the angle θ of B's velocity after the collision with respect to the original direction of A and (b) the speed v'_B of B after the collision?

The system is subject to no net external force, and the conservation of momentum requires that

$$\mathbf{p}_A = \mathbf{p}'_A + \mathbf{p}'_B$$

as shown in Fig. 5-12, where $\mathbf{p}_A$ and $\mathbf{p}'_A$ are the momenta of A before and after the collision, respectively, and $\mathbf{p}'_B$ is the momentum of B after the collision. The total vector momentum is constant through-

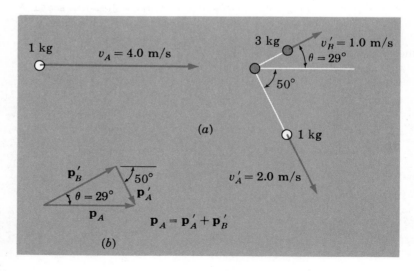

FIG. 5-12. (a) The velocity vectors for a non-head-on collision. (b) The momentum vectors for this collision.

out the collision. Therefore, the algebraic sum of the components of A's and B's momenta along any given direction is also constant. We first choose the original direction of A as the direction for computing momentum components:

$$p_A = p'_A \cos 50° + p'_B \cos \theta \tag{5-6}$$

$(1.0 \text{ kg})(4.0 \text{ m/s}) = (1.0 \text{ kg})(2.0 \text{ m/s}) \cos 50° + (3.0 \text{ kg})(v'_B) \cos \theta$

Equation (5-6) implies that the total linear momentum is unchanged along the forward direction, body B acquiring a momentum component along this direction which is just equal to the component of momentum lost by A.

Now consider the components of momentum along a direction perpendicular to that of A's original motion:

$$0 = p'_A \sin 50° - p'_B \sin \theta \tag{5-7}$$

$0 = (1.0 \text{ kg})(2.0 \text{ m/s}) \sin 50° - (3.0 \text{ kg})(v'_B) \sin \theta$

Before the collision there is no component of momentum along this direction; after the collision the transverse momentum component of A is just balanced by that of B. (There is no component of momentum, before or after the collision, in a direction perpendicular to the plane of vectors $\mathbf{p}'_A$ and $\mathbf{p}'_B$.)

We have two equations, (5-6) and (5-7), and two unknowns, v'_B and θ. We can eliminate v'_B from the two equations by solving for v'_B in (5-7) and substituting in (5-6). The result, after a little algebra, is

$$\theta = 29.4° \approx 29°$$

Using $\theta = 29°$ in (5-6) or (5-7) gives

$$v'_B = 1.0 \text{ m/s}$$

5-7 MOMENTUM CONSERVATION IN ATOMIC COLLISIONS

Momentum conservation applies to collisions of molecules, atoms, and their constituent particles. Much information concerning atomic or subatomic particles can be extracted by applying the conservation of momentum principle to such collisions.

The tracks of electrically charged particles are rendered visible and can be photographed when such particles pass through the superheated liquid of a bubble chamber and bubbles form along the wake. The colliding particles are effectively isolated, and momentum into the atomic collision is exactly equal to the momentum out of the collision.

Such particles as electrons, protons, and neutrons all have linear momentum. Furthermore, electromagnetic radiation, illustrated by radio waves, visible light, and x-rays, also carries linear momentum. (See Sec. 30-5.) (Light exerts a pressure on objects it strikes.) The quantum theory shows that radiation has particlelike aspects, the particles of electromagnetic radiation being called "photons." (See Sec. 36-2.) Each photon carries momentum, and the collision between a photon and a material particle is also governed by the conservation of momentum principle.

SUMMARY

According to the law of inertia, the linear momentum $\mathbf{p} = m\mathbf{v}$ of a single isolated particle is constant. According to the law of mass conservation, the total mass of an isolated system is constant. According to the law of momentum conservation, the total momentum $\Sigma \mathbf{p}$ of a system of interacting particles isolated from external influence is constant in both magnitude and direction.

PROBLEMS

5-1. A 2.0×10^3 kg limousine traveling at 100 km/h collides head-on with a 1.0×10^3 kg car moving in the opposite direction at the same speed. *(a)* If the cars become entangled, what is the speed and direction of the wreckage immediately after impact? *(b)* Sketch momentum-time graphs for the two vehicles both before and after the collision.

5-2. A 1.0×10^4 kg truck traveling at 88 km/h collides with the rear of a 1.0×10^3 kg car moving at 80 km/h in the same direction. If the vehicles stick together, what is the speed and direction of motion of the wreckage immediately after the collision?

5-3. A 3.0 kg rifle, at rest on the smooth surface, fires a 5.0 g bullet at a muzzle speed of 500 m/s. What is the recoil speed of the rifle?

5-4. An object of mass m is projected with an initial velocity v_0 at an angle θ with respect to the horizontal. Write expressions giving the horizontal and vertical momentum components as a function of the time t.

5-5. What is the change in momentum of a 5.00 kg object first moving south at 16.0 m/s and later north at 9.00 m/s?

5-6. A 5.0 g bullet moving horizontally at 400 m/s strikes a 500 g steel block initially at rest. Following the collision the bullet moves in the opposite direction at speed 300 m/s. Sketch momentum-time graphs for the bullet and the block, and give the final speed of the block.

5-7. A 5.0 g bullet moving horizontally at 400 m/s passes through a 500 g block of wood initially at rest on a frictionless surface. The bullet emerges with a speed of 100 m/s. Sketch momentum-time graphs for the bullet and the block, and calculate the final speed of the block.

5-8. A 6 kg object is thrown vertically upward at a speed of 10 m/s. With what speed does the earth (mass, 6×10^{24} kg) recoil?

5-9. An object of mass m is fired vertically upward and comes to rest momentarily at a height h. *(a)* Show that the earth of mass M recoils a distance of $(m/M)h$ if the earth's acceleration is smaller than that of m by a factor m/M. *(b)* How large a mass m achieving a maximum height of 10 m would be required to make the earth (mass, 6×10^{24} kg) recoil 10^{-10} m (approximately the size of an atom)?

5-10. An object initially traveling north at 15 m/s explodes in flight into two pieces of equal mass. One goes east at 10 m/s. What is the velocity of the second piece?

5-11. An object at rest explodes into three pieces of equal mass. One moves east at 20 m/s; a second moves southeast at 30 m/s. What is the velocity of the third piece?

5-12. A helicopter in level flight at 90 km/h drops a 20 kg canister from a height of 30 m above flat terrain. Find the magnitude and direction of the momentum of the canister *(a)* at the moment of release, *(b)* after 2 s, and *(c)* at impact with the ground. Take g to be 10 m/s² (and thereby neglect friction).

5-13. A 1.0 kg steel ball 4.0 m above the floor is released, falls, strikes the floor, and rises to a maximum height of 2.5 m. Find the momentum transferred from the ball to the floor in the collision.

5-14. A star and a planet circling it constitute an isolated system. The planet's momentum is constant in magnitude but changes direction continuously. Show that the star itself must move in a circle, albeit a much smaller one.

5-15. The uranium-238 nucleus is unstable and decays into a thorium-234 nucleus and an α particle. The α particle is emitted with a speed of 1.4×10^6 m/s. What is the recoil velocity of the thorium-234 nucleus, assuming the uranium-238 atom to be at rest at the time of the decay? The thorium-234 and α-particle masses are in the ratio 234 to 4.

5-16. A proton strikes an atom of gold and is thereby scattered from its original direction by an angle of 30°. Its speed after the collision is essentially the same as before. In what direction relative to the original direction of the proton does the gold atom recoil? A gold atom is much more massive than a proton.

5-17. When an electron meets a positron (a particle having the same mass as the electron but carrying a positive, rather than a negative, electric charge), the two particles may annihilate one another, the particles disappearing and electromagnetic radiation appearing in the form of photons. (a) Show that electron-positron annihilation must produce at least two photons. (b) In what relative directions must the two photons travel if the electron and positron are initially at rest?

5-18. An α particle (helium nucleus), originally moving at 1.0×10^7 m/s, collides with the nucleus of a gold atom at rest. Find the recoil speed of the gold nucleus if (a) the α particle comes out of the collision in the opposite direction, with (essentially) its original speed, and (b) the α particle leaves the collision along a direction 60° to its original motion and with its original speed. A gold nucleus has a mass of 197 u; the α particle's mass is 4 u.

5-19. Two girls sit on opposite ends of a sled 6.0 m long initially at rest on frictionless ice. Each girl has a mass of 50 kg; the sled's mass is 30 kg. The girl at one end throws a 4.0 kg object to the other girl so that the object travels horizontally at 5.0 m/s relative to the ice. What is the sled's speed (a) before the second girl catches the object? (b) After she catches the object? (c) Over what distance does the sled move while the object is in flight?

5-20. A pitched baseball is traveling horizontally at 30 m/s at the instant the batter hits it to center field; the ball is hit at an angle of 60° above the horizontal and carries for 90 m in the air. Find (a) the initial *velocity* of the batted ball and (b) the vector impulse imparted to the ball by the bat. Assume that the horizontal direction of the batted ball is exactly opposite to that of the pitched ball and that the trajectory is not influenced by air resistance. Disregard the height above the ground at which the bat-ball impact takes place. The mass of the ball is 160 g; take $g = 10.0$ m/s².

5-21. In a rocket, fuel of mass $-dm$ is ejected to the rear with a constant exhaust velocity of v_E, measured relative to the rocket. At the same time the mass of the rocket changes from m to $m + dm$. (Note that dm is a *negative* quantity.) (a) Show that $m\, dv = -v_E\, dm$, where m and dv are the mass and the change in velocity of the residual rocket, respectively. (b) Use this result to show that $v_b = v_E \ln(M_0/M_b)$, where v_b is the speed of the payload at "burn out." M_0 is the rocket mass at lift-off, and M_b is the payload at burn out—after all of the fuel has been exhausted. (c) Finally, show that this result can be written in the form $(M_0/M_b) = e^{(v_b/v_E)}$. (This discussion is valid in gravity-free space.) This relation, the rocket equation, was first derived in the 1890s by Tsiolkovsky, a self-taught Russian schoolmaster.

5-22. A typical rocket engine produces an exhaust velocity v_E of 2 km/s. If the residual rocket is to achieve orbital speed of 8 km/s, use the results of Prob. 5-21 to find the ratio of the initial mass of the rocket to the mass of the payload.

CHAPTER 6

Center of Mass and Reference Frames

The law of momentum conservation, as well as still other properties of a system of particles yet to be developed, has an even simpler meaning when it is recognized that an entire system of interacting particles may, in effect, be replaced by a single equivalent particle. This is accomplished through the concept of center of mass. We shall also explore the relations which translate physical quantities measured in one reference frame into their corresponding values in a second, moving reference frame. In particular, we shall look at the special convenience of that reference frame in which a system's center of mass is at rest.

6–1 VELOCITY OF THE CENTER OF MASS

According to the law of inertia, the momentum of a single isolated particle is constant. And according to the law of momentum conservation, the momentum of an isolated system of particles is constant. An isolated particle is like an isolated system of particles in that both have constant momentum. This applies quite apart from the details of, or even knowledge of, the motions of the individual particles in the system, which may be in general quite complicated. This suggests that, because of the momentum equivalence of the isolated system and the isolated particle, we may replace the entire system by a single equivalent particle whose momentum is equal to that of the system. We think of this single particle as having a velocity $\mathbf{V}$ and a mass M equal to that of the whole system $M = \Sigma m$, where m is the mass of a particle moving with velocity $\mathbf{v}$ in the system. Then equating the particle's momentum $M\mathbf{V}$ to the total momentum $\Sigma m\mathbf{v}$ of the system, we have

$$MV = \Sigma m\mathbf{v}$$

where $\mathbf{V}$ is the velocity of the *center of mass*. [Note that the individual $\mathbf{v}$'s need not be constant (the particles, interacting with one another, may wander about in the most general fashion); it is the *sum* $\Sigma m\mathbf{v}$ which is constant and therefore also $\mathbf{V}$.]

The equation above then gives the velocity $\mathbf{V}$ of the center of mass as

$$\mathbf{V} = \frac{\Sigma m\mathbf{v}}{M} = \frac{\Sigma m\mathbf{v}}{\Sigma m} \qquad (6\text{-}1)$$

This vector equation may be replaced by the scalar equations for the velocity components (we suppose in this chapter that motion is confined to the XY plane).

and
$$V_x = \frac{\Sigma m v_x}{M} = \frac{\Sigma m v_x}{\Sigma m}$$
$$V_y = \frac{\Sigma m v_y}{M} = \frac{\Sigma m v_y}{\Sigma m} \qquad (6\text{-}2)$$

Example 6-1

A 4-kg object initially at rest explodes into three pieces. A 1-kg piece flies east at 5.0 m/s, and a 2-kg piece flies 60° south of west at 4.0 m/s. What is the velocity of the system's center of mass?

The momentum of the system is initially zero, and it remains zero. Therefore, quite apart from any of the details of what happens after the explosion, we know that the center-of-mass velocity must be initially, and must remain, *zero*.

Example 6-2

A 1-kg particle traveling east at 4 m/s collides with a 3-kg particle initially at rest, and the two stick together after the collision. What is the velocity of the system's center of mass?

Applying Eq. (6-2), we have

$$V_x = \frac{\Sigma m v_x}{\Sigma m} = \frac{(1 \text{ kg})(4 \text{ m/s}) + (3 \text{ kg})(0)}{(1 + 3) \text{kg}}$$

$$V_x = 1 \text{ m/s}$$

$$V_y = \frac{\Sigma m v_y}{\Sigma m} = 0$$

The center-of-mass velocity, 1 m/s to the east, is just the velocity of the composite 4-kg object after the collision, hardly a surprise since the center of mass is used to replace the whole system. Moreover, anticipating the definition of the location of the center of mass, we would think of the center of mass as being at the location of the composite particle after the collision.

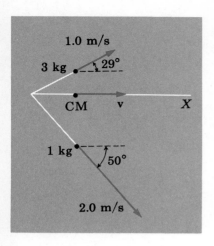

FIG. 6-1. Center-of-mass velocity V.

Example 6-3

A 3-kg particle travels at 1.0 m/s in the direction 29° above the X axis, and a 1-kg particle with a speed of 2.0 m/s at 50° below the X axis. What is the velocity of the center of mass for this isolated system?

The particle velocities are shown in Fig. 6-1. We resolve these velocities into X and Y components and apply equations to find V_x and V_y.

$$V_x = \frac{\Sigma m v_x}{\Sigma m} = \frac{(3.0 \text{ kg})(1.0 \text{ m/s}) \cos 29° + (1.0 \text{ kg})(2.0 \text{ m/s}) \cos 50°}{(3.0 + 1.0) \text{ kg}}$$

$$= 1.0 \text{ m/s}$$

$$V_y = \frac{\Sigma m v_y}{\Sigma m} = \frac{(3.0 \text{ kg})(1.0 \text{ m/s}) \sin 29° - (1.0 \text{ kg})(2.0 \text{ m/s}) \sin 50°}{(3.0 + 1.0) \text{ kg}}$$

$$= 0$$

The center of mass moves along the positive X axis at 1.0 m/s. This is, in fact, the same answer as that for Example 6-2. And it should be, because in these two examples we have computed the velocity of the center of mass before and after the collision discussed earlier in Example 5-5 and shown in Fig. 5-12.

6–2 LOCATION OF THE CENTER OF MASS

If we take the displacement of a system's center of mass to be represented by **R** and the displacement of an individual particle of the system to be **r**, then by analogy with Eq. (6-1) we expect the center-of-mass location to be given by

$$\mathbf{R} = \frac{\Sigma m \mathbf{r}}{\Sigma m} = \frac{\Sigma m \mathbf{r}}{M} \tag{6-3}$$

To check the appropriateness of this relation, we first suppose that all particles of a system have the same location **r**. Then the particles are clustered as a composite particle at **r**, and we expect that this also is the location of the center of mass. So, when all **r**'s are alike, $\Sigma m \mathbf{r}/M = \mathbf{r}\Sigma m/M = \mathbf{r}M/M = \mathbf{r} = \mathbf{R}$, as anticipated.

It is also easy to check that the center-of-mass velocity relation follows from (6-3). By definition, $\mathbf{V} = d\mathbf{R}/dt$, and for each particle, $\mathbf{v} = d\mathbf{r}/dt$. Taking the derivative with respect to time of (6-3) yields

$$\frac{d\mathbf{R}}{dt} = \frac{\Sigma m \, d\mathbf{r}/dt}{\Sigma m}$$

$$\mathbf{V} = \frac{\Sigma m \mathbf{v}}{\Sigma m} \tag{6-1}$$

The scalar equations giving the X and Y coordinates of the center of mass in terms of the X and Y coordinates of the constituent particles are

$$X = \frac{\Sigma mx}{\Sigma m}$$

(6-4)

and
$$Y = \frac{\Sigma my}{\Sigma m}$$

In general the particles of a system may be in motion so that the displacements **R** and **r**, or the coordinates X, Y, x, and y, are not necessarily constant in time. But if the system of particles is a rigid body, then the center of mass has a fixed location relative to the body, and it moves with the body.

To find the center of mass for a continuous distribution of mass, we replace the summation Σmx by the integral $\int x\, dm$, where dm is a small element of mass, small enough so that all atomic particles within dm can be taken as having the same coordinate x but large enough so that the actually discrete atomic particles are equivalent to a continuous mass distribution. If the mass per unit volume, or density, is ρ, the mass element dm may be written $dm = \rho\, dv$, where dv represents the volume element containing mass dm of density ρ. Therefore, the center-of-mass coordinates become

$$X = \frac{\Sigma mx}{\Sigma m} = \frac{\int x\, dm}{\int dm} = \frac{\int \rho x\, dv}{\int \rho\, dv}$$

For constant density ρ throughout the continuous mass distribution,

$$X = \frac{\int x\, dv}{\int dv}$$

(6-5a)

where the limits of the integrals correspond to the physical boundary of the object.

The corresponding relation for the Y coordinate is, of course,

$$Y = \frac{\int y\, dv}{\int dv}$$

(6-5b)

Example 6-4

A 1-kg particle is at the origin, and a 3-kg particle is on the X axis at $x = 8$ m. Where is the system's center of mass?

Since both particles have $y = 0$, $Y = 0$. Using Eq. (6-4), we have

$$X = \frac{\Sigma mx}{\Sigma m} = \frac{(1\text{ kg})(0) + (3\text{ kg})(8\text{ m})}{(1 + 3)\text{ kg}} = 6\text{ m}$$

As Fig. 6-2a shows, the center of mass is 6 m from the 1-kg particle and 2 m from the 3-kg particle; the distances from the center of mass to the two particles is in the inverse ratio of their masses. This result holds in general for a two-particle system: If the center of mass is at the origin so that $X = 0$, then

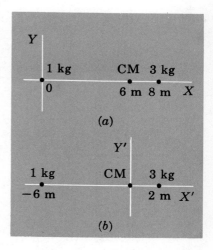

FIG. 6-2. Center of mass for a two-particle system. *(a)* Origin at the 1-kg particle and *(b)* origin at the center of mass.

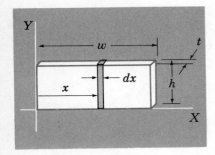

FIG. 6-3. Center of mass of a rectangular plate. An element of mass is contained in the vertical strip of displacement x, width dx, and thickness t.

$$0 = \frac{m_1 x_1 + m_2 x_2}{m_1 + m_2}$$

and

$$\frac{m_1}{m_2} = -\frac{x_2}{x_1}$$

the minus sign implying that the two particles must be on opposite sides of the line joining them and passing through the center of mass. Thus, if we choose a new coordinate origin at the location of the center of mass ($X = 6$ m, but now $X' = 0$), the 1-kg particle has the coordinate $x'_1 = -6$ m and the 3-kg particle the coordinate $x'_2 = +2$ m. See Fig. 6-2b. We find that

$$X' = \frac{\Sigma m x'}{\Sigma m} = \frac{(1 \text{ kg})(-6 \text{ m}) + (3 \text{ kg})(2 \text{ m})}{(1+3) \text{ kg}} = 0$$

and

$$\frac{m_1}{m_2} = -\frac{x'_2}{x'_1} = \tfrac{1}{3}$$

Although the coordinates of the center of mass will depend on the choice of origin, the *location* of the center of mass is always the same *relative to* the positions of the constituent *particles*.

Example 6-5

Find the center of mass of a uniform rectangular plate.

The width is w, the height h, and the thickness t. The coordinate origin is chosen to be at a corner of a rectangle, as shown in Fig. 6-3. To find the X component of the center of mass, we imagine the rectangle to be divided into thin vertical sections, each of width dx. All mass points within any one section have the same X coordinate x. We then sum the contributions from the vertical sections from $x = 0$ to $x = w$. The volume element is $dv = ht\,dx$, and (6-5a) gives

$$X = \frac{\int x\, dv}{\int dv} = \frac{\int_0^w x(ht\, dx)}{htw} = \frac{\int_0^w x\, dx}{w} = \frac{w^2/2}{w} = \frac{w}{2}$$

The center of mass is halfway between the ends. Similarly, the center of mass is halfway up. In short, the center of mass is the center of symmetry of the rectangle.

The center of mass of any uniform symmetrical solid is at the center, or on the line, of symmetry—that is, at the center of a sphere, or spherical shell, or ring, or along the axis of symmetry of a cylinder or a cone. In every instance, if the origin of coordinates is at the center, or line, of symmetry, for each contribution $m_i x_i$ from one side there is a contribution $-m_i x_i$ from the opposite side, yielding, therefore, the center of mass at the center of symmetry, $X = 0$.

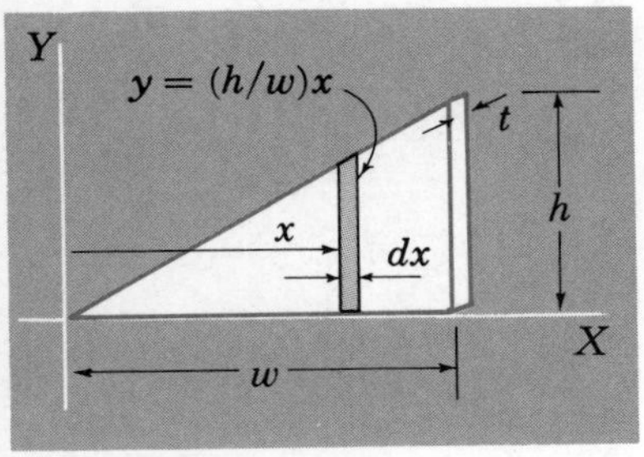

Example 6-6

Find the center of mass of a uniform right triangular plate, height h and width w.

See Fig. 6-4, where the volume element dv has width dx,

FIG. 6-4. Center of mass of a uniform right triangular plate.

height y, and thickness t. The hypotenuse of the triangle is the line $y = (h/w)x$. Therefore, $dv = yt\,dx = (h/w)tx\,dx$, and (6-5) gives

$$X = \frac{\int x\,dv}{\int dv} = \frac{\int_0^w (h/w)tx^2\,dx}{hwt/2} = \tfrac{2}{3}w$$

In similar fashion, $Y = h - \tfrac{2}{3}h = \tfrac{1}{3}h$, again one-third the distance from the square corner.

6–3 REFERENCE FRAMES AND RELATIVE VELOCITIES

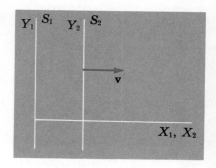

FIG. 6-5. Reference frame S_2 with velocity v with respect to S_1.

Heretofore we have assumed implicitly that the displacements, velocities, momenta, and accelerations of objects were measured with respect to a frame of reference attached to the earth. We wish now to find the relations for translating the values of these quantities as measured in one reference frame to the corresponding values measured in a second reference frame in motion at constant velocity relative to the first. We label the earth's reference frame S_1; quantities measured relative to this frame carry the subscript 1. A second reference frame S_2 moves to the right relative to S_1 at the constant velocity **v**; quantities measured in S_2 are labeled with the subscript 2. For convenience we assume that the origins of S_1 and S_2 coincide at time $t = 0$ and that the two sets of axes are aligned. See Fig. 6-5.

Because system S_2 moves to the right with velocity $+\mathbf{v}$ relative to S_1, it follows that an observer in S_2 would say that S_1 and all objects at rest in this frame move left with velocity $-\mathbf{v}$. Strictly, we can speak *only* of the *relative motion* of any two systems, their relative speed v being simply the separation distance between the two origins divided by the elapsed time. To name S_1 as a frame attached to earth is merely a matter of convenience; S_1 could be any reference frame.

Suppose that at some time t a particle's location is given by radius vector $\mathbf{r}_1$ in S_1 and $\mathbf{r}_2$ in S_2. How these vectors are related is shown in Fig. 6-6, where the displacement of S_2's origin relative to S_1 over the time t is shown as $\mathbf{v}t$:

$$\mathbf{r}_2 = \mathbf{r}_1 - \mathbf{v}t \tag{6-6}$$

This vector equation is equivalent to two scalar equations which give the components of $\mathbf{r}_1$ as the coordinates x_1 and y_1 and those of $\mathbf{r}_2$ as x_2 and y_2.

$$x_2 = x_1 - vt \tag{6-7}$$

$$y_1 = y_2$$

We may rearrange the first equation to find

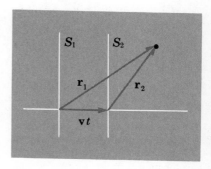

FIG. 6-6. Relative to S_1 the particle's location is $\mathbf{r}_1$; relative to S_2 the particle's location is $\mathbf{r}_2$.

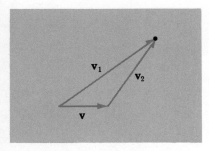

FIG. 6-7. Vector relation for relative velocities: $v_1 = v_2 + v$.

$$x_1 = x_2 + vt \qquad (6\text{-}8)$$

This also follows from another consideration: Since the two reference frames are altogether equivalent, and since S_1 has a velocity $-v$ relative to S_2 if S_2 has a velocity $+v$ relative to S_1, we may interchange subscripts 1 and 2 if we also replace v by $-v$. In this way (6-8) follows from (6-7).

The velocity of the particle as measured in S_1 is $\mathbf{v}_1 = d\mathbf{r}_1/dt$; similarly the velocity of the same particle but now observed in S_2 is $\mathbf{v}_2 = d\mathbf{r}_2/dt$. Therefore, recognizing that $\mathbf{v}$ is a constant and taking the derivative with respect to time of (6-6) yields

$$\mathbf{v}_2 = \mathbf{v}_1 - \mathbf{v}$$
or
$$\mathbf{v}_1 = \mathbf{v}_2 + \mathbf{v} \qquad (6\text{-}9)$$

In words, the instantaneous velocity of the particle relative to S_1 is the velocity of the same particle relative to S_2 plus the velocity of S_2 relative to S_1. See Fig. 6-7. The velocity relations—although simple and (apparently) obvious—merely express in formal mathematical language our everyday, common-sense conceptions of space and time. These relations are not, however, altogether correct. For particle speeds approaching that of light (3×10^8 m/s) they must be supplanted by the more general velocity transformation relations of the special theory of relativity, (Sec. 34-5), which reduce, as they must, to the galilean transformations for the moderate speeds with which we are concerned in classical mechanics.

There is a helpful mnemonic for keeping straight the meaning of the terms and their signs in Eq. (6-9). If we represent the velocity of the particle relative to system 1 by $\mathbf{v}_{p1}$, the velocity of the particle relative to 2 by $\mathbf{v}_{p2}$, and the velocity of system 2 with respect to system 1 by $\mathbf{v}_{21}$, (6-9) becomes

$$\mathbf{v}_{p1} = \mathbf{v}_{p2} + \mathbf{v}_{21}$$

Notice that the "outside" subscripts on the right-hand side correspond to the subscripts of the term on the left. This is illustrated also in Fig. 6-8. The head of each velocity vector is identified with the moving object (or reference frame), and the tail has the label of the reference frame with respect to which the velocity is measured.

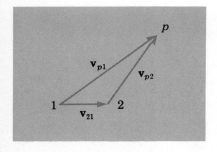

FIG. 6-8. A mnemonic device for relative velocities. The heads and tails are labeled p, 1, and 2 to correspond to the particle, reference frame 1, and reference frame 2, respectively; $\mathbf{v}_{p1}$ gives the velocity of the particle relative to 1, etc.

A simple conclusion follows from (6-9): If a body moves with a constant velocity in one reference frame, then its velocity relative to any other reference frame is also constant (but different), provided the two reference frames have a constant relative velocity. Moreover, a reference frame can always be found in which the body's velocity is zero. For example, if a block slides to the right at 20 m/s relative to the earth, then, with respect to a train traveling 20 m/s to the right, the block *is* (not merely appears to be) *at rest*. Thus, *the states of motion with constant velocity and of rest differ from one another only through the arbitrary choice of a reference frame.*

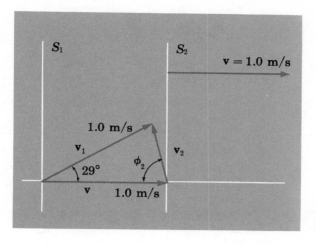

FIG. 6-9. A particle's velocity relative to reference frames S_1 and S_2.

Moreover, if some one reference frame is an inertial frame, then any other reference frame moving with constant velocity relative to it is also an inertial frame. Since the earth is at least an approximate inertial frame, a train moving at constant speed on a straight track, or an elevator ascending or descending at constant speed, is also an inertial frame in that an undisturbed object will maintain a constant horizontal velocity relative to it.

Example 6-7

A particle moves at 1.0 m/s in the direction 29° above the X axis. What is this particle's velocity relative to a reference frame moving along the positive X axis at 1.0 m/s?

The velocity vectors are shown in Fig. 6-9, where the particle's velocity relative to moving frame S_2 is $\mathbf{v}_2$ at an angle ϕ_2 relative to the negative X axis. In symbols

$$\mathbf{v} = \mathbf{v}_1 - \mathbf{v}_2$$

For the X components of velocity vectors we have

$$1.0 \text{ m/s} = (1.0 \text{ m/s}) \cos 29° - v_2 \cos \phi_2$$

where v_2 is the magnitude of $\mathbf{v}_2$. For the Y components of the velocity vectors we have

$$(1.0 \text{ m/s}) \sin 29° = v_2 \sin \phi_2$$

Solving simultaneously for v_2 and ϕ_2 we find that

$$v_2 = 0.51 \text{ m/s}$$

and

$$\phi_2 = 80°$$

Observers in two different inertial frames will measure different velocities for a particle. But they will measure the *same acceleration*. If the acceleration of a particle in S_1 is $\mathbf{a}_1 = d\mathbf{v}_1/dt$ and the acceleration of the same particle in S_2 is $\mathbf{a}_2 = d\mathbf{v}_2/dt$, then taking the time derivative of Eq. (6-9) yields

$$\frac{d\mathbf{v}_1}{dt} = \frac{d\mathbf{v}_2}{dt} + \frac{d\mathbf{v}}{dt}$$

Since the relative velocity **v** between the two inertial frames is a constant, $d\mathbf{v}/dt = 0$, and the above equation becomes

$$\frac{d\mathbf{v}_1}{dt} = \frac{d\mathbf{v}_2}{dt}$$

$$\mathbf{a}_1 = \mathbf{a}_2 \tag{6-10}$$

Thus, if an object has an acceleration (constant or changing) in one reference frame at a given instant, it has, at this same instant, exactly the *same* acceleration in any other reference frame which is moving with a *constant* velocity with respect to the first.

Example 6-8

A boy, standing on a train traveling east at 20 m/s, throws a ball straight up at 10 m/s relative to the train at the moment the train passes a crossing. (*a*) How far from the crossing is the boy when he catches the ball? (*b*) What is the path of the ball as viewed by an observer on the ground?

(*a*) The ball's motion is simplest when viewed from a reference frame traveling with the train. With respect to this frame the boy is at rest, and the ball is seen to rise and descend vertically. The time of flight of the ball can be computed from the equation of motion. With $y_0 = 0$,

$$y = v_0 t + \tfrac{1}{2}at^2$$
$$0 = (10 \text{ m/s})t + \tfrac{1}{2}(-10 \text{ m/s}^2)t^2$$
$$t = 0, \text{ or } 2.0 \text{ s}$$

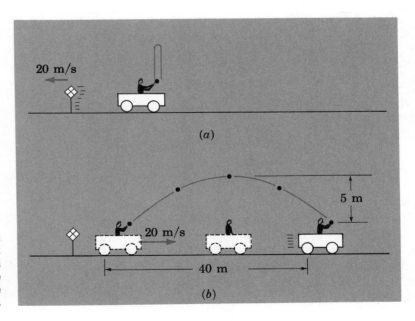

FIG. 6-10. A boy riding in a train throws a ball vertically upward. The motion as seen (*a*) from a reference frame at rest with the train and (*b*) from a reference frame at rest with the earth.

Two seconds elapse from the moment the ball leaves the boy's hand until it is caught.

An observer on ground agrees that the ball is in the air for 2.0 s. During this time, however, the train had advanced a distance of (20 m/s)(2.0 s) = 40 m.

(b) Figure 6-10 shows the path of the ball as viewed from the train and from the ground.

6–4 MOMENTUM CONSERVATION IN THE CENTER-OF-MASS REFERENCE FRAME

Momentum is conserved, and an isolated system's center of mass has a constant velocity in any inertial frame. But one particular inertial frame is of special interest—the *center-of-mass reference frame*—because in it an isolated system's *total momentum* is not only constant but *zero* and the *center-of-mass velocity* not only constant but *zero*. That the center-of-mass frame is indeed the zero-momentum frame follows immediately from Eq. (6-1):

By definition, $\qquad M\mathbf{V} = \Sigma m\mathbf{v}$

If $\qquad\qquad\qquad \mathbf{V} = 0$

then $\qquad\qquad\quad \Sigma m\mathbf{v} = 0$

In words, if the center of mass is at rest, the system's total momentum $\Sigma m\mathbf{v}$ must be zero.

Let us reexamine the collision appearing earlier in Fig. 5-12, both from the point of view of an observer in the laboratory reference frame and now also from the point of view of an observer in the center-of-mass reference frame. In Examples 6-2 and 6-3 we found the center of mass for this system moving at the constant velocity 1.0 m/s to the right in the laboratory, both before and after the collision. Since the two colliding particles had masses of 1 kg and 3 kg, their distances from the center of mass were always in the ratio of 3 to 1, as shown in Example 6-1.

The same collision is also shown in Fig. 6-11 in the reference frame in which the center of mass is at rest. We compute the velocities of the two particles in this reference frame by applying Eq. (6-9), the velocity transformation relation. Indeed, Example 6-7 gives the velocity of the 1-kg mass after the collision in the center-of-mass frame. Once again the two particles maintain a 3 to 1 distance from what is now a fixed center of mass. The particles always move in opposite directions, approaching the center of mass before the collision and receding from it afterward, and the system's total momentum is zero at all times. If we know the motion of one of two interacting particles in the center-of-mass frame, we can immediately find the motion of the other. For example, in the situation above, if we know the location and velocity of the 1-kg

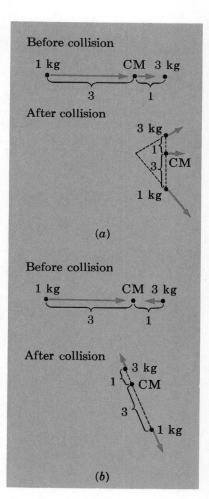

FIG. 6-11. Particle velocities before and after collision as viewed from (a) the laboratory reference frame and (b) the center-of-mass reference frame.

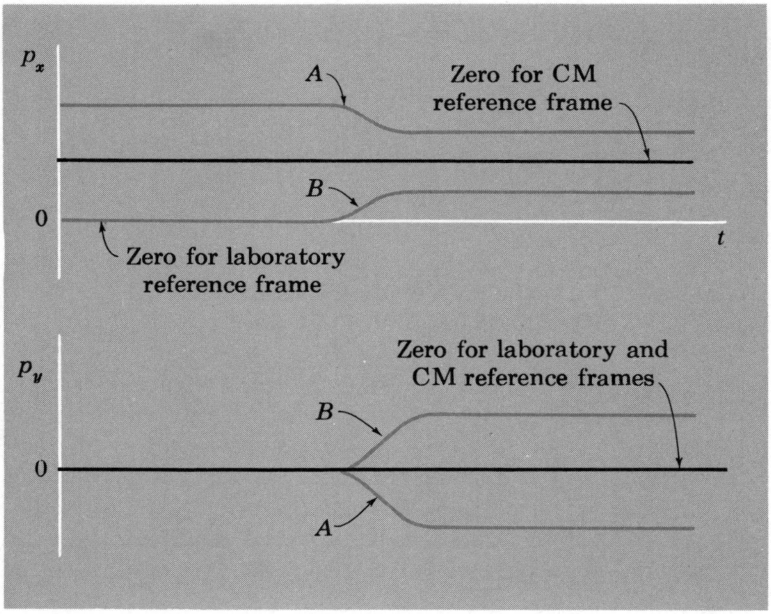

FIG. 6-12. Momentum-time graphs for the X and Y components of the collision shown in Figure 6-11. Note that the momenta for the laboratory and center-of-mass reference frames differ only in the zero for momentum. A represents the 1-kg object, B the 3-kg object.

particle at some instant, then the 3-kg particle must lie on the line connecting the 1-kg particle with the origin at the center of mass, be at one-third the distance from the origin, and move in the opposite direction at one-third the speed.

Consider the momentum-time graphs for the laboratory and center-of-mass reference frames, as shown in Fig. 6-12. The X and Y components of the momenta are plotted separately. The X momentum components of the two particles are different in the two reference frames. In fact, the two reference frames differ simply in the choice of the zero of momentum. Only in the center-of-mass reference frame is the system's total momentum zero. Because the center of mass moves along the X axis, the Y momentum components are the same for the laboratory and center-of-mass reference frames.

SUMMARY

The velocity $\mathbf{V}$ and displacement $\mathbf{R}$ of a system's center-of-mass are given by

$$\mathbf{V} = \frac{\Sigma m \mathbf{v}}{\Sigma m} \qquad (6\text{-}2)$$

$$\mathbf{R} = \frac{\Sigma m \mathbf{r}}{\Sigma m} \qquad (6\text{-}3)$$

The velocity of a particle $\mathbf{v}_1$ relative to reference frame 1 is related to the velocity of the same particle $\mathbf{v}_2$ measured in frame 2 by

$$\mathbf{v}_1 = \mathbf{v}_2 + \mathbf{v} \qquad (6\text{-}9)$$

where $\mathbf{v}$ is the velocity of 2 with respect to 1.

The reference frame for zero total momentum is that in which the system's center of mass is at rest.

PROBLEMS

6-1. As measured in the laboratory, an electron is moving at 1.0×10^6 m/s toward a stationary proton. What is the magnitude and direction of the velocity of the center of mass of the electron–proton system? The electron and proton masses are 9.11×10^{-31} kg and 1.67×10^{-27} kg, respectively.

6-2. Ball A of 2.0-kg mass collides with stationary ball B, whose mass is 6.0 kg. After the collision A is moving northwest at 30 m/s and B is traveling east at 10 m/s. (a) Find the speed and direction of the center of mass of the system. (b) What was the speed of A before the collision? (Assume that A and B form an isolated system.)

6-3. A moving projectile explodes into two fragments with masses in the ratio of three to one. The least massive fragment moves in the Y direction at 200 m/s, and the other piece has a speed of 100 m/s in the X direction. (a) Find the speed and direction of the center of mass. (b) What was the original speed of the projectile? (Assume that the projectile is isolated from external influences.)

6-4. Object A moving east at 10 m/s collides with object B. After the collision A moves in the direction 30° north of east at 10 m/s and B is traveling south at 20 m/s. A and B are of equal mass. (a) Find the speed and direction of the center of mass of the system. (b) What was the original speed and direction of motion for object B?

6-5. The following are the mass, X and Y coordinates, and X and Y components of the velocity $(m; x, y; v_x, v_y)$ of the three particles: (1 kg; 0 m, 2 m; -2 m/s, 0); (2 kg; -4 m, 0 m; 1 m/s, -1 m/s); and (2 kg; 1 m, -2 m; 0 m/s, 4 m/s). Find (a) the X and Y coordinates and (b) the velocity components V_x and V_y of the system's center of mass.

6-6. The mass, X and Y coordinates, and X and Y velocity components $(m; x, y; v_x, v_y)$ of two particles are: (1 kg; 2 m, -2 m; 0 m/s, 4 m/s) and (2 kg; 0 m, 1 m; 2 m/s, 0 m/s). What must be the coordinates and velocity components of a third particle of 2-kg mass if the center of mass of the three-particle system is to be at rest at the origin?

6-7. Find the coordinates of the center of mass of the bracket shown in Fig. P6-7. The bracket is fashioned from a flat metal plate of uniform density. (*Hint:* Subdivide the plate into rectangles, and locate the center of mass of each rectangle. Use the fact that each rectangle's mass is proportional to its area.)

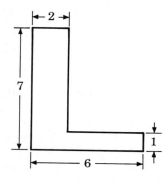

FIG. P6-7

6-8. Show that the center of mass of a uniform triangular plate lies at the intersection of the three lines joining the center of each side with the opposite vertex.

6-9. An airplane is moving 30° east of north with a ground speed of 800 km/h, and the wind is eastward at 100 km/h. What are the airspeed and the compass heading?

6-10. Raindrops are falling vertically at 6.0 m/s, relative to the ground. What is the speed of the drops and their direction relative to the vertical as measured by a man who walks through the rain at 1.5 m/s?

6-11. Hailstones are falling at an angle of 60° with respect to the vertical at a speed of 40 m/s. In what direction and at what speed would a ground observer (moving horizontally) need to travel in order for the hailstones to appear to fall "straight down"?

6-12. A telescope is to capture the light from a distant star directly overhead at the zenith. Because the earth revolves about the sun at the speed 3.0×10^4 m/s and light travels at the finite speed of 3.0×10^8 m/s, the telescope cannot be pointed directly upward, but must be tipped slightly with respect to the vertical. *(a)* If the telescope is at the equator, in which direction, east or west, must it be tipped? (Neglect the earth's rotation.) *(b)* What is the angle of the telescope relative to the vertical? (The phenomenon, known as the *stellar aberration of light,* was first observed by J. Bradley in 1725; Bradley computed the speed of light from the observed aberration angle.)

6-13. A woman rows a boat in still water at 3.0 mi/h. She is to cross a river in which the current is 4.0 mi/h. In what direction should she head, relative to the downstream direction, to cross the river *(a)* in a minimum time, *(b)* with a minimum distance traveled, *(c)* at a minimum speed relative to the river bank?

6-14. A man rows a boat at 4.0 mi/h with respect to still water. He is to cross a river 1.0 mi wide in which the current is 3.0 mi/h. *(a)* If the man continuously heads the boat up river at an angle of 30° to the river bank, how long does it take him to cross the river? *(b)* At what point *down river* will he reach the opposite bank? *(c)* If the man heads the boat to cross the river in the least time, how far down river will he be transported? *(d)* What is then the crossing time? *(e)* The man now heads the boat so as to move the boat the least distance down river. What is this distance? *(f)* What is the corresponding crossing time?

6-15. A boy on a bicycle traveling east at 4.0 m/s tosses a newspaper toward the northeast at 3.0 m/s (as he sees it). What are the speed and direction of the paper as seen by someone standing on the ground?

6-16. Trucks A and B are approaching an intersection, A moving east at 50 km/h and B headed north at 80 km/h. Find the velocity of each truck relative to the other.

6-17. A ball is dropped from the mast of a ship moving at 5.0 m/s. The mast is 15.0 m high. Neglect air effects. *(a)* Where does the ball strike the ship relative to the base of the mast? *(b)* With what speed does it strike the deck relative to an observer on board ship? *(c)* With what speed does it strike the deck relative to an observer on land? *(d)* Through what horizontal distance does the ship move as the ball falls?

  Galileo argued that an object dropped from the mast of a moving ship would strike the deck at the base of the mast, rather than fall behind. The dropped object has zero horizontal velocity relative to the ship or, equivalently, the object and the ship have the same horizontal velocity relative to an observer on land. This argument was used to make plausible the heliocentric cosmology proposed by Copernicus, in which the earth and objects on its surface are imagined to hurtle through space at extraordinarily high speeds by virtue of the earth's revolution about the sun and its rotation about its own axis. An important objection to the Copernican cosmology was this: "If the earth is not at rest, but rather travels at high speeds through space, how is it that objects on the earth's surface are not left behind, that there is no strong wind, that birds can fly serenely through the atmosphere?"

6-18. Two billiard balls approach at equal speeds and collide head on. Each ball recedes from the collision with the same speed as that with which it approached. Describe what happens when one billiard ball in motion strikes head on a second ball which is initially at rest.

6-19. A box sits on a frictionless surface. Inside the box a particle initially moves to the right at constant speed and then strikes the right wall of the box and bounces off to the left. Later it hits the left wall and bounces off to the right. Describe the motion of the box.

6-20. A ping-pong ball hitting a billiard ball at rest bounces back with the same speed, the billiard ball remaining essentially at rest. *(a)* If a billiard ball with speed v hits a ping-pong ball initially at rest, what is the ping-pong ball's speed after the collision? *(b)* If the billiard ball has velocity $\mathbf{v}_b$ before colliding head on with a ping-pong ball of velocity $\mathbf{v}_p$, what is the velocity of the ping-pong ball after collision?

6-21. A space probe approaches the planet Jupiter at speed v_i, which is itself moving in approximately the opposite direction at speed v_J. The probe turns the corner around the planet and is later found moving in nearly the opposite direction to its initial velocity and at the same distance from the planet with the speed v_f. See Fig. P6-21. The mass of the space probe

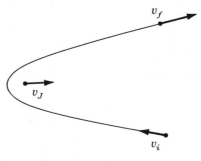

FIG. P6-21

is much less than that of Jupiter, and its collision with the planet is rather like that of a ping-pong ball with a billiard ball. Show that $v_f \approx v_i + 2v_J$; through its interaction with the planet the space probe's speed has increased by *twice* the speed of the planet. This illustrates the "slingshot" effect which is being employed for space probes on a "Grand Tour" of a number of the planets of the solar system.

6-22. Two particles are attached to the ends of a spring of negligible mass. The string is stretched and it, together with the two attached particles, is set in motion along the long dimension of the spring. Each of the two particles executes a complicated motion, and the spring coasts in one direction, all the while contracting and expanding. Show that the two particles have the same velocity at those times when the spring's length is a minimum or a maximum. (*Hint:* Consider the motion from the center-of-mass frame.)

6-23. A particle has a speed v when measured in the laboratory. Its speed is also v when measured in the center-of-mass frame. (*a*) What are the *two* possible speeds of the center-of-mass frame relative to the laboratory? What is the angle between the particle's velocity and the velocity of center of mass relative to the laboratory as measured in (*b*) the laboratory and (*c*) center-of-mass frame?

6-24. An object initially moving east at 10 m/s explodes in flight into two pieces of equal mass. One travels at 40 m/s in the direction 60° north of east. (*a*) What is the velocity of the second piece as measured in the laboratory? (*b*) What is the velocity of the system's center of mass? (*c*) What is the velocity of the first piece and (*d*) second piece as measured from the center-of-mass reference frame?

6-25. A 20-kg mass travels to the right at 15 m/s, and a 10-kg mass travels to the left at 10 m/s. (*a*) What is the total linear momentum of the system of two bodies as viewed from the laboratory? (*b*) What is the velocity of the center of mass of the system? (*c*) What is the velocity of each body relative to the center of mass? (*d*) What is the linear momentum of each of the bodies relative to a reference frame moving with the center of mass?

6-26. A billiard ball traveling initially at speed v collides with another billiard ball at rest. After the collision each ball moves at a speed of $v/\sqrt{2}$ and at an angle of 45° with respect to the direction of incident ball's initial velocity (the angle between the velocities of the two outgoing balls is 90°). (*a*) What is the velocity of the system's center of mass? (*b*) What are the velocities of the two balls before collision in the center-of-mass frame? (*c*) What are the velocities of the two balls after collision in the center-of-mass frame?

6-27. An isolated system consists of a particle with mass m_a and velocity $\mathbf{v}_{a1}$ in the laboratory and a particle of mass m_b and velocity $\mathbf{v}_{b1}$ also in the laboratory. Show that the velocity of m_a relative to the system's center of mass is $\mathbf{v}_{a2} = m_b(\mathbf{v}_{a1} - \mathbf{v}_{b1})/(m_a + m_b)$ and that m_b's velocity relative to the center of mass is $\mathbf{v}_{b2} = -m_a \mathbf{v}_{a2}/m_b = m_a(\mathbf{v}_{b1} - \mathbf{v}_{a2})/(m_a + m_b)$.

6-28. The momentum-conservation law is a universal law of physics because it applies in any reference frame. Observers in different reference frames in relative motion will attribute different values to the velocity and momentum of the same object, but all such observers will agree that in a collision between two otherwise isolated objects the total momentum is constant. An observer in frame S_1 writes for the collision between masses m and M,

$$m\mathbf{v}_1 + M\mathbf{V}_1 = m\mathbf{v}'_1 + M\mathbf{V}'_1$$

where $\mathbf{v}_1$ is m's velocity before and $\mathbf{v}'_1$ its velocity after collision, and similarly for M. Use the velocity transformation relation to change these velocities into those observed in another inertial frame S_2, and show that

$$m\mathbf{v}_2 + M\mathbf{V}_2 = m\mathbf{v}'_2 + M\mathbf{V}'_2$$

where $\mathbf{v}_2$ and $\mathbf{v}'_2$ are the velocities of m before and after the collision, and similarly for M. What is proved thereby is that the law of momentum conservation is invariant under a galilean velocity transformation.

CHAPTER 7

Force

This chapter is concerned with force—its fundamental origins, its relation to linear momentum, and its appearance in the fundamental laws of mechanics. We first consider qualitatively the sources, or origins, of force in the fundamental interactions between particles. Then we define force as the time rate of change of linear momentum and explore the consequences of this definition. The superposition principle for forces, which shows that simultaneous forces add as vectors, lead us to Newton's second law of motion. The impulse of a force is defined, and the impulse-momentum theorem developed. Next, the appearance of equal and opposite action-reaction force pairs, known as Newton's third law of motion, is derived from the momentum-conservation law. Weight is defined, and its relation to mass is given.

In this chapter we lay the foundations of the meaning of force and of Newton's laws, the foundation of classical mechanics.

7–1 THE FUNDAMENTAL ORIGINS OF FORCE

A single particle is said to be acted upon by a force when its momentum *changes;* to test whether a body is subject to an unbalanced, or net, force, one simply notes whether the body departs from motion at constant velocity. A force is most simply illustrated when one has just two interacting particles; each influences the motion of the other, and we can describe this interaction in terms of a force acting on each particle by virtue of the presence of the other. See Fig. 7-1.

There is a variety of forces with which we are familiar: a muscular exertion producing a push or pull, the force of one solid object on another when they come in contact, the gravitational force between the earth and a satellite, the force of a magnet on a

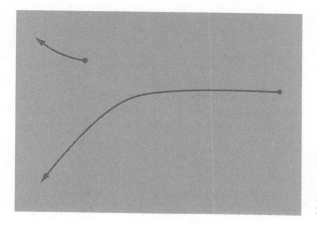

FIG. 7-1. A particle is acted upon by a force when, in interacting with at least one other particle, its momentum changes direction or magnitude.

piece of iron or on another magnet, the electrostatic force of a rubbed insulator, such as a plastic rod, on bits of paper. Despite the diversity of these forces it is a remarkable fact that all of them—indeed all forces in physics now known—can be traced to four basic forces: (1) the universal gravitational force of attraction between any two objects; (2) the so-called weak-interaction force which governs the β decay of certain radioactive particles and other subatomic processes; (3) the force between electric charges, at rest or in motion, which give rise to the electric and magnetic forces; and (4) the very strong nuclear force between the constituents of the atomic nucleus. The weak-interaction and nuclear forces, although important in nuclear structure, have virtually no effect even in atomic structure. Therefore, the only forces that play a role in classical physics are the gravitational, electric, and magnetic forces. *All* other forces in classical physics can be traced ultimately to these.

7-2 FORCE DEFINED

Since a change in a particle's momentum implies that it is being acted upon by a force, the size of the force must be related to the size of the momentum change. But momentum change alone cannot be a proper quantitative measure of force because the time interval over which the momentum change occurs must also enter. A particle's momentum may change slowly or abruptly. A gentle force changes an object's momentum; a strong force produces the same momentum change in a shorter time interval. What matters is the time rate at which the momentum changes. Indeed we define the *force* acting on an object to be the *time rate of change of* the object's *momentum*.†

---

†This definition is justified, not merely because it accords with our common-sense notions of what constitutes a large or small push or pull, but more especially because with force so defined we

If an object's momentum $\mathbf{p} = m\mathbf{v}$ changes by $\Delta \mathbf{p} = \Delta(m\mathbf{v})$ over the time interval Δt, the average force $\overline{\mathbf{F}}$ acting on the object is

$$\overline{\mathbf{F}} = \frac{\Delta \mathbf{p}}{\Delta t} = \frac{\Delta(m\mathbf{v})}{\Delta t} \qquad (7\text{-}1)$$

The instantaneous force (hereafter referred to merely as force) is the average force computed over a vanishingly small time interval; equivalently, force is the derivative with respect to time of momentum:

$$\mathbf{F} = \frac{d\mathbf{p}}{dt} = \frac{d(m\mathbf{v})}{dt} \qquad (7\text{-}2)$$

For a classical particle, whose mass is a constant independent of speed or any other circumstance, we may write

$$\mathbf{F} = \frac{d(m\mathbf{v})}{dt} = m\frac{d\mathbf{v}}{dt} = m\mathbf{a} \qquad (7\text{-}3)$$

The force acting on a particle is equal to simply its mass multiplied by its acceleration.

Force is a vector having the same direction as the momentum change with which it is associated. Clearly, a particle whose speed changes is acted upon by a force because it is accelerated and its momentum changes. A particle which moves in a circle at constant speed is also acted upon by a force because its momentum direction changes continuously. In short, a force acts whenever a particle changes momentum, either magnitude or direction, or both.

From its definition as momentum divided by time, force has dimensions of ML/T^2 and, in the International System, units of kilogram meters per second squared, kg-m/s². This combination of units is known as a *newton* and abbreviated by N.

$$1 \text{ newton} = 1 \text{ N} = 1 \text{ kg-m/s}^2$$

Thus, a force of 1 N causes the momentum of an object to change by 1 kg-m/s in 1 s; equivalently, 1 N imparts an acceleration of 1 m/s² to a 1-kg object.

In the cgs (centimeter-*gram*-second) metric system of units the force unit is the *dyne:* 1 dyn = 1 g-cm/s² = $(10^{-3}$ kg$)(10^{-2}$ m/s²$)$ = 10^{-5} N. In the U.S. Customary System of Units the force unit is the *pound:* 1 lb = 4.448 N.

Example 7-1

A 10-kg body interacts with a second body. The 10-kg body initially moves east at 20 m/s; $\frac{1}{2}$ s later it is traveling north at 15 m/s. What is the average force acting on the 10-kg body during the half-second time interval?

---

find that simultaneous forces add as vectors and such fundamental forces as the gravitational force take on simple mathematical form, as do the related potential-energy functions.

To compute the force, we first find the change in momentum $\Delta\mathbf{p}$. See Fig. 7-2. The momentum vector diagram shows that $\Delta\mathbf{p} = \mathbf{p} - \mathbf{p}_0$ is 250 kg-m/s in the direction 37° north of west. Therefore, from Eq. (7-1), the average force is

$$\overline{\mathbf{F}} = \frac{\Delta\mathbf{p}}{\Delta t} = \frac{250 \text{ kg-m/s}}{0.50 \text{ s}}$$

$$= 500 \text{ N at } 37° \text{ north of west}$$

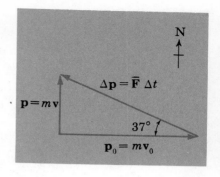

FIG. 7-2

Example 7-2

An electron (mass, 9.1×10^{-31} kg) starts from rest at a negatively charged plate and arrives at a positively charged, parallel plate 2.0 cm away in a time of 1.4×10^{-7} s. If the electron's acceleration is constant throughout the motion, what force acts on it?

The displacement of a particle at constant acceleration starting from rest is

$$x = \tfrac{1}{2}at^2 \quad (3\text{-}7)$$

Therefore, the constant force on the electron is

$$F = ma = \frac{m(2x)}{t^2} = \frac{(9.1 \times 10^{-31} \text{ kg})(4.0 \times 10^{-2} \text{ m})}{(1.4 \times 10^{-7} \text{ s})^2}$$

$$= 1.9 \times 10^{-18} \text{ N}$$

7-3 THE SUPERPOSITION PRINCIPLE FOR FORCES

When an object interacts with a second object, it is subject to a force which is, by definition, equal to the object's time rate of change of momentum. In this section we establish that if an object interacts with *two or more* other objects and is, therefore, subject to several forces simultaneously, the time rate of change of the object's momentum is equal to the *vector sum* of the individual forces acting on it.

Suppose that we have an object of mass m on a frictionless surface and that a spring is attached to it to produce a force on it. (An ordinary helical spring when stretched or compressed produces a force directly proportional to the amount of stretch or compression.) With a first spring attached, the object has an acceleration $\mathbf{a}_1$, so that the spring's force is $\mathbf{F}_1 = m\mathbf{a}_1$. When the first spring is removed and a second one attached, the object undergoes an acceleration $\mathbf{a}_2$ (not necessarily in the same direction as $\mathbf{a}_1$), so that the force of the second spring is $\mathbf{F}_2 = m\mathbf{a}_2$.

Now suppose that both springs are attached to the object, each spring being stretched by the same amount and in the same direction as before. The object is now subject to two simultaneous forces. See Fig. 7-3. Its acceleration is $\mathbf{a}$, where *by experiment*,

$$\mathbf{a} = \mathbf{a}_1 + \mathbf{a}_2$$

both in magnitude and in direction. It follows that

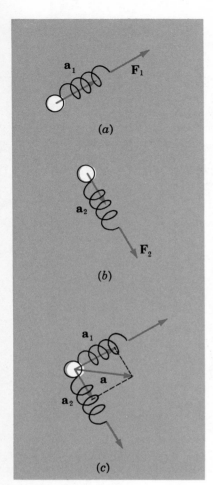

FIG. 7-3. Object acted on by (a) force F_1 alone, (b) force F_2 alone, (c) forces F_1 and F_2 simultaneously.

$$\mathbf{a} = \frac{\mathbf{F}_1}{m} + \frac{\mathbf{F}_2}{m}$$

or

$$\mathbf{F}_1 + \mathbf{F}_2 = m\mathbf{a}$$

The left-hand side of this equation is the *vector sum* of the simultaneously applied forces. Two or more forces acting simultaneously on any object are altogether equivalent to a single force equal to their vector sum. This is the *superposition principle* for forces.

If the vector sum of forces acting on an object is represented by $\Sigma\mathbf{F}$,

$$\Sigma\mathbf{F} = \frac{d\mathbf{p}}{dt} = m\mathbf{a} \tag{7-4}$$

The time rate of change of an object's momentum is equal, in magnitude and direction, to the vector sum of the forces acting on it. This result, which comes from the definition of force and the experimentally determined superposition principle, is known as *Newton's second law of motion*. It is the fundamental principle governing the motions of interacting objects. The next chapter will illustrate its wide applicability.

One simple but important consequence of the force superposition principle is that two or more forces may in combination produce *zero* resultant force. Then the object's momentum is constant and its acceleration zero. For this reason the law of inertia must be stated as follows: A particle has constant momentum when acted upon by zero *resultant* force.

7-4 FORCE PAIRS

Consider two particles with momenta $\mathbf{p}_1$ and $\mathbf{p}_2$ which interact with one another but are otherwise isolated. Their total momentum is constant:

$$\mathbf{p}_1 + \mathbf{p}_2 = \text{constant}$$

Taking the time derivative gives

$$\frac{d\mathbf{p}_1}{dt} + \frac{d\mathbf{p}_2}{dt} = 0$$

or

$$\frac{d\mathbf{p}_1}{dt} = -\frac{d\mathbf{p}_2}{dt}$$

By definition, $d\mathbf{p}_1/dt$ is the force on particle 1 arising from its interaction with particle 2,

$$\mathbf{F}_{2 \text{ on } 1} = \frac{d\mathbf{p}_1}{dt}$$

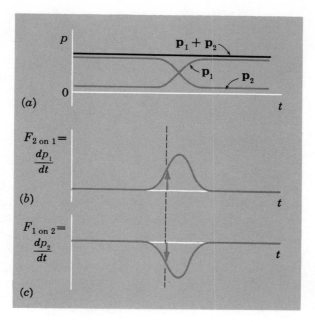

FIG. 7-4. (a) The momentum-time graphs for two colliding objects. (b) Force-time graph for object 1. (c) Force-time graph for object 2.

Similarly, the force on particle 2 arising from 1 is

$$\mathbf{F}_{1 \text{ on } 2} = \frac{d\mathbf{p}_2}{dt}$$

Substituting in the equation above yields

$$\mathbf{F}_{2 \text{ on } 1} = -\mathbf{F}_{1 \text{ on } 2} \tag{7-5}$$

The two forces, often referred to as the *action* and *reaction* forces, are *equal in magnitude* and *opposite in direction*. They act on different objects—one on 1, the other on 2—and consequently they cannot place one of the two interacting objects in equilibrium.

We can see the same thing graphically for two objects interacting in a head-on collision; the momentum-time and the related force-time graphs are shown in Fig. 7-4. Momentum conservation requires that the total momentum of this isolated system remain constant at all times. Momentum gained by object 1 is balanced at all times by an equal momentum loss by object 2: The momentum-time graph for object 2 is like that for 1, but inverted. The force-time graphs are also of the same shape, but the force on 1 is always positive while that on 2 is negative. Moreover, the instantaneous force of 2 on 1 is at each instant equal in magnitude and opposite in direction to the force of 1 on 2, as given by (7-5).

This important result, a consequence of the definition of force and the momentum-conservation law, is known as *Newton's third law of motion*.

Forces occur in pairs only. When a hammer strikes a nail, the force of the hammer on the nail is precisely equal in magnitude

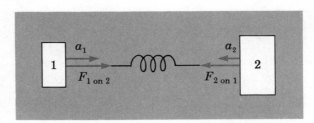

FIG. 7-5. The ratio of the acceleration magnitudes for two interacting objects is equal to the inverse ratio of their respective masses.

to the force of the nail on the hammer. When a ball crashes through a window, the force of the ball on the window is exactly equal to the force of the window on the ball. The force of a rocket on its exhaust particles is exactly equal to the force of the exhaust particles on the rocket. The force of the earth on a falling apple is precisely equal in magnitude to the force of the falling apple on the earth.

We can use Newton's third law to compare masses. Consider two interacting, but otherwise isolated, particles. From (7-5),

$$\mathbf{F}_{2\text{ on }1} = -\mathbf{F}_{1\text{ on }2}$$

Each particle has only one force acting on it and, by Newton's second law,

$$m_1 \mathbf{a}_1 = -m_2 \mathbf{a}_2$$

or

$$\frac{m_1}{m_2} = -\frac{a_2}{a_1} \tag{7-6}$$

The mass ratio equals the inverse acceleration ratio. See Fig. 7-5. If two interacting particles have equal accelerations, their masses are the same.

7–5 WEIGHT

Experiment shows that any object in free flight over short distances near the earth has a constant acceleration **g** toward the earth's center. This implies that any object is acted upon by a constant force arising from its interaction with the entire earth. The phenomenon of an object's being attracted by the earth is called *gravitation*, and the force on the object due to its gravitational interaction with the earth is called its *weight*.

Applying Newton's second law to a freely falling object of mass m, weight **w**, and acceleration **g**, we have (Fig. 7-6)

$$\Sigma \mathbf{F} = m\mathbf{a}$$

$$\mathbf{w} = m\mathbf{g} \tag{7-7}$$

The weight magnitude is simply the mass multiplied by g, the acceleration due to gravity. The direction of the weight is "down," which is to say, toward the earth's center. The acceleration of a

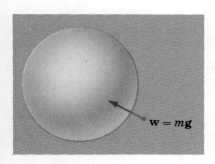

FIG. 7-6. Newton's second law applied to a freely falling object.

freely falling object is, from another point of view, that vector $\mathbf{g} = \mathbf{w}/m$, which gives the gravitational force per unit mass.

The so-called standard acceleration due to gravity, close to the measured value at latitudes of 45° near sea level, is

$$\text{Standard } g = 9.80665 \text{ m/s}^2 = 32.17398 \text{ ft/s}^2$$

Therefore, the weight of a 1-kg object is about 9.8 N.† A small apple weighs about 1 N.

In nonscientific usage, the terms mass and weight are often taken as interchangeable. Mass and weight are, however, quite different quantities.

Weight is a vector; it is a measure of the earth's pull on an object (or of the object's pull on the earth). In magnitude weight is not constant but depends on the magnitude of g, which, although constant at one location near the earth, shows variations with altitude and latitude and is, in fact, zero for objects far from the earth in interstellar space. Thus, the weight of a 1-kg mass is 9.80 N at a location where g is 9.80 m/s²; at another location on earth, where g is 9.79 m/s², the weight is 9.79 N. A 1-kg object, or an object of any mass, is truly weightless when in interstellar space far from other objects.

On the other hand, mass is a scalar; it is an intrinsic property of the object, not dependent on its interaction with others. A force of 1 N must act on a mass of 1 kg to produce an acceleration of 1 m/s², whether the object is at the earth's surface or center, or at any other location.

Suppose we compare the weights of two objects of masses m_1 and m_2 at the same location. Both bodies have the same acceleration $\mathbf{g}$ when falling; therefore,

$$\mathbf{w}_1 = m_1 \mathbf{g} \qquad \mathbf{w}_2 = m_2 \mathbf{g}$$

and

$$\frac{w_1}{w_2} = \frac{m_1}{m_2} \tag{7-8}$$

The mass ratio of two objects at the same location is the same as their weight ratio. Thus, an ordinary gravitational beam balance, which compares the *weights* of two bodies, also indicates, when it is in balance, that the *masses* are equal (see Sec. 5-2).

An object's weight is independent of its state of motion. We can see this by considering an object tossed up in the air: Its acceleration is *always* $\mathbf{g}$, whether it is going up or down or is momentarily at rest at the zenith. Therefore, the gravitational force $\mathbf{w} = m\mathbf{g}$ is likewise independent of the object's motion.

---

†The pound is a force, or weight, unit, not a mass unit. Since the mass of an object having weight w is $m = w/g$, we may write the mass of, say, a 64-lb object as $m = (64 \text{ lb})/(32 \text{ ft/s}^2) = 2 \text{ lb}/(\text{ft/s}^2)$. Sometimes the term *slug* is used for the combination of units lb/(ft/s²), but this can be avoided entirely simply by writing the mass of an object in the U.S. Customary System of Units as its weight divided by g.

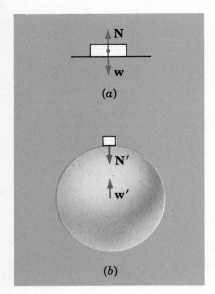

FIG. 7-7 (a) Forces on an object at rest on a horizontal surface. (b) The reaction forces to the forces shown in (a). $N = N'$ and $w = w'$.

For example, when a 1-kg body is at rest on a horizontal surface, it is said to be in equilibrium. There is no net force on it. The earth exerts the force **w** downward; there must be an upward force **N**, equal in magnitude to the weight, as shown in Fig. 7-7. The upward force on the body arising from its interaction with the surface is perpendicular (normal) to the surface; it is termed the *normal force*. In magnitude, $N = w$ *for this situation*. The normal force and weight, although equal and opposite forces, do *not* comprise an action-reaction pair.

Consider the reaction forces in Fig. 7-7b. The weight is the force of earth on block; the reaction force is the pull of block on earth. The normal force is the force of surface on block; the reaction force is the force of block on surface. The only forces that are relevant in describing the state of motion, or rest, *of the block* are the forces acting *on it* — the weight and the normal force. The reactions to these forces do *not* act on the block and are irrelevant to its motion.

7-6 IMPULSE AND MOMENTUM

From Eq. (7-1) the momentum change $\Delta\mathbf{p}$ occurring over a time interval Δt may be written

$$\Delta\mathbf{p} = \overline{\mathbf{F}}\,\Delta t \tag{7-9}$$

where $\overline{\mathbf{F}}$ is the average force over this interval.

Similarly, the incremental momentum change $d\mathbf{p}$ over time dt is

$$d\mathbf{p} = \mathbf{F}\,dt$$

where $\mathbf{F}$ is the instantaneous force. Integrating both sides of this equation from some initial time t_i to a final time t_f yields

$$\int_{\mathbf{p}_i}^{\mathbf{p}_f} d\mathbf{p} = \int_{t_i}^{t_f} \mathbf{F}\,dt$$

$$\mathbf{p}_f - \mathbf{p}_i = \Delta\mathbf{p} = \int_{t_i}^{t_f} \mathbf{F}\,dt \tag{7-10}$$

where $\mathbf{p}_f$ is the final momentum, $\mathbf{p}_i$ the initial momentum, and $\Delta\mathbf{p}$ the net momentum change.

The integral of force over time $\int \mathbf{F}\,dt$ is known as the *impulse* of the force. As (7-10) shows, *impulse equals change in momentum*. This is known as the *impulse-momentum theorem*.

Force is the derivative of momentum with respect to time; momentum change is the integral of force over time, this integral being termed the impulse. Force is a measure of the rate of flow of momentum between interacting objects; impulse is a measure of the net momentum transferred, as shown schematically in Fig. 7-8.

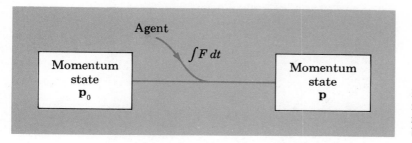

FIG. 7-8. A schematic representation of the transfer of momentum from an agent to a body.

When a force changes with time, it is often useful to speak of the equivalent time-average force $\overline{\mathbf{F}}$. Combining (7-9) and (7-10), we have

$$\text{Impulse} = \int_{t_i}^{t_f} \mathbf{F}\, dt = \overline{\mathbf{F}}\, \Delta t \qquad (7\text{-}11)$$

where $\Delta t = t_f - t_i$. In words, a constant force $\overline{\mathbf{F}}$ has the same impulse and produces the same momentum change as does a time-varying force $\mathbf{F}$ when the forces are related as given in (7-11).

Although the term *impulse* used colloquially usually implies a force which acts only for a very short time, the formal definition of impulse includes any force, however long the period during which it acts. An *impulsive force* usually implies one acting only for a short time, for example, the force of a hammer on a nail.

Impulse has the same units as momentum: kg-m/s or N-s.

It is useful to examine the graphical relations among momentum, time, force, and impulse. Figure 7-9a shows the momentum of an object plotted as a function of time. The momentum is constant before and after the interaction period Δt; during Δt the momentum increases at a constant rate. The average force on the object $\overline{\mathbf{F}} = \Delta \mathbf{p}/\Delta t$ is constant over the time Δt, and otherwise zero, as

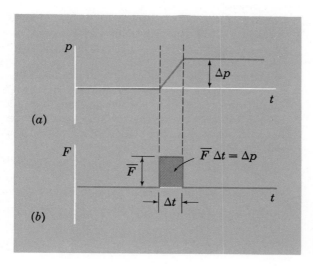

FIG. 7-9. Graphs for a constant force: (a) momentum-time and (b) force-time. The impulse is the area under the force-time graph.

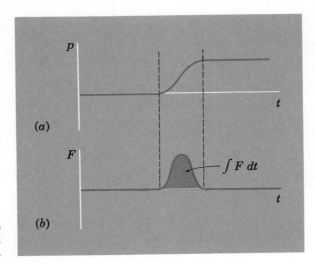

FIG. 7-10. Graphs for variable force: (a) momentum-time and (b) force-time.

shown in Fig. 7-9b. In graphical terms force is the slope of the momentum-time graph. Conversely, the impulse $\bar{\mathbf{F}} \Delta t$ is the area under the force-time graph.

A more realistic situation is portrayed in the momentum-time and corresponding force-time graphs of Fig. 7-10a and b. Here the momentum changes smoothly, and the force rises continuously to a peak and falls back to zero. The instantaneous force $\mathbf{F} = d\mathbf{p}/dt$ is plotted, as derived from the slope of the momentum-time graph. Again impulse is the area under the force-time graph: The area of a slender rectangle is $\mathbf{F}\, dt$, and the entire area under the curve $\int \mathbf{F}\, dt$.

Example 7-3

A 3.0-kg block slides on a frictionless horizontal surface, first moving left at 50 m/s, as shown in Fig. 7-11a. It collides with the spring, compresses it, and is brought to rest momentarily. The body continues to be accelerated to the right by the force of the compressed spring. Finally, the body moves to the right at 40 m/s. The block remains in contact with the spring for 0.020 s. (a) What was the direction and magnitude of the impulse of the spring on the block? (b) What was the spring's average force on the block?

(a) We do not have a record of the force on the block as a function of time; therefore, it is not possible to compute the impulse by finding the area under the instantaneous force-time graph. We do know the change in the block's momentum $\Delta \mathbf{p}$ arising from the impulses. It is

$$\Delta \mathbf{p} = \mathbf{p}_f - \mathbf{p}_i$$

Choosing the right as the positive direction, we have

$$\Delta \mathbf{p} = (3.0 \text{ kg})(+40 \text{ m/s}) - (3.0 \text{ kg})(-50 \text{ m/s}) = +270 \text{ kg-m/s}$$

Note that initial momentum $\mathbf{p}_i$ is *negative*, signifying a momentum to the left. We now know the impulse from Eq. (7-10):

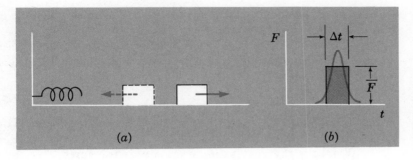

FIG. 7-11. An object initially moving to the left strikes and compresses a spring and then moves to the right. *(b)* Force-time graph for the collision in *(a)*.

$$\text{Impulse} = \Delta \mathbf{p} = 270 \text{ kg-m/s} = 270 \text{ N-s} \quad \text{(to the right)}$$

(b) To find the average force acting on the block during the collision, we use Eq. (7-11),

$$\text{Impulse} = \overline{\mathbf{F}} \Delta t = \int_{t_i}^{t_f} \mathbf{F} \, dt$$

where Δt is now the interaction time 0.020 s. Therefore, we can compute the average force:

$$\overline{\mathbf{F}} = \frac{\text{impulse}}{\Delta t} = \frac{270 \text{ N-s}}{0.020 \text{ s}}$$

$$\overline{\mathbf{F}} = 13{,}500 \text{ N} \approx 1.5 \text{ ton!} \quad \text{(to the right)}$$

A *constant* force of 13,500 N acting over 0.020 s produces the same impulse as that produced by the varying spring force. Since the force of the spring first increases from zero to a maximum, then decreases to zero again, the maximum instantaneous force *exceeds* the equivalent constant force of 13,500 N, as shown in Fig. 7-11*b*. By definition, the equivalent average force $\overline{\mathbf{F}}$ is so chosen that its impulse $\overline{\mathbf{F}} \Delta t$, over the interaction time Δt, is equal to the actual impulse $\int \mathbf{F} \, dt$; that is, the area of the rectangle in Fig. 7-11*b* equals the area under the actual force-time graph (both areas representing the magnitude of the impulse).

SUMMARY

When two particles, 1 and 2, interact, the instantaneous force, $\mathbf{F}_{1 \text{ on } 2}$, of 1 on 2 is defined as the derivative with respect to time of the linear momentum of 2:

$$\mathbf{F}_{1 \text{ on } 2} = \frac{d\mathbf{p}_2}{dt} \quad (7\text{-}2)$$

In its general form Newton's second law of motion for a particle is

$$\Sigma \mathbf{F} = \frac{d\mathbf{p}}{dt} = m\mathbf{a} \quad (7\text{-}4)$$

where $\Sigma\mathbf{F}$ is the vector sum of all forces acting on the particle and $\mathbf{p}$ is its linear momentum.

Newton's third law of motion, a direct consequence of momentum conservation, states that when two particles interact through a mutual force,

$$\mathbf{F}_{2 \text{ on } 1} = -\mathbf{F}_{1 \text{ on } 2} \tag{7-5}$$

Forces occur in pairs: an action force and a reaction force. These forces act on different objects.

Weight is the gravitational force $\mathbf{w}$ of the earth:

$$\mathbf{w} = m\mathbf{g} \tag{7-7}$$

The impulse-momentum theorem is

$$\int_{t_i}^{t_f} \mathbf{F}\, dt = \mathbf{p}_f - \mathbf{p}_i \tag{7-10}$$

where the left-hand side is called the impulse of the force $\mathbf{F}$ and the right-hand side is the resultant change in linear momentum.

PROBLEMS

7-1. Figure P7-1 is a plot of the instantaneous momentum of an object as a function of time. What is the average force on the object over the intervals (a) $t = 0$ to 1 s, (b) $t = 1$ to 2 s, (c) $t = 2$ to 3 s, (d) $t = 3$ to 4 s, (e) $t = 4$ to 5 s, and (f) $t = 5$ to 6 s?

7-2. Figure P7-2 is a plot of the instantaneous force on a 2.0-kg object as a function of time. What is the time average of the force over time intervals (a) $t = 0$ to 1 s, (b) $t = 1$ to 2 s, (c) $t = 2$ to 3 s, (d) $t = 3$ to 4 s, (e) $t = 4$ to 5 s, and (f) $t = 0$ to 5 s?

7-3. The 2.0-kg head of a small sledgehammer is moving at 6.0 m/s when it strikes a spike, driving it into a log; the duration of the impact is 0.0020 s. (a) Find the time average of the impact force, and (b) estimate the distance the spike penetrates into the log as a result of this blow.

7-4. A particle travels southwest at a constant speed of 80 km/h. It is subjected to forces of 10 N northeast and 10 N southeast. What third force must act on the particle?

7-5. What is the magnitude of the average force which changes the direction but not the magnitude of a particle's momentum $m\mathbf{v}$ by 60° in time Δt?

7-6. A particle of mass m travels in a circle of radius r at the constant speed v. What is the magnitude of the time-average force on the particle for (a) one complete cycle, (b) one half cycle, and (c) one quarter cycle?

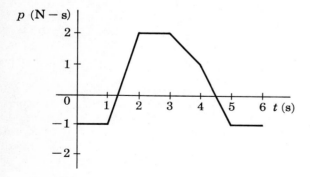

FIG. P7-1

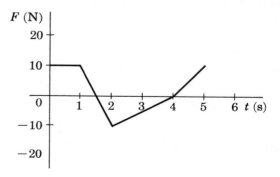

FIG. P7-2

7-7. From a standing start a 2400-lb racing car covers $\frac{1}{4}$ mile in 15.0 s. Find (a) the final speed and (b) the average *horizontal* force between the road surface and each of the two drive wheels. Assume constant acceleration and take $g = 32.0$ ft/s².

7-8. A gun with a barrel 6.0 m long fires a 10.0-kg shell with a muzzle velocity of 1,000 m/s. The shell is capable of penetrating 30 cm of armor plate. Find (a) the average force on the shell while it is moving through the barrel and (b) the average force exerted on the armor by the shell. (Note the ratio of these two forces—and the reason for the difference.)

7-9. A 5.0-g bullet moving at 300 m/s penetrates a 20-cm thick post and emerges at a speed of 100 m/s. What is the average force exerted on the post by the bullet?

7-10. A beam of gas molecules, each with 3.0×10^{-27} kg mass and 360 m/s speed, strikes a wall perpendicular to the surface. The molecules bounce back with the same speed. What is the average force per unit area on the wall if the incident beam contains 2.0×10^{19} molecules/m³?

7-11. A space pistol is used by an astronaut to maneuver in the weightless condition of outer space. The ejection of high-speed gas molecules produces a recoil force in the same fashion as a rocket. (a) What is the average force produced by a space pistol emitting gas molecules at the rate of 14 g/s with a speed of 0.6 km/s? (b) How much does the speed of an astronaut with a mass of 100 kg change if he holds the pistol while it is turned on for 5.0 s?

7-12. A 5.0-kg object dropped from a height of 180 cm above the moon's surface hits in 1.5 s. What is its weight there?

7-13. A 1.0-kg ball is dropped from 1.5 m above a floor. It rebounds to 1.0 m. The ball is in contact with the floor for 1.0×10^{-3} s. What are the direction and magnitude of the average force on the ball (a) while it is in contact with the floor, (b) while it is falling or rising, and (c) over the entire motion? Sketch the vertical force on the ball as a function of time.

7-14. A ball of mass m is dropped from a height h. It makes a number of bounces before coming to rest after a time interval Δt. (a) What is the average force of the ball on the floor over the time Δt? (b) What is the average force of the floor on the ball over the same time interval? Sketch the instantaneous vertical force on the ball as a function of time.

7-15. A 1.0-kg ball is thrown into the air at a speed of 10 m/s at an angle of 30° with respect to the horizontal. (a) What is the impulse on the ball over the first 2.0 s? (b) What is the ball's momentum after 2.0 s? (c) Plot graphs showing the vertical momentum component and force as functions of time.

7-16. The weight of a vessel containing a gas exceeds the weight of the vessel when empty by exactly the weight of the molecules within the vessel. The molecules of the gas are in constant motion, moving in random directions, striking the walls of the vessel, rebounding with equal speed, and being "in contact" with the walls for only a small fraction of the time. Explain, in terms of the impulse and momenta of the molecules, why it is that a weight-measuring device will register exactly the weight of the vessel plus that of its contained molecules.

7-17. A Saturn 5 rocket has a mass of 2.80×10^6 kg, and the first-stage engines generate a total thrusting force of 3.30×10^7 N. If the rocket is launched vertically, (a) what is the initial acceleration and (b) how many seconds does it take for the rocket to move upward a distance equal to its own 110-m length? Assume that $g = 9.80$ m/s².

7-18. A projectile of mass m is launched from the surface of the earth with an initial speed v_0 directed at an angle θ above the horizontal. Using the basic definition of force as given in Eq. (7-2) and the kinematic equations for projectile motion—Eq. (4-10)—show that the instantaneous force on the projectile during its trajectory is (indeed) mg, directed vertically downward.

7-19. A particle with momentum mv strikes a hard surface and rebounds with the same speed. What is the impulse to the struck surface if (a) the particle approaches and recedes along the normal to the surface and (b) the particle's initial and final velocities make an angle θ with respect to the normal?

7-20. A billiard ball initially moving with momentum mv strikes a second ball and is thereby deflected through an angle θ with no change in speed. (a) What is the magnitude of the impulse on either ball? (b) If the balls are in contact for a period Δt, what is the magnitude of the average force on either ball?

7-21. Using the same notation as in Prob. 5-21, the thrust of a rocket engine is $v_E(dm/dt)$. (The *thrust* of a rocket, $v_E(dm/dt)$, is the force of the rocket on the exhaust particles, and therefore it is equal in magnitude to the force of the exhaust particles on the rocket.) At lift-off, the first stage of the Saturn 5 rocket produces a total thrust of 3.3×10^7 N. Find the rate at which fuel is consumed if the exhaust velocity v_E is 2.6 km/s.

CHAPTER 8

Newtonian Mechanics and Its Applications

Classical mechanics is sometimes referred to as newtonian mechanics because of the profound contributions of Sir Isaac Newton. His formulation of classical mechanics was, however, but a part of Newton's scientific work. He discovered the binomial theorem and invented the differential and integral calculus. He made important contributions to the study of light, showing that color was an intrinsic property of light and not of the refracting medium, he devised the reflecting telescope, and he observed the phenomenon known as Newton's rings. He formulated the law of universal gravitation (Sec. 13-1) and thereby established the basic laws of celestial mechanics.

Perhaps Newton's greatest contribution was his three laws of motion, already encountered in Chap. 7. We shall now restate and discuss these laws, and then apply them to various physical problems.

8-1 NEWTON'S LAWS OF MOTION

1. *When an object is subject to no resultant external force, it moves with constant velocity.*   (When $\Sigma \mathbf{F} = 0$, $\mathbf{v} = \mathbf{p}/m =$ constant.)
2. *When an object is subject to one or more external forces, the time rate of change of its momentum is equal (in magnitude and direction) to the vector sum of the external forces acting on it.*   ($\Sigma \mathbf{F} = d\mathbf{p}/dt = m\mathbf{a}$.)
3. *When one object interacts with a second object, the force of the first object on the second is equal in magnitude but opposite in direction to the force of the second object on the first.*   ($\mathbf{F}_{2 \text{ on } 1} = -\mathbf{F}_{1 \text{ on } 2}$.)

Consider some implications of these laws.

1. The first law is Galileo's law of inertia. It defines an inertial frame, a reference system in which an isolated body moves with constant velocity.
2. The second law applies only for observers in inertial frames of reference. The second law provides a procedure for measuring an unknown force; for if a body's motion is recorded in an experiment and its acceleration at each instant is computed, then the resultant force on the body at any instant is simply the acceleration multiplied by the mass. Conversely, the second law permits us to predict in complete detail the future history of a body, provided that the forces acting on it are known; for if one knows the forces and computes the acceleration, he can derive the future displacement and velocity from the present displacement and velocity. Kinematics and dynamics are thus united.

Any force on a particle can be regarded as the sum of two force components, one tangent to the particle's instantaneous velocity and the other at right angles to it.

The resultant instantaneous force on an object can influence its motion in two ways. *(a)* The force component *along* the line of the instantaneous velocity changes the *magnitude* of the velocity but *not* the *direction;* which is to say, a tangential force component changes the object's speed. *(b)* The force component at *right angles* to the velocity changes the *direction* of the instantaneous velocity but *not* the *magnitude;* which is to say, a perpendicular, or radial, force component deflects the object. See Fig. 8-1.

The second law embodies the superposition principle for forces. We may replace a number of forces acting simultaneously on an object by a single resultant force equal to their vector sum.

$$\Sigma \mathbf{F} = m\mathbf{a}$$

This vector equation is, of course, equivalent to three component scalar equations:

$$F_{1x} + F_{2x} + F_{3x} + \cdots = \Sigma F_x = ma_x$$
$$F_{1y} + F_{2y} + F_{3y} + \cdots = \Sigma F_y = ma_y \quad (8\text{-}1)$$
$$F_{1z} + F_{2z} + F_{3z} + \cdots = \Sigma F_z = ma_z$$

where $F_{1x}, F_{2x}, F_{3x}, \ldots$ are the X components of the forces $\mathbf{F}_1, \mathbf{F}_2, \mathbf{F}_3$, ..., respectively, and a_x is the X component of the acceleration $\mathbf{a}$; similarly, for the X and Z components. In solving problems, it is often most convenient to employ the component scalar equations, Eqs. (8-1).

If a particle has zero acceleration, then $\mathbf{a} = 0$ ($a_x = a_y = a_z = 0$). The particle then has a constant velocity (including, of course, the special case of zero velocity, or rest). It is said to be in *translational equilibrium.* By (8-1) we see that a particle can be in translational equilibrium only if the resultant force acting on it is zero or, equivalently, if the sum of the force components along all directions is zero. Thus, equilibrium (object has constant velocity) is a special case of dynamics, and statics (object at rest) is a special case of equilibrium.

The second law is, of course, consistent with the first law, for when no net external force acts on the body, the second law gives $\Sigma \mathbf{F} = m\mathbf{a} = 0$ and the velocity is constant.

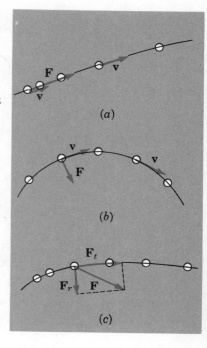

FIG. 8-1. *(a)* A force along the instantaneous velocity changes the object's speed (but not its direction). *(b)* A force at right angles to the instantaneous velocity changes the object's direction (but not its speed). *(c)* Radial (F_r) and tangential (F_t) force components.

3. The third law is a direct consequence of the momentum-conservation law and the definition of force as the time rate of change of the momentum.

Newton's laws are "true" because they are consistent with experiment. They successfully describe the motion of objects as small as molecules (10^{-9} m) and as large as galaxies (10^{21} m). Thus, newtonian mechanics has an enormous range of applicability. Only for the submicroscopic world of the atom and nucleus and for speeds approaching that of light must the classical laws of mechanics be supplanted by the more nearly correct mechanics of the quantum theory and the theory of relativity.

8-2 APPLICATIONS OF NEWTON'S LAWS

It is neither possible nor desirable to follow blindly a prescribed set of rules in solving problems involving Newton's laws. Nevertheless, the following procedure is generally useful.

1. Draw a simple, clear diagram.
2. Choose the object whose motion is to be analyzed. It is often helpful to draw this object, isolated from its surroundings, on a separate diagram. This drawing is called a force diagram, or a free-body diagram.
3. Draw *all* the forces acting on the chosen object, for example, the body's weight, forces applied by objects with which the chosen object is in contact, and noncontact forces (in addition to the weight), such as electric and magnetic forces which may act on the body. (Remember that when we apply Newton's second law, we are concerned only with the *external* forces acting on the chosen object. It is true, of course, that the chosen body exerts forces on its surroundings, but such forces are irrelevant because they do not act *on* the chosen object.) Indicate on the force diagram the magnitude and direction of each force, if known; otherwise, choose symbols to represent the unknown forces.
4. Choose appropriate axes for finding the components of the forces. By a judicious choice of axes one can often simplify the computation of force components. Then resolve each force vector into its components along the chosen axes.
5. Use the component form of Newton's second law, Eqs. (8-1), to solve algebraically for the unknowns. Worked-out examples illustrate these steps.

Example 8-1

What is the nature of the forces exerted by an ideal rope, one which is perfectly flexible, inextensible, and massless? No such rope (or cord, string, or wire) exists. But in many situations a real rope closely approximates the ideal.

We now prove that the tension in the rope is the same at all points along the length of the rope, whether the rope is accelerated or in equilibrium.

We concentrate first on the short segment of cord shown in Fig. 8-2a. The forces on it from the adjoining portions of the cord at the

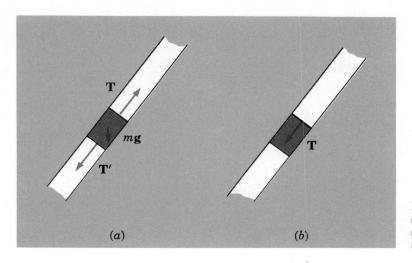

FIG. 8-2. (a) Forces on a small segment of cord. (b) The short segment produces a force T on the adjoining cord.

two ends are labeled **T** and **T'**; its weight is $m\mathbf{g}$. Applying Newton's second law to this effective particle, we have

$$\Sigma \mathbf{F} = m\mathbf{a}$$

$$\mathbf{T} + \mathbf{T}' + m\mathbf{g} = m\mathbf{a}$$

where **a** is the segment's acceleration. But if the mass m is negligibly small, we may discard both the $m\mathbf{g}$ and $m\mathbf{a}$ terms, and the above equation becomes

$$\mathbf{T} = -\mathbf{T}'$$

or in magnitude

$$T = T'$$

The *tension* T has the same magnitude at both ends of the small segment. Moreover, the tension, defined as the force exerted by *any* cord segment on an adjoining segment, is the same at all points along the cord, provided the cord's mass and weight are negligible. Consider Fig. 8-2b, where the force of the small segment on the portion above it is again T. By Newton's third law, the force at the imaginary cut of the short segment on the cord immediately above it is equal in magnitude to the force of the upper cord on the segment.

When a flexible rope passes over a light pulley, the magnitude of the tension is again the same at all points along the rope, and the pulley changes the direction of the force without affecting its magnitude. Hereafter, unless stated otherwise, it will be assumed that any rope is an ideal rope and that therefore the tension has the same magnitude at all points along the rope.

Example 8-2

A 10-N weight is suspended from the center of a massless rope. The rope sags, its two segments making an angle of 10° with the horizontal. What is the tension in the rope segments?

This is a problem in equilibrium. All parts of the system in Fig. 8-3 are at rest, and the resultant force on any one part—on the weight, on the left rope segment, on the right rope segment, on the

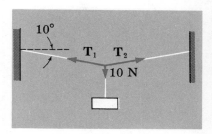

FIG. 8-3

Applications of Newton's Laws   SEC. 8-2   97

left wall, on the right wall, or on the knot—is zero. What part should we choose as our body? We choose the knot, because the forces on it—the tension T_1 by the left rope segment, the tension T_2 by the right rope segment, and the suspended weight W—are the forces we either know or wish to find.

Choosing the X and Y axes as horizontal and vertical, respectively, and resolving each of the tensions into X and Y components, we have, using Eq. (8-1),

$$\Sigma F_x = ma_x$$
$$-T_1 \cos 10° + T_2 \cos 10° = 0$$
$$\Sigma F_y = ma_y$$
$$T_1 \sin 10° + T_2 \sin 10° - W = 0$$

Both a_x and a_y have been set equal to zero because the knot is at rest. The first equation yields

$$T_1 = T_2$$

The magnitude of the tension is the same in the left- and right-hand segments, as we also would have deduced on the basis of symmetry. The second equation then gives, for the magnitude of the tension,

$$T_1 = T_2 = \frac{W}{2 \sin 10°} = \frac{10 \text{ N}}{2 \times 0.174} = 29 \text{ N}$$

The tension *exceeds* the weight. If the angle were smaller, the tension would be larger.

Example 8-3

A 160-lb man stands on a scale in an elevator (Fig. 8-4). The elevator originally moves downward at constant speed, then it decelerates at 4.0 ft/s², and finally comes to rest. What does the scale read before, during, and after its acceleration?

When the man moves with constant velocity, as is the case both before and after the deceleration, the resultant force on him is zero and the scale registers his true weight, 160 lb.

But we are to find the force of the man on the scale while he is decelerating. We do this by finding the magnitude of the equal and opposite force F of the scale on the man. Since his velocity *decreases* in the *downward* direction, his acceleration is *upward*. We make the positive direction that of the acceleration, which is upward. Therefore, writing the man's weight as w and his mass as w/g,

$$\Sigma F = F - w = \frac{w}{g} a$$

or

$$F = w\left(1 + \frac{a}{g}\right) = (160 \text{ lb})\left(1 + \tfrac{4}{32}\right) = 180 \text{ lb}$$

The scale registers 180 lb. The force of the scale on the man (180 lb) exceeds the man's weight (160 lb); it produces a resultant force upward. If the elevator were accelerated downward, the acceleration a above would be negative and the corresponding force F would be less than w.

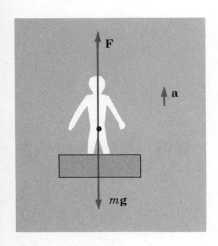

FIG. 8-4. The forces acting on a man on an elevator accelerating upward.

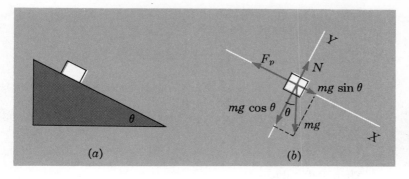

FIG. 8-5. (a) A block sliding on an inclined plane of angle θ. (b) The forces acting on the block, resolved into components parallel and perpendicular to the incline.

Example 8-4

A 2.0-kg block on a perfectly smooth inclined plane of 30° is acted on by a constant force of 15 N upward along the direction of the plane. (a) What is the block's acceleration along the incline? (b) What force along the incline would allow the block to move with constant velocity?

(a) In this problem, as in all problems, it is simplest first to use symbols for all quantities and, only after the problem has been solved in symbols, to substitute the specific numerical quantities. In so doing, one solves, in effect, all problems of that particular type. Moreover, certain quantities may cancel out! Here the block's mass is m, the angle of the incline θ, and $\mathbf{F}_p$ is the upward force parallel to the incline.

The forces acting on the block are shown in Fig. 8-5: the weight $m\mathbf{g}$, the force $\mathbf{F}_p$, and the normal force $\mathbf{N}$ of the incline. The force $\mathbf{N}$ is normal, or perpendicular, to the surface inasmuch as the surface is perfectly smooth and no frictional force acts along the surface.

The X axis is chosen down the incline and the Y axis is perpendicular to it. Then $\mathbf{N}$ is completely along Y, and $\mathbf{F}_p$ is along the negative X axis. We may replace the weight $m\mathbf{g}$, acting vertically downward, by two force components: one of magnitude $mg \cos \theta$ along the negative Y axis and the other of magnitude $mg \sin \theta$ along the positive X axis. (The earth pulls vertically downward on the block; its pull, however, is altogether equivalent to the simultaneous action of the two force components of the weight.)

Applying Newton's second law in the form given in Eq. (8-1), we have

$$\Sigma F_x = ma_x \qquad (8\text{-}1)$$
$$mg \sin \theta - F_p = ma_x \qquad (8\text{-}2)$$
$$\Sigma F_y = ma_y$$
$$N - mg \cos \theta = 0 \qquad (8\text{-}3)$$

We set a_y equal to zero because the block does not accelerate along Y. Equation (8-3) gives the magnitude of the normal force as $N = mg \cos \theta$. The normal force is *not* equal to the weight.

We can solve for a_x from (8-2):

$$a_x = \frac{mg \sin \theta - F_p}{m}$$

Substituting the numerical values, $m = 2.0$ kg, $mg = (2.0$ kg$) \times (9.8$ m/s$^2) = 19.6$ N, and $F_p = 15$ N, yields

$$a_x = \frac{(19.6 \text{ N})(\sin 30°) - 15 \text{ N}}{2.0 \text{ kg}} = -2.6 \text{ m/s}^2$$

The minus sign indicates that the body accelerates along the *negative X* axis, that is, *up* the plane.

(b) If the body is to move with a constant velocity along the incline,

$$a_x = 0$$

Then (8-2) reduces to

$$F_p = mg \sin \theta = (19.6 \text{ N})(\sin 30°) = 9.8 \text{ N}$$

When a 9.8-N force is applied upward along the plane, the block is in equilibrium; it may go up or down the incline at constant speed, or remain at rest.

Example 8-5

A block coasts down a perfectly smooth inclined plane of angle θ. What is the magnitude of its acceleration?

We solve this problem simply by setting $F_p = 0$ in Example 8-4. Then (8-2) becomes

$$mg \sin \theta = ma_x$$

or

$$a_x = g \sin \theta$$

The acceleration is constant and *independent* of the block's mass. If the "incline" is made vertical ($\theta = 90°$), the body falls freely with acceleration g. If the incline is made horizontal ($\theta = 0$), the body is unaccelerated. For intermediate angles, the acceleration is less than g because the factor $\sin \theta$ is less than 1. Motion down a smooth incline is, in effect, slow motion of a freely falling body. Galileo recognized this as a way of "diluting" gravity. He demonstrated that a freely falling body has a constant acceleration by showing that a block sliding freely down an incline had a constant acceleration.

Example 8-6

A mass m_1 on a frictionless incline of angle θ is attached to an inextensible massless cord which passes over a small frictionless pulley to a second mass m_2, as shown in Fig. 8-6a. What is the acceleration of the two masses and the tension in the cord connecting them?

Here we have a system comprised of two coupled objects. Both have the same acceleration magnitude because they are connected by a stretchless cord. To solve this problem, we first consider how Newton's laws apply to the mass m_1 alone and the forces that act on it. See Fig. 8-6b. This problem is just that of Example 8-4 (Fig. 8-5), where the tension **T** replaces force **F**$_p$. Therefore, (8-2) becomes

$$m_1 g \sin \theta - T = m_1 a \qquad (8\text{-}4)$$

where the acceleration a is positive when m_1 accelerates down the incline.

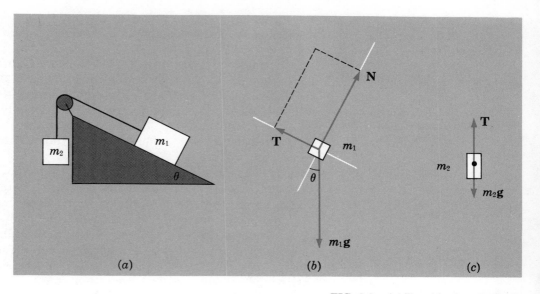

FIG. 8-6. (a) Two blocks attached to a pulley. (b) The forces on m_1. (c) The forces on m_2.

Figure 8-6c shows the forces acting on m_2: the upward tension and the downward weight. The tension **T** of the cord on m_2 is equal in magnitude to the tension **T** on the mass m_1 in Fig. 8-6b (Example 8-1). This tension is *not*, however, equal to the weight m_2g; if it were, the body would not accelerate, but would move with constant velocity. The positive direction (for forces, accelerations, velocities) was chosen as *down* the incline for body m_1. To be consistent and to take into account the fact that the bodies are coupled, we must designate the *positive* direction for body m_2 as upward. Applying Newton's second law to m_2 yields

$$\Sigma F_y = T - m_2 g = m_2 a \qquad (8\text{-}5)$$

Solving (8-4) and (8-5) simultaneously for the unknowns a and T yields

$$a = \frac{(m_1 \sin\theta - m_2)g}{m_1 + m_2} \qquad (8\text{-}6)$$

$$T = \frac{m_1 m_2 g(1 + \sin\theta)}{m_1 + m_2} \qquad (8\text{-}7)$$

Note that a is positive, m_2 accelerating up and m_1 accelerating down the incline, only if $m_1 \sin\theta$ exceeds m_2; otherwise, the acceleration is reversed.

8–3 FRICTION

Whenever two surfaces are in contact and one surface is moved relative to the other, then friction forces come into play. An exact analysis of them is very complicated because it involves interactions between protuberances on the two surfaces and, ultimately,

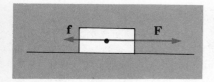

FIG. 8-7. A block subject to two horizontal forces: the applied force F and the friction force f.

the forces between the atoms or molecules on the surfaces. Our discussion is limited to approximate empirical relations found to hold for some surfaces.

Consider a block at rest on a horizontal surface. What happens when we apply a horizontal force to the block (Fig. 8-7)? If the external force is not too large, the block remains at rest. This implies that there is another force acting on the block to maintain its state of equilibrium. This force, called the *static-friction force* $\mathbf{f}_s$, is produced by the surface on which the block rests. As we increase the magnitude of the external force $\mathbf{F}$, the block remains at rest. The magnitude of $\mathbf{f}_s$ must also increase so that $f_s = F$. This situation persists until F reaches a critical value called $f_s(\max)$. For F larger than this maximum static-friction force, the block begins to move and accelerates to the right.

Once in motion, the retarding friction force is *less* than $f_s(\max)$. The friction force now acting is called the *kinetic-friction force* f_k. We will take f_k to be a constant, independent of the speed. (Strictly, f_k decreases with increasing speed.) Thus, once set in motion, the block is subject to a resultant force $F - f_k$ to the right, as shown in Fig. 8-8. If the applied force F were removed, the block would then be subject to an unbalanced force (f_k) alone, and it would be decelerated to rest.

It is found by experiment that, for a given pair of surfaces, the magnitudes of $f_s(\max)$ and f_k are proportional to the normal force N pressing the two surfaces together, and independent of the area in contact or the surfaces' relative velocity. Therefore, by introducing proportionality constants, we may write for the static and kinetic friction forces:

$$f_s \leq \mu_s N \qquad \text{with } f_s(\max) = \mu_s N$$
$$f_k = \mu_k N \tag{8-8}$$

The constants μ_s and μ_k are called the coefficients of static and kinetic friction, respectively. They are dimensionless numbers but depend on the types of surfaces in contact.

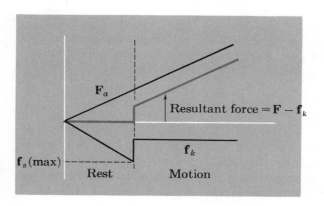

FIG. 8-8. The friction force f and its relation to the applied force F_a for the states of rest and motion.

Typically, $\mu_s > \mu_k$; it takes a larger force to set a body in motion than to maintain it in motion at constant speed. Values of μ are usually, but not necessarily, less than 1.00. The rougher the surfaces, the larger the μ. The inequality sign in (8-8) indicates that the force of static friction can assume any value from zero up to a maximum value of $\mu_s N$; its value is governed by the applied force.

In summary: (1) The frictional force is parallel to the surfaces in contact; (2) the static frictional force is always opposite to the applied force when the surfaces are at rest relative to one another; (3) the kinetic frictional force is opposite to the direction of motion; (4) both the kinetic and maximum static frictional forces are proportional to the normal force.

Example 8-7

A block is released from rest at the top of an inclined plane of angle 30° and length 3.0 m, as shown in Fig. 8-9a. The coefficient of kinetic friction between the block and the inclined plane is $\mu_k = 0.30$. What is the speed of the block when it reaches the bottom of the incline?

The external forces acting on the block during its motion down the plane are shown in Fig. 8-9b. We choose positive X and Y directions as indicated in the figure. Then, by Newton's second law,

$$\Sigma F_y = ma_y = 0$$
$$N - mg \cos \theta = 0$$
$$N = mg \cos \theta \qquad (8\text{-}9)$$
$$\Sigma F_x = ma_x$$
$$mg \sin \theta - f_k = ma_x \qquad (8\text{-}10)$$

By definition [Eq. (8-8)],

$$f_k = \mu_k N$$

From (8-9), this may be written

$$f_k = \mu_k mg \cos \theta$$

Therefore, (8-10) becomes

$$mg \sin \theta - \mu_k mg \cos \theta = ma_x$$

Notice that the mass cancels out, and the acceleration is

$$a_x = g(\sin \theta - \mu_k \cos \theta)$$

Since a_x is constant, we may use the kinematic relation

$$v_x^2 = v_{0x}^2 + 2a_x(x - x_0)$$

If we set $v_{0x} = 0$ and choose $x_0 = 0$, the last two equations yield

$$v_x^2 = 2g(\sin \theta - \mu_k \cos \theta)x$$

Substituting the numerical quantities for v_{0x}, θ, and μ_k and using $g = 9.8$ m/s², we obtain

$$v_x = 14 \text{ m/s}$$

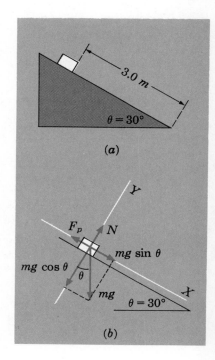

FIG. 8-9

8–4 SYSTEMS OF INTERACTING PARTICLES

A single particle's acceleration is determined by the resultant of all forces acting on it. We shall consider here a system of particles which interact with one another through *internal* forces and are also subject to *external* forces. For such a nonisolated system we show that the acceleration of the system's center of mass is controlled solely by the external forces on the system. In other words, not only can we replace an isolated system by an equivalent single particle at the system's center of mass, but we can do so also when a collection of particles is subject to external forces.

A system of particles 1, 2, 3, . . . is shown in Fig. 8-10; its boundary, shown by a dashed line, is so drawn as to include within all of the particles chosen as part of the system. The forces are of two types: an *internal force*, which arises from one particle in the system acting on another particle *within* the system (the corresponding force vectors are shown lying within the system's boundary), and an *external force*, which arises from one particle within the system being acted upon by another particle *outside* the system (external force vectors penetrate the system's boundary).

Applying Newton's second law to particle 1 alone, we have

$$\mathbf{F}_1 = m_1 \mathbf{a}_1$$

where $\mathbf{F}_1$ is the resultant force on m_1 producing acceleration $\mathbf{a}_1$. Likewise for other particles

$$\mathbf{F}_2 = m_2 \mathbf{a}_2$$
$$\mathbf{F}_3 = m_3 \mathbf{a}_3$$

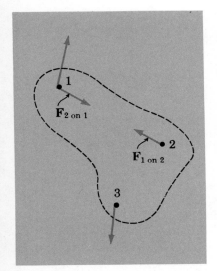

FIG. 8-10. Particles 1, 2, and 3 are within a system whose boundary is shown by a dashed line. External forces on the particles penetrate the system's boundary; internal forces such as $\mathbf{F}_{2\,\text{on}\,1}$ and $\mathbf{F}_{1\,\text{on}\,2}$, lie within the system's boundary.

Adding these equations we have

$$\mathbf{F}_1 + \mathbf{F}_2 + \mathbf{F}_3 + \cdots = m_1 \mathbf{a}_1 + m_2 \mathbf{a}_2 + m_3 \mathbf{a}_3 + \cdots$$

or
$$\Sigma \mathbf{F}_{\text{all forces}} = \Sigma m \mathbf{a} \tag{8-11}$$

The vector sum of the left side of this equation includes *all* forces—both internal and external.

Consider the internal forces $\mathbf{F}_{2\,\text{on}\,1}$ and $\mathbf{F}_{1\,\text{on}\,2}$ acting between particles 1 and 2. From Newton's third law,

$$\mathbf{F}_{2\,\text{on}\,1} = -\mathbf{F}_{1\,\text{on}\,2}$$

or
$$\mathbf{F}_{2\,\text{on}\,1} + \mathbf{F}_{1\,\text{on}\,2} = 0$$

The sum of this pair of internal forces is zero. Indeed, since the internal forces all consist of action-reaction pairs, the sum of all internal forces must be zero.

$$\Sigma \mathbf{F}_{\text{int}} = 0 \tag{8-12}$$

But, since the sum of all forces must be the sum of external forces $\Sigma \mathbf{F}_{\text{ext}}$ together with the sum of the internal forces,

$$\Sigma \mathbf{F}_{\text{all forces}} = \Sigma \mathbf{F}_{\text{ext only}} + \Sigma \mathbf{F}_{\text{int}}$$

$$\Sigma \mathbf{F}_{\text{all forces}} = \Sigma \mathbf{F}_{\text{ext only}}$$

from (8-12). Therefore, (8-11) becomes

$$\Sigma \mathbf{F}_{\text{ext only}} = \Sigma m\mathbf{a} \qquad (8\text{-}13)$$

Now recall that the velocity $\mathbf{V}$ of a system's center of mass is given by

$$M\mathbf{V} = m_1\mathbf{v}_1 + m_2\mathbf{v}_2 + m_3\mathbf{v}_3 + \cdots = \Sigma m\mathbf{v} \qquad (6\text{-}1)$$

where $M = \Sigma m$ is the system's total mass.

Taking the time derivative of the equation above yields

$$M\frac{d\mathbf{V}}{dt} = m_1\frac{d\mathbf{v}_1}{dt} + m_2\frac{d\mathbf{v}_2}{dt} + m_3\frac{d\mathbf{v}_3}{dt} + \cdots = \Sigma m\frac{d\mathbf{v}}{dt}$$

The acceleration of the system's center of mass is $\mathbf{A} = d\mathbf{V}/dt$. Likewise, $\mathbf{a}_1 = d\mathbf{v}_1/dt, \ldots$. Therefore, the last equation can be written

$$M\mathbf{A} = m_1\mathbf{a}_1 + m_2\mathbf{a}_2 + m_3\mathbf{a}_3 + \cdots = \Sigma m\mathbf{a} \qquad (8\text{-}14)$$

Substituting this result into (8-13) we have finally

$$\Sigma \mathbf{F}_{\text{ext only}} = M\mathbf{A} \qquad (8\text{-}15)$$

This remarkable result—that any system's center of mass has an acceleration determined solely by the external forces on the system—results simply from the fact that the internal forces add to zero because they occur in equal and opposite pairs. It is because of (8-15) that we may treat a complicated system, such as a finite-sized block with its myriad of interacting nuclei and electrons, as a mere mass point! It must be emphasized, however, that although we may trace out the motion of a system's center of mass by knowing only the external forces on the constituent particles, to trace out the motion of the individual particles requires a knowledge in detail of the internal forces.

Another way of writing (8-15) is

$$\Sigma \mathbf{F}_{\text{ext}} = \frac{d(M\mathbf{V})}{dt} = \frac{d(\Sigma m\mathbf{v})}{dt} \qquad (8\text{-}16)$$

which is to say, the sum of the external forces equals the time rate of change of the system's total momentum $\Sigma m\mathbf{v}$. If the system is isolated and $\Sigma \mathbf{F}_{\text{ext}} = 0$, then $d(\Sigma m\mathbf{v})/dt = 0$, or $\Sigma m\mathbf{v} = $ constant. A system subject to no net external force has constant total momentum—the momentum-conservation law.

Example 8-8

A bomb thrown into the air explodes in flight into two parts of equal mass. Describe the motion of the center of mass and of the two fragments.

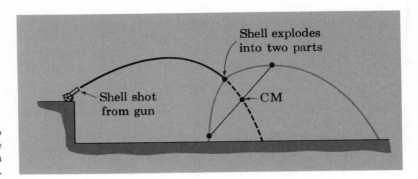

FIG. 8-11. A bomb exploding into two parts of equal mass. The center of mass traces out a parabolic path.

The bomb, which becomes two fragments after explosion, is chosen as our system. The total external force on the system, the weight of the bomb, is constant. Any particle subject to a force of constant magnitude and direction moves in a parabolic trajectory (including the straight-line case). Therefore, from Eq. (8-15), the bomb's center of mass traces out a parabolic path, both before and after the explosion. See Fig. 8-11. Because the two fragments are of equal mass, the center of mass always lies midway between the two fragments, each of which also traces out a parabolic path.

When the first fragment strikes the ground, the system is then subject to another external force, the force of the ground on the fragment. Thereafter, the center of mass no longer follows the same parabolic path.

8–5 UNIFORM CIRCULAR MOTION

An object is accelerated whenever an unbalanced force acts on it. If the resultant force is along the direction of the velocity, the speed changes, but not the direction of the velocity. If the resultant force is always at right angles to the velocity, the direction changes, but not the speed. The body is again accelerated. A body moves in a circle with constant speed when the resultant force is perpendicular to the velocity and is of constant magnitude.

The kinematics of uniform circular motion was treated in Sec. 4-4. The results were these: A particle moving at speed v in a circular arc of radius r has a radial, or centripetal, acceleration $\mathbf{a}_r$ whose direction is toward the center of the circular arc and whose magnitude is v^2/r; equivalently, $a_r = \omega^2 r$, where ω is the angular speed (in *radians* per unit time).

Therefore, if an object is to move in uniform circular motion, the resultant force along the radial direction cannot be zero. Indeed, by Newton's second law it must be

$$\Sigma F_r = ma_r$$

$$\Sigma F_r = \frac{mv^2}{r} = m\omega^2 r \tag{8-17}$$

By the same token, for *uniform* circular motion (constant speed), the resultant *tangential* force is zero: $\Sigma F_t = ma_t = 0$. See Fig. 8-12.

As always, it is essential that we apply Newton's second law from the point of view of an observer in an *inertial frame*. This deserves special emphasis when one deals with objects moving in circles. For example, if one considers the motion of a man in an automobile traveling in a circle, one must view his motion as an observer fixed in the inertial frame of the earth (or one moving with constant velocity relative to it). One cannot properly apply Newton's second law from the point of view of the occupant, as is perhaps psychologically more appealing. Such an observer is in an accelerated, and therefore noninertial, frame; the law of inertia does not hold in such a reference frame.

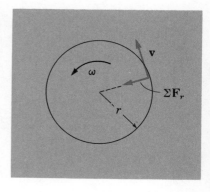

FIG. 8-12. A particle in uniform circular motion under the influence of a resultant radial force ΣF_r.

Example 8-9

An astronaut is riding in a space capsule in a circular orbit at constant speed about the earth. What is the resultant force on the astronaut?

The forces on the astronaut are his weight $m\mathbf{g}$ and possibly a force $\mathbf{N}$ of the seat on him. See Fig. 8-13. These are the only possible forces acting on the astronaut if we observe the motion from an inertial frame of reference. We know that the force $\mathbf{N}$, if it exists, must be along the radial direction, inasmuch as there cannot be a resultant force along the tangential direction since the speed is constant. Applying Newton's second law along the radial direction (taking inward as positive) gives

$$\Sigma F_r = ma_r$$

$$mg - N = \frac{mv^2}{r}$$

where v is the speed of the capsule and r is the radius of the orbit.

From the kinematical analysis of satellite motion in Example 4-2, we found that the radial acceleration is g. Moreover, this acceleration is related to the orbital speed and radius by

$$g = \frac{v^2}{r}$$

Using this result in the equation above yields

$$N = 0$$

While in orbit at constant speed, the spaceship applies *no* force on the astronaut or, for that matter, on any object within the capsule. The only force on the astronaut is his weight. This is the resultant force that permits the astronaut to go in circular orbit.

The astronaut, or any other object in the capsule, is, of course, *not* weightless, the quantities g and mg being nonzero. Because N is zero, all objects in the capsule float about when unattached. For this reason the astronaut experiences *apparent weightlessness*. Paradoxically, the only time a body experiences apparent weightlessness is when the only force acting on it is its weight!

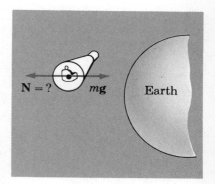

FIG. 8-13. Forces on an astronaut in a satellite orbiting the earth in uniform circular motion.

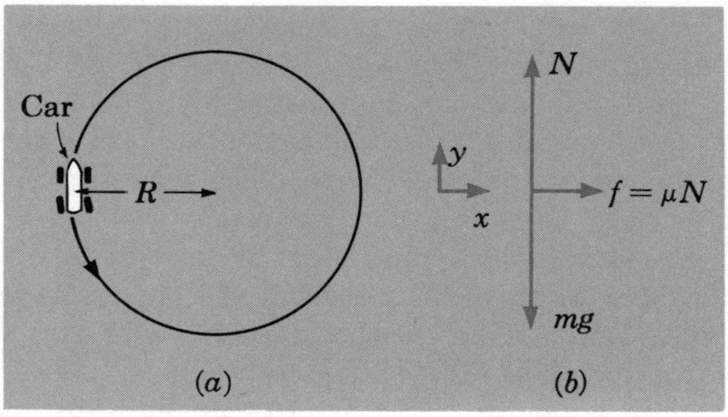

FIG. 8-14. (a) Schematic diagram of racing car traveling in a horizontal circle of radius R. (b) Free-body diagram of forces on car.

Example 8-10

At what maximum speed (in mi/h) can a racing car travel around an unbanked curve 150 ft in radius if the coefficient of static friction between the tires and the road is 0.75? Assume that the car does not skid. See Fig. 8-14.

By Newton's second law

$$\Sigma F_y = ma_y$$
$$N - mg = 0$$
$$N = mg$$

and

$$\Sigma F_x = ma_x$$
$$\mu_s N = \frac{m v_{\max}^2}{R}$$

where we have made use of (8-13); $v_{\max}$ is the maximum speed. In this example the inward radial force is supplied by the force of static friction.

Eliminating N from the two equations above and canceling mass m gives

$$v^2 = \mu_s g R$$

If we take $g = 32$ ft/s² and substitute the numerical values for μ_s and R we find that

$$v = \sqrt{\mu_s g R}$$
$$= \sqrt{0.75 \times 32 \text{ ft/s}^2 \times 150 \text{ ft}} = 60 \text{ ft/s} = 41 \text{ mi/h}$$

Note that the critical speed does not depend upon the car's mass. Although the skidding of the car is independent of its mass and of any of the car's dimensions, the *tipping* of the car as it travels around an unbanked curve depends crucially on the distance between the wheels and the height of the car's center of mass above the road surface.

The resultant force acting on a body in uniform circular motion is sometimes referred to as the *centripetal force*. This designation is mischievous because the centripetal force is not a distinctive type of force. A body may move in a circle under the influence of its weight, the tension of a cord, a friction force, a normal force, or a combination of forces. If the resultant of all forces acting on the body produces circular motion, this resultant is *the* centripetal force.

Suppose that you are an observer on a uniformly rotating platform, such as a merry-go-round. To you an object at rest in this system appears to be in static equilibrium. As a confirmed newtonian you may explain this by saying that the force directed inward toward the center is balanced by a fictitious (or inertial) force directed away from the center. This inertial force, existing only in accelerated and therefore noninertial frames of reference, is called the *centrifugal force*. A centrifugal force also exists for an astronaut circling the earth in a spaceship. From *his* point of view, every object in the spaceship is subject to two forces: the inward pull of the earth and an equal centrifugal force away from the earth. The net force on any object is, in *his* view, zero. Unattached objects float *relative to the astronaut,* and he may conclude that all objects in the spaceship *are weightless*. On the other hand, an inertial observer says that the objects in the spaceship *appear* to be weightless because all such objects have the *same* acceleration.

SUMMARY

Newton's laws of motion may be written:

1. When $\Sigma \mathbf{F} = 0$, then $\mathbf{v} = \mathbf{p}/m =$ constant.
2. $\Sigma \mathbf{F} = d\mathbf{p}/dt = m\mathbf{a}$.
3. $\mathbf{F}_{2 \text{ on } 1} = -\mathbf{F}_{1 \text{ on } 2}$.

For a system of interacting particles

$$\Sigma \mathbf{F}_{\text{ext}} = \frac{d\mathbf{P}}{dt} = M\mathbf{A} \qquad (8\text{-}13)$$

where $\Sigma \mathbf{F}_{\text{ext}}$ is the vector sum of the external forces only, M is the system's total mass, $\mathbf{P}$ its total momentum, and $\mathbf{A}$ the center-of-mass acceleration.

Applied to a particle in uniform circular motion, Newton's second law becomes

$$\Sigma F_r = \frac{mv^2}{r} = m\omega^2 r \qquad (8\text{-}17)$$

where ΣF_r is the resultant radial force.

PROBLEMS

8-1. A particular cord will break if the tension exceeds $T_{\max}$. Find the largest mass *(a)* that can be lifted vertically with acceleration a and *(b)* that can be lowered with acceleration a.

8-2. A 2.0-kg object is suspended from the end of a light cord. The cord is pulled to the side by a horizontal force until the angle between the cord and the vertical is 60°. What are *(a)* the magnitude of the horizontal force, (b) the tension in the lower portion of the cord, and (c) the upper portion of the cord?

8-3. A plumb bob is suspended from one end of a cord which is attached to the ceiling of an aircraft in level flight. The angle between the vertical and the cord is $\theta = 14°$, see Fig. P8-3. Explain this effect quantitatively.

FIG. P8-3

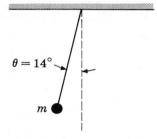

8-4. A cord has a mass per unit length of 1.0 g/cm. If one end of such a cord 1.00 m long is attached to the ceiling and a 1.0-kg object is suspended from the other end, what is the tension in the cord at a distance down from the ceiling of (a) 25 cm, (b) 50 cm, and (c) 75 cm?

8-5. A chain hangs without friction over the top of a right triangle with sides a and b, as shown in Fig. P8-5. What is the ratio of the chain length on the a side to that on the b side when the chain is in equilibrium? [This problem was first solved by S. Stevinus (1548–1620) by considering an entire loop of chain draped over the two upper sides of a triangle and hanging beneath. Stevinus generalized the rule that when

FIG. P8-5

three forces act on a particle in equilibrium, the force vectors form a closed triangle.]

8-6. Identify all the forces acting on the indicated objects under the stated conditions, and identify also the reactions to these forces: (a) a person sitting in a chair, (b) an airplane in a power dive, (c) a child moving on a swing, (d) a phonograph stylus while playing a recording.

8-7. A 1.0-kg object is initially at the origin, traveling west at 5.0 m/s. It is subject to constant forces of 10.0 N northeast and 10.0 N northwest. Find the (a) displacement and (b) velocity of the object after 3.0 s.

8-8. When small oil droplets fall through (nonturbulent) air, they quickly achieve a *terminal velocity*, falling at *constant* speed downward. It is found that the terminal velocity is directly proportional to the cube of the radius of the droplet for droplets of the same liquid. How does the resistive force of the air on the droplet vary with speed?

8-9. A 5.0-kg object is suspended from a cord attached to the ceiling of an elevator initially traveling downward at 3.0 m/s. The elevator comes to rest at constant acceleration in 6.0 s. What is the tension in the cord during this time?

8-10. What is the force of the man in Example 8-3 on the elevator floor when the elevator accelerates *downward* at 2.0 ft/s²?

8-11. The following are the mass, x and y coordinates, v_x and v_y velocity components, and F_x and F_y force components of three particles, given as $(m; x, y; v_x, v_y; F_x, F_y)$, (1.0 kg; 0 m, 0 m; 2.0 m/s, −3.0 m/s; 4.0 N, 2.0 N), (2.0 kg; 4.0 m, 0 m; −4.0 m/s, 4.0 m/s; −6.0 N, 2.0 N), and (2.0 kg; 0 m, 4.0 m; −5 m/s, 0 m/s; −4.0 N, 0 N). What is the acceleration of this system's center of mass?

8-12. Blocks A and B, of masses 5.0 and 3.0 kg, respectively, are in contact on a smooth horizontal surface, as shown in Fig. P8-12. A force of 16 N is applied to block A. What are (a) the acceleration of the masses, (b) the force of B on A, and (c) the force of A on B?

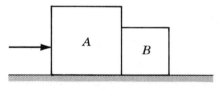

FIG. P8-12

8-13. The three blocks in Fig. P8-13 have masses of 1.0, 2.0, and 3.0 kg and move upward along a frictionless 60° incline. A 60-N force is applied upward along

FIG. P8-13

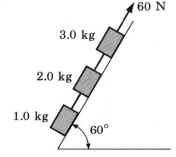

110

the incline to the uppermost block. Take g to be 10 m/s². What is the tension in the massless cord between (a) the upper and middle blocks and (b) the lower and middle blocks? (c) What is the acceleration of the blocks?

8-14. A block coasts down a rough incline of angle θ at constant speed. What is the coefficient of kinetic friction between the block and the surface of the incline?

8-15. The breaking strength of a steel cable is 1.0×10^4 N. If one pulls horizontally with this cable, what is the maximum horizontal acceleration which can be given to a 4000-kg object resting on a rough horizontal surface if the coefficient of kinetic friction is 0.15? Assume that $g = 10.0$ m/s².

8-16. Because of friction, a 20-kg block traveling in a straight line along a horizontal plane surface slows from a speed of 10.0 m/s to 1.0 m/s in 4.0 s. Find (a) the average force over this time interval, and (b) the coefficient of kinetic friction.

8-17. A crate rests on the rough floor of a truck. What is the maximum horizontal acceleration which can be given to the truck before the crate starts to slip? The coefficient of static friction is 0.75.

8-18. A constant 10.0-N force is applied to a 2.00-kg object at rest on a horizontal surface. As a result, the object is accelerated and reaches a speed of 12.0 m/s after it has moved a distance of 24.0 m. Find (a) the coefficient of friction and (b) the time interval for the 24.0-m displacement. Assume that $g = 10.0$ m/s².

8-19. Starting from rest a wooden block slides down a 60° inclined plane, moving 6.0 m in 1.5 s. What is the coefficient of sliding friction for the block-plane interface?

8-20. An object of mass m is constrained to move in a horizontal circle on a flat, friction-free surface by a cord with a breaking strength of 100 N. What is the limiting value of m if the object is to move at 600 rev/min in a circle 2.0 m in radius?

8-21. A small block is a distance r from the center of a turntable. The coefficient of static friction between the block and the turntable surface is μ_s. What is the maximum angular velocity of the turntable if the block is not to slide?

8-22. A 10.0-kg object is acted upon by a single force having a *constant* magnitude for 6.0 s; the object maintains constant speed during this time. What simple path satisfies these conditions?

8-23. An automobile travels at constant speed v around a curve of radius r. At what angle with respect to the horizontal would the curve have to be banked if there were to be no friction force (up or down the incline) of the road surface on the car when it takes the curve?

8-24. A ball is suspended from the rim of a rotating wheel by means of an inextensible cord. The wheel is spinning about a vertical axis and the ball moves radially outward from this axis so that the cord is at angle θ with respect to the vertical. Show that the angular speed of the wheel is given by

$$\omega = \sqrt{\frac{g \tan \theta}{R + l \sin \theta}}$$

where g is the acceleration of gravity, R the radial distance from the upper end of the cord to the rotation axis, and l the cord length. See Fig. P8-24.

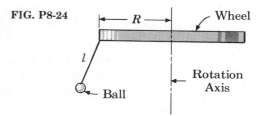

FIG. P8-24

8-25. A 1,000-kg automobile is driven at a *constant speed* of 15 m/s along a hilly road; see Fig. P8-25. The radii of curvature at the points a, b, and c are 40, 30, and ∞ m, respectively. (a) Find the normal force of the road on the automobile at these three points. (b) What would happen if the car were to travel at a constant speed of 30 m/s?

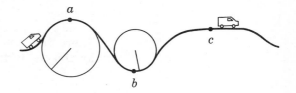

FIG. P8-25

8-26. When a 1.5-V battery is connected across two parallel metal plates separated by 1.0 cm, electrons (starting from rest) are accelerated to a speed of 7.3×10^5 m/s as they cross the gap. What is the ratio of the electric force on an electron to the gravitational force on it? (In atomic physics the principal forces are of electric origin, and gravitational forces between the particles are entirely negligible.)

8-27. An earth satellite of mass m in a circular orbit of radius R_0 travels at speed v_0. Find the magnitude and direction of the additional force needed to maintain the same orbit at speed (a) $2v_0$ and (b) $\tfrac{1}{2}v_0$.

CHAPTER 9

Work and Kinetic Energy

In this chapter we introduce the concepts of work and kinetic energy, which are important because they lead ultimately to the conservation of energy principle.

9–1 KINETIC ENERGY DEFINED

Consider a constant force F acting on a particle of mass m along the direction of its displacement Δx. See Fig. 9-1. The particle's speed goes from an initial value v_i to a final speed v_f over this interval.

Newton's second law yields

$$F = ma$$

where the constant acceleration a is related to the displacement Δx and the initial and final speeds through the kinematic equation

$$2a\,\Delta x = v_f^2 - v_i^2 \tag{3-10}$$

Elimination of the acceleration from the two equations above results in a relation which can be written in the form

$$F\,\Delta x = \tfrac{1}{2}mv_f^2 - \tfrac{1}{2}mv_i^2 \tag{9-1}$$

The constant force multiplied by the displacement over which it acts equals the change in the quantity $\tfrac{1}{2}mv^2$. We shall return to a consideration of the left side; we first concentrate on the right side and the quantity $\tfrac{1}{2}mv^2$, which is defined as the *kinetic energy* of a particle:

$$\text{Kinetic energy} = K = \tfrac{1}{2}mv^2 \tag{9-2}$$

Kinetic energy is a scalar quantity. It is, as we shall see, as significant as the momentum vector $m\mathbf{v}$, which also charac-

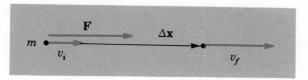

FIG. 9-1. A constant force F acts over the displacement Δx in the same direction, increasing the speed of a particle from v_i to v_f.

terizes a particle in motion. From its definition, kinetic energy has dimensions of ML^2/T^2, either mass multiplied by the square of speed or, as shown by (9-1), force multiplied by distance. Thus, the kinetic energy of a particle may be expressed in units of kg-m²/s², or its equivalent N-m. For convenience the unit *joule,* abbreviated J, is used for kinetic energy rather than the compound units.† Thus, by definition,

$$1 \text{ joule} = 1 \text{ J} = 1 \text{ kg-m}^2/\text{s}^2 = 1 \text{ N-m}$$

It follows from (9-2) that a 2-kg particle moving at 1 m/s has a kinetic energy of $K = 1$ J.

The kinetic energy of a particle may also be expressed in terms of the magnitude of its linear momentum $mv = p$:

$$K = \tfrac{1}{2}mv^2 = \frac{(mv)^2}{2m} = \frac{p^2}{2m} \tag{9-3}$$

9-2 WORK DEFINED

In Fig. 9-1 we had a *constant* force acting *along the direction of motion* of a particle and Eq. (9-1) resulted. What if the force is not constant nor along the direction of motion? As we shall see, we then get a relation identical with (9-1) on the right side, but with a slightly modified left side.

First consider the situation in which a resultant force **F** acts on a particle but not along its direction of motion. See Fig. 9-2. The angle between the force **F** and the small displacement $d\mathbf{r}$ of the particle is θ. We may resolve the force **F** into a component $F \cos \theta$ along $d\mathbf{r}$ and a component $F \sin \theta$ perpendicular to the particle's velocity. The tangential force component $F \cos \theta$ changes the particle's speed but not its direction, while the radially inward force component $F \sin \theta$ changes the particle's direction of motion but not its speed. In terms of kinetic energy, *only a force component along the direction of a particle's displacement can change its kinetic energy.*

Thus, applied to the situation shown in Fig. 9-2, Eq. (9-1) becomes

$$F \cos \theta \, dr = \tfrac{1}{2}mv_f^2 - \tfrac{1}{2}mv_i^2 \tag{9-4}$$

---

†In the cgs system kinetic energy is expressed in ergs, where 1 erg = 1 dyn-cm = 1 g-cm²/s² = 10^{-7} J. The energy unit in the U.S. Customary System is the foot-pound, where 1 ft-lb = 1.356 J. See Appendix II for other energy units and their conversion factors.

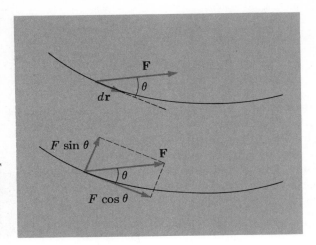

FIG. 9-2. (a) Resultant force F acting on a particle undergoing a displacement $d\mathbf{r}$. (b) The force's component along the displacement is $F\cos\theta$ and at right angles to the displacement it is $F\sin\theta$.

We may properly use the relation above without further modification if the displacement $d\mathbf{r}$ is sufficiently small that the magnitude of the resultant force and the acceleration may be taken as constant.

The quantity appearing on the left side of (9-4) is, by definition, the *work* done by the force **F** over the displacement $d\mathbf{r}$:

$$\Delta W = \text{work by } \mathbf{F} \text{ over } d\mathbf{r} = F\cos\theta\, dr \qquad (9\text{-}5)$$

In the terminology of work and kinetic energy, (9-4) then says that the *work done by a resultant force on a particle equals the change in its kinetic energy*. This is the *work–kinetic-energy theorem*. Note that if $\theta = 90°$, with the force at right angles to the particle's velocity, there is no change in its kinetic energy. Further, if $\theta > 90°$, so that the force component is opposite in direction to the particle's displacement, the particle is slowed, its kinetic energy decreased, and the work done by the force is negative.

Recalling that the scalar product $\mathbf{A} \cdot \mathbf{B}$ of two vectors is simply $AB\cos\theta$, where θ is the angle between **A** and **B** (Sec. 2-6), we may express (9-5) more compactly as follows:

$$\text{Work done by } \mathbf{F} \text{ over } d\mathbf{r} = F\cos\theta\, dr = \mathbf{F} \cdot d\mathbf{r}$$

and therefore (9-4) may be written

$$\mathbf{F} \cdot d\mathbf{r} = \tfrac{1}{2}mv_f^2 - \tfrac{1}{2}mv_i^2 \qquad (9\text{-}6)$$

It is easy to show that a particle's kinetic energy is equal to the work done by the particle in coming to rest. The work-energy theorem concerns the work done *by* the resultant force *on* a particle. But such a resultant force exists only if the particle in question interacts with at least one other particle. We may lump all such sources of the resultant force on the particle with the term *external agent*. Thus, an external agent does work on a particle to change its kinetic energy. By Newton's third law, if the agent produces a force on the particle, the particle produces an equal but opposite force on the agent, and this force by the particle may do work on the agent.

Consider a particle with an initial kinetic energy K_i which is acted upon by a retarding force and brought to rest. From the work-energy theorem.

$$W_{i \to f} = K_f - K_i = 0 - K_i$$
$$K_i = -W_{i \to f}$$

The work done by the external agent on the particle is $W_{i \to f}$; the work done by the particle on the agent is $-W_{i \to f}$. Therefore, the equation above shows that a particle's *kinetic energy* is the *work done by the particle in coming to rest*.

9-3 THE GENERAL WORK–ENERGY THEOREM

As given above, we have seen that the work done by a force over an *infinitesimal* displacement $d\mathbf{r}$ equals the particle's kinetic energy change over this displacement.

What about a path of *finite* length and one in which the resultant *force* on the particle may *change* both in magnitude and direction from point to point? Consider Fig. 9-3, where a particle moves along a path which is subdivided into a number of small straight-line segments, $d\mathbf{r}_1, d\mathbf{r}_2, d\mathbf{r}_3, \ldots$. The resultant force at these points is $\mathbf{F}_1, \mathbf{F}_2, \mathbf{F}_3, \ldots$, respectively, the segments being taken so small that the force can be regarded as locally constant.

The work done over the first segment is

$$\mathbf{F}_1 \cdot d\mathbf{r}_1 = \tfrac{1}{2} m v_{f1}^2 - \tfrac{1}{2} m v_{i1}^2$$

where v_{i1} is the speed with which the particle enters the segment 1 and v_{f1} is the speed with which it leaves.

Similarly, for other segments,

$$\mathbf{F}_2 \cdot d\mathbf{r}_2 = \tfrac{1}{2} m v_{f2}^2 - \tfrac{1}{2} m v_{i2}^2$$
$$\mathbf{F}_3 \cdot d\mathbf{r}_3 = \tfrac{1}{2} m v_{f3}^2 - \tfrac{1}{2} m v_{i3}^2$$

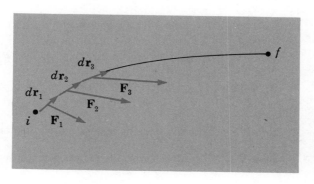

FIG. 9-3. A particle traces out a general path from *i* to *f*. The resultant force is $\mathbf{F}_1$ over the infinitesimal displacement $d\mathbf{r}_1$, $\mathbf{F}_2$ over $d\mathbf{r}_2$, etc.

Now we recognize that the speed v_{f_1} with which the particle leaves segment 1 is just the speed v_{i2} with which it enters segment 2: $v_{f1} = v_{i2}$. Similarly, $v_{f2} = v_{i3}$, etc. If we add together the last work equations, the result is

$$\mathbf{F}_1 \cdot d\mathbf{r}_1 + \mathbf{F}_2 \cdot d\mathbf{r}_2 + \mathbf{F}_3 \cdot d\mathbf{r}_3 + \cdots = \tfrac{1}{2}mv_f^2 - \tfrac{1}{2}mv_i^2$$

Note that, because of the equality of exit and entrance speeds, the only terms surviving on the right-hand side are the kinetic energy $\tfrac{1}{2}mv_i^2$ with which the particle enters the first segment and the kinetic energy $\tfrac{1}{2}mv_f^2$ with which the particle emerges from the last segment. Writing the above relation more compactly, we have

$$\sum_i^f \mathbf{F} \cdot d\mathbf{r} = \tfrac{1}{2}mv_f^2 - \tfrac{1}{2}mv_i^2$$

or, if the summation is indicated by an integration,

$$\int_i^f \mathbf{F} \cdot d\mathbf{r} = \tfrac{1}{2}mv_f^2 - \tfrac{1}{2}mv_i^2 \tag{9-7}$$

or
$$W_{i \to f} = K_f - K_i \tag{9-7a}$$

This is the work–kinetic-energy relation in its most general form: *The total work done on a particle by the resultant force on it between some starting point i and ending point f is the change in the particle's kinetic energy between these two end points.* This relation applies quite apart from the particular path followed, so long as the local component of the force along the direction of the path is used in evaluating contributions to the total work $\int_i^f \mathbf{F} \cdot d\mathbf{r}$ done by a resultant force. Moreover, $\mathbf{F}$ may vary from point to point along the path in both direction and magnitude.

We complete our formal analytical treatment of work with two useful theorems:

I The work done by the *resultant* force $\mathbf{F}$ may be expressed in a useful alternate form in terms of the work done by the individual forces on the particle whose resultant is $\mathbf{F}$. Suppose that $\mathbf{F} = \mathbf{F}_1 + \mathbf{F}_2 + \mathbf{F}_3 + \cdots$ Then

$$\int_i^f \mathbf{F} \cdot d\mathbf{r} = \int_i^f (\mathbf{F}_1 + \mathbf{F}_2 + \mathbf{F}_3 + \cdots) \cdot d\mathbf{r}$$

or
$$\int_i^f \mathbf{F} \cdot d\mathbf{r} = \int_i^f \mathbf{F}_1 \cdot d\mathbf{r} + \int_i^f \mathbf{F}_2 \cdot d\mathbf{r} + \int_i^f \mathbf{F}_3 \cdot d\mathbf{r} + \cdots \tag{9-8}$$

$W_{i \to f}$(by resultant of 1, 2, 3, ...)
$$= W_{i \to f}(\text{by } 1) + W_{i \to f}(\text{by } 2) + W_{i \to f}(\text{by } 3) + \cdots \tag{9-8a}$$

But $\int_i^f \mathbf{F}_1 \cdot d\mathbf{r}$ is the work done between points i and f by the individual force $\mathbf{F}_1$, $\int_i^f \mathbf{F}_2 \cdot d\mathbf{r}$ the work done between the same two end points by individual force $\mathbf{F}_2$, etc. Therefore, (9-8) shows that *the work done by the resultant force on a particle is equal to the sum of the works done by each force contributing to the resultant force over the same path.* This result often simplifies the computation of the work done by a resultant force and the corresponding increase in a particle's kinetic energy.

II The scalar product $\mathbf{A} \cdot \mathbf{B}$ may be expressed in terms of the rectangular components of $\mathbf{A}$ and $\mathbf{B}$ as follows:

$$\mathbf{A} \cdot \mathbf{B} = A_x B_x + A_y B_y + A_z B_z \qquad (2\text{-}7)$$

where the components carry algebraic signs.

If we apply this result to the force vector $\mathbf{F}$ with rectangular components F_x, F_y, and F_z and the incremental displacement vector $d\mathbf{r}$ with rectangular components dx, dy, and dz, we have

$$\mathbf{F} \cdot d\mathbf{r} = F_x\, dx + F_y\, dy + F_z\, dz$$

and
$$\int_i^f \mathbf{F} \cdot d\mathbf{r} = \int_i^f F_x\, dx + \int_i^f F_y\, dy + \int_i^f F_z\, dz \qquad (9\text{-}9)$$

In words, *the work done by force $\mathbf{F}$ is equal to the sum of the works done by its rectangular components separately.*

Example 9-1

A particle of mass m falls from rest and traverses a vertical distance y. *(a)* What is the work done by the resultant force on m? *(b)* What is the particle's final speed?

(a) The particle is subject to a single constant force mg, its weight, in the downward direction. Its displacement is also y downward. Therefore the total work done is

$$\int \mathbf{F} \cdot d\mathbf{r} = mgy$$

(b) Since the work done by the resultant force is equal to the change in kinetic energy, initially zero, we have

$$W_{i \to f} = K_f - K_i$$
$$mgy = \tfrac{1}{2}mv^2 - 0$$

or
$$v = (2gy)^{1/2}$$

a result which we could have arrived at simply from the kinematics of falling objects.

Example 9-2

A 2-kg object is raised vertically upward a distance of 1.5 m by a man who holds the object in his hand. The object travels at constant speed. What is the work done by *(a)* the resultant force on the object, *(b)* the man, and *(c)* the force of gravity? See Fig. 9-4.

(a) If the object travels at constant speed, the resultant force on it must be zero. The downward weight mg is just balanced by a force of equal magnitude from the man's hand. Since the *resultant* force is zero, it does *zero work*, and, of course, the object's kinetic energy does not change.

(b) The man's hand produces a force of magnitude $mg = (2.0 \text{ kg})(9.8 \text{ m/s}^2) = 19.6$ N on the object. This force is upward, and

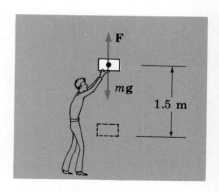

FIG. 9-4. A man raises an object vertically at constant speed.

the object travels 1.5 m upward. Therefore, with the force and displacement along the same (upward) direction, the work done by the man's hand is $F \Delta x = (19.6 \text{ N})(1.5 \text{ m}) = 29.4 \text{ J}$.

(c) The weight of the object is downward and of magnitude 19.6 N, whereas the object is displaced upward. Therefore, the work done by this force is *negative* [$\theta = 180°$ in Eq. (9-5)] and is $F \cos \theta \, \Delta x = -(19.6 \text{ N})(1.5 \text{ m}) = -29.4 \text{ J}$.

Note that the results from the three parts are consistent with the general theorem [Eq. (9-8)]: The work by the resultant force equals the sum of work done by the individual forces whose sum is the resultant force. Here we have:

Work by resultant force = work by hand + work by gravity

$$0 = 29.4 \text{ J} + (-29.4 \text{ J})$$

Example 9-3

A block slides down a perfectly smooth surface with initial speed v_i. What is its speed at a point a distance y vertically beneath the starting point?

Figure 9-5 shows that the forces acting on the particle are its weight $m\mathbf{g}$ and the normal force $\mathbf{N}$ of the surface on the block. (That the surface is perfectly smooth is reflected in the fact that the force of the surface is, in fact, perpendicular to the surface, with no component along the surface.) The work done by the resultant force on the block between the two end points is the change in the block's kinetic energy:

$$\int_i^f \mathbf{F} \cdot d\mathbf{r} = \tfrac{1}{2}mv_f^2 - \tfrac{1}{2}mv_i^2$$

But here $\mathbf{F} = m\mathbf{g} + \mathbf{N}$, and we can compute the work done by the resultant force by adding the work by $m\mathbf{g}$ alone to that done by $\mathbf{N}$ alone. Since the block slides along the plane, the normal force $\mathbf{N}$ is always at right angles to the displacement $d\mathbf{r}$, so that $\mathbf{N} \cdot d\mathbf{r} = 0$. This leaves the work done by $m\mathbf{g}$, which, for simplicity, we compute by adding the work done separately by its rectangular components.

$$m\mathbf{g} \cdot d\mathbf{r} = F_x \, dx + F_y \, dy$$

The weight $m\mathbf{g}$ is vertical, so that $F_x = 0$ and $F_y = -mg$ (downward). Over an incremental downward displacement dy,

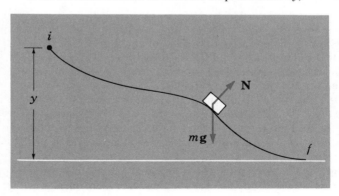

FIG. 9-5. A block sliding on a smooth surface from *i* to *f* is subject to two forces: the normal force N and its weight $m\mathbf{g}$.

$$\int_i^f m\mathbf{g} \cdot d\mathbf{r} = \int_0^{-y} -mg\,dy = mgy$$

Therefore, the total work done by the resultant force is just mgy, and the work-energy theorem yields

$$mgy = \tfrac{1}{2}mv_f^2 - \tfrac{1}{2}mv_i^2$$

or a final speed, $$v_f^2 = v_i^2 + 2gy$$

which is independent not only of the mass m but also of *any* details concerning the shape of the surface leading from the starting to ending point. It follows that *all* objects starting from rest and sliding on frictionless surfaces connecting two points whose vertical separation is the same reach the *same* final speed. The various paths may differ in the total time required to traverse the particular path, but the final speeds are the *same*.

Example 9-4

An object is thrown at an initial speed v_0 at an angle θ relative to the horizontal. With what speed does it arrive at a point a vertical distance y down from the starting point? See Fig. 9-6.

The only force acting on the object in flight is its weight $m\mathbf{g}$, and as we have seen in Example 9-3, the work done by $m\mathbf{g}$ over a displacement having a downward vertical component y is mgy. Therefore, the work-energy theorem yields

$$mgy = \tfrac{1}{2}mv_f^2 - \tfrac{1}{2}mv_0^2$$

or $$v_f^2 = v_0^2 + 2gy$$

Note that the angle θ does not enter into the final result. An object of any mass tossed in any direction, upward or downward from the horizontal, will change its speed by the same amount when traversing a vertical distance y.

One special case is of special interest: that with $y = 0$. Then the result above gives $v_f = v_i$, and the total work done by gravity is zero. Here we have negative work done by the weight as the object rises ($m\mathbf{g}$ downward, displacement upward) and positive work of the same amount done by the weight as the object falls back again ($m\mathbf{g}$ downward, displacement also downward); their sum is zero.

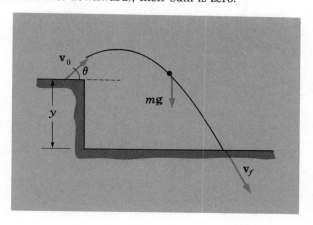

FIG. 9-6. An object thrown with initial velocity v_0 from an elevation y.

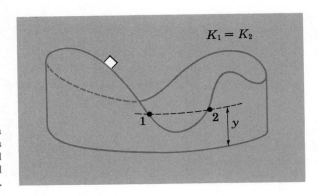

FIG. 9-7. An object sliding on a perfectly smooth roller coaster has the same kinetic energy at all points on the same horizontal level.

Indeed, whenever the only work-producing force is an object's weight—an object thrown into the air, or sliding on a frictionless surface, for example—the *total work done by the force of gravity over any closed path which brings the object* back to the same vertical position is *zero*. This follows from the fact that a downward displacement results in positive work done by gravity, an upward displacement corresponds to negative work done by gravity, and a horizontal displacement, with the force at right angles to the displacement, involves zero work done. Thus, an object sliding on a perfectly smooth roller coaster as in Fig. 9-7, will return to its starting point with precisely the same kinetic energy as that with which it started the complete cycle, and the particle's kinetic energy at all points at the same horizontal level will be the same.

Example 9-5
A 2.0-kg block is projected up a 30° rough incline at an initial speed of 2.0 m/s. It comes to rest momentarily and then slides down the incline, reaching the starting point with a speed of 1.1 m/s. (*a*) What is the total work done by the force of friction? (*b*) How far up along the incline does the block travel if the coefficient of kinetic friction between the block and surface is 0.30?

(*a*) Figure 9-8 shows the forces acting on the block while it is traveling *up* the incline; the force of friction $\mathbf{f}_k$ is then *down* the incline, opposite in direction to the block's velocity. On the return trip, the velocity is down and the friction force up the incline.

We know that the work done on the block by the resultant force on it is the change in the block's kinetic energy:

$$W(\text{by resultant } \mathbf{F}) = K_f - K_i$$
$$= \tfrac{1}{2}mv_f^2 - \tfrac{1}{2}mv_i^2$$
$$= \tfrac{1}{2}(2.0 \text{ kg})[(1.1 \text{ m/s})^2 - (2.0 \text{ m/s})^2]$$
$$= -2.8 \text{ J}$$

But we also know that, since the resultant force is $\mathbf{F} = m\mathbf{g} + \mathbf{N} + \mathbf{f}_k$,

$$W(\text{by resultant } \mathbf{F}) = W(\text{by } m\mathbf{g}) + W(\text{by } \mathbf{N}) + W(\text{by } \mathbf{f}_k)$$

where each of the terms is to be evaluated from an initial state with

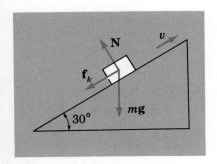

FIG. 9-8. Forces on a block sliding up a rough incline.

the block at the bottom to a final state with the block again at the bottom. As was shown in Example 9-4, the work done by the weight $m\mathbf{g}$ over a round trip is zero; W(by $m\mathbf{g}$) = 0. Since the normal force $\mathbf{N}$ is always perpendicular to the displacement, W(by $\mathbf{N}$) = 0. We are left with W(by $\mathbf{f}_k$), so that

$$W(\text{by resultant } \mathbf{F}) = W(\text{by } \mathbf{f}_k) = -2.8 \text{ J}$$

The friction force here, as always, does *negative* work.

(b) As shown in Example 8-7, the magnitude of the friction force is

$$f_k = \mu_k mg \cos \theta$$

If the total distance traveled by the block up the incline is s, the friction force acts over a total distance $2s$ as the block completes the round trip. The force $\mathbf{f}_k$ is opposite to the displacement for both the upward and downward segments, so that it does negative work throughout, and we have

$$-(2s)(\mu_k mg \cos \theta) = -2.8 \text{ J}$$

$$s = \frac{2.8 \text{ J}}{2\mu_k mg \cos \theta} = \frac{2.8 \text{ J}}{2(0.30)(2.0 \text{ kg})(9.8 \text{ m/s}^2)(0.866)}$$

$$= 0.27 \text{ m}$$

9-4 WORK DONE BY A SPRING

If an ordinary helical spring is not stretched or compressed too much, it will return to its undeformed, or equilibrium, configuration when the deforming forces are removed. For small deformations the force F_s produced by the spring along its length is given by

$$F_s = -kx \qquad (9\text{-}10)$$

where x is the displacement of the spring from its unstretched length and k is a constant.† See Fig. 9-9a.

The minus sign appears in (9-10) because the spring's force is always opposite in direction to the displacement. When $x = 0$ and the spring is relaxed, $F_s = 0$ (Fig. 9-9b). When the spring is stretched (Fig. 9-9c) and x is positive, the force produced by the spring is negative. Likewise, when the spring is compressed and x is negative, as in Fig. 9-9d, the force F_s is positive. In every instance the spring applies a force to an object attached to it which tends to restore this object and the spring to an equilibrium configuration, and the spring is said to apply a *restoring force*. The spring constant, or force constant, k is a measure of a spring's stiffness; the larger the k, the stiffer the spring.

In Chap. 14 we shall consider in detail the motion of an object attached to a spring. Here we are concerned with the work

---

†This relation is known as Hooke's law after its discoverer, Robert Hooke (1635–1703).

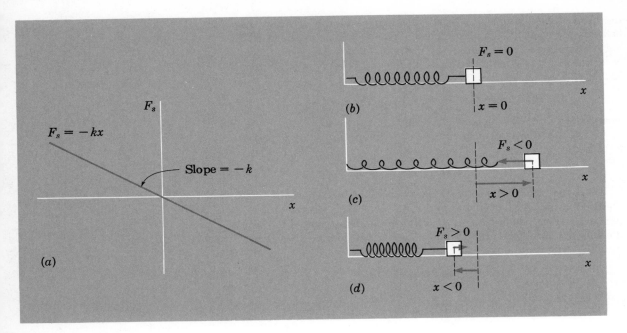

FIG. 9-9. Force characteristics of a spring. (a) The spring force F_s as a function of its displacement x. (b) The spring in its relaxed state. (c) The spring stretched by a displacement x to the right and with a spring force F_s to the left. (d) The spring compressed with a displacement x to the left and the spring force F_s to the right.

done by a stretched or compressed spring, a simple example of work done by a force which varies with displacement.

Suppose that a spring is initially compressed, as in Fig. 9-9d; it applies a force F_s to the right to a block pressed against the spring. We wish to find the work done by the spring as its right end goes from the position $-x$ to zero. Note that the force and displacement are both to the right, so that the angle θ between the resultant force $\mathbf{F}$ and the displacement $d\mathbf{r}$ is zero. Applying the general definition of work, we then have

$$W_{i \to f} = \int_i^f \mathbf{F} \cdot d\mathbf{r} = \int_{-x}^0 F_s\, dx = \int_{-x}^0 (-kx)\, dx$$
$$W = \tfrac{1}{2}kx^2 \tag{9-11}$$

In general the integral $\int F\, dx$ is the area under the curve of F plotted against x. For this particular curve—a straight line—the area is $\tfrac{1}{2}kx^2$. In general, the area under a force-displacement curve is equal to the work done by the force. See Fig. 9-10.

A compressed spring does *positive* work in the amount $\tfrac{1}{2}kx^2$ as it expands by an amount x to its relaxed length; it imparts kinetic energy to a block attached to its end. Similarly, a stretched spring does *positive* work in the amount $\tfrac{1}{2}kx^2$ as it contracts by an amount x to its relaxed length; here again the spring force and displacement are in the same direction, and a block attached to the spring gains kinetic energy. On the other hand, the spring does *negative* work while being stretched or compressed; then the spring force and the applied displacement are in opposite directions, and a block at-

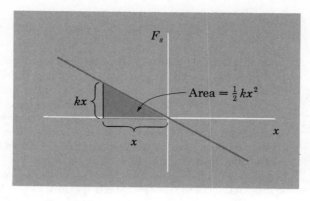

FIG. 9-10. The work done by a spring expanding from $-x$ to 0 is the area under the force-displacement graph, $\frac{1}{2}kx^2$.

tached to the spring and initially in motion loses kinetic energy. In any event, the *magnitude* of the work done by a spring in going from $\pm x$ to zero or from 0 to $\pm x$ is $\frac{1}{2}kx^2$ [$\frac{1}{2}kx^2 = \frac{1}{2}k(-x)^2$]; the sign of the work done by the spring is determined by whether force and displacement are parallel or antiparallel, or equivalently, whether the spring speeds up or slows down an object attached to it.

Example 9-6

A 25-N force compresses a spring by 4.0 cm. A 1.0-kg object is attached to the spring when it is compressed by 4.0 cm and then released from rest. (a) What is the object's speed when the spring is relaxed? (b) Where is the object when it first comes to rest again? (c) What is the net work done by the spring when the object has returned to its starting point?

(a) As the spring expands, it does positive work in the amount $\frac{1}{2}kx^2$, where here x is 4 cm. The object's kinetic energy goes from an initial zero to a final $\frac{1}{2}mv^2$. Therefore, from the work-energy theorem,

$$W_{i \to f} = K_f - K_i$$

$$\tfrac{1}{2}kx^2 = \tfrac{1}{2}mv^2 - 0$$

or

$$v = \left(\frac{k}{m}\right)^{1/2} x$$

The spring's force constant is

$$k = -\frac{F_s}{x} = -\frac{25 \text{ N}}{-4.0 \times 10^{-2} \text{ m}} = \frac{10^4}{16} \text{ N/m}$$

so that the speed at $x = 0$ is

$$v = \sqrt{\frac{k}{m}}\, x = \sqrt{\frac{(10^4/16) \text{ N/m}}{1.0 \text{ kg}}}\, (4.0 \times 10^{-2} \text{ m}) = 1.0 \text{ m/s}$$

(b) After it passes the equilibrium position, the object is slowed by the spring force which now acts in a direction opposite to the object's velocity. The spring does negative work until the object is brought momentarily to rest. Now the work done by the spring is $-\frac{1}{2}kx^2$, the object's *initial* kinetic energy is $\frac{1}{2}mv^2$, and its final kinetic energy is zero. For this segment of the object's motion, we have

$$W_{i \to f} = K_f - K_i$$
$$-\tfrac{1}{2}kx^2 = 0 - \tfrac{1}{2}mv^2$$

But this is precisely the same relation as that for part *(a)*, so that the object comes to rest with the spring extended 4.0 cm, or the same amount as that by which it was initially compressed.

(c) We know that after the object has come to rest at the farthest extension of the spring, it will then be returned to the equilibrium position by a force in the same direction as its displacement; the spring does positive work, again in the amount $\tfrac{1}{2}kx^2$, as the object gains kinetic energy which is, at $x = 0$, again $\tfrac{1}{2}mv^2 = \tfrac{1}{2}kx^2$. In the final segment of the cycle, the spring again does negative work and brings the object to rest at its starting point, $x = -4.0$ cm.

We can find the total amount of work done by the spring over one complete cycle simply by adding, with proper attention to signs, the work done in each of the four segments:

$$W_{\text{complete cycle}} = W_{-x \to 0} + W_{0 \to x} + W_{x \to 0} + W_{0 \to -x}$$
$$= \tfrac{1}{2}kx^2 - \tfrac{1}{2}kx^2 + \tfrac{1}{2}kx^2 - \tfrac{1}{2}kx^2$$
$$= 0$$

Over a round trip the net work done by the spring force is *zero*. This result is, in fact, independent of how a spring is brought from some initial configuration to an identical final configuration: Since the direction and magnitude of the spring force depends only on the displacement $x(F_s = -kx)$, the net work done by the spring as it retraces its path back to the starting point is zero. Over any small displacement dx where the force is F_s, the work done has magnitude $F_s\, dx$; on the return trip the force at that location is identical in magnitude and direction but the direction of the displacement dx is reversed, so that the work, again of magnitude $F_s\, dx$, has the opposite sign. If the net work is zero for any set of equal incremental displacements in opposite directions, it must be zero for a complete cycle. See Fig. 9-11.

9-5 POWER

The work done by a force and the kinetic energy acquired by a particle do not involve the time t explicitly. For practical considerations, however, it is often important to know not only the amount of work done by an agent but also how rapidly the agent can do it. It is useful, therefore, to define a quantity called the *power*, which represents the time rate at which work is done.

The average power $\overline{P}$ over the time interval Δt is defined as the energy transferred $W_{i \to f}$ divided by the time Δt over which the transfer takes place:

$$\overline{P} = \frac{W_{i \to f}}{\Delta t}$$

The instantaneous power P is the time rate of doing work:

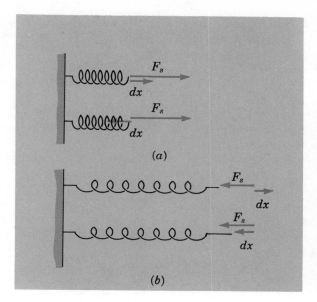

FIG. 9-11. (a) A compressed spring does work of equal magnitude but *opposite* sign for equal incremental displacements in opposite directions. (b) A stretched string does work of equal magnitude but *opposite* sign for equal incremental displacements in opposite directions. In both (a) and (b) the force is equal in magnitude and direction, whereas the incremental displacements are equal in magnitude but opposite in direction.

$$P = \frac{dW}{dt} \qquad (9\text{-}12)$$

These relations can be written in different forms by recalling the definition of work in terms of force and displacement:

$$W_{i \to f} = \int \mathbf{F} \cdot d\mathbf{r}$$

However, $d\mathbf{r} = \mathbf{v}\, dt$, where $\mathbf{v}$ is the instantaneous velocity, and we can write

$$W_{i \to f} = \int (\mathbf{F} \cdot \mathbf{v})\, dt = (\overline{\mathbf{F} \cdot \mathbf{v}})\, \Delta t$$

where $(\overline{\mathbf{F} \cdot \mathbf{v}})$ is the time average of the scalar product of the instantaneous force $\mathbf{F}$ and velocity $\mathbf{v}$ over the time interval Δt. Using the last equation in the definitions of P and $\overline{P}$, we have:

$$\overline{P} = (\overline{\mathbf{F} \cdot \mathbf{v}})$$

$$P = \mathbf{F} \cdot \mathbf{v} \qquad (9\text{-}13)$$

Appropriate units for power are joules per second. This combination of units is termed the watt, abbreviated W. Therefore,

$$1 \text{ watt} = 1 \text{ W} = 1 \text{ J/s} = 1 \text{ kg-m}^2/\text{s}^3$$

SUMMARY

The work-energy theorem:

$$\int_i^f \mathbf{F} \cdot d\mathbf{r} = \tfrac{1}{2}mv_f^2 - \tfrac{1}{2}mv_i^2 \qquad (9\text{-}7)$$

where $\mathbf{F}$ is the resultant force on a particle of mass m going from initial speed v_i to find speed v_f and $d\mathbf{r}$ is an incremental displacement along the particle's path.

The work done by force $\mathbf{F}$ from initial state i to final state f is

$$W_{i \to f} = \int_i^f \mathbf{F} \cdot d\mathbf{r} \qquad (9\text{-}5)$$

Work is computed from the component $F \cos \theta$ along the displacement.

A particle's kinetic energy is

$$K = \tfrac{1}{2}mv^2 \qquad (9\text{-}2)$$

The force exerted by a spring is given by

$$F = -kx \qquad (9\text{-}10)$$

where x is the displacement of the spring from its unstretched length and k is the spring constant.

The work done by a spring (as it expands from $-x$ to 0, or contracts from $+x$ to 0) is

$$W = \tfrac{1}{2}kx^2 \qquad (9\text{-}11)$$

Power is the time rate of doing work:

$$P = \frac{dW}{dt} = \mathbf{F} \cdot \mathbf{v} \qquad (9\text{-}12),(9\text{-}13)$$

PROBLEMS

9-1. Particles A and B have equal kinetic energies, and the mass of A is twice that of B. What is the ratio of their momenta?

9-2. The kinetic energy of particle A is five times that of particle B, and the momentum of A is four times the momentum of B. What is the ratio of their masses?

9-3. A woman lowers a 300-N weight through a vertical distance of 90 cm at a constant speed of 20 cm/s. What is the work done by (a) the woman, (b) the earth, and (c) the resultant force?

9-4. A man (a) raises a 20-kg suitcase from the floor to a point 75 cm above the floor in 2.0 s, (b) then holds it in this position for 4.0 h, and finally (c) places the suitcase back of the floor in 2.0 s. What is the work done by the man on the suitcase for each of the three parts above? What is the net work done by the man on the suitcase for the entire sequence?

9-5. A man (a) stretches an initially relaxed spring (with a force constant of 200 N/m) so that he pulls its two ends farther apart by a distance of 20.0 cm, (b) then holds it in this position for 10.0 minutes, and (c) finally returns it to its initial state in 5.0 s. What is the work done by the man on the spring for each of the three parts above? (d) What is the net work done by the man on the spring for the entire sequence?

9-6. For a 5.0-s interval an unbalanced force is applied to a 2.0-kg object which is initially at rest, giving it a kinetic energy of 100 J. What are (a) the force and

(b) the final momentum? (c) How far has the object moved?

9-7. A billiard ball of mass m sliding initially with speed v collides head on with an identical second ball and is brought to rest. What is the work done by the incident ball on the struck ball?

9-8. A 0.20-g particle initially traveling left at 50 cm/s is acted upon by a constant force to the right of 4.0 N. What is the particle's kinetic energy when it is 50 cm to the right of its starting point?

9-9. A girl pulls with a rope attached to a 12-kg box moving at constant speed over a rough horizontal surface. The rope makes an angle of 30° with respect to the horizontal, and the girl applies a force of 180 N. The box moves 2.0 m along the horizontal. What is the work done by (a) the resultant force on the box, (b) the force of the girl on the box, (c) the normal force of the surface on the box, (d) the force of gravity on the box, and (e) the friction force on the box?

9-10. A man applies a force along an incline at angle θ relative to the horizontal so that an object of mass m on this frictionless surface moves up along the incline at constant speed. The object advances a distance s along the incline. What is the work done on the object by (a) the man, (b) the normal force of the surface, and (c) the force of gravity on the object?

9-11. A constant braking force **F** is applied to an object moving with an initial velocity $\mathbf{v}_0$, stopping it in a distance D and in a time interval $\Delta\tau$. What are the corresponding (a) distance and (b) time intervals required for the same force to stop the same object moving initially at a speed larger by a factor of three.

9-12. (a) A particle of mass m moves in a circle of radius r at the constant speed v_i. How much work is done on the particle by the force on it when it travels half way around the circle? (b) Now the particle's speed, initially v_i, is increased by a factor of three as it travels half way around the circle. What is the total work done on the particle by the resultant force on it?

9-13. A particle of mass m hangs from a string of length L. Then the particle is pulled to the side with the string remaining taut until the angle of the string with the vertical is θ. How much work is done (a) by a person moving the particle from its initial to its final location, (b) by the force of gravity, and (c) by the tension in the string?

9-14. Figure P9-14 shows the force on a particle of mass 0.50 kg moving along the positive X axis. The particle's velocity at $x = 0$ is 2.0 m/s in the positive X direction. What is the particle's kinetic energy at (a) $t = 1.0$ s, (b) $t = 2.0$ s, and (c) $t = 3.0$ s?

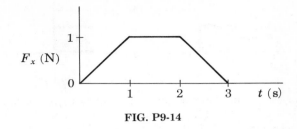

FIG. P9-14

9-15. A 0.40-kg ball is thrown vertically upward with a speed of 8.0 m/s. What is the ball's kinetic energy when it is (a) 2.0 m above the ground and traveling upward, (b) 2.0 m above the ground and traveling downward, and (c) back at the starting point?

9-16. A 2.0-kg block coasts down a 30° incline at constant speed traveling a total distance of 50 cm along the incline. What is the work done on the block by (a) the normal force, (b) the force of friction, (c) the force of gravity on the block, and (d) the resultant force on the block?

9-17. The magnitude of the attractive force between two unlike electric charges of magnitudes q_1 and q_2 separated by a distance r is given by kq_1q_2/r^2. How much work must be done to increase the separation of the two charges from r_1 to r_2?

9-18. A 5.0-kg object is thrown vertically upward at a speed of 2.0 m/s. From the time when it first rises until it returns to its starting point, what is (a) the net work done by the force of gravity and (b) the net impulse by the force of gravity?

9-19. A force acts on a particle initially at rest. Its impulse is 4.0 kg-m/s, and the work done by the force over the same interval is 2.0 J. What is the particle's mass?

9-20. A golfer drives a 45-g ball at 45° above the horizontal so that it travels for 100 m (measured along the ground). Find (a) the change in momentum of the ball and (b) the work done by the golf club. (c) What is the magnitude of the average force on the ball during impact if contact lasts for 1.0×10^{-3} s? Assume $g = 10.0$ m/s².

9-21. Initially a 2.0-kg block of wood is moving horizontally over a rough surface at 3.0 m/s; the block is brought to rest after moving 5.0 m. How much power would be required to keep it moving over the same surface at a constant speed of 3.0 m/s?

9-22. A pitcher throws a 160-g baseball into a 1,440-g absorber block which is initially at rest on a rough horizontal surface. The collision is perfectly inelastic—the ball and block move together—and the impact

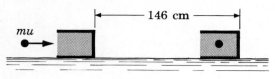

FIG. P9-22

moves the block (and ball) forward a distance of 146 cm, see Fig. P9-22. Find the speed of the pitched ball if the coefficient of kinetic friction is 0.40.

9-23. A spring with a force constant of 30 N/m is stretched from its relaxed length by 6.0 cm, then allowed to return to its relaxed length, then compressed 6.0 cm, and finally allowed to expand 3.0 cm. What is the net work done by the spring?

9-24. A spring with force constant k is compressed an amount x and two objects, each of mass m, are placed at its ends. Then the objects are released from rest. (a) What is the work done by the spring as it expands to its relaxed length? (b) What is the kinetic energy of each object afterward?

9-25. Two springs having force constants k_1 and k_2 have the same length when relaxed. Both springs are fixed at their left ends and pulled together at their right ends by a single applied force, as shown in Fig. P9-25. (a) What is the work done when the two springs are stretched through a distance x? (b) What is the force as a function of x?

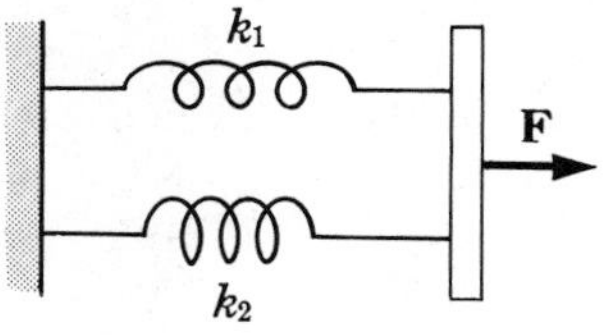

9-26. A racing car travels at constant speed along a horizontal roadway at 45 m/s. The engine is producing energy at the rate of 450 kW ($\approx$600 hp). (a) What is the equivalent total resistive force on the racer? (b) What is the horizontal force, along the direction of motion, of the road surface on the car?

9-27. A 3.0-kg object is thrown vertically upward at a speed of 29.4 m/s. What is the rate (in watts) at which the force of gravity does work on this object (a) 1.0 s, (b) 2.0 s, and (c) 3.0 s after it has been projected?

9-28. Show that it is impossible to do work at constant power on a body initially at rest.

9-29. A small sphere of mass 300 g is attached to a string 1.0 m long. The other end of the string is attached to the ceiling, and the small sphere is displaced until the string makes an angle of 60° with respect to the vertical and is then released from rest. (a) Find the work done by the force of gravity and by the tension in the string during the time the angle goes from 60 to 0°. (b) What is the kinetic energy and speed of the object at its lowest position? (c) What is the tension in the string at the lowest position?

9-30. The earth of mass 6.0×10^{24} kg moves in a nearly circular orbit at constant speed about the sun in 1 yr $\approx 3.16 \times 10^7$ s. During a $\frac{1}{2}$-yr period (a) what is the net work done by the sun on the earth? (b) What is the net impulse? (The sun-earth distance is 1.5×10^{11} m.)

FIG. P9-25

CHAPTER 10

Potential Energy and Energy Conservation

Up to this point we have recognized one type of energy only—the kinetic energy an object has by virtue of its motion. In this chapter we introduce the concept of potential energy, so that we may speak of the conservation of energy. We consider in some detail the potential energy associated with two important systems in which a conservative force acts: the earth and an object near it, and two masses attached to a spring. We treat the general conservation of energy principle and reexamine collisions in terms of energy conservation.

10-1 POTENTIAL ENERGY OF A SPRING

Momentum is an important physical quantity because, for any isolated system, momentum is conserved. We have seen further that kinetic energy is a significant property of an object in motion. Is kinetic energy conserved?

To answer this question we consider in some detail a simple situation: the head-on collision between two objects which interact through a spring attached to one of them. For simplicity we suppose that the two objects, not necessarily of the same mass, collide in a reference frame in which the system's total momentum is zero (the center-of-mass reference frame).

As Fig. 10-1 shows, *(a)* the objects first approach at constant velocities, *(b)* a repulsive force acts on each as the spring is compressed and later expands to its relaxed length, and the objects reverse their directions of motion, and *(c)* then the two objects separate, again at constant velocities. While the spring is compressed and a force acts on each object, both the momentum mv and kinetic energy $\frac{1}{2}mv^2$ of each object are changing.

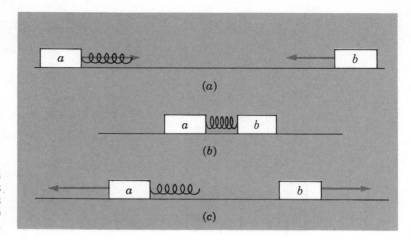

FIG. 10-1. Objects a and b in collision; they *(a)* approach at constant velocities, *(b)* interact through a spring, and *(c)* recede from one another.

The momenta p_a and p_b of the two objects, together with the system's total momentum $p = p_a + p_b$, is plotted as a function of time in Fig. 10-2a. The *total* momentum p is constant (zero) before, after, and *during* the interaction, in accord with the law of momentum conservation.

What about the kinetic energy? Each object's kinetic energy, K_a and K_b, and their sum, $K = K_a + K_b$, the total kinetic energy for the system, are plotted as a function of time in Fig. 10-2b. (Although both a and b have the same momentum magnitude, their kinetic energies are not the same because their masses differ—$K = p^2/2m$.) We see that the total kinetic energy is the same before and after the interaction, but it decreases during the interaction and, in fact, drops to zero at that instant when both objects have been brought momentarily to rest. Although unchanged before and after the collision, kinetic energy is *not* conserved throughout the collision.

First let us see why the kinetic energy out of the collision must equal the total kinetic energy into the collision. The force of the spring does negative work on the objects in slowing them to rest, and then positive work in speeding them. We know from the discussion in Sec. 9-4 that the *net* work done *by* a spring which is deformed but then returned to its initial configuration is *zero*. Therefore, from the work-energy theorem, there can be no overall change in the objects' kinetic energy over the entire cycle. In short, the properties of a spring, together with the work-energy theorem, require that the objects emerge from the interaction with the same kinetic energy as they had initially.

Even though the isolated system's initial and final total kinetic energies are the same, we cannot say that kinetic energy is conserved during collision because, clearly, the kinetic energy decreases during the collision. But we know from the arguments above that we will get back *all* of the kinetic energy that is tem-

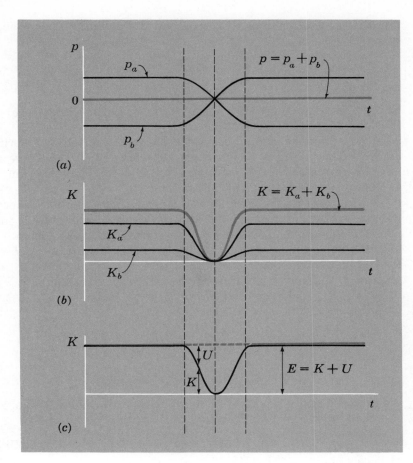

FIG. 10-2. For the situation shown in Fig. 10-1, *(a)* the momenta of the separate objects and the total momentum $p = p_a + p_b$ of the system plotted as a function of time, and *(b)* the kinetic energies of the separate objects and the total kinetic energy $K = K_a + K_b$ of the system plotted as a function of time. *(c)* The system's potential energy U is so defined that $E = K + U$ is a constant.

porarily "lost" while the objects are interacting. The kinetic-energy debt is sure to be repaid. If this credit in kinetic energy is certain to be honored, then the amount by which the total kinetic energy drops from its initial (or final) value may be regarded as potentially kinetic energy, or *potential energy* for short. Thus, we may define a potential energy for the system of two objects and spring to be that quantity U which when added to $K = K_a + K_b$ ensures that the total is constant. If we label the system's *total energy E*, we may then write for an isolated system (Fig. 10-2c):

$$E = K + U = \text{constant} \qquad (10\text{-}1a)$$

If K_1 and U_1 are the kinetic and potential energies of the system at one time and K_2 and U_2 their respective values at some second time, then the constancy of E implies that

$$K_1 + U_1 = K_2 + U_2 \qquad (10\text{-}1b)$$

Put into a different form, we so define the system's potential energy that for every change ΔK in the system's kinetic energy there is a corresponding change ΔU in the potential energy such that

$$\Delta K + \Delta U = 0$$

or
$$\Delta U = -\Delta K \qquad (10\text{-}1c)$$

Thus, when K decreases (increases), U must increase (decrease) by the same amount to ensure that $\Delta E = 0$.

In summary, the properties of the force exerted by a spring are such that we may *invent potential energy so that the system's total energy is conserved.* Equations (10-1a to c) are alternative expressions of the *law of energy conservation.*

Having established that we can associate potential energy with a deformed spring, let us find the expression for the spring potential energy U_s in terms of the spring's stiffness constant k and the displacement x of one end relative to its position when relaxed. In being compressed from its equilibrium shape by an amount x, the spring does negative work. As shown in Sec. 9-4, the work $W_{0 \to -x}$ done *by* the spring is then $-\tfrac{1}{2}kx^2 = W_{0 \to -x}$. It follows from the work-energy theorem that

$$W_{i \to f} = \Delta K$$

and from (10-1c),

$$\Delta U = -\Delta K = -(-\tfrac{1}{2}kx^2) = \tfrac{1}{2}kx^2$$

We take the spring's potential energy to be zero when it is relaxed ($x = 0$ and also $F_s = 0$). Therefore,

$$U_s = \tfrac{1}{2}kx^2 \qquad (10\text{-}2)$$

The potential energy of a stretched or compressed spring varies as the square of x.

Example 10-1

A 1.0-kg object coasts along a smooth horizontal surface at 1.0 m/s and strikes a spring with $k = 25$ N/m whose right end is firmly attached to a very massive second object. See Fig. 10-3. What is the maximum amount by which the spring is compressed?

The spring has its maximum compression when the initially moving object has been brought to rest momentarily. Taking 1 to represent the state of affairs before collision and 2 the state when the spring has its maximum compression and the object is at rest, we have, using Eq. (10-1b),

$$K_1 + U_1 = K_2 + U_2$$
$$\tfrac{1}{2}mv^2 + 0 = 0 + \tfrac{1}{2}kx^2$$

(We may ignore the kinetic energy of the massive object because it remains essentially at rest throughout.)

$$x = \left(\frac{m}{k}\right)^{1/2} v = \left(\frac{1.0 \text{ kg}}{25 \text{ N/m}}\right)^{1/2} (1.0 \text{ m/s}) = 0.20 \text{ m}$$

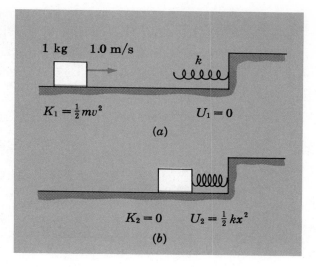

FIG. 10-3

Of course, when the object is at rest, the spring force continues to act to the left, and the 1.0-kg object is accelerated to the left. In the language of energy, after achieving its maximum compression, the spring loses potential energy, and the light object at its end gains an equal amount of kinetic energy. Thus, the 1.0-kg object leaves the spring traveling left at 1.0 m/s, with the same speed and kinetic energy as that with which it entered the collision.

10–2 GRAVITATIONAL POTENTIAL ENERGY

In the previous section we saw that it is possible to associate potential energy with a system of two objects interacting through a spring. Now our system consists of the earth and a ball thrown vertically upward into the air, and our task is to see whether a gravitational potential energy may be associated with this system and, if so, what its form must be. (We suppose that the ball covers a distance small compared with the earth's radius, so that the force mg on it, which from Newton's third law is also the magnitude of the force on the earth, is constant.)

We again have, in effect, a collision, but it is now one in which the two objects interact continuously with one another and the velocity of each changes continuously. Our considerations here will, nevertheless, parallel exactly those given for the spring. Whereas the spring produced a repulsive force on the interacting objects, we now have interaction through an attractive force.

We suppose that the motion of the earth (very slight indeed) and the ball thrown away from it are viewed from a reference frame in which the system's total momentum is zero (the center-of-mass reference frame). As the ball moves upward, the earth recoils downward. Both ball and earth come to rest momentarily

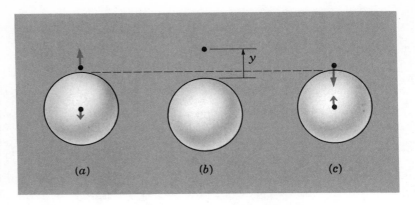

FIG. 10-4. A ball interacting with the earth: *(a)* The ball is thrown vertically upward, and the earth recoils downward, *(b)* the ball and earth are momentarily at rest when their separation distance is a maximum *y*, and *(c)* the ball returns to meet the earth's surface.

at the same time in our reference frame, and since they continue to attract one another, the ball falls downward and the earth upward as they meet once again. See Fig. 10-4. The coordinate *y* gives the ball's vertical displacement *relative to the earth's surface*.

Figure 10-5 shows the momentum along the *Y* axis of each object as a function of time, together with the system's total momentum. (The lines on the momentum-time graph are straight because the gravitational force on both ball and earth is assumed to be constant.) Here too momentum is conserved, the system's total momentum being zero at *all* times, as shown in Fig. 10-5*a*. And again the kinetic energy is not constant. The ball's kinetic energy first decreases as it rises and comes to rest at its highest point, and then increases as the ball falls. The earth's kinetic energy, very much smaller than the ball's because of the large mass ratio, also decreases to zero and then increases. We see in Fig. 10-5*b* that the total kinetic energy of ball and earth returns to its initial value when the ball strikes the ground, and the system has the same configuration ($y = 0$) as initially. Again this happens because the force mg acting on the ball and a force of the same magnitude acting on the earth does *no net work over the round trip*. Therefore, there is no net change in kinetic energy.

Here again a potential energy may be assigned to the system since the kinetic-energy loss is certain to be repaid. As before, the gravitational potential energy U_g is to be defined in such a way as to ensure that

$$E = K + U_g = \text{constant}$$

where $K = K_b + K_e$, strictly the sum of the kinetic energies of ball and earth. See Fig. 10-5*c*.

As the ball rises a distance *y* above the earth, the downward force mg on it is opposite to its upward displacement, and the work done on it by the gravitational force is $-mgy$.†

---

†Strictly, as the ball rises h_1 in the center-of-mass reference frame, the work on it is $-mgh_1$. And as the earth falls a distance h_2, again in the center-of-mass frame, the equal force on it does

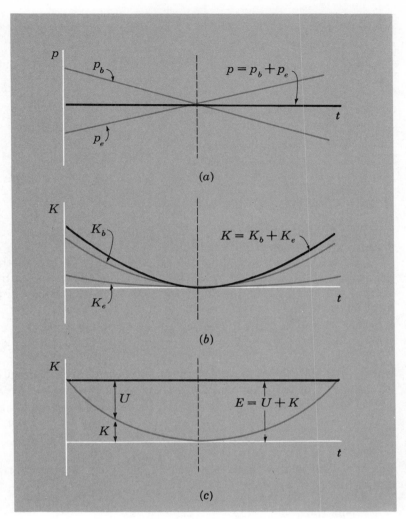

FIG. 10-5. For the situation shown in Fig. 10-2, *(a)* the momenta of the separate objects and the total momentum $p = p_b + p_e$ of the system plotted as a function of time, and *(b)* the kinetic energies of the separate objects and the total kinetic energy $K = K_b + K_e$ of the system plotted as a function of time. *(c)* The system's potential energy U is so defined that $E = K + U$ is a constant.

$$W_{i \to f} = -mgy = \Delta K$$

But it is required that

$$\Delta U_g = -\Delta K = -(-mgy) = mgy$$

If we take the zero for U_g to correspond to $y = 0$, then

$$U_g = mgy \tag{10-3}$$

It must be emphasized that (10-3) gives the gravitational potential energy only for small changes in the separation distance between the earth and an object with which it interacts. The general rela-

---

work $-mgh_2$. The total work done is then

$$-mgh_1 - mgh_2 = -mgy$$

where $y = h_1 + h_2$.

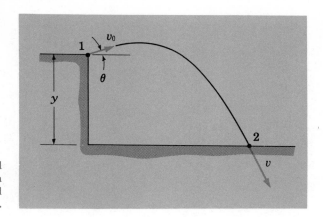

FIG. 10-6. An object projected with a speed v_0 at point 1 has a speed v at point 2, a vertical distance y below 1.

tion for gravitational potential energy, applicable for all separation distances, is developed in Sec. 13-5.

It is easy to show that the kinetic energy of the earth is negligibly small compared with that of a light object, like a ball. Taking the earth's kinetic energy to be $K_e = \frac{1}{2}MV^2$ and that of the light object to be $K_b = \frac{1}{2}mv^2$, we have for their ratio:

$$\frac{K_e}{K_b} = \frac{\frac{1}{2}MV^2}{\frac{1}{2}mv^2} = \frac{m}{M}\left(\frac{MV}{mv}\right)^2$$

But by momentum conservation the magnitudes of the momentum of the light object and earth must always be equal: $MV = mv$. Therefore,

$$\frac{K_e}{K_b} = \frac{m}{M}$$

For $m = 1$ kg, and with $M \approx 10^{25}$ kg, $K_e/K_b \approx 10^{-25}$, which is truly negligible.

Example 10-2

An object of mass m is thrown with an initial speed v_0 at angle θ relative to a horizontal plane at an elevation y above the earth's surface. With what speed v does the object strike the earth?

This is just Example 9-4, worked earlier by applying the work-energy theorem. Now we apply an equivalent principle, that of energy conservation. We take 1 and 2 to designate the initial and final states, as shown in Fig. 10-6. From energy conservation,

$$K_1 + U_1 = K_2 + U_2$$
$$\tfrac{1}{2}mv_0^2 + mgy = \tfrac{1}{2}mv^2 + 0$$

or
$$v = (v_0^2 + 2gy)^{1/2}$$

Note again that the final speed is independent of the object's mass and of the projection angle θ. The great utility of the energy-conservation principle is that we need not be concerned with what happens in detail at all the intermediate stages but may instead deal

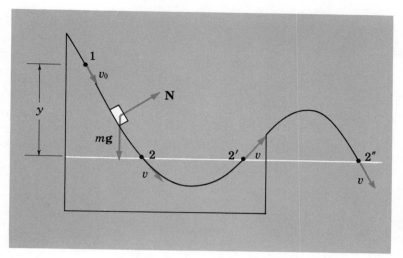

FIG. 10-7. An object sliding on a smooth surface has a speed v_0 at point 1. Its speed at points 2, 2', and 2", all a vertical distance y below point 1, is v.

only with the end points. Energy conservation does not, however, give us all available information about a system. For example, in this example we cannot find the time of flight or the direction of the final velocity from energy considerations alone.

Example 10-3

Suppose that the object in the previous example, instead of being projected at speed v_0 to fly through the air and strike the ground a vertical distance y below, is now sliding at initial speed v_0 on a smooth surface, as in Fig. 10-7. We wish to find its speed v at points a distance y vertically beneath the starting point.

The only significant difference from the previous example is that, in addition to the object's weight, we now have a normal force N which constrains the object to remain on the surface. But the normal force is always at right angles to the object's displacement along the surface. It does no work, and no potential energy can be associated with it. Therefore, exactly the same analysis applies here, and we have at once that $v = (v_0^2 + 2gy)^{1/2}$ at points 2, 2', and 2" in Fig. 10-7.

Example 10-4

A small object is attached to the end of a taut cord of negligible mass as shown in Fig. 10-8. Its speed is initially v_0, and it swings downward in a vertical circle until its vertical displacement is down by y. What is the object's speed at this lower point?

Again we have, in effect, the same problem as in the two previous examples. The string's tension T is always at right angles to the particle's displacement along the circular arc, so that it does no work and a potential energy cannot be associated with this constraint force. And again the force of gravity (weight) is accounted for by the gravitational potential energy. The speed is, once more, $v = (v_0^2 + 2gy)^{1/2}$.

Notice in the figure that, after swinging downward, the string is interrupted by a post, so that a shortened segment of string now

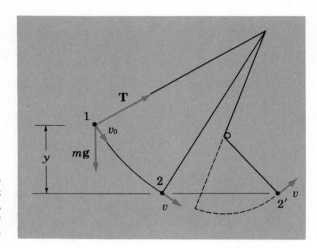

FIG. 10-8. A particle attached to a taut string has speed v_0 at point 1. Its speed at points 2 and 2', both a vertical distance y below point 1, is v.

turns about it. The force of this post does no work on the system—it does not act over a displacement—and there is then no influence on the system's total energy. Consequently, the particle at the string's end will again have the same speed and kinetic energy for all points along a horizontal line.

Example 10-5

A 1.0-kg block is pressed against, but not attached to, a light spring having a stiffness constant of 1.0×10^3 N/m. When the spring has been compressed 7.0 cm, the block is released. The block slides along a frictionless surface and up an incline, as shown in Fig. 10-9. What maximum height y does the block achieve?

After the spring has been compressed and released, the total energy content of the system remains constant. First, potential energy of the spring $\frac{1}{2}kx_1^2$ is transformed into kinetic energy $\frac{1}{2}mv_2^2$ of the sliding block; then the block's kinetic energy becomes gravitational potential energy mgy_3 as the block comes to rest at the highest point:

$$\tfrac{1}{2}kx_1^2 = \tfrac{1}{2}mv_2^2 = mgy_3$$

$$y_3 = \frac{kx_1^2}{2mg} = \frac{(1.0 \times 10^3 \text{ N/m})(7.0 \times 10^{-2} \text{ m})^2}{2(1.0 \text{ kg})(9.8 \text{ m/s}^2)} = 0.25 \text{ m}$$

This example illustrates the usefulness of the potential-energy concept and the conservation principle following from it. We merely equate the total energy of the system at the end points taking no concern for such details of the motion as the shape of the incline or the velocity at intermediate points. We could, of course, have arrived at the same final result by applying Newton's second law directly. However, this would have been mathematically difficult and extremely tedious, for it would have required the following procedure. From the initial force acting on the block (and this force *varies* with position), one computes the initial instantaneous acceleration; from this acceleration one computes the instantaneous velocity at a time dt later; from this instantaneous velocity one then computes the corresponding displacement; knowing the new displacement, one finds

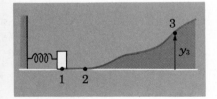

FIG. 10-9. A block is pressed against a compressed spring at point 1, slides past point 2 on the horizontal level, and comes finally to rest at point 3, where the vertical displacement is y_3.

the new instantaneous force. The cycle is repeated over and over until finally one arrives at the position for which the velocity is zero. This procedure requires, of course, a detailed knowledge of the shape of the incline.

Example 10-6

A spring of stiffness constant k is attached to a ceiling. A block of mass m is attached to the lower end of the spring while the spring is unstretched, and then the block is released at this point. See Fig. 10-10. What is the distance, say $2A$, below the point of release at which the block comes momentarily to rest?

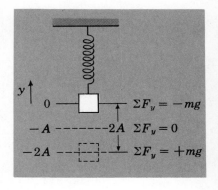

FIG. 10-10. An oscillating block subject to both spring and gravitational forces. The spring is unstretched at $y = 0$; it has its maximum extension at $y = -2A$.

As the block descends, the gravitational potential energy decreases, and the potential energy of the spring increases. We take the vertical displacement y (positive upward) of the block, to be measured relative to the block's initial position; y then represents the elongation of the spring also. The total energy is initially *zero*, and it must remain zero at every instant thereafter. In symbols,

$$E = K + U_g + U_s$$
$$0 = K + mgy + \tfrac{1}{2}ky^2$$

At the lowest point, $y = -2A$ and $K = 0$. Therefore,

$$0 = 0 + mg(-2A) + \tfrac{1}{2}k(-2A)^2$$

or
$$A = \frac{mg}{k} \quad \text{and} \quad 2A = \frac{2\,mg}{k}$$

We leave it to the reader to show that at $y = -A$ the resultant force on the block is zero, and that at $y = -2A$ the resultant force on the block has the magnitude mg in the *upward* direction, as shown in Fig. 10-10. Upon being released from rest at $y = 0$, the block oscillates between $y = 0$ and $y = -2A$, with an equilibrium position at $y = -A$.

10-3 GENERAL DEFINITION OF POTENTIAL ENERGY

We have already arrived at the relations for potential energy associated with particles interacting by gravity and by means of a spring. In both instances the potential energy was so defined that the system's total energy remains constant. Now we wish to find a general relation for potential energy in terms of the force acting between a pair of particles.

Consider a system consisting of two particles, a and b, which interact with one another with a force whose magnitude and direction depend only on the relative positions of the two objects and on their intrinsic properties. To simplify the discussion we assume that, whereas particle a has finite mass and undergoes a displacement when acted on by particle b, particle b's mass is infinite so that its velocity (and kinetic energy K_b) remains zero when interacting with a. See Fig. 10-11.

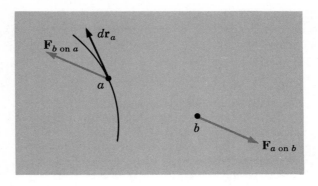

FIG. 10-11

The interaction force on object a—labeled $\mathbf{F}_{b\text{ on }a}$—causes a to undergo a displacement $d\mathbf{r}_a$ and thereby does work on it, changing its kinetic energy by ΔK_a, where

$$\Delta K_a = \mathbf{F}_{b\text{ on }a} \cdot d\mathbf{r}_a$$

The change in the kinetic energy of the two-particle system is then

$$\Delta K = \Delta K_a + \Delta K_b$$
$$= \Delta K_a$$

where we have made use of the fact that $\Delta K_b = 0$ because b's mass is assumed to be infinite.

If the total energy of the system is to remain constant, then

$$\Delta U = -\Delta K = -\Delta K_a = -\mathbf{F}_{b\text{ on }a} \cdot d\mathbf{r}_a$$

where ΔU is the change in the system's potential energy. In going from some *i*nitial state i to a *f*inal state f the potential energy change is

$$U_f - U_i = -\int_i^f \mathbf{F}_{\text{int}} \cdot d\mathbf{r} \qquad (10\text{-}4)$$

where we have replaced $d\mathbf{r}_a$ by $d\mathbf{r}$ and have let $\mathbf{F}_{\text{int}}$ represent $\mathbf{F}_{b\text{ on }a}$, a force which is *internal* to the system.

In words, the change in a system's potential energy is the *negative* of the work done by the interaction force between the two particles from an initial to a final configuration, and this work is computed by taking the displacement of one particle relative to the other.

A potential energy may be defined only if the work done by the internal force is zero when the particles are brought back to their original relative positions. We symbolize this by writing

$$\oint \mathbf{F} \cdot d\mathbf{r} = 0 \qquad (10\text{-}5)$$

where the circle is added to the integral sign to emphasize that the work is computed around *any closed path*. A force satisfying this condition is, like the spring force or force of gravity, one for which a potential energy may be defined and the energy-conservation principle applied. It is denoted a *conservative force*.

10-4 PROPERTIES OF POTENTIAL ENERGY

1. *Potential energy is a property of a system of interacting bodies as a whole.* One cannot speak of the potential energy of a single object. When, for example, a light ball falls toward the earth, it is perhaps natural to speak of the "ball" as losing potential energy, inasmuch as it is the ball rather than the earth that gains most of the kinetic energy; however, it is the earth-ball *system* that gains kinetic energy as the two objects approach one another, and it is the *system* that loses potential energy.

2. Because one is always concerned only with differences in potential energy, *the choice of the zero of potential energy is arbitrary*. If the force associated with the potential energy is constant, as in the case of the gravitational force on an object close to the earth, we may choose any convenient horizontal level (usually the lowest, or *ground* level) as the zero for gravitational potential energy. If, on the other hand, the force varies with displacement, as in the case of the spring, it is customary to choose the zero of potential energy at that displacement for which the force is zero (thus, both $U_s = \frac{1}{2}kx^2$ and $F_s = -kx$ are zero at $x = 0$). There is also, in fact, an arbitrariness in the choice of the zero for kinetic energy: A body always has zero kinetic energy relative to the reference frame in which it is at rest.

    Both kinetic and potential energies are scalar quantities but, unlike kinetic energy, which must always be positive, potential energy may be either positive or negative.

3. Whereas kinetic energy is energy of motion, *potential energy is energy of position* or, more properly, *energy of relative separation of interacting particles*. A spring has potential energy when it is deformed by being stretched or compressed, and it retains this potential energy as long as it remains deformed. If we clamp a compressed spring, the spring's potential energy is locked in. No matter how long a time elapses, the spring is potentially able to do work, and if later we release the clamp, this stored energy is released.

    When a spring is deformed, the atoms that comprise it must change their separation distances. Any potential energy is, in fact, related ultimately to the relative separation of particles.

4. *A potential energy can be defined only for a conservative force.* The requirement that the line integral of the interaction force around a closed path be zero for a conservative force,

Conservative force: $$\oint \mathbf{F} \cdot d\mathbf{r} = 0$$

implies that for each element of positive work done by the interaction force there must be a corresponding negative work somewhere on the return trip. For example, the gravitational force is conservative because when an object is displaced vertically upward the force of gravity does negative work, but when the same object is displaced downward through the same distance the force does positive work in the same amount. A horizontal displacement, for which the force of gravity is at right angles to the displacement, involves no work by the gravity force.

Conservative forces, of which the gravitational force and spring force are examples, depend only on the separation distances between the interacting particles, but not on their velocities or on the time.

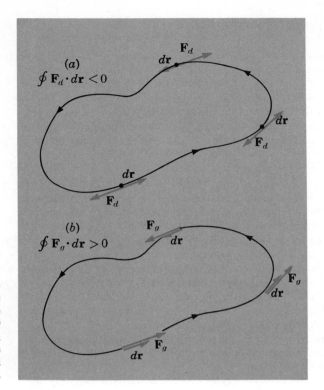

FIG. 10-12. Nonconservative forces: *(a)* An energy-dissipative force F_d is opposite to the displacement and $\oint \mathbf{F}_d \cdot d\mathbf{r} < 0$ around a closed loop; *(b)* an energy-generative force F_g is along the displacement and $\oint \mathbf{F}_g \cdot d\mathbf{r} > 0$ around a closed loop.

For a *nonconservative force* the work done around a closed path is not zero, and no potential energy can be associated with such a force:

Nonconservative force: $\qquad \oint \mathbf{F} \cdot d\mathbf{r} \neq 0 \qquad$ (10-6)

Suppose a man pushes an object around a closed path on a rough horizontal surface. The force of friction is always opposite in direction to the object's displacement, and the friction force always does *negative* work. See Fig. 10-12. For such an *energy-dissipative force* as friction,

$$\oint \mathbf{F}_d \cdot d\mathbf{r} < 0$$

Now consider the work done by the hand in the above situation. The hand's force is along the displacement, so as to balance out the friction force, and this force always does *positive* work. Around a closed path such an *energy-generative force* produces net work, and

$$\oint \mathbf{F}_g \cdot d\mathbf{r} > 0$$

Considering now both the friction and hand forces, we see that the object returns to its starting point with no net change in kinetic energy because the negative work done by the frictional (energy-dissipative) force is just balanced by the positive work done by the hand (energy-generative) force.

5. *The energetics of a motion may be portrayed usefully by a graph of potential energy as a function of displacement.* Consider the plot of the

gravitational potential energy U_g versus the displacement y, corresponding to an object falling freely under the influence of gravity (or sliding on a smooth surface or swinging from a taut string). Because $U_g = mgy$ and the potential energy is proportional to the height y, a plot of U_g versus y is simply a straight line of positive slope mg, as shown in Fig. 10-13. If we draw a horizontal line to represent the total energy of the system $E = K + U_g$, then the vertical segment between the y axis and the curve represents the potential energy, and the vertical segment between the curve and the line for constant E represents the kinetic energy.

The slope of a curve of potential energy vs. displacement has an interesting significance. For the gravitational potential, the slope dU_g/dy is

$$\frac{d}{dy} U_g = \frac{d}{dy} mgy = mg$$

But
$$F_g = -mg$$

Therefore,
$$F_g = -\frac{dU_g}{dy}$$

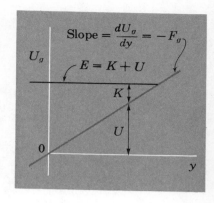

FIG. 10-13. Plot of the gravitational potential energy U_g versus the vertical displacement y for a constant gravitational force F_g. The force is the negative slope of the potential-energy–displacement curve: $F_g = -dU_g/dy$.

The force is the negative derivative of the potential energy with respect to displacement. Therefore, the force is a measure of the steepness, or *grade,* of the slope. For this reason a conservative force may be referred to as the negative *gradient* of the potential energy.

A similar relation can be written for the spring force:

$$\frac{d}{dx} U_x = \frac{d}{dx} \tfrac{1}{2}kx^2 = kx$$

But
$$F_s = -kx$$

Therefore,
$$F_s = -\frac{dU_s}{dx}$$

Again the negative of the slope of potential energy vs. displacement gives the force. The potential-energy curve for a spring is a parabola, as shown in Fig. 10-14. The motion is *bound,* the oscillating particle being confined by the "potential well" to the region between $x = -A$

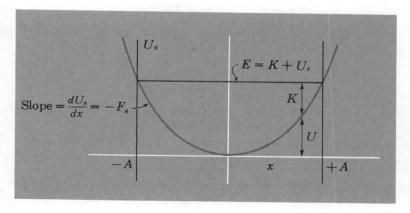

FIG. 10-14. Plot of a spring's potential energy U_s versus its displacement x. The curve is a parabola. The object's motion is restricted to displacements between $x = \pm A$, for which the kinetic energy $K = E - U_s$ is positive.

Properties of Potential Energy   SEC. 10-4

and $x = +A$. Outside this region the kinetic energy would be negative—an impossibility. When the body is displaced from the equilibrium position at $x = 0$, a restoring force acts to return it to the lowest point in the potential-energy valley. This is in contrast to the *unbound* motion of the falling body in Fig. 10-13, where the kinetic energy is positive for *all* points lower than that for which the body has zero kinetic energy.

The gravitational and spring forces are two special examples of a general relation giving the components of a conservative force in terms of the corresponding potential energy. If one has a potential energy $U(x, y, z)$ depending on the coordinates x, y, and z, the components of the force F_x, F_y, and F_z are given by†

$$F_x = -\frac{\partial U}{\partial x} \qquad F_y = -\frac{\partial U}{\partial y} \qquad F_z = -\frac{\partial U}{\partial z} \qquad (10\text{-}7)$$

10-5 THE CONSERVATION OF ENERGY LAW

Systems with Conservative Forces. When an isolated system is comprised of particles that interact by strictly conservative forces, the total energy E of the system is constant. By definition, in fact,

$$E = K + U = \text{constant}$$

where K is the sum of the kinetic energies of the particles $a, b, c, \ldots$ in the system.

$$K = K_a + K_b + K_c + \cdots$$

and U is the sum of the potential energies between each pair of interacting particles,‡

$$U = U_{ab} + U_{bc} + U_{ac} + \cdots \qquad (10\text{-}8)$$

The particles of the system may lose kinetic energy; if they do, the potential energy then must increase to keep the total energy constant. The *i*nitial energy E_i of a conservative system is exactly equal to the *f*inal energy E_f at any later time:

Conservative system: $\qquad E_i = E_f \qquad (10\text{-}9)$

If this system is no longer isolated from its surroundings and an external agent does work $W_{i \to f}$ on the system, then

$$W_{i \to f} = E_f - E_i$$

---

†This follows from the definition of potential energy in Eq. (10-4):

$$\Delta U = -\mathbf{F} \cdot d\mathbf{r} = -(F_x dx + F_y dy + F_z dz)$$

In computing the partial derivative, $\partial U/\partial x$, one takes y and z to be constant.

‡The additivity of potential-energy terms for pairs of particles follows from the superposition principle for forces (Sec. 7-3): If $\mathbf{F} = \mathbf{F}_{ab} + \mathbf{F}_{bc} + \mathbf{F}_{ac} + \cdots$, then

$$-\int \mathbf{F} \cdot d\mathbf{r} = -\int \mathbf{F}_{ab} \cdot d\mathbf{r} - \int \mathbf{F}_{bc} \cdot d\mathbf{r} - \int \mathbf{F}_{ac} \cdot d\mathbf{r} - \cdots$$

and $\qquad U = U_{ab} + U_{bc} + U_{ac} + \cdots$

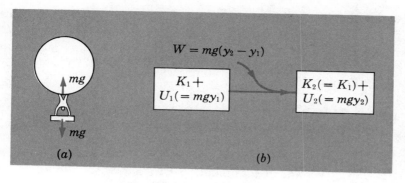

FIG. 10-15. (a) An external agent increases the separation distance between a block and the earth by applying a force of magnitude mg to both. (b) The change in the potential energy of the earth-block system is equal to the work $W = mg(y_2 - y_1)$ done by the external agent in separating the objects at constant velocity.

where the energy content of the system has been increased by $W_{i \to f}$. Conversely, if the system does work on an external agent, the energy content of the system decreases.

Suppose, for example, that a man raises a block of weight mg vertically from y_1 to y_2. The man interposes himself between the earth and the block, pushing on each to increase their separation, as shown in Fig. 10-15. He, as the external agent, does work on the system, thereby increasing its total energy content.

Imagine that the man raises the block at constant velocity. Since the resultant force on it must be zero, the man must apply a force of magnitude mg to the block (and also to the earth). Neither the kinetic energy of the block nor the earth changes. Therefore, the potential energy of the earth-block system must then have increased by $mg(y_2 - y_1)$. But $mg(y_2 - y_1)$ is the work done by the man. Here, and in general, the work done by an external agent in changing the separation distance between interacting objects *at constant velocity* is equal to the change in the system's potential energy. This idea is shown schematically in Fig. 10-15. As another example, a man pulling on the two ends of a spring to elongate it by x does work $\frac{1}{2}kx^2$; the spring's potential energy increases by $\frac{1}{2}kx^2$.

Systems with Nonconservative Forces. It is impossible to construct any large-scale system of objects in which *all* frictional and other nonconservative forces are absent. Perpetual-motion machines are impossible. Consider an isolated system having nonconservative internal forces. When particles in such a system lose kinetic energy, the potential energy of the system does *not* increase in the same amount. Consequently, the initial total energy of an isolated system with nonconservative forces always *exceeds* the final energy. The total energy *decreases* with time in a nonconservative system:

Nonconservative system: $\quad E_i > E_f \quad$ (10-10)

Energy is *not* conserved, if by "energy" is meant the kinetic energy of *large-scale* objects one can see and the potential energy one can identify with a *discernible* change in their relative separation. Thus, if energy *is* to be conserved in a system with nonconservative forces, a new form, or perhaps several new forms, of energy must be named, so that the inequality above can be replaced by an equation such as

$$E_i = E_f + E_{\text{non-}m} \tag{10-11}$$

where $E_{\text{non-}m}$ represents the sum of all forms of what might be called *nonmechanical* energy—that is, energy different from the kinetic and potential energies of large-scale bodies.

One of the greatest discoveries in classical physics was that (10-11) agrees with experiment. As we shall see in later chapters, physicists found that nonmechanical forms of energy exist in that they can be measured in meaningful ways, their amounts being such as to make the *total energy content* of an isolated system constant. When the nonconservative force is friction, $E_{\text{non-}m}$ is (mostly) heat, or thermal energy. But still other forms of energy are recognized: the energy associated with electromagnetic radiation (as illustrated by light or radio waves), chemical energy (which can properly be called *atomic* energy), nuclear energy (which is sometimes improperly called *atomic energy*), and still others.

Experiments in all branches of physics are consistent with the *general conservation of energy law, that the total energy content of an isolated system is constant.* By an "isolated" system is now meant a collection of objects on which no work is done and into (or out of) which neither thermal energy nor radiation flows. Energy may be converted from one form to another, but it is not created or destroyed. Thus, energy conservation ranks, with the conservation laws of mass and of linear momentum, as one of the truly fundamental principles of physics. (In the theory of special relativity, the separate conservation laws of mass and energy are combined into a single conservation law, that of mass-energy, which is thus of an even greater simplicity and generality.) Physicists' confidence in the universality of energy conservation has always been vindicated by experiment.

The recognition that various forms of energy can be delineated, that energy appears as thermal energy, or as *radiation,* as well as kinetic energy and potential energy, was indeed a remarkable discovery. But even more remarkable perhaps has been the realization, coming mostly in the last 50 years, that so-called nonmechanical forms of energy are, after all, just kinetic and potential energy on a submicroscopic scale. It is now known that thermal energy is, in fact, kinetic and potential energy associated with the disordered motion of atoms or molecules; the energy of electromagnetic radiation is the kinetic energy of particle-like photons traveling at the speed of light; chemical energy is traceable to the kinetic energy of subatomic particles and the electric potential energy of their interaction; nuclear energy is the kinetic energy of the nuclear constituents and the potential energy associated with their interaction through the nuclear force. Macroscopically, many forms of energy must be delineated; submicroscopically—at the level of the smallest particles known to physicists—one needs only kinetic and potential energy. At the submicroscopic level, *all* energy is either kinetic energy or potential energy because all the fundamental forces between particles are *conservative forces*.

Example 10-7

A billiard ball collides with a second billiard ball of the same mass but initially at rest. What is the angle between the directions of motion of the two balls after the collision, assuming that the total kinetic energy K is the same before and after the collision? Such a collision, in which total *kinetic* energy before the collision equals the total *kinetic* energy after the collision, is said to be *perfectly elastic*.† (Although interacting objects are deformed during an elastic collision, they spring back to their original shapes after the collision.)

From the conservation-of-momentum law we have

$$m\mathbf{v}_1 = m\mathbf{v}'_1 + m\mathbf{v}'_2$$

where m is the mass of either ball, $\mathbf{v}_1$ and $\mathbf{v}'_1$ are the velocities before and after the collision, respectively, of the ball originally in motion, and $\mathbf{v}'_2$ is the velocity after the collision of the ball originally at rest. We may write this equation as

$$\mathbf{v}_1 = \mathbf{v}'_1 + \mathbf{v}'_2$$

The velocity vectors form a triangle, as shown in Fig. 10-16a; θ is the angle between velocities after the collision, as shown in Fig. 10-16b.

The fact that this is a perfectly elastic collision requires that

$$\tfrac{1}{2}mv_1^2 = \tfrac{1}{2}mv'^2_1 + \tfrac{1}{2}mv'^2_2$$

or

$$v_1^2 = v'^2_1 + v'^2_2$$

The last equation implies, through the pythagorean theorem, that the velocity vectors form a *right* triangle. Therefore, the angle θ between $\mathbf{v}'_1$ and $\mathbf{v}'_2$ is 90° for any non-head-on collision. Fig. 10-17 is a cloud-chamber photograph of such a collision.

When a light ball strikes a massive ball in an elastic head-on collision, as in the case of a ping-pong ball striking a billiard ball, the light ball bounces back with the same speed as that with which it approached the massive ball, the massive ball remaining at rest. On the other hand, when a massive ball strikes a light ball originally at rest in an elastic head-on collision, the massive ball continues to move forward with essentially unchanged speed and the light ball is set in motion. Although the light ball's speed exceeds that of the massive ball, its kinetic energy is very much less. An object can lose *all* its kinetic energy in a collision only when it strikes head on a stationary object of *equal* mass. If it strikes head on either a more massive ob-

---

†It is useful to classify collisions according to whether or not the total *kinetic* energy of the objects going into the collision equals, exceeds, or is less than the total *kinetic* energy of the objects emerging from the collision. The terminology is as follows:
Elastic collisions: Total K before = total K after
Inelastic collisions: Total K before > total K after
Explosive collisions: Total K before < total K after

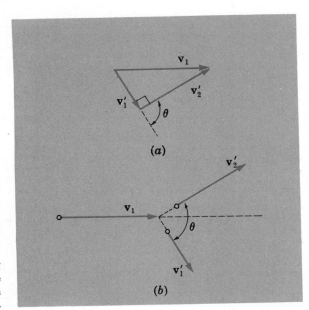

FIG. 10-16. (a) Velocity vectors for the non-head-on collision between two particles of equal mass. Momentum conservation requires that $\mathbf{v}_1 = (\mathbf{v}_1' + \mathbf{v}_2')$. (b) Velocity vectors corresponding to part (a); the angle between the two velocities after the collision is θ.

ject or a less massive object, only a fraction of its kinetic energy is transferred to the struck object.

This result is of importance when one is concerned with the problem of reducing the kinetic energies of the high-speed neutrons produced in a nuclear reactor. The neutrons are slowed down (or *moderated*) when they collide with, and transfer kinetic energy to, nuclei within the material of the nuclear reactor. This material (or *moderator*) must then consist of particles whose mass is not greatly different from that of the neutron, for the neutrons will be slowed

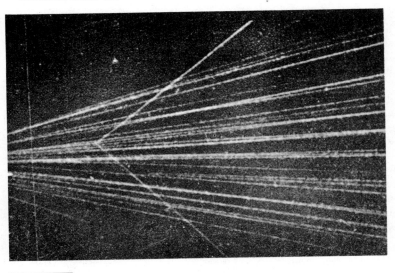

FIG. 10-17. Cloud chamber† photograph showing the elastic collision of an α particle (helium nucleus) with an α particle initially at rest. The angle between the paths of the emerging particles is 90°. (From *An Atlas of Typical Expansion Chamber Photographs,* 1954, Pergamon Press, Inc., Elmsford, N.Y.)

†Charged particles passing through a supersaturated vapor in a cloud chamber produce streaks of condensed droplets along their paths which are analogous to the contrails *(con*densation *trails)* left by jet aircraft and rockets moving at high altitudes.

down quickly, in a relatively small number of collisions, only if they collide elastically with nuclei of comparable mass. For this reason, the deuteron nuclei in heavy water are suitable, the deuteron mass being only twice the neutron mass. (Ordinary water, with hydrogen nuclei, is unsuitable because neutrons produce a nuclear reaction with protons.)

Example 10-8

A bullet of mass m initially traveling at speed v is shot at and becomes imbedded in a block of mass M originally at rest on a smooth horizontal surface. (a) What is the speed of the block (and bullet) after the collision? (b) If the block is suspended by a thin vertical massless rigid rod pivoted at the top, and is hit by the bullet, as in the example just given, through what vertical distance would the block travel before coming to rest?

(a) The bullet and the block are subject to no net horizontal external force; therefore, the conservation of momentum requires that

$$m\mathbf{v} = (m + M)\mathbf{V}$$

where $\mathbf{V}$ is the block's velocity after the collision. Therefore,

$$\mathbf{V} = \frac{m}{m + M}\mathbf{v} \qquad (10\text{-}12)$$

One can measure the bullet's speed v indirectly through knowing the masses m and M and measuring the lower speed V.

The collision is said to be *completely inelastic* in that the two interacting objects are at rest with respect to one another after the collision. A nonconservative force of friction acts in stopping the bullet, energy is dissipated, and the kinetic energy out of the collision is less than the kinetic energy into the collision.

(b) The bullet strikes a block suspended by a rod of length L, as shown in Fig. 10-18. If the bullet comes to rest in the block, its speed is reduced from v to V in a short distance, and the time interval during which the bullet is decelerated and during which the block acquires the speed V is very short indeed. The block is acted on by an impulsive force for a time interval that is much less than the time required for the block and rod to swing upward and come to rest at a height y above the lowest point. Thus, the complete motion consists of two parts: (a) the short time interval during which the bullet is brought to rest in the block and during which the block acquires the speed V without rising appreciably and (b) the much longer time interval during which the block swings to its highest point. During interval 1, momentum is conserved (no *external* force on the system of bullet and block), but kinetic energy is not conserved (a nonconservative *internal* force acts); during interval 2, mechanical energy is conserved (the block is subject only to conservative forces), but momentum is not conserved (the bullet-block system is now subject to an unbalanced external force).

During interval 2,

$$\tfrac{1}{2}(m + M)V^2 = (m + M)gy$$

$$V = \sqrt{2gy}$$

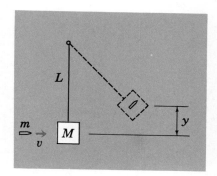

FIG. 10-18. A ballistic pendulum. A bullet of mass m and velocity v strikes and becomes imbedded in a block of mass M attached to the end of a light rod of length L. The block rises a vertical distance y in coming to rest.

During interval 1, from (10-12),

$$v = \frac{m+M}{m}\sqrt{2gy}$$

A bullet's speed can be measured with a device of this sort, known as a *ballistic pendulum,* simply by measuring the height y and masses m and M.

Example 10-9
A 5.0-g projectile moving horizontally at 1.0×10^3 m/s strikes a 1.0-kg block of wood at rest on a frictionless horizontal surface. The projectile penetrates the block, emerging with a velocity of 40 m/s in the forward direction. *(a)* What is the final velocity of the block? *(b)* What is the ratio R between the final total kinetic energy and the initial total kinetic energy?

(a) From the conservation of momentum law we have

$$m_p \mathbf{u}_p + m_b \mathbf{u}_b = m_p \mathbf{v}_p + m_b \mathbf{v}_b$$

Where m_p, $\mathbf{u}_p$, and $\mathbf{v}_p$ are the mass, velocity before collision, and velocity after collision of the *projectile.* Similarly, m_b, $\mathbf{u}_b$, and $\mathbf{v}_b$ are the analogous parameters for the *block.* Solving this equation for $\mathbf{v}_b$ we find

$$\mathbf{v}_b = \mathbf{u}_b + (m_p/m_b)(\mathbf{u}_p - \mathbf{v}_p)$$

$$v_b = 0 + \frac{5 \times 10^{-3} \text{ kg}}{1.0 \text{ kg}} (1.0 \times 10^3 \text{ m/s} - 40 \text{ m/s})$$

$$= 4.8 \text{ m/s}$$

(b)
$$R = \frac{\tfrac{1}{2} m_p v_p^2 + \tfrac{1}{2} m_b v_b^2}{\tfrac{1}{2} m_p u_p^2 + \tfrac{1}{2} m_b u_b^2} = \frac{m_p v_p^2 + m_b v_b^2}{m_p u_p^2 + m_b u_b^2}$$

$$= \frac{(5.0 \times 10^{-3} \text{ kg})(40 \text{ m/s})^2 + (1.0 \text{ kg})(4.8 \text{ m/s})^2}{(5.0 \times 10^{-3} \text{ kg})(1.0 \times 10^3 \text{ m/s})^2 + 0}$$

$$= 0.0062 = 0.62\%$$

Therefore more than 99 percent of the initial kinetic energy has been dissipated in what is clearly an inelastic ($R < 1$) collision.

SUMMARY

The total energy E of a system of particles consists of the sum of the particles' kinetic energies K and the potential energy U of their interaction:

$$E = K + U \qquad (10\text{-}1)$$

The gravitational potential energy U_g for a particle of mass m which has undergone a vertical displacement y near the surface of the earth is

$$U_g = mgy \qquad (10\text{-}3)$$

The potential energy U_s of a spring of stiffness constant k, stretched or compressed by a distance x, is

$$U_s = \tfrac{1}{2}kx^2 \qquad (10\text{-}2)$$

A potential energy can be defined only for a conservative force—the net work done around a closed path being zero for such a force. Then the difference in potential energy between two states i and f is defined to be

$$U_f - U_i = -\int_i^f \mathbf{F} \cdot d\mathbf{r} \qquad (10\text{-}4)$$

The conservation of energy law: The total energy of any isolated system is constant. If the particles within the system interact through conservative forces, the sum of kinetic and potential energies is constant with time; if nonconservative dissipative forces act, the sum of the *macroscopic* kinetic and potential energies decreases with time.

PROBLEMS

10-1. A 5.0-kg block falls from a height of 3.0 m onto a vertical spring. Find the distance that the spring will be compressed if the spring constant is 1.0×10^3 N/m. Assume $g = 10$ m/s².

10-2. A 1.0-kg object is thrown into the air from ground level with a speed of 16.0 m/s at an angle of 37° above the horizontal. Using energy conservation, find (a) the maximum height, (b) the object's minimum kinetic energy, and (c) its speed when 2.0 above the ground.

10-3. An object is pressed against one end of a spring whose other end is attached to a solid wall, and then released from rest. What is the ratio of the speed of the object when it leaves the spring to its speed when the spring has expanded halfway to its relaxed length?

10-4. An object is brought to rest by compressing a spring with one end fixed. Suppose that the spring compression is to be doubled. By what factor must the object's momentum increase?

10-5. In 1970 the United States consumed 6.45×10^{16} Btu of energy. (a) What mass of water could be heated from 32°F to 212°F by this quantity of energy? (b) Find the edge length of a cube which could contain this mass of water. Consult Appendix II for appropriate conversion factors. (The volume of Lake Michigan is estimated to be about 5×10^{12} m³.)

10-6. By 1980 total energy consumption in the United States is expected to reach 1.05×10^{20} J/year. (a) What is the corresponding rate of power consumption in megawatts? (b) Express this projected rate of energy consumption in megawatt-hours/year. (One megawatt-hour is the total energy produced by a one megawatt source operating for one hour.)

10-7. To pull a bowstring from its undisplaced position requires a force of 200 N when the middle of the string is pulled back 60 cm. The force is closely proportional to the displacement of the center point on the string. If the bow shoots a 40-g arrow vertically upward, what is the maximum height achieved by the arrow.

10-8. A particle is suspended at one end of a taut string whose upper end is fixed in position (a simple pendulum). The string's length is 12.5 m, and the particle passes through the lowest point at a speed of 7.0 m/s. What is the angle between the string and the vertical when the particle is momentarily at rest?

10-9. Two particles with a 2 to 1 mass ratio are pressed against the ends of a spring and released from rest. What is the ratio of the kinetic energy of the light particle to that of the more massive particle after the spring has expanded to its relaxed length?

10-10. Show that in Example 10-6, (a) at $y = -A$, $\Sigma F_y = 0$ and at $y = -2A$, $\Sigma F_y = +mg$. (b) Choose a new origin for vertical displacement such that $y' = y + A$, and show that $\Sigma F_y = -ky'$.

10-11. An object of mass m is attached to one end of a spring of stiffness constant k and is initially at rest on a horizontal frictionless surface. A second object also of mass m collides head on with the free end of the spring

151

with a speed v. Find (a) the system's total momentum, (b) the system's total energy, (c) the velocity of the center of mass, (d) the system's total momentum relative to the center of mass, (e) the system's total energy relative to the center of mass, (f) the maximum compression of the spring.

10-12. A 2.0-kg block rests on a rough surface ($\mu = 0.40$) and is held in contact with one end of a spring ($k = 800$ N/m) which has been compressed by 25 cm. The other end of the spring is fixed. If the block is released (a) how far will it move over the surface and (b) what will its speed be when it has traveled half that distance?

10-13. A block is released from rest to slide on the frictionless surface shown in Fig. P10-13. (a) What is the minimum height y if the block is to make it to the top of the hill with radius of curvature r? (b) What is the maximum height if the block is not to leave the surface at the top of the hill?

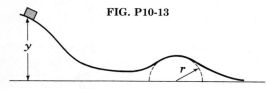

FIG. P10-13

10-14. A block slides down a frictionless surface and makes a loop-the-loop in a circle of radius r. At what minimum height y relative to the bottom of the loop should the block be started from rest in order that it may make the loop while remaining in contact with the surface? See Fig. P10-14.

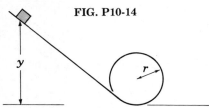

FIG. P10-14

10-15. In 1970 the total energy consumed in the United States was 6.8×10^{19} J $\equiv E_T$. According to Einstein, mass and energy are different forms of the same thing—and the mass equivalent to energy E is given by $M_{eq} = E/c^2$, where c is the speed of light in vacuum ($c = 3.0 \times 10^8$ m/s). (a) What is the mass-equivalent of E_T? (b) In a uranium *fission* reactor just 0.090 percent of the mass of the "burned" fuel is converted into energy. At this rate, what mass of uranium would be needed to produce E_T? (c) In a nuclear *fusion* reactor which uses heavy hydrogen isotopes (deuterium and tritium) 0.38 percent of the mass of the burned fuel is converted into energy. On this basis, what mass of fusion reactants would be required to produce energy E_T?

10-16. The sun radiates energy at the rate of 3.8×10^{26} J/s. (The energy is produced by transforming mass into energy, as discussed in the introduction to Problem 10-15.) (a) At what rate is the sun's mass being converted into energy? (b) If this rate of conversion of mass into energy were maintained constant for the next 1.0 billion years, what fraction of the sun's mass would be transformed? (The sun's mass is 2.0×10^{30} kg.)

10-17. A 4.0-kg block and a 1.0-kg block are attached to opposite ends of a massless cord 2.0 m long. The cord is hung over a small frictionless and massless pulley a distance of 1.5 m from the floor, with the 1.0-kg block initially at floor level. See Fig. P10-17. Then the blocks are released from rest. What is the speed of either block when the 4.0-kg block strikes the floor?

FIG. P10-17

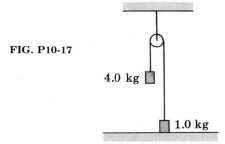

10-18. Initially a 1,000-kg automobile is moving along a horizontal section of road at 90 km/h. Sighting a small rise in the road, the driver switches off the ignition and coasts up and over the crest, which rises 15 m above the level of the original section of the road. The radius of curvature of the road at the crest is 60 m. (a) What is the speed of the car when it is at the crest? (b) With what force does the car press down on the road surface at this instant? Assume $g = 10$ m/s² and neglect frictional losses.

10-19. A 10-kg block is thrust up a 30° inclined plane with an initial speed of 12 m/s. The block travels 8.0 m along the plane, stops, and then slides back to the bottom. Assume $g = 10$ m/s². (a) Compute the coefficient of friction between the block and the surface of the plane, and (b) find the speed of the block when it returns to the bottom of the incline.

10-20. The electric potential energy between two like charges q_1 and q_2 separated by a distance r is given by $U = kq_1q_2/r$, where k is a constant. What is the electric force between the charges?

10-21. A 2.0-kg object is released from rest to slide down a rough surface. At a point 3.0 m vertically downward from its starting point the object's speed is 5.0 m/s. How much energy was dissipated through friction with the surface?

10-22. A 5.0-kg object dropped from a height of 3.0 m bounces back to a height of 1.5 m. How much energy is dissipated in the collision with the floor?

10-23. A 4.0-kg object falls through a viscous liquid at the constant speed of 0.75 m/s. At what rate (in watts) is energy being dissipated?

10-24. A 25-g bullet moving vertically upward strikes a 1,475-g wood block suspended from a massive framework by means of a spring. As a result of the impact the block is driven upward 30.0 cm, compressing the spring. See Fig. P10-24. Find (a) the original speed of the bullet, (b) the fraction of the bullet's kinetic energy transferred to the block, and (c) the fraction of the block's kinetic energy absorbed by the spring. The spring constant is 600 N/m; assume $g = 9.80$ m/s².

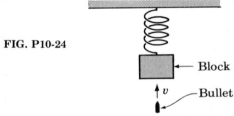

FIG. P10-24

10-25. A particle of mass m traveling initially at speed v collides head on with and sticks to a second particle of equal mass initially at rest. What is the system's total kinetic energy (a) before and (b) after the collision? (c) What fraction of the initial kinetic energy is dissipated in the collision? (d) With what initial speed must the two particles move toward each other in order that the same amount of energy be dissipated in the collision?

10-26. An object initially at rest is struck by a second object, and the two stick together after the collision. (a) If the composite object moves at one-third the speed of the incident object, what is the ratio of struck object's mass to that of the incident object? (b) What fraction of the incident object's kinetic energy is dissipated in the collision?

10-27. The gravitational potential energy between two point-masses m_1 and m_2 separated by a distance r is given by $U = Gm_1m_2/r$, where G is a constant. What is the gravitational force between the masses?

10-28. An object of mass m is attached to one end of an inextensible cord and is whirled in a circular orbit in a vertical plane. Show that the tension in the cord can be written as

$$T = mv_0^2/R + (3\cos\theta + 2)mg$$

where v_0 is the speed of the object at the top of its path, R is the radius of the circle, θ is the angle between the downward vertical and the cord, and g is the acceleration of gravity. Assume that the mass of the cord is negligible and that there are no frictional losses.

10-29. A particle of mass m_1 makes an elastic head-on collision with a particle of mass m_2 initially at rest. Show that the fraction of the kinetic energy of m_1 transferred to m_2 in the collision is $4m_1m_2/(m_1+m_2)^2$ and that this fraction is a maximum for $m_1 = m_2$.

10-30. In a perfectly elastic collision† a 1.0-kg object traveling east at 3.0 m/s collides head on with a 2.0-kg object traveling west at 1.5 m/s. (a) What are the velocities of the two objects after collision? (b) What are the velocities of both objects, before and after collision, relative to the system's center of mass?

10-31. A 2.0-kg ball moving at 3.0 m/s strikes head on a 1.0-kg ball moving in the same direction at 2.0 m/s. After the collision the 2.0-kg ball moves in the opposite direction at 2.0 m/s. (a) What is the velocity of the 1.0-kg ball? (b) Is the collision elastic, inelastic, or explosive?†

10-32. If a group of identical billiard balls are arranged in a straight line, each ball just touching its neighbors, and another billiard ball strikes the first in the line head on, it is found that the last ball in the line is set in motion with a speed identical with that of the originally moving ball. Similarly, if two billiard balls moving together at the same speed strike the line of balls at rest, the last two balls in the line are set in motion with the same speed as that of the two balls striking the line. Show that this behavior is consistent with the conservation laws of energy and momentum for head-on elastic collisions.†

10-33. Every time the radioactive nucleus radium (^{226}Ra) decays (explodes) into the nucleus radon (^{222}Rn) and an α particle (^{4}He), a total energy of 4.87 MeV is released (1 MeV = 1.6×10^{-13} J). What is the kinetic energy of (a) the α particle and (b) the radon nucleus? (The superscripts to the chemical symbol give the relative masses of the particles.)

---

†It is useful to classify collisions according to whether or not the total *kinetic* energy of the objects going into the collision equals, exceeds, or is less than the total *kinetic* energy of the objects emerging from the collision. The terminology is as follows:
Elastic collisions: Total K before = total K after
Inelastic collisions: Total K before > total K after
Explosive collisions: Total K before < total K after

CHAPTER 11

Angular Momentum

The linear momentum $m\mathbf{v}$ of a particle, or of a collection of particles, is important because the total vector momentum of either an isolated particle, or an isolated system of particles, is constant. In short, momentum is conserved. Moreover, if linear momentum changes, we may relate the time rate of change of momentum to the resultant force.

Angular momentum is similar. It is constant for an isolated system—there is an angular-momentum–conservation law. And when the angular momentum changes, we may relate the time rate of change of angular momentum to the resultant torque. But even if a particle *is* acted upon by a net force, we shall see that its angular momentum may, nevertheless, remain constant. Therefore, angular momentum is a fundamental physical quantity which, because of its constancy under appropriate conditions, simplifies the description and analysis of many physical situations.

As the term angular momentum suggests, spinning objects have angular momentum. We shall first consider qualitatively some aspects of the angular momentum of a rotating object, then see how a single particle may have angular momentum attributed to it, and finally derive the formal relations governing changes in angular momentum.

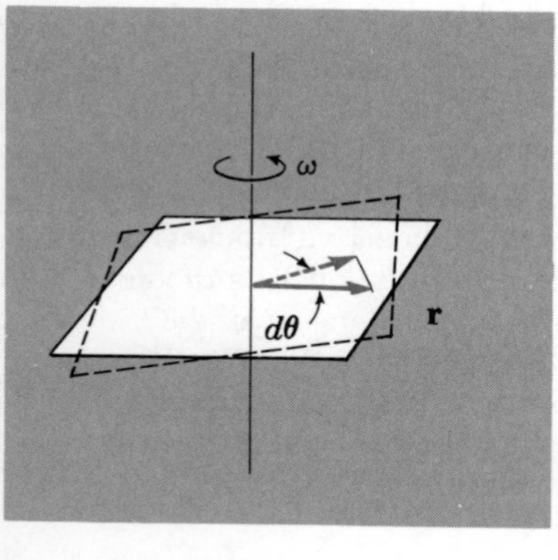

11–1 ANGULAR VELOCITY AS A VECTOR†

Consider the rigid plate in Fig. 11-1, which is undergoing a rotation about an axis normal to the plane of the plate. The radius vector $\mathbf{r}$ from the axis to some point in the plane undergoes an

FIG. 11-1. Radius vector r attached to a plate undergoing rotation about its normal at angular velocity ω undergoes an angular displacement $d\theta$ in time dt.

---

†This section deals only with the vector properties of angular velocity. Other aspects of rotational kinematics are treated in Sec. 12-1.

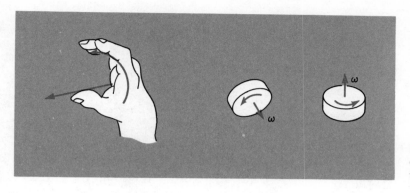

FIG. 11-2. The right-hand rule convention, giving the direction of the angular velocity vector.

angular displacement $d\theta$ in the time interval dt as the plate turns. By definition the angular velocity of **r**, or of the entire plate, is

$$\omega = \frac{d\theta}{dt} \qquad (4\text{-}14)$$

By convention a counterclockwise rotation is taken as positive and a clockwise rotation as negative. Thus, the magnitude of ω gives the angle turned through per unit time; its sign gives the sense of rotation. There is an axis associated with every rotation, and two directions are associated with every straight line representing a rotation axis—one along the line in one sense, and the other opposite to it. With two signs (+ for counterclockwise, − for clockwise) associated with the rotation sense, and with two directions associated with a rotation axis, we may use the two directions of the axis to specify the two rotation senses. All we need is a convention.

The rule is this: If the curled fingers of the *right* hand correspond to the rotational sense, the outstretched thumb points along the direction of the vector representing angular velocity. See Fig. 11-2. The magnitude of **ω** gives the angular speed; its direction specifies the rotation axis and the sense of rotation. For example, counterclockwise rotation in the plane of the paper corresponds to **ω** out of the paper.

11-2 SOME QUALITATIVE FEATURES OF ANGULAR MOMENTUM

Because the mathematical apparatus used for discussing angular momentum is somewhat formidable, it is useful to discuss, in advance of the formal proofs, some qualitative features of the angular momentum not only of a spinning object but also of a single particle.

The following assertions will all be proved in sections that follow. The angular momentum of a spinning symmetrical object, like a right circular cylinder, is a vector in the same direction as the angular velocity of the object. For a given object and axis of

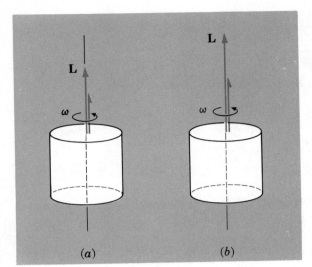

FIG. 11-3. *(a)* For a cylinder spinning about its symmetry axis the angular momentum vector **L** is in the same direction as the angular velocity vector **ω**. *(b)* When the angular velocity increases, the angular momentum increases proportionately.

rotation, the magnitude of the angular momentum is directly proportional to its angular velocity. Angular momentum is conserved, the total angular momentum for an isolated system being constant in magnitude and direction. The recognition of angular momentum as a vector allows certain otherwise difficult problems to be treated in a relatively straightforward manner. For example, see Prob. 11-25 for an analysis of the motion of a top.

Consider the cylinder shown in Fig. 11-3 rotating about its axis of symmetry. As its rotation speed is increased, the angular-velocity vector **ω** grows and so does the vector **L** representing its angular momentum. In other words, the magnitude of the angular momentum increases or decreases when the angular velocity increases or decreases. Suppose now that a bullet of mass m and velocity **v** is shot into a freely pivoted cylinder, misses hitting the rotation axis by an amount $r_\perp$, and becomes *imbedded* in it. See Fig. 11-4. In Fig. 11-4a the cylinder acquires angular velocity and angular momentum in the upward direction; in Fig. 11-4b, where the bullet hits the cylinder on the other side, the cylinder acquires angular velocity and angular momentum in the downward direction; and in Fig. 11-4c, where the bullet strikes along a line passing through the rotation axis, the cylinder is not set into rotation and thereby gains zero angular momentum.

We may attribute angular momentum to a particle because it can change a cylinder's spin rate and therefore its angular momentum. If the bullet and cylinder constitute an isolated system whose total angular momentum is constant, then we must conclude that in part *(a)* of Fig. 11-4 the incident particle had upward angular momentum; in *(b)* the particle's initial angular momentum was downward; and in *(c)* the particle's angular momentum was zero. Note that in associating a direction with the angular momentum of a single particle, we are not indicating the direction in which

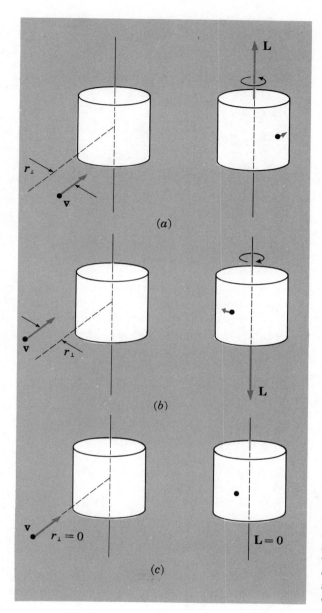

FIG. 11-4. A bullet shot into a freely pivoted cylinder. (a) The bullet misses hitting the symmetry axis by $r_\perp$, and the cylinder acquires angular momentum L upward; (b) the bullet misses by $r_\perp$ on the other side, and the cylinder acquires angular momentum L downward; (c) with $r_\perp = 0$, $L = 0$.

the particle is traveling (the particle's velocity **v** does that); rather, the direction of the angular momentum of a particle indicates the direction of the angular velocity **ω** and angular momentum **L** acquired by a cylinder initially at rest. As a further example consider Fig. 11-5. Here a cylinder is initially spinning with an upward angular momentum. When the bullet strikes and becomes imbedded in the cylinder, the cylinder is brought to rest. Thus, with the final angular momentum zero, we must infer that the particle alone had downward angular momentum, which when

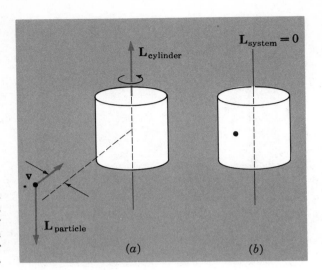

FIG. 11-5. *(a)* A particle with downward angular momentum hits an initially spinning cylinder with upward angular momentum and *(b)* the system's total angular momentum is zero.

added to the equal magnitude of upward angular momentum of the initially spinning cylinder, yields zero total angular momentum.

As the above illustrations indicate, we may assign angular momentum to a single particle. Further, both the magnitude and the direction of a particle's angular momentum depend not only on the particle's mass m and velocity $\mathbf{v}$ but also on the perpendicular distance $r_\perp$. When $r_\perp = 0$, $\mathbf{L}$ is also zero.

11-3 TORQUE AS TIME RATE OF CHANGE OF ANGULAR MOMENTUM

Changes in the linear momentum $m\mathbf{v}$ of a single particle are governed by the relation

$$\mathbf{F} = \frac{d(m\mathbf{v})}{dt}$$

where $\mathbf{F}$ is the resultant force on the particle. In general $\mathbf{F}$ and $\mathbf{v}$ are not in the same direction.

The radius vector $\mathbf{r}$ locates the particle relative to some arbitrary origin O, as shown in Fig. 11-6, here chosen for convenience to lie in the plane of $\mathbf{v}$ and $\mathbf{F}$. The vectors $\mathbf{r}, \mathbf{v},$ and $\mathbf{F}$ will, in general, all vary with time.

We take the cross product (Sec. 2-6) of $\mathbf{r}$ with both sides of the equation above:

$$\mathbf{r} \times \mathbf{F} = \mathbf{r} \times \frac{d(m\mathbf{v})}{dt} \quad (11\text{-}1)$$

Now consider the following derivative:

$$\frac{d}{dt}(\mathbf{r} \times m\mathbf{v}) = \left(\frac{d\mathbf{r}}{dt} \times m\mathbf{v}\right) + \left(\mathbf{r} \times \frac{d(m\mathbf{v})}{dt}\right) \quad (11\text{-}2)$$

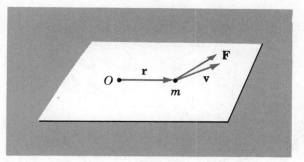

FIG. 11-6. A particle of mass m and velocity v is subject to a resultant force F when its displacement relative to point O is r.

(The rule for the derivative of a cross product is exactly analogous to that for the derivative of an algebraic product, as may be verified in detail by taking the derivative of the components of a cross product.)

Since vector **r** locates the particle at every instant, its time derivative is the particle's velocity, $d\mathbf{r}/dt = \mathbf{v}$. Therefore,

$$\frac{d\mathbf{r}}{dt} \times m\mathbf{v} = \mathbf{v} \times m\mathbf{v} = 0$$

the cross product between any two *parallel* vectors, such as **v** and $m\mathbf{v}$, being zero.

The first term on the right side of (11-2) is then zero, and (11-1) becomes

$$\mathbf{r} \times \mathbf{F} = \frac{d}{dt}(\mathbf{r} \times m\mathbf{v}) \tag{11-3}$$

This is the fundamental relation upon which all else in this and the following chapter is based.

The *torque* τ of force **F** relative to the origin O (relative to which **r** is measured) is, by definition,

$$\text{Torque} = \tau = \mathbf{r} \times \mathbf{F} \tag{11-4}$$

Relative to the origin O the angular momentum **L** of the particle with linear momentum $m\mathbf{v}$ is, by definition,

$$\text{Angular momentum} = \mathbf{L} = \mathbf{r} \times m\mathbf{v} \tag{11-5}$$

Therefore, (11-3) can be written

$$\tau = \frac{d\mathbf{L}}{dt} \tag{11-6}$$

In words, *torque equals time rate of change of angular momentum*. The torque is the cross product of **r** and the resultant force on the particle; the particle's angular momentum is the cross product of the *same* **r** and the particle's linear momentum.

The direction and magnitude of the torque $\tau = \mathbf{r} \times \mathbf{F}$ are illustrated in Fig. 11-7. The torque τ is perpendicular to the plane

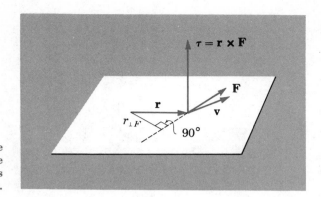

FIG. 11-7. The torque on the particle is $\tau = \mathbf{r} \times \mathbf{F}$. In magnitude $\tau = Fr_{\perp F}$, where $r_{\perp F}$ is perpendicular to F (*not* v).

of **r** and **F**. Its direction is given by the right-hand rule for cross products (turn **r** into **F** through the smaller angle with the fingers of the right hand; the right-hand thumb points in the direction of $\mathbf{r} \times \mathbf{F}$). The magnitude of $\mathbf{r} \times \mathbf{F}$ is the product of F and $r_{\perp F}$, the component of **r** perpendicular to **F**; that is,

$$\tau = r_{\perp F} F \tag{11-7}$$

The direction and magnitude of a particle's angular momentum $\mathbf{r} \times m\mathbf{v}$ are illustrated in Fig. 11-8. The direction of $\mathbf{r} \times m\mathbf{v}$ is perpendicular to the plane of **r** and $m\mathbf{v}$; its sense is given by the right-hand rule. (When **r**, $m\mathbf{v}$, and **F** all lie in a single plane, as will be the case for almost all our examples in this chapter, $\boldsymbol{\tau}$ and **L** are either parallel or antiparallel, both being perpendicular to the plane.) The magnitude of $\mathbf{r} \times m\mathbf{v}$ is equal to mv multiplied by the component $r_{\perp v}$ of **r** perpendicular to the velocity **v** (or momentum $m\mathbf{v}$). Thus,

$$L = r_{\perp v} mv \tag{11-8}$$

Note particularly that $r_{\perp F}$, the component of **r** perpendicular to the force **F**, and $r_{\perp v}$, the component of **r** perpendicular to the momentum $m\mathbf{v}$, are *not* generally the same.

When $\boldsymbol{\tau}$ and **L** are along the *same* line, (11-3) reduces to the algebraic equation

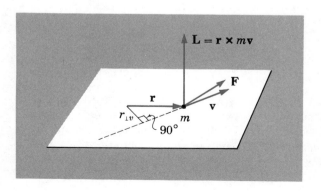

FIG. 11-8. The particle's angular momentum $\mathbf{L} = \mathbf{r} \times m\mathbf{v}$. In magnitude $\mathrm{L} = mvr_{\perp v}$, where $r_{\perp v}$ is perpendicular to v (*not* F).

$$r_{\perp F} F = \frac{d}{dt} r_{\perp v} mv \qquad (11\text{-}9)$$

The dimensions of angular momentum are ML^2T^{-1}; those of torque are ML^2T^{-2}. Appropriate units for angular momentum are kg-m²/s, or its equivalent, J-s. Torque is expressed in the units m-N. (Torque is the cross product of displacement and force; work is the dot product of displacement and force. Although their dimensions are alike, they are different physical concepts, and it is customary to express torque units as m-N and work units as N-m, or J.)

Example 11-1

A 1.0-kg particle at the location 3.0 m along the positive X axis, otherwise denoted (3.0 m, 0, 0), has a velocity of 2.0 m/s along the positive Y direction and is subject to a resultant force of 4.0 N at an angle of 53° with respect to the positive X axis and lying in the XY plane. What is the particle's angular momentum and the torque of the force on it *(a)* relative to the coordinate origin (0, 0, 0) and *(b)* relative to point (0, −4.0 m, 0)? See Fig. 11-9a.

(a) The origin of the radius vector **r** is the coordinate origin, so that **r** extends along the positive X direction. Since **r** is at right angles to **v**, $r_{\perp v} = 3.0$ m, and the magnitude of the particle's angular momentum is

$$L = r_{\perp v} mv = (3.0 \text{ m})(1.0 \text{ kg})(2.0 \text{ m/s})$$

$$= 6.0 \text{ kg-m}^2/\text{s}$$

The direction of **L** is, from the right-hand rule for cross products, in the positive Z direction.

As Fig. 11-9b shows, the component of **r** perpendicular to **F** is $r_{\perp F} = (3.0 \text{ m}) \sin 53°$. (In general the magnitude of the torque is $rF \sin \theta$, where θ is the angle between **r** and **F**.) Therefore, the torque magnitude is

$$\tau = r_{\perp F} F = (3.0 \text{ m})(0.80)(4.0 \text{ N})$$

$$= 9.6 \text{ m-N}$$

The direction of τ is also along the positive Z axis.

(b) With angular momentum and torque computed relative to the point (0, −4.0 m, 0), the magnitude and direction of the radius vector **r** to the particle are changed. See Fig. 11-9c. The magnitude of **r** is now 5.0 m, and it makes an angle of 53° with respect to the positive X axis.

With $r_{\perp v} = r \sin 37°$, the magnitude of the angular momentum is

$$L = r_{\perp v} mv = (5.0 \text{ m})(0.60)(1.0 \text{ kg})(2.0 \text{ m/s})$$

$$= 6.0 \text{ kg-m}^2/\text{s}$$

Both the direction and magnitude of the angular momentum **L** are unchanged compared with part *(a)*.

Since the radius vector **r** is parallel to the force **F**, their cross

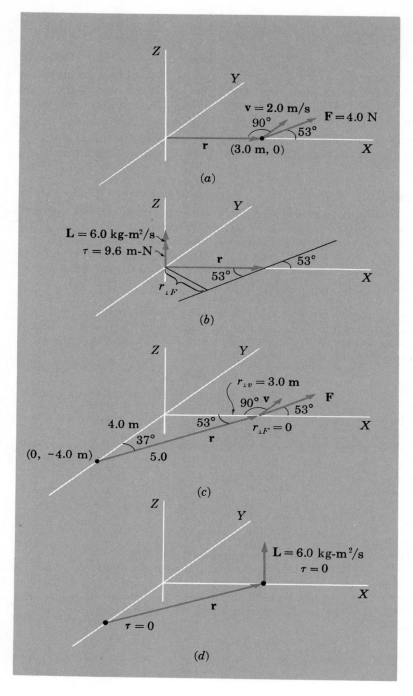

FIG. 11-9. Computation of the torque and the angular momentum of a particle for two different origins.

product $\mathbf{r} \times \mathbf{F}$ is zero. The torque on the particle relative to point $(0, -4.0 \text{ m}, 0)$ is zero. See Fig. 11-9d.

As this example illustrates, it is meaningless to specify a particle's angular momentum or the torque on a particle, unless one also specifies the point relative to which $\mathbf{r} \times m\mathbf{v}$ and $\mathbf{r} \times \mathbf{F}$ are measured. Both the magnitude and direction of $\mathbf{L}$ and of $\boldsymbol{\tau}$ depend upon this choice.

11-4 A PARTICLE WITH CONSTANT ANGULAR MOMENTUM

We have shown that for a single particle

$$\mathbf{r} \times \mathbf{F} = \frac{d(\mathbf{r} \times m\mathbf{v})}{dt} \quad \text{or} \quad \boldsymbol{\tau} = \frac{d\mathbf{L}}{dt} \qquad (11\text{-}3), (11\text{-}6)$$

In itself this relation says nothing more than is already contained in Newton's second law. We may, however, use this alternative expression to inquire into the circumstances in which the torque $\boldsymbol{\tau} = \mathbf{r} \times \mathbf{F}$ on a particle is *zero,* so that $d\mathbf{L}/dt = 0$ and the particle's angular momentum $\mathbf{L} = \mathbf{r} \times m\mathbf{v}$ must then be constant:

$$\mathbf{L} = \mathbf{r} \times m\mathbf{v} = \text{constant} \quad \text{when } \boldsymbol{\tau} = \mathbf{r} \times \mathbf{F} = 0$$

The torque of a force on a single particle may be zero under three circumstances: (1) $\mathbf{r} = 0$, (2) $\mathbf{F} = 0$, or (3) with neither $\mathbf{r}$ nor $\mathbf{F}$ zero, $\mathbf{r} \times \mathbf{F} = 0$ because $\mathbf{r}$ is parallel to $\mathbf{F}$ or antiparallel to $\mathbf{F}$.

The case $\mathbf{r} = 0$ corresponds to the particle's momentarily being at the origin of $\mathbf{r}$, a trivial situation which deserves no further comment.

When $\mathbf{F} = 0$, we have effectively an isolated particle. We know that the particle's momentum $m\mathbf{v}$ is then constant, and we now know in addition that its angular momentum $\mathbf{r} \times m\mathbf{v}$ is also constant. See Fig. 11-10, where a particle travels in a straight line at constant speed. The radius vector changes direction and magnitude from $\mathbf{r}_1$ to $\mathbf{r}_2$ to $\mathbf{r}_3$, etc., but the component at right angles to $\mathbf{v}$ remains constantly $r_{\perp v}$. The angular-momentum vector $\mathbf{L}$, at right angles to $\mathbf{r}$ and $\mathbf{v}$, is unchanged in magnitude and direction. Note especially that the particle does *not* advance along the direction of $\mathbf{L}$; indeed, nothing—no thing—is moving in the direction of the angular-momentum vector.

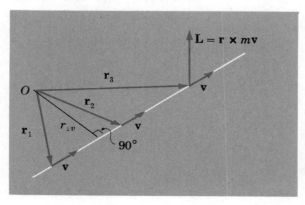

FIG. 11-10. A particle with constant velocity has constant angular momentum relative to point O. The radius vector changes from r_1 to r_2 to r_3, but $r_{\perp v}$ is constant.

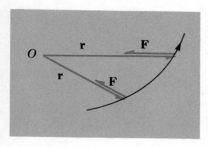

FIG. 11-11. A particle subject to a force F which is always directed toward the origin at O and opposite to the radius vector r.

The third situation in which the torque $\mathbf{r} \times \mathbf{F}$ is zero—that in which $\mathbf{r}$ and $\mathbf{F}$ are in either the same or opposite directions (and the angle between them is either 0 or 180°)—will illustrate, in part, the usefulness of the whole idea of angular momentum and torque.

When $\mathbf{r} \times \mathbf{F} = 0$, $d\mathbf{L}/dt = 0$, and

$$\mathbf{L} = \mathbf{r} \times m\mathbf{v} = \text{constant} \qquad (11\text{-}10)$$

The angular momentum of a particle is constant if the torque of the force on it remains zero. So here we have the case of a *non*isolated particle, one on which a force of varying magnitude and direction may act; yet if this force always points away from or toward the origin of $\mathbf{r}$, the particle has a physical property—angular momentum—which does not change direction or magnitude. The particle is isolated from any torque.

Suppose that we choose the force center as origin for the radius vector $\mathbf{r}$. Then the force always points toward or away from that origin. See Fig. 11-11, where a particle is attached to one end of an elastic band whose other end is fixed in position at the origin O. Then the attractive force produced by the stretched band points at all times toward a force center at O; $\mathbf{r}$ and $\mathbf{F}$ are always in opposite directions; and the cross product $\mathbf{r} \times \mathbf{F}$ is always zero.

Since the particle's angular momentum is constant, the direction of $\mathbf{L} = \mathbf{r} \times m\mathbf{v}$ remains perpendicular to the plane of $\mathbf{r}$ and $\mathbf{v}$. Indeed, the constancy of $\mathbf{L}$ requires that the particle travel in a *single* plane. And even though $\mathbf{r}$ and $\mathbf{v}$ may change both magnitude and direction, the magnitude of the angular momentum is constant:

$$r_{\perp v} mv = \text{constant} \qquad (11\text{-}11)$$

where the quantity $r_{\perp v}$ is measured from the same origin at O and is always perpendicular to the velocity $\mathbf{v}$.

A force between two particles whose direction is along the line connecting the two particles is called a *central force*. Examples are the repulsive electric force between two particles of like electric charge, the attractive force between two particles of unlike electric charge, or the gravitational force between any two particles. See Fig. 11-12. When a central force acts between a pair of par-

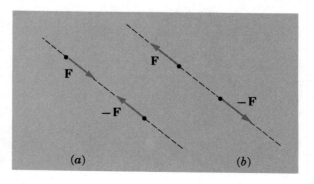

FIG. 11-12. A pair of particles interacting through (a) an attractive central force and (b) a repulsive central force.

ticles, and we choose the origin for computing torques and angular momentum at the location of a fixed particle, then the angular momentum of the other particle is constant. In summary, *a particle subject to a central force has constant angular momentum* when the angular momentum is measured relative to the force center.

Not only is the angular momentum constant for a particle subject to a central force, so too is the rate at which the radius vector to the orbiting particle sweeps out area. Consider Fig. 11-13, where in the short time interval dt the particle moves with velocity $\mathbf{v}$ through a displacement $\mathbf{v}\,dt$. The area dA swept out by the radius vector in the time interval dt is shown shaded in Fig. 11-13; it is half the area of a parallelogram with sides $\mathbf{r}$ and $\mathbf{v}\,dt$. But the area of the entire parallelogram is $|\mathbf{r} \times \mathbf{v}\,dt|$, so that

$$dA = \tfrac{1}{2}|\mathbf{r} \times \mathbf{v}\,dt|$$

But by definition,

$$\mathbf{L} = \mathbf{r} \times m\mathbf{v} = m(\mathbf{r} \times \mathbf{v})$$

Therefore, in magnitude

$$dA = \frac{1}{2}\frac{L}{m}\,dt$$

and the rate dA/dt at which the radius vector sweeps out area is

$$\frac{dA}{dt} = \frac{L}{2m}$$

If the force on the particle is *central* and the angular momentum $\mathbf{L}$ is therefore constant,

$$\frac{dA}{dt} = \frac{L}{2m} = \text{constant} \qquad (11\text{-}12)$$

Conversely, if one finds from observation that a particle's motion is such that the radius vector sweeps out equal areas in equal time intervals, then one knows that the particle's angular momentum is constant and the force on it is *central*. Astronomical observations show that this is true for each planet of the solar system in its motion around the sun; therefore, the gravitational force between the sun and each planet must lie along the line connecting them. More about this in Chap. 13.

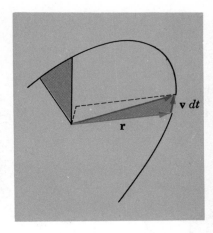

FIG. 11-13. The area dA (shaded) swept out by the radius vector $\mathbf{r}$ in time dt is equal to $dA = \tfrac{1}{2}|\mathbf{r} \times \mathbf{v}|\,dt$.

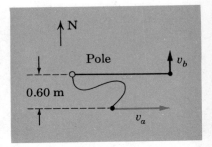

FIG. 11-14. A body attached to one end of a slack cord is traveling east from point a. The other end of the cord is tied to a fixed pole. At point b the cord is taut and the body is traveling north.

Example 11-2

A small object attached to a 2.0-meter cord slides on a frictionless horizontal surface. The other end of the cord is fixed to a pole. Initially the cord is loose and the object travels east at 4.0 m/s, as shown in Fig. 11-14. Later the body is found to be traveling north with the cord taut. What is its speed at that time?

No resultant force acts on the object when the cord is slack; when the cord is taut it exerts a central force. Therefore, the angular momentum of the object relative to the pole is constant, and Eq. 11-11 applies:

$$mr_{a\perp}v_a = mr_{b\perp}v_b$$

$$v_b = \frac{r_{a\perp}}{r_{b\perp}} v_a = \frac{0.60 \text{ m}}{2.0 \text{ m}} (4.0 \text{ m/s}) = 1.2 \text{ m/s}$$

11–5 ANGULAR MOMENTUM OF A RIGID BODY

We shall be concerned exclusively with two simple types of rotating rigid objects: (1) a sheet of any shape rotating about an axis perpendicular to the sheet or (2) an extended symmetrical object rotating about a symmetry axis or about an axis parallel to a symmetry axis. Under these circumstances the object's angular-momentum vector is in the same direction as its angular-velocity vector. In general, however, **L** is not parallel to **ω**. (See Prob. 11-6.)

The total angular momentum of a collection of particles is simply the vector sum of the angular momenta of individual particles. When each particle's position is fixed relative to the other particles, as in the case of a rigid body, then all particles have the same angular velocity about the rotation axis, and the total angular momentum of the rigid body takes a particularly simple form.

Consider the *i*th particle of the symmetrical rotating rigid body in Fig. 11-15. Its mass is m_i, its velocity $\mathbf{v}_i$, and the radius vector $\mathbf{r}_i$ specifies its location relative to the rotation axis. Since the particle travels in a circle of radius r_i, the velocity $\mathbf{v}_i$ is perpendicular to $\mathbf{r}_i$, and the particle's angular momentum $\mathbf{L} = \mathbf{r}_i \times m_i\mathbf{v}_i$ is along the direction of the angular-velocity vector **ω**, parallel to the rotation axis. In magnitude the particle's angular momentum is

$$L_i = r_i m_i v_i$$

But the particle's speed v_i around the circle of radius r_i is given by $v_i = r_i\omega$ [Eq. (4-15)], and the relation above may be written

$$\mathbf{L}_i = m_i r_i^2 \boldsymbol{\omega}$$

a *vector* equation in which it is recognized that $\mathbf{L}_i$ is parallel to **ω**.

To find the rigid body's total angular momentum, we sum over all i particles in the body, recognizing that all particles have the same angular velocity **ω**.

$$\mathbf{L} = \Sigma m_i r_i^2 \boldsymbol{\omega} \tag{11-13}$$

The quantity $\Sigma m_i r_i^2$ by which the angular velocity must be multiplied to yield the angular momentum is called the object's *moment of inertia I* about the rotation axis:

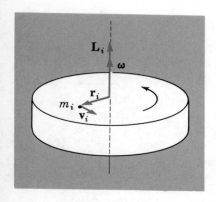

FIG. 11-15. The angular momentum $\mathbf{L} = r_i m_i v_i$ of particle i of a rigid body rotating at angular velocity ω.

$$I = \Sigma m_i r_i^2 \qquad (11\text{-}14)$$

Equation (11-13) can then be written as

$$\mathbf{L} = I\boldsymbol{\omega} \qquad (11\text{-}15)$$

It must be emphasized that the relation above applies only for a rigid body *spinning* symmetrically about a *fixed* axis. In form it is analogous to the relation for the linear momentum of a particle: $\mathbf{p} = m\mathbf{v}$. But whereas the mass is an intrinsic property of particle, and has a constant value which is independent of the state of motion of the particle, the moment of inertia $I = \Sigma m_i r_i^2$ depends not only on the object's mass but also on the distribution of the mass relative to the rotation axis.†

When all of a rigid body's mass is at the same distance R from the rotation axis, as for the particle, ring, and cylindrical shell of Fig. 11-16a to c, the moment of inertia is particularly simple. With $r_i = R$ for all particles, $I = \Sigma m_i R^2 = MR^2$. For more complicated mass distributions, a detailed computation is required. Procedures for computing moments of inertia for objects of various shapes are given in some detail in Appendix VI. In this chapter we shall concentrate on the basic properties of angular momentum and restrict our considerations to objects with simple shapes and moments of inertia: those shown in Fig. 11-16, including a right circular cylinder of uniform density, mass M, and radius R, spinning about its symmetry axis, for which $I = \frac{1}{2}MR^2$ (proved in detail in Appendix VIII.)‡

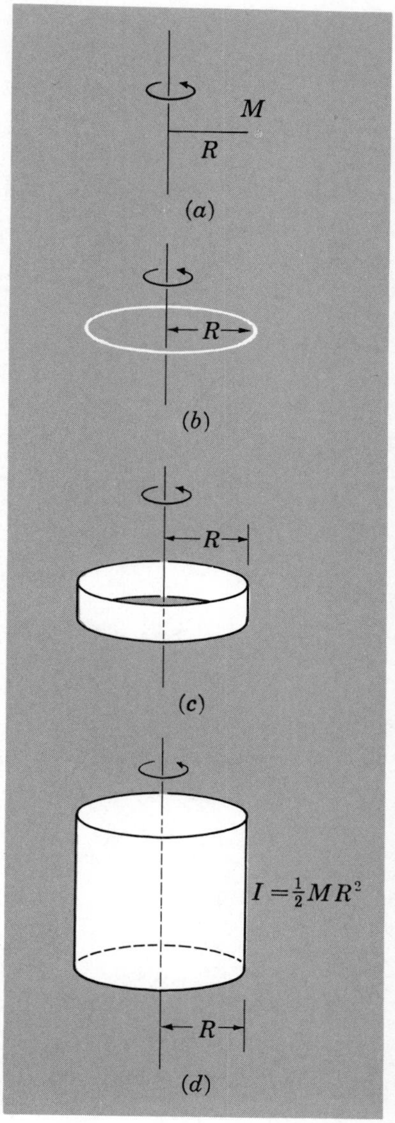

11-6 KINETIC ENERGY OF A SPINNING RIGID BODY

The total kinetic energy of any collection of particles is

$$K = \Sigma K_i = \Sigma \tfrac{1}{2} m_i v_i^2$$

If all particles in the system belong to a rigid body undergoing rotation at the angular velocity ω about a fixed axis, then for each particle $v_i = r_i \omega$. Then the total kinetic energy may be written

$$K = \Sigma K_i = \Sigma \tfrac{1}{2} m_i v_i^2 = \Sigma \tfrac{1}{2} m_i (r_i \omega)^2$$
$$= \tfrac{1}{2}(\Sigma m_i r_i^2)\omega^2$$
$$K = \tfrac{1}{2} I \omega^2 \qquad (11\text{-}16)$$

where we have used the fact that every particle in a rigid body rotating about a fixed axis has the *same* angular velocity ω. Equation (11-16) gives the kinetic energy of a rigid object spinning about

FIG. 11-16. *(a)* A point particle, *(b)* a circular ring, and *(c)* a cylindrical shell, each of mass M and radius R, have a moment of inertia $I = MR^2$ with respect to the axis shown. *(d)* The moment of inertia of a cylinder rotating about its symmetry axis is $I = \tfrac{1}{2}MR^2$.

---
†The angular momentum $\mathbf{r} \times m\mathbf{v}$ is sometimes referred to as the *moment of momentum,* and the torque $\mathbf{r} \times \mathbf{F}$ as the *moment of force.* It is then natural that the quantity I, which plays the role analogous to the mass, or inertia, be known as the *moment of inertia.*
‡Worked examples and problems at the end of the chapter involve only the mass configurations in Fig. 11-16.

a *fixed* rotation axis. When the rotation axis moves, as in the case of a rolling cylinder, (11-16) no longer gives the total kinetic energy of the particles in motion. An additional term must be added to the kinetic-energy relation which represents the kinetic energy of the system's center of mass.†

The relation for the kinetic energy of a spinning object $K = \frac{1}{2}I\omega^2$ is analogous to that for the kinetic energy of a single particle $K = \frac{1}{2}mv^2$, where moment of inertia I replaces inertial mass m and angular velocity ω replaces linear velocity v.‡

Example 11-3

A cylindrical shell of radius 20 cm and mass 1.0 kg is rotating about its symmetry axis at the rate of 33 rev/min. What is the object's *(a)* angular momentum and *(b)* kinetic energy.

(a) Relative to the rotation axis, the angular momentum is

$$L = I\omega = MR^2\omega$$
$$= (1.0 \text{ kg})(0.20 \text{ m})^2(33 \text{ rev/min})(2\pi \text{ rad/rev})(1 \text{ min/60 s})$$
$$= 0.138 \text{ kg-m}^2/\text{s}$$

(b) With the kinetic energy of a rotating object given by $K = \frac{1}{2}I\omega^2$ and $L = I\omega$, we may express the kinetic energy as

$$K = \tfrac{1}{2}I\omega^2 = \frac{(I\omega)^2}{2I} = \frac{L^2}{2I}$$

so that for a cylindrical shell of mass M and radius R

$$K = \frac{L^2}{2MR^2} = \frac{(0.138 \text{ kg-m}^2/\text{s})^2}{2(1.0 \text{ kg})(0.20 \text{ m})^2}$$
$$= 0.238 \text{ J}$$

11-7 THE LAW OF ANGULAR–MOMENTUM CONSERVATION

We are concerned here with a collection of particles, not necessarily parts of a rigid body, which interact with one another and are also subject to external forces. Particle 1 has mass m_1, velocity $\mathbf{v}_1$, and the resultant force on it is $\mathbf{F}_1$. The radius vector from some arbitrarily chosen origin to particle 1 is $\mathbf{r}_1$. Then the resultant torque on particle 1 is $\boldsymbol{\tau}_1 = \mathbf{r}_1 \times \mathbf{F}_1$, and its angular momentum is $\mathbf{L}_1 = \mathbf{r}_1 \times m_1\mathbf{v}_1$. Similarly for particles 2, 3, etc., with, however, the *same* origin for all radius vectors, angular momenta, and torques. See Fig. 11-17.

---

†For a proof see R. T. Weidner and R. L. Sells, *Elementary Classical Physics*, 2d ed., sec. 12-7, Allyn and Bacon, 1973.
‡For a comprehensive listing of analogous mechanical quantities in translational and rotational motion, see Table 12-1.

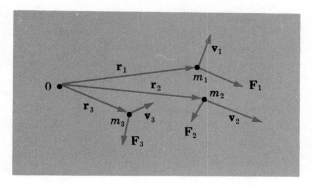

FIG. 11-17. A collection of particles subject to forces. Particle 1 of mass m_1 and velocity v_1 has a displacement r_1 relative to the origin O and is subject to a resultant force F_1. Likewise for particles 2, 3, ...

Applying Eq. (11-6) to each of the particles in the system, we have

$$\tau_1 = \frac{d\mathbf{L}_1}{dt}$$

$$\tau_2 = \frac{d\mathbf{L}_2}{dt}$$

. . . .

Adding all these equations yields

$$\tau_1 + \tau_2 + \tau_3 + \cdots = \frac{d\mathbf{L}_1}{dt} + \frac{d\mathbf{L}_2}{dt} + \frac{d\mathbf{L}_3}{dt} + \cdots$$

The resultant of *all* torques is

$$\Sigma \tau_{\text{all torques}} = \tau_1 + \tau_2 + \tau_3 + \cdots$$

and the total vector angular momentum of the system is

$$\mathbf{L} = \mathbf{L}_1 + \mathbf{L}_2 + \mathbf{L}_3 + \cdots$$

Therefore, the equation above for the sum of the torques may be written

$$\Sigma \tau_{\text{all torques}} = \frac{d\mathbf{L}}{dt} \qquad (11\text{-}17)$$

Now consider two representative particles 1 and 2 in the system. They interact with one another by the internal forces $\mathbf{F}_{2 \text{ on } 1}$ and $\mathbf{F}_{1 \text{ on } 2}$, where

$$\mathbf{F}_{2 \text{ on } 1} = -\mathbf{F}_{1 \text{ on } 2}$$

by Newton's third law. See Fig. 11-18.

We wish to consider the torques on particles 1 and 2 arising from mutual interaction when the interaction force is a *central* force, so that $\mathbf{F}_{2 \text{ on } 1}$ and $\mathbf{F}_{1 \text{ on } 2}$ are not merely in opposite directions but also lie along the line connecting particles 1 and 2. As Fig. 11-18 shows, the perpendicular distance $r_{\perp F}$ from the origin to the

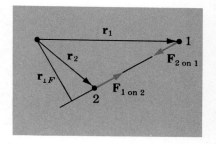

FIG. 11-18. Particles 1 and 2 interacting through a central force. The torques of both forces have the same moment arm $r_{\perp F}$, and the resultant torque is zero.

The Law of Angular-Momentum Conservation SEC. 11-7

common line of the two internal forces is the *same* for both forces. Therefore, the magnitude of the torque $\boldsymbol{\tau}_{2\text{ on }1} = \mathbf{r}_1 \times \mathbf{F}_{2\text{ on }1}$ equals the magnitude of $\boldsymbol{\tau}_{1\text{ on }2} = \mathbf{r}_2 \times \mathbf{F}_{1\text{ on }2}$, since the magnitudes of the forces are alike. But the directions of the two internal torques are opposite; applying the right-hand rule for cross products, we see that $\boldsymbol{\tau}_{2\text{ on }1}$ is into the paper, while $\boldsymbol{\tau}_{1\text{ on }2}$ is out of the paper. The internal torques between particles 1 and 2 are equal and opposite, and their sum is zero:

$$\boldsymbol{\tau}_{2\text{ on }1} + \boldsymbol{\tau}_{1\text{ on }2} = 0$$

But if this applies for particles 1 and 2 interacting by a central force, it applies for *all* pairs of particles interacting mutually through central forces. Thus, the sum of the *internal* torques is zero.

$$\Sigma \boldsymbol{\tau}_{\text{internal torques}} = 0 \tag{11-18}$$

Now, the vector sum of all torques on the system of particles is merely the sum of the external torques (those arising from external forces) and the internal torques (from internal *central* forces):

$$\Sigma \boldsymbol{\tau}_{\text{all torques}} = \Sigma \boldsymbol{\tau}_{\text{external torques}} + \Sigma \boldsymbol{\tau}_{\text{internal torques}} \tag{11-19}$$

Using (11-18) and (11-19) in (11-17), we have finally

$$\Sigma \boldsymbol{\tau}_{\text{ext}} = \frac{d\mathbf{L}}{dt} \tag{11-20}$$

The vector sum of the *external* torques only is equal to the time rate of change of the system's total angular momentum. This is *the* fundamental relation between angular momentum and torque. [It is similar to Eq. (8-16), $\Sigma \mathbf{F}_{\text{ext}} = d\mathbf{P}/dt$, where $\Sigma \mathbf{F}_{\text{ext}}$ is the resultant external force on any system and $\mathbf{P}$ is the system's total linear momentum.] Chapter 12 is devoted, in the main, to illustrating its application to rigid bodies in rotation.

If some system of particles is completely isolated, with no external forces acting upon it, then clearly the resultant external torque on the system is zero, and the system's angular momentum $\mathbf{L}$ does not change magnitude or direction. But even if a system is not isolated, and external forces *do* act upon it, the system's total angular momentum $\mathbf{L}$ will be constant, provided that the resultant *torque* of the external forces is zero. From (11-20), if

$$\Sigma \boldsymbol{\tau}_{\text{ext}} = 0 \quad \text{then} \quad \frac{d\mathbf{L}}{dt} = 0$$

or

$$\mathbf{L} = \mathbf{L}_1 + \mathbf{L}_2 + \mathbf{L}_3 + \cdots = \text{constant} \tag{11-21}$$

This is the *law of conservation of angular momentum*: Any system of particles interacting through central forces and subject to no net external *torque* has constant angular momentum.

Example 11-4

A bullet of mass m traveling with a speed v is shot into the rim of a right circular cylinder of radius R and mass M, as shown in Fig. 11-19. The cylinder has a fixed horizontal rotation axis and is originally at rest. What is the angular speed of the cylinder after the bullet has become imbedded in it?

Our system is the bullet and cylinder. We choose a point on the cylinder's axle for computing torques and angular momentum. Although a force clearly acts on the cylinder at its axle, the torque of this force is zero. Consequently, the system of bullet and cylinder is subject to no external torque (we neglect the torque due to the weight of the bullet), and its total angular momentum is constant.

The magnitude of the bullet's angular momentum before striking the cylinder is Rmv, R being the perpendicular distance between the point on the axle and the direction of **v**. After the bullet is captured, the total moment of inertia is that for the cylinder alone ($\tfrac{1}{2}MR^2$) plus that for the bullet (mR^2): $I = \tfrac{1}{2}MR^2 + mR^2$.

Equating total angular momentum before collision to that after, we have

$$Rmv = (\tfrac{1}{2}M + m)R^2\omega$$

where ω is the cylinder's angular velocity, or

$$\omega = \frac{mv}{(\tfrac{1}{2}M + m)R}$$

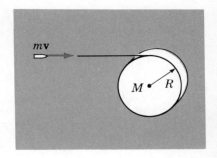

FIG. 11-19. A bullet of mass m and velocity v is shot into the rim of a cylinder of mass M and radius R.

Example 11-5

A phonograph *r*ecord having a moment of inertia I_r relative to an axis through its center is dropped onto a *t*urntable having a moment of inertia I_t. The turntable is *i*nitially spinning with the angular speed ω_i as shown in Fig. 11-20. *(a)* What is the final angular speed of the turntable and record after the record has come to rest on the turntable? *(b)* What fraction of the initial kinetic energy is dissipated?

Our system is chosen to be the record and the turntable. Assuming that the turntable spins about a frictionless axle, there is no resultant external torque on the system, and its total angular momentum remains constant. We label the *f*inal angular speed ω_f. Then angular-momentum conservation requires that

$$I_t\omega_i = (I_r + I_t)\omega_f$$

$$\omega_f = \frac{I_t}{I_r + I_t}\omega_i$$

Note that *internal* torques act within the system. The record produces a frictional torque on the turntable, slowing it, and the turntable produces a frictional torque on the record, speeding it up. The sum of these two torques is zero at every instant of time.

(b) The initial kinetic energy is

$$K_i = \tfrac{1}{2}I_t\omega_i^2$$

and the final kinetic energy is

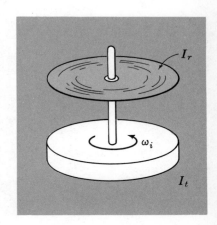

FIG. 11-20. A phonograph record of moment of inertia I_r is dropped onto a turntable of moment of inertia I_t initially spinning with angular velocity ω_i.

$$K_f = \tfrac{1}{2}(I_r + I_t)\omega_f^2$$

Their ratio is, with the use of the relation for ω_f above,

$$\frac{K_f}{K_i} = \frac{\tfrac{1}{2}(I_r + I_t)\omega_f^2}{\tfrac{1}{2}I_t\omega_i^2} = \frac{I_t}{I_r + I_t}$$

Since K_f/K_i is always less than 1, some of the initial kinetic energy is always dissipated in what might be called an "inelastic rotational collision." It is altogether analogous to an ordinary inelastic collision in which two objects stick together after collision; the system's linear momentum is constant, but its kinetic energy decreases. Here too the kinetic energy decreases because a nonconservative, frictional force acts on the two spinning objects, but now it is the system's total angular momentum which remains constant.

SUMMARY

The angular momentum **L** of a particle with linear momentum $m\mathbf{v}$ relative to the origin of the radius vector **r** is

$$\mathbf{L} = \mathbf{r} \times m\mathbf{v} \qquad (11\text{-}5)$$

The torque $\boldsymbol{\tau}$ of a force **F** relative to the origin of the radius vector **r** is

$$\boldsymbol{\tau} = \mathbf{r} \times \mathbf{F} \qquad (11\text{-}4)$$

For a single particle the torque of the resultant force on the particle is equal to the time rate of change of the particle's angular momentum:

$$\boldsymbol{\tau} = \frac{d\mathbf{L}}{dt} \qquad (11\text{-}6)$$

When a particle is subject to a central force, its angular momentum relative to the force center is constant. Equivalently, the radius vector from the force center to the orbiting particle sweeps out equal areas in equal times:

$$\frac{dA}{dt} = \frac{L}{2m} \qquad (11\text{-}12)$$

When a rigid body rotates symmetrically about a fixed axis at the angular velocity $\boldsymbol{\omega}$, its spin angular momentum is

$$\mathbf{L} = I\boldsymbol{\omega} \qquad (11\text{-}15)$$

and its kinetic energy is

$$K = \tfrac{1}{2}I\omega^2 \qquad (11\text{-}16)$$

where the object's moment of inertia I relative to the rotation axis is

$$I = \Sigma m_i r_i^2 \qquad (11\text{-}14)$$

For any system of particles interacting with one another through central forces,

$$\Sigma \tau_{ext} = \frac{d\mathbf{L}}{dt} \qquad (11\text{-}20)$$

where $\mathbf{L}$ is the system's total angular momentum and $\Sigma \tau_{ext}$ is the resultant external torque on the system.

According to the law of angular-momentum conservation, a system's total angular momentum is constant when the resultant torque on the system is zero.

PROBLEMS

11-1. A pitcher throws a 160-g baseball across the plate with a speed of 30 m/s. At the instant of release the ball is 60 cm from his shoulder joint. What are the magnitude and *direction* of the angular momentum (relative to the shoulder) given to the ball? Assume that the ball's trajectory is confined to a vertical plane.

11-2. What are the magnitude and direction of the angular momentum of a 1.0-kg particle relative to the coordinate origin if its X and Y coordinates and velocity components in the XY plane $(x, y; v_x, v_y)$ are: *(a)* (3.0 m, 2.0 m; 6.0 m/s, 2.0 m/s); *(b)* (−2.0 m, 1.0 m; 1.0 m/s, 3.0 m/s); and *(c)* (2.0 m, −3.0 m; 3.0 m/s, 4.0 m/s).

11-3. A 2.0-kg particle has an angular momentum of 12.0 J-s in the positive Z direction relative to the coordinate origin when it is traveling in the negative Y direction with a speed of 4.0 m/s. Where is the particle at this instant?

11-4. A 2.0-kg particle is initially at the location $x = 10.0$ m and $y = 0$ m. The particle is subject to a constant force of 6.0 N in the negative Y direction. Relative to the point (4.0 m, 0 m), *(a)* what is the particle's angular momentum and *(b)* the torque on the particle, both as functions of time? *(c)* Show for this particular example that torque equals time rate of change of angular momentum.

11-5. A system consists of two particles each of 1.0-kg mass. One particle travels north at a constant speed of 4.0 m/s; the second particle travels south at the same constant speed in the same horizontal plane, the separation distance between the lines of the two velocities being 3.0 m. What are the direction and magnitude of this system's total angular momentum relative to *any* point?

11-6. Figure P11-6 shows two particles of mass m attached to the ends of a massless rigid rod of length L and oriented at a constant angle θ relative to the Z

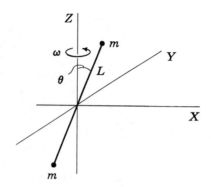

FIG. P11-6. An example of a system whose angular velocity ω is *not* parallel to its angular momentum **L**.

axis. The rod rotates at the angular velocity ω along the positive Z direction. (The rotation axis is *not* along the symmetry axis of the rod.) What are the direction and magnitude of the system's angular momentum at the instant when the rod lies in the XZ plane, as shown in the figure? (Note that the angular-momentum vector is *not* along the direction of the angular-velocity vector.)

11-7. Show that, relative to the X, Y, and Z axes, the components of the angular momentum of a particle of mass m passing the point x, y, and z with velocity components v_x, v_y, and v_z are:

$$L_x = m(yv_z - zv_y)$$
$$L_y = m(zv_x - xv_z)$$
$$L_z = m(xv_y - yv_x)$$

11-8. A particle is subject to a central attractive force from a fixed force center. Show that unless the particle is aimed at some instant directly at the force center, it will never reach the force center.

11-9. A particle of mass m is thrown into the air with an initial velocity v_0 at an angle θ relative to the horizontal. What is the particle's angular momentum and the torque due to the particle's weight, both relative to the point where the particle was projected, when *(a)* the particle is at its highest point and *(b)* when the particle is at the same vertical elevation as initially?

11-10. According to the quantum theory, all orbital angular momenta occur in integral multiples of 1.1×10^{-34} J-s (Planck's constant divided by 2π). How many such quanta of angular momentum does an electron (mass, 9.1×10^{-31} kg) moving in a circle of 5.3×10^{-11} m radius at a frequency of 6.6×10^{15} Hz have? (These are circumstances for a hydrogen atom in its lowest energy state according to the Bohr atomic theory.)

11-11. A simple pendulum consists of a 1.0-kg ball suspended from one end of a cord 1.0 m long. At the maximum of its swing, the cord makes a 60° angle with the vertical. Find the maximum value of the angular momentum of the ball relative to the center of its arc. Assume $g = 10$ m/s².

11-12. A 1.0×10^3 kg earth satellite in a circular orbit at an altitude of 100 km completes one orbit in 86.4 min. Find *(a)* the orbital speed and *(b)* the orbital angular momentum of the satellite relative to the earth's center. The earth's radius is 6.37×10^6 m.

11-13. A long flexible cord is wound around a heavy cylinder which is free to rotate on a horizontal axle coincident with its symmetry axis. Initially the cylinder, which is 30.0 cm in diameter, is stationary. A steady 20.0-N force is applied to the cord—this force being tangential to the cylinder and perpendicular to the axle. As a result, a 12.0-m length of cord is unwound in 10.0 s. Find *(a)* the moment of inertia of the cylinder and *(b)* its mass (assuming uniform density). *(c)* What is the final angular speed of the cylinder?

11-14. A mass m slides in circle of radius r on a smooth horizontal surface. The mass is attached to a string that passes through a hole at the center of the circle, the other end of the string being attached to a mass M, as shown in Fig. P11-14. *(a)* What is the speed of the

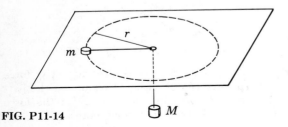

FIG. P11-14

mass m? *(b)* What is the angular momentum of m? *(c)* If the mass M is pulled down slowly through a vertical distance y, what is then the speed of m? *(d)* How much work was done in pulling the mass M downward in part *(c)*?

11-15. A diatomic molecule can be thought of as a dumbbell-like structure in which particles of masses m_1 and m_2 are attached to a massless rod of length r. *(a)* Show that the angular momentum of a molecule rotating about an axis through the center of mass and perpendicular to the interatomic axis is given by $L = \mu r^2 \omega$, where μ, the so-called reduced mass, is $m_1 m_2 \div (m_1 + m_2)$ and ω is the angular speed. *(b)* Show that the rotational kinetic energy is given by $K = \frac{1}{2}\mu r^2 \omega^2 = L^2/2\mu r^2$.

11-16. Consider the angular momentum of two particles each of mass m on diametrically opposite sides of a ring of radius R, as shown in Fig. P11-16. The two

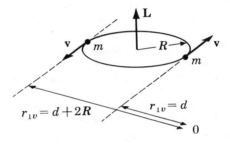

FIG. P11-16

particles move in opposite directions, each with speed v, in a circle of radius R. Relative to the arbitrary origin O, the nearer particle has $r_{\perp v} = d$, and an angular momentum magnitude dmv; its angular momentum is *down*. *(a)* Show that the angular momentum magnitude of the farther particle is $(d + 2R)mv$, and *(b)* show that the net angular momentum of the pair of particles is

$$L = R(2m)v \quad \text{(up)}$$

Note that distance d does not appear. The magnitude and direction of the angular momentum of this pair of symmetrically located particles depends only on their mass m, distance r from the rotation axis, and speed v. Since any symmetrical object rotating about its symmetry axis can be considered as a collection of pairs of particles in the fashion of Fig. P11-16, the angular momentum of such a spinning object—its *spin angular momentum*—is an intrinsic property of the object; the spin angular momentum is *independent* of the point relative to which angular momentum is measured.

It can be shown that for any system of particles, the total angular momentum can be expressed as the sum of two parts: the angular momentum of all particles measured relative to an origin at the center of mass, $L_{\text{relative to CM}}$, and what might be called the angular momentum of the center of mass, $L_{\text{of CM}}$.

$$L_{\text{total}} = L_{\text{relative to CM}} + L_{\text{of CM}}$$

The latter term is $L_{\text{of CM}} = \mathbf{R} \times M\mathbf{V}$, where $\mathbf{R}$ is the radius vector from the chosen origin to the system's center of mass, M is the mass of the entire system, and $\mathbf{V}$ is the velocity of the center of mass.

For a symmetrical object spinning about an internal axis and also moving in translation motion,

$$L_{\text{relative to CM}} = L_{\text{spin}}$$

and the remaining angular momentum, arising from the orbiting of the object as whole, is termed its *orbital angular momentum*, L_{orbital}.

$$L_{\text{of CM}} = L_{\text{orbital CM}} = \mathbf{R} \times M\mathbf{V}$$

11-17. Calculate (*a*) the spin angular momentum and (*b*) the orbital angular momentum of the earth. The earth's spin is associated with its daily rotation about the north-south polar diameter and its orbital motion is simply its annual turn around the sun. (Consult Tables 13-1 and 13-2 for astronomical data.)

11-18. A 60-kg ice skater grasps a rope attached to a relatively thin, fixed post. Initially the rope is 8.0 m long, and the ice skater travels in a circle 8.0 m in radius at a speed of 1.0 m/s. As the skater continues, the rope becomes wrapped around the post until its length to the skater is 2.0 m. (*a*) What is then the skater's speed? (*b*) If the rope were attached by a loop to the post, so that it did not become wrapped around the post, and if the length of rope between the skater and the post were reduced slowly by the skater's pulling on it, what would be the skater's speed when the rope length was 2.0 m?

11-19. When its temperature rises, a freely rotating body increases all its linear dimensions by 3 parts in 10^5. By what fraction is its angular speed of rotation reduced?

11-20. Two electrically charged particles interact by a *central* repulsive force for which the magnitude of the potential energy is given by K/r, where K is a constant and r is the distance between the two particles. In Fig. P11-20 we see a charged particle of small mass m moving initially with speed v_i when it is far from a second charged particle of essentially infinite mass (the force center). If there were no repulsive force to deflect it, the light particle would miss hitting the massive one

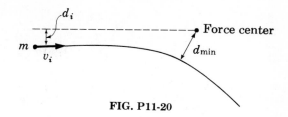

FIG. P11-20

head on by a distance of d_i. What is the minimum separation distance $d_{\min}$ between the two particles?

11-21. One of the two coaxial disks is initially rotating with an angular speed of 120 rev/min. Its moment of inertia is 10.0 kg-m². The second disk, having a moment of inertia of 50.0 kg-m² and initially at rest, is then pressed against the rotating disk in the fashion of a clutch. (*a*) What is the final common angular speed of the two disks? (*b*) How much energy is dissipated by the clutch? (*c*) Does the total amount of energy dissipated depend upon whether the disks are brought together slowly or quickly?

11-22. The conservation of angular momentum law is sometimes demonstrated with a freely rotating horizontal platform on which a toy electric train runs on a track near the circumference. Suppose that a train with a mass of 8.0 kg runs around a track of 0.75 m in radius on a platform whose moment of inertia relative to the vertical axle at the center is 3.0 kg-m². The train and the platform are initially at rest. Then the train is set in motion, its final speed relative to the tracks being 0.20 m/s. (*a*) What is the angular speed of the platform relative to the ground? (*b*) What is the angular momentum of the platform relative to the ground? (*c*) What is the angular momentum of the train relative to the ground? (*d*) What is the angular speed of the platform relative to the ground after the motor on the train has been shut off and the train has come to rest relative to the tracks? (*e*) Is angular momentum conserved if the train's wheels slip on the tracks?

11-23. A particle of mass m is attached to one end of a spring of relaxed length l_0 and force constant k. The spring's other end is fixed in position. The particle slides over a frictionless horizontal surface. See Fig. P11-23. Initially, the spring is relaxed, and the particle's velocity $\mathbf{v}_1$ is at right angles to the long axis of the spring. At some later time the particle reaches point 2: here the spring's length is $l_0 + x$, and the particle's velocity is $\mathbf{v}_2$, the direction between $\mathbf{v}_2$ and the long axis of the spring being θ. Apply the conservation laws for energy and angular momentum to show that

$$v_2 = (v_1^2 - kx^2/m)^{1/2}$$

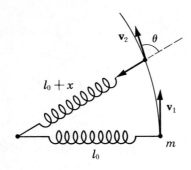

FIG. P11-23

and $\quad \sin \theta = \dfrac{v_1}{(1 + x/l_0)(v_1^2 - kx^2/m)^{1/2}}$

11-24. A 50-g bullet moving horizontally at 300 m/s penetrates an upright 10.0-kg wood cylinder and emerges with a speed of 60 m/s. As a result, the cylinder—initially at rest—is set in rotation about its (vertical) axis of symmetry, see Fig. P11-24. The radius of the cylinder is 15.0 cm and the bullet's trajectory is 12.0 cm to one side of the axis. Find the angular speed of the cylinder after the bullet has emerged from it.

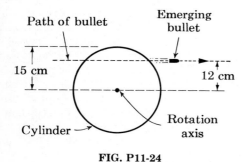

FIG. P11-24

11-25. A top is set spinning about its symmetry axis, which is initially in a horizontal plane. The top is supported at one end of its axis by a vertical post. See Fig. P11-25, where the instantaneous spin angular momentum is labeled $\mathbf{L}_0$. *(a)* Identify the direction of $\boldsymbol{\tau}$, the instantaneous gravitational torque exerted on the top; $\boldsymbol{\tau}$ is computed relative to the point of support of the top. (Note that $\boldsymbol{\tau}$ is *always* at right angles to $\mathbf{L}_0$.) *(b)* Draw a vector diagram (in the horizontal plane) which displays the *change* in angular momentum, $\Delta \mathbf{L}$, which occurs over some short time interval Δt. *(c)* Show that $\Delta \mathbf{L} = \boldsymbol{\tau} \Delta t$ and that the relative directions of $\mathbf{L}_0$ and $\Delta \mathbf{L}$ are such as to indicate that the spinning top as a whole rotates in a horizontal plane about its point of support. This motion is known as *precession*. *(d)* Use the geometrical relationships contained in your diagram to show that the angular frequency of precession is given by $\omega_p = \lim\limits_{\Delta t \to 0} \Delta \phi / \Delta t = \tau / L_0$.

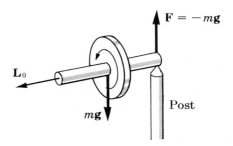

FIG. P11-25

11-26. A woman walking east carries a suitcase in which a massive object is rotating freely on pivots attached to the suitcase. The spinning object's angular velocity is in the west direction. What happens to the suitcase if the woman turns *(a)* north and *(b)* south? (Do not attempt this problem without first understanding Prob. 11-25.)

CHAPTER 12

Rotational Dynamics

In this chapter we develop the rotational dynamics of rigid bodies. It is simply a matter of applying the basic relations concerning angular momentum and torque to objects which do not undergo deformation. In addition to arriving at the rotational form of Newton's second law, we derive the rotational form of the work-energy theorem, define the center of gravity, and consider the general conditions for rigid-body equilibrium.

12–1 ROTATIONAL KINEMATICS

Since all rotations to be discussed in this chapter will be about axes perpendicular to the plane of the paper, it is sufficient to consider the angular velocity as an algebraic quantity, without concern for its vector properties (Sec. 11-1). By definition the angular velocity ω of *all* particles of a perfectly rigid body undergoing rotation is

$$\omega = \frac{d\theta}{dt} \qquad (4\text{-}14), (12\text{-}1)$$

the time rate of change of angular displacement. Positive ω means counterclockwise rotation, negative ω clockwise. Appropriate units for ω are rad/s.

If a rotating object's rate of rotation changes, it undergoes an angular acceleration. The instantaneous angular acceleration α is defined as the time rate of change of angular velocity:

$$\alpha = \frac{d\omega}{dt} \qquad (12\text{-}2)$$

A positive angular acceleration implies that the rotational velocity is increasing in the counterclockwise sense (or, equivalently, decreasing in the clockwise sense). For $\alpha = 0$, the body rotates at constant ω. Angular acceleration is given in units of rad/s². Angular velocity and angular acceleration have the dimensions T^{-1} and T^{-2}, respectively.

Equations (12-1) and (12-2) relating the angular velocity ω and acceleration α to the angular displacement θ and the time t are altogether analogous to the relations between the linear displacement x, linear velocity v, and linear acceleration a. Therefore, we can arrive at the kinematic equations describing rotational motion at a *constant* angular acceleration from the linear kinematic equations for *constant* linear acceleration. We simply replace:

$$x \quad \text{by} \quad \theta$$

$$v = \frac{dx}{dt} \quad \text{by} \quad \omega = \frac{d\theta}{dt}$$

$$a = \frac{dv}{dt} \quad \text{by} \quad \alpha = \frac{d\omega}{dt}$$

Thus, for *constant angular acceleration*, the rotational kinematic equations for a rigid body about a fixed axis follow directly from the analogous translational kinematic equations for constant linear acceleration, as shown below.

	Constant linear acceleration a	Constant angular acceleration α	
(3-6)	$v = v_0 + at$	$\omega = \omega_0 + \alpha t$	(12-3)
(3-7)	$x = x_0 + v_0 t + \tfrac{1}{2}at^2$	$\theta = \theta_0 + \omega_0 t + \tfrac{1}{2}\alpha t^2$	(12-4)
(3-9)	$x - x_0 = \dfrac{(v_0 + v)t}{2}$	$\theta - \theta_0 = \dfrac{(\omega_0 + \omega)t}{2}$	(12-5)
(3-10)	$v^2 = v_0^2 + 2a(x - x_0)$	$\omega^2 = \omega_0^2 + 2\alpha(\theta - \theta_0)$	(12-6)

Of course one can derive Eqs. (12-3) to (12-6) from the definitions, Eqs. (12-1) and (12-2), by integration in the same fashion as shown in Sec. 3-5. The initial ($t = 0$) angular displacement is θ_0, and the initial angular velocity is ω_0.

Although all particles of a rigid body will have turned through the same angle θ and have, at any instant, the same ω and α, their *linear* speeds and accelerations along the circular arcs in which they travel differ according to their relative distances from the axis of rotation.

The *tangential* speed v_t of a particle a distance r from the rotation axis of an object rotating with angular velocity ω is

$$v_t = r\omega \qquad (4\text{-}15),(12\text{-}7)$$

When ω changes with time, so does the speed v_t. We find the *tangential* acceleration a_t by taking the time derivative of (12-7):

$$\frac{dv_t}{dt} = r\frac{d\omega}{dt}$$

or
$$a_t = r\alpha \qquad (12\text{-}8)$$

In using (12-7) and (12-8) for computation, it is important to recognize that ω must be given in radians per unit time and α in radians per time squared.

Any particle in motion in a circle always has a radially inward acceleration a_r, where

$$a_r = \omega^2 r = \frac{v_t^2}{r} \qquad (4\text{-}16)$$

so that the total acceleration of a particle moving in a circle, comprised of a tangential component a_t and a perpendicular radial component a_r, has the magnitude $(a_t^2 + a_r^2)^{1/2}$.

Example 12-1

A wheel initially spinning at 60 rev/min is brought to rest in 10 s.
(a) What is the wheel's angular acceleration, assumed to be constant?
(b) How many turns does the wheel make in coming to rest?

(a) We have from Eq. (12-3)

$$\omega = \omega_0 + \alpha t$$

Here $\omega_0 = (60 \text{ rev/min})(2\pi \text{ rad/rev})(1 \text{ min}/60 \text{ s}) = 2\pi \text{ s}^{-1}$, and $\omega = 0$ at $t = 10$ s.

$$\alpha = -\frac{\omega_0}{t} = -\frac{2\pi \text{ s}^{-1}}{10 \text{ s}} = -\tfrac{1}{5}\pi \text{ s}^{-2}$$

The minus sign indicates that the spin rate of the rotating object is decreasing.

(b) To find the total angular displacement θ, we use Eq. (12-5), taking θ_0 to be zero.

$$\theta - \theta_0 = \frac{(\omega_0 + \omega)t}{2}$$

$$\theta = \tfrac{1}{2}(2\pi \text{ s}^{-1})(10 \text{ s})\frac{1 \text{ turn}}{2\pi \text{ rad}} = 5.0 \text{ turns}$$

12-2 NEWTON'S SECOND LAW FOR ROTATION

As shown in Chap. 11, the angular momentum L and kinetic energy K of a solid object rotating symmetrically at angular velocity ω are

$$L = I\omega \qquad (11\text{-}15),(12\text{-}9)$$

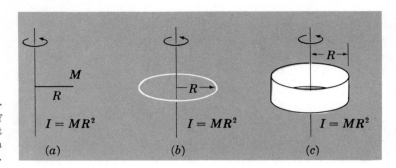

FIG. 12-1. Moments of inertia for several simple shapes, all of constant density: *(a)* a point particle, *(b)* a circular ring, *(c)* a cylindrical shell.

$$K = \tfrac{1}{2}I\omega^2 \qquad (11\text{-}16),(12\text{-}10)$$

where the moment of inertia I is, by definition,

$$I = \Sigma m_i r_i^2 \qquad (11\text{-}14),(12\text{-}11)$$

We sum the contributions from each particle of mass m_i at a distance r_i from the rotation axis to find a rigid body's moment of inertia. So the moment of inertia depends not only on the object's mass but also on the distribution of mass relative to the line chosen as axis of rotation. Clearly, then, even for a single object, the moment of inertia depends upon the choice of the rotation axis.

For the three simple shapes shown in Fig. 12-1*a* to *c,* where *all* mass is at the *same* distance R from the axis,

$$I = MR^2$$

The moment of inertia I_{CM} of any object with respect to an axis passing through its center of mass is related in a simple way to the moment of inertia I with respect to another parallel axis:

$$I = I_{CM} + Md^2 \qquad (12\text{-}12)$$

where M is the total mass and d is the distance between the two parallel axes. See Fig. 12-2. This is the so-called parallel-axis theorem for moments of inertia.† As (12-12) shows, the moment of inertia about an axis through the center of mass is a minimum; for any other parallel axis, clearly $I > I_{CM}$. See Appendix VIII for detailed procedures for use in computing the moment of inertia.

Any change in the position of a rigid body is a combination of its translation through space and rotation about an axis. The *translational motion* of the object's center of mass is controlled by the *external forces* acting on it:

$$\Sigma \mathbf{F}_{ext} = M\mathbf{A} \qquad (8\text{-}15),(12\text{-}13)$$

where M is the total mass and $\mathbf{A}$ the acceleration of the center of mass.

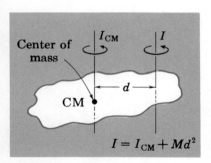

FIG. 12-2. Illustration of the parallel-axis theorem. The moment of inertia about an axis through the center of mass is I_{CM}; the moment of inertia I about a parallel axis, separated a distance d, is $I_{CM} + Md^2$, where M is the object's mass.

---

†For a detailed proof see R. T. Weidner and R. L. Sells, *Elementary Classical Physics,* 2d ed., sec. 12-7, Allyn and Bacon, 1973.

As was shown in Sec. 11-7, the resultant external torque on a system of particles determines the time rate of change of that system's total angular momentum **L**:

$$\Sigma \boldsymbol{\tau}_{\text{ext}} = \frac{d\mathbf{L}}{dt} \qquad (11\text{-}20), (12\text{-}14)$$

Recall that the internal forces and internal torques (for central forces) add to zero.

Now, if the system consists of a rigid body, its angular momentum† is

$$\mathbf{L} = I\boldsymbol{\omega} \qquad (11\text{-}15)$$

and if the object is truly rigid, every particle within it remaining at a fixed location relative to every other particle, the moment of inertia I is a *constant*. Therefore,

$$\frac{d\mathbf{L}}{dt} = I\frac{d\boldsymbol{\omega}}{dt} = I\boldsymbol{\alpha}$$

and (12-14) may be written

$$\Sigma \boldsymbol{\tau}_{\text{ext}} = I\boldsymbol{\alpha} \qquad (12\text{-}15)$$

Although (12-15) is a vector equation, we shall use its algebraic form for torques, axes of rotation, and angular accelerations all perpendicular to the plane of the paper. It must be emphasized that, in applying (12-15), the torques, the moment of inertia, and the angular acceleration are all measured relative to the axis of rotation.

Recall that the torque $\boldsymbol{\tau}$ of the force $\mathbf{F}$; $\boldsymbol{\tau} = \mathbf{r} \times \mathbf{F}$, has a magnitude given by

$$\tau = r_{\perp F} F \qquad (11\text{-}7)$$

where F is the magnitude of the force and $r_{\perp F}$ is the perpendicular distance from the origin of **r**, or axis of the torque, to the line of action of **F**. The term *moment arm*, or *lever arm*, is sometimes used to denote $r_{\perp F}$. As Fig. 12-3 shows, the torque magnitude may also be written as $\tau = rF \sin\theta$, where θ is the angle between **r** and **F**. We take counterclockwise torques to be positive and clockwise torques to be negative.

We may certainly use *Newton's second law of motion for rotation*, $\Sigma \tau_{\text{ext}} = I\alpha$, when the rotation axis is fixed in an inertial frame. The same relation is valid when the rotation axis moves at constant velocity. What about the situation in which a rotating rigid body is *accelerating* in its translational motion? Curiously,

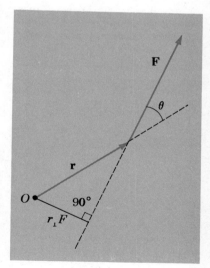

FIG. 12-3. The magnitude of the torque of the force F relative to the origin O is $\tau = r_{\perp F} F = rF \sin\theta$.

---

†As remarked before in Chap. 11, we are restricting our discussion to rigid objects whose angular momentum is *parallel* to their angular velocity. This implies that the object is a sheet of any shape which rotates about an axis perpendicular to the sheet or an extended symmetrical object which rotates about an axis of symmetry.

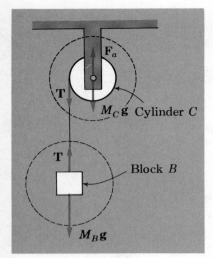

FIG. 12-4. Forces acting on a cylinder and a block. The cylinder is free to turn about its symmetry axis. The block descends, unwinding the cord — which was wrapped around the cylinder — and spinning the cylinder. The dashed lines indicate the boundaries of the two objects considered individually.

here again we *may* properly apply the same relation if the torques, moment of inertia, and angular acceleration are all measured relative to the accelerating body's *center of mass*. To be sure, Newton's second law, including now (12-15), applies only in an inertial frame. But the sum of the inertial forces is equivalent to a single inertial force acting on the system's center of mass, and this force has zero torque relative to the center of mass.

Example 12-2

Block B of mass M_B is attached to a massless inextensible cord which is wrapped around a uniform right circular cylinder C. The cylinder is free to rotate about an axle through its symmetry axis. The axle is fixed in position, suspended from an overhead support. See Fig. 12-4. (*a*) What is the linear acceleration of the block? (*b*) What is the tension in the cord?

(*a*) The block moves downward in response to the forces on it and the cylinder rotates as the cord unwinds. First we isolate each of the two bodies, as indicated by the dashed loops in Fig. 12-4, and then show *all* of the forces on each object.

Applying Newton's second law to the translational motion of the block yields

$$\Sigma \mathbf{F}_{\text{ext}} = M_B \mathbf{a}$$

Block B $\qquad M_B g - T = M_B a \qquad$ (12-16)

We have chosen downward as the positive direction of the block's acceleration. This choice requires that the positive direction for the angular acceleration of the cylinder be counterclockwise (see Fig. 12-4).

Next we apply Newton's second law in its rotational form to the motion of the cylinder C. The forces on C are the tension $\mathbf{T}$ (which is tangent to the cylinder), the force $\mathbf{F}_a$ exerted by the axle on C, and the weight $M_C \mathbf{g}$ of C. The tension $\mathbf{T}$ is the only force on C which produces a torque about the symmetry axis:

$$\Sigma \tau_{\text{ext}} = I\alpha$$

Cylinder C: $\qquad RT = I\alpha$

where R is the moment arm for T.

The moment of inertia of the cylinder about its center is

$$I = \tfrac{1}{2} M_C R^2$$

Because the rope does not slip on the cylinder, the tangential acceleration of a point on its rim has the same acceleration magnitude a as the cord and the block. Therefore

$$\alpha = \frac{a}{R}$$

Substituting the relations for I and α into the torque equation above yields

$$RT = \tfrac{1}{2} M_C R^2 \frac{a}{R}$$

$$T = \tfrac{1}{2} M_C a \qquad (12\text{-}17)$$

Note that the cylinder radius cancels out.

Equations (12-16) and (12-17) can now be solved simultaneously for a and T. After very simple algebra, we find

$$a = \frac{M_B}{M_B + \frac{1}{2}M_C} g = \frac{g}{1 + M_C/2M_B}$$

(b) The tension T is then

$$T = \frac{\frac{1}{2}M_B M_C}{M_B + \frac{1}{2}M_C} g = \frac{M_B g}{1 + 2M_B/M_C}$$

Note that $a < g$ and $T < M_B g$, where $M_B g$ is the weight of the block.

Example 12-3

A rope is wrapped around the circumference of a uniform circular disk of radius R and mass M. The other end of the rope is attached to the ceiling, and the disk descends with the rope unwinding from it. See Fig. 12-5. What is (a) the acceleration of the cylinder's center and (b) the tension in the cord?

The forces on the disk are its weight $M\mathbf{g}$ and the upward rope tension $\mathbf{T}$. Choosing the downward direction as positive, we have for Newton's second law applied to the translational motion:

$$\Sigma \mathbf{F}_{\text{ext}} = M\mathbf{A}$$

$$Mg - T = Ma$$

where $a = A$ is the acceleration of the disk's center of mass, located along its symmetry axis.

We may apply Newton's second law for rotational motion to the *accelerating* disk if the rotation axis passes through the object's center of mass, as it does in Fig. 12-5.

$$\Sigma \tau_{\text{ext}} = I\alpha$$

where the cylindrical disk's moment of inertia relative to its symmetry axis is $I = \frac{1}{2}MR^2$, and α is the angular acceleration of the spinning cylinder. Now the acceleration of the rope relative to the cylinder's center at the point where it leaves the cylinder is $a = R\alpha$, which is the same quantity, a, in magnitude as the acceleration of the disk's center relative to the ceiling. The only torque relative to the center arises from the tension, and it has the magnitude RT. Substituting these quantities into the equation above, we have

$$RT = \frac{1}{2}MR^2 \frac{a}{R}$$

or

$$T = \frac{1}{2}Ma$$

But from above we had

$$Mg - T = Ma$$

so that

$$a = \tfrac{2}{3}g \quad \text{and} \quad T = \tfrac{1}{3}Mg$$

Note that the disk's radius R cancels out.

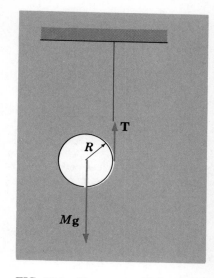

FIG. 12-5. Forces on a disk about which a rope has been wrapped, with one end attached to a ceiling.

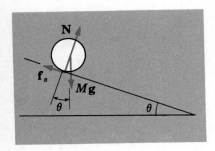

FIG. 12-6. The forces acting upon a cylinder rolling *up* an incline. Note that the force of static friction f_s (*up* the plane) is in the *same* direction as the velocity of the cylinder's center of mass.

Example 12-4

A cylinder of mass M and radius R is projected up an inclined plane of angle θ. The cylinder rolls without slipping. The initial speed of its center of mass is v_0. What is the acceleration of the cylinder's center of mass?

In Fig. 12-6 are shown the forces acting on the cylinder: its weight $M\mathbf{g}$, the normal force $\mathbf{N}$, and the force of friction $\mathbf{f}_s$. The tails of the force vectors are placed at the points where the forces are applied to the cylinder; the computation of the torques necessitates this. Notice that the *static* friction force is *up* the incline, in the *same* direction as the cylinder's motion.

The cylinder has a translational acceleration down the incline and an angular acceleration about its center of mass. Choosing up-the-incline as the positive direction for translation, counterclockwise as the corresponding positive sense for rotation, and applying Newton's second law to the translational motion gives

$$\Sigma \mathbf{F}_{\text{ext}} = M\mathbf{a}_{\text{CM}}$$

Along the incline: $\quad f_s - Mg\sin\theta = Ma_{\text{CM}}$

Perpendicular to the incline: $\quad N - Mg\cos\theta = 0$

Because the cylinder rolls but does not slip, it is always at rest momentarily at its point of contact with the plane, and a *static* force of friction acts on the cylinder where

$$f_s \leq \mu_s N$$

In considering the rotational motion, the torques, moment of inertia, and angular acceleration must all be determined relative to an axis of rotation passing through the center of mass. The forces $\mathbf{N}$ and $M\mathbf{g}$ both pass through the axis, yielding no torques; the resultant torque on the cylinder is produced solely by the friction force $\mathbf{f}_s$ having a moment arm R. Therefore, Newton's second law for rotation yields

$$\Sigma \tau_{\text{ext}} = I_{\text{CM}} \alpha_{\text{CM}}$$

$$-Rf_s = I_{\text{CM}} \alpha_{\text{CM}}$$

But $\quad I_{\text{CM}} = \tfrac{1}{2} MR^2$

and $\quad \alpha_{\text{CM}} = \dfrac{a_{\text{CM}}}{R}$

The last relation requires justification. First we note that the acceleration a_{CM} of the cylinder's center relative to its instantaneous point of contact with the plane is equal in magnitude to the acceleration of the point of contact relative to the cylinder's center. Therefore, if we view the motion from a reference frame moving with the cylinder's center, and if the cylinder rolls without slipping, then the point of contact has the *same* tangential acceleration as the cylinder's rim, and $a_{\text{CM}} = R\alpha_{\text{CM}}$.

Using the relations for I_{CM} and α_{CM} in the equation above, we obtain

$$-Rf_s = \tfrac{1}{2}MR^2 \frac{a_{\text{CM}}}{R}$$

$$-f_s = \tfrac{1}{2}Ma_{\text{CM}}$$

Substituting this result into the first equation gives

$$-\tfrac{1}{2}Ma_{\text{CM}} - Mg\sin\theta = Ma_{\text{CM}}$$

$$a_{\text{CM}} = -\tfrac{2}{3}g\sin\theta$$

The acceleration does *not* depend on the cylinder's mass or radius. The center of mass of *all* uniform solid cylinders, whatever their masses or radii, roll up (or down) an incline of a given angle with the *same* linear acceleration. If, however, solid uniform cylinders were replaced by cylindrical shells, the corresponding linear acceleration magnitudes would be found to be $a_{\text{CM}} = \tfrac{1}{2}g\sin\theta$ and the result for solid spheres would be $a_{\text{CM}} = \tfrac{5}{7}g\sin\theta$. The acceleration depends on the character of the mass distribution, or shape, but not on the mass or dimensions.

12–3 WORK AND KINETIC ENERGY IN ROTATIONAL MOTION

We first find an expression giving the work done by the torque $\boldsymbol{\tau}$ of an external force $\mathbf{F}$ when a rigid body rotates an angle $d\theta$ about a fixed axis. See Fig. 12-7. The force is applied at a distance r from the axis of rotation. We resolve the force into the components $\mathbf{F}_\parallel$ and $\mathbf{F}_\perp$ which are, respectively, parallel and perpendicular to the radius vector $\mathbf{r}$. At the point of application of the force, the object is displaced through a circular arc of length ds, where $ds = r\,d\theta$. Inasmuch as $\mathbf{F}_\parallel$ is always perpendicular to the displacement $d\mathbf{s}$, the force component $\mathbf{F}_\perp$ alone does work:

$$dW = \mathbf{F}\cdot d\mathbf{s} = F_\perp\,ds = F_\perp r\,d\theta$$

But $F_\perp r$ is the external torque τ of the force; therefore,

$$dW = \tau_{\text{ext}}\,d\theta \qquad (12\text{-}18)$$

If several external torques act simultaneously on the body, then τ_{ext} represents their sum. The work done by the resultant external torque in producing an angular displacement $d\theta$ is $\tau_{\text{ext}}\,d\theta$, or $\boldsymbol{\tau}_{\text{ext}}\cdot d\boldsymbol{\theta}$. The work is positive when the sense of rotation of the torque and the angular displacement are alike. Then energy is transferred *to* the object, and its rotational kinetic energy *increases*. On the other hand, the work is negative when a braking torque acts.

We find the power, the rate of doing work, by taking the time derivative of (12-18):

$$\frac{dW}{dt} = \tau_{\text{ext}}\frac{d\theta}{dt}$$

$$P = \tau_{\text{ext}}\omega = \boldsymbol{\tau}_{\text{ext}}\cdot\boldsymbol{\omega} \qquad (12\text{-}19)$$

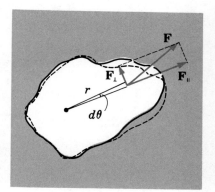

FIG. 12-7. A rigid body undergoing an angular displacement $d\theta$ under the influence of force F applied at r. The perpendicular and parallel components of F are $F_\perp$ and $F_\parallel$, respectively.

From the work-energy theorem, the work done on any object by external forces is equal to the change in the object's kinetic energy:

$$W_{i \to f} = K_f - K_i$$

Since the kinetic energy K of a body having a moment of inertia I and an angular speed ω about a fixed axis is

$$K = \tfrac{1}{2}I\omega^2 \tag{11-16}$$

the work-energy theorem yields

$$W_{i \to f} = \int_i^f \tau_{\text{ext}}\, d\theta = (\tfrac{1}{2}I\omega_f^2 - \tfrac{1}{2}I\omega_i^2) \tag{12-20}$$

The most general motion of a rigid body is that in which it undergoes simultaneously a translational motion and a rotational motion. The total kinetic energy K can always be expressed as the sum of two contributions:†

$$K = \tfrac{1}{2}Mv_{\text{CM}}^2 + \tfrac{1}{2}I_{\text{CM}}\omega_{\text{CM}}^2 \tag{12-21}$$

The first term represents the translational motion of the center of mass, $\tfrac{1}{2}Mv_{\text{CM}}^2$, where M is the total mass and v_{CM} is the speed of the center of mass. The second term represents rotational motion about the center of mass, $\tfrac{1}{2}I_{\text{CM}}\omega_{\text{CM}}^2$, where I_{CM} is the moment of inertia about an axis passing through the center of mass and ω_{CM} is the angular speed of the object about this axis.

Using Eq. (12-11) the rotational term becomes

$$\tfrac{1}{2}I_{\text{CM}}\omega_{\text{CM}}^2 = \tfrac{1}{2}(\Sigma m_i r_i^2)\omega_{\text{CM}}^2 = \Sigma(\tfrac{1}{2}m_i v_i^2)$$

where $v_i = r_i \omega_{\text{CM}}$ is the tangential speed of m_i relative to the center of mass. This leads to an alternate form of (12-21):

$$K = \tfrac{1}{2}Mv_{\text{CM}}^2 + \Sigma(\tfrac{1}{2}m_i v_i^2) \tag{12-22}$$

You have probably perceived the parallelism between the dynamical quantities in rotation and the corresponding quantities in translational dynamics. Table 12-1 lists the corresponding quantities.

The advantages of using energy considerations to solve problems in rotational dynamics are the same as for single particles: One deals with *scalar,* rather than vector, quantities; only the *end points,* not the detailed intermediate states, of some motion need be considered, since the total energy of an isolated system with conservative forces is constant. It should not be thought that the kinetic energy $\tfrac{1}{2}I\omega^2$ of a rotating object is somehow different from kinetic energy expressed as $\tfrac{1}{2}mv^2$; rather, $\tfrac{1}{2}I\omega^2$ is merely an economical way of expressing $\Sigma \tfrac{1}{2}m v_i^2$ for rotation. Indeed, there is

---

†A detailed proof of (12-21) and (12-22) is given in R. T. Weidner and R. L. Sells, *Elementary Classical Physics,* 2d ed., sec. 12-7, Allyn and Bacon, 1973.

Table 12–1

	Translation		Rotation	
Linear momentum	$\mathbf{p} = m\mathbf{v}$	Angular momentum	$\mathbf{L} = \mathbf{r} \times m\mathbf{v}$
Force	$\mathbf{F}$	Torque	$\boldsymbol{\tau} = \mathbf{r} \times \mathbf{F}$
Mass	m	Moment of inertia	$I = \Sigma m_i r_i^2$
Newton's law	$\Sigma \mathbf{F}_{\text{ext}} = d\mathbf{P}/dt$ or $\Sigma \mathbf{F}_{\text{ext}} = M\mathbf{A}$	Newton's law	$\Sigma \boldsymbol{\tau}_{\text{ext}} = d\mathbf{L}/dt$ or $\Sigma \boldsymbol{\tau}_{\text{ext}} = I\boldsymbol{\alpha}$
Work	$\int \mathbf{F} \cdot d\mathbf{r}$	Work	$\int \boldsymbol{\tau} \cdot d\boldsymbol{\theta}$
Kinetic energy	$\tfrac{1}{2}mv^2$	Kinetic energy	$\tfrac{1}{2}I\omega^2$
Power	$\mathbf{F} \cdot \mathbf{v}$	Power	$\boldsymbol{\tau} \cdot \boldsymbol{\omega}$

really no new physics in this chapter; it is merely an application of the basic angular momentum and torque relations developed in the last chapter, which are themselves merely the consequences of Newton's laws and the recognition of central forces.

Example 12-5

Block B of mass M_B is attached to a cord which is, in turn, wrapped around a uniform right circular cylinder of radius R. The cylinder is free to turn on an axle through its symmetry axis. The block is released from rest a distance y above the floor. As the block descends, the cord unwinds and the cylinder spins counterclockwise. (*a*) With what speed does the block hit the floor? (*b*) What is the angular velocity of rotation of the cylinder at that instant?

(*a*) The situation here is that of Example 12-2 and Fig. 12-4. It is easy to find the speed by applying the energy-conservation principle. When the block has descended through the distance y, its gravitational potential energy changes by $M_B g y$. In the process the block acquires a translational kinetic energy $\tfrac{1}{2}M_B v_B^2$ and the cylinder a rotational kinetic energy $\tfrac{1}{2}I_0 \omega^2$, where I_0 is the moment of inertia of the cylinder about its symmetry axis. Energy conservation requires that

$$M_B g y = \tfrac{1}{2}M_B v_B^2 + \tfrac{1}{2}I_0 \omega^2$$

where

$$I_0 = \tfrac{1}{2}M_B R^2$$

A point on the rim of the cylinder moves at the same speed as the cord and the block. Therefore the angular speed of the cylinder is

$$\omega = \frac{v_B}{R}$$

The first equation then becomes

$$M_B g y = \tfrac{1}{2}M_B v_B^2 + \tfrac{1}{2}(\tfrac{1}{2}M_C R^2)\left(\frac{v_B}{R}\right)^2$$

$$v_B = \sqrt{\frac{M_B}{\frac{1}{2}M_B + \frac{1}{4}M_C}\, gy}$$

Note that v_B is independent of the cylinder radius R. The result obtained here is consistent with that contained in Example 12-2.

(b) From above,

$$\omega = v_B/R = \sqrt{\frac{M_B}{\frac{1}{2}M_B + \frac{1}{4}M_C}\frac{gy}{R^2}}$$

12-4 CENTER OF GRAVITY

The *center of gravity* of a rigid body is defined as that point about which the resultant of the gravitational torques on all the particles comprising the rigid body is zero. It is easy to show that the center of gravity has the same location as the center of mass, provided (as is typically the case) that the body is in a *uniform* gravitational field—that is, that the acceleration due to gravity **g** is the same throughout the body.

Consider the simple situation with a uniform beam suspended at its center of mass. We compute the torques acting on two segments of the beam, each of mass m, located at equal distances from the center of mass. See Fig. 12-8. The weight of each segment is $m\mathbf{g}$, and with respect to an axis of rotation passing through the center of mass, the moment arms of the two torques are equal. One torque is clockwise, the other counterclockwise: Their sum is zero. The gravitational torques occur in equal but opposite pairs for other segments symmetrically located with respect to the center of mass. Thus, the resultant of the gravitational torques, relative to an axis of rotation through the center of mass, is zero. The center of gravity coincides with the center of mass. For any object that is small compared with the radius of the earth, **g** has the same magnitude and direction at all points within the object; then the center of gravity and center of mass are coincident, both being fixed relative to the body.

The existence of the center of gravity, fixed relative to the body, leads to an important consequence. We may imagine that all of the mass of an extended rigid body is concentrated at the center of gravity and that a *single* force equal to the weight of the body acts downward at this point. A rigid body under the influence of gravity alone may be put into equilibrium by a single upward force equal to the weight and applied so as to pass through the center of gravity. Then, the resultant force and resultant torque are both zero, and the object is in both translational and rotational equilibrium.

If a rigid body is suspended from a cord, the center of gravity must lie along a vertical line passing through the point of support.

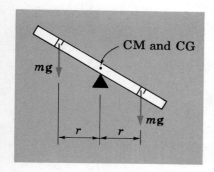

FIG. 12-8. The forces acting on two symmetrically located mass elements of a thin rod rotating about a horizontal axis through the center of mass. The resultant gravitational torque of each pair of masses is zero.

Furthermore, if the center of gravity is *below* the point of support (Fig. 12-9a), the center of gravity will rise for any rotational displacement about this point of support. An object so displaced will be subject to a restoring torque, tending to lower the center of gravity. Such a situation corresponds to *stable equilibrium*. On the other hand, if a rigid object is supported by an upward external force equal to the weight, passing through the center of gravity, which now lies *above* the point of support (Fig. 12-9b), the equilibrium is unstable. Any rotational displacement about the point of support will *lower* the center of gravity, and the resultant torque on the body will tend to increase the rotational displacement. This is *unstable equilibrium*.

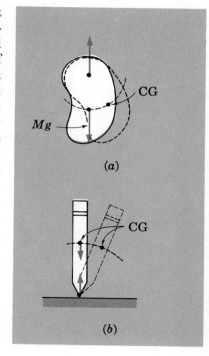

FIG. 12-9. (a) An example of stable equilibrium: An object's point of support is *above* the center of gravity (CG). (b) An example of unstable equilibrium: The objects' point of support is *below* the center of gravity.

Example 12-6

A 6-in pencil, initially standing on end, falls over (Fig. 12-9b). With what speed does the eraser strike the horizontal surface, assuming that the pencil point does not move?

This problem is best solved by energy-conservation methods. The pencil is a uniform rod of mass M and length L. The pencil's center of gravity is at its center (originally a vertical distance $L/2$ above the surface). The pencil's weight $M\mathbf{g}$ can be regarded as a single force applied at the center of gravity. In falling, gravitational potential energy decreases by $Mg(L/2)$, the pencil's center of gravity having descended $L/2$. Therefore, upon striking the surface, the increase in the rotational kinetic energy of the pencil about the fixed point equals the loss in potential energy:

$$\tfrac{1}{2}I\omega^2 = \frac{MgL}{2}$$

where I is the moment of inertia of the pencil and ω is the angular speed, both with respect to an axis of rotation passing through fixed point. If v represents the linear speed of the eraser upon striking, then

$$v = \omega L$$

The moment of inertia of a uniform rod about an axis at an end (Appendix VIII) is

$$I = \tfrac{1}{3}ML^2$$

Substituting the relations for I and ω in the energy equation above yields

$$\tfrac{1}{2}(\tfrac{1}{3}ML^2)\left(\frac{v}{L}\right)^2 = \frac{MgL}{2}$$

$$v = \sqrt{3gL} = \sqrt{3(32 \text{ ft/s}^2)(0.50 \text{ ft})} = 6.9 \text{ ft/s}$$

If the eraser had been detached and fallen freely from rest through the same vertical distance, its speed upon striking the surface would have been only $\sqrt{2gL} = 5.7$ ft/s!

12-5 EQUILIBRIUM OF A RIGID BODY

When the vector sum of the external forces on a body is zero, the linear acceleration of its center of mass is zero. The center of mass then maintains a constant linear velocity. When the vector sum of the external torques about the body's center of mass is zero, the angular acceleration is zero, and it maintains a constant angular velocity about its center of mass. When both these conditions are fulfilled, the object is said to be in equilibrium. If, furthermore, the linear and angular velocities are initially zero, the body remains at rest and is said to be in static equilibrium. Thus, equilibrium is but a special case of dynamics, and statics is a special case of equilibrium.

$$\Sigma \mathbf{F}_{ext} = 0 \quad \text{or} \quad \Sigma F_x = 0, \Sigma F_y = 0, \text{ and } \Sigma F_z = 0$$

and $\quad \Sigma \boldsymbol{\tau}_{ext} = 0 \quad$ or $\quad \Sigma \tau_x = 0, \Sigma \tau_y = 0, \text{ and } \Sigma \tau_z = 0$

where F_x, F_y, and F_z and τ_x, τ_y, and τ_z represent, respectively, the force and torque components along the X, Y, and Z axes.

We shall treat only those situations in which the forces lie entirely in the XY plane; the torques then lie entirely along the Z axis. Then equilibrium occurs only if

$$\Sigma F_x = 0$$
$$\Sigma F_y = 0 \tag{12-23}$$
$$\Sigma \tau_z = 0$$

Problems involving the equilibrium of rigid, or nearly rigid, bodies are of great importance in many branches of engineering. For example, when an engineer designs a bridge, he computes the forces acting on each member and chooses the materials and dimensions of the structural components such that these parts can withstand the forces.

If a rigid body is at rest, and therefore in static equilibrium, the resultant torque about an axis through the center of mass must clearly be zero. It is then not difficult to prove the following theorem: If a rigid body is in static equilibrium, the resultant external torque is zero *about any axis*. (See Prob. 12-29 for a proof.)

Example 12-7

A uniform beam, weighing 100 lb and making an angle of 30° with respect to the horizontal, is hinged at its lower end; a horizontal wire is attached to its upper end, as shown in Fig. 12-10. What is the tension in the wire and the direction and magnitude of the force of the hinge on the beam?

First we remark that, in a problem such as this, one can solve for only three unknowns. All forces lie in a single plane, and the equilibrium conditions lead to *three* scalar equations [Eqs. (12-23)].

The three external forces on the beam are shown in Fig. 12-10a. They are: the weight of the beam acting at the center of gravity, a distance $L/2$ from either end; the force of the hinge on the beam, replaced for convenience by its vertical and horizontal components $\mathbf{F}_v$ and $\mathbf{F}_h$; and the tension in the wire, $\mathbf{T}$. Translational equilibrium requires that

$$\Sigma F_x = 0$$
$$F_h - T = 0$$
$$\Sigma F_y = 0$$
$$F_v - 100 \text{ lb} = 0$$

A convenient choice of axis of rotation is one passing through the hinge. This choice is made, not primarily because the hinge represents a natural point of rotation, but rather because the forces $\mathbf{F}_v$ and $\mathbf{F}_h$ have lines of action through this point, thereby making their torques zero relative to this axis. For this axis, the 100-lb force produces a clockwise (negative) torque with moment arm $(L/2) \times \cos 30°$; the force $\mathbf{T}$ has a counterclockwise (positive) torque with moment arm $L \sin 30°$. Rotational equilibrium then requires that

$$\Sigma \tau_z = 0$$
$$-\frac{L}{2} \cos 30°(100 \text{ lb}) + L \sin 30°(T) = 0$$

Solving simultaneously for F_h, F_v, and T in the equations given above yields

$$F_h = 86.6 \text{ lb} \qquad F_v = 100 \text{ lb} \qquad T = 86.6 \text{ lb}$$

Notice that the length L of the beam does *not* enter.

We can readily compute the magnitude F and direction θ with respect to the horizontal of the force of the hinge on the beam:

$$F = \sqrt{F_h^2 + F_v^2} = \sqrt{86.6^2 + 100^2} \text{ lb} = 132 \text{ lb}$$
$$\tan \theta = \frac{F_v}{F_h} = \frac{100}{86.6} = 1.15$$
$$\theta = 49°$$

The force of the hinge on the beam is *not* along the length of the beam. This is also obvious from Fig. 12-10b, where it is seen that the three forces on the beam are concurrent, their lines of action intersecting at a single point. Indeed, this result obtains whenever an object subject to *three* forces is in equilibrium. Clearly, we can choose an axis of rotation passing through the point of intersection of the lines of action of two forces; the torques of these forces are then zero about this axis. But if the resultant torque is to be zero, the torque of the third force must be zero relative to the chosen axis, and its line of action must also pass through the intersection point.

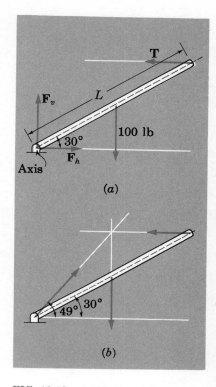

FIG. 12-10. (a) External forces acting on a uniform beam. The force on the hinge at the left end is replaced by its vertical and horizontal components F_v and F_h. (b) The three forces on the beam interact at a common point. Note that the force on the beam's left end is *not* along the direction of the beam.

SUMMARY

The motion of a rigid body is governed by the external forces and external torques acting upon it:

$$\Sigma \mathbf{F}_{ext} = M\mathbf{A} \qquad (12\text{-}13)$$

$$\Sigma \tau_{ext} = I\alpha \qquad (12\text{-}15)$$

where M is the object's total mass, $\mathbf{A}$ the acceleration of its center of mass, I its moment of inertia relative to the axis chosen for computing torques, and $\alpha = d\omega/dt$ its angular acceleration.

The work dW done by a torque τ turning an object through an angular displacement $d\theta$ is

$$dW = \tau\, d\theta \qquad (12\text{-}18)$$

The total kinetic energy of any system of particles may be written as

$$K = \tfrac{1}{2} M v_{CM}^2 + \Sigma \tfrac{1}{2} m_i v_i^2 \qquad (12\text{-}22)$$

where m_i is the mass of particle i, v_i its speed relative to the system's center of mass, M the mass of the entire system, and v_{CM} the system's center-of-mass speed. In effect, the total kinetic energy consists of the particles' kinetic energy relative to the center of mass plus the kinetic energy of the center of mass relative to the observer. For a rigid body this relation reduces to

$$K = \tfrac{1}{2} M v_{CM}^2 + \tfrac{1}{2} I \omega^2 \qquad (12\text{-}21)$$

where I and ω are computed relative to the center of mass.

The center of gravity is that point about which the resultant gravitational torque is zero; in a uniform gravitational field it coincides with the center of mass.

A rigid body is in translational and rotational equilibrium with the resultant force on it is zero and the resultant torque about any axis is zero.

PROBLEMS

12-1. In 12.0 s an automobile engine accelerates from an idling speed of 600 rev/min to a final speed of 6000 rev/min. (a) What is the angular acceleration, assuming that it is constant? (b) How many revolutions does the engine make in this time interval?

12-2. A flywheel, initially spinning at 3000 rev/min, slows to 2400 rev/min in 5.0 min. (a) Find its angular acceleration in rad/s². (b) How many revolutions will it turn through before coming to rest?

12-3. The second hand of a clock has a radius of 7.00 cm. What are the magnitudes of (a) the linear velocity, (b) the linear acceleration, (c) the angular velocity, and (d) the angular acceleration, of the tip of the second hand?

12-4. A fan initially rotating clockwise at 180 rev/min is found rotating at 180 rev/min in the counterclockwise sense 10 s later. Assuming it to be constant, what is the magnitude of the fan's angular acceleration?

12-5. An ultracentrifuge of the type used in biochemistry spins a sample at 60,000 rev/min in a circle 6.5 cm in radius. The rotor is magnetically supported and spins in a vacuum (1.0×10^{-7} atmospheres). Friction is so small that—when "coasting"—the rotational speed is reduced by just 1 rev/s after 10 hours! *(a)* At operating speed what is the magnitude of the radial acceleration to which the sample is exposed? *(b)* What is the angular acceleration of the rotor once the driving motor is turned off? *(c)* How much time would it take for the rotor to come to a stop if this rate of deceleration remains constant?

12-6. Prof. J. W. Beams made measurements of the tensile strength of steel by employing rotating steel balls suspended in vacuum and supported by a magnetic field; the driving torque was supplied by a rotating electromagnetic field. In one such experiment a tiny sphere 0.30 mm in diameter reached a rotational speed of 48×10^6 rev/min? Find *(a)* the linear speed and *(b)* the radial acceleration (in units of g) of a point on the surface of the sphere which is 0.15 mm from the spin axis.

12-7. Show that the magnitude of the linear acceleration of a point moving in a circle of radius r with angular velocity ω and angular acceleration α is given by $a = r(\omega^4 + \alpha^2)^{1/2}$.

12-8. A truck with wheels 120 cm in diameter accelerates from rest at the constant rate of 2.0 m/s² without slipping. *(a)* What is the angular acceleration of a wheel relative to its axle? *(b)* After 3.0 s what is the speed relative to the ground of the point on a wheel in contact with the ground? *(c)* What is the speed relative to the ground of the highest point on a wheel?

12-9. The *mean solar day* is the time required for the earth to make one complete rotation relative to the sun (that is, the time interval elapsing between two successive appearances of the sun overhead at the same longitude). The *mean sidereal day* is the time required for the earth to make one complete rotation relative to the fixed stars. The solar day is 24 hr long. Show that the approximate length of the sidereal day is 23 hr, 56 min, 4 s.

12-10. *(a)* Show that the moment of inertia about the center of symmetry of a uniform right circular cylindrical shell of inner and outer radii R_1 and R_2, respectively, and mass M is $(M/2)(R_1^2 + R_2^2)$. *(b)* Show that the moment of inertia about a diameter of a thin spherical shell of radius R and mass M is $\frac{2}{3}MR^2$. (*Hint:* Use the relations giving the moments of inertia of a solid cylinder and a solid sphere. The moment of inertia for a solid cylinder is developed in Appendix VIII and that for a sphere is quoted in Prob. 12-11.)

12-11. Verify that the moment of inertia of a uniform solid sphere of mass M and radius R about a diameter is $\frac{2}{5}MR^2$. (*Hint:* Imagine the sphere to be subdivided into circular disks, and sum the moments of inertia of these disks.)

12-12. What is the ratio of the earth's spin angular momentum to its orbital angular momentum measured relative to the sun? (Earth radius, 6.4×10^6 m; earth–sun distance, 1.5×10^{11} m.)

12-13. Find *(a)* the rotational kinetic energy of the earth and *(b)* the kinetic energy of the earth in its revolution about the sun. The earth's mass is 6.0×10^{24} kg; other relevant quantities are given in Prob. 12-12.

12-14. A meterstick is pivoted at one end. It is released from rest with the meterstick vertically above the pivot point. *(a)* At what angle relative to the vertical is the stick's angular acceleration a maximum? *(b)* What is the angular acceleration for this orientation?

12-15. Two noncollinear forces having equal magnitudes but opposite directions comprise a *couple*. Show that *(a)* the resultant force of a couple is zero and that *(b)* the magnitude of the resultant torque of a couple for *any* axis of rotation perpendicular to the plane of the two forces is Fd, where F is the magnitude of either force and d is the distance separating the lines of action of the two forces.

12-16. Show that any number of forces acting on a rigid body are equivalent to: a single force equal to their resultant, acting at the center of mass of the body, and a single couple causing rotation about the center of mass equal to the resultant torque of the forces about the center of mass. (See Prob. 12-15 for the definition of *couple*.)

12-17. A small object attached to a 1.0-m length of string swings in a vertical plane as a simple pendulum. The object is released from rest when the string is horizontal. What is the angular speed of the string *(a)* when the string is vertical and *(b)* when the string makes an angle of 30° with the vertical?

12-18. A meterstick with a mass of 100 g has one end on a rough floor and makes an angle of 45° with the horizontal. *(a)* What is the gravitational torque on the meterstick about a horizontal axis perpendicular to it and passing through the pivot? *(b)* What is the angular acceleration of the meterstick about this axis at the instant it is released from this position?

12-19. A man pulls with a constant force F on a rope wrapped around a cylinder with a fixed rotation axis. A length L of rope is withdrawn and the cylinder, initially at rest, then has an angular velocity ω. What is the cylinder's moment of inertia (in terms of F, L, and ω) relative to its rotation axis?

12-20. Consider the situation described in Example 12-4 and shown in Fig. 12-6. Find the distance d that the cylinder travels up the plane before coming (momentarily) to rest.

12-21. Consider the situation described in Example 12-4 and shown in Fig. 12-6. If the cylinder is to roll without slipping, show that $\mu_s \geq \frac{1}{3}\mu_k$, where μ_s and μ_k are, respectively, the coefficients of *s*tatic and *k*inetic friction.

12-22. A solid cylinder of mass M and radius R rolls without slipping on a rough horizontal surface and is accelerated to the right by a constant force to the right of magnitude F applied at the cylinder's symmetry axis. Show that the force of static friction on the cylinder is to the left and of magnitude $F/3$ and that the cylinder's center of mass has an acceleration of magnitude $\frac{2}{3}F/M$.

12-23. Show that *(a)* the linear acceleration of the center of mass of *all* right circular cylindrical shells rolling without slipping down an incline of angle θ is $\frac{1}{2}g \sin \theta$ and that *(b)* the acceleration of *all* rolling solid spheres of uniform density is $\frac{5}{7}g \sin \theta$. (Refer to Appendix VIII if necessary.)

12-24. Initially a 16-cm pencil stands vertically, with its point on a perfectly smooth horizontal surface. When the pencil is released, it tips over and its point slides along the surface. What is the separation distance between the point at which the pencil first rested and the point on the surface where the eraser strikes?

12-25. A solid cylinder, a solid sphere, and a circular loop—all with the same radius and mass—roll without slipping with the same kinetic energy. Which object has *(a)* highest, *(b)* intermediate, and *(c)* lowest speed for its center of mass? (If necessary, refer to Appendix VIII.)

12-26. What is the ratio of the kinetic energy of rotation about the center of mass to the kinetic energy of translational motion of the center of mass for any uniform cylinder which rolls without slipping?

12-27. A 60-kg solid cylinder of 50-cm radius is mounted with its fixed axis horizontal. *(a)* What is the cylinder's moment of inertia? *(b)* A rope wrapped tightly around the cylinder is unwound when a constant force of 6.0 N is applied to the rope; what is the torque of the rope on the cylinder? *(c)* What is the angular acceleration of the cylinder? *(d)* What is its angular speed after 5.0 s? *(e)* How much work is done when 1.0 m of rope is unwound from the cylinder? *(f)* What is the cylinder's final kinetic energy after 1.0 m of rope is unwound?

12-28. A 5.0-kg cubical box with edge length 20 cm has its center of gravity at its geometric center. It rests initially on a horizontal floor and then is tipped along an edge until its base makes an angle of 30° with the surface. Then the box is released from rest, and it rotates about the fixed edge in contact with the floor. *(a)* What is the initial angular acceleration of the box? *(b)* With what speed does the far edge strike the floor?

12-29. If an object is in equilibrium, then the sum of the torques on the object is zero, not merely relative to the object's center of mass, but relative to *any point* chosen for computing torques. Figure P12-29a shows a rigid body in equilibrium under the action of three coplanar forces ($\mathbf{F}_1$, $\mathbf{F}_2$, and $\mathbf{F}_3$), the points of application of these forces relative to an axis of rotation through the center of mass being designated by the radius vectors $\mathbf{r}_1$, $\mathbf{r}_2$, and $\mathbf{r}_3$. Since the body is in static equilibrium, the sum of the external forces is zero ($\mathbf{F}_1 + \mathbf{F}_2 + \mathbf{F}_3 = 0$), and the sum of the external torques about the center of mass CM is also zero [($\mathbf{r}_1 \times \mathbf{F}_1$) + ($\mathbf{r}_2 \times \mathbf{F}_2$) + ($\mathbf{r}_3 \times \mathbf{F}_3$) = 0]. Now consider a new axis through point P, where $\mathbf{d}$ is the radius vector from P to the center of mass. This is shown in Fig. P12-29b, where $\mathbf{R}_1$ is the radius vector from P to the point of application of $\mathbf{F}_1$: $\mathbf{R}_1 = \mathbf{r}_1 + \mathbf{d}$. Similar relations hold for $\mathbf{R}_2$ and $\mathbf{R}_3$. Substitute the equations for $\mathbf{r}_1$, $\mathbf{r}_2$, and $\mathbf{r}_3$ in terms of $\mathbf{R}_1$, $\mathbf{R}_2$, and $\mathbf{R}_3$ into the torque equations about the center-of-mass to show that the sum of the external torques about point P is zero also.

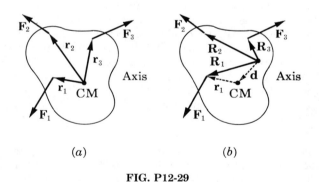

FIG. P12-29

12-30. The following are the force components and points of application (F_x, F_y; x, y) of two forces acting on a rigid body: (0 N, 10.0 N; 0 m, 2.0 m) and (3.0 N, −6.0 N; 8.0 m, 0 m). What are the force components and the X coordinate of the point of application of a third force, applied at $y = 0$, which places the rigid body in equilibrium for both translational and rotational motion?

12-31. The human forearm contains two major bones (the *radius* and the *ulna*) and articulates with the single large bone in the upper arm (the *humerus*) at the elbow joint. The *biceps* muscle is attached to the humerus and to the radius; when the biceps contracts, the angle between upper and low arms (at the elbow) is *decreased*. (For completeness, *extension* of the forearm is effected by contraction of the *triceps,* which is located in the upper arm to the rear of the bone — with the biceps in front.) In this problem we will investigate the forces involved in holding a load in the hand — with the forearm horizontal and the upper arm vertical. See Fig. P12-31, where the dimensions given are roughly correct for an adult. Assume that the forearm is horizontal and that its mass is 2.50 kg, and take the load to be 10.0 kg. The angle at the elbow is 90°, and the center line of the biceps is at 10° to the vertical. Assume $g = 9.80$ m/s². Find (a) T_B, the tension in the biceps, (b) F_H, the compressive force of the humerus on the forearm at the elbow, and (c) the angle between F_H and the vertical. (The forearm is a class III lever — characterized by the fact that *effort* force is applied at a point *between* the fulcrum (or pivot) and the load. In class III levers the effort force *always* exceeds the load — but the load travels farther and faster than does the point at which the effort force is applied. Most of the large bones in the body which are mobile are class III levers.)

12-32. A door 8.00 ft high and 4.00 ft wide weighs 90 lb. Two hinges, H_1 and H_2, support this weight; each supports half of the total. See Fig. P12-32. Find the horizontal and vertical components of the forces on the door by the hinges. Assume that the center of gravity of the door is at its geometrical center.

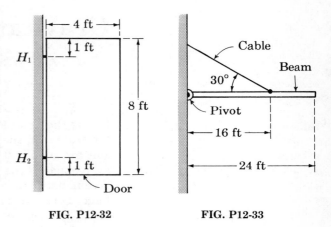

FIG. P12-32   FIG. P12-33

12-33. A uniform horizontal beam is attached to a vertical wall by a free pivot at one end; the beam is 24 ft long and weighs 1200 lb. The beam is supported by a cable attached to it at a point which is 16 ft from the wall, and the other end of the cable is fixed to the wall so that the cable makes an angle of 30° with the horizontal. The cable will break if the tension in it exceeds 2100 lb. See Fig. P12-33. (a) What are the direction and magnitude of the force on the beam at the pivot? (b) How far out along the beam can a 160-lb man walk before the cable breaks?

12-34. A uniform rod is placed such that its lower end rests on a rough floor and its upper end rests against a perfectly smooth wall. The angle between the rod and the floor is θ. Show that the minimum coefficient of static friction between the rod and the floor is $1/(2 \tan \theta)$.

12-35. A lawn roller of uniform density, radius R, and weight w is to be raised over a curb of height h, where $h < R$. The roller's handle is attached to an axle along the roller's axis of symmetry. (a) What are the magnitude and direction of the minimum force required to pull the roller over the curb? (b) How much work is done in raising the roller?

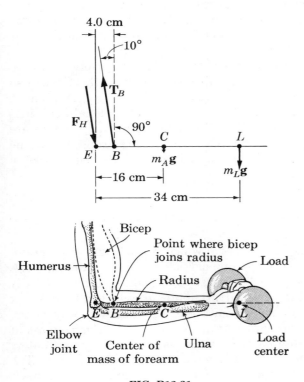

FIG. P12-31

CHAPTER 13

Gravitation

It was the genius of Isaac Newton to recognize that objects falling toward earth, the moon circling the earth, and planets orbiting the sun are all illustrations of a universal attraction acting between all particles—the force of gravity. Indeed, the gravitational interaction between any pair of particles, together with the electric force between electrically charged particles and the magnetic force between electrically charged particles in motion, are the only fundamental forces operating for separation distances above those at the atomic level.

In this chapter we shall see how the law of universal gravitation is arrived at and its direct demonstration in the Cavendish experiment. We discuss the acceleration due to gravity as a measure of the gravitational field and the gravitational field lines which represent this field. Kepler's laws of planetary motion are related to the gravitational force. The general relation for the gravitational potential energy is derived, and the energetics of satellite motion is treated.

13-1 THE LAW OF UNIVERSAL GRAVITATION

The intrinsic property of a particle which is the origin of its gravitational interaction with a second particle is its *gravitational mass*. In other words, two particles attract one another gravitationally because each has the physical attribute gravitational mass. It should be understood that, despite the fact that we speak of the gravitational *mass,* this property is not to be taken as equivalent to the particle's *inertial mass* m_i, which is the property that enters in momentum, $p = m_i v$, and in Newton's second law, $F = m_i a$. The term gravitational *charge* is used instead of gravitational mass to emphasize that it is quite distinct from the inertial mass. Then, just as two particles exert an electric force on one another because each has electric charge, we may say in parallel fashion that a gravitational force exists between two particles because each has a gravitational charge.

Gravitational mass is defined and measured through a simple procedure. Suppose that particles 1 and 2 with respective gravitational masses of m_{g1} and m_{g2} are brought, in turn, to the same location with respect to a third object, such as the earth. The gravitational forces on each, F_{g1} and F_{g2}, are measured; that is, the weights of the two objects are found to be F_{g1} and F_{g2}. Then, *by definition,* the ratio of respective gravitational masses (or gravitational charges) is equal to the ratio of the corresponding forces:

$$\frac{F_{g1}}{F_{g2}} = \frac{m_{g1}}{m_{g2}} \qquad (13\text{-}1)$$

Suppose now that particles 1 and 2 are dropped at the same location and their accelerations a_1 and a_2 are measured (as in the legendary experiment at the Tower of Pisa, or in the actual experiment performed on the moon during the Apollo 15 mission in 1971). It is found that both particles have the same acceleration. For particles at the earth's surface,

$$a_1 = a_2 = g$$

When we apply Newton's second law to the motion of each particle and recall that it is the *inertial* mass that must be multiplied by the acceleration to yield the resultant force, we have

$$F_{g1} = m_{i1} a_1 = m_{i1} g$$

and

$$F_{g2} = m_{i2} a_2 = m_{i2} g$$

In taking the ratio of the two equations above and using (13-1), we find that

$$\frac{m_{g1}}{m_{g2}} = \frac{m_{i1}}{m_{i2}}$$

By experiment the ratio of the gravitational masses of two particles is equal to the ratio of their inertial masses. This result, although simple, must be regarded as an extraordinary coincidence of nature, not predictable on theoretical grounds alone. After all, two particles with identical electric charges need not, and in general do not, have identical inertial masses; for example, the proton and positron have precisely equal electric charges, but their inertial masses differ by a factor of 1,836. Suffice it to say that a variety of experiments have established that an object's gravitational mass is proportional to its inertial mass, the most recent to within 1 part in 10^{11}.

Since any particle's gravitational mass is directly proportional to its inertial mass, we may use the standard 1-kg object not only as the basis for measuring inertial mass but also as the basis of gravitational mass. Then, a 1-kg object has an inertial mass of 1 kg and a gravitational mass also of 1 kg. For this reason we may drop the adjectives, inertial and gravitational, and refer hereafter simply to a particle's mass:

$$m_i = m_g \tag{13-2}$$

How does the gravitational force between a pair of particles depend on their respective masses? If particles 1 and 2 interact with one another through gravitation, then by the definition of gravitational mass we know that the force on particle 1 must be proportional to its mass m_1, and similarly for particle 2.

$$F_{2 \text{ on } 1} \propto m_1$$

and

$$F_{1 \text{ on } 2} \propto m_2$$

But we also know, from Newton's third law, that the magnitudes of these forces are the same: $F_{1 \text{ on } 2} = F_{2 \text{ on } 1}$. It follows, then, that the magnitude of the gravitational force on either 1 or 2 must be proportional to both m_1 and m_2.

$$F \propto m_1 m_2 \tag{13-3}$$

What is the direction of the gravitational force between two particles at rest in otherwise empty space? It *must* lie along the line connecting the two particles, since this is the *only line* defined by such a pair of particles. If the force did not lie along this line, it would mean that one could distinguish some one direction in empty space (different from that of the connecting line) from another, in contradiction of the assumed isotropic character of empty space. As we shall see, experiment confirms that the gravitational force is a *central force*. See Fig. 13-1.

Thus far we have established that the gravitational force between a pair of particles is central and directly proportional to the product of the respective masses. How does the magnitude of the force vary with the distance r between a pair of particles? Newton's analysis showed that $F \propto 1/r^2$, the force varying inversely as the square of the separation distance. Combining this result with that of (13-3), we may then write

$$F \propto \frac{m_1 m_2}{r^2}$$

or

$$F = \frac{G m_1 m_2}{r^2} \tag{13-4}$$

where G is a universal constant. By experiment the magnitude of the universal gravitational constant G is

$$G = 6.673 \times 10^{-11} \text{ N-m}^2/\text{kg}^2$$

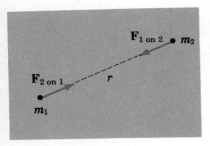

FIG. 13-1. Two point masses attract one another by a central force.

CHAP. 13 Gravitation

the units assigned to G being those which yield the force in newtons when the masses are in kilograms and the distance in meters.

It follows directly from (13-4) that two 1-kg particles separated by 1 m attract one another gravitationally with a force of a mere 6.7×10^{-11} N. The gravitational force is sizeable only if at least one of the two interacting objects has a very large mass—for example, the earth.

The gravitational force is the weakest of the fundamental forces in physics; yet it is the only fundamental force readily experienced. The attractive electric force between a proton and an electron is 10^{39} times larger than the attractive gravitational force between them! The reason why the much stronger electric force is not encountered in ordinary experience is simply that all large-scale objects ordinarily are electrically neutral; they contain equal amounts of positive and negative charge. The still stronger nuclear force, which acts between the particles in the nuclei of atoms, is never evident in ordinary experience because this force is effective only over a very short range (10^{-15} m).

One argument used by Newton to prove that the gravitational force is inverse square goes as follows. At the earth's surface the gravitational force on a particle of mass m is its weight mg, where $g = 9.8$ m/s². Now, if the gravitational force varies as $1/r^2$, so does the acceleration. Then, for example, at six times the distance from the earth's center to its surface, a particle's acceleration would be $(\frac{1}{6})^2 (9.8$ m/s²$)$, and at 60 times the earth's radius the earth's gravitational acceleration would be only $(\frac{1}{60})^2(9.8$ m/s²$) = 2.7 \times 10^{-3}$ m/s². To test whether the gravitational force and acceleration have been diluted by the factor $(\frac{1}{60})^2$ at this distance from earth requires that the acceleration of a test object at that location be measured. But the moon is 60 times the earth's radius away from the earth's center, and its acceleration toward the earth can be computed. Since the moon orbits the earth in a circle of radius r (3.8×10^8 m) and period T (27 days), its acceleration toward earth is

$$a = \omega^2 r = \left(\frac{2\pi}{T}\right)^2 r = \left(\frac{2\pi}{27 \text{ days}}\right)^2 (3.8 \times 10^8 \text{ m})$$

$$= 2.7 \times 10^{-3} \text{ m/s}^2$$

or exactly the magnitude required for an inverse-square force.

Newton's analysis, as given above, is based on an important assumption: The earth, certainly not a particle when it acts on a small object at its surface, is equivalent in its gravitational effects to a particle located at its center. Indeed, Newton invented the integral calculus to carry out the proof, given in detail in Problem 13-28, of the following assertions: If a *spherical shell* of mass M and radius R interacts with a particle of mass m *outside* the shell, the gravitational force between them is given by GmM/r^2, where r is the distance from the shell's *center* to mass m. On the other

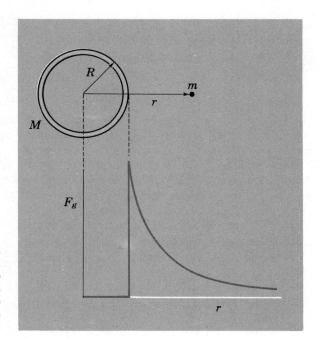

FIG. 13-2. Gravitational interaction between a spherical shell and a mass point. The gravitational force is inverse-square for the particle outside the shell and zero for the particle inside.

hand, if the point mass is located at any point *inside* the shell, the gravitational force on it is exactly *zero!* See Fig. 13-2. Insofar as gravitational influence goes, the shell is equivalent to a point mass M at its center when particle m is outside, and the shell is gravitationally nonexistent when the particle is inside. Inasmuch as a solid sphere consists simply of the wrapping of many shells of different radii, like an onion, in interacting with external bodies a solid sphere is equivalent in its gravitational effects to a particle of equal mass at its center. Thus, two nonoverlapping spheres interact as two particles, provided that the density of each depends only on the distance from the center.

Example 13-1

Compute the ratio of the sun's mass to the earth's mass, given only the radii of the moon's and the earth's orbits (3.84×10^8 m and 1.49×10^{11} m) and their respective periods (27.3 days and 365 days).

Let M_s, M_e, and M_m represent the respective masses of the sun, earth, and moon. Furthermore, let r_{es} represent the radius, and T_{es} the period, of the earth's orbit about the sun, and r_{me} and T_{me} the radius and period of the moon's orbit about the earth. Then we may write for the forces F_{es} on the earth due to the sun and F_{me} on the moon due to the earth:

$$F = ma = m\omega^2 r$$

$$F_{es} = \frac{GM_e M_s}{r_{es}^2} = M_e \frac{4\pi^2 r_{es}}{T_{es}^2}$$

$$F_{me} = \frac{GM_m M_e}{r_{me}^2} = M_m \frac{4\pi^2 r_{me}}{T_{me}^2}$$

Dividing the first equation by the second equation yields

$$\frac{M_s}{M_m}\left(\frac{r_{me}}{r_{es}}\right)^2 = \frac{M_e}{M_m}\frac{r_{es}}{r_{me}}\left(\frac{T_{me}}{T_{es}}\right)^2$$

Cancelling out the moon's mass M_m and rearranging terms gives

$$\frac{M_s}{M_e} = \left(\frac{T_{me}}{T_{es}}\right)^2 \left(\frac{r_{es}}{r_{me}}\right)^3 = \left(\frac{27.3 \text{ days}}{365 \text{ days}}\right)^2 \left(\frac{1.49 \times 10^{11} \text{ m}}{3.84 \times 10^8 \text{ m}}\right)^3$$

$$\frac{M_s}{M_e} = 3.3 \times 10^5$$

The sun is 330,000 times more massive than the earth. This calculation of the sun-earth mass ratio, which does not require a knowledge of the numerical value of the constant G, was first made by Newton.

13-2 THE CAVENDISH EXPERIMENT

Without knowing the magnitude of the universal gravitational constant G, Newton showed that an inverse-square universal force of gravity accounted for the motions of the planets about the sun and of the "falling" of the moon and an apple toward the earth. That two objects of ordinary size attract one another gravitationally was first shown in an historic experiment of extraordinary delicacy by Henry Cavendish in 1798, performed more than a hundred years after Newton had formulated the law of gravitation. The Cavendish experiment is significant, not only because it demonstrated directly that a gravitational force acts between two ordinary objects, but also because it enabled the numerical value of the constant G to be computed.

Cavendish secured two small spheres, each of mass m_1, to the opposite ends of a light rigid rod and suspended the rod at its midpoint by a light elastic (quartz) fiber to form a torsion balance as shown in Fig. 13-3. Two large lead spheres, each of mass m_2, were then placed near the masses m_1, the distance between m_1 and the nearby m_2 being r. In such an arrangement each lead sphere attracts the mass m_1 close to it, and a clockwise torque then acts on the rod (looking downward). (The gravitational attraction between one lead sphere and the more distant mass m_1 is ignorable: The separation distance is much greater, and the torque has a smaller lever arm.) The clockwise torque rotates the rod. As the rod rotates, the wire twists. This deformation produces a counterclockwise restoring torque which, for small angles of twist, is proportional to the twist angle θ. Equilibrium is reached when the gravitational torque equals (in magnitude) the restoring torque.

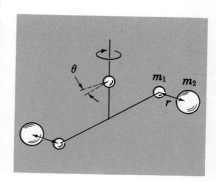

FIG. 13-3. Schematic representation of the Cavendish experiment.

By measuring the restoring torque (which is proportional to θ), one finds the attractive force between m_1 and m_2. This force, together with the masses m_1 and m_2, and their separation r, allows the constant G to be computed from Eq. (13-4).

The small angle θ, through which the rod is turned by the gravitational torque, is most readily measured by attaching a mirror to the fiber, shining a beam of light from a distant fixed light source on the mirror, and measuring the deflection of the light beam reflected from the mirror as the rod and mirror turn. The proportionality constant between the restoring torque and the angular displacement θ is determined by measuring the fiber's period of oscillation with the rod and attached spheres suspended from it. (This combination has a known moment of inertia.)

Example 13-2

Compute the earth's mass.

We can write the gravitational force F_g on an object of mass m at the surface of the earth in two equivalent ways:

$$F_g = mg$$

or

$$F_g = \frac{Gmm_e}{r_e^2}$$

where, in the second relation, the weight is attributed to the gravitational pull of the earth of mass m_e and radius r_e.

Equating the two expressions for the weight gives

$$mg = \frac{Gmm_e}{r_e^2}$$

or

$$m_e = \frac{gr_e^2}{G} = \frac{(9.8 \text{ m/s}^2)(6.4 \times 10^6 \text{ m})^2}{6.7 \times 10^{-11} \text{ N-m}^2/\text{kg}^2}$$

$$= 6.0 \times 10^{24} \text{ kg}$$

Cavendish is said to have measured the earth's mass in his famous experiment, inasmuch as a knowledge of the constant G permits the earth's mass to be calculated.

From a knowledge of the earth's mass and radius one can compute its average density ρ:

$$\rho = \frac{m_e}{\text{volume}} = \frac{m_e}{\frac{4}{3}\pi r_e^2}$$

$$= \frac{6.0 \times 10^{24} \text{ kg}}{\frac{4}{3}\pi (6.4 \times 10^6 \text{ m})^3} = 5.5 \text{ g/cm}^3$$

The mean density of the earth, 5.5 times that of water, exceeds that of material at the earth's surface. One must conclude, then, that the earth's interior consists largely of material of high density, most probably metals.

13-3 THE GRAVITATIONAL FIELD

Since we write the weight of an object as $m\mathbf{g}$, the acceleration due to gravity $\mathbf{g}$ gives the gravitational force per unit mass, or the *gravitational field,* for any object attracted by the earth:

$$\mathbf{g} = \frac{\mathbf{F}_g}{m}$$

The magnitude of $\mathbf{F}_g$ can be written

$$F_g = \frac{Gmm_e}{r^2}$$

where m_e is the earth's mass and r is the distance of the mass m from the earth's center. Therefore,

$$g = \frac{Gm_e}{r^2} \tag{13-5}$$

The acceleration due to gravity falls off inversely as the square of the distance from the center of the earth. At an altitude of 32 km, g is 1.0 percent less than its magnitude at the surface.

As we shall see in Chap. 14, the magnitude of $\mathbf{g}$ can be measured to high precision (1 part in 10^6) by using various types of pendula. Such measurements show that g differs at various points on the earth's surface. There are several causes: (1) the difference in altitude; (2) the ellipsoidal, rather than spherical, shape of the earth (because of the equatorial bulge, objects at low latitudes are farther from the earth's center than objects at the poles); (3) local deposits in the earth, which influence the value of g (and sometimes indicate the presence of oil or minerals); and (4) the rotation of the earth, which causes the effective g to differ from the true g. At 45° latitude and at sea level, g is close to the so-called standard acceleration, 9.80665 m/s².

The magnitude of $\mathbf{g}$ is governed by (13-5); its direction is always toward the earth's center. The gravitational field $\mathbf{g}$ may be portrayed by vectors, as shown in Fig. 13-4a. There is a still simpler way of mapping the gravitational effect of the earth on a body near it; see Fig. 13-4b. Here lines are drawn radially inward, indicating that the gravitational force is attractive. If one surrounds the mass toward which the lines point by an imaginary sphere, the number of *gravitational field lines* penetrating the imaginary spherical surface is the *same* for *any* radius. However, because a sphere's surface area varies directly as the square of its radius ($A = 4\pi r^2$), the number of field lines penetrating a *unit* area on the imaginary sphere falls off inversely as the square of the radius. Thus, radial lines, spaced uniformly in three dimensions, may be used to indicate the magnitude as well as the direction of $\mathbf{g}$. The magnitude of $\mathbf{g}$ is proportional to (or, by an appropriate choice

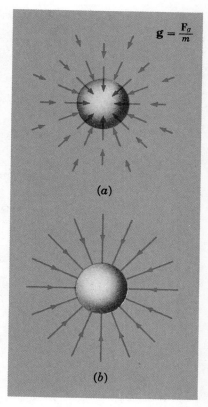

FIG. 13-4. (a) The gravitational intensity g represented by vectors. (b) The gravitational intensity represented by gravitational field lines.

of units, equal to) the number of gravitational field lines passing through a unit area oriented perpendicular to the lines.

A *uniform* gravitational field is one in which the field lines are uniformly spaced and parallel, **g** being constant in both magnitude and direction. For example, over a small volume near the surface of the earth, the gravitational field is approximately uniform.

We may describe the gravitational interaction between two masses through the field concept as follows: The first mass establishes a *gravitational field* in space. The second mass, finding itself in this gravitational field, is acted upon by a gravitational force. Thus, the interaction between particles 1 and 2 may be viewed as taking place in two steps: (1) Particle 1 establishes a gravitational field at all points in space, including the location of particle 2; and (2) particle 2 is subject to a force by virtue of its presence in this field. Conversely, particle 2 establishes a field which acts on particle 1.

The gravitational-field concept is useful as a means of visualizing, through gravitational field lines, the interaction between particles. But it has a much more fundamental significance when one inquires into the propagation of gravitational effects. For example, suppose that star 1 is somehow suddenly displaced away from star 2, situated a distance of 1 light-yr from it. Does star 2 *immediately* feel a weaker gravitational force on it? The answer, physicists believe, is no—it would take 1 yr, the time for light to travel from one star to the other, for the reduced gravitational force to reach and act upon the distant star. In other words, it is believed that gravitational fields propagate through space at the speed of light (exactly in analogy to the propagation of electromagnetic waves, also at the speed of light, and representing the interaction between electrically charged particles). Because gravitational effects are extremely feeble compared with electric effects, it requires a catastrophic astronomical event to generate a sizeable gravitational wave and an extremely sensitive device to detect it. Nevertheless, recent experiments have revealed some evidence, still not conclusive, for the existence of gravitational waves.

13-4 KEPLER'S LAWS OF PLANETARY MOTION

Within our solar system are a number of examples of relatively light objects moving in orbit about a massive object (see Table 13-1). The moon moves in a nearly circular orbit, and many artificial satellites travel in elliptical paths about the earth. Twelve moons encircle the planet Jupiter. The planets trace out elliptical orbits about the massive sun. It was the analysis of planetary motion that led Newton to the law of gravitation. He proved that the

Table 13-1

Body	Mass (earth masses)†	Mean diameter (kilometers)
Sun	333,400	1,391,000
Mercury	0.0549	5,140
Venus	0.8073	12,620
Earth	1.0000	12,760
Mars	0.1065	6,860
Jupiter	314.5	143,600
Saturn	94.07	120,600
Uranus	14.40	53,400
Neptune	16.72	49,700
Pluto	—	—
Moon	0.01228	3,476

†Mass of earth: 5.975×10^{24} kg.

astronomical observations were consistent with a universal attractive force, which varied inversely as the square of the distance between the interacting bodies and whose direction was along the line connecting them.

Observations of a light celestial body moving in the vicinity of a much more massive celestial body show that:

1. The path of the light body is a conic section—an ellipse, a parabola, or an hyperbola—depending on the injection speed. The massive body is at one focus.
2. The area swept out per unit time by the radius vector from the massive body to the light body is constant.
3. For elliptical orbits, the square of the period is proportional to the cube of the semimajor axis.

These empirical relations, which summarize the motions of planets in the solar system, were first recognized by Johannes Kepler (1571–1630); they are known as *Kepler's three laws of planetary motion*. Of course, Kepler's original formulation of these laws was restricted to the motion of the then known planets of the solar system. Newton's great contribution lay in proving that Kepler's laws were the logical consequences of a universal inverse-square gravitational force.

Kepler's First Law. Figure 13-5 shows an elliptical orbit with the force center at one focus. The semimajor axis is a, the semiminor axis b. At its closest and farthest distances from the sun, a planet is at the *perihelion* and *aphelion* positions, respectively; the corresponding terms for a satellite about the earth are *perigee* and *apogee* (*helios*, sun; *geo*, earth). In general, the gravitational force has a tangential component F_t along the path, which changes the magnitude of the planet's velocity but not its direction, and a radial

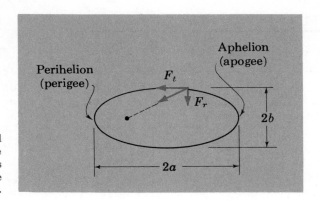

FIG. 13-5. Elliptical path traced out by a planet about a sun. The tangential and radial components of the gravitational force are F_t and F_r, respectively.

component F_r, which changes the direction of the planet's velocity but not its magnitude. Qualitatively, it is clear from Fig. 13-5 that, as the planet *recedes* from the force center, the gravitational force *slows* the planet and that, as the planet approaches the force center, the gravitational force has a tangential component *along* the velocity and the planet's speed *increases*.

For the special case of a circular orbit, closely approximated by the planets of the solar system (except Pluto) and by the moons, the two foci of the ellipse coalesce into the center of the circle. Then the force is always at right angles to the velocity ($F_t = 0$), the speed of the planet is constant, and $a = b = r$ (where r is the radius of the circle).

The gravitational force of one planet on the sun is precisely equal, by Newton's third law, to the force of the sun on the planet. Consequently, the force on the sun changes with time in precisely the same fashion as the force on the planet. If the planet moves in an ellipse, so does the sun. See Fig. 13-6. Indeed, both objects move in ellipses about their common center of mass, which remains at rest. Their mass ratio m_p/m_s is in the inverse ratio of their respective distances r_p and r_s from the center of mass: $m_p/m_s = r_s/r_p$. The total linear momentum of the system is, of course, zero: The planet and sun have momenta of equal magnitudes in opposite directions.

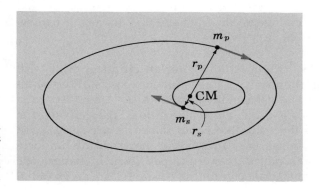

FIG. 13-6. Both planet and sun trace out elliptical paths about their center of mass. The total linear momentum of the system is zero.

CHAP. 13 Gravitation

If a sun is very massive compared with a planet, as in the case of the solar system, the center of mass is close to the center of the sun. The sun is then nearly at rest, and the total kinetic energy of the sun-planet system is equal to the kinetic energy of the planet alone. In addition, the total orbital angular momentum of the system is essentially that of the planet alone.

Kepler's Second Law. First recall a proof demonstrated in Sec. 11-4. When a particle is acted upon by a *central* force—one which lies along the line joining to a fixed force center—the particle's orbital angular momentum relative to the force center is constant. Moreover, dA/dt, the rate at which the area is swept out by the radius vector from the force center to the moving particle of mass m, is proportional to the particle's angular momentum L; that is,

$$\frac{dA}{dt} = \frac{L}{2m} \quad (11\text{-}12)$$

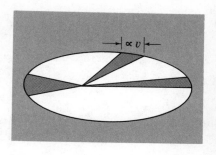

FIG. 13-7. Kepler's second law illustrated. The three shaded areas, representing the area swept out by the radius vector from the force center to the orbiting particle in equal times, are all equal.

Astronomical observations show that dA/dt is a constant for planets and comets about the sun, and for moons about a planet. This implies that the gravitational force is a *central* force, acting along the radius vector **r**. Figure 13-7 shows the area swept out by the radius vector from the sun to a light body for different locations relative to the sun. Note that the body's speed is greatest when it is closest to the force center, and conversely.

If v_p and v_a are the planet's speeds at the *p*erihelion and *a*phelion positions, respectively, and r_p and r_a are the corresponding distances from the sun, the conservation of angular momentum principle requires that

$$mr_p v_p = mr_a v_a$$

The direction of the angular momentum is perpendicular to the plane of the orbit and is derived from the right-hand rule.

Kepler's Third Law. It can be shown that the period T of a planet moving under the influence of an inverse-square force in an elliptical orbit is given by

$$T^2 = \frac{4\pi^2 a^3}{GM} \quad (13\text{-}6)$$

where a is the semimajor axis and M is the mass of the force center. For a given force center, the square of the period is proportional to the cube of the semimajor axis. This is Kepler's third law of planetary motion.

The periods and semimajor axes of the sun's planets (including Halley's comet), some of the satellites of the earth, and the several moons of Jupiter are given in Table 13-2. These data are portrayed graphically in Fig. 13-8.

Table 13–2

Satellites	Semimajor length a (m)	Period of revolution (s)
Planets of the sun		
Mercury	5.79×10^{10}	7.60×10^{6}
Venus	1.08×10^{11}	1.94×10^{7}
Earth	1.49×10^{11}	3.16×10^{7}
Mars	2.28×10^{11}	5.94×10^{7}
Jupiter	7.78×10^{11}	3.74×10^{8}
Saturn	1.43×10^{12}	9.30×10^{8}
Halley's comet	2.69×10^{12}	2.39×10^{9}
Uranus	2.87×10^{12}	2.66×10^{9}
Neptune	4.50×10^{12}	5.20×10^{9}
Pluto	5.91×10^{12}	7.82×10^{9}
Some moons of the earth		
Sputnik I	6.94×10^{6}	5.77×10^{3}
Explorer I	7.81×10^{6}	6.87×10^{3}
Telstar I	9.66×10^{6}	9.48×10^{3}
The Moon	3.84×10^{8}	2.36×10^{6}
Moons of Jupiter		
I (Io)	4.20×10^{8}	1.53×10^{5}
II (Europa)	6.70×10^{8}	3.06×10^{5}
III (Ganymede)	1.07×10^{9}	6.18×10^{5}
IV (Callisto)	1.88×10^{9}	1.44×10^{6}
V	1.81×10^{8}	4.30×10^{4}
VI	1.14×10^{10}	2.16×10^{7}
VII	1.17×10^{10}	2.24×10^{7}
VIII	2.35×10^{10}	6.38×10^{7}
IX	2.40×10^{10}	6.43×10^{7}
X	1.17×10^{10}	2.33×10^{7}
XI	2.25×10^{10}	6.35×10^{7}

We see from the figure that, for a given force center, astronomical observations are summarized by the following relation between the period and semimajor axis:

$$T^2 = Ka^3$$

where, by (13-6) the constant K is inversely proportional to the mass M of the force center. That is, $K = 4\pi^2/GM$.

It is easy to show that, for the special situation in which the elliptical orbit becomes a circle (and a becomes r), (13-6) is a necessary consequence of the inverse-square character of the gravitational force. A particle moving in uniform circular motion has a radial acceleration $\mathbf{a}_r$ of magnitude

$$a_r = \omega^2 r = \frac{4\pi^2 r}{T^2} \qquad (4\text{-}16)$$

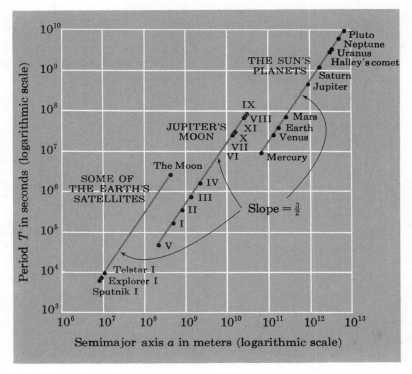

FIG. 13-8. Kepler's third law illustrated for three planetary systems: the sun and its planets, the earth and its satellites, the planet Jupiter and its moons. The logarithm of the period is plotted against the logarithm of the semimajor axis. All three lines have the same slope, $3/2$. Therefore, each system obeys the relation $\log T = 3/2 \log a + 1/2 \log K$, which is equivalent to $T^2 = Ka^3$, K being a different constant for each system.

where ω is the angular speed and T is the period of the motion. If the radial force arises from the gravitational attraction of the force center, then

$$\Sigma \mathbf{F} = m\mathbf{a}$$

$$\frac{GmM}{r^2} = \frac{m\,4\pi^2 r}{T^2}$$

or

$$T^2 = \frac{4\pi^2 r^3}{GM}$$

This is precisely (13-6) with a replaced by r. Whereas Kepler's *second* law requires only that the gravitational force be a *central* force, Kepler's *third* law requires, in addition, that $F_g \propto r^{-2}$.

13-5 GRAVITATIONAL POTENTIAL ENERGY

The gravitational force between two particles depends only on their separation distance, not on their velocities. It is, therefore, a conservative force, and one may associate with it a gravitational potential energy (see Sec. 10-3). Heretofore, we considered the special case in which the gravitational force (and therefore also the acceleration due to gravity) was constant. Then, the difference

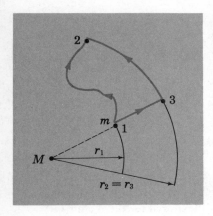

FIG. 13-9. The gravitational potential-energy difference is computed for mass m at distances r_1 and r_2 from the mass M.

in gravitational potential energy for a particle of mass m displaced between two points separated by a vertical distance Δy was $\Delta U_g = mg\,\Delta y$. We wish now to find the more general expression for gravitational potential energy, taking into account the fact that the gravitational force between two particles varies with their separation distance.

By definition, the potential-energy difference $U_f - U_i$ between the points i and f is given by

$$U_f - U_i = -\int_i^f \mathbf{F} \cdot d\mathbf{r} = -W_{i \to f} \qquad (10\text{-}4),(13\text{-}7)$$

Thus, the gravitational potential difference is equal to the negative of the work done by the gravitational force when a particle is moved from some *i*nitial to some *f*inal point along *any path*.

Consider two point masses m and M, where M remains fixed. Because the gravitational force $\mathbf{F}_g$ between them is conservative, the work done on mass m as it is moved from point 1 to point 2 (see Fig. 13-9) is independent of the path from 1 to 2.

Whether we take the mass m along a complicated path from 1 to 2, or along the path from 1 to 3 (along which $\mathbf{F}_g$ is parallel to the displacement) and then from 3 to 2 (along which $\mathbf{F}_g$ is at right angles to the displacement), the *total work done* is the *same*. Thus, we can find the gravitational potential-energy difference $U_2 - U_1$ by computing the work $W_{1 \to 3}$ done along the path 1 to 3 and the work $W_{3 \to 2}$ done along the path 3 to 2. From (13-7),

$$U_2 - U_1 = -W_{1 \to 2} = -(W_{1 \to 3} + W_{3 \to 2}) \qquad (13\text{-}8)$$

Along the circular arc from 3 to 2, the gravitational force $\mathbf{F}_g$ is always perpendicular to the displacement $d\mathbf{r}$; thus, the work done $W_{3 \to 2}$ in moving the mass m along this path is zero: $W_{3 \to 2} = 0$.

Along the radial path from 1 to 3 the gravitational force $\mathbf{F}_g$ is opposite in direction to the displacement $d\mathbf{r}$, the angle θ between them being $180°$. Therefore,

$$W_{1 \to 3} = \int_{r_1}^{r_3} \mathbf{F}_g \cdot d\mathbf{r} = \int_{r_1}^{r_3 = r_2} F_g \cos\theta\, dr$$

$$= \int_{r_1}^{r_2} \frac{GMm}{r^2}(-1)(dr) = GMm\left(\frac{1}{r_2} - \frac{1}{r_1}\right)$$

Substituting this result in (13-8) gives

$$U_2 - U_1 = -W_{1 \to 2} = GMm\left(\frac{1}{r_1} - \frac{1}{r_2}\right) \qquad (13\text{-}9)$$

It is customary to choose the gravitational potential energy to be zero at the point where the force is zero—namely, at $r = \infty$. Putting $U_1 = U_\infty = 0$ for $r_1 = \infty$, and dropping the subscript 2, we have

$$U_r = -\frac{GMm}{r} \qquad (13\text{-}10)$$

The negative sign in (13-10) appears because the zero for potential energy corresponds to $r = \infty$ and because the gravitational force is attractive. As two masses are brought closer together, their potential energy *decreases* (becomes more negative). We can see that the gravitational potential energy must be negative by considering two particles initially infinitely far apart and at rest. The particles' kinetic energy is zero, and their potential energy is also zero. So the system's total energy is zero, and it must, by energy conservation, remain zero. As the particles attract one another, their kinetic energy increases; therefore, the system's potential energy must decrease and become negative to ensure that the total energy remains zero.

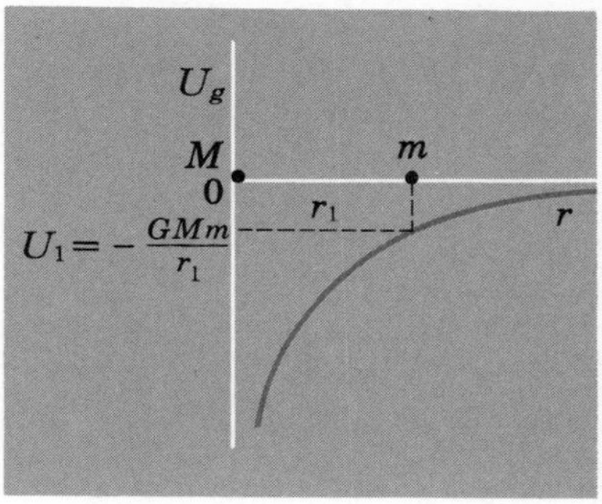

Figure 13-10 shows the gravitational potential energy between two point masses M and m as a function of their separation distance r.

Equation (13-10) expresses the potential energy of two *point* particles. It gives, as well, the potential energy between two spherical objects. Just as the gravitational *force* between two spherical shells can be computed by regarding the entire mass of each shell as concentrated at a point at its center, so too the gravitational *potential energy* between two nonoverlapping spherical shells is computed by assuming the masses of the shells to be concentrated at their respective centers.

The gravitational potential energy of a pair of particles m_1 and m_2 separated by r_{12} is $-Gm_1m_2/r_{12}$. What is the total potential energy for a system consisting of three or more particles? It is not difficult to show that it is merely the sum of the respective potential energies for each pair of particles. Thus, with particles m_1, m_2, and m_3,

$$U_g = U_{12} + U_{13} + U_{23} = -\frac{Gm_1m_2}{r_{12}} - \frac{Gm_1m_3}{r_{13}} - \frac{Gm_2m_3}{r_{23}} \quad (13\text{-}11)$$

where r_{13} is the distance between particles 1 and 3, and r_{23} that between 2 and 3.†

Example 13-3

An object of mass m close to the earth's surface is raised through a vertical height Δy. Show that, if Δy is small compared with the earth's radius r_e, the general relation for gravitational potential energy reduces to the familiar relation $\Delta U = mg\,\Delta y$.

From Eq. (13-9),

$$\Delta U_g = U_2 - U_1 = Gm_e m \left(\frac{1}{r_1} - \frac{1}{r_2}\right)$$

We choose $r_1 = r_e$ and $r_2 = r_e + \Delta y$. See Fig. 13-11. If $\Delta y \ll r_e$, we can write $-1/r_2$ as

FIG. 13-10. Gravitational potential energy U_g as a function of the separation distance r between point particles of masses m and M.

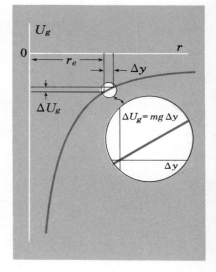

FIG. 13-11. For objects close to the earth, the gravitational potential-energy difference is given by $\Delta U_g = mg\,\Delta y$.

---

†For a proof of (13-11) see R. T. Weidner and R. L. Sells, *Elementary Classical Physics,* 2d ed., sec. 13-5, Allyn and Bacon, 1973.

$$-\frac{1}{r_2} = -\frac{1}{r_e + \Delta y} = -\frac{1}{r_e}\left[\frac{1}{1+\Delta y/r_e}\right]$$

$$= -\frac{1}{r_e}\left[1 - \frac{\Delta y}{r_e} + \left(\frac{\Delta y}{r_e}\right)^2 - \cdots\right] \approx -\frac{1}{r_e} + \frac{\Delta y}{r_e^2}$$

Then the equation above becomes

$$\Delta U_g = Gm_e m \left[\frac{1}{r_e} + \left(-\frac{1}{r_e} + \frac{\Delta y}{r_e^2}\right)\right] = m\frac{Gm_e}{r_e^2}\Delta y$$

But, by definition,

$$g = \frac{Gm_e}{r_e^2} \tag{13-5}$$

Therefore, $\quad\Delta U_g = mg\,\Delta y$

Example 13-4

Consider a rocket which carries a payload of mass m. We wish to launch the payload from the surface of a planet of mass M and radius R so that its speed is v after it has escaped from the planet's gravitational field. (*a*) What is the necessary launch speed? (*b*) What is the minimum launch speed needed to break free of the planet?

(*a*) As a simplifying assumption, consider the planet to be at rest. Then the total energy of the m-M system is conserved:

$$\tfrac{1}{2}mv_0^2 - \frac{GMm}{R} = \tfrac{1}{2}mv^2 - \frac{GMm}{D}$$

where v_0 is the launch speed and $D \to \infty$ so that $(GMm/D) \to 0$. The necessary launch speed is then

$$v_0 = \sqrt{\frac{2GM}{R} + v^2}$$

(*b*) The minimum launch speed corresponds to the case where $v = 0$:

$$(v_0)_{\min} = \sqrt{\frac{2GMm}{R}}$$

This minimum launch speed for escape is called the *escape velocity* v_e:

$$v_e = (v_0)_{\min}$$

If the payload is launched with speed v_e, the total energy is zero and the trajectory can be shown to be a parabola. In the more general case for escape ($v_e > 0$), the trajectory is an hyperbola. For a particle of *any* mass, the escape velocity from the earth's surface is approximately 11 km/s.

Example 13-5

A spacecraft is in circular orbit as a satellite about the earth close to its surface. By how much must its speed increase so as to be put into a new elliptical orbit about the earth with its greatest distance from the earth's surface equal to the earth's radius?

See Fig. 13-12. In the elliptical orbit, the earth's center is at one focus. The satellite's speed at the perigee position is v_p, and its distance from the earth's center is r_p, where $r_p = r_e$, the earth's radius. When at its greatest distance from earth in the apogee position, the satellite's speed is v_a and its distance from the earth's center is $r_a = 2r_e$.

Applying the energy-conservation principle at the perigee and apogee positions, we have for the system's total energy

$$K_p + U_p = K_a + U_a$$

$$\tfrac{1}{2}mv_p^2 - \frac{Gmm_e}{r_p} = \tfrac{1}{2}mv_a^2 - \frac{Gmm_e}{r_a}$$

where m_e is the earth's mass.

The satellite's angular momentum is also constant when measured relative to the force center, and applying angular-momentum conservation to the perigee and apogee positions (at both positions the radius vector and velocity vector are perpendicular) requires that

$$r_p m v_p = r_a m v_a$$

Eliminating v_a from the two equations above and solving for v_p yields:

$$v_p = \left[2\left(\frac{Gm_e}{r_p}\right)\left(\frac{r_a}{r_a + r_p}\right) \right]^{1/2}$$

But the speed v_c of a satellite in circular orbit near the earth's surface is $v_c = (Gm_e/r_e)^{1/2} = (Gm_e/r_p)^{1/2}$, and $r_a = 2r_p$. Substituting these values into the equation above gives $v_p = (2/\sqrt{3})v_c = 1.15 v_c$. The satellite's speed must increase by 15 percent.

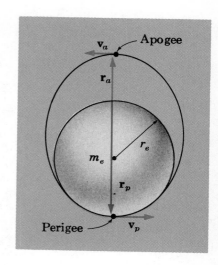

FIG. 13-12. A spacecraft initially in circular orbit near the earth's surface speeds up to go into an elliptical orbit.

Example 13-6

With what minimum speed must a particle of mass m be projected from the surface of a uniform sphere A if it is to reach the surface of an identical sphere B? Spheres A and B both have mass M and radius R; their centers are separated by the distance D. See Figure 13-13a.

The particle is more strongly attracted by A than by B in the first half of the trip, and more strongly attracted by B than by A in the second half of the trip. If projected with too low a speed, the particle travels outward from A, comes to rest, and then falls back to the starting point without reaching the midpoint. But if the particle is projected with a sufficiently large speed from A toward B so that it reaches the point midway between the two spheres—a location at which the two spheres attract it equally but in opposite directions—then the particle will gain speed in the second half of its trip and arrive at the surface of sphere B.

The same situation may be described in terms of energy: The particle loses kinetic energy and the system's potential energy increases as the particle moves from A to the midpoint, and conversely for the second portion of the trip. The critical launch speed $v_{\min}$ is that for which the particle's kinetic energy becomes zero at the midpoint. To find $v_{\min}$ most directly we apply energy conservation,

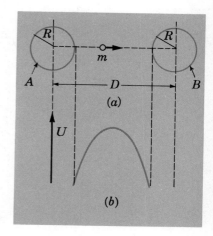

FIG. 13-13. (a) Spheres A and B, with a particle of mass m projected from the surface of A and intended to reach the surface of B. (b) The system's gravitational potential energy as a function of x, the distance from the center of A to the particle. Spheres A and B are assumed to be solid so that $U(x)$ is not plotted inside either sphere.

Gravitational Potential Energy SEC. 13-5

equating the system's total energy with the particle at the launch position to the total energy with the particle at the midpoint.

The total energy of the system for the particle at the surface of A may be written:

$$E = K + U$$

Particle on surface of A:
$$= \tfrac{1}{2}mv_{\min}^2 + \left(-\frac{GmM}{R}\right) + \left(-\frac{GmM}{D-R}\right)$$

where the second term on the right is the gravitational potential energy of particle m interacting with sphere A of mass M at a distance R from its center, while the third term is the potential energy between m and sphere B with the particle initially at a distance $D-R$ from this sphere's center.

With the particle at the midpoint and momentarily at rest, the system's total energy may be written:

Particle at midpoint:
$$E = K + U = 0 + \left(-\frac{GmM}{\tfrac{1}{2}D}\right) + \left(-\frac{GmM}{\tfrac{1}{2}D}\right)$$

At this location the particle is at the same distance $\tfrac{1}{2}D$ from both spheres. (We have ignored the gravitational potential energy of sphere A interacting with sphere B in both equations above, because this potential energy remains unchanged.)

Equating the two relations above for the total energy and solving for $v_{\min}$, we have

$$\tfrac{1}{2}mv_{\min}^2 - \frac{GmM}{R} - \frac{GmM}{D-R} = -\frac{4GmM}{D}$$

and finally, after some algebra,

$$v_{\min} = \sqrt{\frac{2GM}{R}\frac{1-(4R/D)(1-R/D)}{1-R/D}}$$

Note that the projection speed does not depend on the particle's mass m.

It is instructive to examine the curve, shown in Fig. 13-13b, in which the gravitational potential energy of particle m interacting with spheres A and B is plotted as a function of the location of the particle. We see that the potential energy rises to a maximum at the midpoint; the particle must have sufficient initial kinetic energy to surmount the "potential hill" if it is to reach the distant sphere. The same sort of potential hill arises, for example, when a spacecraft is to travel from the earth's surface to the moon. In this instance the highest point in the potential hill is much closer to the moon than to the earth (because of the large difference in earth and moon masses.) Furthermore, the height of the potential hill to be surmounted in an earth-to-moon trip is far greater than that for a moon-to-earth trip (again because of the large mass difference); for this reason a relatively large rocket must be used to launch a spacecraft from earth-to-moon, while a much more modest rocket suffices for the return trip.

SUMMARY

The universal central attractive force of gravitation between any two particles with masses m_1 and m_2 separated by a distance r has the magnitude

$$F_g = \frac{Gm_1 m_2}{r^2} \qquad (13\text{-}4)$$

where G is a constant. This relation gives, as well, the force between two spherical shells with centers separated by r. The gravitational force on a point mass in the interior of a spherical shell of mass is zero.

The gravitational field $\mathbf{g}$ is the gravitational force per unit gravitational mass, $\mathbf{F}_g/m$; it may be represented by gravitational field lines.

In their general form, Kepler's laws are written: (1) A particle under the influence of a central attractive inverse-square force moves in a path which is a conic section; (2) the radius vector from the force center to the particle sweeps out equal areas in equal times; (3) for elliptical orbits, the square of the period is proportional, for a given force center, to the cube of the ellipse's semimajor axis.

The gravitational potential energy between two masses M and m separated by r is

$$U_r = -\frac{GMm}{r} \qquad (13\text{-}10)$$

PROBLEMS

See Tables 13-1 and 13-2 for astronomical data.

13-1. Three particles, each of mass m, are placed at the vertices of an equilateral triangle of edge length d. What is the magnitude of the resultant gravitational force on any one mass?

13-2. At what separation distance would the gravitational force between two protons equal the weight of a proton at the surface of the earth? The mass of a proton is 1.7×10^{-27} kg and its nominal "radius" is of the order of 10^{-15} m.

13-3. According to the Bohr model of the hydrogen atom (1913), the atom can be represented as a single electron moving at 2.2×10^6 m/s in a circular orbit about the proton; the orbit radius is 5.3×10^{-11} m. (a) If the centripetal force were supplied by gravitation alone, what would be the proton mass? (Assume the accepted value of the gravitational constant.) (b) From the alternate point of view, what would be the gravitational "constant" if the proton mass were actually 1.67×10^{-27} kg (the accepted value)?

13-4. The period of a certain pendulum differs by 0.10 percent between the base and the peak of a mountain. The period of any pendulum varies inversely as the square root of g. What is the height of the mountain?

13-5. At what distance from the earth's center would a satellite in circular orbit above the earth's equator revolve around the earth with the same angular speed as that of the earth's spin about its axis and thereby appear to be stationary as viewed from earth?

13-6. Before Cavendish first determined the numerical value of G, the absolute masses of the sun and planets could not be calculated from the universal law of gravitation. One could, however, find the ratio of the masses of any two bodies, each of which has an orbiting satellite. As an example, calculate, without using the numerical value of G, the ratio of Jupiter's mass to the earth's mass from the following observations:

	Radius of orbit	Period
Earth-moon	2.4×10^5 mi	27 days
Jupiter-moon	4.2×10^5 mi	3.5 days

13-7. Once nuclear "burning" is no longer possible, a star such as the sun will evolve into a *white dwarf*—a highly condensed star with a normal stellar mass compressed within the dimensions typical of an earth-sized planet. The star Sirius is in fact two stars bound together by their mutual gravitational attraction. Sirius B, the less bright of the pair, is a white dwarf with a mass equal to that of the sun; however, its radius is only 2.2 percent that of the sun. Find *(a)* the mass density and *(b)* the acceleration of gravity at the surface of Sirius B.

13-8. When a massive star nears the end of its existence it explodes; such an exploding star is called a *supernova*. One result of this explosion is the creation of a *neutron star*. A typical neutron star has a radius of 10 km and a mass equal to that of the sun. Find *(a)* the average mass density and *(b)* the surface gravitational field strength—the value of g at the surface—of such a star. *(c)* Compare this result with the acceleration of gravity at the surface of the sun.

13-9. The magnitude of g is less at the equator than at the poles because of two factors: the earth's spin and the equatorial bulge. The distance from the center of the earth to the equator is 13.5 mi more than from the center to the poles. Assuming, for simplicity, that the earth is a perfect sphere of radius 3,960 mi, compute the fractional decrease in g for *(a)* a point 13.5 mi *above* the earth's surface and *(b)* a point 13.5 mi *below* the earth's surface (neglect the earth's spin).

13-10. Compute the period of a spacecraft's circular orbit close to the surface of the moon.

13-11. Particles with masses M and $3M$ are separated by a distance d. At what distance from the $3M$ mass is the gravitational field from the two masses zero?

13-12. Two objects of masses m and $2m$, respectively, are a distance d apart. *(a)* What must be the ratio of their speeds and the directions of their velocities if each mass is to rotate in a circle about the system's center of mass, which remains at rest? *(b)* What is the minimum energy required to separate completely these two masses?

13-13. Upon reentry into the earth's atmosphere, a satellite loses energy because of atmospheric drag. The drag force is initially small compared with the radial force on the satellite, and the satellite can be assumed to move in an approximately circular orbit. Show that upon reentry the satellite actually *speeds up*, rather than slows down.

13-14. A binary star system consists of two stars held together by mutual gravitational attraction and rotating about the center of mass of the system. Show that $M_{sys}T^2 = D^3$, where M_{sys} is the mass of the binary system in solar mass units, T is the orbit period in years, and D is the separation distance in astronomical units (AU). One astronomical unit equals the mean value of the earth-sun distance. (*Hint:* The period of the earth's orbit is one year. One solar mass unit $M_\odot$ equals the mass of the sun.)

13-15. Express the potential energy of an object of mass m at the surface of the earth in terms of the weight of the object and the earth's radius. (Use Eq. 13-10.)

13-16. Suppose that the moon were at rest at its present distance from the earth, rather than orbiting it. With what speed would it strike the earth? (Take the earth to be infinitely massive relative to the moon.)

13-17. A projectile is fired vertically upward from the earth's surface at a speed of 6.0 km/s. *(a)* What is its maximum distance from the earth's center? *(b)* In what direction with respect to the vertical would the projectile have to be fired to be placed in a stable orbit?

13-18. What is the escape velocity from the moon?

13-19. Two particles each of mass M are initially separated by a distance d. They are released from rest and approach one another through gravitational attraction. What is the speed of each particle when their separation distance is $\frac{1}{2}d$?

13-20. Consider an object of mass m which is initially at rest on the surface of the earth. If m is lifted vertically through a height h, the *approximate* change in the gravitational potential energy of the earth–object system is $U_{appr} = mgh$, where g is the acceleration of gravity at the surface. *(a)* Express U_{ex}, the *exact* value of the potential energy change, in terms of U_{appr} and earth radius R_e. *(b)* Use this result to show that the relative error in using the approximation is h/R_e.

13-21. Show that the escape velocity from the earth's surface is equal to or greater than 11 km/s.

13-22. What is the escape velocity from the surface of Jupiter?

13-23. A satellite is observed to orbit a planet close to the planet's surface with a period T. Show that the average density of the planet is $3\pi/GT^2$.

13-24. Find the minimum energy necessary to remove the earth from the gravitational influence of the sun.

13-25. Show that the kinetic energy required to launch an object in a circular orbit close to the earth's surface is just one-half of the kinetic energy required for the object to escape from the earth.

13-26. Halley's comet, which moves in a highly eccentric elliptical orbit about the sun, has been observed in the vicinity of the earth every 75.5 yr since the year 87 B.C. Its perihelion distance is 8.9×10^{10} m. (a) What is the greatest distance between Halley's comet and the sun? (b) What is the comet's maximum speed?

13-27. At what percentage of the distance from the earth's to the moon's center is the peak in the gravitational potential-energy "hill" of the earth and moon, as "seen" by some third object?

13-28. Proof that, when interacting with an external mass, a solid sphere of uniform mass density ρ is gravitationally equivalent to a *particle* of the same mass located at the geometrical center of the sphere: Since a solid sphere can be considered as being composed of a continuum of concentric spherical shells, it is sufficient to show that the gravitational potential energy between a thin, spherical shell of mass M_s and an external mass m is given by $U = -GmM_s/r$, where r is the distance between m and the center of the shell. Also, show that the net gravitational force between such a shell and a point mass interior to it is *zero*. Consider a thin, spherical shell of mass M_s, radius R, thickness ΔR, and uniform mass density ρ; the many mass elements dM composing the shell are at various distances s from the external mass m. See Fig. P13-28. According to Eq. 13-10, the gravitational potential energy between one mass element dM and the point mass m is given by $-Gm\,dM/s$, where s is the distance between m and dM. All points on the particular ring whose axis is the line between m and the center of the shell are at the same distance s from the mass m. The radius of this ring is $R \sin \theta$, its width is $R\,d\theta$, and its thickness is ΔR. (a) Show that the gravitational potential energy between this ring and m is given by $dU = -Gm[\rho(2\pi R^2 \Delta R) \sin \theta\, d\theta]/s$. (b) Using the law of cosines applied to the triangle with sides s, r, and R and yielding $s^2 = R^2 + r^2 - 2Rr \cos \theta$, show that $\sin \theta\, d\theta/s = ds/rR$. (c) Making use of this change of variable, from θ to s, show that dU becomes $dU = (Gm/r)[\rho(2\pi R \Delta R)]ds$. (d) Next, show that the gravitational potential energy between m and the entire shell is given by

$$U = -\frac{Gm}{r}[\rho(2\pi R \,\Delta R)]\int_{r-R}^{r+R} ds = -\frac{GmM_s}{r}$$

where M_s is the mass of the entire shell. (e) Finally, show that the gravitational potential for a point interior to the spherical shell is a constant and therefore that the gravitational force on a particle anywhere inside a spherical shell is zero.

13-29. The earth has an equatorial bulge, and its spin axis is not perpendicular to the plane of its orbit about the sun. See Fig. P13-29. Because the gravitational force decreases with distance, the sun's pull on the near protuberance of the earth is greater than its pull on the far protuberance. These forces are *not* collinear; consequently, the sun applies a resultant gravitational *torque* on the spinning earth, the direction of the torque being perpendicular to the earth's spin angular momentum. Thus, the earth precesses, like a top. (a) What is the direction of the gravitational torque of the sun on the earth (with respect to the arrangement shown in Fig. P13-29)? (b) What is the direction of the angular-velocity vector representing the precession of the earth's spin axis?

This phenomenon, first explained by Newton, is known as the *precession of the equinoxes*. The period of the precession is about 26,000 yr. One consequence of this very slow precession is that the North Star (Polaris), toward which the earth's spin axis presently points, will, as time goes on, no longer be found along this axis.

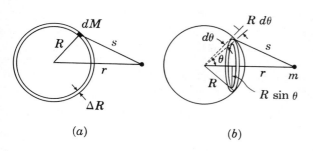

FIG. P13-28

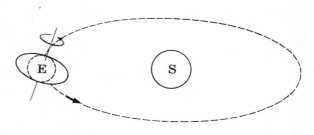

FIG. P13-29

CHAPTER 14

Simple Harmonic Motion

In this chapter we develop the basic dynamics of simple harmonic motion and illustrate this type of motion with a number of examples. We also consider briefly the damped oscillator and forced oscillations. Simple harmonic motion is characteristic of all objects subject to a linear restoring force. Many situations in physics illustrate this behavior: a mass attached to a spring, an object attached to a torsion rod, a simple or compound pendulum, the atoms of any solid, the atoms of a molecule, the current in a conductor carrying an alternating current, the electric or magnetic fields in a simple electromagnetic wave, the pressure variations in a sound wave. Indeed, *any particle undergoing small oscillations about a point of stable equilibrium executes simple harmonic motion.*

14-1 DYNAMICS AND KINEMATICS OF SIMPLE HARMONIC MOTION

The force F_s exerted by a spring stretched or compressed by a small amount x from its equilibrium position is given by

$$F_s = -kx \qquad (9\text{-}10),(14\text{-}1)$$

where the minus sign implies that the force is opposite to the displacement and k is the spring's force constant.†

The potential energy U_s of a deformed spring is

$$U_s = \tfrac{1}{2}kx^2 \qquad (10\text{-}2),(14\text{-}2)$$

---

†Equation (14-1) is sometimes referred to as *Hooke's law*, after Robert Hooke (1635–1678), a contemporary of Newton.

Actually, the spring serves as a prototype of the situation arising in many areas of physics in which a particle is subject to a *linear restoring force* and, therefore, the system's *potential energy varies as the square of the particle's displacement* from the equilibrium position. Under these circumstances the particle executes *simple harmonic motion*.

For a particle of mass m attached to one end of a spring whose other end is fixed, Newton's second law gives

$$\Sigma F_x = ma_x$$

$$-kx = m\frac{d^2x}{dt^2}$$

where d^2x/dt^2 is the particle's acceleration a_x along x.

Rearranging the equation above, we have

$$\frac{d^2x}{dt^2} + \frac{k}{m}x = 0$$

Putting the ratio k/m equal to the constant ω_0^2, whose meaning we shall shortly elucidate,

$$\omega_0^2 \equiv \frac{k}{m} \qquad (14\text{-}3)$$

the equation above becomes

$$\frac{d^2x}{dt^2} + \omega_0^2 x = 0 \qquad (14\text{-}4)$$

Any physical situation which leads to a second-order differential equation of motion of the form of (14-4) implies simple harmonic motion.

It is easy to show that a solution of (14-4) is

$$x = A\cos(\omega_0 t + \delta) \qquad (14\text{-}5)$$

where A and δ are constants.†

Taking the first and second time derivatives of the displacement x to find the particle's velocity $v_x = dx/dt$ and acceleration $a_x = d^2x/dt^2$, we have

$$v_x = \frac{dx}{dt} = -\omega_0 A \sin(\omega_0 t + \delta) \qquad (14\text{-}6)$$

and

$$a_x = \frac{d^2x}{dt^2} = -\omega_0^2 A \cos(\omega_0 t + \delta) = -\omega_0^2 x \qquad (14\text{-}7)$$

Comparing (14-7) and (14-4), we see that (14-5) is indeed a solution.

---

†The solution may be given equivalently as $x = B\sin(\omega_0 t + \epsilon)$, or $x = C\cos\omega_0 t + D\sin\omega_0 t$, where B, ϵ, C, and D are all constants.

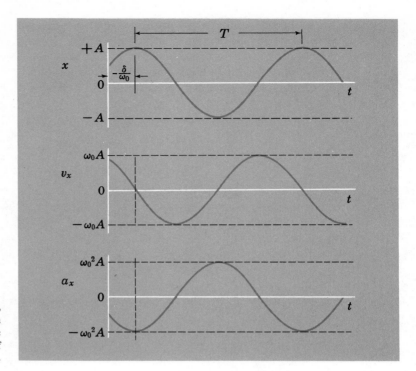

FIG. 14-1. *(a)* Displacement, *(b)* velocity, and *(c)* acceleration of a particle in simple harmonic motion plotted as a function of time.

The displacement x, velocity dx/dt, and acceleration d^2x/dt^2 all *vary sinusoidally with time*, as shown in Fig. 14-1. The displacement x takes on values ranging from $-A$ to $+A$; the maximum displacement A is known as the *amplitude* of the oscillation. Notice from the figure that x, v, and a do not reach their maxima at the same time; the maximum in the velocity comes one-quarter of a full cycle before the maximum in the displacement, and the acceleration maximum is one-quarter cycle ahead in time of the velocity maximum.

The *phase constant* δ is determined by the displacement and velocity at the starting time $t = 0$. In (14-5) and (14-6), with $t = 0$, we have $x(t = 0) = A \cos \delta$ and $v(t = 0) = -\omega_0 A \sin \delta$. Thus, $\tan \delta = -v(t = 0)/\omega_0 x(t = 0)$.

The *period T* of the oscillation is, by definition, the *time* interval for the oscillating particle to complete *one full cycle*. Thus, if the time t increases by one period to $t + T$, then the argument of the cosine in (14-5) must have increased by one cycle, or 2π rad:

$$\omega_0(t + T) + \delta = (\omega_0 t + \delta) + 2\pi$$

or
$$T = \frac{2\pi}{\omega_0} \qquad (14\text{-}8)$$

The *frequency f*,† giving the *number of oscillations per unit*

---

†The appropriate unit for frequency is s⁻¹ or, now more commonly, the Hertz (abbreviated Hz), where 1 Hz = 1 s⁻¹.

time, must be the reciprocal of the period:

$$f = \frac{1}{T} = \frac{\omega_0}{2\pi} \tag{14-9}$$

The quantity ω_0 whose magnitude determines the frequency and period of the oscillations is sometimes referred to as the *angular frequency*. The subscript zero is to emphasize that the oscillations take place in the absence of damping (*zero* damping force).

Substituting (14-3) into (14-8) and (14-9), we have for the period and frequency of a particle attached to spring:

$$T = \frac{1}{f} = 2\pi \left(\frac{m}{k}\right)^{1/2} \tag{14-10}$$

Note especially that the period and frequency do *not* depend on the amplitude A. The time for one oscillation depends on the particle's inertia (measured by m) and the spring's stiffness (measured by k) but not on the maximum displacement of the particle from the equilibrium position. We might say that the particle's inertia makes it overshoot the equilibrium position, while the spring's restoring force binds the particle to the equilibrium position. The competing inertial and elastic effects are reflected in (14-10): T increases with m, but decreases with k.

Simple harmonic motion can be defined kinematically not merely as the sinusoidal variation with time of the displacement but also as the *projection along a diameter of uniform circular motion*. Consider Fig. 14-2, where a particle moves counterclockwise in a circle of radius A with an angular velocity of magnitude ω_0. The angle θ, measured in radians, which the radius vector makes with the X axis is $\theta = \omega_0 t + \delta$, where $\theta = \delta$ at $t = 0$. Therefore, the X component of the particle's circular motion is

$$x = A \cos \theta = A \cos (\omega_0 t + \delta) \tag{14-5}$$

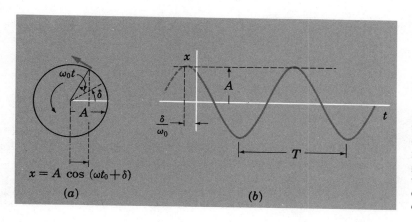

FIG. 14-2. (*a*) Representation of the uniform circular motion corresponding to simple harmonic motion. (*b*) Displacement-time graph for simple harmonic motion with $x = A \cos (\omega_0 t + \delta)$. The phase constant is δ, angular frequency ω_0, amplitude A, and period T.

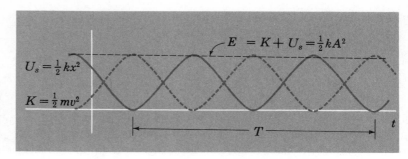

FIG. 14-3. The variation with time of the kinetic energy K and the elastic potential energy U_s for a simple harmonic oscillator. Note that the total energy $E = K + U_s$ is constant.

14–2 ENERGETICS OF SIMPLE HARMONIC MOTION

The oscillating particle's kinetic energy K is, using (14-6),

$$K = \tfrac{1}{2}mv^2 = \tfrac{1}{2}m\omega_0^2 A^2 \sin^2(\omega_0 t + \delta) = \tfrac{1}{2}kA^2 \sin^2(\omega_0 t + \delta)$$

where we have used (14-3) in the last step.

The potential energy of the spring is, from (14-2) and (14-5),

$$U_s = \tfrac{1}{2}kx^2 = \tfrac{1}{2}kA^2 \cos^2(\omega_0 t + \delta)$$

Therefore the total energy E of the oscillating system is

$$E = K + U = \tfrac{1}{2}kA^2[\sin^2(\omega_0 t + \delta) + \cos^2(\omega_0 t + \delta)]$$

$$E = \tfrac{1}{2}kA^2 \tag{14-11}$$

The energy of a simple harmonic oscillator is proportional to the *square of the amplitude*.

The total energy of an undamped simple harmonic oscillator is constant. In the absence of dissipative forces, the oscillations persist indefinitely with undiminished amplitude. Both the kinetic and the potential energies vary sinusoidally with time, as shown in Fig. 14-3, but their sum E is constant. The oscillating mass has a maximum kinetic energy when it passes through the equilibrium position ($x = 0$), and the spring has a maximum potential energy when the body is momentarily at rest at the amplitude position ($x = \pm A$).

Figure 14-3 shows the potential energy U_s as a function of time. It is also worthwhile, since simple harmonic motion is repetitive, to examine the potential energy as a function of the displacement. Figure 14-4 shows U_s as a function of x; $U_s = \tfrac{1}{2}kx^2$, a parabola, following (14-2).

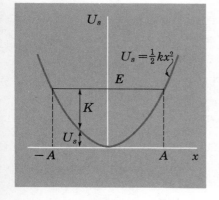

FIG. 14-4. Elastic potential energy U_s as a function of displacement x for a simple harmonic oscillator. The motion is restricted to the region between the amplitude positions $x = \pm A$, for which the kinetic energy K is positive.

Example 14-1

A 2.0-kg block rests on a horizontal frictionless surface, attached to the right end of a spring whose left end is fixed. See Fig. 14-5. The block is displaced 5.0 cm to the right from its equilibrium position and held motionless at this position by an external force of 10.0 N. (a) What is the spring's force constant? (b) The block is then released. What is the period of the block's oscillations? What are

(c) the kinetic energy of the block and (d) the potential energy of the spring at the time $t = \pi/15$ s?

(a) From Eq. (14-1),

$$k = -\frac{F}{x} = -\frac{-10.0 \text{ N}}{0.050 \text{ m}} = 200 \text{ N/m}$$

(b) The period is given by (14-10).

$$T = 2\pi \sqrt{\frac{m}{k}} = 2\pi \sqrt{\frac{2.0 \text{ kg}}{200 \text{ N/m}}} = \frac{\pi}{5} \text{ s}$$

(c) The kinetic energy is

$$K = \tfrac{1}{2}mv^2 = \tfrac{1}{2}m\omega_0^2 A^2 \sin^2(\omega t_0 + \delta) = \tfrac{1}{2}kA^2 \sin^2(\omega t_0 + \delta)$$

Since the motion is started from rest at the amplitude position $A = 0.050$ m, we know that $\delta = 0$. Therefore, at $t = \pi/15$ s, the equation above becomes

$$K = \tfrac{1}{2}(200 \text{ N/m})(0.050 \text{ m})^2 \sin^2 \left[\frac{2\pi(\pi/15 \text{ s})}{\pi/5 \text{ s}}\right]$$

$$= (0.25 \text{ N-m}) \sin^2 120°$$

$$= 0.19 \text{ J}$$

(d) We can find the spring's potential energy by using the relation $U_s = \tfrac{1}{2}kA^2 \cos^2(\omega t_0 + \delta)$. But, we can equally well find the spring's potential energy at $t = \pi/15$ s by subtracting the block's kinetic energy at this instant from the total energy $E = \tfrac{1}{2}kA^2$ of the block-spring system:

$$U_s = \tfrac{1}{2}kA^2 - K$$

$$= \tfrac{1}{2}(200 \text{ N/m})(0.050 \text{ m})^2 - (0.19 \text{ J}) = 0.06 \text{ J}$$

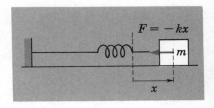

FIG. 14-5. A simple harmonic oscillator. A mass m attached to a spring of force constant k, subject to a restoring force $F = -kx$, where x is the mass's displacement from its equilibrium position.

14-3 THE SIMPLE PENDULUM

Consider a particle of mass m attached to the lower end of a massless cord of length l whose upper end is fixed. See Fig. 14-6. This system constitutes a *simple pendulum*.

The angle of the cord with the vertical is θ. We assume in what follows that the angle θ is always small, so that the particle, when released from the side with the string taut, undergoes oscillations of small amplitude. The forces on the particle are the cord tension **T** and the particle's weight $m\mathbf{g}$. The resultant force along the circular arc of radius l in which the particle moves is of magnitude $mg \sin \theta$, in a direction opposite to the displacement θ. If θ is sufficiently small, the particle's displacement along the circular arc can be closely approximated by a horizontal displacement x from the equilibrium position, where $x \approx \theta l$. Thus, for small θ,

$$\frac{x}{l} \approx \sin \theta \approx \theta$$

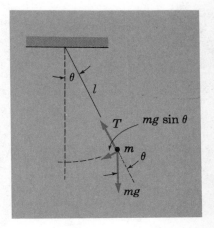

FIG. 14-6. A simple pendulum. The resultant force on the particle m along the circular arc is $mg \sin \theta$.

so that, in applying Newton's second law, we have

$$\Sigma F_x = ma_x$$

$$-mg \sin \theta = m \frac{d^2x}{dt^2}$$

$$-mg \frac{x}{l} \approx m \frac{d^2x}{dt^2}$$

or

$$\frac{d^2x}{dt^2} + \frac{g}{l} x = 0$$

The equation of motion above is of exactly the same form as Eq. (14-4), the general equation for simple harmonic motion, where

$$\omega_0^2 = \frac{g}{l}$$

Therefore, the period of a simple pendulum *for small amplitude* is, using (14-8),

$$T = \frac{1}{f} = 2\pi \left(\frac{l}{g}\right)^{1/2} \tag{14-12}$$

The period depends only on the length l and acceleration g but is independent of the amplitude (whether of x or of θ) of the oscillation.†

A pendulum is a suitable timing device, since it is *isochronous:* Its period is (nearly) *independent of its amplitude.* The simple pendulum provides a simple and precise basis for measuring g: One measures the period of a pendulum of known length and applies (14-12).

The period of a simple pendulum does not depend on the particle's mass. All simple pendula of the same length oscillate at the same rate. We have, however, made an important assumption in arriving at this result; namely, that the particle's inertial mass m_i and gravitational mass m_g are equal (see Sec. 13-1). As Newton found by experiment, pendula of the same length but with different masses and materials do indeed have the same period. This implies that $m_i = m_g$.

Example 14-2

Show that the period T of a simple pendulum of length l and mass m *must,* from considerations of the dimensions of these quantities alone

---

†The general relations giving the period for an arbitrary angular amplitude θ_m (in radians) is

$$T = 2\pi \sqrt{l/g} \left[1 + \tfrac{1}{4}\sin^2 \frac{\theta_m}{2} + \tfrac{9}{64}\sin^4 \frac{\theta_m}{2} + \cdots \right]$$

With $\theta_m = 30°$, $T = 2\pi \sqrt{l/g}$ (1.017); thus, even for this large angle of swing, the actual period is greater than that given by the simple relation in Eq. (14-12) by less than 2 percent.

(*dimensional analysis*), have the functional dependence: $T \propto (l/g)^{1/2}$, or equivalently, $T = b(l/g)^{1/2}$, where b is a dimensionless constant.

The quantities which might a priori be imagined to determine the pendulum's period T are (with dimensions in parentheses): $m(M)$, $l(L)$, and $g(LT^{-2})$.

Since neither l nor g involves the mass dimension, the period must be *independent of mass*. The acceleration g alone has dimensions of time, and it must appear as $g^{-1/2}$ to yield for T the required dimensions $(T)^1$. But if $T \propto g^{-1/2}$, we have the dimension $(L)^{-1/2}$ also appearing. To ensure that the period has no length dimension, the length l must appear as $l^{1/2}$ with dimensions $(L)^{1/2}$. Therefore,

$$T \propto \left(\frac{l}{g}\right)^{1/2}$$

or, in terms of dimensions,

$$(T)^1 = \left(\frac{L}{LT^{-2}}\right)^{1/2} = (T)^1$$

The proportionality above may be written as an equation:

$$T = b\left(\frac{l}{g}\right)^{1/2}$$

The value of the dimensionless constant b [2π from our detailed dynamical analysis, Eq. (14-12)] cannot be determined from dimensional analysis. The power of this procedure is that, without concern with the details, or even with the physical principles, in some instances we can arrive at the required functional dependence simply by considering the dimensions of physical quantities.

14-4 SMALL OSCILLATIONS AND SIMPLE HARMONIC MOTION

It is easy to show that *any particle undergoing small oscillations about a point of stable equilibrium executes simple harmonic motion.* If a particle has an equilibrium position which is stable, then it must be subject to a force toward the equilibrium position when displaced in one direction or the other. Said differently, if the particle is in motion away from its equilibrium position, its kinetic energy must decrease. In consequence, the system's potential energy must increase to maintain a constant total energy. Thus, the potential energy U must rise for either a positive or negative displacement x from equilibrium, as shown in Fig. 14-7. In the vicinity of the potential-energy minimum (the equilibrium position), the potential energy can be approximated by a parabola†

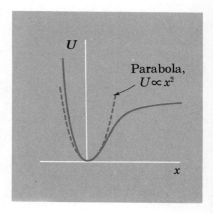

FIG. 14-7. Potential energy as a function of displacement for a particle in stable equilibrium. The solid line is the potential energy for an atom bound in a molecule; the dashed line, which approximates the actual potential-energy curve near its minimum, is parabolic.

---

†We can write the potential energy U close to the minimum as an expansion in the displacement x:

$$U(\text{near } x = 0) = a_0 + a_1 x + a_2 x^2 + a_3 x^3 + \cdots$$

where $a_0, a_1, a_2, \ldots$ are constants. Since the zero for potential energy is arbitrary, and we

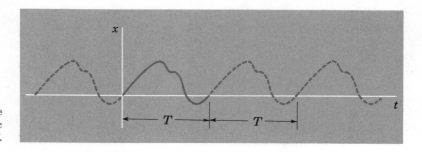

FIG. 14-8. A displacement-time graph for one example of periodic motion with period T.

with a vertical symmetry axis, and we can write:

$$U \propto x^2 \qquad \text{for small } x$$

But a potential energy varying as the square of the displacement from equilibrium implies, as we have seen, a linear restoring force. A particle displaced a small distance from $x = 0$ will, then, undergo simple harmonic motion.

Simple harmonic motion is one important example of *periodic motion*. By periodic motion is meant repetitive motion. The *period T* is defined in general as the smallest time interval such that, if the motion is known during any interval of duration T, we can find the motion for *all times* merely by shifting the displacement-time curve to the right or left an integral number of periods. The periodic motion illustrated by Fig. 14-8 is clearly *not* simple harmonic motion. The general mathematical theorem known as the *Fourier theorem* implies the following important property. *Any periodic motion* can be regarded as the *superposition*, or summation, *of a number of simple harmonic motions*, which may differ in amplitude, frequency, and phase.

Example 14-3

Find the period T of a simple pendulum without resorting to Newton's second law. The strategy will be to write the potential energy of a simple pendulum oscillating with small amplitude in the form

$$U = \tfrac{1}{2}kx^2 \qquad (14\text{-}2)$$

in order to find the equivalent "force constant" k. Then the period can be found immediately through the relations

$$\omega_0^2 = \frac{k}{m} \qquad (14\text{-}3)$$

and

$$T = \frac{2\pi}{\omega_0} \qquad (14\text{-}8)$$

Relative to the equilibrium position — the low point in its path —

---

may choose $U = 0$ when $x = 0$, the constant $a_0 = 0$. If a_1 were not zero, its contribution to U would make U rise for positive x and fall for negative x, in contradiction to the fact that U must rise in both directions. Thus, $a_1 = 0$ for an equilibrium point. For sufficiently small x, then, the leading term is $a_2 x^2$, and $U \approx a_2 x^2$. For *stable* equilibrium we must have $a_2 > 0$.

the gravitational potential energy of the suspended mass m is

$$U = mgh = mgl(1 - \cos\theta)$$

where θ is the angle which the cord makes with the vertical, l is the length of the cord, and g is the acceleration of gravity. See Fig. 14-9.

The cosine of any angle θ (in radians) can be written as a series expansion:

$$\cos\theta = 1 - \frac{\theta^2}{2!} + \frac{\theta^4}{4!} - \cdots$$

If θ is small we may discard all but the first two terms in the above approximation, and the general relation for the potential energy becomes

$$U = \tfrac{1}{2}mgl\theta^2$$

As in Sec. 14-3 the particle's displacement along the circular arc can be closely approximated by x, its horizontal displacement from equilibrium, where $x = l\theta$, so that the equation above may be written

$$U = \tfrac{1}{2}mgl\left(\frac{x}{l}\right)^2 = \frac{\tfrac{1}{2}mg}{l}x^2$$

The standard form for the potential energy of a simple harmonic oscillator is

$$U = \tfrac{1}{2}kx^2$$

Comparing these equations we discover that the effective value of the "spring constant" for the simple pendulum is

$$k = \frac{mg}{l}$$

Using the relations for ω_0^2 and T given above we have finally

$$\omega_0^2 = \frac{k}{m} = \frac{mg/l}{m} = \frac{g}{l}$$

$$T = \frac{2\pi}{\omega_0} = 2\pi\left(\frac{l}{g}\right)^{1/2} \tag{14-12}$$

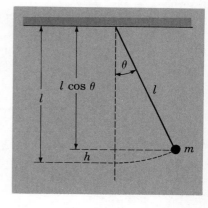

FIG. 14-9. A simple pendulum consisting of a particle of mass m suspended from a cord of length l. The particle is a distance $h = l(1 - \cos\theta)$ above its lowest point when the cord makes angle θ with the vertical.

Example 14-4

Imagine that a small hole is drilled diametrically through the earth. A particle travels along this tunnel. Show that it executes simple harmonic motion for an earth of uniform mass density.

The situation is shown in Fig. 14-10a. We imagine a particle of mass m to be just at the *outside* of a sphere of radius r and *inside* a spherical shell with inner and outer radii of r and r_e, respectively, where r_e is the radius of the earth.

We know that there is *no* gravitational force on m arising from the *surrounding* shell (Prob. 13-28). Therefore, the resultant gravitational force on m arises solely from the sphere of radius r:

$$F_g = -\frac{GMm}{r^2}$$

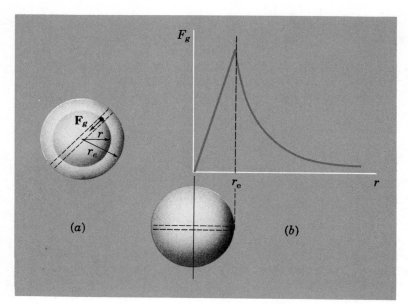

FIG. 14-10. (a) Gravitational force on a particle traveling in a tunnel through the earth. (b) Gravitational force as a function of the separation distance between a particle and the earth's center.

where M is the mass of this sphere, and the minus sign indicates that the force is inward.

If we *assume* that the earth has a *uniform* density ρ, we may write

$$M = \rho \tfrac{4}{3}\pi r^3$$

Using the relation for M in the equation for F_g given above, we obtain

$$F_g = -\frac{G(\rho \tfrac{4}{3}\pi r^3)m}{r^2} = -Kmr$$

where K is a constant. The gravitational force on the particle is proportional to its distance r from the earth's center.

When the particle is at the earth's surface, $r = r_e$, and the force F_g on the particle is mg. Therefore,

$$mg = Kmr_e$$

where g is the acceleration due to gravity at the earth's surface.

Eliminating K from the last two equations gives, finally,

$$F_g = -\frac{mg}{r_e} r$$

The gravitational force is proportional to the particle's displacement r from the earth's center and is therefore of the form $F = -kx$, where the equivalent "stiffness constant" is $k = mg/r_e$. The particle executes simple harmonic motion with a period

$$T = 2\pi \left(\frac{m}{k}\right)^{1/2} = 2\pi \sqrt{\frac{m}{mg/r_e}} = 2\pi \left(\frac{r_e}{g}\right)^{1/2} \quad (14\text{-}10)$$

The relation for T is precisely that for a satellite orbiting the earth close to its surface, with $T \approx 85$ min. That the particle in the diametrical earth tunnel should have exactly the same period as that of an orbiting satellite follows also from the fact that the projections along the tunnel of the satellite's resultant force, displacement, velocity, and acceleration are equal to the corresponding quantities for the particle in simple harmonic motion.

14–5 DAMPED OSCILLATIONS AND RESONANCE

When a particle is subject to a linear restoring force only, the system's energy remains constant, and the oscillatory motion persists indefinitely with undiminished amplitude. However, if there is a small opposing force—a damping force—the energy is no longer constant, but decays exponentially, as shown in Fig. 14-11a. Furthermore, the displacement is then no longer a sinusoidal function of time, and the particle oscillates with continuously decreasing amplitude. See Fig. 14-11b. Finally, if there is damping, the resistive force slows the motion and the period increases—corresponding to a lower frequency.

When set into motion, a damped oscillator loses energy and eventually comes to rest. How can its energy be maintained constant? One must do work on the system, so that the energy fed into the system compensates for the energy dissipated. Energy is fed into the oscillator only when *positive* work is done by some external agent, that is, when the agent pushes in the *same* direction as the oscillator moves. For example, one can keep a damped oscillator, such as a playground swing, oscillating with constant amplitude by pushing it (at the right time) once each cycle. Then the frequency of the pushes will equal the natural frequency of the swing, and positive work will be done on the swing each time the agent pushes.

In general, if a damped oscillator is driven by an external oscillating force whose frequency can be varied, the oscillator shows the greatest response—that is, the largest oscillator amplitude—when the frequency of the driving force is at (or near) the natural frequency of the oscillator alone. More explicitly, for an oscillator with small damping the amplitude A_ω of the driven oscillator is a maximum when the external driving frequency ω_e is equal to ω_0. See Fig. 14-12. This resonance behavior—strong response at the condition when the frequency of the driving force matches the natural oscillation frequency of the oscillator alone—characterizes not only mechanical oscillators, but any system (including electrical oscillators, atoms, or molecules) for which there are one or more characteristic oscillation frequencies. The resonance is sharp when the damping is small, the oscillator responding substantially to the driving force only over the narrow band in which the natural and driving frequencies are nearly equal. On the other hand, if

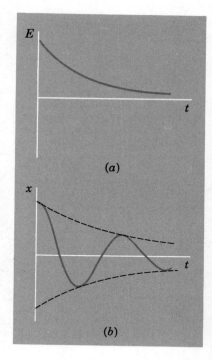

FIG. 14-11. (a) Total energy E as a function of time for a damped oscillator. (b) Displacement-time graph for a damped oscillator.

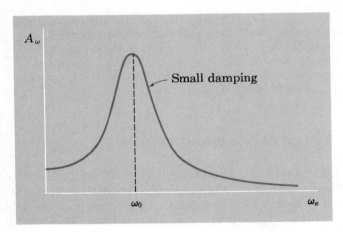

FIG. 14-12. The amplitude A_ω of the forced motion of a damped oscillator as a function of the frequency of the external driving force. Resonance occurs when the driving frequency approaches the natural frequency of the oscillator. The resonance peak is narrow for small damping.

the damping is large, the resonance is broad; then the oscillator is influenced less by the external driving force, but it responds over a broader band of frequencies.

The forced oscillator has a constant displacement amplitude for a given frequency ω_e. That is to say, the total energy of the oscillating mass and spring remains constant with time. Since energy is transferred out of the system continuously, through damping, it follows that the driving force feeds energy into the system at the same rate. The resonance frequency represents that frequency at which energy is most readily transferred from the external agent to the oscillating system.

Resonance phenomena appear in many areas of physics: in resonating mechanical devices, in electric circuits, in atomic and molecular structure. In fact, resonance occurs whenever an oscillator is acted upon by a second driving oscillator at the same frequency.

SUMMARY

When a particle of mass m is subject to a restoring force of the form

$$F = -kx \qquad (14\text{-}1)$$

and with an associated potential energy

$$U = \tfrac{1}{2}kx^2 \qquad (14\text{-}2)$$

the differential equation of motion has the form

$$\frac{d^2x}{dt^2} + \omega_0^2 x = 0 \qquad (14\text{-}4)$$

where the constant $\omega_0^2 = k/m$.

The displacement then varies sinusoidally with time:

$$x = A \cos(\omega_0 t + \delta) \qquad (14\text{-}5)$$

where A is the amplitude, δ is the phase constant, and the angular frequency ω_0 is related to the period T and frequency of the oscillation by $\omega_0 = 2\pi/T = 2\pi f$.

A simple harmonic oscillator's total energy E is

$$E = \tfrac{1}{2}kA^2 \qquad (14\text{-}11)$$

PROBLEMS

14-1. A 4.0-kg mass is oscillating with a frequency of 4.0 Hz and an amplitude of 2.0 cm. The clock reads zero when the mass is at the equilibrium position, with the particle moving in the $+X$ direction, and the displacement x is given by $x = A \cos(\omega_0 t + \delta)$. Find (a) the phase constant δ, (b) the force constant k, and (c) the maximum velocity of the mass. Also, find the sign and magnitude of the (d) displacement, (e) acceleration, and (f) velocity when the clock time is 0.20 s.

14-2. An object vibrates in simple harmonic motion with an amplitude of 6.0 cm and a frequency of 5.0 Hz. Find the (a) acceleration and (b) speed of the object when it is 4.0 cm from its equilibrium position. (c) What is the angular frequency in rad/s of the corresponding uniform circular motion?

14-3. A 100-g plumb bob is suspended from a vertical spring. When it is displaced vertically downward from its equilibrium position and released, the bob oscillates in simple harmonic motion with a frequency of 6.0 Hz. (a) Find the force constant of the spring. (b) If a 1.00-kg mass is substituted for the plumb bob, what is the new oscillation frequency?

14-4. A particle oscillating in simple harmonic motion travels a distance of 20 cm during the time of one complete cycle. The maximum acceleration is 4.0 cm/s². What is the frequency of oscillation?

14-5. A typical atom in a solid at ordinary temperatures undergoes simple harmonic motion with an amplitude of 10^{-10} m and a frequency of 10^{13} Hz. What are (a) the maximum speed of the atom and (b) its maximum acceleration?

14-6. A particle executes simple harmonic motion according to the relation $x = A \cos(\omega_0 t + \delta)$. The particle's initial ($t = 0$) displacement and velocity are x_0 and v_0. Show that the amplitude A of the oscillation is given by $A = [x_0^2 + (v_0/\omega_0)^2]^{1/2}$, and that the phase constant δ is given by $\tan \delta = -v_0/\omega_0 x_0$.

14-7. A simple harmonic oscillator has a frequency f. Show that the oscillator's kinetic energy and potential energy both oscillate at the frequency $2f$.

14-8. When a mass m is attached to one end of a spring whose second end is fixed in position, the mass oscillates at the frequency f. If two particles, each of mass m, are attached one to each end of the same spring and then set in oscillation, what is the frequency?

14-9. A 2.0-kg mass is supported on a smooth horizontal plane and attached to two identical springs, as shown in Fig. P14-9. The mass is oscillating in simple harmonic motion with a 6.0-s period and an amplitude of 10.0 cm. Take right as positive, left as negative, and set time at zero when the mass is at the equilibrium position and moving to the right. Find (a) the equivalent force constant and (b) the magnitudes of the maximum values of the velocity and the acceleration. (c) Also, find the velocity and acceleration when the displacement is -6.0 cm.

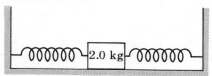

FIG. P14-9

14-10. An automobile is lowered $\tfrac{1}{4}$ in when a 160-lb man seats himself at the wheel. When the car (and driver) hits a bump in the road, it oscillates up and down with a period of 2 s and an amplitude of 4 in. (Assume that $g = 32$ ft/s².) Find (a) the force constant of the suspension system, (b) the weight of the car, (c) the maximum value of the kinetic energy associated with this oscillation, and (c) the angular frequency in rad/s of the corresponding uniform circular motion.

14-11. A coin rests on a table vibrating in the vertical direction with a frequency of 1,000 Hz. What is the maximum amplitude of the vibrations if the coin is not to lose contact with the surface?

14-12. A particle of mass m is located at the midpoint of a vertical wire of length L, and the wire is under tension T. See Fig. P14-12. (a) Show that if the particle is displaced slightly to the side and released it will vibrate horizontally with simple harmonic motion. Find (b) the force constant and (c) the frequency of the vibratory motion.

FIG. P14-12

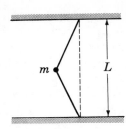

14-13. An ordinary helical spring with force constant k is cut in half to produce two springs of half the original length. What is the force constant of one of these shorter springs?

14-14. A spring of stiffness k with one end fixed and with a mass m attached to the other end rests on a frictionless table, and the mass oscillates along the horizontal. Then the spring is oriented along the vertical, with its upper end fixed and the mass m hanging from the other end stretching the spring when in the equilibrium position. The mass is pulled down further and released. What is the ratio of the period of oscillations in the first to that in the second arrangement?

14-15. (a) Show that uniform circular motion can be considered as the superposition of simple harmonic motions with the same amplitude and frequency but 90° out of phase along two mutually perpendicular lines. (b) Show that simple harmonic motion along a line can be considered as the superposition of two uniform circular motions having the same radius and angular frequency but rotating in opposite senses.

14-16. A pen is attached to the lower end of a pendulum, and the pen writes on a sheet of paper drawn at constant speed along a horizontal plane at right angles to the plane of oscillation of the pendulum. Show that the trace on the paper is a sine or cosine curve. (Conversely, that the trace is sinusoidal is a demonstration that the motion is simple harmonic.)

14-17. By dimensional analysis deduce the functional dependence of the acceleration of a particle moving in a circle of radius r with constant angular velocity ω.

14-18. A particle of 0.10-kg mass is in an environment for which the potential energy is given by $U = (4.0 \times 10^3 \text{ J/m}^2)x^2$, where x is the particle's displacement from its equilibrium position. What is the particle's oscillation frequency?

14-19. Show that the superposition of two simple harmonic motions along the X and Y axes, respectively, having *unequal* amplitudes and one lagging behind the other by one quarter of a cycle, yields an elliptical path, rather than a circular path, whose axis of symmetry is along the X or Y axis.

14-20. A particle at the end of a long taut string (a simple pendulum) is displaced slightly from the vertical and released with a small velocity perpendicular to its initial displacement. (a) Show that the particle's path is, in general, an ellipse (neglecting the small vertical motion). (b) Does a radius vector from the center of the ellipse to the particle sweep out equal areas in equal times? (c) Is the square of the period of the particle's motion proportional to the cube of the semi-major axis of the ellipse?

14-21. A simple pendulum of mass m and length l executes simple harmonic motion with horizontal amplitude A. Show that the maximum tension in the support cord is given by $T = [1 + (A/l)^2]mg$, where g is the acceleration of gravity.

14-22. A particle slides across the bottom of a perfectly smooth hemispherical bowl of radius of curvature R. (a) Show that the situation is completely analogous to that of the simple pendulum and that, for small amplitude oscillations, the particle executes simple harmonic motion. (b) Show that the period of oscillation is given by $2\pi(R/g)^{1/2}$.

14-23. A simple pendulum 1.0 m in length is attached to the ceiling of an elevator. What is the period of the pendulum when the elevator (a) moves upward at constant speed, (b) moves upward with a constant acceleration of 2.5 m/s², and (c) accelerates downward at the rate 9.8 m/s²?

14-24. A simple pendulum is designed to oscillate with a period of exactly 2 s when it is at the Naval Research Laboratory in Washington, D.C., where the acceleration of gravity is $g_W = 9.8011$ m/s². Suppose that this device were transported to Anchorage, Alaska, where $g_A = 9.8218$ m/s². (a) Find the *change* in the period at Anchorage. (b) What would be the cumulative error in one true 24-hour day? (Hint: Use the binomial expansion.)

14-25. What is the maximum speed of a particle dropped into a diametrical tunnel through the center of the earth?

14-26. By dimensional analysis deduce the functional dependence of the period of a satellite orbiting the earth close to its surface in terms of the earth's mass M, radius R, and the universal gravitational constant G.

14-27. The reciprocal of the force constant of a mechanical system is called its *compliance*. What is the effective value (in grams) of the dynamic mass of a phonograph arm and stylus if the dynamic compliance of the stylus is 30×10^{-6} cm/dyne and the arm-stylus system resonates at 10 Hz?

CHAPTER 15

Waves on a String

The study of waves is crucially important in physics. Such phenomena as sound and light are wave phenomena. Even more importantly, from the point of view of the quantum theory, particles exhibit wave characteristics in *all* phenomena at the level of atomic and nuclear physics.

In this and the next chapter we shall be concerned with mechanical waves, waves that involve the coupling together of a series of particles and which have their origin in the elastic properties of the medium transmitting them. First we shall deal exclusively with a simple type of wave, a wave traveling in one dimension along a string. Many of the results we find for this familiar type of wave behavior can be carried over, essentially unchanged, to more complicated types of mechanical waves.

15–1 BASIC WAVE BEHAVIOR

Consider a uniform perfectly flexible string, under tension and attached at its right end. Suppose that the left end is suddenly displaced laterally and then returned to its initial position. A *wavepulse* is produced, which travels to the right at constant speed with unchanged shape, as shown in Fig. 15-1. Each particle along

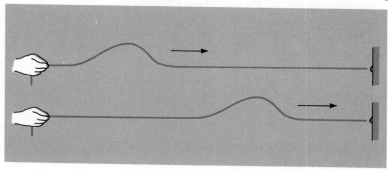

FIG. 15-1. A wavepulse traveling along a stretched string.

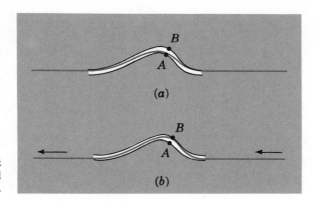

FIG. 15-2. A tube with a taut string *(a)* at rest and *(b)* reeled through at a very high speed.

the string undergoes, in turn, a transverse motion and returns to its initial position. The shape of the wave as a whole moves to the right.

This is a commonplace observation. Yet, on further thought, the phenomenon of wave propagation is truly remarkable in that each of the string's particles "knows enough" to return precisely to its initial position, while the shape of the wavepulse is unaltered as the disturbance travels along the string. Actually, the behavior of mechanical waves can be deduced quite directly from Newton's laws and the ideas of reference frames, as we shall see.

We now suppose that a flexible string of constant linear density ρ (mass per unit length) and under constant tension F_t is threaded through a smooth tube, as shown in Fig. 15-2. Clearly, if the taut string is at rest with respect to the tube, it will touch the tube at point A. On the other hand, if the string, still under tension F_t, is pulled to the left through the tube at a uniform high speed, the string's inertia will cause it to strike the tube at the higher point, B. There must be an intermediate speed at which the string can be reeled through the tube, touching neither A nor B. Let us find it.

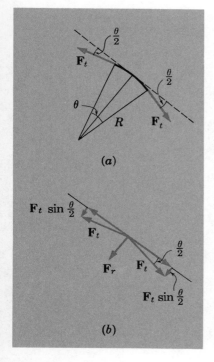

FIG. 15-3. *(a)* Forces on a small string segment. *(b)* The resultant force is $F_r = 2F_t \sin(\theta/2)$.

We concentrate on the small string segment shown in Fig. 15-3a. This segment is assumed to be so short that we may regard it as a circular arc of radius of curvature R, the angle θ, subtended by the segment about the center of curvature, being small. Now, if the speed c of this segment is just right, it will not touch the tube, and the only external forces acting on it will be the tension forces at the ends, both of magnitude F_t. These forces are tangent to the ends of the segment, since the string is assumed to be perfectly flexible. As Fig. 15-3b shows, the resultant force $\mathbf{F}_r$ on the segment points to the center of curvature and has the magnitude $2F_t \sin(\theta/2)$. Applying Newton's second law along the radial direction yields

$$\Sigma \mathbf{F} = m\mathbf{a}$$
$$2F_t \sin \frac{\theta}{2} = \rho R \theta \frac{c^2}{R}$$

where we have used the fact that the segment of length $R\theta$ and mass $\rho R\theta$ travels with constant speed c in a circular arc of radius R and therefore undergoes a radial acceleration c^2/R. Since θ is small, $\sin(\theta/2) \approx \theta/2$. Then the equation above reduces to

$$F_t = \rho c^2$$

$$c = \sqrt{\frac{F_t}{\rho}} \qquad (15\text{-}1)$$

When the string is drawn through the tube at the speed $\sqrt{F_t/\rho}$, the small string segment does not touch the tube at A or at B. Indeed, *no* segment of the string touches the tube, inasmuch as the speed c is independent of R and θ. Thus, if the string is reeled through the tube at just the right speed, the tube may be removed and the string will maintain its shape, each particle of string following the tube's shape as it passes from right to left. Then, *any shape*, once established, will stand in place as long as the speed of the string is that given by (15-1).

Note the assumptions: The string is uniform and perfectly flexible, the transverse displacements are small enough so that the tension is the same at all points along the string, and there are no sharp corners, which would preclude the assumption that $\sin(\theta/2) \approx \theta/2$.

Now let us view this behavior from another reference frame, one in which the *string* (to the left and right of the disturbance) remains *at rest*. If we travel to the left at the same speed c as does the string relative to the tube, then in this new reference frame we see a wave disturbance, unchanged in shape, moving to the right at the speed c, as in Fig. 15-1.

Our derivation for the wavespeed did not depend on the direction in which the string was moving through the tube. Thus, if the string were reeled through the tube with the speed c in the opposite direction, the wave disturbance would be seen to travel to the left in the reference frame in which the string is at rest.

We can express these results more formally. First recall the galilean coordinate transformation relations, Eqs. (6-7), for two reference frames in relative motion. The coordinate x measured in a reference frame traveling to the left (along the $-X$ axis) at speed c, relative to a reference frame in which the coordinate is x', is given by $x = x' + ct$ or $x' = x - ct$. (The time t is zero in both frames at the instant their origins coincide.) By the same token, if our moving reference frame travels to the right at speed c, and x again represents the X coordinate in this "moving" frame, then $x' = x + ct$. Now let $y = F(x')$ represent the shape of the disturbance (and of the tube), whatever it may be, as seen in the reference frame of the tube. Then, if the *string* travels to the *left* at speed c relative to the wavepulse, the *wavepulse* must travel to the *right* as seen from the reference frame of the string. The shape† is given, in general, by

Wave to *right*: $\qquad y = F(x - ct) \qquad$ (15-2a)

Wave to *left*: $\qquad y = F(x + ct) \qquad$ (15-2b)

In wave motion one must clearly distinguish between two speeds: the *wavespeed c*, the speed with which the *shape* travels along the X direction, and the transverse velocity $\partial y/\partial t$ at any fixed x, which gives the *velocity of a particle of the string*. Whereas the wavespeed $c = \sqrt{F_t/\rho}$ is constant for a given string (a given ρ and F_t), the particle speed is not.

Example 15-1

A uniform 10-m string having a mass of 490 g is attached at its upper end and has a mass of 50 kg suspended from its lower end. The lower end of the string is suddenly displaced horizontally. How long does it take for the wavepulse which is produced to travel to the upper end?

Denoting the string length by L, we can write the time t for the pulse to travel this length as $t = L/c$. From Eq. (15-1), $c = \sqrt{F_t/\rho} = \sqrt{Mg/(m/L)}$, where M is the suspended mass and m is the mass of the string. Therefore,

$$t = \frac{L}{c} = \frac{L}{\sqrt{Mg/(m/L)}} = \sqrt{\frac{mL}{Mg}} = \sqrt{\frac{(0.490 \text{ kg})(10 \text{ m})}{(50 \text{ kg})(9.8 \text{ m/s}^2)}} = 0.10 \text{ s}$$

We have ignored the 1 percent difference in tension between the top and bottom string ends, arising from the nonzero weight of the string itself. Strictly, the wavespeed at the top, where the tension is largest, exceeds that at the bottom by 0.5 percent.

15-2 THE SUPERPOSITION PRINCIPLE AND INTERFERENCE

Suppose that two wave disturbances traveling in opposite directions along the same string are allowed to "collide." The results of observation are shown in Fig. 15-4. As the two pulses merge, the resultant disturbance at each point along the string and at any instant of time, is found merely by adding algebraically, or superposing, the separate wave disturbances. By the *principle of superposition*, the resultant transverse displacement is just the algebraic sum of the displacements of the individual waves. Thus, any wave will pass through another wave with *unchanged* shape. Said differently, there is no wave-wave interaction; each wave carries away from the "collision" precisely the energy and transverse linear momentum it carried into the "collision."

---

†For a wave to the right, the argument of F must remain unchanged; if t increases (later time), then x must increase (farther to the right), so that $x - ct$ is constant.

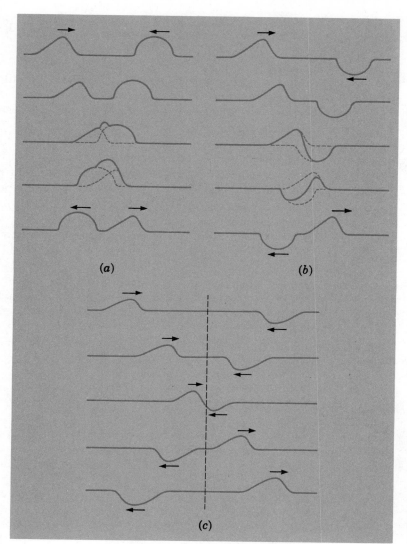

FIG. 15-4. The "collision" of wavepulses traveling in opposite directions: *(a)* constructive interference, *(b)* destructive interference. *(c)* As the pulses interfere, a single point on the string remains undisplaced at all times. Each group shows snapshots taken at equal time intervals, time increasing downward.

The superposition principle for wave displacements reminds us of another superposition principle, that for forces. In fact, wave superposition results from force superposition. According to the principle of superposition for forces, two or more forces acting simultaneously on an object are equivalent to a single force which is their vector sum. Now, in the case of transverse waves along a string, an accelerating force arises when the string is deformed, the string deformation, as measured by the curvature of a small segment, being proportional to the deforming force. As long as this proportionality holds, superposing forces is equivalent to superposing displacements. Thus, for mechanical waves, superposition applies only for *small* deformations (here, small string displacements). If the string were displaced great distances, or were deformed with sharp corners, or if the tension were not constant, the approximations made in deriving Eq. (15-1) would no longer apply. Then the principle of superposition would be inapplicable. In all that follows we shall assume that the wave disturbances are small enough for the superposition principle to hold strictly.

The superposing of separate wave displacements, to arrive at the resultant displacement, is known as *interference*. (The term is an unhappy one, since the separate waveforms really do not interfere with one another.) When two waves both have positive (or both negative) y displacements, as in Fig. 15-4a, the magnitude of their resultant displacement is greater than that of each wave separately. This is called *constructive interference*. On the other hand, when two waves of opposite y displacements are superposed, as in Fig. 15-4b, the magnitude of the resultant displacement is less than that of each wave separately, and the waves are said to show *destructive interference*.

One rather special case of wave interference is shown in Fig. 15-4c. Here two waves, one to the left, the other to the right, have identical shapes but are inverted both up-down and left-right. We see that when the two waves interfere, there is a single point along the string at which the resultant displacement *always* remains zero.

15-3 REFLECTION OF WAVES

When one wave collides with a second wave on the same string, nothing happens to either wave. What happens when a wave collides with a boundary, a point at which the medium propagating the wave (here, the string) changes?

First imagine a string attached firmly to a very massive wall, as in Fig. 15-5. Since the wall cannot move, we may describe this situation formally by saying that the displacement y must always be zero at the point where the string joins the wall. Now, if a wavepulse is propagated to the right, we find it reflected to the left from the boundary. After reflection, the shape is reversed *left-right;* the initial leading edge of the pulse is still the leading edge after reflection. In addition, the sign of the wave is reversed; that is, the wave is inverted *up-down*, a positive transverse displacement becoming negative upon reflection, and conversely.

One can give a physical basis for this behavior. When the leading edge of the wave disturbance arrives at the boundary, the tension of the string produces an upward force on the infinitely massive, and therefore essentially immovable, wall. By Newton's third law, the wall applies an equal downward force on the string. This force is, in fact, of greater magnitude then that of the force applied by an adjoining segment of string in the absence of the wall,

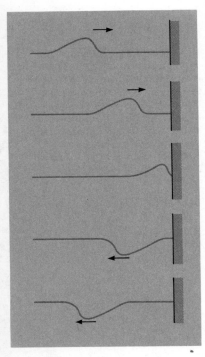

FIG. 15-5. Reflection of a wavepulse at an infinitely massive boundary.

because the string, being fixed vertically at the boundary, undergoes a larger change in curvature at that point. The force of the wall on the string is so great that it does not merely return it to $y = 0$; the wall pulls so hard on the string that the string is brought below the line $y = 0$. Thus, an inverted wave to the left is generated.

We now imagine that the end of the string, rather than being tied down, is perfectly free to move in the transverse direction. For definiteness, we suppose that the string is terminated with a small massless ring that can slide freely along a smooth vertical post. The results of a reflection are shown in Fig. 15-6. Again there is a left-right shape reversal, as the direction of propagation is changed from right to left. There is, however, *no* change in the sign of the waveshape. An upright incident wave is reflected as an upright wave. A simple physical basis also can be given for reflection from a free boundary. Since there is no string to the right of the free end to provide a downward force component through the tension, the string overshoots as the disturbance reaches the end.

We have treated two extreme cases of reflection: a string attached to an infinitely massive second medium and a string attached to a massless second medium. Now consider the more general case in which one string with linear density ρ_1 is connected to a second string of linear density ρ_2. The tension F_t is the same in both strings. If $\rho_2 > \rho_1$, we see from Eq. (15-1) that the wavespeed c_1 in the first string exceeds the wavespeed c_2 in the second string. Figure 15-7a shows what happens when a wavepulse incident from the left with speed c_1 encounters the boundary. The wave is partially transmitted into the second medium and partially reflected back into the first medium. Here, the boundary moves laterally as the waves reach it. As we would expect, the transmitted wave undergoes no change in sign; on the other hand, the reflected wave is reversed. Moreover, the transmitted waveshape is compressed longitudinally by virtue of the decreased speed c_2. That the reflected wave is reversed in sign follows from the behavior found earlier for reflection from an infinitely massive second medium. The second medium here is not infinitely massive, but its inertia is greater than that of the first. Consequently, a reversal in polarity occurs upon reflection. As Fig. 15-7b shows, the polarity is not reversed, in either the transmitted or reflected wave, for an incident wave which travels from the more massive string into the less massive string, as we might expect from Fig. 15-6.

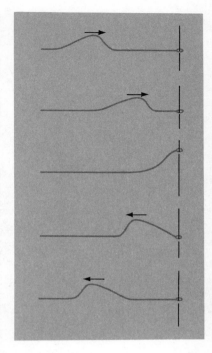

FIG. 15-6. Reflection of a wavepulse from a free end.

15-4 SINUSOIDAL WAVES

Thus far our discussion of waves on a string has been quite general. We have not been concerned with the specific waveshape, that is, with the mathematical relation giving the displacement y of a particle on the string as an explicit function of time t and position x.

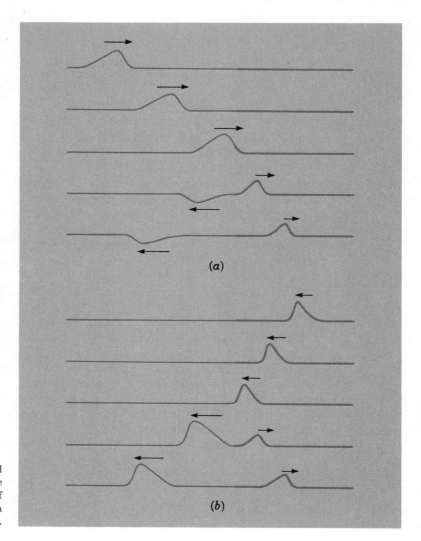

FIG. 15-7. Reflection and transmission of a wavepulse incident upon (a) a string of greater linear density and (b) a string of lesser linear density.

Here we do this for the special case of continuous sinusoidal waves generated by a simple harmonic oscillator. The special importance of sinusoidal waves is not merely that their generators are common but, more especially, that *any* waveshape can be regarded as the superposition of sinusoidal waves of various frequencies.

Suppose that the left end ($x = 0$) of a very long string under tension is moved up and down in simple harmonic motion with amplitude A and frequency f (and period T, where $T = 1/f$). Then (Sec. 14-1) the transverse displacement at this point is given by

$$y(x, t) = y(0, t) = A \sin \omega t$$

where $\omega = 2\pi f = 2\pi/T$. From the arguments of Sec. 15-1 we know that the displacement of any other point on the string to the right will show exactly the same variation with time, but with a delay, or phase lag, δ:

$$y(x, t) = A \sin (\omega t - \delta)$$

Since the wave disturbance travels along the X direction at a constant speed, the phase lag δ is proportional to x, and we may write

$$\delta = kx \tag{15-3}$$

where k is known as the *wave number* (for reasons soon to be seen). The transverse displacement $y(x, t)$ can then be written

Wave to *right*: $\qquad y(x, t) = A \sin (\omega t - kx) \qquad$ (15-4a)

This is one important form of the equation for a traveling sinusoidal wave. It gives the transverse displacement y as a function of both the coordinate x along the direction of propagation and the time t for a wave traveling to the right. If the wave were to travel to the left, the displacement at point x would *lead*, rather than lag, the displacement at $x = 0$. Then the equation for the traveling wave would be written

Wave to *left*: $\qquad y(x, t) = A \sin (\omega t + kx) \qquad$ (15-4b)

The displacement y varies sinusoidally with time for every point along the string. Equations (15-4) also show that, for any time t, y varies sinusoidally with x; that is, a snapshot of a wave generated by a simple harmonic oscillator is a sine or a cosine, as shown in Fig. 15-8.

Points along the wave having the same displacement and *transverse* velocity are in the same phase; for example, the dots of Fig. 15-8. The distance between any two such adjacent points is known as the *wavelength*, λ. Thus, if x changes by λ, the phase must change by 2π or, from (15-3),

$$k\lambda = 2\pi$$

$$k = \frac{2\pi}{\lambda} \tag{15-5}$$

The wave number k gives the number of wavelengths per 2π phase change; that is, k is the rate of change in phase with distance.

Using the definition $\omega = 2\pi/T$, we may write (15-4a) in another useful form:

$$y(x, t) = A \sin 2\pi \left(\frac{t}{T} - \frac{x}{\lambda} \right) \tag{15-6}$$

Equation (15-6) applies for a wave traveling to the right; as before, a wave to the left is represented by the same equation, but a plus sign replaces the minus sign.

Clearly, a change in time t of the duration of one period T is equivalent to a change in space along X of one wavelength λ. That

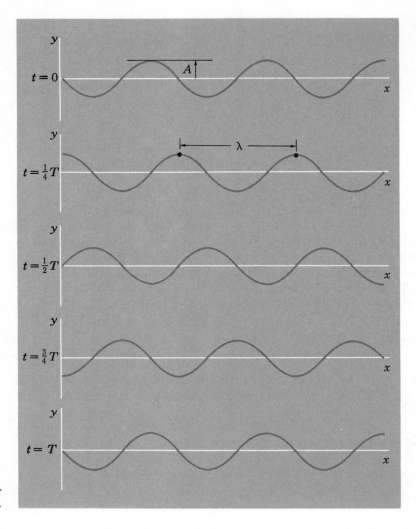

FIG. 15-8. A sinusoidal wave for several different times.

is, in the time T, during which the wave generator and each particle on the string complete one cycle, the wavelength advances a distance λ. The wavespeed c is then

$$c = \frac{\lambda}{T} \tag{15-7}$$

or, since $1/T = f$,
$$c = f\lambda \tag{15-8}$$

An alternate form of this relation is $c = (2\pi f)(\lambda/2\pi) = \omega/k$. Since the phase of oscillation advances along the propagation direction at the same rate as does the waveshape, it is appropriate to call the wavespeed the *phase speed*.

The phase speed for a wave on a string depends only on the elastic and inertial properties of the string (the tension and linear density), not on the form of the wave. It follows, then, from (15-8) that high-frequency oscillations generate waves of relatively short wavelength, while low-frequency oscillations generate long wavelengths, the product of f and λ always remaining constant.

Using the alternate form of (15-8), $c = \omega/k$, the equation for the traveling wave, (15-4a) can be written in still another form:

$$y = A \sin (\omega t - kx) = A \sin k(ct - x)$$

Because $\sin(-\theta) = -\sin \theta$, we have

$$y(x, t) = -A \sin k(x - ct) \qquad (15\text{-}9)$$

again for a wave propagated along positive X. We note that the quantities x, c, and t appear in (15-9) in the combination $x - ct$, as required by the very general arguments leading to (15-2).

The three forms of the equation for a traveling sinusoidal wave given in (15-4), (15-6), and (15-9) are equivalent. They differ only as to which of the quantities k, λ, ω, T, f, and c appear. We have assumed, however, that the displacement y is zero when $x = 0$ and $t = 0$. To allow for an initial displacement at $x = 0$ and $t = 0$, we need merely incorporate a phase constant ϕ, writing

$$y(x, t) = A \sin [k(x - ct) - \phi]$$

Example 15-2

A sinusoidal wave is generated along a rope that has a linear density of 70 g/m and is under a tension of 10 N from a transverse simple harmonic oscillator at the point $x = 0$. The oscillator executes 4.0 oscillations per second with an amplitude of 2.0 cm. (a) What is the wavespeed? (b) What is the wavelength? (c) Assuming that the oscillator is at the upper amplitude position at time $t = 0$, write the equation for the traveling sinusoidal wave as a function of x and t.

(a) From Eq. (15-1),

$$c = \sqrt{\frac{F_t}{\rho}} = \sqrt{\frac{10 \text{ N}}{0.070 \text{ kg/m}}} = 12 \text{ m/s}$$

(b) From (15-8),

$$\lambda = \frac{c}{f} = \frac{12 \text{ m/s}}{4.0 \text{ Hz}} = 3.0 \text{ m}$$

(c) The transverse displacement y is *not* zero initially; therefore, we must include a phase constant ϕ:

$$y(x, t) = A \sin [k(x - ct) - \phi]$$

Since $y = A$ at $t = 0$ and $x = 0$,

$$A = A \sin (-\phi) = -A \sin \phi$$

or

$$\phi = -\frac{\pi}{2}$$

Then

$$y(x, t) = A \sin \left[k(x - ct) + \frac{\pi}{2} \right]$$

or
$$y(x, t) = A \cos k(x - ct)$$

Recalling that $k = 2\pi/\lambda$, and substituting the numerical values of λ, A, and c into the last equation, we have

$$y(x, t) = (2.0 \times 10^{-2}) \cos \frac{2\pi}{3} (x - 12t)$$

where y and x are in meters and t is in seconds.

15–5 STANDING WAVES

When two or more sinusoidal waves traveling in the *same* direction at the same speed are superposed, the resultant waveform travels with the wavespeed of the component waves. Now consider the superposition of two sinusoidal waves of the same wavelength and amplitude traveling at the same speed but in *opposite* directions. The results are shown graphically in Fig. 15-9, where the two traveling waves, together with their resultant, are shown for a succession of times.

When the amplitude of the resultant exceeds that of either of the two component waves, we have an example of *constructive interference*. Conversely, whenever the amplitude of the resultant is smaller than either of the two waves taken separately, we have a case of *destructive interference*.

First, we note that there are certain times at which the two waves fall exactly on top of one another, namely, at $t = 0$ or $t = \frac{1}{2}T$, where T is the period. There are other times at which the waves give complete destructive interference, namely, at $t = \frac{1}{4}T$, $\frac{3}{4}T$, etc. Figure 15-9 shows, further, that there are certain *points* along the string at which the two traveling waves always interfere destructively. These points, at which the string never undergoes a displacement, are known as nodal points, or *nodes*. The string has its largest amplitude of oscillation at locations, known as *loops*, midway between adjoining nodes. Clearly, *adjacent nodes are separated by one half-wavelength* ($\frac{1}{2}\lambda$); likewise *adjacent loops are separated by* $\frac{1}{2}\lambda$. The resultant oscillating disturbance is called a *standing wave*, or *stationary wave*. These terms are appropriate, since no resultant waveform is seen traveling left or right, and no energy is transferred left or right. The particles of the string oscillate, but the pattern stands still in the sense that the loops and nodes are fixed along the X axis.

How can one produce two similar sinusoidal waves traveling in opposite directions and thus obtain standing waves? One way is to place transverse simple harmonic oscillators, or wave generators, at opposite ends of a string. Even more simply, one can reflect an incident sinusoidal wave at a boundary to produce a reflected wave traveling in the opposite direction. As we saw in Sec. 15-3, any waveform is reflected completely at an infinitely massive boundary.

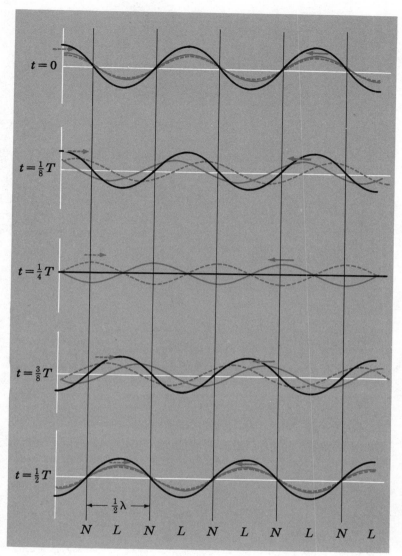

FIG. 15-9. Standing waves. The resultant waveform (heavy black line) for two waves traveling in opposite directions for a succession of times. Adjacent nodes *(N)* and adjacent loops *(L)* are separated by one-half wavelength.

With respect to the incident wave, the reflected wave is inverted up-down and reversed left-right. When one changes the sign of *y* for a sinusoidal wave, inverting it up-down, every wave crest becomes a wave trough, and conversely; that is, the *phase* of the wave is *shifted* 180°. Thus, when a sinusoidal wave on a string is reflected from a hard boundary, it undergoes a phase shift of 180° relative to the incident wave. (By the same token a wave on a string reflected from an altogether "soft" boundary, as in Fig. 15-6, is *not* shifted in phase.) One can, therefore, generate a standing-wave pattern on a string by oscillating one end laterally while keeping the other end fixed, as shown in Fig. 15-10. This effect is strikingly demonstrated by using a tuning fork or other mechanical oscillator; the

FIG. 15-10. An arrangement for demonstrating standing waves on a string. One string end is attached to a vibrating tuning fork; the other end is attached to a weight and hung over a pulley.

oscillations are typically so rapid that one sees only the envelope of the standing-wave pattern.

Standing waves can, in fact, be produced when *both* ends of the string are fixed. The standing-wave pattern must, however, fit between the two ends of the string; that is, the string length L must be an integral multiple of half-wavelengths, if the *boundary conditions* at the string ends, $x = 0$ and $x = L$, are to be satisfied. The allowed wavelengths and frequencies for standing waves on a string fixed at both ends are given by

$$n\frac{\lambda}{2} = L \qquad \text{where } n = 1, 2, 3, \ldots \qquad (15\text{-}10a)$$

$$f = \frac{c}{\lambda} = n\frac{\sqrt{F_t/\rho}}{2L} \qquad (15\text{-}10b)$$

where we have used Eq. (15-1) for the wavespeed c.

The lowest frequency, called the *fundamental* frequency, occurs for $n = 1$. We designate this frequency f_1. Then standing waves can be produced for the frequencies

$$f = nf_1 = f_1, 2f_1, 3f_1, \ldots$$

all allowed frequencies being integral multiples of the fundamental, as shown in Fig. 15-11. The lowest frequency is often denoted the

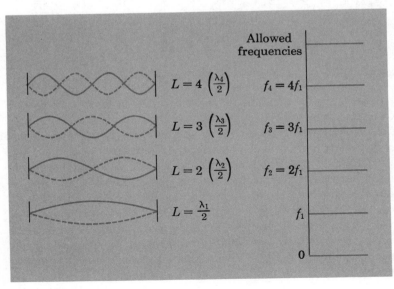

FIG. 15-11. Allowed oscillations for a string attached at both ends.

first harmonic. The second harmonic is $2f_1$, the third harmonic is $3f_1$, etc. The term *harmonic* is used only when the characteristic frequencies are *integral* multiples of the fundamental frequency.

An alternative argument leading to the conditions for standing waves is the following. Imagine a wave incident from the left on a boundary. The reflected wave may be considered to be the incident wave "folded" backward to the left. This folded, or reflected, wave then strikes the left boundary, is folded once more, and again proceeds to the right. Now, this twice-reflected wave will destructively interfere with the initial wave to the right (really, its own tail end) unless the distance ($2L$) it has traveled in one round trip between the boundaries is exactly an integral multiple of *whole* wavelengths. See Fig. 15-12. The wave will then, so to speak, constructively interfere with itself, provided $2L = n\lambda$, or $n(\lambda/2) = L$, as in (15-10a). (Although a 180° phase change occurs upon reflection from a hard boundary, the argument is unchanged, since *two* such phase shifts are made in one round trip.) Thus, a standing wave may be thought of as a traveling wave of infinite extent folded upon itself an infinite number of times, the wavelength being such as to ensure that the boundary conditions are met, namely, that an integral multiple of half-wavelengths fits between the ends.

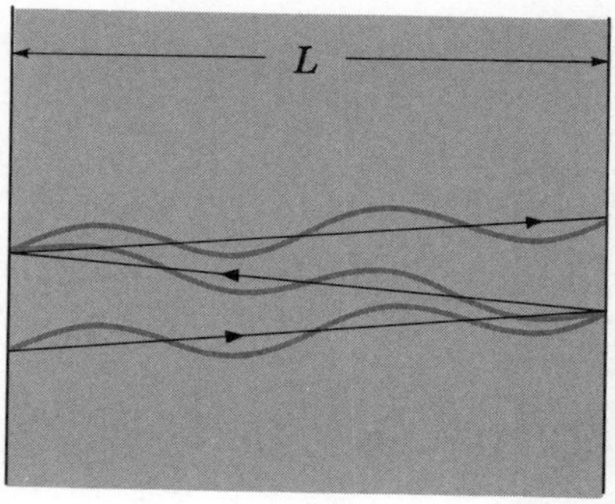

FIG. 15-12. A traveling wave undergoing multiple reflections at boundaries.

According to the superposition principle, several waves may exist simultaneously on a string without their disturbing one another in any way. The waves pass through one another without interaction. Thus, two or more standing waves of *different* frequencies and wavelengths may exist simultaneously on a single string with fixed ends. See Fig. 15-13 which shows the resultant wave pattern on a string for simultaneous excitation of the first and second harmonics. Conversely, *any* periodic disturbance on a

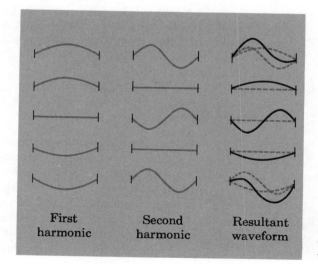

FIG. 15-13. Wave pattern for a succession of times corresponding to the simultaneous excitation of the first and second harmonics.

string — any oscillations of a string which persist in time — must consist of one or more standing waves. On the other hand, any wave disturbance which is not at one of the allowed frequencies — any oscillation which does not correspond to one of the allowed *modes* of oscillation — must, through the arguments of the last paragraph, quickly die out, because such a wave destructively interferes with itself, the energy being dissipated at the boundaries. Thus, when a string is struck or bowed, as in the piano or violin, it can oscillate simultaneously in one or more of the allowed frequencies. Typically, the amplitude of the fundamental exceeds that of higher harmonics. The higher harmonics are sometimes referred to as "overtones"; they have a frequency "over" that of the fundamental. For a string with fixed ends, the second harmonic is the first overtone, etc.

Example 15-3

A string 2.0 m long is held fixed at both ends. It is found that if a wavepulse is generated near one end, it takes 0.10 s for the pulse to travel one round trip to the far end and return. What are the allowed frequencies of oscillation for this string?

The wavespeed c is the distance traveled by the wavepulse divided by the elapsed time, $(4.0 \text{ m})/(0.10 \text{ s}) = 40$ m/s. The allowed standing-wave patterns are those for which the string length L is an integral multiple of one half-wavelength: $L = n(\lambda/2)$. Thus,

$$f = \frac{c}{\lambda} = \frac{nc}{2L} = \frac{n(40 \text{ m/s})}{2(2.0\text{m})}$$

$$= n(10 \text{ Hz}) = 10, 20, 30, \ldots, \text{Hz}$$

We can arrive at the same result through another argument: With the string oscillating in its fundamental mode, a sinusoidal wave will travel one round trip between the boundaries in a time of one period, because only then will it constructively interfere with itself in further reflections. Thus, the period of the first harmonic is 0.10 s and its frequency is 10 Hz. The higher harmonics are integral multiples of 10 Hz. Note that string length and wavespeed do *not* enter here.

15-6 RESONANCE

First recall some results concerning the phenomenon of resonance, given in Sec. 14-5. There we considered the simple case of a single mass attached to a spring, damped by a force opposing the motion, and driven by an external oscillator of variable frequency. With relatively small damping and no driving force, the mass oscillates at its natural frequency with a slowly decreasing amplitude. The oscillations die out. When the driving force is applied, the mass always oscillates at the same frequency as the driving force, but its oscillation amplitude reaches a sharp maximum when the frequency of the applied force is equal to the oscillator's natural frequency. That is, energy is transferred from the driver to the oscillator at a maximum rate, and *resonance* is achieved, when the oscillator is driven at its natural frequency.

Now consider a string fixed at its ends. It has many natural frequencies of oscillation, not just one. It has, in fact, an infinite number of characteristic frequencies and characteristic, or normal, modes of oscillation. The resonant frequencies are given by Eq. (15-10b); they are the frequencies corresponding to the existence of standing waves. All oscillations of a string are, of course, damped, either by friction with the surrounding air or by internal friction arising from the stretching of the string. If a string is excited by an external driving force, for example, by variations in the air pressure from a sound wave, it may exhibit resonance; the oscillation amplitude will be large if the exciting frequency is one of the string's natural frequencies. Note that, although there exist an infinite number of resonant frequencies for a given string, these frequencies are not distributed continuously, but rather are in the ratio of integers, as shown in Fig. 15-11.

The fact that waves on a string, or other waves trapped between reflecting boundaries, have a set of discrete frequencies and characteristic modes of oscillation has an important analogy in the quantum theory. Indeed, the stability and structure of atoms, molecules, and nuclei can be understood on the basis of the wave properties one must attribute to such particles as electrons, protons, and neutrons.

Example 15-4

A violin with a string 31.6 cm long, of linear density 0.65 g/m, is placed near a loudspeaker fed by an audio-oscillator of variable frequency. It is found that as the frequency of the sound waves reaching the violin string is varied continuously over the range of 500 to 1,500 Hz, the string is set in oscillation only at the frequencies 880 and 1,320 Hz. What is the tension in the string?

The violin string, fixed at both ends, oscillates in resonance at its characteristic frequencies, which are in the ratio of the integers 1, 2, 3, ... The ratio of the resonance frequencies here is $\frac{1320}{880} = \frac{3}{2}$. Therefore, the two resonances correspond to the string oscillating in its second and third harmonics. (If the two harmonics were, for example, the fourth and sixth, their ratio would again be $\frac{3}{2}$, but one would then find, in addition, a resonance at the fifth harmonic, 1,100 Hz.) Consequently, the fundamental, or first harmonic is 440 Hz; this corresponds to "A" above "middle C" on the concert scale. The length of the string for this frequency is one half-wavelength:

$$L = \frac{\lambda}{2} = \frac{c}{2f} = \frac{\sqrt{F_t/\rho}}{2f}$$

$$F_t = 4f^2 L^2 \rho = 4(440 \text{ Hz})^2 (0.316 \text{ m})^2 (6.5 \times 10^{-4} \text{ kg/m}) = 50 \text{ N}$$

SUMMARY

The general form of the function F describing a wave, traveling along the X axis at the wavespeed c, is

$$y = F(x \mp ct) \tag{15-2}$$

where the minus and plus signs refer, respectively, to wave propagation in the positive and negative X directions. For a transverse wave on a stretched uniform string the wavespeed is

$$c = \sqrt{\frac{F_t}{\rho}} \tag{15-1}$$

where F_t is the tension in the string and ρ is its linear density.

A sinusoidal wave, generated by a simple harmonic oscillator of amplitude A, angular frequency ω, and period T, and traveling along the positive X axis, may be represented in any of the following equivalent forms:

$$y = A \sin(\omega t - kx) \tag{15-4}$$

$$y = A \sin 2\pi \left(\frac{t}{T} - \frac{x}{\lambda}\right) \tag{15-6}$$

$$y = -A \sin k(x - ct) \tag{15-9}$$

where the wavelength λ, the distance between two nearest points with the same phase, is related to the wave number k by $k = 2\pi/\lambda$. The product of the frequency f and wavelength λ is a constant, c, the wavespeed, or phase speed.

$$c = f\lambda \tag{15-8}$$

The superposition principle governs wave interference: The resultant wave disturbance is simply the vector sum of the component disturbances. There are no wave-wave interactions in linear systems. Superposition of transverse waves on a string requires that the transverse displacements be small. Constructive and destructive interference refer, respectively, to situations in which the resultant disturbance is greater than or less than the component disturbances. At a boundary, or change in medium for wave propagation, the incident wave is partially transmitted and partially reflected. Reflection from a more massive second medium results in a 180° phase change at the boundary; reflection from a less massive second medium produces no phase change at the boundary.

An example of wave interference is standing waves which, for a string of length L fixed at both ends, lead to the allowed frequencies $nc/2L$.

PROBLEMS

15-1. Find the speed of a transverse wave on a 75-cm length of cord when the tension in the cord is 320 N. The mass of the cord is 120 g.

15-2. A transverse wave travels at a speed of 400 m/s in a steel wire with linear density of 0.10 g/cm. What is the tension in the wire?

15-3. A transverse wavepulse is propagating from one end of a taut string to the other. The applied tension in the string is suddenly doubled. Describe what happens to the *(a)* shape, *(b)* amplitude, and *(c)* speed of the wave.

15-4. A triangular wavepulse has the same slope (magnitude) of 5.0×10^{-3} along the leading and trailing edges (see Fig. P15-4). The wave moves to the right at 5.0 m/s. What is the transverse particle speed *(a)* along the leading edge and *(b)* along the trailing edge?

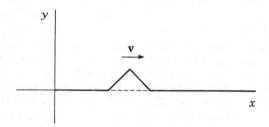

FIG. P15-4

15-5. A rope hangs vertically under its own weight. Show that the wavespeed at a distance y from the lower end is given by $\sqrt{gy}$. Note that the wavespeed is independent of the rope's linear density; hence, this relation applies even if the rope is *not* uniform.

15-6. Figure P15-6 shows a snapshot of the waveshape $y(x)$ at one instant of a pulse's traveling to the right. Sketch curves showing the transverse *(a)* displacement, *(b)* velocity, and *(c)* acceleration, all as a function of time.

FIG. P15-6

15-7. Two strings with linear densities ρ_1 and ρ_2 are joined together and are under the same tension. *(a)* If a wave has a wavelength of 2.0 cm in string 1, what is the wavelength in string 2? *(b)* If the frequency is 100 Hz in string 1, what is the frequency in string 2? *(c)* What is the wavespeed of the string with density ρ_1 and *(d)* the string with density ρ_2?

15-8. Figure P15-8 shows a sinusoidal traveling wave for three different times. What are *(a)* the (minimum) wavespeed, *(b)* the amplitude, *(c)* the frequency, and *(d)* the wavelength? *(e)* In what direction does the wave travel? *(f)* Write an equation for this traveling wave.

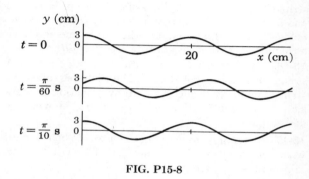

FIG. P15-8

15-9. A traveling transverse wave is described by the equation $y = (2.0 \times 10^{-3}) \sin(10x + 200t)$, where all quantities are in SI units. *(a)* What are the amplitude, wavelength, and frequency of the wave? *(b)* What is the velocity of the wave? *(c)* What is the transverse particle velocity at $x = (\pi/10)$ m and $t = (\pi/160)$ s?

15-10. A transverse wave with an amplitude of 5.0 mm travels along the negative X axis at a speed of 120 m/s; the wavelength is 6.0 m. Find the following quantities: *(a)* frequency in Hz, *(b)* angular frequency, *(c)* period T, and *(d)* wavenumber k. Using SI units, write equations for this wave *(e)* in the form of Eq. (15-4) and *(f)* in the form of Eq. (15-6).

15-11. Consider the sinusoidal wave on a rope which is described in Example 15-2. *(a)* Find the magnitude of the maximum transverse linear momentum of a small segment of rope 1.0 mm long. *(b)* What is the magnitude of the maximum resultant force on this segment?

15-12. A requirement for the applicability of the superposition principle is that the wavespeed be large compared with the transverse speed of a particle. Show that, for a sinusoidal wave, this requirement implies that the amplitude be small compared with the wavelength.

15-13. Two wavepulses traveling in opposite directions completely annul each other at one instant. What has become of the energy carried by each wavepulse?

15-14. What is the relative change in the tension of a particular violin string when the fundamental frequency is changed from 435 Hz to 440 Hz? (Concert A is 435 Hz according to International Pitch; according to Standard Pitch, Concert A = 440 Hz.)

15-15. If the fundamental frequencies of piano strings A and D are to be in the ratio $f_D/f_A = \frac{3}{2}$, what is the required ratio in their tensions? Assume that the two strings are made from sections of the same piece of wire and that their lengths are in the ratio $L_D/L_A = \frac{5}{6}$. (Two musical notes which sound at the same time and have a frequency ratio of $\frac{3}{2}$ constitute a *perfect fifth*.)

15-16. Initially the fundamental frequency of the A string of a violin is 444 Hz. The violinist tunes the string to 440 Hz (Concert Pitch) by adjusting the tension. By what fraction must the string's tension be decreased?

15-17. A string fastened at both ends is to be resonant with wavelengths of 0.16 and 0.18 m. What is the minimum string length?

15-18. A string fixed at its ends oscillates in the normal modes corresponding to its first and third harmonics. Both modes have the same amplitude. Sketch the resultant wave pattern for several instants of time in the fashion of Fig. 15-13.

15-19. What tension in the violin string of Example 15-4 would produce resonances in the string of frequencies 880, 1,100, and 1,320 Hz when the audio-oscillator is varied continuously over the range of 700 to 1,500 Hz?

15-20. When attached at both ends, a string under a tension of 400 N and having a linear density of 1.0×10^{-2} kg/m is resonant at the frequency of 150 Hz. The next highest frequency at which the string resonates is 200 Hz. (*a*) What is the string's length? (*b*) What are the two harmonics?

15-21. The lower end of a 25-cm-long vertical rod is clamped at the center and attached to an oscillator which oscillates with small amplitude in the horizontal direction. The upper end of the rod is free. The lowest frequency at which resonance occurs is 1,000 Hz. (*a*) With what speed do transverse waves propagate through the rod? (*b*) At what frequency will the next resonance occur?

15-22. An oscillating string is damped by a force which is proportional to the transverse velocity of the string. Show that the higher-frequency modes of a struck or plucked string are damped out more rapidly than the fundamental. (This effect can be heard, particularly in bass notes of such a stringed instrument as the harpsichord: The tone quality changes with time, the fundamental becoming relatively more important as the overtones are quickly damped.)

CHAPTER 16

Compressional Waves

This chapter deals with compressional elastic waves through a solid, liquid, or gas. There is a close parallel between the wave characteristics developed in the last chapter for transverse waves on a stretched string and those for longitudinal waves. Happily, all the phenomena discussed earlier—superposition, reflection, standing waves, resonance—apply to compressional waves, essentially without modification. The chief difference is this: The property describing the wave disturbance is now not a transverse displacement but rather a longitudinal displacement, a density variation, or a pressure variation.

16-1 LONGITUDINAL WAVES

Figure 16-1 shows a simple structure which serves as a prototype for the propagation of longitudinal elastic waves, a number of identical masses coupled together by springs to form a chain. Each spring follows Hooke's law for small elongations or compressions.
When any one mass is displaced longitudinally, the springs attached to it are stretched or compressed; these deformations produce a force not only on the displaced mass but also on the neighboring masses. Thus, neighboring masses are set in motion, and when the mass at the left end of the chain is suddenly displaced to the right, a compressional disturbance is found to be propagated to the right along the chain. Each mass undergoes, in turn, a longitudinal displacement and then returns to its equilibrium position, as a *longitudinal*, or *compressional*, wave travels along the chain.

FIG. 16-1. A simple model for longitudinal waves.

Energy is transported longitudinally—along the line of the mass's motion—as the kinetic energy of the masses and the elastic potential energy of the deformed springs. In addition, a traveling longitudinal wave carries linear momentum *along* the direction of wave propagation. (Recall that a transverse wave on a stretched string carries linear momentum at right angles to the direction of energy propagation.)

The essential conditions for the existence of longitudinal waves are that the *medium,* in this case the chain of masses and springs, possess *inertia* and be *elastically deformable.* Clearly, if there were no coupling between masses (if the springs were infinitely weak), displacing one mass would not influence neighboring springs or masses, and a wave would not be produced. On the other hand, if the medium were perfectly rigid (if the springs were infinitely stiff), displacing one mass would cause *all* the other masses to be displaced *simultaneously.* Again, a wave traveling at finite speed would not be produced. With an elastic medium, however, an external force initially deforms and sets in motion only a portion of the medium, and this deformed portion, through its coupling with adjoining portions, sets them in motion.

For simplicity, suppose that one end of the chain in Fig. 16-1 is moved longitudinally in simple harmonic motion. All the masses will eventually be set in simple harmonic motion with the same amplitude and frequency as that of the source, each oscillating mass having a phase lag relative to the source which is proportional to its distance from it. We have a longitudinal sinusoidal wave. How do we represent such a wave graphically and describe it analytically?

The coordinate x gives the equilibrium location of each mass in the chain relative to, say, the left end. Let us use y to represent the *longitudinal* displacement of a particle with *its* equilibrium position along the X axis, the direction of wave propagation. Figure 16-2a shows the Y displacements of the particles along the chain,

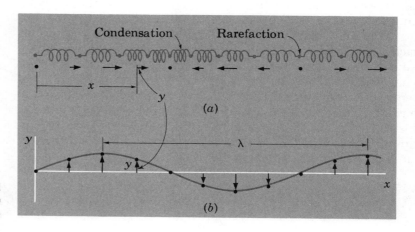

FIG. 16-2. *(a)* A sinusoidal longitudinal wave. *(b)* The corresponding waveshape.

CHAP. 16 *Compressional Waves*

and Fig. 16-2b is a plot of y against x, both for a sinusoidal wave. Just as for a transverse wave on a stretched string, the function $y(x)$ for any given time t is a sinusoidal wave. This waveshape travels to the right at the wavespeed, or phase speed, c, and we may represent it analytically by any of the relations [Eqs. (15-4), (15-6), or (15-9)] describing a traveling sinusoidal wave. For example,

$$y = A \sin(\omega t - kx) \qquad (15\text{-}4a),(16\text{-}1)$$

where A is the longitudinal oscillation amplitude, $\omega = 2\pi f$, and $k = 2\pi/\lambda$. As before, the wavelength λ is the distance between adjoining oscillators having the same phase, and the phase velocity is given by $c = f\lambda$.

It must be emphasized that Fig. 16-2b, which graphs the waveshape $y(x)$, is *not* a snapshot of the longitudinal wave at any instant, as was the case of a transverse wave on a string. We may, however, use exactly the same mathematical expressions to describe longitudinal waves, if it is understood that the particle displacement y is now a *longitudinal* displacement. The regions in Fig. 16-2a (a snapshot) in which the particles are crowded together, the regions of greatest density, are known as *condensations*. Regions in which the particles have their greatest relative separation, regions of minimum density, are called *rarefactions*. Rarefactions and condensations travel in the direction of wave propagation at the wavespeed c. In fact, a longitudinal wave in a deformable medium may be described as a disturbance for which the propagated property, or wave function, is the density variation of the medium.

Our model for a longitudinal wave consists of discrete masses coupled by identical springs. A still simpler physical arrangement for longitudinal waves is that of a single long stretched helical spring. Here the mass is distributed continuously throughout the length of the spring, rather than concentrated in regularly spaced masses. The displacement y then gives the shift from its equilibrium position of any turn of the spring, and a longitudinal wave is seen as compressed or elongated segments traveling along the spring.

16-2 COMPRESSIONAL WAVESPEEDS

The molecules of a solid or liquid are subject to intermolecular forces which, for small displacements, obey Hooke's law, the intermolecular restoring force being proportional to the molecule's displacement from equilibrium position (Sec. 14-4). Thus, the simple model of masses and springs shown in Fig. 16-1 represents quite faithfully an actual deformable solid or liquid. The discrete masses are now to be thought of as atoms or molecules, and the springs correspond to the interatomic or intermolecular forces. For a three-dimensional medium any one molecule is coupled to *all* molecules that immediately surround it.

The wavespeed c for a compressional wave is given by

$$c = \sqrt{B/\rho} \qquad (16\text{-}2)$$

where ρ is the mass per unit volume. The *bulk modulus* B is defined as

$$B = \frac{\Delta p}{-\Delta V/V} \qquad (16\text{-}3)$$

where $-\Delta V/V$ is the relative *decrease* in volume due to added pressure Δp.† For a typical solid, $B \approx 10^{11}$ N/m² and $\rho \approx 10^4$ kg/m³. Using (16-2), we find a typical compressional wavespeed to be 3×10^3 m/s. For a frequency of 10^3 Hz, the wavelength is 3 m.

Equation (16-2) is of the general form: Wavespeed = $\sqrt{\text{elastic property/inertial property}}$. All types of mechanical waves follow a relation of this type. It is also illustrated, by our earlier formula $c = \sqrt{F_t/\rho}$, Eq. (15-1), for the wavespeed of a transverse wave on a stretched string, where the elastic and inertial properties are, respectively, the string's tension and linear density.

A compressional wave may be described as a *displacement* wave; it may also be described as a *pressure* wave. It can be shown that the pressure variations are 90° out of phase with respect to the displacement variations; that is, the pressure is greatest (and least) at those points where the particles undergo no displacement. This is evident from Fig. 16-2b: the condensations and rarefactions—the regions of respective maximum and minimum pressure (and density)—occur at those locations where the particles are temporarily at their equilibrium positions.

Compressional waves through a gas are, of course, responsible for sounds that reach the ear. In an ideal gas we do not have a collection of particles coupled together in the fashion of the molecules of a solid or liquid; the molecules are in random thermal motion, interacting only during collisions. There are no elastic forces acting constantly. But, if one portion of a gas is suddenly compressed, and gas molecules are thereby crowded together in a condensation, the molecules will, because of the higher pressure, tend to move to adjoining portions of the gas where the pressure and density are lower. A compressional wave for a gas, then, corresponds to density changes, in exactly the same fashion as for solids and liquids, superposed on the random molecular motion.

For sound waves through a gas—waves having frequencies in the audible range, roughly 20 to 20,000 vibrations per second—the oscillations in gas density take place so rapidly, and the thermal conductivity of the gas is so low, that there is not enough time for the thermal energy to flow from a condensation to a rarefaction, from a region of high density and temperature to one of low density and temperature. The process is said to be *adiabatic*.

---

†The concept of *pressure* is defined and discussed in Sec. 17-3.

For an adiabatic process in an ideal gas Eq. (16-2) can be written in the form

$$c = \sqrt{\gamma \frac{RT}{w}} \qquad (16\text{-}4)$$

where R is the gas law constant, T is the absolute temperature, w is the molecular "weight," and γ is a constant for any particular gas.†

The speed of sound through an ideal gas varies directly as the square root of the absolute temperature T. For air at 0°C, $c = 331$ m/s.

16–3 SUPERPOSITION, REFLECTION, AND STANDING WAVES

These phenomena applied to elastic waves are altogether similar to the corresponding effects found earlier in transverse waves on a stretched string. The wave property simply is different: Now it is a longitudinal displacement, or a pressure or density change, rather than a transverse displacement. Two compressional waves interfere constructively when the density or pressure is enhanced by superposition of the separate changes, and conversely for destructive interference.

Suppose that a deformable medium is terminated by an infinitely hard and massive second medium. When a compressional wave reaches the boundary, particles there cannot be displaced longitudinally. Since the wave cannot be propagated into the second medium, the wave is completely reflected. The longitudinal displacement y must be zero at the boundary; therefore, a *displacement node* occurs at the boundary between a deformable medium and a nondeformable medium. Recall that the displacement and pressure variations are 90° out of phase. Thus, if y is zero at the boundary, Δp must be a maximum there: A *pressure loop,* or antinode, exists at the boundary. This follows also from the fact that the particles will tend to pile up at the boundary, thereby producing a high density and pressure at this point. On the other hand, if the deformable medium is terminated with a less dense medium, the displacement y will be a maximum, or loop, and the pressure Δp will be a minimum, or node, at this boundary. As before, a wave undergoes a 180° change in phase when reflected from a boundary leading to a second medium in which the wavespeed is less, whereas no phase change occurs for waves traveling in the reverse direction, from the second to the first medium.

Standing waves can exist in a medium propagating a wave disturbance when waves of the same wavelength travel in opposite

---

†See Sec. 17-5 for a discussion of the parameters R and T.

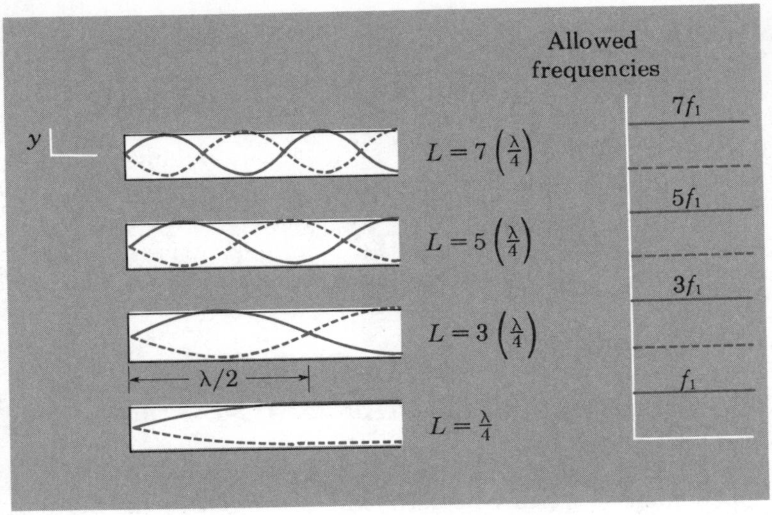

FIG. 16-3. Allowed oscillation modes and characteristic frequencies for sound waves in a pipe closed at one end and open at the other. Note that the standing wave patterns show the longitudinal displacements, rather than the pressure variations.

directions. The resultant disturbance is a standing wave, with alternate nodes and loops. Adjacent loops and adjacent nodes are separated by half-wavelengths.

Example 16-1

A sound wave is propagated through a pipe closed at one end. What are the allowed oscillation modes and characteristic frequencies?

The boundary conditions are as follows: At the closed end there is a displacement node or pressure loop; at the open end there is a displacement loop or pressure node. The allowed oscillation modes, or standing-wave patterns, are shown in Fig. 16-3 (the actual displacement loop lies somewhat beyond the open end of the pipe). Note particularly that the quantity plotted here is the longitudinal particle displacement y; if one were to plot Δp, the loops and nodes would be reversed.

The length L of the tube is an odd multiple of $\frac{1}{4}\lambda$; $L = n(\lambda/4)$, where $n = 1, 3, 5, \ldots$. The allowed frequencies are $f = c/\lambda = n(c/4L) = nf_1 = f_1, 3f_1, 5f_1, \ldots$. The overtones are *odd* harmonics of the fundamental frequency f_1. There may, of course, be two or more characteristic oscillation modes existing simultaneously, and any disturbance persisting in time must consist of a superposition of allowed modes.

If the length of the closed pipe is 0.50 m, roughly the length of a clarinet, the fundamental frequency is

$$f_1 = \frac{c}{4L} = \frac{344 \text{ m/s}}{4(0.50 \text{ m})} = 172 \text{ Hz}$$

roughly at the center (logarithmically) of the audible range.

Example 16-2

A sound wave is propagated through a pipe open at both ends. What are its allowed oscillation modes and characteristic frequencies?

The boundary conditions are now these: A displacement loop,

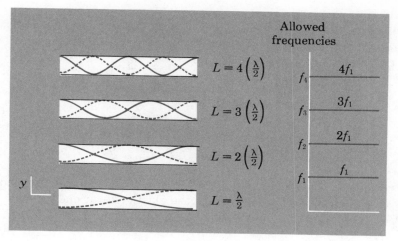

FIG. 16-4. Allowed oscillation modes and characteristic frequencies of sound waves in a pipe open at both ends. The wave pattern shows the *longitudinal* displacement of the gas particles.

or pressure node, must exist at each of the two open ends. The allowed oscillation modes are shown in Fig. 16-4. The tube length is always an integral multiple of $\frac{1}{2}\lambda$; the allowed frequencies consist of *all* harmonics of the fundamental.

16-4 SOUND AND ACOUSTICS

Mechanical vibrations over the audible frequency range, from 20 to 20,000 Hz (a factor of 10^3), and with intensities lying between 10^{-16} and 10^{-4} W/cm², are perceived by a typical human ear as sound. (The eye detects electromagnetic vibrations only over a frequency range of factor 2, or one octave.) Nonperiodic variations in the air pressure correspond to noise. Oscillations at frequencies in the ratios of simple integers are recognized as musical tones.

By the *pitch* of a musical tone is meant the lowest, or fundamental, frequency. The relation of perceived pitch to fundamental frequency is not precise, however. The *loudness* of the tone is a measure of the sound intensity. The tonal *quality*, by which the ear can distinguish, for example, between a violin and a clarinet playing the same note, is a measure of the relative amplitudes of the overtones excited simultaneously with the fundamental oscillation.

Musical instruments usually involve arrangements whereby the characteristic overtones are harmonics, or integral multiples of the fundamental frequency. This is illustrated by oscillating strings in string instruments and oscillating air columns in pipes, such as in the woodwind and brass instruments. Since the wavelength of sound through air at 400 Hz, roughly the center of the audible range, is about 0.8 m, wind instruments are expected to be, and are found to be, roughly this size.

The acoustic spectrum is shown in Fig. 16-5. Note that

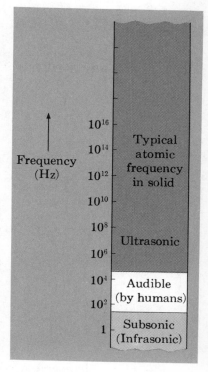

FIG. 16-5. The acoustic spectrum.

Sound and Acoustics SEC. 16-4

259

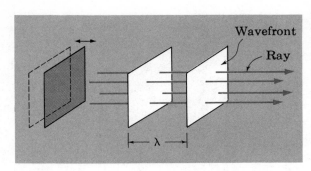

FIG. 16-6. Plane wave source and plane wavefronts.

frequencies alone are given; the wavelengths depend on the medium (the wavespeed) through which the elastic vibrations travel.

16–5 WAVEFRONTS AND RAYS

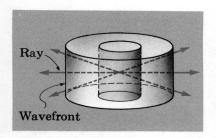

FIG. 16-7. Line source and cylindrical wavefronts.

The simplest configuration for wave propagation in two or three dimensions is shown in Fig. 16-6. Here the wave source is spread over a plane, as in the case of an oscillating flat diaphragm generating sound waves, and the waves progress to the right in a single direction. We indicate the direction in which waves radiate from the source by straight lines, or *rays*. A *wavefront* is defined as a surface on which all points have the same phase of oscillation. The wavefronts of a plane wave source consist of planes at right angles to the rays. The perpendicular distance between adjoining wavefronts in the same phase is the wavelength. Surfaces of constant phase, or wavefronts, advance from the source at the *phase speed*, or wavespeed, c. A line source generates cylindrical wavefronts centered about it, as shown in Fig. 16-7. Rays again intercept wavefronts at right angles. Figure 16-8 shows spherical wavefronts in three dimensions emanating from a point source. In general, rays are always perpendicular to the associated wavefronts.

Any small portion of a cylindrical or spherical surface is very closely a plane; that is, over a small solid angle, a cylindrical or spherical wavefront closely approximates a plane wavefront.

16–6 INTENSITY VARIATION WITH DISTANCE FROM SOURCE

The intensity, or energy flux, of any wave is defined as the energy passing per unit time through a unit cross-sectional area. That is, the intensity I is the power P per unit transverse area A:

$$I = \frac{P}{A}$$

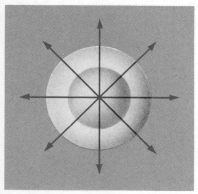

FIG. 16-8. Point source and spherical wavefronts.

We wish to find how the intensity of a wave varies with distance from the source for three simple configurations: plane, cylindrical, and spherical wavefronts. In a plane wave there is no divergence of the rays; the wavefronts are parallel planes; and the same energy from the source passes through every transverse area. Therefore, the intensity is *constant,* independent of distance from the source, assuming no absorption of energy by the medium.

For cylindrical waves the energy carried by the wave is diluted in space as the expanding cylindrical wavefronts increase their surface area with time. Consider one angular segment, shown in Fig. 16-9. The energy passing through the area A_1 with radius r_1 is exactly the same as the energy passing later through the area A_2 with radius r_2. Clearly, $A_1 = r_1 \theta h$ and $A_2 = r_2 \theta h$, or $A_1/A_2 = r_1/r_2$. Since the intensity varies inversely with the transverse area and $A = r\theta h$, the intensity of a cylindrical wave varies inversely as the distance r from the line source.

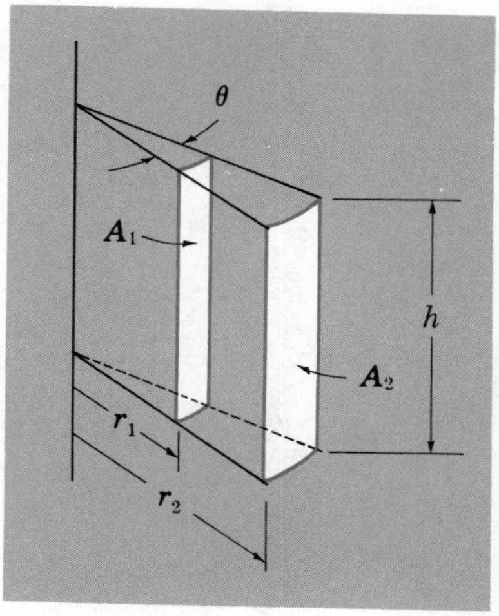

In the case of a spherical wave propagated from a point source, all the energy within the small cone having angles θ_1 and θ_2 passes, in turn, through areas A_1 and A_2, as shown in Fig. 16-10. But $A_1 = r_1^2 \theta_1 \theta_2$ and $A_2 = r_2^2 \theta_1 \theta_2$; that is, the surface area of a given solid angle varies directly as the square of the sphere's radius. Thus, the intensity of waves from a point source varies inversely as the square of the distance from the source. In summary, when no wave energy is absorbed by the medium,

FIG. 16-9. Angular segment for cylindrical wavefronts.

Plane waves: $\quad\quad\quad\quad I = \text{constant}$

Cylindrical waves: $\quad\quad I \propto \dfrac{1}{r}$ $\quad\quad\quad\quad\quad\quad$ (16-5)

Spherical waves: $\quad\quad\; I \propto \dfrac{1}{r^2}$

Example 16-3

A point source radiates isotropically at the rate of 10 kW. What is the intensity of the radiation at a distance of 10 m from the source?

The intensity I at a radius r from a point source is given by the power P through a spherical area $A = 4\pi r^2$:

$$I = \frac{P}{A} = \frac{P}{4\pi r^2} = \frac{(1.0 \times 10^4 \text{ W})}{4\pi (10 \text{ m})^2} = 8.0 \text{ W/m}^2$$

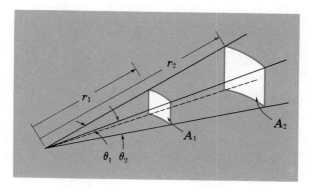

FIG. 16-10. Angular segment for spherical wavefronts.

Intensity Variation with Distance from Source SEC. 16-6

16–7 HUYGENS' PRINCIPLE

Given one wavefront for a progressing wave, how does one find a future wavefront? A remarkably simple geometrical procedure, *Huygens' principle,* may be used to chart the progress of a wavefront through a medium. Devised by C. Huygens (1629–1695) in 1678, this principle asserts that each point on an advancing wavefront may be regarded as a new point source generating spherical *Huygens' wavelets* in the forward direction of wave propagation. Thus, to find the wavefront at a time t later, one draws circular arcs of radius ct centered at points along the wavefront; the new wavefront at time t is merely the envelope of these wavelets; see Fig. 16-11. Thus a plane wavefront generates another plane wavefront; a spherical wavefront generates another spherical wavefront, of larger radius, about the same point source.

Huygens' construction is a geometrical procedure, not a physical method. Clearly, if one is to find a wavefront at some *future* time, one must draw the envelope of the wavelets along the *leading,* rather than the trailing, side of the wavefront. Huygens' method gives only the *possible* wavefronts at some future time; it does not give the distribution of energy over these wavefronts.

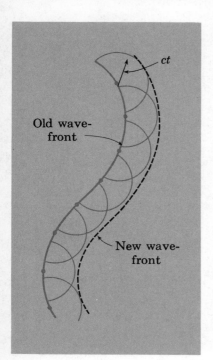

FIG. 16-11. Huygens' construction for advancing wavefronts.

SUMMARY

In a compressional or longitudinal wave disturbance the particles oscillate along the same direction as that in which the energy and linear momentum of the wave travel. The superposition principle applies when the elastic deformations follow Hooke's law.

The pressure variations for a longitudinal wave through a compressible medium are 90° out of phase with the longitudinal particle displacement.

Wavefronts are surfaces of constant phase. Rays indicate the direction of wave propagation and are always at right angles to the wavefronts. Plane, line, and point sources generate, respectively, plane, cylindrical, and spherical wavefronts.

The variation with the distance of the intensity, the power per transverse area of a wave, follows the rules:

Plane waves: $\qquad I = \text{constant}$

Cylindrical waves: $\qquad I \propto \dfrac{1}{r}$ \qquad (16-5)

Spherical waves: $\qquad I \propto \dfrac{1}{r^2}$

In using Huygens' principle to construct future wavefronts, one imagines each point on the present wavefront to act as a point source of Huygens' wavelets, the future wavefront being the envelope of such wavelets.

PROBLEMS

16-1. If the speed of the longitudinal wavepulse shown in Fig. P16-1 is 6.0 km/s, what is the maximum longitudinal *particle* speed?

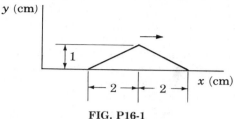

FIG. P16-1

16-2. The volume of a certain solid shrinks by 2 parts in 10^6 when it is subject to an external hydrostatic pressure of 1 atm. The density of the solid is 8.0 g/cm^3. What is the speed of a longitudinal wave through this material?

16-3. In the sonar method of measuring distances under water a sound pulse travels from the generator to a distant object and is reflected back to the location of the generator. The distance is computed by measuring the time interval between the transmission and reception of the pulse. Taking the bulk modulus of water as 2.0×10^9 N/m^2 and the density of water as 1.0×10^3 kg/m^3, what is the time interval for a sonar pulse striking an object 50 m distant from the transmitter?

16-4. A compressional earthquake tremor travels through the earth's crust with a wavespeed of 8.0 km/s. Find the average value of the bulk modulus of the crustal rocks if their mean density is 3.3 g/cm^3. (These values characterize the crust at a depth of 50 km.)

16-5. An observer fires a shot and hears its echo from a surface 150 m distant 0.860 s later. What is the temperature of the air, assumed to be an ideal gas?

16-6. What is the wavelength in air (at 20°C) of the limits of the audible sound spectrum?

16-7. What is the frequency of the lowest frequency sound which can produce a resonance by virtue of reflection from the opposite walls of a room if the walls are 6.0 m apart?

16-8. The left and right ears of a person are separated by 15 cm. (a) What is the difference in arrival time at the two ears of a sound pulse originating from a location 45° between the front and side directions? (b) For what frequency of acoustic oscillation would this difference in arrival time equal the period of oscillation?

16-9. (a) Use differential calculus to show that in an ideal gas the differential rate of change of wavespeed with temperature is given by $(d/dT)(c) = c/(2T)$.

(b) Find the rate of change with temperature of the speed of sound at 0°C.

16-10. (a) Show that an alternate form of Eq. (16-2) is given by $c = (dp/d\rho)^{1/2}$, where p is the pressure and ρ the mass density. (b) Use this equation to find the speed of sound at the *tropopause* (lower boundary stratosphere), where $p = 0.223\,p_0$ and $\rho = 0.282\rho_0$. At ground level nominal pressure and density are $p_0 = 1.013 \times 10^5$ N/m^2 and $\rho_0 = 1.292$ kg/m^3.

16-11. A pipe, open at both ends, is set in resonance only at the frequencies of 800 and 1,000 Hz when an exciting audio-oscillator is varied in frequency over the range from 700 to 1,100 Hz. What is the length of the pipe?

16-12. A pipe, initially open at both ends, oscillates at the frequency of 840 Hz (not necessarily its fundamental frequency). When one end is closed, its oscillation frequency is 210 Hz. What is the minimum pipe length satisfying these conditions?

16-13. When a sound generator sweeps through all frequencies between 1,000 Hz and 2,000 Hz, resonance is excited in a hollow tube at just two frequencies — 1,400 Hz and 1,800 Hz. The 45-cm-long tube is open to the air at one end and closed at the other. Find (a) the fundamental resonance frequency for the tube, (b) the speed of sound in air, and (c) the air temperature. (At 0°C, the speed of sound in air is 331 m/s.)

16-14. A particular organ pipe is in tune at 20°C. Over what temperature range will the frequency remain within 1 percent of the "true" value? Assume that changes in the dimensions of the pipe can be neglected.

16-15. The speed of sound is measured with a tube, open at one end and fitted at the other end with a movable piston. Resonance oscillations are excited by a tuning fork of frequency 880 Hz placed near the open end. It is found that the sound intensity exhibits peaks when the piston is at the positions 9.2, 28.3, and 47.2 cm. What is the speed of sound?

16-16. A 12-in 33⅓-rpm phonograph record can reproduce sound with frequencies of up to 16 kHz. (a) What is the maximum separation distance in the record groove (near the beginning of the record) between the oscillations corresponding to the highest frequency? (b) What is the corresponding separation distance on a magnetic tape played at the speed of 1⅞ in/s?

16-17. Since the ear is sensitive to sound over an enormous range of intensities, from 10^{-16} to 10^{-4} W/cm^2, it is convenient to measure the loudness of sound on a logarithmic scale. The sound level in *decibels* (db) of an intensity I is given by $10 \log_{10}(I/I_0)$, I_0 being the minimum detectable intensity, 10^{-16} W/cm². What is the noise level in decibels of (a) ordinary conversation, which corresponds to 10^{-10} W/cm², and (b) the threshold of pain, 10^{-4} W/cm²?

16-18. In the musical *diatonic scale,* the "fifth" has a frequency which is $\frac{3}{2}$ times that of the tonic ("G," 396 Hz, as compared with "C," 264 Hz, for example). Indeed, all the tones in the major scale are in the ratio of simple integers. In the *well-tempered scale,* on the other hand, any two adjoining tones of the 12 half-tones comprised in one octave differ in frequency by the factor $2^{1/12} = 1.0595$. (a) What is the ratio of the fifth (the seventh half-tone) to the tonic on the well-tempered scale? (b) By what fraction do the fifths on the diatonic and well-tempered scale differ?

16-19. When the acoustic spectrum of a soprano musical instrument playing a single note is analyzed, it is found that *all* harmonics of the played note are present in varying amounts. Show that the instrument *cannot* be a clarinet (which can be taken as approximating a cylindrical tube open at the bell end and effectively closed at the mouthpiece end.)

16-20. What is the minimum power output of an isotropic point source of sound that will produce an intensity of 10^{-16} W/cm² (the minimum threshold of hearing) at the ear of a listener 1.0 km distant from the source?

16-21. A camera with an aperture of 25 mm diameter ("$f/2$" for 50-mm focal-length lens) takes a photograph of a 50-W point light source 100 m distant with an exposure time of $\frac{1}{100}$ s. What is the energy of the light incident upon the film?

16-22. From the viewpoint of an observer on earth, the sun acts as an isotropic source of radiation. What fraction of the total energy radiated by the sun is intercepted by the earth (and its atmosphere)? In this calculation the earth can be regarded as a circular disc 6.4×10^6 m in radius; the earth-sun distance is 1.5×10^{11} m. Assume that none of the solar radiation is absorbed while in transit from sun to earth.

16-23. Relative to a reference frame attached to a source of frequency f and moving at the speed v_s through a medium transmitting an elastic wave at the speed c, the pressure variations are written as

$$\Delta p = \Delta p_0 \sin 2\pi \left(ft - \frac{x'}{\lambda'}\right)$$

where x' is the coordinate in the moving frame and λ' is the wavelength. Since the speed of the waves traveling along the positive X' axis relative to this moving frame is $c - v_s = f\lambda'$, we may write

$$\Delta p = \Delta p_0 \sin 2\pi \left(ft - \frac{fx'}{c - v_s}\right)$$

Use the galilean coordinate transformations (Sec. 6-3) to write the corresponding equation for the pressure variations as measured by an observer at rest in the medium, and show thereby that the frequency measured by the fixed observer when the source recedes from him is given by $f/(1 + v_s/c)$.

CHAPTER 17

Temperature and the Macroscopic Properties of an Ideal Gas

This is the first of three chapters on the general topic of heat, including the kinetic theory of gases and the laws of thermodynamics. We set forth here not only a preview of the contents of this chapter but also a brief outline of the succeeding chapters as well.

Historically, the subject of heat developed quite separately from that of mechanics. Heat was first thought to be a substance (called *caloric*), and the units devised for measuring the amount of heat (calories) were quite distinct from those in mechanics. Moreover, the temperature of a body was not thought to be related in any simple way to its mechanical properties. The single most profound discovery concerning heat was the recognition, emerging about 1840, that *heat and temperature* — indeed, *all thermal phenomena — can be interpreted in mechanical terms*.

The key is the molecular theory of gases. In this theory, ordinary newtonian mechanics is applied to the microscopic motions of molecules of a gas. One finds that thermal phenomena are merely mechanical phenomena taking place on a microscopic scale, that thermal energy is simply the disordered mechanical energy of atoms and molecules, that temperature is a measure of the translational kinetic energy of gas molecules, and that heat is nothing more than a thermal-energy transfer caused by a temperature difference.

We shall not follow the historical development of ideas concerning heat and thermodynamics. Instead, our strategy will be as follows. In this chapter we shall first consider the measurement of temperature, the meaning of thermal equilibrium, and the task of finding reliable thermometers. Several varieties of thermometers will be discussed. Then we shall investigate the properties of an ideal gas, the general-gas law, ideal-gas thermometers, and several types of transformations taking place in the state of a gas.

We turn then (Chap. 18) to the molecular model, or kinetic theory, of the simplest of all many-particle systems—an ideal gas. Here we apply mechanics to molecular collisions, deduce the general-gas law from a mechanical model, and interpret temperature in terms of molecular behavior. We shall interpret the meanings of the words *heat, internal energy,* and *specific heat* in terms of the molecular behavior and discuss the first and second laws of thermodynamics.

Next (Chap. 19), we extend the molecular model to more complicated systems—solids and liquids. We interpret the thermal behavior of solids and liquids, and the meanings of heat, temperature, and the first and second laws of thermodynamics for these systems.

17–1 TEMPERATURE AND THE ZEROTH LAW OF THERMODYNAMICS

The most rudimentary conception of temperature arises from our sense of touch. We can tell by touch, although only qualitatively and roughly, whether two bodies have the same temperature or whether one is relatively hot or cold compared with the second. What are needed are a means of asserting that two bodies have the *same* temperature and a quantitative measure of *difference* in temperature. That is, we need a thermometer.

Suppose that we bring two bodies together, place them in contact, isolate them (that is, insulate them) from external influences—and wait. After a sufficiently long time has elapsed, the bodies are said to be in *thermal equilibrium,* and the property the two bodies then have in common—whatever differences there may be in their size, mass, material, and past history—is *temperature*. Two bodies in thermal equilibrium have the same temperature, by definition.

Suppose now that we have three bodies: vessel A, vessel B, and a thermometer. The thermometer is first placed in vessel A. After these two objects have come to thermal equilibrium, we read their common temperature on the thermometer. (A thermometer reads not only its own temperature but also that of a body with which it is in equilibrium.) Suppose that the thermometer is then placed in vessel B and we find the same final temperature as that of vessel A. The thermometer has the same temperature as that of vessels A and B because it has achieved thermal equilibrium with each in turn. Do A and B have the same temperature? Are A and B in thermal equilibrium? According to the *zeroth law of thermo-*

dynamics, they are. That is, *of bodies A and B are separately in thermal equilibrium with a third body C, then A and B are in thermal equilibrium with each other.* The zeroth law of thermodynamics, which is fundamental to all thermodynamics, is so named because it has priority over the first and second laws of thermodynamics.

17–2 THERMOMETRY

Just as a physicist defines length as the quantity that one measures with a meterstick, and time as the quantity that one reads on a clock, so too one must first say that temperature is simply what one reads on a thermometer. But what is a thermometer? *A thermometer is any device having some measurable physical property that changes with the degree of relative hotness or coldness of the body.* For example, the change in the volume of a liquid with temperature is the basis of the common liquid-in-glass thermometer. We shall consider this type of thermometer in some detail, to see how the temperature scales are defined and the thermometers calibrated, and also to see some important problems in thermometry that apply to all thermometers.

Our liquid-in-glass thermometer consists of a spherical bulb of glass filled to the brim with the dense liquid mercury. If the temperature is then raised, the mercury expands more rapidly than the glass container. Some of the liquid overflows. One can, in principle, read the temperature by measuring the amount of mercury overflow. A simpler procedure, however, is to attach a cylindrical stem of uniform cross section to the bulb so that, when the mercury overflows from the bulb into the stem, one can measure this overflow by noting the position of the meniscus along the stem.

Next we calibrate the thermometer. The thermometer is placed in a mixture of ice and water under standard atmospheric pressure. We mark the position of the meniscus and label it "0." By definition, the *ice point* at standard pressure has a temperature of 0° on the *Celsius* (also known as the *centigrade*) temperature scale. The thermometer is then brought into thermal equilibrium with a mixture of water vapor and boiling water, again under standard atmospheric pressure. We make a second mark at the new position of the meniscus and label it 100°C. By definition, the *steam point* of water at standard pressure is 100° Celsius. The calibration is completed when we add uniformly spaced graduations between 0 and 100°C, to place 100 graduations (centigrade) between the two chosen *fixed points*. See Fig. 17-1. We can extend the scale adding graduations of the same size below the 0° and above the 100° marks.

In the familiar *Fahrenheit* scale of temperature, the ice point is defined as 32°F and the steam point as 212°F. Thus, 180 grad-

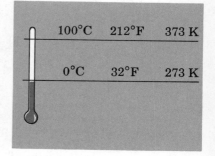

FIG. 17-1. Definition of the fixed points for the Celsius, Fahrenheit, and Kelvin temperature scales.

uations separate the two fixed points on the Fahrenheit scale, as against 100 graduations on the Celsius scale. Not only are the temperature readings different at the fixed points on the two scales, but the sizes of the degrees differ as well. As Fig. 17-1 shows, any temperature t_C on the Celsius scale is related to the same temperature t_F registered on the Fahrenheit scale, by

$$\frac{t_C}{180} = \frac{t_F - 32}{180} \tag{17-1}$$

Although there is no upper limit on temperature, the lowest possible temperature is $-273.15°C$.†

How do we know that the mercury in our thermometer expands uniformly between 0 and 100°C? We don't know—in fact, can't know—since we can test for uniform thermal expansion only with a thermometer. Suppose that we construct two thermometers using *different* liquids, one with mercury and the second with alcohol (colored with a pink dye). We calibrate this second thermometer in the same fashion as the first, using the two fixed points of water and uniformly spaced graduations.

How do the readings on the two thermometers compare when we bring both thermometers into thermal equilibrium with a third body? At the two fixed points the thermometers agree perfectly. They must, of course. But, if we compare the readings at other temperatures, we find that the two readings, although nearly the same, do *not* agree precisely. See Fig. 17-2. Thus, the behavior of the thermometer depends on the *thermometric substance* (here,

---

†In order to distinguish temperature *differences* from temperatures, it is customary to represent temperature *differences* as Celsius degrees (C°) rather than as degrees Celsius (°C). Thus, 60°C differs from 40°C by 20 C°. Furthermore, 100 C° = 180 F°, although 100°C is *not* the same as 180°F. In general, the relative sizes of degrees on the two scales is given by 1 C° = $\frac{9}{5}$ F°.

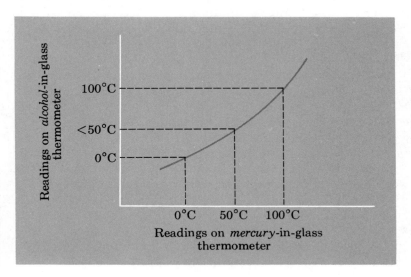

FIG. 17-2. Readings of a uniformly graduated alcohol-in-glass thermometer against the corresponding readings for a mercury-in-glass thermometer. (The departure from linearity is greatly exaggerated.)

Table 17-1

Thermometer and physical property or effect	Temperature range
Magnetic properties of paramagnetic salts	−273 to −272°C
Pressure of helium vapor in equilibrium with liquid helium	−272 to −269°C
Resistance thermometer (electrical resistance of substances)	−261 to 600°C
Thermocouple (electromotive force of two wires of dissimilar materials, the junctions maintained at different temperatures)	−250 to 1000°C
Liquid-in-glass thermometers (difference in expansion properties)	−196 to 500°C
Gas thermometers (constant pressure or constant volume, ideal gas of very low density)	$> \approx 300°C$
Optical pyrometer (visible electromagnetic radiation from body to be measured—the color of an incandescent filament is compared with that of the body)	$> 600°C$

mercury or alcohol). We cannot, at this stage, say which of the two thermometers is more nearly correct, but only that the thermal expansion of mercury differs from that of alcohol.

Thermometers may be based on physical effects other than thermal expansion. Some properties that change with temperature are the electrical resistance (resistance thermometer), the electromotive force developed by a pair of unlike wires with their junctions at different temperatures (thermocouple), the color of light emitted from a solid, or from a gas, as in the case of a star (pyrometer), and the pressure of a gas (gas thermometer). Table 17-1 lists several different types of thermometers, the physical property or effect on which each is based, and the temperature ranges over which each is most commonly used.

17-3 PRESSURE IN A GAS

The meaning of pressure is most easily illustrated through a definition in which one specifies the laboratory operations to be used in measuring it. Consider the simple pressure-measuring device in Fig. 17-3. It consists of an evacuated cylinder and a tightly fitted, weightless piston attached to the cylinder by a spring. When in a vacuum, no net force acts on the piston, and the spring assumes its relaxed length.† If the spring system is immersed in a fluid, the

---

†Although spring devices of the sort shown in Fig. 17-3 ordinarily are not used in measuring

pistonhead is acted upon by a compressive force arising from the fluid, and the spring is compressed until the inward force of the fluid is balanced by the outward force of the compressed spring. Knowing the force constant of the spring, one can compute the force of the fluid on the pistonhead simply by measuring the amount by which the spring has been compressed (and by applying $F = -kx$).

Let ΔF represent the magnitude of the force on the pistonhead, as measured by the spring, and ΔA the area of the piston. Then the pressure p of the fluid is, by definition,

$$p = \lim_{\Delta A \to 0} \frac{\Delta F}{\Delta A} \qquad (17\text{-}2)$$

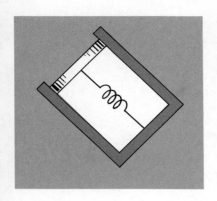

FIG. 17-3. A simple pressure-measuring device.

Note that we take the limit of the ratio of $\Delta F/\Delta A$ as the area of the piston is made very small. This defines the pressure at a *point* within the fluid. If ΔA is sizeable, $\Delta F/\Delta A$ gives the *average* pressure over the chosen area.

One finds by experiment that, at any one point within a fluid, the compression of the spring is the *same* for *all* orientations of the spring device. Pressure is a *scalar* quantity, not a vector. If one chooses a surface of area ΔA, a compressive force $\Delta F = p\,\Delta A$ acts perpendicular to the surface.

Inasmuch as the forces within a nonviscous fluid can only be those of compression, the spring shown in Fig. 17-3 is always compressed, never elongated. The lowest possible pressure it registers is zero—a vacuum. We have been discussing the so-called absolute pressure, whose zero corresponds to a vacuum. It is sometimes convenient, however, to speak of the pressure of a fluid *relative* to the atmospheric pressure p_0 which arises from the earth's atmosphere of nitrogen and oxygen. This relative pressure, called the *gauge pressure* p_g, is related to the absolute pressure p and atmospheric pressure p_0 by

$$p_g = p - p_0 \qquad (17\text{-}3)$$

An ordinary pressure gauge, used, for example, to measure the pressure of an inflated automobile tire, indicates the gauge pressure: If p_g is positive, the tire is inflated; if $p_g = 0$, the tire is flat; and if p_g is negative, the tire is at least partially evacuated.

Standard atmospheric pressure is defined as

$$1 \text{ atm} = 1.013250 \times 10^5 \text{ N/m}^2$$

Atmospheric pressure has its origin in the gravitational pull of the earth on the gases surrounding it. In fact, atmospheric pressure (≈ 10 N/cm²) represents the weight of all the air in a column 1.0 cm² in cross section, extending from the earth's surface to the top of the

---

pressure, the *aneroid barometer* operates on the same basic principle. This device employs an evacuated flexible metal chamber, one of whose walls is displaced by pressure changes. The elasticity of the walls replaces the spring, and the displacement of the flexible wall indicates the absolute pressure.

atmosphere. Since the air in this column weighs 10 N, the total mass of air in it is approximately 1 kg.

By definition, pressure is $\Delta F/\Delta A$. Now consider a vertical column within a liquid, where A is the horizontal cross section and h the height of the column. The pressure difference Δp between the top and bottom of the column is mg (the weight of the liquid within the column) divided by A:

$$\Delta p = \frac{mg}{A}$$

The mass m of the liquid within the column is $\rho A h$, where ρ is the mass density of the liquid and Ah is the volume of the column. Then

$$\Delta p = \frac{\rho A h g}{A} = \rho h g \qquad (17\text{-}4)$$

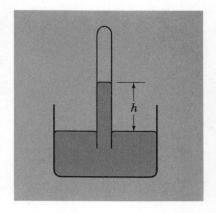

FIG. 17-4. A closed-tube manometer, or barometer.

A common barometer is shown in Fig. 17-4. The liquid used is usually mercury because at room temperature (1) it is liquid, (2) it has a high density, (3) it is chemically stable, and (4) it has a relatively low vapor pressure. The tube is closed at the upper end, and a near vacuum exists within this region above the liquid surface. Thus, the pressure existing at the outside liquid surface, equal to that within the tube at the same elevation, is $\rho g h$, where h is the height of the mercury column. Standard atmospheric pressure corresponds to a mercury height of exactly 760 mm (at a temperature of 27°C). Therefore,

1 atm = $\rho g h$ = $(13.595 \times 10^3 \text{ kg/m}^3)(9.80665 \text{ m/s}^2)(0.760000 \text{ m})$

= $1.013250 \times 10^5 \text{ N/m}^2$

An alternative unit commonly used to describe the pressure of a gas is the *torr,* where 1 torr $\equiv$ 1 mm Hg.

Since the density ρ of the liquid and the gravitational acceleration g are essentially constants, one may measure pressures directly in units of centimeters of mercury.

17-4 THE CONSTANT–VOLUME GAS THERMOMETER

Can one devise a thermometer whose readings are independent of the thermometric substance? Yes, if one uses a gas of very low density to construct a *constant-volume gas thermometer.* A simple type is shown in Fig. 17-5. *The pressure of the gas* is the physical quantity we use *to register the temperature* of the gas. The temperature of the gas must be high enough, and its pressure low enough, for the gas to have a very low density; that is, the gas must not be close to the condition under which it condenses into a liquid.

Our calibration procedure is just like that used with the

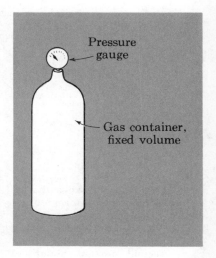

FIG. 17-5. A rudimentary form of a constant-volume gas thermometer.

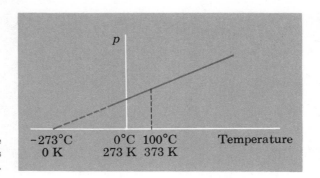

FIG. 17-6. A pressure-temperature graph for a constant-volume gas thermometer.

liquid-in-glass thermometers. Maintaining the gas volume constant, we record the pressures at the ice point and at the steam point. Then, *assuming* the pressure to vary linearly with temperature, we mark the pressures corresponding to 0 and 100°C on graph paper, and draw a *straight* line through the two points, to establish temperatures between 0 and 100°C. By extrapolation, we extend to temperatures beyond this range, as shown in Fig. 17-6. (The drawing of a *straight* line corresponds here to the marking of *uniform* graduations on the stem of the mercury thermometer.)

We have not indicated what gas is to be placed in the constant-volume gas thermometer. Experiment shows that it does *not* matter! All very low-density gases show exactly the same behavior. Two thermometers with different low-density gases agree not only at the fixed points but also at *all* other temperatures when we assume a linear pressure-temperature relation. See Fig. 17-7. This is, of course, why a gas thermometer is a good thermometer.

Figure 17-6 implies a temperature scale in which the pressure p_t is directly proportional to the temperature. When we extrapolate the straight line to low temperatures, we find that the pressure is zero when the Celsius temperature is −273.15°. More properly, if the behavior of the gas at elevated temperatures persisted unchanged to the lowest temperatures (which it does not, because gases liquefy), its pressure would be zero at −273.15°C. The

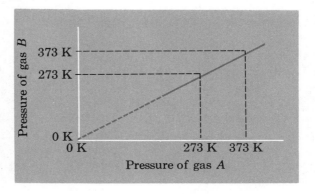

FIG. 17-7. Pressure (that is, absolute temperature) for one constant-volume gas thermometer against the corresponding pressure (temperature) for a second constant-volume gas thermometer. Both gas densities are very low. Compare with Fig. 17-2.

temperature −273.15°C provides a *possible* lower limit of temperature. It is, therefore, convenient to introduce a new temperature scale with its zero at this point. The *Kelvin temperature scale* is so chosen that its zero is at −273.15°C, the size of the Kelvin degree being the same as the Celsius degree. Kelvin temperatures are represented by T. Then, through the *constant-volume gas thermometer*, the *Kelvin temperature*† is given by

$$T = \frac{p_T}{p_{273}} 273.15 \text{ K} \tag{17-5}$$

with
$$T = t_c + 273.15°$$

where p_{273} is the pressure of the gas at the ice point.

We have referred to the ice and steam points of water under standard atmospheric pressure as the bases of temperature-scale determinations. We could equally well choose the zero of the Kelvin scale and the triple point of water as the two fixed points to define the Kelvin scale. The *triple point* of water, which occurs at a pressure of 4.58 torr and a temperature of 273.16 K, corresponds to that unique condition under which water can exist simultaneously in the liquid, solid, and vapor states.

Zero on the Kelvin scale, the so-called absolute zero, *is* the lowest possible temperature. That this is indeed the case must be proved through thermodynamic arguments; it is suggested by, but *not* deduced from, Fig. 17-6. We shall see that the Kelvin, or *absolute*, scale of temperature has a simple interpretation in terms of the mechanics of molecular motion. Suffice it to say here that the independence of the thermometric substance found for low-density gases is not entirely unexpected. In low-density gases the molecules are very distant from one another and interact by intermolecular forces only during the small periods of time that they collide.

17-5 THE GENERAL–GAS LAW

We are concerned here with an *ideal,* or *perfect,* gas. Any gas behaves as an ideal gas if its density is very low. This requires, in effect, that the gas's temperature be relatively high and its pressure relatively low, so that the gas is not close to the conditions under which it condenses into a liquid. Most gases at room temperature and atmospheric pressure can be regarded as ideal gases.

The general-gas law is concerned with the relationship among the macroscopic measurable properties of an ideal gas: the absolute temperature T, the pressure p, the volume V, and the mass

---

†The writing of Kelvin temperatures without a degree mark, as in 273.15 K, is officially sanctioned. More properly, Eq. (17-5) defines the Kelvin temperature only for extremely low-density gas thermometers.

m. It is customary to specify the mass m indirectly through the number n of moles of gas. By definition

$$n = \frac{m}{w} \tag{17-6}$$

where w is the molecular weight† of the gas molecules. If the mass m is given in grams, then n is the number of gram moles. Thus, 1 g mol of any material contains w grams of that material. For example, oxygen has a molecular weight of 32; therefore, 1 g mol of oxygen consists of 32 g.

A common form of the general-gas law, based on experiment with low-density gases, is

$$pV = nRT \tag{17-7}$$

where R is the *universal gas constant*.

It is an experimental fact that 1 g mol of an ideal gas at standard temperature (273.15 K) and standard atmospheric pressure $(1.013250 \times 10^5 \text{ N/m}^2)$‡ occupies a volume of 22.415×10^{-3} m³ ≈ 22.4 liters. Therefore, from (17-7),

$$R = \frac{pV}{nT}$$

$$= \frac{(1.013250 \times 10^5 \text{ N/m}^2)(22.415 \times 10^{-3} \text{ m}^3)}{(1 \text{ g mol})(273.15 \text{ K})}$$

$$= 8.314 \text{ J/g mol-K°}$$

The numerical value of R depends, of course, on the units used for p and V. With p in atmospheres and V in liters,

$$R = 0.08207 \text{ liter-atm/g mol-K}$$

and if energy is expressed in units of the gram calorie, where 1 g cal = 4.1840 J,

$$R = 1.986 \text{ g cal/g mol-K}$$

(We shall see in Chap. 19 that the calorie has a simple meaning in terms of the thermal properties of water.)

An alternate form of the general-gas law, (17-7), is

$$\frac{p_1 V_1}{m_1 T_1} = \frac{p_2 V_2}{m_2 T_2} \tag{17-8}$$

where subscripts 1 and 2 denote two different states of the gas.

For a constant volume and mass of gas, (17-8) becomes

$$\frac{p_1}{T_1} = \frac{p_2}{T_2} \quad \text{for constant } V \text{ and } m$$

---

†Molecular "weight" is, strictly, molecular *mass* relative to one atom of carbon 12 having mass 12.
‡Conditions of standard pressure and standard temperature are often abbreviated by STP.

The pressure is inversely proportional to the absolute temperature, the basis of the constant-volume gas thermometer (Sec. 17-4).

For constant pressure and mass, (17-8) reduces to

$$\frac{V_1}{T_1} = \frac{V_2}{T_2} \quad \text{for constant } p \text{ and } m$$

The proportionality of volume and temperature may be used as the basis of a constant-pressure gas thermometer; the relation is sometimes referred to as the law of J. L. Gay-Lussac and J. A. C. Charles, its discoverers.

Boyle's law, applying for a constant temperature and mass of ideal gas, is

$$p_1 V_1 = p_2 V_2 \quad \text{for constant } T \text{ and } m$$

An alternative form, using the mass density $\rho = m/V$, is

$$\frac{p_1}{\rho_1} = \frac{p_2}{\rho_1}$$

Example 17-1

What is the density of helium under STP?

We compute ρ directly by recalling that, for a pressure of 1 atm and a temperature of 0°C, the volume of 1 g mol of any gas is 22.415×10^{-3} m³. Since helium is *monatomic* and has an atomic weight of 4.003,

$$\rho = \frac{4.003 \times 10^{-3} \text{ kg}}{22.415 \times 10^{-3} \text{ m}^3} = 0.1786 \text{ kg/m}^3$$

Example 17-2

A tank having a volume of 30 liters contains nitrogen (N_2) gas at 20°C at a gauge pressure of 3.00 atm. The tank's valve is opened momentarily, and some nitrogen escapes. After the valve is closed and the gas has returned to room temperature, the tank's pressure gauge reads 2.40 atm. How much nitrogen leaked out?

We designate the number of moles of gas and the pressure before and after the leak by subscripts 1 and 2, respectively. Then, from Eq. (17-7),

$$n_1 - n_2 = \frac{p_1 V}{RT} - \frac{p_2 V}{RT} = \frac{V}{RT}(p_1 - p_2)$$

$$n_1 - n_2 = \frac{(30 \text{ liters})(0.60 \text{ atm})}{(0.082 \text{ liter-atm/g mol} - \text{K})(293 \text{ K})} = 0.75 \text{ g mol}$$

Since nitrogen has an atomic weight of 14, 1 g mol has a mass of 28 g. The mass escaping is, then,

$$m = nw = (0.75 \text{ g mol})(28 \text{ g/g mol}) = 21 \text{ g}$$

17-6 CHANGES IN THE STATE OF A GAS

Several types of change in the state of a gas are particularly simple, inasmuch as one variable in the equation of state remains constant. Each process is briefly described below, for the simple case of an ideal gas.

In an *isothermal process* the *temperature* is constant. Lines of constant T, corresponding to $pV =$ constant, are called *isotherms*. The pressure increases in an isothermal compression and decreases in an isothermal expansion. See Fig. 17-8a.

An *isobaric process* is one in which the *pressure* remains constant. We see from Fig. 17-8b that when a gas expands isobarically, its temperature rises; and conversely.

In an *isovolumetric process* the *volume* remains constant. Figure 17-8c shows that the temperature of a gas in a container of fixed size drops as its pressure falls.

An *adiabatic process* is one in which there is *no heat* into or out of the gas. A gas will undergo an adiabatic process under two simple circumstances: when the container is a heat insulator, which isolates the gas thermally from its surroundings, and when a process takes place so rapidly that there is not enough time for energy to be transferred to or from the gas through the container walls. An adiabatic process is described for an ideal gas by the relation

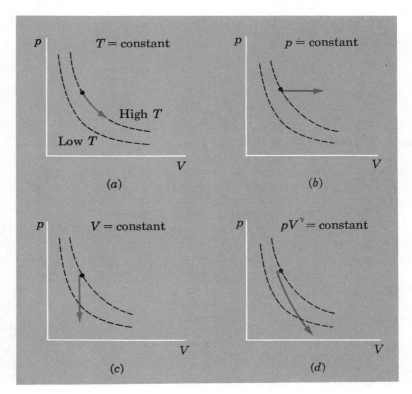

FIG. 17-8. Several important changes in the state of a gas. The dotted lines represent two isotherms. *(a)* Isothermal, *(b)* isobaric, *(c)* isovolumetric, and *(d)* adiabatic processes.

$$pV^\gamma = \text{constant} \qquad (17\text{-}9)$$

where γ is a constant. The numerical value of γ depends on the nature of the gas, but it is always greater than 1. Therefore, adiabatic lines on a pV diagram are more steeply inclined than isotherms, for which pV is constant. Figure 17-8d shows that the temperature falls in an adiabatic expansion.

SUMMARY

After a sufficiently long time has elapsed, a collection of objects isolated from external influence achieves thermal equilibrium. The property that all objects in the collection then have in common is temperature. According to the zeroth law of thermodynamics, if A and C are in thermal equilibrium, and B and C are also in thermal equilibrium, then A and B are in thermal equilibrium.

Pressure p is defined as

$$p = \frac{\Delta F}{\Delta A}$$

where ΔF is the force normal to a plane element of area ΔA. The pressure difference Δp between two points in a liquid separated by vertical distance h is

$$\Delta p = \rho h g \qquad (17\text{-}4)$$

The equation of state for any low-density gas, the so-called general-gas law, is

$$pV = nRT \qquad (17\text{-}7)$$

where n is the number of moles of gas and R is the universal gas constant (8.314 J/g mol–K). At STP 1 g mol of any ideal gas occupies a volume of 22.4 liters. An ideal gas may be used as the basis for a thermometer registering the absolute, or Kelvin, temperature T, which is related to the Celsius temperature t_C by $T = t_C + 273.15°$.

PROBLEMS

17-1. (a) At what temperature will the reading on a Celsius thermometer be the same as the reading on a Fahrenheit thermometer? (b) At what temperature will the Kelvin and Celsius temperatures be the same to within 0.10 percent?

17-2. William Rankine, a nineteenth-century Scottish engineer, devised a temperature scale in which 1 R° = 1 F° with the absolute zero $t_R = 0°$R. Find the equation which relates Rankine temperature t_R to Celsius temperature t_C.

17-3. Suppose that we define a temperature scale based on the melting and boiling points of the element mercury (Hg). We define this new scale as follows: $t_H = 0°$H $= -38.9°$C (melting point of mercury) and $t_H = 600°$H $= 356.9°$C (boiling point of mercury). Find (a) 1 H° in terms of 1 C° and (b) t_H in terms of t_C.

17-4. An automobile tire is inflated to a gauge pressure (excess over atmospheric pressure of 15 lb/in²) of 30 lb/in² (often referred to, improperly, as "30 lb") when the temperature is 68°F. What is the gauge pressure when the temperature of the tire and the air within has risen to 90°F? Assume for simplicity that the tire does not expand.

17-5. What is the total mass of the earth's atmosphere (atmospheric pressure, 1.0×10^5 N/m²; earth's radius, 6.4×10^6 m)? (Neglect the variation in g with elevation above the earth's surface.)

17-6. If the earth's atmosphere had the same density at all distances above the earth's surface as at the surface (1.29×10^{-3} g/cm³), what would be the height of the atmosphere when the pressure at the earth's surface is 1 atm? (Neglect the variation in g with elevation.)

17-7. An ideal gas is contained in an upright right circular cylinder, into which a tightly fitting piston, free to slide vertically without leaking gas, has been placed so that it floats in equilibrium. Show that the distance from the bottom of the cylinder to the lower face of the piston is directly proportional to the absolute temperature of the gas within.

17-8. A gas occupies a volume of 6.0 liters when its temperature is 30°C and its pressure is 760 torr. What volume must this gas occupy when its temperature has been raised to 100°C and its pressure increased to 240 cmHg?

17-9. (a) Find the isobaric rate of change of volume with temperature for an ideal gas at temperature T. (b) Evaluate $[(1/V)(dV/dT)]_p$ – the volume coefficient of thermal expansion at constant pressure – when the temperature is 273.15 K. (An *isobaric process* is one in which the *pressure* is constant.)

17-10. A meteorological balloon is designed to become fully inflated at an altitude of 10 km where the temperature is −50°C and the pressure is 2.64×10^4 N/m². What volume V_0 of helium should be placed in the balloon at sea level where the temperature is 15°C and the pressure is 1 atmosphere? Express V_0 as a percentage of V_F, the volume of the gas when the balloon is fully inflated.

17-11. At STP the density of chlorine gas is 3.214 g/liter. Calculate the mean value of the molecular weight of chlorine gas (Cl_2).

17-12. Compute the density of air at STP, assuming the air to consist of N_2 and O_2 in the ratio of 5:1 by weight. The molecular weights of nitrogen and oxygen molecules are 28 and 32, respectively. (According to the law of partial pressures, the pressure of each of the two gases is independent of the presence of the other.)

17-13. Molecular hydrogen is contained in a spherical balloon of a 1.0-m diameter. The mass of the balloon alone is 1.0 kg. How much hydrogen is contained if the balloon just floats in the atmosphere?

17-14. Show that for an ideal gas an adiabatic process can be described by the relation $TV^{\gamma-1}$ = constant.

17-15. In a diesel engine the fuel is ignited by air which has been heated by adiabatic compression — there is no spark plug. Find (a) the temperature and (b) pressure at the end of the compression stroke under the following assumptions: the adiabatic compression ratio is 16 to 1; intake air is at 27°C and 1 atm pressure; $\gamma = 1.40$ for air under these conditions.

17-16. Show that the speed of a compressional wave in an ideal gas can be written in the form $c = (\gamma p/\rho)^{1/2}$. (*Hint:* Compare differential forms of Eq. (17-9) and of the bulk modulus B.)

17-17. Use the equation developed in Problem 17-16 to find the speed of sound in air at an altitude of 10 km, where the pressure is 0.261 atm, the density 0.413 kg/m³, and the temperature −56.9°C. (At 15°C and 1 atm, the air density is 1.225 kg/m³ and sound travels at 340 m/s.)

17-18. Consider an ideal gas confined within a cylindrical container closed at one end by a fixed wall and at the other by a moveable piston. (a) Show that the force on the piston is $F = pA$, where p is the gas pressure and A is the piston area. (b) As the gas volume increases by dV, show that the work done on the piston by the gas is $dW = p\,dV$. See Fig. P17-18a. (c) Final-

FIG. P17-18

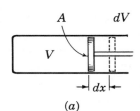

(a)

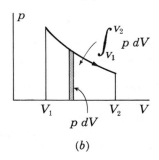

(b)

ly, as the volume goes from V_1 to V_2, show that the total work done is

$$W_{1\to 2} = \int_{V_1}^{V_2} p\, dV$$

See Fig. P17-18b. Thus, the work done by an expanding gas is equal in magnitude to the area under the curve of the line describing the expansion on a p-V diagram.

17-19. (a) Make use of the equation developed in Problem 17-18 to show that the work done by an ideal gas as it expands *isobarically* from volume V_1 to volume V_2 is given by $W_{1\to 2} = p(V_2 - V_1)$. (b) Use the equation of state to show that the work done by an ideal gas when it expands *isothermally* from V_1 to V_2 is $W_{1\to 2} = nRT \ln(V_2/V_1)$.

17-20. (a) Use the equation developed in Problem 17-18 to show that the work done by a gas expanding adiabatically from the state (p_1, V_1) to the state (p_2, V_2) is $(p_1 V_1 - p_2 V_2)/(\gamma - 1)$. (b) Does the temperature rise or fall in this process? (c) Suppose that, starting from the same initial state, the gas expands isothermally to the same final pressure p_2 as in the adiabatic transformation. Is more work or less work done by the gas?

CHAPTER 18

The Kinetic Theory and the Fundamental Concepts of Thermodynamics

In this chapter we consider a gas to be composed of a large number of microscopic particles. Through the kinetic theory, we shall relate the microscopic properties of gas molecules — their speeds, diameters, masses, kinetic energies, and their number — to the macroscopic behavior of ideal gases. We shall thereby be led to a mechanical interpretation of the concepts of temperature and heat. We shall identify the thermal energy of a gas with the disordered energy of the molecules and consider the first and second laws of thermodynamics.

18–1 MOLECULAR SIZE

Recall first that 1 g mol of any element represents the element's atomic weight in grams. Thus, 1 g mol of *any* pure substance contains the same number of atoms or molecules, called *Avogadro's number*, N_A. By experiment,

$$\text{Avogadro's number} = N_A = \frac{6.02217 \times 10^{23}}{\text{g-mol}}$$

Thus, 2 g of H_2, 32 g of O_2, and 4 g of He all contain 6×10^{23} molecules. Avogadro's number gives, as well, the number of atoms per gram mole for solids existing as pure elements; for example, 197 g of solid gold (atomic weight, 197) contains N_A atoms, and 2 g of solid H_2 contains $2N_A$ atoms or N_A molecules.

We wish to find the approximate size of a typical molecule and the average distance between neighboring molecules in a gas.

Consider oxygen. At STP 1 g mol (32 g) is known to occupy a volume of 22.4 liters = 22.4×10^{-3} m³. The gas is, of course, easily compressed.

If we lower the temperature of this oxygen to 90 K while maintaining the pressure constant, the gas will condense into a nearly incompressible liquid. In the liquid state the 32 g of oxygen is found to have a volume smaller than the original gaseous volume by a factor of 700. Since the volume of liquid oxygen can be reduced appreciably only by relatively large external pressures, we can assume that in the liquid state the oxygen molecules are "in contact" with one another. Every molecule exerts strong repulsive intermolecular forces on its neighboring molecules when compressive external forces are applied to the liquid.

Let us compute the volume associated with each oxygen molecule. We know that in the liquid state, 1 molecule of oxygen occupies a volume of $(22.4 \times 10^{-3}$ m³$)/(700)(6.023 \times 10^{23}) = 53 \times 10^{-30}$ m³. We can find the approximate diameter d of the molecule through the relation $d^3 \approx 53 \times 10^{-30}$ m³, or $d \approx 4 \times 10^{-10}$ m. Thus, the diameter of an oxygen molecule is close to 4×10^{-10} m = 4 Å (the *angstrom* unit of distance, abbreviated Å, is defined as 1 Å = 10^{-10} m). All molecules comprised of no more than a few atoms are found to have diameters of the order of a few angstroms.

We have computed the molecular diameter by using precisely the same procedure that one might employ to find the diameter of a large-sized sphere, such as a billiard ball. The diameter of any one of identical billiard balls is the distance between the centers of a pair of such balls in contact with one another; that is, the diameter is that separation distance below which the force between the balls becomes strongly repulsive. In like fashion, we have found the molecular diameter to be the separation distance between adjacent molecules marking the onset of a strongly repulsive intermolecular force.

Knowing the approximate size of an oxygen molecule (4×10^{-10} m), we can also find the average distance between oxygen molecules under STP. The available volume per molecule in the gaseous state is 700 times larger than in the liquid state. Thus, the average distance between molecular centers is $(700)^{1/3} \approx 9$ times the distance between molecular centers in liquid oxygen. For oxygen at STP, the molecules are, therefore, separated on the average by 9 molecular diameters, so that a low-density gas may be regarded as mass points in a vacuum.

18–2 THE KINETIC THEORY APPLIED TO AN IDEAL GAS

We list below the assumptions of the kinetic theory of gases. These assumptions are amply justified, not only because they lead to a

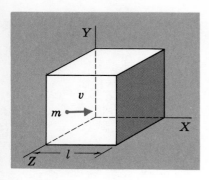

FIG. 18-1. Molecule of mass m and speed v striking the YZ wall of a cubical container of edge length l.

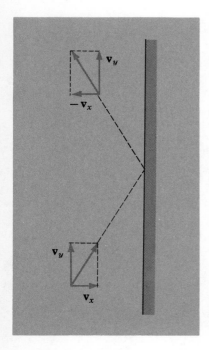

FIG. 18-2. A molecule collides obliquely with a container wall. Only the velocity component at right angles to the surface changes.

successful interpretation of the macroscopic properties of ideal gases in terms of the microscopic properties of gas molecules but also because the microscopic properties are verified directly by experiment.

1. The molecules of a gas are, on the average, separated by distances that are large compared with the molecular diameters. In fact, for ideal gases we assume the *molecules* to *have a negligible volume* compared with the volume of the container of the gas.
2. The molecules are in constant *random motion*. Because the number of molecules is enormous, we assume that this motion is utterly chaotic: The molecules move in all directions with equal probability and with a variety of speeds. The center of mass of the gas as a whole remains at rest, the *total* momentum being zero.
3. The collisions of the molecules with one another and with the walls are *perfectly elastic.*
4. Between collisions the molecules are free of forces and move with a constant speed.† A collision takes place when one molecule is within the short-range intermolecular force of a second molecule. The *duration of any collision is assumed to be very small compared with the time between collisions.* Thus, although a pair of molecules will lose kinetic energy and gain potential energy during a collision, the potential energy can be ignored because a molecule spends a negligible fraction of its time in collisions.
5. *Newtonian mechanics applies* to molecular collisions.

We wish to compute the pressure (a macroscopic property) of an ideal gas in terms of the microscopic properties of the gas molecules: N, the number of molecules in the container; m, the mass of each molecule; and v, a typical molecular speed. The pressure of the gas is attributed to the bombardment of the container walls by molecules. The N molecules are taken to be in a cubical box of edge length l.

Consider the collision of a molecule with the right-hand YZ wall of the cubical container of edge length l in Fig. 18-1. The molecule approaches the wall with velocity components v_x, v_y, v_z and collides elastically with the wall, rebounding with the same speed. Its v_y and v_z components are not changed, while the v_x component becomes $-v_x$ after the collision. See Fig. 18-2. Therefore, in this collision the X component of the molecule's momentum changes by $\Delta(mv) = (-mv_x) - (+mv_x) = -2mv_x$, and the container wall has its momentum changed by $+2mv_x$. The time interval Δt between two successive collisions with the right-hand wall is $\Delta t = 2l/v_x$, since the molecule travels a total distance $2l$ along the X axis over a round trip at the speed v_x. (You can easily convince

---

†Strictly, the molecular speeds are constant between collisions only if the molecules move in gravity-free space. All molecules in a gas at the earth's surface are subject to a constant acceleration **g** downward. But we shall see that the change in velocity arising from gravity is small compared with the molecular velocities and can be ignored. Stated differently, the change in the kinetic energy of a molecule arising from its rise or fall in a gravitational field is negligible compared with the kinetic energy of a typical molecule.

yourself that this time interval is not altered by collisions with top, bottom, and side walls.)

The average force F_{av} on the wall over the time between successive collisions is, then, from Newton's second law,

$$F_{av} = \frac{\Delta(mv)}{\Delta t} = \frac{2mv_x}{2l/v_x} = \frac{mv_x^2}{l} \quad (18\text{-}1)$$

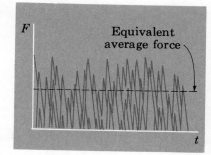

FIG. 18-3. Force on container wall, arising from discrete molecular collisions, as a function of time. The equivalent average force is constant.

Each time a molecule strikes the container wall it imparts an impulsive force to it; the total force is the resultant of a very large number of discrete and abrupt molecular collisions. See Fig. 18-3. The molecular speeds vary, and the molecules strike the wall from different directions; consequently, the individual impulses differ in size. But because the number of molecules is large, the total force ΣF_{av} on the wall is essentially constant. Dividing this total force by the wall area l^2 yields the pressure p on the wall:

$$p = \frac{\Sigma F}{l^2} = \frac{F_1 + F_2 + F_3 + \cdots}{l^2}$$

which becomes, with the use of (18-1),

$$p = \frac{m(v_{x1}^2 + v_{x2}^2 + v_{x3}^2 + \cdots)}{l^3} \quad (18\text{-}2)$$

where the subscripts 1, 2, 3, etc., refer to the various individual molecules.

The *speeds* of the molecules are distributed over a continuous range of values. So are the X components of the velocities. If $\overline{v_x^2}$ represents the average of the squares,

$$\overline{v_x^2} = \frac{v_{x1}^2 + v_{x2}^2 + v_{x3}^2 + \cdots}{N} \quad (18\text{-}3)$$

then (18-2) may be written

$$p = \frac{Nm\overline{v_x^2}}{l^3} \quad (18\text{-}4)$$

The averages for the Y and Z velocity components are given by relations analogous to (18-3), and the average of the squares of the molecular speeds is, by definition,

$$\overline{v^2} = \frac{v_1^2 + v_2^2 + v_3^2 + \cdots}{N}$$

where, for any one particle,

$$v_1^2 = v_{x1}^2 + v_{y1}^2 + v_{z1}^2$$

It then follows that

$$\overline{v^2} = \overline{v_x^2} + \overline{v_y^2} + \overline{v_z^2} \quad (18\text{-}5)$$

The Kinetic Theory Applied to an Ideal Gas   SEC. 18-2

Now, if the molecules move in truly random directions, one direction is like any other, so that

$$\overline{v_x^2} = \overline{v_y^2} = \overline{v_z^2}$$

Equation (18-5) can then be written

$$\overline{v^2} = 3\overline{v_x^2} \quad \text{or} \quad \overline{v_x^2} = \tfrac{1}{3}\overline{v^2}$$

and (18-4) becomes

$$pV = \tfrac{1}{3}Nm\overline{v^2} \tag{18-6}$$

where $V = l^3$.

It is useful to introduce the *root-mean-square speed*, v_{rms}, which is the square *root* of the *m*ean of the *s*quares of the molecular speeds. By definition,

$$v_{\text{rms}} \equiv \sqrt{\overline{v^2}}$$

and (18-6) may be written

$$pV = \tfrac{1}{3}Nmv_{\text{rms}}^2 \tag{18-7}$$

Through the analysis above we have related the macroscopic properties of a gas (the pressure p and volume V) to the microscopic molecular properties N, m, and v_{rms}. Equation (18-7) can be written more compactly by introducing the mass density of gas ρ, where ρ is the total mass Nm of molecules divided by the volume V they occupy, $\rho = Nm/V$.

$$p = \tfrac{1}{3}\rho v_{\text{rms}}^2 \tag{18-8}$$

This equation permits us to compute the rms speed of the molecules by knowing merely the pressure and density of the gas. For example, under conditions of STP, molecular hydrogen has a density

$$\rho = \frac{2.016 \text{ g}}{22.4 \text{ liters}} = 9.00 \times 10^{-2} \text{ kg/m}^3$$

and a pressure $p = 1.013 \times 10^5$ N/m². From (18-8),

$$v_{\text{rms}} = \sqrt{\frac{3p}{\rho}} = \sqrt{\frac{3(1.013 \times 10^5 \text{ N/m}^2)}{9.00 \times 10^{-2} \text{ kg/m}^3}} = 1.84 \times 10^3 \text{ m/s}$$

A typical hydrogen molecule at STP has a speed of nearly 2 km/s ≈ 4,000 mi/h! This does not mean that, in 1 s, such a molecule will travel along a straight line for 2 km, nor for that matter that it will travel directly from one end of a small box to the opposite end (it makes frequent collisions with other molecules). What (18-8) *does* say is that, between collisions, a typical hydrogen molecule at STP travels at a speed of 2 km/s.

18-3 THE MEANING OF TEMPERATURE AND INTERNAL ENERGY

It is now easy to relate the macroscopic concept of the temperature of an ideal gas, as defined through the general-gas law,

$$pV = nRT \qquad (17\text{-}7), (18\text{-}9)$$

to the average kinetic energy of the molecules comprising the gas. From the kinetic theory we have

$$pV = \tfrac{1}{3}Nmv_{\text{rms}}^2 = \tfrac{1}{3}Nm\overline{v^2} \qquad (18\text{-}7), (18\text{-}10)$$

The total number of molecules N is the product of the number of moles n and the number of molecules per mole N_A (Avogadro's number); that is, $N = nN_A$.

Equating (18-9) and (18-10), we obtain

$$\tfrac{1}{2}m\overline{v^2} = \frac{3}{2}\frac{R}{N_A}T$$

The *average translational kinetic energy* of a molecule of an *ideal gas* is *directly proportional* to the *absolute temperature* of the gas. From the point of view of the kinetic theory, temperature is simply a measure of the average *translational* kinetic energy per molecule in an ideal gas. If gases followed the general-gas law to very low temperatures—which they do not—the molecules would be at rest at the absolute zero of temperature.

The ratio of the universal gas constant R to Avogadro's number N_A is called the *Boltzmann constant k*:

$$\text{Boltzmann constant} = k = \frac{R}{N_A} = 1.381 \times 10^{-23} \text{ J/K}$$

The Boltzmann constant may be regarded as the universal gas constant *per molecule*.

Our equation above can then be written:

Average translational kinetic energy per molecule

$$= \tfrac{1}{2}m\overline{v^2} = \frac{3}{2}\frac{R}{N_A}T = \tfrac{3}{2}kT \qquad (18\text{-}11)$$

This equation has profound implications. For instance, suppose that we bring together two gases, initially at different temperatures, within a single container. Molecules of the hotter gas will transfer energy through intermolecular collisions to the molecules of the cooler gas until thermal equilibrium is established at a single final temperature. Equation (18-11) implies that, at any fixed temperature, *all* molecules, whatever their masses, have the *same* average translational kinetic energy. Therefore, in a mixture of different gases, the molecules of small mass move at relatively high speeds, and massive molecules move at low speeds.

Another consequence of (18-11) is this: For a given gas (given mass m) the rms speed is directly proportional to the square root of the absolute temperature. The molecular speed depends on neither the pressure nor the density singly, but on their ratio p/ρ which, as can be seen from the general-gas law, Eq. (18-9), is proportional to T.

Since the potential energy between molecules during a collision may be ignored, the average total energy per particle is, at least for monatomic gases, simply the average translational kinetic energy. For liquids, solids, and nonideal gases, where the intermolecular potential energy is not negligible, the relation between temperature and particle energy is more complicated. Nevertheless, the relation $\frac{1}{2}m\overline{v^2} = \frac{3}{2}kT$ between the *translational* kinetic energy and the temperature, applies in all cases.

Example 18-1

What is the rms molecular speed for hydrogen and oxygen molecules at (a) 300 K and (b) 30,000 K?

(a) At the end of Sec. 18-2 we found the rms speed for hydrogen molecules at 0°C to be 1.84 km/s. From Eq. (18-11), the rms speed for a fixed molecular mass is

$$v_{\text{rms}} \propto \sqrt{T}$$

Therefore, the rms speed of hydrogen molecules at 300 K is

$$v_{\text{rms}} = (1.84 \text{ km/s})\sqrt{\tfrac{300}{273}} = 1.93 \text{ km/s}$$

We can easily compute the rms speed for O_2 at 300 K by recognizing that molecules of all gases at the same temperature have the same average kinetic energy. Thus,

$$\tfrac{1}{2}m_H v_H^2 = \tfrac{1}{2}m_O v_O^2$$

where the subscripts H and O refer to hydrogen and oxygen, respectively. Then, for O_2 at (300 K),

$$v_O = v_H \sqrt{\frac{m_H}{m_O}} = (1.93 \text{ km/s})\sqrt{\frac{2.016}{32.00}} = 0.484 \text{ km/s}$$

(b) To find the speeds at 30,000 K, we simply recognize that, since the temperature is up by a factor of 100, the energies increase by the factor 100, and the speeds by factor 10, over the corresponding values of 300 K. Therefore,

$$v_{\text{rms}} \text{ for } H_2 \text{ at } 30{,}000 \text{ K} = (1.93 \text{ km/s})(10) = 19.3 \text{ km/s}$$

$$v_{\text{rms}} \text{ for } O_2 \text{ at } 30{,}000 \text{ K} = (0.484 \text{ km/s})(10) = 4.84 \text{ km/s}$$

The term *internal energy* is used to designate the total energy content of a system. Thus, if the system consists of a spring with attached masses, the internal energy U of the system is the sum of the kinetic energies of the masses and the potential energy of the spring. If the system is an ideal gas composed of N particles† in thermal equilibrium, its internal energy U is just the average kinetic energy per particle, $\frac{3}{2}kT$, multiplied by the number of particles N; that is,

$$U = \tfrac{3}{2}NkT = \tfrac{3}{2}nRT \qquad (18\text{-}12)$$

As (18-12) implies, the internal energy of an ideal gas is a function of the absolute temperature only.

The internal energy of an ideal gas arises from the *disordered* motion, or molecular chaos, of many small particles. Although the molecules have kinetic energy, the total linear momentum of the system as a whole is *zero*. Figure 18-4 shows two systems, both with the same total kinetic energy; in Fig. 18-4a the motion of the particles is *ordered*, and the total momentum of the system is just N times the momentum of each particle; in Fig. 18-4b the motion of particles is *disordered*, and the center of mass of the system remains at rest.

Thus, the internal energy of a system of particles may be either ordered or disordered (or a combination of the two). When the motion is ordered, it is manifest macroscopically as recognizable kinetic and potential energy. When the motion is disordered, as in an ideal gas, one "sees" no energy on a macroscopic scale, although it exists microscopically. This disordered, chaotic, internal energy of a system is called *thermal energy*.

Thermal energy is measured by an observer at rest with respect to the system's center of mass. Suppose that we are in motion relative to a container of gas. The speeds and kinetic energies of the molecules are then higher than their values for the container at rest. But we can *not* find the temperature of the gas to have increased simply because we are in motion relative to the container. So, the disordered thermal energy of gas molecules must be measured relative to the system's center of mass.

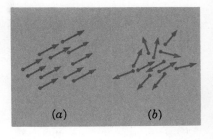

FIG. 18-4. *(a)* Molecules with the same velocity display *ordered* motion. *(b)* Molecules with randomly distributed velocities display *disordered* motion.

18-4 THE MEANING OF HEAT AND THE FIRST LAW OF THERMODYNAMICS

Here we are concerned with the ways in which the internal energy of a gas can change. Assume that we have a gas in thermal equilibrium with its container; the temperature of the gas is the same as that of the container. As time goes on, the total internal energy, and therefore the temperature, of the gas remains unchanged. The molecular collisions are perfectly elastic. This implies that, when a molecule strikes a wall, it rebounds with the same kinetic energy.

---

†For simplicity we here assume that the molecules are monatomic. Then no energy is associated with the vibration of atoms within the molecule, and there is no kinetic energy of rotation of the molecule as a whole.

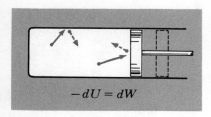

FIG. 18-5. The internal energy dU decreases and work dW is done by the gas when, on the average, a molecule rebounds from the piston at a lower speed.

But the container wall is not a perfectly smooth, hard surface. It consists of atoms bound to one another in a solid, all oscillating continuously. When a molecule strikes a wall, it hits an atom which is, in general, moving, and the gas molecule may leave the wall with a greater or lesser kinetic energy. For thermal equilibrium, then, the *average* energy of molecules rebounding from the wall equals the *average* energy of molecules striking the wall.

There is only one way in which the internal thermal energy of a gas in a container can change: The molecules must, on the average, either gain or lose kinetic energy in their collisions with atoms in the container walls. (We suppose, for simplicity, that the molecules are monatomic.) We may distinguish *two modes of energy transfer* to or from the gas molecules: *work* and *heat*.

Suppose that a gas is in a container with a movable, freely sliding piston, as shown in Fig. 18-5. Molecules bombard the piston and walls. Suppose, further, that the pressure outside the container is zero. Then the piston is moved outward by the molecular bombardment; that is, the contained gas does work on the piston, and the piston gains ordered kinetic energy.

As the piston moves outward, the gas temperature (and pressure) falls. We imagine that the container walls are always maintained at the same temperature as the gas. Then any molecule striking a fixed container wall (excluding the moving piston) will rebound, on the average, with unchanged kinetic energy. But what about the molecules hitting the piston? Each does work on the piston in striking it. Each molecule rebounds with a reduced speed (and reduced kinetic energy), as shown in Fig. 18-5. Consequently, the total internal energy of the gas is lowered, the decrease in internal energy, $-dU$, being exactly equal to the work done by the gas on the piston:

$$-dU = dW$$
or
$$0 = dU + dW$$

(Note that dU and dW are *algebraic* quantities.) In this process some of the disordered thermal energy of the gas molecules is converted into the ordered kinetic energy of the piston.

The process described above can, of course, be reversed. Then an external agent compresses the gas, the walls again being maintained at the same temperature as the gas. With the piston displaced inward, work is done *on* the gas, the molecules leave the piston wall with a *larger* kinetic energy, and the total internal energy of the gas is *increased*.

Is it possible to change the internal energy of the gas, even if no work is done? To answer this, we now imagine that the container is initially at a higher temperature than the gas within; that is, we start with a cool gas in a hot container. The piston is locked in position. The atoms in the wall are oscillating energetically, and when a gas molecule strikes the wall, it rebounds, on the average, with a *higher* kinetic energy. See Fig. 18-6. This energy-transfer process, which has its origin in the temperature difference between the gas and the container, is called *heat*. The gas has been heated because there is a difference in temperature between it and the container. The internal energy of the gas increases by dU. This gain in internal energy is equal to the thermal energy dQ transferred into the system by heat; that is,

With no work: $\qquad dQ = dU$

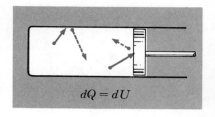

FIG. 18-6. The gas is heated dQ and its internal energy increased dU when, on the average, a molecule rebounds from the container wall at a higher speed.

(Again, dU and dQ are *algebraic* quantities.)

This process may also be reversed. With the gas initially at a higher temperature than the container walls, a typical molecule *loses* energy in striking it. Then the internal energy of the gas changes by $-dU$, the amount of heat transferred *into* the gas being $-dQ$. Again, $dQ = dU$.

Heat and work both describe energy-transfer processes. *Work* is identified with the *macroscopic* displacement of an object having *ordered* kinetic and potential energy; no temperature difference need be involved. *Heat*, on the other hand, is identified with the transfer of thermal energy, energy of *disordered* motion, arising from a temperature difference. When a system is heated, work is done, to be sure, but it occurs on a microscopic scale and in disordered fashion. It involves the interaction between individual atoms and molecules, and it can be described macroscopically in terms of a temperature difference.

Heat may be expressed in any units of work or energy. By tradition, heat is often given in terms of the energy units *calories* where 1 cal = 4.18 J.

On a microscopic scale, where one deals with individual particles, there is no such thing as heat, or "thermal energy." The kinetic energy of individual particles changes, because one particle does work on a second particle. On the other hand, heat and thermal energy have meaning only for large numbers of particles. They allow us to preserve and use the energy-conservation principle without regard for the detailed motions of the many interacting particles.

Unfortunately, the term *heat* is sometimes used not only to describe an energy-transfer process itself but also to denote the thermal-energy content of a system. But one cannot properly speak of the *heat* "content" of a system any more than one can speak of the "work content" of a system. Moreover, it is misleading to describe heat as a "form" of energy or to speak of the "flow" of heat. These phrases are left over from the notion, now rejected, that heat (or "caloric") is a substance. They are, unhappily, thoroughly entrenched in the parlance of thermodynamics.

Figure 18-5 shows a process in which there is no heat; work is done and the internal energy changes. Figure 18-6 shows a process in which no work is done; there the gas is heated and the internal

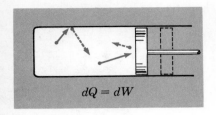

FIG. 18-7. The gas is heated dQ, does work dW, but has no change in its internal energy when, on the average, a molecule rebounds with increased speed from the container and with corresponding decreased speed from the piston.

energy changes. Now consider the situation of Fig. 18-7, in which the gas is heated and does work without, however, changing its internal energy. Energy flows into the system through heat dQ; the same amount flows out, as work dW done on the piston:

With constant internal energy: $dQ = dW$

The internal energy of the gas (therefore, its temperature) is unchanged, and the following occurs: On a molecular scale a typical molecule gains kinetic energy when striking a fixed (hot) wall but loses kinetic energy when striking (and pushing) the piston. (Since heat requires a temperature difference, the container walls must, strictly, be at an infinitesimally higher temperature than the gas. The "direction" of temperature difference always gives, so to speak, the direction of net thermal-energy flow.)

Finally, consider a process involving changes in all three—the heat, internal energy, and work done by the system. If heat in exceeds work out—that is, if the thermal energy into the system exceeds the ordered energy leaving the system—the internal energy will increase. The relation governing such a process is

$$dQ = dU + dW \qquad (18\text{-}13)$$

This is the *first law of thermodynamics*. It is simply the conservation of energy principle expressed in its most general form. It recognizes transfer of energy through either work or heat, and it includes in the internal energy of the system *all* forms of energy, whether ordered or disordered.

Note the convention for the signs in (18-13): $+dQ$ represents the thermal energy *entering* the system through heating, $+dW$ represents the energy *leaving* the system by virtue of work done *by* the system, and dU is the change in internal energy of the system. Figure 18-8 is a symbolic representation of the first law.

Although we have arrived at the first law of thermodynamics by considering the microscopic behavior of molecules in an ideal gas, this relation is of complete generality. We shall later consider its applications to systems of liquids and solids and its complete confirmation in experiment.

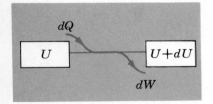

FIG. 18-8. Schematic representation of the first law of thermodynamics. Energy enters the system as heat dQ, leaves as work dW, and the internal energy increases by dU.

Example 18-2

One half of an insulated container holds 1 g mol of helium gas at 0°C; the other half holds 1 g mol of neon gas at 100°C. The insulating partition between the two halves is then removed, the gases mix, and thermal equilibrium is achieved. What is the final temperature of the mixture?

Initially the total internal energy of the helium molecules is equal to their total number of N_A multiplied by the initial kinetic energy $\tfrac{3}{2}kT$ per molecule, or $\tfrac{3}{2}N_A kT = \tfrac{3}{2}RT = \tfrac{3}{2}R(273 \text{ K})$. Similarly, the total initial internal energy of the neon molecules is $\tfrac{3}{2}RT = \tfrac{3}{2}R(373 \text{ K})$. After the molecules are mixed and achieve thermal equilibrium, each of the two types of gas will have an internal energy of $\tfrac{3}{2}RT_f$, where T_f is the *f*inal temperature. Therefore, by energy conservation,

$$\tfrac{3}{2}R(273\text{ K}) + \tfrac{3}{2}R(373\text{ K}) = (2)\tfrac{3}{2}RT_f$$

$$T_f = \frac{273 + 373}{2}\text{ K} = 323\text{ K} = 50°\text{C}$$

18-5 DISORDER AND THE SECOND LAW OF THERMODYNAMICS

The first law of thermodynamics, nothing more than the energy-conservation principle applied to a system comprised of a very large number of molecules, is *not* sufficient to account for the thermal behavior of gases. For example, when we mix a hot gas with a cold gas, we know that the final equilibrium temperature lies *between* the two initial temperatures. We do *not* find the hot gas at a higher final temperature and the cold gas at a lower final temperature, although this possibility is not ruled out by energy conservation. An additional fundamental law of physics, the *second law of thermodynamics*, operates here.

One very general formulation of the second law is this: *An isolated system, free of external influence, will, if it is initially in a state of relative order, always pass to states of relative disorder until it eventually reaches the state of maximum disorder.* We shall discuss qualitatively the ideas embodied in this statement, as illustrated by the behavior of gas molecules.

From the microscopic point of view, *all* that ever happens in a gas is this: Molecules collide. In each collision the molecules interchange energy and momentum in such a way that each of these quantities is conserved. *All* general conclusions concerning the behavior of the gas can be drawn from an analysis of the molecular collisions.

Consider the collision shown in Fig. 18-9a. Here a molecule with a high speed strikes, and transfers energy and momentum to, a molecule with a low (zero) speed. In this particular collision the energies and speeds of the two molecules have become more nearly equalized. This is, of course, just one of many possible collisions. Another is shown in Fig. 18-9b. Here two molecules, both initially in motion, collide in such a way that one molecule is brought to rest while the other molecule leaves the collision with all of the energy originally shared by the two. To assure ourselves that this is, in fact, a possible collision, one consistent with both momentum and energy conservation, we need merely note that it is simply that of Fig. 18-9a run backward in time: If we saw the first collision in a motion picture, the second would be seen by running the film backward, all molecular velocities now being reversed. *On the microscopic scale, all processes*—which is to say, all collisions—*are reversible.*

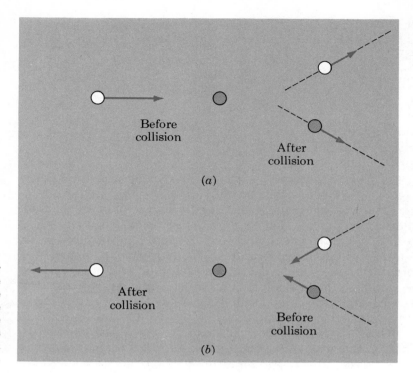

FIG. 18-9. Two molecular collisions. In collision *(a)* the molecular energies become more nearly equal; in collision *(b)* the molecular energies become more nearly unequal. Collision *(b)* is merely collision *(a)* run backward in time.

We have seen that molecular speeds can become more nearly equalized or they can become more *un*equalized, through a collision. That is, in molecular collisions, energy may be transferred from fast to slow particles, or from slow to fast ones. In an ordinary gas there are frequent collisions, and any one molecule will change velocity frequently, moving in a zigzag path. We wish to consider here how the individual molecular kinetic energies of a large number of molecules will change when the particles interact through intermolecular collisions. The detailed analysis of such processes lies in the area of *statistical mechanics,* which combines the laws of newtonian mechanics with the rules of statistics to predict the most probable behavior of collections of particles.

Consider a gas of N monatomic molecules confined to a volume V and having a constant total internal energy, thus a constant translational kinetic energy. What shape can we expect for the energy-distribution curve? That is, what curve will give the number of molecules N_E as a function of the kinetic energy E per particle? There are, of course, many possibilities consistent with the conservation laws of energy and of momentum; three are shown in Fig. 18-10. In Fig. 18-10*a* all molecules have essentially the same kinetic energy; for this distribution all molecules move at the same speed, and the state of the system is a highly ordered one. Figure 18-10*b* shows another possible energy distribution; in this case there are two peaks in the curve. Finally, Fig. 18-10*c* shows the experimentally observed distribution for any gas in thermal equilibrium, the so-called *maxwellian distribution*.†

Now suppose that we start with the gas molecules in the highly ordered state indicated by Fig. 18-10a. If we wait long enough for many collisions to have occurred (perhaps 10^{-5} s), the distribution becomes that of Fig. 18-10c. In fact, *any* initial distribution of molecular energies becomes the maxwellian distribution when the molecules have been allowed to transfer energy in collisions and achieve molecular chaos. The total energy is so distributed that the system achieves the state of maximum possible disorder, namely, the state of thermal equilibrium at temperature T. Indeed, one can define a temperature for a collection of gas molecules *only* when the molecules have achieved thermal equilibrium; then, the average kinetic energy per molecule is, of course, $\frac{3}{2}kT$. One cannot speak of a temperature for nonequilibrium distributions like those in Fig. 18-10a and b, even though all three distributions have the same average kinetic energy per molecule.

A collection of molecules achieves thermal equilibrium, whatever the initial distribution of molecular energies, because the equilibrium state represents the *most probable* state. The basic assumption of statistical mechanics in arriving at this result is the following: *All microscopic states are equally probable,* such states being occupied according to pure chance. There are, of course, the additional requirements that the total energy of all molecules be constant and that the total momentum of all molecules be zero.

By a microscopic state is meant a state corresponding to a specific position and momentum for each and every molecule of a gas. There are available so many more microscopic states representing disorder or near disorder than representing order that the most probable macroscopic state is that of maximum disorder; indeed, it is a near certainty. Similarly, one is not likely to find the cards of a thoroughly shuffled pack in such an ordered state that the first four cards are aces, simply because the number of disordered arrangements greatly exceeds the number of arrangements for which the first four cards are aces. It is *possible* that molecules of a gas initially in equilibrium would, at some later time, all be found moving with exactly the same speed, as in Fig. 18-10a, but such a possibility is so overwhelmingly unlikely as to be virtually impossible. An isolated system will pass from a state of relative order to one of relative disorder because the disordered state is, in fact, the most probable state.

Consider the following example. Oxygen gas is held in the left half of an insulated container, the gas being initially in thermal equilibrium at the temperature T_l. See Fig. 18-11a. An equal number of helium molecules is in the right half of the container, separated from the left by a removable partition. The helium gas

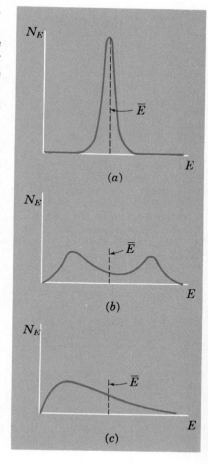

FIG. 18-10. Three possible molecular-energy distributions, all with the same average molecular energy $\overline{E}$. (a) All molecules have nearly the same energy. (b) Two peaks in the energy-distribution distribution for a gas in thermal equilibrium. (c) The maxwellian energy distribution for a gas in thermal equilibrium.

---

†First *derived* by J. C. Maxwell, using the basic assumptions of statistical mechanics.

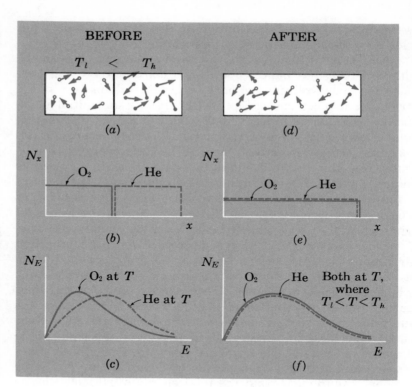

FIG. 18-11. Oxygen (O_2) molecules at a *low* temperature T_l and helium (He) molecules at a *high* temperature T_h in a container with a partition. The molecules in the container *(a)* before, and *(d)* after, the partition is removed. The spatial distribution of the molecules *(b)* before, and *(e)* after, mixing. The molecular energy distribution *(c)* before, and *(f)* after, mixing.

is initially in thermal equilibrium at the temperature T_h, where $T_h > T_l$. The distributions of the two kinds of molecules, in *space* and in *energy,* are shown in Fig. 18-11b and c, respectively. Suppose that we remove the partition and wait until a final equilibrium state is reached. What will the distributions of the molecules in space and energy now be?

Before the partition is removed, there is a degree of spatial order: All oxygen molecules are in the left half, all helium molecules in the right half. There is, moreover, disorder in energy among the oxygen molecules at T_l; similarly, there is disorder in energy among the helium molecules at the temperature T_h. After the partition is removed, the oxygen molecules are free to move into the right half and the helium molecules into the left. Intermolecular collisions occur, the gases become thoroughly mixed, and in a short time we find that both the oxygen and the helium molecules are uniformly distributed throughout the container. Thus, the final distribution of both kinds of molecules is one of maximum disorder, both in space and in energy.

By the second law, the energy distribution will, after the partition is removed, correspond to maximum disorder in energy. Inasmuch as the helium molecules are initially at a higher temperature than the oxygen molecules, the probability that energy will be transferred in a collision from a helium molecule to an oxygen molecule is greater than for energy transfer in the reverse sense. Thus, thermal energy will go from the hot helium gas to the cooler oxygen gas until the two reach a common temperature intermediate between T_l and T_h. They will then have common space and energy distributions, as shown in Fig. 18-11c and f.

Is it possible that, at some later time, we shall again find the oxygen molecules all on the left side at T_l and the helium molecules all on the right side at temperature T_h? Yes. Is it probable? No. It is extraordinarily improbable because there are very many more microscopic states corresponding to the molecules distributed uniformly in space and having a maxwellian energy distribution than there are states corresponding to the relatively high degree of order in which the two gases are separated in the container and at different temperatures.

We saw earlier that individual collisions between particles are reversible in time: We cannot tell whether a moving-picture film portraying an intermolecular collision is being run forward or backward. This is *not* the case when one deals with large numbers of particles. On the macroscopic scale, processes are essentially irreversible, since a system moves inexorably from states of relative order to disorder. From our experience, we *can* tell at once when a moving picture of some ordinary large-scale phenomenon, such as an exploding bomb, is run backward. Thus, the second law of thermodynamics implies a directionality of time. At the macroscopic level, time's arrow points to the future. Order turns to disorder; ordered energy is degraded into disordered or thermal energy. So too, the direction of thermal-energy flow—the "direction" of heat, as it were—is from the higher- to the lower-temperature body, it being thereby ensured that two isolated bodies, initially at different temperatures, achieve a common final temperature.

SUMMARY

In the kinetic theory it is assumed that molecules of negligible size are in random motion and make perfectly elastic collisions with one another and with the container walls, the duration of any one collision being short compared with the time elapsing between collisions. This theory relates the pressure p and volume V of an ideal gas to the number N of molecules, and molecular mass m, and the root-mean-square molecular speed v_{rms}:

$$pV = \tfrac{1}{3} N m v_{rms}^2 \qquad (18\text{-}7)$$

The average translational kinetic energy of a molecule in a gas in thermal equilibrium at temperature T is $\tfrac{3}{2}kT = \tfrac{1}{2}mv_{rms}^2$.

The internal energy U of a monatomic gas of N molecules arising from translational molecular motion is

$$U = N(\tfrac{3}{2}kT) \qquad (18\text{-}11)$$

The first law of thermodynamics can be written

$$dQ = dU + dW \qquad (18\text{-}13)$$

where dQ is the heat entering the system, dU the change in the system's internal energy, and dW the work done by the system on its surroundings. This law is merely the conservation of energy principle generalized to include the transfer of energy by heat, which is the transfer of disordered energy arising from a temperature difference.

The molecules of a gas in thermal equilibrium are distributed over a large range of speeds. The root-mean-square speed v_{rms} is related to temperature T by $\tfrac{1}{2}mv_{\text{rms}}^2 = \tfrac{3}{2}kT$. The average translational kinetic energy per molecule, $\tfrac{3}{2}kT$, is the same for all gases at the same temperature T.

In terms of microscopic molecular behavior, the second law of thermodynamics can be stated as follows: Any system free of external influence will always pass from states of relative order to states of relative disorder, until it reaches the state of maximum disorder, thermal equilibrium. Then the state of maximum disorder, consistent with energy and momentum conservation, corresponds to the most probable state.

PROBLEMS

18-1. Most materials of low atomic weight have densities of the order of a few grams per cubic centimeter. Compute the approximate size of a molecule in such a material.

18-2. Show that the (approximate) atomic diameters d_1 and d_2 of elements "1" and "2" in the form of solids are related in the following way:

$$\frac{d_2}{d_1} = \left(\frac{A_2/\rho_2}{A_1/\rho_1}\right)^{1/3}$$

where A and ρ refer to the atomic weight and the mass density, respectively. (Although the magnitudes of A range from 1 to about 240 and those of ρ from 0.075 to 22.48, the A/ρ ratio is restricted to a range of about 5 to 1, and all elements have quite closely the *same* atomic diameter.)

18-3. A certain beam consists of particles each of mass m and traveling in the same direction with speed v. The beam is incident normally on a surface, the number of particles striking it per unit time being N. What is the average force on the surface if all particles collide with it elastically?

18-4. Listed below are the speeds, in meters per second, of 10 molecules: 2, 4, 6, 6, 8, 9, 9, 10, 11, 15. Compute (a) the average speed and (b) the root-mean-square speed for these molecules.

18-5. The weight of a vessel containing gas exceeds the weight of the same vessel evacuated; a balance indicates a weight for a filled vessel equal to the sum of the weight of the evacuated vessel and weight of the gas alone. Explain why this is so, even though any one molecule of the gas is, according to the kinetic theory, moving freely in space except during collisions of negligible duration. Use the following considerations: A vessel of height h contains one molecule of mass m moving vertically up and down and making perfectly elastic collisions with the top and bottom walls. Its speed is v as it moves upward immediately after striking the lower wall, and it is constantly accelerated downward at the rate g. Show that the average force of the molecule on the bottom wall exceeds that on the top wall by mg; that is, the average force of the molecule on the container is just its weight.

18-6. Show that the following are consequences of the kinetic theory of gases. (a) Avogadro's law: Under the same conditions of temperature and pressure, equal volumes of gas contain equal numbers of molecules. (b) Dalton's law of partial pressures: When two or more gases which do not interact chemically are present together in the same container, the total pressure is the sum of partial pressures contributed independently by each of the several gases.

18-7. What is the rms speed of helium gas molecules at 4.216 K? Helium gas is monatomic, with a molecular weight of 4.00. (Liquid helium boils at 4.216 K when the pressure is one atmosphere.)

18-8. What is the ratio of the rms speed of oxygen molecules in air to the rms speed of nitrogen molecules in air at any temperature?

18-9. A thermal neutron is one whose average kinetic energy is equal to that of a molecule of gas at 300 K. What is the rms speed of a thermal neutron (mass, 1.67×10^{-27} kg)?

18-10. The fragments from uranium atoms which have undergone nuclear fission have an average kinetic energy of 1.1×10^{-11} J. What would be the approximate temperature of a gas consisting of such fission fragments?

18-11. Show that the rms speed of a gas of molecular weight w at the absolute temperature T is given by $v_{\text{rms}} = \sqrt{3RT/w}$.

18-12. The speed of sound through an ideal gas is directly proportional to the rms speed of the gas molecules. What is the ratio of the speed of sound through helium gas to the speed of sound through oxygen gas, both assumed to be ideal, at the same temperature?

18-13. What is the total translational kinetic energy of 1 g mol of any ideal gas at 300 K?

18-14. Find the speed at which the kinetic energy of a 1,000-kg automobile would equal the thermal energy content of 1.0 kg of helium gas at STP. The atomic weight of helium, a monatomic gas, is 4.00 g.

18-15. An ideal gas in a perfectly insulated container is compressed when a piston is displaced 2.0 cm by a force of 100 N. (a) By how much is the internal energy of the gas changed? (b) Does its temperature rise or fall?

18-16. (a) For an ideal gas show that $pw = \rho RT$, where w is the atomic weight of the gas. (b) Use this relation to find the air density at an altitude of 20 km, where the temperature and pressure are $-56°C$ and 0.0540 atm, respectively. The average molecular weight of air is 29.0.

18-17. At the top of the photosphere, the visual surface of the sun, the temperature, pressure, and mass density are 4,500 K, 1.0×10^{-2} atm, and 2.8×10^{-8} g/cm³, respectively. Find the average mass of a particle in the solar atmosphere at this level. (In this computation, assume that the gases in the solar atmosphere can be treated as though they were ideal gases.)

18-18. A *diffusion pump* can easily maintain a pressure of 1.0×10^{-6} torr at 273 K. (The *torr* is a unit often used to characterize low pressures; 1 torr = 1 mm Hg.) (a) If the residual gas is helium, find the corresponding number density and intermolecular spacing. The atomic (and molecular) weight of helium gas is 4.00. (b) Find the ratio of these values to the corresponding quantities for helium gas at STP.

18-19. A container with a volume of 1,000 cm³ holds 10^8 molecules. What is the probability of (a) finding one particular molecule located within a certain 1-cm³ region of the container, (b) finding all 10^8 molecules within this region, (c) finding a particular molecule to be moving to the right, (d) finding all molecules moving to the right, and (e) finding all molecules located within a 1-cm³ region and moving to the right?

CHAPTER 19

Thermal Properties of Solids and Liquids

In this chapter we extend the basic thermodynamic concepts—temperature, thermal energy, heat, and the first and second laws of thermodynamics—from the simple ideal-gas systems to the more complicated solid and liquid systems. We consider the fundamental experiments of Joule, which established the first law of thermodynamics, together with specific heats and latent heats. Finally, we treat modes of thermal-energy transfer: conduction and radiation.

19-1 SOLIDS AND LIQUIDS AS THERMAL SYSTEMS

Ideal gases are simple thermal systems. At low densities all gases follow the same equation of state, the general-gas law; the absolute temperature of any ideal gas is merely a measure of the average molecular translational kinetic energy, and one may relate the microscopic molecular behavior to the macroscopic properties of a gas directly through a mechanical model, the kinetic theory

What is the fundamental reason that gases show such relatively simple behavior? It is this: Apart from perfectly elastic intermolecular collisions of negligible duration, a molecule in a low-density gas is always in force-free motion, traveling along a straight line at constant speed. Intermolecular collisions serve merely to maintain molecular chaos and thermal equilibrium; otherwise, one need not (for ideal gases) be concerned with interactions between the molecules.

Solids and liquids are different. In these states, the molecules are *not* separated, on the average, by distances which are large compared with the range of the intermolecular force. Indeed, neighboring molecules interact continuously in the solid and liquid states. As a consequence, the total energy of the particles is *not* merely the sum of their kinetic energies. One must take into account, in addition, the intermolecular *potential* energies. Let us, then, inquire into the meaning of thermal energy for the solid and liquid states.

We first recall that, for a monatomic ideal gas in thermal equilibrium, the only contribution to the thermal energy is the disordered *translational* kinetic energy of the molecules. The total linear momentum of all the molecules is zero. For diatomic or polyatomic molecules, there may be additional contributions to the system's total energy. If a molecule rotates, the kinetic energy of rotation is a part of the gas's thermal energy. This *rotational* kinetic energy is disordered because the molecules rotate about axes oriented in all directions; that is, the total angular momentum of all the rotating molecules is zero. Moreover, if the molecules also undergo vibrational motion, the energy associated with the molecular oscillations contributes to the gas's thermal energy. The *vibrational* energy—now a combination of both kinetic energy and potential energy—is disordered because the vibrational axes are randomly oriented.

Now consider the contributions to the disordered, or thermal, energy for a solid. A solid may be thought of as a collection of atoms bound together, each atom oscillating about an equilibrium position at any finite temperature. In addition, some of the least tightly bound atomic electrons may wander throughout the solid. Except for extremely low temperatures, the translational kinetic energy of these electrons is not important thermodynamically, for reasons explained by the quantum theory. Therefore, for a solid, the thermal energy consists (almost entirely) of the kinetic and potential energy of atoms vibrating about equilibrium positions.

It is important to distinguish again between *ordered* and *disordered* energy. Suppose that all the atoms of a solid had the same velocity. Then the solid as a whole would be in motion in the direction of the velocity vectors, and the kinetic energy of the body would be *ordered*. See Fig. 19-1a. Likewise, if the body were at rest and a pair of external forces were applied to the solid to stretch it, the atoms would increase their separation distances along the direction of the applied forces and the body as a whole would have *ordered* elastic potential energy. See Fig. 19-1b.

The *thermal* energy of a solid is something else. It consists partly of the kinetic energy of atomic vibrations; this *kinetic energy* is *disordered* by virtue of the variety of atomic velocities, both in magnitude and direction, as shown in Fig. 19-1c. The resultant linear momentum associated with this disordered energy is zero. Furthermore, there is thermal energy associated with the intermolecular potential energy. This *potential energy* is *disordered* (Fig. 19-1d); the atoms have random displacements, from their equilibrium positions, that differ in magnitude and in direction, and

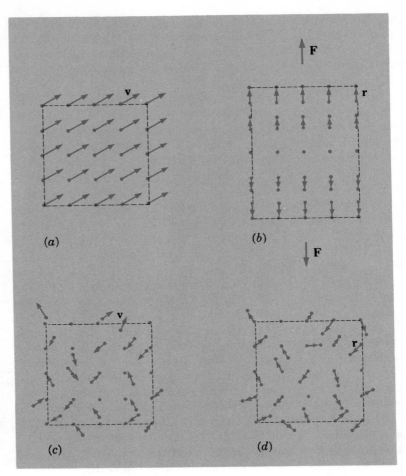

FIG. 19-1. Ordered-vs.-disordered energy for the particles of a solid. *(a)* The body as a whole is in motion and has *ordered kinetic energy*, inasmuch as all atoms have the *same* velocity. *(b)* The body as a whole is stretched and has *ordered potential energy*, inasmuch as the atoms are displaced along the common direction of the applied forces. *(c) Disordered kinetic energy:* The vectors show the atomic velocities, which are distributed at random in magnitude and direction. *(d) Disordered potential energy:* The vectors show the atomic displacements, which are distributed at random in magnitude and direction.

the solid as a whole shows no net deformation (apart from thermal expansion). In short, the random oscillations of atoms in a solid are the origin of the solid's thermal energy (with a relatively minor contribution from the "free" electrons within the solid).

In like fashion, the thermal energy of a liquid is associated with disordered molecular kinetic and potential energy. In this state, however, the molecules are not bound to an equilibrium position, and the disordered translational kinetic energy of the molecules also contributes to the thermal energy.

19-2 THE FIRST LAW OF THERMODYNAMICS

We wish to apply the first law of thermodynamics to liquids and solids:

$$\Delta Q = \Delta U + \Delta W \tag{18-13}$$

Here ΔU is the change in the internal energy of the system. Both

ΔQ and ΔW describe energy-transfer processes. The heat ΔQ measures the disordered energy entering the system by virtue of a temperature difference between the system and its surroundings, whereas the work ΔW represents ordered energy transferred from the system to its surroundings.

To establish the validity of the first law, one need only show that the change in a system's internal energy depends on the *net amount* of energy entering the system, but *not on whether* the internal-energy change arises *by work or by heat*, or some combination of the two energy-transfer modes.

The classic experiments of J. Joule (1818–1889) did just this. Consider the situation shown in Fig. 19-2a in which a liquid has its internal energy changed by heating. Here an electric generator (with whose details we need not be concerned) is run by a descending weight. The generator sends an electric current through the coil of wire, and the hot wire heats the liquid. Assuming that the generator's internal energy does not change, the work $\Delta W = mg\,\Delta y$ done by the weight mg descending a distance Δy is equal in magnitude to the heat ΔQ into the liquid, and ΔQ is also the change ΔU_{liq} in the liquid's internal energy.

$$\Delta W = \Delta Q = \Delta U_{\text{liq}}$$

For the situation shown in Fig. 19-2b, energy enters the liquid as work, not heat. Now the descending weight turns a stirrer immersed in the liquid. Initially, the liquid is set into rotational motion by the stirrer, and it acquires macroscopic rotational kinetic energy. But the liquid soon comes to rest because of internal friction, and its ordered rotational kinetic energy is converted into disordered thermal energy. The work ΔW done by the descending weight equals the work done by the stirrer on the liquid, which is, in turn, equal to the increase ΔU in the liquid's internal energy.

$$\Delta W = \Delta U_{\text{liq}}$$

The change in the liquid's internal energy is manifest as a change ΔT in its temperature. Joule's experiments showed that the amount of this temperature change was the same for arrangements in Fig. 19-2a or b—that is, energy added either by heat or by work—if the work ΔW done by the descending weight was the same.

The experiments also showed that the liquid's temperature change ΔT is proportional to the work ΔW; and for a constant ΔW, ΔT is inversely proportional to the mass m of liquid. In summary,

$$\Delta W \propto m\,\Delta T$$

which, from the two equations above, is equivalent to

$$\Delta U_{\text{liq}} = cm\,\Delta T \qquad (19\text{-}1)$$

where c is a proportionality constant whose magnitude depends

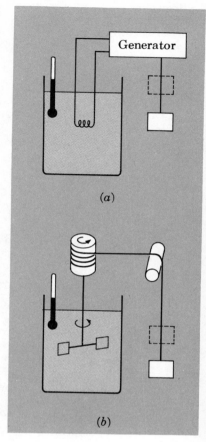

FIG. 19-2. The Joule experiment: (a) A descending weight runs an electric generator, which operates an electric heater, which in turn raises the temperature of a liquid; (b) a descending weight runs a stirrer, which stirs the liquid and raises its temperature.

The First Law of Thermodynamics SEC. 19-2

on the nature of the liquid. Equation (19-1) implies that the change in the thermal energy of a liquid is directly proportional to both the liquid's mass m and the temperature change ΔT. We postpone consideration of the values of c and of the units to the next section.

The temperature of a liquid can be changed by heating it, or by doing work on it and by changing internal ordered kinetic energy or potential energy into internal thermal energy. When the temperature changes, the internal thermal-energy change is proportional to the mass and temperature difference, irrespective of the mode by which energy enters the system.

The same result is found for solids. One may, for example, change the temperature of a metal bar by heating it, by doing work on it in repeatedly flexing it or hammering it, or by bringing it to rest in an inelastic collision. Once again, (19-1) relates the change in internal thermal energy to the solid's mass and temperature rise.

19–3 SPECIFIC HEATS AND HEAT UNITS

Experiments of the sort described in the last section show that 4.18 J of work or heat is required to raise the temperature of 1 g of water by 1 C°. It is convenient to give a special name, the *calorie,* to this amount of energy. By definition (U.S. National Bureau of Standards),

$$1 \text{ cal} = 4.18400 \text{ J}$$

Strictly, the amount of energy required to change the temperature of water 1 C° differs slightly, according to the initial temperature of the water; therefore, the calorie is, very closely, the thermal energy that changes the temperature of 1 g of water from 14.5 to 15.5°C. This energy unit is sometimes referred to as the *gram calorie.* The *kilogram calorie,* also known as the "large" calorie and abbreviated kcal, is 10^3 g cal. One kilogram calorie raises the temperature of 1 kg of water 1 C°.†

It must be emphasized that the calorie, although used most often in measuring heat and thermal energy, is merely another energy unit. One may perfectly well express the kinetic energy of a baseball in calories.

By the *specific heat* (capacity) c of a substance is meant the thermal energy required to change the temperature of a unit mass of material 1°; that is,‡

$$c = \frac{\Delta U}{m \, \Delta T} \tag{19-2}$$

This relation is simply (19-1) solved for the constant c.

---

†The name *British thermal unit* (Btu) is given to the energy required to raise the temperature of 1 lb of water 1 F°.
‡Strictly, for gases we must define the specific heat as $c = \lim_{\Delta T \to 0} (\Delta Q / m \, \Delta T)$. But $\Delta Q \approx \Delta U$ for solids and liquids inasmuch as very little ordered work is done by them when they are heated.

Table 19–1

Substance	Temp. (°C)	Spec. heat (cal/g-C°)	Molar spec. Heat (×R)
Gases†			
Helium	−260–3000	1.25	2.49
Water vapor	100–120	0.48	4.33
Solids			
Aluminum	20	0.217	2.92
Copper	−263	0.00086	0.005
Copper	20	0.093	2.94
Copper	1000	1.10	3.48
Iron	20–100	0.113	3.00
Lead	20–100	0.031	3.22
Glass	20–100	0.199	
Ice	−10–0	0.55	
Liquids			
Mercury	0–100	0.033	3.35
Water	0	1.009	
	15	1.000	
	100	1.006	

†Specific heats of gases are at constant pressure.

The specific heat of water is 1.00 cal/g-C°. It is found that to raise the temperature of 1 g of mercury by 1 C° requires 0.033 cal; therefore, the specific heat of mercury is 0.033 cal/g-C°.

Table 19-1 lists the specific heats for a number of materials. For liquids and solids the specific heats at constant volume and at constant pressure are so nearly alike that one need not ordinarily distinguish between them. This is so because liquids and solids are highly incompressible and the work done in expanding is negligible. It should be noted that the specific heat of a liquid or solid is *not* independent of temperature. For example, copper has a specific heat of 0.093 cal/g-C° at 20°C, but its specific heat at −263°C is only 0.00086 cal/g-C°. Indeed, the specific heats of all substances approach zero at the absolute zero of temperature. Thus, in general the specific heat c is not a constant, and (19-2) applies only when the temperature difference ΔT is sufficiently small. Nevertheless, the specific heats of most materials at temperatures not very far from room temperature (300 K) may be properly regarded as constants.

The term *heat capacity* (more properly, *thermal-energy capacity*) is used to denote the thermal energy required to change the temperature of an object by 1°. Thus, the heat capacity of an object of mass m and specific heat c is equal to $\Delta U/\Delta T = cm$.

When thermal energy enters or leaves a substance, in addition to (or instead of) a change in temperature, there may well be a change in state. *Latent heat* must be associated with this change in state. For example, to melt ice at 0° C into water at 0° C requires that 80 cal/g be *added* at atmospheric pressure; conversely, freezing water at 0° C into ice at 0° C requires that 80 cal/g be extracted from the water.

Example 19-1

A 40-g block of copper originally at 20°C is dropped into an insulated container holding 200 g of water originally at 80°C. What is the temperature of copper and water after thermal equilibrium has been achieved?

The copper block is heated by the water, initially at the higher temperature. The thermal-energy content of the copper is increased and the thermal-energy content of the water is decreased, both by the same amount, until both substances reach a common temperature. The increase in the internal energy of copper is, from Eq. (19-2),

$$\Delta U_{copper} = cm\,\Delta T_{copper} = (0.093 \text{ cal/g-C°})(40 \text{ g})(t_f - 20°C)$$

where t_f is the final temperature. Similarly, the decrease in the internal energy of the water is

$$\Delta U_{water} = cm\,\Delta T_{water} = (1.00 \text{ cal/g-C°})(200 \text{ g})(80 \text{ C°} - t_f)$$

Since energy is conserved, $\Delta U_{copper} = \Delta U_{water}$, and we have

$$(0.093)(40)(t_f - 20°C) = (1.00)(200)(80 - t_f)$$

$$t_f = 79°C$$

The temperature of the water decreases by only 1 C°, not only because the mass of water exceeds that of the copper but more especially, because the specific heat of water greatly exceeds that of copper. As may be seen from Table 19-1, the specific heat of water is relatively high; as solids or liquids go, water is very hard to heat.

The procedure of this example can be used to measure the specific heat of a substance. The masses, initial temperatures, and final equilibrium temperature are easily measured, and the unknown specific heat can then be computed. This experimental procedure is known as *calorimetry*.

Example 19-2

A block of copper, originally at 20°C and at rest, is dropped from a height of 40 m. The block strikes and comes to rest on an insulated surface. What is the final temperature of the copper, assuming no thermal energy losses?

In Example 19-1 the internal energy and temperature of the copper were increased because the copper had energy transferred to it through heat. In this example the copper is *not* heated; that is, the copper does *not* change temperature by virtue of being in contact with a hotter material. Here the block (and earth) lose gravitational potential energy, and the block gains macroscopic kinetic energy as it falls. Upon striking the surface, *work* is done on the block. Its internal energy increases as the block becomes deformed, and internal friction changes ordered energy into disordered, or thermal, energy.

The original gravitational potential energy $mg\,\Delta y$ has become internal thermal energy $cm\,\Delta T$. Therefore,

$$mg\,\Delta y = cm\,\Delta T$$

$$\Delta T = \frac{g\,\Delta y}{c} = \frac{(9.8 \text{ m/s}^2)(40 \text{ m})(1 \text{ kg}/10^3 \text{ g})}{(0.093 \text{ cal/g-C°})(4.18 \text{ J/cal})} = 1.0 \text{ C°}$$

The copper's final temperature is 21°C.

Although *all* of the work done on a gas, liquid, or solid may be transformed into thermal energy, the converse is not true. A fundamental limitation on the conversion of thermal energy into useful work arises, for example, in considerations of an ideal *heat engine*, the term used to designate any device which, when operating over a complete cycle that returns it to its initial state, transforms heat from a high-temperature reservoir into work on an external object. Even if such a device were completely frictionless and operated in the most efficient of all conceivable cycles (the Carnot cycle), according to the second law of thermodynamics only a portion of the heat entering the engine from a hot reservoir can be extracted as useful work done by the engine over a complete cycle, the remaining thermal energy necessarily being discarded (unused) into a lower-temperature heat reservoir. Consequently, any thermal-energy source, such as chemical or nuclear fuels, and the devices used to convert thermal energy at high temperatures into useful work, are subject to a fundamental limitation in efficiency. These results follow from very general arguments in formal thermodynamics, which typically employ the concept of *entropy*, a quantitative measure of the relative disorder of a system. Just as the relative disorder of any isolated system must increase with time, the second law of thermodynamics may also be formulated in terms of increasing entropy of an isolated system.†

19-4 THERMAL CONDUCTION

Heat is that energy-transfer process which takes place by virtue of a temperature difference. When a hot object is in contact with a cold object, thermal energy flows from the hot to the cold one. Such a flow can take place within a single object: Thermal energy flows from one region at a high temperature to an adjoining region at a lower temperature, if the temperature difference is maintained between the two points. This thermal-energy transfer (often termed, redundantly, a "heat-transfer process") is *thermal conduction*.

The thermal energy of a solid consists mostly of the vibration of the atoms about their equilibrium positions. Adjoining atoms

---

†See R. T. Weidner and R. L. Sells, *Elementary Classical Physics*, 2d ed., vol 1, chap. 21, Allyn and Bacon, 1973.

in a solid interact; as one atom oscillates, its influences the motion of a neighboring atom. If one region of a solid is at a higher temperature than an adjoining region, the amplitudes (and energies) of the atomic oscillations are greater at the hot regions. Thermal energy is then transferred from the hot to the cold regions by the coupling between neighboring oscillators.

If, in addition to vibrating atoms, a solid also has free electrons, as in the case of metals, these free electrons also contribute to the thermal conduction. That the free electrons play a significant role in thermal conductivity is shown by the fact that good thermal conductors, such as metals, are usually also good electrical conductors. Free electrons are the origin of electric currents; these electrons also transfer thermal energy.

For the quantitative aspects of thermal conduction, consider the situation shown in Fig. 19-3. Here a rod of uniform cross section A is surrounded by an insulating material, so that no heat leaks into or out of the rod through its sides. The *hot* left end is maintained at a constant high temperature T_h while the *cold* right end is maintained at a constant lower temperature T_c. The thermal energy entering the rod per unit time from the hot reservoir (that is, the heat rate into the rod) is dQ/dt. At a steady-state condition, this is also the heat per unit time leaving the right end. In fact, dQ/dt represents the thermal energy crossing the area A per unit time at *any* point along the rod. The *net* heat entering the rod, or entering any small volume of the rod, is *zero;* therefore, the rod's internal energy remains *constant*. The rod, a thermal conductor, acts merely as a "heat pipe" between the hot and cold reservoirs, degrading thermal energy by sending it to a lower temperature. This behavior continues, of course, only as long as the two ends are maintained at T_h and T_c. When $T_h = T_c$, then $dQ/dt = 0$.

We concentrate on the temperature drop dT occurring across a thin section of thickness dx. The quantity dT/dx, called the *temperature gradient,* measures the temperature change per unit displacement along the direction of heat flow. If x is taken as

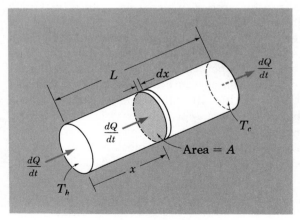

FIG. 19-3. Thermal conduction through a rod of length L and cross section A, having its left and right ends at the constant hot and cold temperatures T_h and T_c, respectively. The heat rate through any cross section is dQ/dt.

increasing along the direction of dQ/dt, the temperature gradient dT/dx is *negative;* that is, the temperature *drops* as x increases.

Experiment shows that the heat rate $R \equiv dQ/dt$ is related to the temperature gradient dT/dx by

$$R = \frac{dQ}{dt} = -KA\frac{dT}{dx} \qquad (19\text{-}3)$$

where A is the cross-sectional area through which the thermal energy flows. The quantity K is a positive constant, called the *thermal conductivity*. It is characteristic of the material of the thermal conductor. Equation (19-3) then shows that, for a slice of infinitesimal thickness, the heat rate varies directly as the cross-sectional area and the temperature gradient.

If dT/dx is constant along the length of the material, then the temperature differential dT can be written as $(T_h - T_c)$ and elementary displacement dx can be replaced by the finite length L. Note that R, K, and A are all constants. Then

$$R = \frac{dQ}{dt} = \frac{KA(T_h - T_c)}{L} \qquad (19\text{-}4)$$

Note that, according to (19-4), the temperature drops *uniformly* along the rod. The differential form of the thermal-conduction equation given in (19-3) is more general than that given in (19-4). Equation (19-3) can be applied to *all* shapes of conductors, not merely to uniform rods. (One can imagine the conductor to consist of a collection of infinitesimally thin sheets for each of which the differential form holds exactly.)

Measured values of the thermal conductivity K for various materials are given in Table 19-2. A good thermal conductor has a high K value; a low K characterizes a poor thermal conductor, or a good insulator.

Table 19–2

Substance	Temp. (°C)	Thermal conduct. (mW/cm-K)
Conducting		
Aluminum	−190 to 30	2.1×10^3
Copper	−160	4.5×10^3
Copper	18	3.9×10^3
Copper	100	3.8×10^3
Silver	18	4.2×10^3
Lead	18	3.5×10^2
Water	20	5.9×10^1
Insulating		
Concrete		8
Cork		0.4
Glass, wool	50	0.4
Ice	0	21
Air	0	0.25

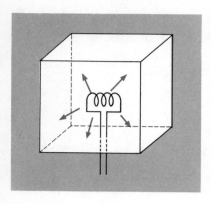

FIG. 19-4. An electric heater inside an insulating box.

Example 19-3

An electric heater operating at 200 W is placed in the interior of a cubical box constructed of insulating material. See Fig. 19-4. The edge length of the box is 20 cm, and the thickness of each side is 1.0 cm. After the heater has been on for a sufficiently long time, dynamic thermal equilibrium is achieved; the interior surfaces of the box remain at the constant temperature 60°C while the exterior surfaces are at the temperature 20°C. (*a*) What is the rate of thermal-energy flow out of the box? (*b*) What is the thermal conductivity of the material of which the box is constructed?

(*a*) The thermal energy from the electric heater passing into the walls of the box must equal the thermal energy leaving the box. Thus, the rate of thermal-energy flow *out* of the box is 200 W.

(*b*) The total area A can be taken as the area of one side, $(0.20 \text{ m})^2$, multiplied by the number of sides (6) of the cube. The thickness L is 0.010 m. Since the edge length of a side is large compared with the thickness of a side, we may properly neglect heat flow along the edges. Thus, Eq. (19-4) gives

$$K = \frac{(dQ/dt)L}{A(T_h - T_c)} = \frac{(200 \text{ W})(0.010 \text{ m})}{6(0.20 \text{ m})^2(40 \text{ C}°)} = 2.1 \text{ mW/cm-°C}$$

Good thermal insulators, such as asbestos, have thermal conductivities of this order.

19–5 THERMAL–ENERGY TRANSFER THROUGH RADIATION

Suppose that a hot object is placed in a completely evacuated container at a low temperature. The hot object and the container are not directly in contact; they are connected only by a thin thread which acts as an almost perfect thermal insulator; see Fig. 19-5. Under these circumstances thermal energy cannot be transferred effectively by conduction. Nor can convection, the bulk transport of heated gas or liquid, occur, inasmuch as the chamber is completely evacuated. Yet, if one waits for a sufficiently long time, two such objects are found by experiment to achieve the *same* final temperature, intermediate between the two initial temperatures. One must conclude that still another thermal-energy transfer process operates here. This process is the *emission and absorption of electromagnetic radiation.*

A thoroughgoing analysis of radiation requires an understanding of the theory of electromagnetism. We merely summarize the most pertinent properties of electromagnetic radiation as they relate to thermal-energy transfer. Whenever an electrically charged particle accelerates (or decelerates), it loses energy by emitting electromagnetic radiation which propagates through empty space at the unique speed of 3.0×10^8 m/s, that of light.†

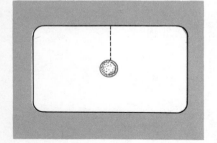

FIG. 19-5. A hot object suspended from an insulating thread in the interior of an evacuated enclosure.

---

†It is interesting to compare the speeds associated with the three thermal-energy transfer processes. Radiation travels at the speed of light ($\approx 10^8$ m/s), thermal-conduction energy is propa-

The reverse process of emission is the absorption of electromagnetic radiation. In absorption electromagnetic fields produce forces on electrically charged particles, thereby transferring energy to the particles. Since matter consists of electrically charged particles, all materials radiate and absorb electromagnetic energy.

Now consider again the situation in Fig. 19-5, in which two objects are at different initial temperatures. The hot object emits *and absorbs* electromagnetic waves. So does the cold one. We find, however, that the hot object loses thermal energy and its temperature falls, while the cold object gains thermal energy and its temperature rises. Since the hot object has a net loss of energy and the cold one a net gain, one must conclude that the hot object emits more radiation than it absorbs and that the cold one absorbs more radiation than it emits. Experiment shows that, although the *absorption process is temperature-independent, emission is temperature-dependent*. For emission, the higher the temperature of a body the greater the rate at which it emits energy.

The total electromagnetic energy emitted per unit time from a unit area of a blackbody (defined below) at the absolute temperature T is found to be directly proportional to T^4. This is the *Stefan-Boltzmann law*, which can be written as

$$P = \sigma T^4 \tag{19-5}$$

where P is the power radiated per unit area and σ is a universal constant equal to 5.67×10^{-8} W/m²-K⁴.

Suppose that some body remains at a fixed temperature. Then it must emit at precisely the same rate as it absorbs. An object is black when it absorbs all visible light striking it, reflecting none, and it is white if it reflects all visible light striking it, absorbing none. In thermodynamics a *blackbody* is defined as an object which absorbs *all* electromagnetic radiation, visible and nonvisible, that impinges on it. Therefore, a blackbody is a *perfect radiator*, as well as a *perfect absorber*, of electromagnetic waves. By the same token, a poor radiator is a poor absorber.

Paradoxically, a blackbody need not appear black; it may, in fact, appear bright. For example, the sun can be considered a blackbody because it absorbs essentially all radiation striking it. A still better blackbody consists merely of a hole leading to the interior of an enclosure made of *any* material; electromagnetic waves entering through the hole undergo reflections inside until all entering radiation is absorbed. Equation (19-5) gives the total electromagnetic power emitted per unit area for *any* blackbody at temperature T. (Recall that the absorbed power per unit area is independent of the temperature of the *absorbing* body.)

We return once again to the situation shown in Fig. 19-5.

---

gated at the speed of sound ($\approx 10^4$ m/s in a typical solid), and convection occurs at a very slow rate, corresponding to the bulk transport of a fluid (≈ 1 m/s).

Both the suspended hot object and the cold enclosure eventually reach thermal equilibrium at some common final temperature. Until this temperature is reached, the hot body emits more radiant energy than it absorbs, and the cold body absorbs more radiant energy than it emits. Because energy is associated with an electromagnetic field in space, we see that there are *three* contributions to the total thermal energy for this system: from the suspended object, from the enclosure, and from the electromagnetic radiation which fills the space between them. The electromagnetic radiation represents energy which has been emitted but not yet absorbed. The radiation within an enclosure which acts as a blackbody at a temperature T may be said to be in thermal equilibrium with the body and is known as *thermal radiation* at the temperature T.

SUMMARY

The thermal energy for solids consists mainly of the disordered kinetic and potential energy of atomic oscillations. For liquids, there is, in addition, disordered translational kinetic energy.

When a solid or liquid undergoes a temperature change ΔT, its internal energy changes by

$$\Delta U = cm\,\Delta T \qquad (19\text{-}1)$$

where m is the mass and c is a constant, characteristic of the material, called the specific heat.

Latent heat is the thermal energy associated with a change in state of a substance with no change in temperature.

The heat rate dQ/dt through a sheet of cross-sectional area A is given by

$$\frac{dQ}{dt} = -KA\,\frac{dT}{dx} \qquad (19\text{-}3)$$

where dT is the temperature change across the sheet, of thickness dx, and the thermal conductivity K is a constant characteristic of the material.

Thermal energy may be transferred by the emission or absorption of electromagnetic radiation. The absorption process is temperature-independent, but the emission of radiation from a blackbody, a perfect radiator (and absorber), is governed by the Stefan-Boltzmann relation,

$$P = \sigma T^4 \qquad (19\text{-}5)$$

where P is the power radiated from a unit surface area of the blackbody at the temperature T and σ is a universal constant.

PROBLEMS

19-1. An aluminum ball is dropped from rest a distance of 3.0 m above an insulating surface. It bounces a number of times and finally comes to rest. By how much has its temperature been increased?

19-2. A 5,000-g block of copper at 20°C is dropped into 3,000 g of water at an initial temperature of 100°C. What is the temperature after thermal equilibrium has been achieved? (Assume that the copper-water system is completely isolated.)

19-3. A 1,000-g metal sphere at 20°C is placed in 1,000 g of water at 80°C which is contained in a 200-g copper vessel insulated against heat loss to the outside. (Such an insulated container is called a *calorimeter*; it is used to measure the specific heats of materials.) Initially the copper is in thermal equilibrium with the water. When thermal equilibrium is re-established, the temperature stabilizes at 72.0°C. Find the specific heat of the metal sphere.

19-4. Find the speed at which the kinetic energy of a 1,000-kg automobile would equal the thermal energy required to heat 1.0 kg of water from 0°C to 100°C.

19-5. In a Joule experiment, of the sort shown in Fig. 19-2, a mass of 50 kg falls through a distance of 1.0 m twenty times. The heated liquid is water and its temperature rises 5.0 C°. What is the mass of the water?

19-6. A lead block moving initially at 3.0 m/s strikes and sticks to an identical lead block initially at rest on a perfectly smooth surface. By how much does the temperature of the composite block rise?

19-7. Two identical blocks moving at equal speeds in opposite directions on a perfectly smooth surface collide, stick together, and come to rest. The temperature of each rises by Δt by virtue of the inelastic collision. What would be the temperature rise if one of the two blocks had been initially at rest?

19-8. Pellets of lead initially at 100°C are placed on a block of ice of the same mass initially at 0°C. What fraction of the ice becomes water?

19-9. What length of an aluminum rod 1.0 cm in radius, and insulated except at the ends, will conduct heat at the rate of 2.0 W between hot and cold reservoirs differing in temperature by 100 C°?

19-10. A long insulated cylindrical bar of silver 5.0 mm in diameter conducts heat along its axis at the rate of 5.0 W. What is the temperature gradient along the bar?

19-11. A windowless room is insulated on all sides (floor, ceiling, and walls) by a 4.0-inch thickness of glass wool, for which the thermal conductivity is $K = 4.0 \times 10^{-1}$ mW/cm-K°. What is the power rating of a heater operating continuously within the room which will maintain the temperature inside the room at 20°C (68°F) if the outside temperature is 0°C (32°F)? The walls are 8.0 ft high and the floor is 15 ft × 20 ft in area.

19-12. The 200-ft² ceiling of a room is insulated from an attic by a 6.0-inch layer of glass wool for which $K = 4.0 \times 10^{-4}$ W/cm-K°. If the attic temperature is 50°C (122°F) and the room temperature is to be maintained at 20°C (68°F), how much energy per hour is transported through the insulation by conduction? Assume that the entire temperature drop is maintained across the glass wool.

19-13. In calculating the heat conduction through a pane of glass it is essential to take into account the fact that a thin film of air adheres to the glass surfaces which are exposed to the atmosphere. As a result, the actual heat path consists of a glass layer sandwiched between two air films. Once dynamic equilibrium has been established, the heat rate R is the same through each of the three layers:

$$R = \frac{dQ}{dt} = K_a A \frac{\Delta T_a}{d_a} = K_g A \frac{\Delta T_g}{d_g} = K_a A \frac{\Delta T_a}{d_a}$$

where K, d, and ΔT are, respectively, the thermal conductivity, thickness, and temperature difference across the various layers. The subscripts a and g refer to *a*ir and *g*lass, respectively. The quantity A is the cross-sectional area of the heat path. (*a*) Show that the total temperature drop across the three-layer sandwich is given by

$$\Delta T = \Delta T_a + \Delta T_g + \Delta T_a = \Delta T_g + 2\,\Delta T_a$$
$$= \frac{R}{A}\left(\frac{1}{K_g/d_g} + \frac{2}{K_a/d_a}\right) = \frac{R}{A}\left(\frac{1}{\Gamma_g} + \frac{2}{\Gamma_a}\right)$$

where we have assumed that the two air films are identical in all of their characteristics. Note that, by definition, $\Gamma_a = K_a/d_a$ and $\Gamma_g = K_g/d_g$. (*b*) Show that for the air-to-glass sandwich the heat rate is given by

$$R_{aga} = \left[\frac{\Gamma_a \Gamma_g}{\Gamma_a + 2\Gamma_g}\right] A\,\Delta T = \left[\frac{\Gamma_a}{\Gamma_a + 2\Gamma_g}\right] K_g A \frac{\Delta T}{d_g}$$
$$= \left[\frac{\Gamma_a}{\Gamma_a + 2\Gamma_g}\right] R_g$$

where R_g is the heat rate which would result if the air film were absent. For air the *film coefficient* Γ_a is about 1.0 mW/cm²-K°, and for window glass 3.2 mm thick

$$\Gamma_g = \frac{K_g}{d_g} = \frac{8.0 \text{ mW/cm-K°}}{0.32 \text{ cm}} = \frac{25 \text{ mW}}{\text{cm}^2\text{-K°}}$$

(c) Evaluate the ratio R_{aga}/R_g and show that it is about 2 percent. This indicates that two air layers effectively improve the insulating properties of the glass by a factor of about 50.

19-14. Two thermal conductors, identical in length, cross-sectional area, and material, are connected between two thermal reservoirs at constant but different temperatures. First the conductors are connected end-to-end (in series) between the reservoirs; then both are connected directly between the two reservoirs (in parallel). What is the ratio of the rate of heat flow in the second case to that in the first?

19-15. A cylindrical pipe of length L and inner and outer radii r_1 and r_2 conducts heat radially outward at the constant rate $R = dQ/dt$, both the inner and outer cylindrical surfaces being maintained at constant temperatures. Show that the temperature difference between the inner and outer surfaces is given by $(R/2LK) \ln (r_2/r_1)$, where K is the thermal conductivity of the material of the pipe. [*Hint:* Apply Eq. (19-3) to a cylindrical shell of thickness dr, and integrate from r_1 to r_2.]

19-16. The white dwarf star 40 Eridani B is known to have the following characteristics: effective surface temperature $T = 2.7\ T_\odot$, luminosity $L = 1.2 \times 10^{-2}\ L_\odot$, and mass $M = 0.427\ M_\odot$; where $T_\odot = 5.76 \times 10^3$ K, $L_\odot = 3.8 \times 10^{26}$ W, $M_\odot = 2.0 \times 10^{30}$ kg, and $R_\odot = 7.0 \times 10^8$ m are the effective temperature, luminosity, mass, and radius of the sun. (The luminosity of a star is the rate at which it radiates energy.) Assuming that stars radiate as though they were perfect blackbodies, find (a) the radius and (b) the mean density of 40 Eridani B.

19-17. A certain blackbody with a surface area of 0.20 m² has a constant temperature of 1000 K. (a) What is the *total* power radiated by the blackbody? (b) If the blackbody absorbs radiation at the rate of 5,000 W, what is the *net* power radiated by the blackbody?

19-18. The temperature of an open hearth furnace used to make steel is optimal at 1850 K. In a particular installation the furnace temperature is monitored by means of a *radiation pyrometer*, an instrument which measures the radiant energy escaping through an aperture in the furnace wall per unit time. (a) Find the rate at which radiant energy escapes from the aperture when its area is 20 mm² and the temperature is 1850 K. (b) Find the minimum and maximum rates of flow of radiant energy if the operating temperature is to be maintained to within 3 percent of 1850 K.

19-19. The temperature of the filament of an incandescent lamp bulb is 2400 K. The filament may be regarded as a blackbody. (a) If the power of the lamp bulb is 100 W, what is the surface area of the filament? (b) If the same lamp is operated at 200 W, what is then the temperature of the filament?

19-20. (a) Show, from the Stefan-Boltzmann law, that the power radiated from a 1-cm² surface of a blackbody at the Kelvin temperature T is $(T/648)^4$ W. (b) At what temperature will a blackbody radiate 1.0 W/cm²?

19-21. A spherical shell 10 cm in radius acts as a blackbody and absorbs all of the 2.0 MW radiation falling continuously on it. The shell's temperature remains constant. What is this temperature?

19-22. The sun may be regarded as a blackbody with a temperature of about 6000 K. The sun's radius is 0.70×10^6 km. (a) What is the total power radiated by the sun? (b) Assuming that the radiated power passing through a unit area placed at right angles to the direction of propagation of the radiation varies inversely as the square of the distance from the sun, what is the power through an area, in square centimeters, at a distance of 1.5×10^8 km, the distance of the earth from the sun? (c) The observed radiation intensity at the earth's surface is 0.13 W/cm². What fraction of the radiation emitted from the sun is intercepted by the earth's atmosphere?

CHAPTER 20

The Electric Force between Charges

All the known forces in physics arise from four fundamental interactions: the strong nuclear force, the electromagnetic force, the weak interaction force, and the gravitational force. The nuclear and the weak interaction forces are of importance only within the atomic nucleus, in certain collisions between nuclei, and in the decays of unstable elementary particles. The familiar gravitational force is important only when the mass of one of two interacting objects is comparable to that of a planet. This leaves the *electromagnetic force*. With the exception of the force due to gravity, all the forces of ordinary experience—the restoring force of a stretched string, the normal force of a floor acting upward on a person, the force between colliding automobiles—indeed, all the forces acting between the atomic nucleus and its surrounding electrons, or between atoms in molecules, are ultimately electromagnetic in origin. The electromagnetic forces dominate all interactions of systems from the size of atoms to that of planets.

We use the word electromagnetism to emphasize that electric and magnetic phenomena are not separate or unrelated. Both electrical and magnetic effects are a consequence of a property of matter called *electric charge*. Although electromagnetic effects have a venerable history, we shall not chronicle the development of this important branch of classical physics.

20–1 ELECTRIC CHARGE

The distinctively new concept introduced in electromagnetism is that of *electric charge*. We know that when any two dissimilar nonmetallic objects are brought into intimate contact with one another—for example, by rubbing glass with silk—and then separated from one another, they show a mutual attraction, exceeding by far their gravitational attraction. Such objects are said to be electrically charged, and the presence of electric charge on each object is responsible for their interaction by an electric force. The reader is assumed to be familiar with this effect and the many qualitative experiments which establish the following fundamental facts concerning these electric interactions: *(a)* There exist *two kinds of charge, (b) like charges repel,* and *(c) unlike charges attract* (Fig. 20.1).

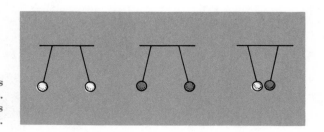

FIG. 20-1. Charged objects suspended from insulating strings. Like charges repel, unlike charges attract.

In this and the next several chapters we shall deal with electrostatics, the science of electric charges at rest. Besides repelling or attracting by the electrical force, charges in motion interact by the so-called magnetic force. Here we shall deal only with situations in which the charges' velocities are so much smaller than the speed of light (3.0×10^8 m/s) that the magnetic force between the charges is negligible compared with the electric force between them.

All electrical phenomena arise from the fact that the fundamental elementary particles of physics may have the property of electric charge. Thus, the electron has a negative charge, the proton a positive charge, and the neutron a zero charge. The use of the algebraic signs + and − to denote the two kinds of charge is appropriate, since combining equal amounts (to be defined precisely below) of positive and negative charges results in a zero electric force acting on an object. Quantization of charge will be discussed in Sec. 20-5. As we know, the nucleus of an atom, consisting of protons and neutrons bound together within a volume whose length dimensions are never much greater than 10^{-14} m, is surrounded by electrons which are bound to it. An atom is electrically neutral as a whole when the number of electrons surrounding the nucleus equals the number of protons in the nucleus.

Atoms are closely packed in solids, their nuclei being separated from one another by distances of the order of 10^{-10} m. In *conductors,* of which metals are examples, most of the electrons are bound to and remain with their parent nuclei, but approximately one electron per atom may be a *free electron.* A free (or conduction) electron, although bound to the conductor, may wander throughout the interior of the conducting material and can easily be displaced within the conductor by external electric forces. In *insulating,* or *dielectric,* materials on the other hand, *all* atomic electrons are *bound,* to a greater or lesser degree, to their parent nuclei. Electrons are removed from or added to an insulating material only with the expenditure of energy. Examples of common conductors are metals, liquids having dissociated ions (electrolytes), the earth, and the human body. Good insulators are very often transparent materials: plastics, glass, and a vacuum, which is a perfect insulator. The best electrical conductors are better than the worst conductors (or best insulators) by enormous factors, up to 10^{20}, which is a number that will be given precise quantitative meaning in Chap. 24. Lying between these extremes are the so-called semiconductors, whose conductivity is intermediate between conductors and insulators. Examples of semiconductors are germanium and silicon. In semiconductors only a very small fraction of the electrons are free. The number of conduction electrons in a semiconductor may be changed by heating the material, by shining light on it, or by applying a very strong external electric field.

Electrostatic effects may be understood on the basis of the atomic model and of the properties of conductors and dielectrics. When two unlike dielectrics are rubbed together, some electrons at the interface between the materials will leave the material to which they would be less tightly bound for the material to which they would be more tightly bound, because systems always go to states of lower energy.

Upon separation one object now carries excess electrons and is negatively charged, while the other object has a deficiency of electrons and is positively charged. When a large-scale object is said to be positively charged, its electrical neutrality has been disturbed by its having lost electrons; similarly, a negatively charged object is an object with excess electrons. "Charging" an object consists simply of adding or subtracting electrons from it. When one type of charge is produced on an ordinary object, the other type must appear in equal amounts on a second object. The charging of any large-scale body results from the separation of charged particles (see Fig. 20-2). Since a charged body has acquired or lost electrons, we sometimes speak of the charge *on* a body. Of course the body acquires (or loses) not only the charge of electrons added to (or removed from) it but also the mass of the added (or removed) electrons. However, the additional mass is usually so trivial as to be negligible.

It is not proper to speak of the charge *on* an electron (or on any other elementary particle for that matter). Electric charge is not something that can be added to or removed from an electron. An electron without charge does not exist. Since electric charge is, like mass, an intrinsic property of an electron, we speak of the electric charge *of* an electron.

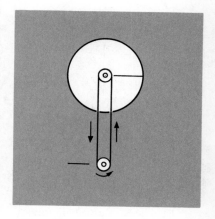

FIG. 20-2. Schematic diagram for a Van de Graaff generator, a device for separating charge. The moving belt continuously carries charge from sharp points near it at the lower roller to points on the interior of a spherical conductor.

Electric Charge   SEC. 20-1

20-2 COULOMB'S LAW

Here we establish the quantitative aspects of the electric interaction between charges. The electric force is often referred to as the coulomb force, named after C. A. de Coulomb (1736–1806) who in 1785 found the electric force to vary as the inverse square of the separation distance.

We shall consider point-charges. A point-charge is a group of one *or more* elementary charged particles confined to a region of space which is small compared with any other dimensions with which we might be dealing, such as the separation distance between two point-charges. A single elementary particle best exemplifies the concept of point-charge, but even here the charge has finite size. The stability of a charged particle of nonzero size against the strong mutual repulsion of its parts is not understood on the basis of present-day fundamental physical theory. We must simply say, for example, that the charge of an electron is confined to a very small volume, leaving the question of what holds it together as an important one not yet answered.

The electric force between point-charges at rest is found to lie along the line connecting them. Thus, the coulomb force is a *central force*. Indeed, it could not be otherwise, for between two points at rest in empty isotropic space the only unique direction is the line between them.

The coulomb force varies *inversely as the square of the distance r* between two point charges:

$$F \propto \frac{1}{r^2}$$

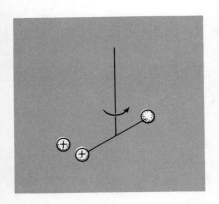

FIG. 20-3. Cavendish torsion balance for measuring the variation in the coulomb force with the distance between two charged objects.

This was confirmed, at least approximately, in experiments by Coulomb and, later, by Cavendish, who used a torsion balance (see Fig. 20-3). The restoring torque of the twisted thin fiber is proportional to the angle of twist.† Thus, one can measure the force of attraction or repulsion between small charged objects of known separation by measuring the angle through which the rod attached to the fiber is displaced. Such experiments with charged objects are similar to the Cavendish experiment, in which the gravitational force between two small objects is measured (Sec. 13-2). Although experiments with a torsion balance can establish that the exponent of r in $F \propto 1/r^n$ is 2 to within a few percent, other experiments (to be described in Sec. 22-5) show by indirect means that the exponent is precisely 2 (to within a few parts in 10^9).

Scattering experiments, in which positively charged particles are shot at atomic nuclei, show that the inverse-square coulomb force holds down to dimensions of about 10^{-14} m, the size of the very small, positively charged nuclei of atoms. Indeed, the coulomb

---

†R. T. Weidner and R. L. Sells, *Elementary Classical Physics,* 2d ed., sec. 14-4, Allyn and Bacon, 1973.

force operates between the protons *within* a nucleus. The electric repulsion between protons competes with the strong nuclear force of attraction among protons and neutrons. In the heavier atoms, in which there are many protons, the coulomb repulsion is responsible for instabilities which can lead to the ejection of helium nuclei in α decay or the splitting of the nucleus in nuclear fission. Furthermore, indirect experiments involving the interaction of electrons and muons with atomic nuclei show that Coulomb's law is valid down to distances of about 10^{-16} m.

One important property of the coulomb force is that it is a *conservative* force (Sec. 10-4). It depends only on the separation distance between two charged particles, not on the time or on the velocities of the particles (provided the velocities are small compared with the velocity of light). As a consequence, one may associate a potential energy with the electric interaction between charged particles.

The gravitational and electric forces are similar in several respects: Both are *central, conservative, inverse-square forces*. Therefore, many of the concepts developed for gravitation—the gravitational field, the gravitational potential energy, the energetics of particles interacting under the gravitational force (Chap. 13)—are equally applicable to the electric force. But there are also emphatic differences. For one thing, there are two types of electric charge, but only one type of gravitational charge (or gravitational mass). Electric charges may attract or repel; gravitational charges attract only. Another important difference is in the relative magnitudes of the electric and gravitational interactions. The electric force is immensely larger than the gravitational force; for example, between an electron and a proton the electric attraction is 10^{39} times greater than the gravitational attraction.

At the atomic level the gravitational force is altogether trivial compared with the electric force. On the other hand, because the negative and positive charges of the elementary particles (the constituents of ordinary objects) are equal in magnitude and the electric force between them so great, ordinary objects are generally electrically neutral. Thus, the electric force, operating internally at the atomic scale, is not manifest for large-scale objects whereas the gravitational force is.

With only two point-charges one can learn how the electric force varies with distance, that it is an inverse-square force. On the other hand, with only two point-charges we cannot determine whether one charge is large and the other small. To learn something about the magnitudes of electric charges requires at least *three* charges. Suppose that small charged objects 1 and 2 are brought in turn to the same distance from a charged object 3 (see Fig. 20-4). We find that the electric force on 1 due to 3 is $\mathbf{F}_1$ and that on 2 due to 3 is $\mathbf{F}_2$. Then, *by definition,* the relative magnitudes of the charges q_1 and q_2 are

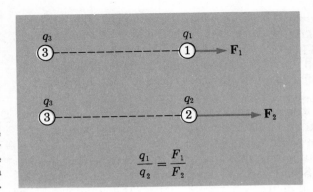

FIG. 20-4. The relative magnitude of two charges q_1 and q_2 is, by definition, the ratio F_1/F_2 of the respective forces arising from a third charged object.

$$\frac{q_1}{q_2} = \frac{F_1}{F_2} \tag{20-1}$$

The ratio of the respective forces is the ratio of the charge magnitudes. (This is, of course, precisely what is done in comparing gravitational masses: If two objects at the same location on earth have weights in the ratio of 2 to 1, their respective gravitational masses are in the ratio of 2 to 1.)

With the choice of some arbitrarily chosen standard electric charge, we can measure any other charge magnitude in units of the standard charge by means of Eq. (20-1).

There is a very simple procedure for dividing the charge on a conductor into halves. Suppose that one has two identical spherical conductors, one initially charged and the other uncharged. If the two spheres are brought into contact, then by symmetry the charge will divide equally, and after separation each will have exactly half the charge. This procedure can be extended, of course, to change the charge on a conductor by factors of 4, 8, ... It was used by Coulomb. The halving procedure cannot, however, be continued indefinitely; eventually one reaches that point where a single extra electron resides on a conductor, and the charge of one electron cannot be divided.

Equation (20-1) gives the charge ratio q_1/q_2 for two charges in terms of the force ratio F_1/F_2 for the two charges interacting, in turn, with a *third* charge. We determine thereby the ratio of the charges: q_1/q_2. Now suppose that these two charges interact with one another only, as shown in Fig. 20-5. From (20-1) it follows that the electric force on q_1 is proportional to q_1. By the same token, the force on q_2 is proportional to q_2. Now, if the forces acting on the two interacting charges are equal and opposite, in accordance with Newton's third law, the magnitude of force **F** acting on either must be proportional to both q_1 and q_2 as well as inversely proportional to r^2; or

$$F = \frac{kq_1q_2}{r^2} \tag{20-2}$$

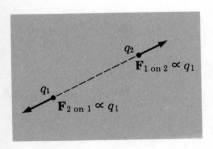

FIG. 20-5. Two interacting charged objects. The force on each charged object is proportional to the magnitude of its charge.

where k is a constant whose value depends on the choice of units for charge and distance.

If we use the radius vector $\mathbf{r}_{12}$ to designate the position of charge q_2 relative to q_1, then we may write (20-2) in vector form as

$$\mathbf{F}_{1 \text{ on } 2} = \frac{kq_1q_2}{r_{12}^2} \frac{\mathbf{r}_{12}}{r_{12}} = -\mathbf{F}_{2 \text{ on } 1} \qquad (20\text{-}3)$$

where $\mathbf{r}_{12}/r_{12}$ is a unit vector whose direction is from q_1 to q_2. See Fig. 20-6. Note that the direction of $\mathbf{F}_{1 \text{ on } 2}$ is given correctly if we use the algebraic signs for q_1 and q_2: The force is positive (that is, repulsive) when both q_1 and q_2 have the same sign, and it is negative (that is, attractive) when the signs are different.

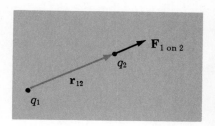

FIG. 20-6. Relation between the force $\mathbf{F}_{1 \text{ on } 2}$ on charge q_2 and the radius vector $\mathbf{r}_{12}$ from charge q_1 to q_2.

Thus far we have treated the coulomb force between only two interacting point-charges at rest. Suppose that a charge q_3 is in the presence of two other charges q_1 and q_2, as shown in Fig. 20-7. Experiment shows that the force on q_3 is just the *vector sum* of the separate forces on it from q_1 and q_2. That is, the *superposition principle* of forces holds for the coulomb force. Said differently, the force between any two charges is independent of the presence of other charges: To find the resultant force, we merely add the individual forces as vectors. The superposition principle, although simple, is *not* self-evident; it is a result of observation for electric interactions. As we shall see, its consequences are many and important.

20-3 ELECTRIC UNITS

The numerical value of the constant k appearing in the Coulomb-law relation [Eq. (20-2)] depends, of course, on the units chosen for F, q, and r. In scientific work the *I*nternational *S*ystem of units—which is metric—is preferred. In the *rationalized mksa* (meter-kilogram-second-ampere) *system of units,* which we shall use exclusively henceforth, the "mechanical" units are those of the SI (or mks) system—meter for length, newton for force, joule for energy, etc.— and the unit of charge is the *coulomb* (abbreviated C).

With F in newtons, r in meters, and q in coulombs, the constant k of (20-2) has the value

$$k \equiv 8.98755 \times 10^9 \text{ N-m}^2/\text{C}^2 \qquad (20\text{-}4)$$

Thus, two point-charges, each of 1 C and separated by 1 m, exert an electric force of approximately 9.0×10^9 N on one another. One coulomb of charge is, as charges go, an enormous amount. The net charge on a laboratory device of ordinary size might be about 10^{-7} C = 0.10 μC. Two point-charges of this magnitude and separated by 1 cm repel one another with a force of about 1 N.

The definition of the coulomb is a bit complicated. First, by definition, a net charge of 1 C passes through the cross section of

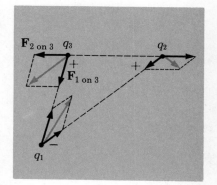

FIG. 20-7. Three interacting point-charges. The principle of superposition applies for coulomb forces.

an electric conductor when an electric current of 1 A (ampere) exists in this conductor for 1 s. The ampere is defined in turn by the magnetic force between two current-carrying conductors (Sec. 27-4).

For most computations it will suffice to round off the value for k, taking it to be 9.0×10^9 N-m²/C². In the SI/system† the constant k is written

$$k \equiv \frac{1}{4\pi\varepsilon_0} \tag{20-5}$$

where ε_0, the *electric permittivity of the vacuum,* has the value

$$\varepsilon_0 = 8.85435 \times 10^{-12} \text{ C}^2/\text{N-m}^2 \tag{20-6}$$

We may then write Coulomb's law in the form

$$F = \frac{1}{4\pi\varepsilon_0} \frac{q_1 q_2}{r^2} \tag{20-7}$$

We shall use both the constant k and ε_0, the choice being dictated by which of these will lead to the mathematically simpler relation.

In the *rationalized* unit system the factor 4π appears in (20-7) but not in certain other fundamental relations, such as Gauss' law, Eq. (22-3).

The electron's charge, whose magnitude is that of any other charged elementary particle, is designated by $-e$; its magnitude is

$$e = 1.602192 \times 10^{-19} \text{ C}$$

The direct measurement of the electronic charge e, as made in the Millikan experiment, is discussed in Sec. 21-5.

The electron's charge is extremely small. One microcoulomb, a typical charge by laboratory standards, corresponds to an excess or deficiency of 6×10^{12} electrons. Thus, we may ordinarily assume electric charge to be infinitely divisible and continuous and ignore its essential "graininess." We may, for example, imagine a negatively charged surface to have charge spread continuously over it, rather than concern ourselves with the actual finite number of electrons, acting as point-charges, which reside on it. We do the same thing in dealing with the mass of an ordinary object: We imagine it to be infinitely divisible, although the atomic nature, or graininess, of all matter implies that the mass is always an integral multiple of the mass of one atom or molecule.

When electrons are transferred to or from a laboratory object being charged, the difference in mass is trivial. Thus, when a body acquires charge of 1 μC (or 6×10^{12} electrons), its mass changes by only $(9.1 \times 10^{-31}$ kg/electron$)(6 \times 10^{12}$ electrons$) = 5 \times 10^{-18}$ kg $= 5 \times 10^{-9}$ μg.

---

†Other than the mksa system, the only system now in common use is the so-called *gaussian cgs system.* The names of the electrical units, the forms of the fundamental laws of electromagnetism, and conversion factors for the gaussian system are given in Appendix III.

The simplest device for detecting electric charge is the electroscope, in which the angular separation of light conducting leaves (which mutually repel one another) is a measure of the charge on the electroscope (Fig. 20-8). The electroscope is not, however, a precise or easily calibrated instrument. The electrometer is a charge-measuring device of this type. It is basically an electroscope that has been so calibrated that the angular displacement of a charged conductor suspended from a torsion fiber measures the charge. Another way of measuring charge is to measure electric current over a period of time or to measure the potential of a charged conductor of simple geometry (Sec. 23-8).

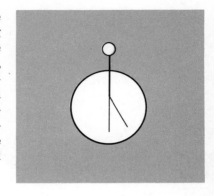

FIG. 20-8. A simple type of electroscope.

Example 20-1
What is the ratio of electric to gravitational attractive forces between a proton and electron?

The gravitational force [Eq. (13-4)] is $F_g = Gm_p m_e/r^2$, and the electric force is $F_e = ke^2/r^2$ where e is the charge magnitude of both particles. The ratio of the forces for *any* separation distance r is:

$$\frac{F_e}{F_g} = \frac{ke^2}{Gm_p m_e}$$

$$= \frac{(9.0 \times 10^9 \text{ N-m}^2/\text{C}^2)(1.6 \times 10^{-19} \text{ C})^2}{(6.67 \times 10^{-11} \text{ N-m}^2/\text{kg}^2)(1.67 \times 10^{-27} \text{ kg})(9.11 \times 10^{-31} \text{ kg})}$$

$$= 10^{39}$$

20–4 CHARGE CONSERVATION

Experimentally it is found that electric charge is conserved. According to the law of conservation of charge, the net charge, or *algebraic sum of the charges, in any isolated system is constant.* This is illustrated very simply by the processes in which *two* neutral objects become charged: Electrons are transferred from one to the other, and the result is one object with positive charge and the second with an equal amount of negative charge. *No violation of charge conservation has ever been observed.*

Electric charge conservation does *not* imply, however, that electric charge can be neither created nor destroyed, but only that the creation of positive charge must be accompanied by the creation of an equal amount of negative charge. An important example is the phenomenon of *pair production*. See Sec. 36-5. When a sufficiently energetic photon, an electrically neutral particle of electromagnetic radiation, enters a closed container, as in Fig. 20-9, it may be annihilated and in its stead two particles appear, an electron with charge $-e$ and a positron with charge $+e$. The *net* charge within the container has not changed.

The positron, identical with an electron in all respects except the sign of its charge (and the consequences of the difference in

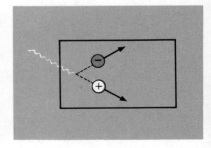

FIG. 20-9. A photon enters a closed chamber and produces an electron-positron pair.

sign), is called the *antiparticle* of the electron. The electron and positron are but one example of a particle-antiparticle pair. Other examples are the proton and antiproton (charges of $+e$ and $-e$, respectively) and more exotic elementary particles such as the π^- meson and the π^+ meson. Just as particle-antiparticle pairs may be created, a particle and its antiparticle may *annihilate* each other, producing two or more photons, or pairs of other particles. No matter what processes take place within a system—whether charge transfer between objects in contact, nuclear transformations, creation of matter, or annihilation of particles—the total charge is always conserved. With the conservation of electric charge, the *classical* list of conservation laws is complete. They are: the conservation of linear momentum (Sec. 5-6), angular momentum (Sec. 11-7), mass (Sec. 5-4), energy (Sec. 10-5) and, now, electric charge.

20–5 CHARGE QUANTIZATION

The net electric charge of any object is just the algebraic sum of the charges of the elementary particles comprising it. *All* the many elementary particles now known, although they may differ greatly in mass and other properties, have just one of *three*† *possible charge values:*

$$+e, \quad 0, \quad \text{or} \quad -e$$

where e is the magnitude of the charge of an electron. For example, the charges of the electron and positron are $-e$ and $+e$, respectively; and the charges of the three kinds of π mesons, the π^+, π^0, and π^-, are $+e$, 0, and $-e$, respectively. The neutron and its antiparticle, the antineutron, have charges of exactly zero. Why the charged fundamental particles of physics have only the charge magnitude e is not known on any more fundamental basis than that it is a well-established experimental fact.

Since every charged object is nothing more than a collection of elementary particles, the only possible values of the total charge Q of any object are given by

$$Q = \pm Ne \qquad \text{where } N = 0, 1, 2, \ldots \tag{20-8}$$

Electric charge is said to be *quantized:* It appears in *integral* positive and negative *multiples* of the charge of the electron, and no others; see Fig. 20-10.

The discreteness, or granularity, of electric charge is evident only through rather subtle experiments, simply because most charged objects have a charge that is very much larger than e; that is, the integer N in (20-8) is typically very much larger than 1. But

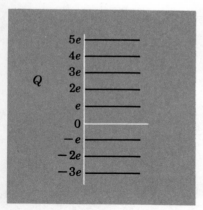

FIG. 20-10. Charge quantization. The only possible values of any charge Q are integral multiples of the electronic charge e.

---

†The very short-lived exotic particle $\Delta(1236)$ has the charge $+2e$.

the unique charge of the electron is shown directly in the fundamental experiments of Millikan (Sec. 21-5). Charge quantization is implicit in the chemical idea of atomic number, where this integer gives the total number of electrons (or protons) and, hence, the total negative (or positive) charge in a neutral atom. For example, the elements $_1$H, $_2$He, $_3$Li have atomic numbers of 1, 2, and 3, respectively. Charge quantization is also implied in the chemical idea of valence, where again one assigns a positive or negative integer.

SUMMARY

The electric, or coulomb, force between point-charges is given by

$$F = \frac{kq_1q_2}{r^2} \qquad (20\text{-}2)$$

where $k = 1/4\pi\varepsilon_0 = 9.0 \times 10^9$ N-m^2/C^2. The unit of charge in the SI system is the coulomb. The electric force is a central conservative force, and the superposition principle for forces applies to the coulomb interaction.

Electric charge is conserved: In any isolated system, the net charge is constant. Electric charge is quantized: Any observed charge is an integral multiple, positive or negative, of the electronic charge $e = 1.60 \times 10^{-19}$ C.

PROBLEMS

20-1. Show that the magnitude of the total charge for one mole of electrons is 96,487 C. (This particular value of electric charge, 96,487 C—equal to the product of Avogadro's number and the electronic charge—is known as the *Faraday constant*.)

20-2. Two small identical metal spheres are originally uncharged. How many electrons must be transferred from one sphere to the other if the spheres are to attract one another with a 1.0-N force when they are separated by 10 cm?

20-3. A charge of +0.30 µC is 10.0 cm from a charge of −2.70 µC. At what (noninfinite) point relative to the +0.30-µC charge can any third positive charge be placed so that the resultant electric force on it is zero?

20-4. Consider the five charges in Fig. P20-4, equally spaced 10 cm apart along a straight line. The two end charges of +2.0 µC each are held fixed. Find the magnitude and sign of the charges q_1 and q_2 such that

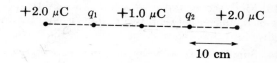

FIG. P20-4

the resultant force on each of the three middle charges is zero.

20-5. What is the distance between an electron and a proton for which the coulomb force of attraction is equal to the gravitational force of attraction between them when their separation distance is 5.3×10^{-11} m? (Note that in the Bohr model of the hydrogen atom, the electron-proton distance is 5.3×10^{-11} m.)

20-6. Two identical spherical conductors initially are charged with *unlike* charges. When they are separated by a distance r, the force on each conductor is F. Then the two conductors are brought in contact and separated again to r. The force on each is again found to have the magnitude F. What is the ratio of the two initial charges?

20-7. In a Cavendish experiment measuring the gravitational attractive force between two objects of masses 1.0 kg and 100 kg, respectively, what is the maximum stray excess charge each object can have if the electrostatic force is to be less than 10^{-3} the gravitational force? Assume equal charges on each object.

20-8. Four charges, each of charge $+Q$, are placed at the four corners of a square of edge length L. Find the magnitude of the force on any one charge arising from the other three.

20-9. When they are 4.0 cm apart two charged small spheres attract each other with a force of 180 N. Find the sign and magnitude of each of the two charges if their algebraic sum is $+4.0$ μC.

20-10. A charge of $+2.0$ μC is located at $(-4.0 \text{ cm}, 0)$ in the XY plane and a second charge of -3.0 μC at $(0, +3.0 \text{ cm})$. A third charge Q is now placed at $(+3.0 \text{ cm}, 0)$. (a) Find the value of Q such that the X component of the resultant electric force on the -3.0-μC charge is zero; (b) find Q such that the Y component of the force on the -3.0-μC charge is zero.

20-11. Three charges, each of charge $+Q$, are placed at the corners of an equilateral triangle of edge length L. Find the point(s) inside the triangle at which a fourth charge would experience no electric force.

20-12. Find the distance at which the electrostatic force of repulsion between two electrons is 1.0 N.

20-13. Find the distance at which the electrostatic force of repulsion between two protons is equal to the weight of a 1.0-kg mass at the surface of the earth.

20-14. Suppose that both the sun and the earth were negatively charged, the net charge of each being the charge of 1.0 gram of electrons. (a) Find the magnitude of this charge. (b) Find the electrostatic force between earth and sun under these circumstances. (The earth-sun distance is 1.5×10^{11} m.)

20-15. Point-charges $q_A = +1.0$ μC, $q_B = -2.0$ μC, and $q_C = +4.0$ μC are fixed in place at the vertices of an isosceles triangle with edge lengths $AB = AC = 5.0$ cm and $BC = 8.0$ cm. Find the magnitude of the force on each charge; indicate the force directions on an appropriate diagram.

20-16. Point-charges q_1, q_2, and q_3 are fixed in position at the vertices of an equilateral triangle 6.0 cm on a side. We are given that $F_{12} = -10$ N, $F_{13} = +7.5$ N, and $F_{23} = -30$ N, where F_{12} is the magnitude and sign (+, repulsion; −, attraction) of the electrostatic force between charges q_1 and q_2, and so forth. Find the magnitudes and signs of the three charges, taking q_1 to be positive.

20-17. Two small spherical conductors, each of mass 10 g, are hung from two insulated 50-cm-long strings which are attached to a common point at their upper ends. One of the conductors is given an initial charge Q. It is then brought in contact with the other conductor, initially electrically neutral. The two conductors now repel one another and come to rest with the angle between the strings being 74°. What was the initial charge Q on the one conductor?

20-18. Two small 6.0-g spheres carry equal electric charges. Each sphere is suspended from the same point by an insulating filament 13 cm long. In the equilibrium condition the spheres are separated by 10 cm. Find the magnitude of the electric charge on each sphere.

20-19. A particle of mass m and charge $-q$ is located midway between two other fixed point-charges, each of charge $+q$. When the particle in the middle is displaced by a small amount transverse to the line joining the two fixed particles, it oscillates in simple harmonic motion. Find the oscillation frequency.

20-20. Assuming that the sun, earth, and moon each has a net electric charge, could you explain their observed motions in terms of electrical interactions rather than gravitational interactions?

20-21. An electron in a hydrogen atom orbits the proton in a circle of radius 0.53 Å. (a) What is the force on the electron? (b) What is the speed of the electron? (1 Å = 10^{-10} m.)

20-22. Two electrons rotate about a proton in the same circular orbit, the electrons being diametrically opposite each other. If the radius of the orbit is 0.53 Å, what is the angular velocity of the electrons' orbital motion?

20-23. Imagine that an electron moves about a proton in a circular orbit 1.0×10^{-10} m in radius. (a) What is the frequency of the electron's orbital motion? (b) What would be the orbit frequency if the charges of the two particles were "turned off" and the electron's orbit were maintained solely by means of the gravitational interaction? (Electron mass, 9.1×10^{-31} kg, proton mass, 1.7×10^{-27} kg.)

20-24. Find the coulomb forces on each of two nuclear fragments of $^{146}_{56}$Ba and $^{90}_{36}$Kr formed in the nuclear fission of $^{236}_{92}$U when, just after their formation, their centers are separated by 1.3×10^{-14} m.

20-25. The particles in atomic nuclei, protons and neutrons, attract each other by the very strong nuclear force. This force shows saturation; that is, any one nuclear particle will interact only with those neighboring particles lying within the range of the nuclear force. The coulomb force, on the other hand, although it falls off with distance, has an infinite range; moreover, it acts between every pair of charged particles. In which type of nucleus, a light one or a heavy one, is the coulomb repulsion more important relative to the nuclear attraction?

20-26. (a) Suppose that overnight all positively charged particles became negatively charged particles and all negatively charged particles became positively charged particles. Could you tell? (b) Suppose that the magnitude of e changed overnight by a factor of 10^{10}. Could you tell?

20-27. What evidence is there that the total charge in the universe, if not exactly zero, is only a small fraction of the total negative or positive charge?

20-9

A —4cm— B   A + B = 4μC

$$F = 180 N = k\frac{A(B)}{.04^2} \qquad A = \frac{3.2 \times 10^{-11}}{B}$$

$$\frac{180(.04)^2}{9 \times 10^9} = AB = 3.2 \times 10^{-11}$$

$$B + \frac{3.2 \times 10^{-11}}{B} = 4 \times 10^{-6} \qquad A = 4 \times 10^{-6}$$

$$4 \times 10^{-6} - A = B$$

$$\frac{4 \times 10^{-6} \pm \sqrt{(4 \times 10^{-6})^2 - 4(3.2 \times 10^{-11})}}{2} \qquad A(4 \times 10^{-6} - A) = 3.2 \times 10^{-11}$$

$$4 \times 10^{-6} A - A^2 - 3.2 \times 10^{-11} = 0$$

$$B^2 + 3.2 \times 10^{-11} = 4 \times 10^{-6} B \qquad A^2 - 4 \times 10^{-6} A + 3.2 \times 10^{-11} = 0$$

CHAPTER 21

The Electric Field

The electric field is a useful concept for computations involving the coulomb force. Electric field lines, which represent this field, are useful in visualizing the coulomb interaction. In this chapter we compute the electric field for several simple charge distributions.

21-1 ELECTRIC FIELD DEFINED

Consider a number of point-charges, such as q_a, q_b, and q_c in Fig. 21-1, *fixed* in position. We introduce still another charge q_t, a test charge, at position P and find the resultant electric force $\mathbf{F}_t$ acting on it by superposing the individual electric forces from the charges q_a, q_b, and q_c. Now suppose that we replace this test charge by another one of the same sign but twice its magnitude. The electric force of each of the three fixed charges on q_t in Fig. 21-1 is proportional to q_t; therefore, the force on a charge $2q_t$ placed at point P will be twice the force on the charge q_t. If the resultant electric force on a unit positive charge placed in the region of a collection of fixed charges is known, the electric force on *any* other point charge placed at the same position is easily found: The force is proportional to the magnitude of the test charge; for a positive charge, the direction of the force is the same as that for the unit positive charge but it is *reversed* for a negative charge.

It is useful to define the *electric field* $\mathbf{E}$ for any point in space as *the resultant electric force per unit positive charge*.

$$\mathbf{E} = \frac{\mathbf{F}}{q} \qquad (21\text{-}1)$$

Knowing $\mathbf{E}$ at some point, we may then find the electric force $\mathbf{F}$ acting on any charge q_t at the same point through the relation

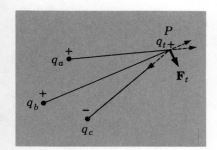

FIG. 21-1. Test charge q_t is acted on by the resultant electric force $\mathbf{F}_t$ arising from the charges q_a, q_b, and q_c.

$$\mathbf{F} = q_t \mathbf{E} \tag{21-2}$$

If we want the electric field due to the source charges (q_a, q_b, and q_c in Fig. 21-1) to remain unchanged, it is essential that the source charges remain fixed in position. Only then will the test charge, when introduced, not cause these charges to be displaced by the electric force it produces on them, thereby altering the field. Thus, to compute a stationary field $\mathbf{E}$ from one or more source charges, we imagine these charges to remain fixed in position; alternatively, we imagine the test charge q_t to be so very small that it will not appreciably affect the locations of the other charges. $\mathbf{E}$ represents a vector field in space whose direction and magnitude give the force on a unit positive charge before any test charge is introduced.

Sometimes the quantity labeled $\mathbf{E}$ is called the electric field "strength" or the "intensity" of the electric field. We shall refer to it hereafter simply as the electric field. The units of electric field are newtons per coulomb. We shall see (Sec. 23-4) that equivalent units are volts per meter.

The electric field at a distance r from a single point-charge q is easily found from Coulomb's law. The force between q and the test charge q_t is

$$F = k\frac{qq_t}{r^2}$$

Therefore,
$$E = \frac{F}{q_t} = \frac{kq}{r^2} \tag{21-3}$$

We may write this relation in vector form by using the radius vector $\mathbf{r}$ to denote the position of the test charge relative to the origin located at the charge q:

$$\mathbf{E} = \frac{kq}{r^2}\frac{\mathbf{r}}{r} \tag{21-4}$$

where ($\mathbf{r}/r$) is a unit vector in the direction from q to q_t. The magnitude of the electric field from a single point-charge varies inversely as the square of the distance from it. Since electric forces superpose as vectors, the electric fields contributed by individual point-charges are added as vectors to give the resultant field.

The electric field concept gives a new way of looking at the interaction between two charges q_a and q_b. We can say that charge q_a acts as the source of, or creates, an electric field $\mathbf{E}_a$ which surrounds it and that the charge q_b, immersed in this field, is then subject to an electric force $\mathbf{F}_{a \text{ on } b}$. That is,

$$\mathbf{E}_a = \frac{kq_a}{r^2}\frac{\mathbf{r}}{r} \quad \text{and} \quad \mathbf{F}_{a \text{ on } b} = q_b \mathbf{E}_a$$

Similarly, we can interpret the electric force of q_b on q_a by saying that q_b generates the electric field $\mathbf{E}_b = (kq_b/r^2)(\mathbf{r}/r)$ and that q_a immersed in the field $\mathbf{E}_b$ then experiences the force $\mathbf{F}_{b \text{ on } a} = q_a \mathbf{E}_b$.

This two-stage process, the production of the field by one charge and the response to the field by the second charge, may seem at first sight to be a pedantic matter. But there is, in fact, real physical justification for the field concept, that is, for visualizing electric interactions as taking place *via* the electric field. For one thing, when q_a is moved, so that its separation distance from q_b is changed, q_b does *not* feel a different force *instantaneously*. Rather, q_b continues to experience the original force (therefore, electric field) for the time required for light to travel from a to b. That is, disturbances in the electric field arising from accelerated charges are propagated at the finite speed of light. This is no mere accident. As we shall see, light consists of electric (and magnetic) fields traveling through space. Thus, an electric field may become detached from the electric charge generating it. Such electric (and magnetic) fields, or electromagnetic waves, are physically real in the sense that one may attribute to them such "mechanical" properties as energy, linear momentum, and angular momentum.

Example 21-1

Two point-charges, one of +36 μC and one of −36 μC, are separated by 8.0 cm. Find the electric field at point P in Fig. 21-2.

We may compute the electric field arising from each of the two charges and add these fields to find the resultant by using Eq. (21-3), or we may equivalently find the resultant electric force on a charge of +1 C located at point P.

The magnitude of the electric field $\mathbf{E}_+$ due to the +36-μC charge is

$$E_+ = \frac{kq}{r^2} = k\,\frac{36 \times 10^{-6}\text{ C}}{(6 \times 10^{-2}\text{ m})^2} = k(1.00 \times 10^{-2}\text{ C/m}^2)$$

The direction of $\mathbf{E}_+$ is *away* from the positive charge along a line connecting point P with the charge. In like fashion, the magnitude of the field $\mathbf{E}_-$ from the negative charge is

$$E_- = \frac{kq}{r^2} = k\,\frac{36 \times 10^{-6}\text{ C}}{(10 \times 10^{-2}\text{ m})^2} = k(0.36 \times 10^{-2}\text{ C/m}^2)$$

the direction of $\mathbf{E}_-$ being *toward* the negative charge.

We add $\mathbf{E}_+$ and $\mathbf{E}_-$ as vectors to find the resultant field E with components E_x and E_y at P. We have

$$E_x = k(0.36 \times 10^{-2}\text{ C/m}^2)\left(\frac{8\text{ cm}}{10\text{ cm}}\right) = k(0.29 \times 10^{-2}\text{ C/m}^2)$$

$$E_y = k\left[(1.00 \times 10^{-2}) - (0.36 \times 10^{-2})\left(\frac{6\text{ cm}}{10\text{ cm}}\right)\right](\text{C/m}^2)$$

$$= k(0.78 \times 10^{-2}\text{ C/m}^2)$$

The magnitude of $\mathbf{E}$ is

$$E = \sqrt{E_x^2 + E_y^2} = k(0.83 \times 10^{-2}\text{ C/m}^2)$$

$$= (9.0 \times 10^9\text{ N-m}^2/\text{C}^2)(0.83 \times 10^2\text{ C/m}^2) = 7.5 \times 10^7\text{ N/C}$$

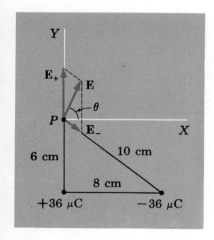

FIG. 21-2

and its direction θ relative to the X axis is

$$\theta = \tan^{-1} \frac{E_y}{E_x} = \tan^{-1} \frac{0.78}{0.29} = 70°$$

21–2 ELECTRIC FIELD LINES

One may use a number of vectors, as in Fig. 21-3a, to represent the electric field of a point-charge (taken to be positive in this example). The vectors are all radially outward, and their magnitude is chosen to be inversely proportional to the square of the distance from the charge. An equivalent way of mapping this electric field is shown in Fig. 21-3b, where a number of uniformly spaced, outwardly directed, *electric-field lines* are shown radiating from the point-charge. Electric-field lines are sometimes called electric lines of force. Clearly, the direction of the continuous lines corresponds to the direction of **E**. Moreover, the number of such lines passing through a small area of fixed size oriented at right angles to the electric field lines varies inversely as the square of the distance from the point-charge. This follows from the inverse-square character of the electric interaction: If the number of lines from a point-charge is fixed, then one knows the number (in three dimensions) passing through a spherical surface ($4\pi r^2$ in area) centered at the charge, and the number of lines through a transverse area varies inversely as the square of the distance. Since the electric field produced by a point-charge also varies inversely as the distance squared, we see that the magnitude of the electric field at a point is proportional to the number of lines through a small area transverse to the electric field lines.

For the electric field or more complicated charge distributions one may always draw lines to give the directions of **E**, but it is not obvious that the density of lines *always* gives the magnitude of the

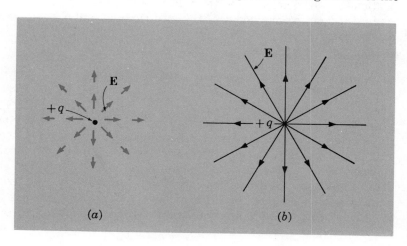

FIG. 21-3. The electric field of a point-charge represented *(a)* by E-vectors and *(b)* by electric field lines.

field. That this is indeed the case for any charge distribution will be proved in the next chapter. At the moment we shall merely note that electric field lines are a very useful means of visualizing the electric field. The electric *field* is real, or as real as any other physically measurable property; the *lines* which represent the field are a useful *fiction*.

Two charge configurations and their associated electric fields are shown in Fig. 21-4. Electric lines originate from positive charges and terminate on negative charges. Where the lines are crowded together, the field is strong. Where the lines are equally spaced and parallel, the field is uniform, or constant. When viewed from a great distance any object carrying charge of a *single sign* will have the field of a point-charge; that is, $E \propto 1/r^2$.

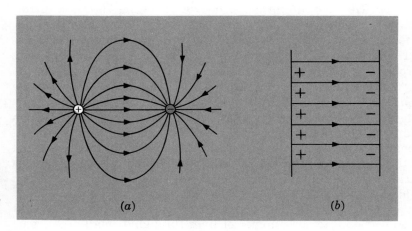

FIG. 21-4. Electric field configurations for *(a)* equal unlike point-charges and *(b)* two parallel, uniformly charged plates of opposite sign.

It must be emphasized that the electric field lines give the direction of the *force* acting upon a unit *positive* charge introduced into the field. The field lines do *not* portray the *paths* of a charged particle released in an electric field. (Of course the path may be *computed* from a knowledge of the electric field and of the particle mass.)

21–3 ELECTRIC FIELDS FOR THREE SIMPLE GEOMETRIES

Coulomb's law applies only to *point-charges*. But one can compute the electric field for any charge distribution by taking the vector sum of the individual electric field vectors contributed by the point-charges which comprise the distribution—where Coulomb's law applies to these point-charges considered individually. In special situations characterized by high geometric symmetry, it is possible to achieve considerable simplification of the computation through the use of Gauss' law (see Chap. 22). There are three simple continuous charge distributions—a point-charge, a line of charge, and a surface of charge—for which one can readily find how **E** varies with position. Since more complicated distributions can frequently be thought of as superpositions of a number of simple distributions, knowing the electric field configurations for these three cases permits one to know, at least qualitatively, the fields for such complicated charge distributions.

Point-charge. We have already solved this problem. The electric field is radially outward from a positive point-charge and varies with the distance r from the charge, following Eq. (21-3):

Point-charge: $$E \propto \frac{1}{r^2}$$

Line of Charge. Here we imagine electric charge to be distributed uniformly along the length of an infinitely long wire, a cylinder of negligible cross-sectional area. We designate the charge per unit length as the linear charge density λ, in coulombs per meter. We recognize that a line of charge can never actually be infinitely long, so that what we find is the electric field for a finite length of charged wire (whether a conductor or a thin dielectric rod) at points which are much closer to the wire than they are to either end.

It is easy to establish the direction of the electric field from symmetry considerations alone: **E** must be radially outward (inward) from a line of positive (negative) charge, the electric lines of force being straight and lying in a plane perpendicular to the line. Only such a direction is consistent with the fact that the infinite line is symmetrical to the left and right at any point. If **E** made some angle other than 90° with the wire, there would be a difference between left and right.

Using Gauss' law it is easy to show (see Example 22-2, Sec. 22-2) that the magnitude of the resultant electric field due to an infinite line of charge is given by

$$E = \frac{2k\lambda}{r} \qquad (21\text{-}5)$$

where r is the perpendicular distance from the infinite line. The electric field is proportional to the linear charge density and falls off with r to the *first* power:

Line of charge: $$E \propto \frac{1}{r}$$

Curiously, we find that the electric field from an infinite line of charge, one having an *infinite* total charge, is *finite*.

Infinite Surface of Charge. Here we consider the electric field produced by a uniformly charged infinite plane surface. The constant surface charge density, the charge per unit area (in coulombs per square meter) is represented by σ. On the basis of symmetry considerations alone, we can immediately say that the electric field must be perpendicular to the surface. It is outward along the normal for a positively charged sheet and inward along the normal for a negatively charged sheet. This follows from the fact that there is but one direction uniquely associated with a plane, that of the normal.

Once again Gauss' law can be used to show (see Example 22-3, Sec. 22-2) that the electric field due to an infinite surface with charge density σ is

$$E = \frac{\sigma}{2\varepsilon_0} \qquad (21\text{-}6)$$

The electric field is *constant* in magnitude and independent of the distance from the surface. A uniformly charged infinite plane surface produces a uniform field along the normal to the plane:

Surface of charge: $E = $ constant

If our uniformly charged surface is imagined to be an infinitesimally thin sheet of charge, then a uniform electric field of magnitude $E = \sigma/2\varepsilon_0$ is produced on *both sides* of the sheet, as shown in Fig. 21-5a. Of course, **E** is away from the surface in both directions for a positively charged sheet and toward the surface in both directions for a negatively charged sheet. Now, if one has two parallel charged sheets with *opposite* surface charges, both of the same magnitude σ, as in Fig. 21-5b, the resultant field between the sheets has equal contributions, both in magnitude and direction, from the two surfaces, and the resultant field is $E = \sigma/\varepsilon_0$; outside this region the fields from the two surfaces cancel and $E = 0$ (see Fig. 21-5c). Such a pair of parallel sheets of opposite charge provides a simple way of establishing a uniform electric field in space. If the two oppositely charged sheets are conductors, the pair constitutes a *capacitor* (see Sec. 23-7).

FIG. 21-5. *(a)* Electric field of a single uniformly charged sheet; *(b)* electric fields of two parallel uniformly charged sheets of opposite sign; *(c)* resultant electric field of part *(b)*.

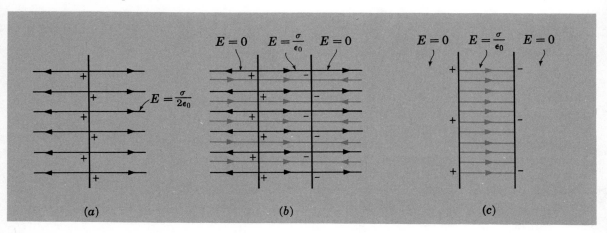

Example 21-2

A beam of electrons, each having an energy of 10 eV (1 eV = 1.6 × 10^{-19} J) enters the region between two oppositely charged parallel plates separated by 10 cm. The electrons enter through a hole in the positively charged plate and travel initially along the direction of the electric field; see Fig. 21-6. What is the minimum magnitude of the electric field which will stop the electrons before they hit the second plate?

The electric force on the electrons is opposite to their initial velocity. The electrons are, therefore, subject to a constant retarding force which can bring them to rest and then accelerate them backward toward the hole. In bringing particles of initial kinetic energy K to rest in a distance d, a constant electric force F does work in the amount Fd. Therefore,

$$K = Fd = eEd$$

or

$$E = \frac{K}{ed} = \frac{(10 \text{ eV})(1.6 \times 10^{-19} \text{ J/eV})}{(1.6 \times 10^{-19} \text{ C})(0.10 \text{ m})} = 100 \text{ N/C}$$

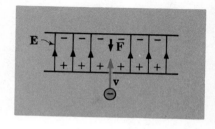

FIG. 21-6. Electron acted upon by a retarding force in a uniform electric field.

Note the conversion from electron volts to joules. Since the magnitude of the charge of the electron is the same number (not by accident, as we shall see), the computation is rendered particularly simple. Note, further, that the mass of the charged particle does *not* enter. A 10-eV chlorine ion, Cl^-, would be stopped by the same electric field. Of course, such an ion would be traveling at a much lower speed initially.

21-4 ELECTRIC FIELDS AND CONDUCTORS

For a conductor in electrostatic equilibrium the electric field, both within the conductor and at its exterior, is very simple: At any point inside the conductor's surface, **E** = 0; at the conductor's surface, **E** is always perpendicular to the surface; see Fig. 21-7. Let us prove these assertions.

Imagine that a net charge is somehow placed in the interior of a solid conductor. Then an electric field is established momentarily within the conductor. This field acts on the free electrons within the conductor, and they redistribute themselves throughout the conductor's volume until they are no longer acted upon by a resultant electric field and finally come to rest. Since the conduction electrons are highly mobile, the state of electrostatic equilibrium is achieved very rapidly indeed. Now if a free electron in the interior of the conductor is subject to *no* net electric force, the electric field in the interior of the conductor must be exactly *zero*. This means that there can be no net charge within the conductor. Where has the net charge added to the conductor gone? It must, of course, be found residing on the conductor's surface, with equal charges of opposite sign having moved so as to cancel the original net charge configuration.

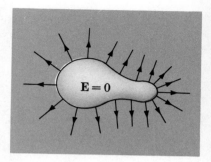

FIG. 21-7. Electric field for a charged conductor. Within the conductor E = 0; at the outer surface E is perpendicular to the surface.

For electrostatic equilibrium the net charge on a conductor appears at rest on its outer surface. If any one charge on the surface is at rest, there can be no electric force acting on it along the surface, inasmuch as such a force would cause the charge to move along the surface. Thus, the electric field at the conductor surface must be *perpendicular* to the surface at all points. In terms of electric field lines we can say that (1) there are no electric field lines within a conductor and (2) electric field lines originate from charges at the surface and are always perpendicular to it (Fig. 21-7).

Recall that the electric field of any uniformly charged infinite plane is perpendicular to the plane and proportional to the surface charge density σ [Eq. (21-6)]. Any very small portion of a conductor's surface closely approximates a flat plane and, as we have seen, the electric field is perpendicular to its surface. Since $E \propto \sigma$, the charge on a conductor's surface is most dense at those points where the electric field is large. Thus, the electric field lines emanating from a charged conductor indicate how the charge is distributed over the surface: Charge is concentrated where the electric lines are most dense.

The behavior of charges placed on a nonconductor is quite different. A charge on an insulator—on its surface or in its interior—does *not* move, because the electrons in the insulator are *not* free to redistribute themselves. In general, an electric field *will* exist within a nonconductor; the charge will *not* necessarily be found on the surface only; and the electric force lines at the surface will *not* necessarily be at right angles to the surface.

21-5 THE MILLIKAN EXPERIMENT

The charge of the electron was first measured directly in the historic experiments of R. A. Millikan (1868–1953) beginning in 1909. These experiments involve observations of electrically charged oil drops in a uniform electric field. The strategy of the experiment is this: One balances the force of an electric field on a single electron, or a small number of electrons residing on the droplet with a second force, the weight of the very small droplet of oil (oil does not evaporate readily). The essential parts of the experimental arrangement are shown in Fig. 21-8. Oil droplets from an atomizer fall into the region between two horizontal parallel plates through a hole in the top plate. If the plates are uncharged, there is no electric field and a drop initially accelerates downward because of the gravitational force. It very quickly reaches a constant terminal speed, however, because it falls under the influence of two forces, its weight $\mathbf{F}_g$ downward and a resistive force $\mathbf{F}_r$ upward. The resistive force, which arises from the molecular bombardment by air molecules, is proportional to the speed v_0 of the oil drop. By observing the motion of an individual oil droplet viewed through

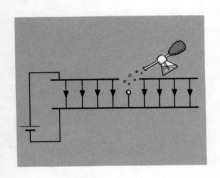

FIG. 21-8. Schematic of the Millikan oil-drop experiment.

a microscope, the speed of the droplet can be measured by timing its fall. Thus, in the absence of an electric field,

$$\mathbf{F}_g + \mathbf{F}_r = 0 \quad \text{where } \mathbf{F}_r = -K\mathbf{v}_0$$

The droplet is too small to permit its weight to be measured directly, but by measuring $\mathbf{v}_0$ and knowing the constant resistive coefficient K, the droplet weight $\mathbf{F}_g$ can be determined.

When the vertical electric field is turned on, the oil droplet, if charged, is subject to an additional electric force $\mathbf{F}_e = q\mathbf{E}$, where q is the charge on the drop and $\mathbf{E}$ is the constant electric field acting either up or down (we shall later see that, for a uniform field, $E = \Delta V/d$, where ΔV is the easily measured potential difference between the charged plates and d is their separation distance). Again, if the charged droplet is initially subject to an unbalanced force, it very quickly reaches a constant terminal velocity under the influence of the resistive force $\mathbf{F}_r$. When this terminal velocity is achieved, the resultant force on the droplet is again zero, so that

$$\mathbf{F}_g + \mathbf{F}_r + \mathbf{F}_e = 0$$

where $\quad \mathbf{F}_e = q\mathbf{E} \quad$ and $\quad \mathbf{F}_r = -K\mathbf{v} \qquad (21\text{-}7)$

One measures the magnitude and direction of the droplet's constant velocity $\mathbf{v}$. The constant K is known, $\mathbf{E}$ is easily measured, and $\mathbf{F}_g$ was determined in the observations without the electric field. Thus, one can determine the magnitude and sign of the charge q by using (21-7). Moreover, by exposing the drop to a short burst of ionizing x-rays, one can change the charge q on the drop (without, of course, having changed the drop's weight $\mathbf{F}_g$).

The results of experiments involving many different droplets and charges are these:

$$q = ne$$

where $\quad n = 0, 1, 2, \ldots \quad$ and $\quad e = 1.60 \times 10^{-19}$ C

Electric charge is quantized. The Millikan experiment showed that the only observed charge is always an integral multiple of nature's basic quantum of charge, the elementary charge e.

SUMMARY

The electric field $\mathbf{E}$ at any point in space is the resultant force per unit positive test charge:

$$\mathbf{E} = \sum \frac{\mathbf{F}}{q} \qquad (21\text{-}1)$$

One may map the magnitude and direction of an electric field with electric lines of force. The electric field varies with position for three simple geometries, as follows:

Point-charge $\quad E = \dfrac{kq}{r^2} \propto \dfrac{1}{r^2} \quad$ (21-3)

Line of charge: $\quad E = \dfrac{2k\lambda}{r} \propto \dfrac{1}{r} \quad$ (21-5)

Surface of charge: $\quad E = \dfrac{\sigma}{2\varepsilon_0} \propto \text{constant} \quad$ (21-6)

The linear charge density is λ; the surface charge density is σ.

The charge on a charged conductor in equilibrium resides on the conductor's outer surface. The electric field is perpendicular to the surface outside the conductor and zero within the conductor.

PROBLEMS

21-1. Stationary charges q_1 and q_2 are fixed in position and separated by a distance of 10 cm. Find the point (or points) where the electric field is zero under the following conditions: (a) $q_1 = -9\ \mu\text{C}$, $q_2 = +4\ \mu\text{C}$; and (b) $q_1 = +9\ \mu\text{C}$, $q_2 = +4\ \mu\text{C}$.

21-2. (a) Two point charges, one of $+1.0\ \mu\text{C}$ and the other of $-4.0\ \mu\text{C}$, are separated by 10 cm. Find the points along the line passing through the two charges at which the electric field is zero. (b) Solve part (a) if the negative charge is $-40\ \mu\text{C}$.

21-3. Find the electric field at point P in Fig. 21-2 if the $+36\ \mu\text{C}$ charge is replaced by a $-36\ \mu\text{C}$ charge.

21-4. A square of edge length L has charges of $+q, -q, +q,$ and $-q$ on its northeast, northwest, southwest, and southeast corners, respectively. (a) What are the direction and magnitude of the electric field at the center of the square? (b) What is the field at the center with the two first-named charges interchanged?

21-5. Sketch electric lines of force for the following configurations: (a) two equal like point-charges q separated by a distance d; (b) an equilateral triangle with charges $+q, +q,$ and $-q$ at its corners; (c) two concentric thin nonconducting shells of radii R and $2R$, respectively, each with charge $+q$ uniformly spread over the spherical shell.

21-6. A point-charge $-q$ is located at $x = -d/2, y = 0$, and a point-charge $+2q$ is located at $x = +d/2, y = 0$. (a) Sketch the graphs of the X and Y components of the resultant electric field, E_x and E_y, along the Y axis. (b) Find the values of y at which E_x is a maximum along the Y axis. (c) Find the values of y at which E_y is a maximum along the Y axis.

21-7. Four equal charges of magnitude Q are located at the vertices of an equilateral tetrahedron of edge length a. (The figure is a pyramid bounded by four equilateral triangles.) Show that the electric field at the location of any one charge (arising from the other three) is given by

$$E = \tfrac{3}{4}\sqrt{13}\,\dfrac{kQ}{a^2}$$

21-8. A particle with charge q and mass m moves in uniform circular motion about an infinitely long, straight wire with uniform charge density λ of sign opposite to that of the particle. Show that the speed of the orbiting particle does *not* depend on the particle's distance from the wire and that the unique speed is given by $(2k\lambda q/m)^{1/2}$.

21-9. An electron travels in a circle around a uniformly charged wire at a speed of 2.0×10^4 m/s. What is the linear charge density of the wire?

21-10. Initially an electron is moving with velocity v_0 parallel to a uniform electric field **E**. Write relations giving (a) the time Δt and (b) the distance Δx the electron travels before coming (temporarily) to a halt.

21-11. What is the ratio of electron speed to proton speed if—starting from rest—an electron and a proton each move through a distance of 10 cm aided by an electric field $E = 1.0 \times 10^3$ N/C $= 1.0$ kV/m?

21-12. The X and Y axes are each uniformly charged with a linear charge density λ. Find the magnitude and direction of the resultant electric field at any point along the line $y = 2x$. Use the superposition principle of electric fields which assumes that the resultant field is merely the sum of the two separate fields of the two lines of charges.

21-13. A uniformly charged ring of radius r has a total charge q. (a) Show that at a distance x along the symmetry axis of the ring the magnitude of the electric field is $kqx(r^2 + x^2)^{-3/2}$. (b) Sketch the electric lines of force for a uniformly charged ring.

21-14. When moving close to a single, uniformly charged flat surface, a hydrogen ion H^+ is subject to a constant force of 4.0×10^{-15} N. What is the magnitude of the surface charge density?

21-15. An electron and a proton, each starting from rest, are accelerated by the same constant electric field of magnitude 5.0×10^4 N/C. Find the distance and time for each particle to acquire a kinetic energy of 500 eV.

21-16. An electron gun shoots a narrow beam of electrons horizontally into the region between two oppositely charged horizontal parallel plates. The uniform electric field produced by the parallel plates is 500 N/C, and the average distance between neighboring electrons in the beam is 1.0×10^{-5} m. Compare the electric force between an electron and the electric field of the plates with (a) the electric force between neighboring electrons and (b) the gravitational force between an electron and the earth.

21-17. In a cathode-ray (electron beam) oscilloscope a narrow beam of electrons is projected horizontally between a pair of oppositely charged horizontal parallel plates. After passing through the plates, the electrons move in free flight and strike a distant fluorescent screen (see Fig. P21-17). Show that $\tan \theta = eEx/2K_0$, where E is the magnitude of the constant field between the plates and K_0 is the initial kinetic energy of an electron. Note that for constant K_0, $\tan \theta \propto x$; therefore, ignoring the small vertical deflection of the beam during its passage between the plates, the Y deflection on the screen is proportional to the electric field E.

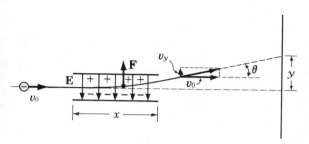

FIG. P21-17

21-18. In Fig. P21-17 an electron enters the region between the plates with an initial horizontal velocity of 3.0×10^5 m/s. The electric field between the horizontal plates is 0.50 N/C, and the horizontal distance along either plate is 3.0 cm. The distant screen is 20 cm from the nearest ends of the plates. (a) What is the electron's vertical displacement upon leaving the plates? (b) What is the direction of the electron's velocity at that time? (c) What is the displacement of the electron on the screen relative to its position when the electric field between the parallel plates is zero?

21-19. Two sheets, one conducting and the other dielectric, are charged with the same constant surface charge density. Hence, the electric field for both configuration is the same. When a charged object is brought to the same distance from first one surface and then the other, it is found that the electric force on the object is *not* the same in the two cases. Explain.

21-20. Two parallel charged sheets with opposite charge densities $+\sigma$ and $-\sigma$ are a distance d apart. (a) Find the electric field due to the positive charged sheet at the site of the negative charged sheet. (b) What is the force on a unit area of the negative charged sheet?

21-21. Starting from rest at a position near a uniformly charged conducting sheet, a proton reaches a kinetic energy of 1.60×10^{-16} J after moving (in vacuum) through a distance of 10 cm. Find (a) the magnitude of the electric field and (b) the surface charge density on the sheet.

21-22. A 2.0-g sphere hangs from an insulating thread mounted between the vertical plates of a charged capacitor. See Fig. P21-22. Find the surface charge density on each of the capacitor plates if the thread makes an angle of 45° to the vertical. The sphere carries an electric charge of $+1.0$ μC.

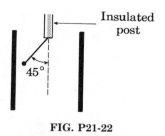

FIG. P21-22

21-23. In a Millikan experiment it is observed that an oil droplet is balanced and at rest between the plates when an electric field of 1.0×10^5 N/C exists and there is an excess charge on the droplet of $+1.6 \times 10^{-19}$ C. (a) What is the direction of the electric field? (b) What is the mass of the oil droplet?

21-24. Refer to Prob. 21-23. When there is no electric field, the droplet is observed to fall at the constant speed of 0.10 mm/s. *(a)* Determine the value of the resistive constant K. *(b)* With what velocity (magnitude and direction) does the droplet move when it has an excess of 5 electrons? *(c)* When it has a deficiency of 10 electrons?

21-25. Refer to Prob. 21-23. When there is no electric field, the droplet falls at the constant speed of 0.10 mm/s. With what velocity will it move if it is deficient by one electron and the electric field is 2.0×10^5 N/C?

21-26. An *electric dipole* consists of charges q and $-q$ separated by a distance $2a$. See Fig. P21-26. *(a)* Show that the electric field at a point on the perpendicular bisector of the dipole axis at a distance r is given by

$$E = \frac{kp}{r^3} \frac{1}{(1 + a^2/r^2)^{3/2}}$$

where $p = q(2a)$ is the *electric dipole moment*—the product of the magnitude of one of the charges and the distance between them. *(b)* What is the direction of the electric field vector found in part *(a)*? *(c)* Show

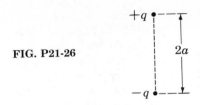

FIG. P21-26

that the electric field at a point along the dipole axis at a point a distance r from its center—where $r \gg a$—is given by

$$E = \frac{2kp}{r^3} \frac{1}{(1 - a^2/r^2)^2}$$

(d) What is the direction of the electric field vector found in part *(c)*? Note that in both cases $E \propto 1/r^3$ at large distances such that $a/r \to 0$. Whereas the electric field from a single point-charge (a monopole) varies inversely as the *square* of r, the electric field from a dipole varies (for large r) inversely as the *cube* of r.

CHAPTER 22

Gauss' Law

Gauss' law is an alternative formulation of Coulomb's law; it expresses in a simple and altogether general way the fact that the electric force between point-charges is strictly inverse-square. For charge distributions of high geometrical symmetry Gauss' law readily leads to the electric field produced by a variety of charge distributions. Conversely, knowing the electric field emerging from any closed surface, one may deduce the net charge within the surface. When applied to an electric conductor, Gauss' law implies that the net static charge on any conductor must lie completely on the outside surface. The experimental observation that this is indeed the case is the most precise demonstration of the coulomb-force law.

22–1 ELECTRIC FLUX

Since Gauss' law is a statement about electric flux, we first define this quantity. Imagine a *closed* surface of arbitrary shape (a so-called gaussian surface) with the electric field penetrating this surface. The surface element of area dS shown in Fig. 22-1 is sufficiently small to be considered perfectly flat. The orientation of this surface element is defined by the direction of the outwardly drawn normal. Then the vector $d\mathbf{S}$ is assigned a direction along the *outward* normal to the surface element, its magnitude being the area of the element. The electric field $\mathbf{E}$, which may arise from one or more point-charges or from a continuous charge distribution, may be assumed to be constant over the very small area dS. The angle between $\mathbf{E}$ and $d\mathbf{S}$ is θ. By definition, the electric flux $d\phi_E$ through this surface element is

$$d\phi_E = E \cos\theta \, dS$$

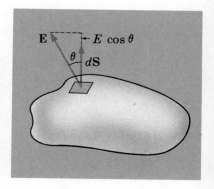

FIG. 22-1. Computing the electric flux through the element of surface $d\mathbf{S}$, whose direction is that of the outward normal to the surface.

That is, the electric flux is the component, $E \cos \theta$, of the electric field parallel to $d\mathbf{S}$ (or perpendicular to the surface) multiplied by the area dS. We may express this more compactly by using the dot product (Sec. 2-5) of the vectors $\mathbf{E}$ and $d\mathbf{S}$. Then we have

$$d\phi_E = \mathbf{E} \cdot d\mathbf{S} \tag{22-1}$$

If the electric field is directed outward from the gaussian surface, that is, if the angle θ is less than 90°, the flux $d\phi_E$ is positive; if the angle θ is greater than 90°, the electric field penetrates the gaussian surface in the inward direction and the flux is negative. When the electric field lies in the plane of the surface element $\theta = 90°$, the flux through the surface is *zero*. Since the gaussian surface may be chosen, as we shall see, to be of any shape, we will often so choose the shape that either $\mathbf{E}$ is parallel to $d\mathbf{S}$ and the flux is simply $d\phi_E = E\, dS$ or $\mathbf{E}$ is perpendicular to $d\mathbf{S}$ and the flux is $d\phi_E = 0$.

To find the *total electric flux* ϕ_E over an entire closed gaussian surface, we merely sum up the contributions from all surface elements, taking into account, of course, the variations in $\mathbf{E}$ both in magnitude and direction from one point on the surface to another.

$$\phi_E = \oint d\phi_E = \oint \mathbf{E} \cdot d\mathbf{S} \tag{22-2}$$

The flux is called the surface integral of $\mathbf{E}$ over the closed gaussian surface. The circle in the integral sign indicates that the integration must be taken over the entire closed gaussian surface. At first sight, evaluation of such an integral might appear formidable. However, we shall see that computing the electric flux can be remarkably simple when the geometrical arrangement of charge is simple.

Example 22-1

Compute the electric flux through a closed surface in a uniform electric field, the surface being that of a cylinder whose axis is the electric field direction, as shown in Fig. 22-2.

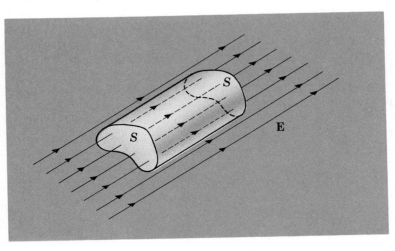

FIG. 22-2. The total electric flux through a closed surface in a uniform electric field is zero.

There is no contribution to ϕ_E along the curved surface, since **E** is perpendicular to $d\mathbf{S}$. On the other hand, the electric force lines are perpendicular to both end surfaces of the cylinder, going in at one end and out at the other. The inward lines produce a flux $-ES$; the outward lines, a flux $+ES$. The total flux $\oint \mathbf{E} \cdot d\mathbf{S}$ is zero.

22-2 GAUSS' LAW

First consider a single point-charge q (assumed positive for definiteness) surrounded by a spherical gaussian surface of radius r_1 centered at the point-charge; see Fig. 22-3. We can easily find the total electric flux through this surface, since the electric field, radially outward, is perpendicular to the surface at all points. The magnitude of the electric field $\mathbf{E}_1$ at a distance r_1 from a point-charge is given by

$$E_1 = \frac{kq}{r_1^2} \qquad (21\text{-}3)$$

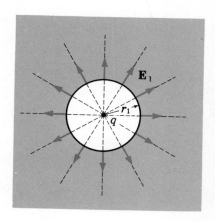

FIG. 22-3. The electric field E_1 at a distance r_1 from a point-charge q.

The surface area of the sphere is $4\pi r_1^2$. Therefore, the electric flux is

$$\phi_E = \oint \mathbf{E} \cdot d\mathbf{S} = \frac{kq}{r_1^2} 4\pi r_1^2 = 4\pi kq$$

$$\phi_E = \frac{q}{\varepsilon_0} \qquad (22\text{-}3)$$

Note that the distance r_1 cancels out. As the radius of the sphere is increased, dilution of the electric field from a point-charge is exactly matched by the dilation in the sphere's surface area. This holds, of course, only because the coulomb interaction is precisely inverse-square.

We have found, following (22-3), that the electric flux from a point-charge through a sphere encircling it is simply the charge q divided by the permittivity constant ε_0. This simple relation holds in general: *The total electric flux ϕ_E through any closed surface arising from any number of electric charges is q/ε_0, where q is the net charge enclosed by the surface.* This is Gauss' law. Let us prove it.

First we imagine a single point-charge q to be surrounded by a surface of arbitrary shape, as shown in Fig. 22-4. We concentrate on the flux through the surface element $d\mathbf{S}_2$ at a distance r_2 from the point-charge. The electric field here is

$$E_2 = \frac{kq}{r_2^2} \qquad (22\text{-}4)$$

The surface element is so chosen that it subtends the same solid angle as does the surface element dS_1 lying on the concentric sphere

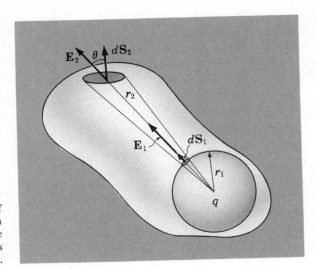

FIG. 22-4. A point-charge q surrounded by a volume of arbitrary shape and by a concentric sphere. The electric flux through the two shaded areas is exactly the same.

of radius r_1 which is centered at q; that is, a single cone drawn outward from the charge intercepts the surface elements dS_1 and dS_2. We wish to show that the flux is the same through these two surface elements.

First, we note that $\mathbf{E}_2$ and $d\mathbf{S}_2$ are not parallel in general. Therefore the flux through this surface element is

$$d\phi_2 = \mathbf{E}_2 \cdot d\mathbf{S}_2 = E_2 \cos\theta \, dS_2 \tag{22-5}$$

The area dS_2 is larger than dS_1, not only because it lies farther from the charge q but also because dS_2 is inclined relative to $\mathbf{E}_2$. (Note that both $\mathbf{E}_1$ and $\mathbf{E}_2$ are directed radially outward from q.) In fact,

$$\frac{dS_2 \cos\theta}{dS_1} = \left(\frac{r_2}{r_1}\right)^2 \tag{22-6}$$

where the factor $(r_2/r_1)^2$ arises from the difference in distances, and the factor $\cos\theta$ arises from the inclination of dS_2 to the direction of $\mathbf{E}$.

Substituting (22-4) and (22-6) in (22-5) yields

$$d\phi_2 = \frac{kq}{r_2^2} \cos\theta \, \frac{(r_2/r_1)^2 \, dS_1}{\cos\theta} = \frac{kq}{r_1^2} \, dS_1$$

But the electric flux $d\phi_1$ through the surface element dS_1 is

$$d\phi_1 = E_1 \, dS_1 = \frac{kq}{r_1^2} \, dS_1$$

and therefore $\qquad d\phi_2 = d\phi_1$

The electric flux through the two matched surface elements is the *same*.

To find the flux through a closed surface of arbitrary shape, we merely extend this procedure, choosing matched pairs of surface elements, until all of the arbitrary surface (and, therefore, all of the spherical surface) is accounted for. It follows that the total electric flux through an arbitrary surface is precisely the same as the electric flux through a sphere centered on the point-charge, and therefore (22-3) applies to any point-charge and any closed arbitrarily shaped gaussian surface surrounding the charge.

When the gaussian surface encloses a positive charge, the electric flux is positive and the electric lines of force pass outward through the surface. On the other hand, an enclosed negative charge gives a negative flux, the electric lines of force passing inward through the gaussian surface.

If the point-charge is *outside* the closed surface, as in Fig. 22-5, the total flux through the surface is zero since as many electric field lines penetrate into the surface as emerge outward from it.

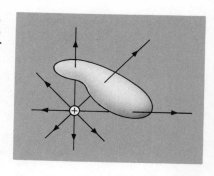

FIG. 22-5. The total electric flux through a closed surface *not* enclosing a net charge is zero.

Thus far we have dealt with only a single point-charge within any closed surface. Now suppose that we have a number of charges $q_a, q_b, \ldots$ within some arbitrarily shaped closed surface. To find the total electric flux through the surface, we first compute the flux through the surface from each charge.

$$\oint \mathbf{E}_a \cdot d\mathbf{S} = \frac{q_a}{\varepsilon_0} \qquad \oint \mathbf{E}_b \cdot d\mathbf{S} = \frac{q_b}{\varepsilon_0} \qquad \text{etc.}$$

Adding these relations into a single equation gives

$$\oint (\mathbf{E}_a + \mathbf{E}_b + \cdots) \cdot d\mathbf{S} = \frac{q_a + q_b + \cdots}{\varepsilon_0} \tag{22-7}$$

This addition (*under* the integral sign) *is* permitted. Recall that the surface was assumed to be the same for each of the point-charges. Moreover, we know that electric forces follow the superposition principle and the resultant electric field is $\mathbf{E} = \mathbf{E}_a + \mathbf{E}_b + \cdots$. Similarly, from the right-hand side of (22-7), the total *net* charge q enclosed by the surface is $q = q_a + q_b + \cdots$. Equation (22-7) may then be written as

$$\phi_E = \frac{q}{\varepsilon_0} \tag{22-3}$$

We have proved Gauss' law: The total electric flux ϕ_E through *any* closed surface is simply the *net* charge q *enclosed* by the surface divided by ε_0. There were two crucial assumptions in the proof: (1) The electric force varies as $1/r^2$, and (2) electric forces follow the superposition principle.

Applied to charge distributions in general, Gauss' law involves a surface integral which may be difficult to evaluate. But for uniform charge distributions of high geometrical symmetry—a

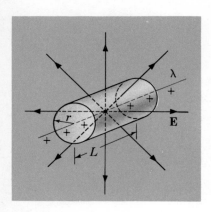

FIG. 22-6. A cylindrical gaussian surface used for computing the electric field from an infinitely long line of charge.

line of charge or a surface of charge, for example—it is a simple, almost trivial, matter to compute the integral. The key is this: Since we are free to choose a gaussian surface of any shape, we choose a shape which matches the symmetry of the charge distribution within.

Example 22-2

Using Gauss' law, find the electric field about a uniformly charged infinite wire.

For this simple geometric configuration we know that the electric field is radially outward (for positive charge), perpendicular to the wire. If this were not so, it would imply that either space or the charge distribution is not uniform—contrary to fact. Since we know the direction of **E**, we choose our gaussian surface so that it will be simple to find the electric flux through it. Its shape is a right circular cylinder of radius r and length L whose axis coincides with the wire, as shown in Fig. 22-6. The flux through the ends of the cylinder is zero, since the electric field is in the plane of the ends. The flux through the cylindrical surface of total area $2\pi rL$ is simply $E(2\pi rL)$ because **E** is always parallel to $d\mathbf{S}$ and by symmetry the magnitude of **E** is the same for all points on the cylindrical surface. Therefore, the *total* flux through the gaussian surface is

$$\phi_E = \oint \mathbf{E} \cdot d\mathbf{S} = E(2\pi rL)$$

From Gauss' law, $\quad \phi_E = E(2\pi rL) = \dfrac{q}{\varepsilon_0} = \dfrac{\lambda L}{\varepsilon_0}$

We have used the fact that the total charge q enclosed by the cylinder is λL, where λ is the linear charge density—the charge per unit length. This gives

$$E = \frac{\lambda}{2\pi\varepsilon_0 r}$$

Example 22-3

Find the electric field from a uniformly charged infinite plane sheet, using Gauss' law.

On the basis of symmetry arguments like those used in Example 22-2 above, we know that the field must be perpendicular to the surface on *both* sides. Note that our surface is imagined to be an infinitesimally thin sheet of charge, *not* a conductor. We again choose the gaussian surface as a cylinder, this time with the axis perpendicular to the surface, as shown in Fig. 22-7. This time the electric field lines lie parallel to the cylindrical surface, and there is no flux through this surface. Over the end surfaces the electric field is perpendicular to and outward from the ends, and the flux through each end is EA, where A is the surface area of an end. The *total* charge enclosed by the gaussian surface is σA, where σ is the surface charge density—the charge per unit area. Therefore, from Gauss' law we have

$$\phi_E = \frac{q}{\varepsilon_0}$$

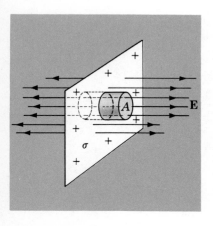

FIG. 22-7. The cylindrical gaussian surface used for computing the electric field from a uniformly charged sheet.

$$2EA = \frac{\sigma A}{\varepsilon_0}$$

$$E = \frac{\sigma}{2\varepsilon_0}$$

22-3 ELECTRIC FIELD LINES

In Chap. 21 it was asserted that one may map the electric field of any charge distribution with electric field lines which originate on positive charges, terminate on negative charges, and are continuous in between. Furthermore, the electric field lines denote the direction of the electric field at any point by their tangent and the magnitude of the electric field by their number through a transverse area. Gauss' law gives us proof of these assertions. Indeed, Gauss' law may be regarded as nothing more than a mathematical statement of the concept of electric field lines.

The electric flux through a surface element $d\mathbf{S}$ is $d\phi_E = \mathbf{E} \cdot d\mathbf{S}$. If this surface is oriented perpendicular to the electric field, then $E = d\phi_E/dS$; that is, the electric field is the flux per unit transverse area. Now, if the electric field is represented by lines such that an electric field of 1 N/C is represented by 1 electric line per unit transverse area, we see that the electric flux is nothing more than the number of field lines passing through a transverse gaussian surface.

We may always construct the gaussian surface so that electric lines enter or leave it at right angles to the surface. Then the total flux ϕ_E is simply the net number of field lines leaving the gaussian surface. If no charge is enclosed, $q = 0$ and hence $\phi_E = 0$, by Gauss' law; then the net number of lines into the gaussian surface equals the number out. Inasmuch as the closed gaussian surface can be made as small as we wish, in a region where there are no electric charges the electric lines are continuous. If, on the other hand, a net charge is enclosed by the gaussian surface, the net flux through it is not zero, and the number of field lines, $\phi_E = q/\varepsilon_0$, is directly proportional to the charge which acts as the source of the field. That is to say, the electric field lines originate and terminate on charges, the number of lines being directly proportional to the charge.

Given the electric field lines at all points on any closed surface, we can immediately say something about the charge within: If the same number of lines go in as come out, there is no net charge inside; if more lines emerge from the surface than go in, the surface encloses positive net charge (and conversely for negative charge); see Fig. 22-8. We cannot say how charge is distributed within the surface or how charge is located outside the surface; we can only say what the net enclosed charge is.

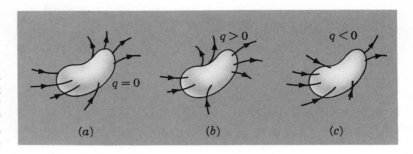

FIG. 22-8. By counting the net number of electric field lines out of a closed surface, one may deduce the net electric charge q within the surface: (a) $q = 0$, (b) $q > 0$, (c) $q < 0$.

22-4 SPHERICALLY SYMMETRIC CHARGE DISTRIBUTIONS

First we find the electric field from a spherical shell of charge. Since the charge is assumed to be spread uniformly over the spherical surface, there can be no preferred direction in space. Consequently, the electric field must be radial (outward for positive charge, inward for negative charge). To find the electric field outside the shell, we apply Gauss' law to a spherical gaussian surface surrounding the charged shell and concentric with it, as shown in Fig. 22-9a.† By symmetry, the electric field has constant magnitude at every point on the gaussian sphere, and the electric flux through the spherical surface is

$$\phi_E = E(4\pi r^2)$$

By Gauss' law we know that this flux is q/ε_0, or

$$E(4\pi r^2) = \frac{q}{\varepsilon_0} \qquad (22\text{-}8)$$

where q is the total net charge on the shell. The electric field is radial and has, from (22-8), the magnitude

$$E = \frac{q}{4\pi\varepsilon_0 r^2}$$

We recognize this equation as being identical with that found earlier for the electric field at a distance r from a *point-charge* [Eq. (21-3)]. Thus, the electric field *exterior* to a spherically symmetric shell of charge is precisely the same as that produced by the same total charge located at the center of the shell.

What about the field on the interior of the shell? We find this easily by taking the gaussian surface to be a concentric spherical surface *inside* the shell of charge, as in Fig. 22-9b. Since no charge is enclosed by this surface, the electric flux through it must be

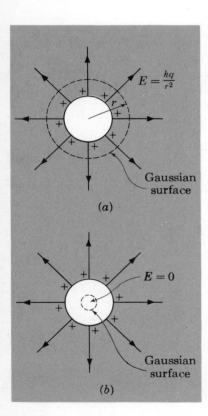

FIG. 22-9. A cross-sectional view of a spherical shell of charge (solid circle) and the spherical gaussian surfaces (dashed circles) used to compute the electric field (a) outside and (b) inside the shell of charge.

†Since the gravitational force between point masses is also inverse square and formally analogous to the Coulomb electric force, there is a corresponding Gauss' law for gravity. The results found here for electric charge apply equally well to the analogous distributions of mass.

zero. But if the shell is truly spherically symmetric, the electric field must be radial and of the same magnitude at all points on the gaussian surface. Equation (22-8) then gives $E(4\pi r^2) = 0$, which implies $E = 0$ at all points within the shell. The only spherically symmetric electric field consistent with the requirement of zero net flux is *zero field*. Thus, there is no electric field in the interior of the spherical shell of charge and no electric force on any test charge placed within the shell. A spherical shell of charge is, from an external point of view, electrically equivalent to a point-charge; from an internal point of view, it is electrically nonexistent.

It is a simple matter to find the electric field for a dielectric solid sphere (*not* a conductor) if one knows the results for a single shell. If the distribution of charges throughout a dielectric sphere depends only on the distance to the center (that is, if the charges have spherical symmetry relative to the center), then one simply imagines the solid sphere to consist of a series of concentric shells. From the outside, the electric field is just that due to a point-charge at the center whose magnitude is the net charge of the sphere. From the inside at a distance r from the center, the resultant field at any point is that due only to those shells whose radius is less than r. At the center, the electric field is zero.

One interesting case is that in which a sphere has constant charge density, the charge per unit volume ρ (in C/m³) being the same at all points throughout the sphere. Taking the radius of the sphere to be R, we know that for any distance $r > R$, the electric field is that of a point-charge at the center. What is the interior electric field with $r < R$? The only charge that matters is the charge *inside* the radius r. The total charge within r is $\rho(\frac{4}{3}\pi r^3)$ and the electric field of this charge is like that due to a point-charge at the center. Therefore,

$$E = \frac{kq}{r^2} = \frac{k\rho(\frac{4}{3}\pi r^3)}{r^2} = (\tfrac{4}{3}\pi k\rho)r$$

The resultant field within the sphere of charge is proportional to the distance r. If the shell has a positive charge, the field is radially outward; if negative, inward.

The electric field for a uniformly charged sphere is shown in Fig. 22-10. Outside the surface, $E \propto 1/r^2$; inside the surface, $E \propto r$.

22–5 GAUSS' LAW AND CONDUCTORS

In what follows we shall speak of the electric field within a conductor. Let us first be clear about what this means. If the electric field is measured over a volume of space which is small compared with the distance between particles on an atomic scale, we see nothing but point-charges; so the field within any material, whether conductor or nonconductor, changes rapidly as we approach and

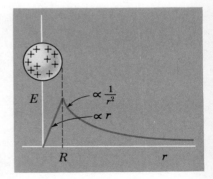

FIG. 22-10. Electric field as a function of distance from the center of a uniformly charged dielectric sphere.

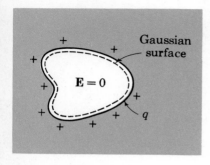

FIG. 22-11. Electrical conductor with a gaussian surface chosen to lie just inside the conductor's outside surface.

recede from these point-charges. For our purposes we need not consider such small volumes, and by the electric field within a material we shall mean the average force acting on a unit positive charge over a volume that is large enough to include many charged particles, but small enough to allow us to speak of the electric field at a "point."

As we have seen in Sec. 21-4, the electric field within an insulated charged conductor is exactly zero. By using Gauss' law we can immediately deduce an important consequence of this fact. Imagine a gaussian surface to be located just inside the outer surface of a charged conductor of arbitrary shape, as shown in Fig. 22-11. The conductor may be a solid piece of metal or one with interior cavities. Since the boundary of the gaussian surface lies on the interior of the conductor, the electric field at every point on the gaussian surface is zero; consequently, the electric flux through the surface is zero. From Gauss' law, if $\phi_E = 0$, then $q = 0$, and there is *no* (net) charge inside the surface of a charged conductor. Thus, atoms within the interior of the conductor are neutral, while any excess or deficiency of electrons occurs in a thin layer of atoms at the conductor's surface. We saw earlier that the electric field within a *spherically symmetric* shell of charge is zero. Our present result for *conductors* is more general: The interior field is zero for *any* shape and for *any* charges on the exterior surface.

The external electric field near the conductor's surface is always perpendicular to the surface at that point. This is always the case for a conductor in electrostatic equilibrium; for, if the electric field were not along the normal to the surface, there would be a component lying along the surface. This component would then act on free electric charges at the surface of the conductor and change the charge distribution. Since the charge distribution on a conductor in equilibrium is fixed, the electric field is always normal to the surface.

The assertion that the net electric charge within a conductor, quite apart from external electric fields or charges on the surface, is always zero, is a consequence of Gauss' law. If we test this assertion by experiment, we test the correctness of Gauss' law, which is to say that we test Coulomb's law and the inverse-square variation in the electric interaction between point-charges. If the net charge within a conductor were found to be just slightly different from zero, the exponent of r in Coulomb's law would differ from 2 by some very small amount. Such experimental attempts at seeking a net charge inside a conducting shell, which are the most sensitive tests of the exponent in Coulomb's law, have been made by many physicists, including Benjamin Franklin, Michael Faraday, and Henry Cavendish. The most recent and precise tests, made by Plimpton and Lawton in 1936, show that in the relation $F = kq_1q_2/r^n$, the exponent n differs from 2 by no more than 1 part in 10^9.

A demonstration of zero net charge inside a conductor is the

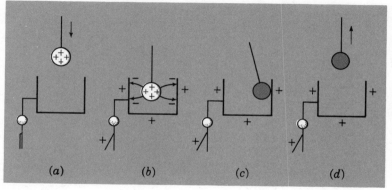

FIG. 22-12. Stages in the transfer of charge from a charged object to the outside of a hollow conductor, whose state of charged is indicated by the attached electroscope. (a) The charged object is outside. (b) The charged object is within the conductor, and equal and opposite charges are found on the inside and outside of the conductor. (c) When the object is touched to the conductor's inside, it annuls the charge on the conductor's inner surface and the exterior charge is unchanged. (d) The object, now electrically neutral, is removed.

famous Faraday ice-pail experiment shown in Fig. 22-12. The conductor is an ice pail, and the charge on its outer surface, indicated by the charge on the attached electroscope, is initially zero. A positively charged object is introduced into the interior of the conductor. Negative charges appear on the inner surface and positive charges of the same magnitude on the outer surface of the pail. We may say that the charged object "induces" charges on the inner and outer surface of the conductor. Then the object is touched to the inside of the conductor and annuls all the induced charge on this inner surface (Fig. 22-12c). Finally, when the object is removed, the electroscope shows the same charge as in Fig. 22-12b. The object itself is found to have zero charge. The experiment shows that the charge originally carried by the object has been transferred entirely to the exterior of a conductor.

Another important example of zero field inside a conductor is in the effects of external charges. Any static electrical effect external to a conductor is not felt by charges within the conductor. Therefore, a closed conductor (or one with only small holes leading to the interior) acts as a perfect shield. This is of practical importance in that electrical apparatus placed within a metallic screen (which is effectively a closed conducting surface) is shielded from any external electrical influence.

Example 22-4

What is the electric field between two parallel oppositely charged conductors, each having a surface charge density of magnitude σ, as shown in Fig. 22-13? (This is the configuration of a parallel-plate capacitor, of which we will have more to say in Sec. 23-8.)

Consider the situation shown in Fig. 22-13. At any point not close to the edges of the plates, where a "fringing" of the electric field occurs, the electric field will, by symmetry, be uniform and perpendicular to the charged surfaces.

As a Gaussian surface we choose a small cylinder whose axis

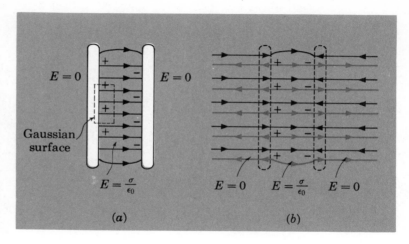

FIG. 22-13. (a) Two oppositely charged parallel conductors and a gaussian surface for evaluating the electric field between the plates in terms of the surface charge density. (b) The electric fields from two oppositely charged parallel infinitesimal sheets; the color lines represent the field of the positive charges, the black lines the field of the negative charges.

is parallel to the normal to the surface; one end of this gaussian surface is inside the conductor and the other end is between the two conductors. The electric field **E** is constant in both magnitude and direction over the chosen surface area of magnitude A. Of course the surface charge density σ is constant over the surface *within* the gaussian cylinder.

The total charge enclosed by the gaussian surface is $q = \sigma A$; therefore, from Gauss' law we have

$$\phi_E = \frac{q}{\varepsilon_0}$$

$$\oint \mathbf{E} \cdot d\mathbf{S} = \frac{\sigma A}{\varepsilon_0}$$

The electric flux is zero within the conductor, since **E** is zero there. The flux is also zero along the curved surface of the cylinder, since the electric lines lie in this surface. The only contribution to ϕ_E is through the outside end of the cylinder, where $\phi_E = \oint \mathbf{E} \cdot d\mathbf{S} = EA$. Therefore, the relation above becomes

$$EA = \frac{\sigma A}{\varepsilon_0}$$

$$E = \frac{\sigma}{\varepsilon_0} \tag{22-9}$$

where σ is the charge density on *one* of the two conducting plates. Note that the electric field at the surface of a conductor is *twice* that found near the surface of a uniformly charged sheet of infinitesimal thickness (Example 22-3). To see that there is no contradiction here, consider Fig. 22-13b, where the electric fields are produced on both sides of two sheets of charge of opposite sign. The field from *one* sheet is $E = \sigma/2\varepsilon_0$. In the region between the two sheets both fields act to the right, giving a resultant field $E = \sigma/2\varepsilon_0 - (-\sigma/2\varepsilon_0) = \sigma/\varepsilon_0$ in agreement with our result above for a conductor. Moreover, to the left of the left sheet and to the right of the right sheet the resultant field is zero, just as required for the interior of a conductor.

SUMMARY

The electric flux $d\phi_E$ through an outwardly directed surface element dS is defined as

$$d\phi_E = \mathbf{E} \cdot d\mathbf{S} \qquad (22\text{-}1)$$

According to Gauss' law, which is an alternative formulation of Coulomb's law, the total electric flux ϕ_E through any closed surface is proportional to the net charge q enclosed by the surface:

$$\phi_E = \oint \mathbf{E} \cdot d\mathbf{S} = \frac{q}{\varepsilon_0} \qquad (22\text{-}3)$$

Gauss' law justifies the use of electric field lines to represent the magnitude and direction of the electric field.

Insofar as the external electric field is concerned, a spherically symmetric charge distribution is equivalent to a point-charge of the same magnitude at its center.

When applied to conductors, Gauss' law implies that no net charge will be found within a conductor, all the net charge residing on the outside surface. The experimental observation that the charge is indeed on the surface of a conductor confirms Gauss' law; that is, it shows that the electric force between point-charges is strictly inverse-square. A closed conducting shell, or screen, produces perfect electrical shielding in its interior; the electric field within is always exactly zero, quite apart from the external electric fields or the electric charges on the surface.

The electric field just outside a conductor's surface is proportional to the surface charge density at that point:

$$E = \frac{\sigma}{\varepsilon_0} \qquad (22\text{-}9)$$

PROBLEMS

22-1. The term "flux" for the quantity $\int \mathbf{E} \cdot d\mathbf{S}$ implies that something is flowing through the chosen surface. Historically, the term "flux" first arose in connection with hydrodynamics, in which the relevant field vector is not the electric field $\mathbf{E}$, but the velocity $\mathbf{v}$ of particles in a fluid. Show that the flux of a velocity field $\int \mathbf{v} \cdot d\mathbf{S}$ is, in fact, equal to the volume of fluid passing through the chosen surface per unit time.

22-2. It is stated in the text that Gauss' law applies to any arbitrarily chosen surface. Is a surface with a hole through it—for example, a doughnut—an acceptable Gaussian surface?

22-3. Use Gauss' law to show that electric field lines in charge-free space cannot cross. (*Hint:* Suppose, for the sake of argument, that two electric field lines *do* cross. Choose a Gaussian surface such that the intersection point of the electric field lines lies on the Gaussian surface.)

22-4. Assume that a uniform electric field of 1.5 N/C in the horizontal direction exists between the oppositely charged parallel plates shown in Fig. 22-13. What is the electric flux through a flat surface bounded by a circle of radius 2.0 cm, the flat surface being completely in the uniform electric field and making an angle of 30° with respect to the field direction?

22-5. Consider a cube of edge length L. What can you say about the outward electric flux through each face of the cube if a positive point-charge Q is located (*a*) at the center of the cube, (b) inside the cube, but very near the center of one face, (c) outside the cube, but very near the center of one face?

22-6. Consider the electric field due to a proton (charge, $+1.6 \times 10^{-19}$ C) at the center of the hydrogen atom. What is the electric flux through a gaussian sphere of radius 0.53×10^{-8} m (size of hydrogen atom) due to the enclosed proton?

22-7. Suppose that we arbitrarily represent an electric field of magnitude 1.0 N/C by drawing one electric field line per square centimeter. How many electric lines would originate from a $+Q$ charge?

22-8. An infinitely long dielectric solid cylinder with radius R has a constant volume charge density ρ (C/m³). What is the electric field as a function of the distance r from the axis for (a) $r < R$ and (b) $r > R$?

22-9. A particle of mass m and charge $-q$ moves diametrically through a small hole in a uniformly charged sphere of radius R and volume charge density ρ (C/m³). Find the frequency of the particle's simple harmonic motion.

22-10. The nucleus of the $^{20}_{10}$Ne atom is 7.60×10^{-15} m in diameter and is composed of 20 *nucleons* — 10 electrically neutral neutrons and 10 protons, each with an electrical charge $+e$. In turn, the nucleus is enclosed in two concentric electron shells; the inner shell is 2.0×10^{-11} m in radius and contains 2 electrons and the outer shell is 2.0×10^{-10} m in radius, being filled with the remaining 8 electrons. Find the electric field at the following radial distances from the center of the nucleus: (a) $r = 1.0 \times 10^{-12}$ m, (b) $r = 1.0 \times 10^{-10}$ m, and (c) $r > 2.0 \times 10^{-10}$ m. (It is proper to assume that all of the charge distributions are spherically symmetric. The pre-subscript to the chemical symbol gives the electric charge of the nucleus in units of the electronic charge e; the pre-superscript gives the total number of protons and neutrons.)

22-11. A net charge of $+2.0$ μC is added to an initially uncharged spherical conducting shell of 6.0-cm radius. What is the magnitude of the electric field (a) inside the shell, (b) just outside the surface of the shell, and (c) 12.0 cm from the center of the shell?

22-12. Show that Gauss' law may be written as $\oint \mathbf{E} \cdot d\mathbf{S} = 1/\varepsilon_0 \int \rho \, dV$, where ρ is the local electric charge density and the volume integral on the right side of the equation above is evaluated over the volume enclosed by the Gaussian surface.

22-13. A solid sheet of dielectric material of thickness t and infinite length and width has a uniform positive charge density ρ (C/m³) throughout the sheet. (a) Use Gauss' law to find the electric field as a function of the distance x from the center of the sheet, where $-t/2 < x < t/2$. (b) If a small hole is drilled through the sheet and a particle of mass m and charge $-q$ is released from rest on one side of the sheet, what is the frequency of the particle's motion?

22-14. A metallic spherical shell of inner radius 5.0 cm and outer radius 10.0 cm has a charge of -2.0 μC. Find the electric field at the following distances from the center of the spherical shell: (a) 3.0 cm, (b) 6.0 cm, (c) 12 cm, and (d) 24 cm.

22-15. Two thin, concentric cylindrical dielectric shells have radii of 4.0 and 6.0 cm and respective linear charge densities of $+2.0$ and -3.0 μC/m. Find the electric field at the following distances from the center of the cylinders: (a) 3.0 cm, (b) 5.0 cm, and (c) 7.0 cm.

22-16. Charge is to be distributed throughout a solid dielectric sphere of radius R such that the magnitude of the electric field inside the sphere is constant. (a) How must the volume charge density ρ vary with distance r from the center of the sphere? (b) Is such a charge distribution physically possible?

22-17. An electron is subject to a constant electric force when it is between two parallel, oppositely charged conducting plates. The plates, 1.0 cm apart, are given a surface charge density such that an electron, initially at rest at the negative plate, takes 1.0×10^{-8} s to reach the positive plate. What is the surface charge density on either plate? The electron charge and mass are 1.6×10^{-19} C and 9.1×10^{-31} kg, respectively.

22-18. The X and Y axes are both uniformly charged with a positive linear charge density λ. Consider a gaussian surface which is cubical in shape with edge length L and centered at the origin. (a) What is the total flux through this surface? (b) Find the fraction of the total flux through each of the six faces of the cube.

22-19. A uniform electric field exists between two parallel, oppositely charged conducting plates. The magnitude of the surface charge density on each plate is σ. A third conducting plate is now placed between the two charged plates and oriented parallel to the two plates. The free charges within the neutral middle plate quickly rearrange themselves so that the electric field is zero everywhere within this plate. Find the surface charge density on both sides of the middle plate (a) by using Gauss' law and (b) by applying the relation $E = \sigma/2\varepsilon_0$ giving the electric field from an infinitesimally thin sheet of charge.

22-20. Three concentric, conducting, spherical shells carry charges $+3Q$ (on the smallest), $-Q$, and $-3Q$ (on the largest). Find the charges on the inner and outer surfaces of each of these three spherical shells.

22-21. A small, hollow, conducting spherical shell having zero net charge is placed between two parallel conducting plates of surface charge densities $+\sigma$ and $-\sigma$. (a) What is the electric field within the shell? (b) Sketch the electric field lines within and around the conducting shell. (c) Where on the surface of the conducting sphere is the surface charge density largest and smallest?

22-22. A charge Q is introduced at the center of an initially uncharged spherical, hollow conductor of radius R. (a) Plot the electric field as a function of distance from the center of the sphere. (b) Now a second charge, $-Q$, is placed outside the conductor. What part of the field, if any, is affected? (c) Does the charge on the interior experience a force arising from the charge on the outside? (d) Does the charge on the outside experience a force arising from the charge on the inside? (e) Is there a violation of Newton's third law?

22-23. Consider a hollow dielectric cylinder with inner radius a and outer radius b; the cylinder is charged, having a charge density which varies with the radius: $\rho(r) = \rho_0 r^n$, where ρ_0 is a constant and $n = 0, 1, 2, \ldots$. (a) Show that the electric field is given by

$$r < a: E = 0; \quad a \leq r \leq b: E(r) = \frac{\rho_0}{\varepsilon_0 r}\left[\frac{r^{n+2} - a^{n+2}}{n+2}\right]$$

$$b < r: E(r) = \frac{\rho_0}{\varepsilon_0 r}\left[\frac{b^{n+2} - a^{n+2}}{n+2}\right]$$

(b) What are the dimensions and the units for ρ_0?

CHAPTER 23

Electric Potential, Capacitance, and Dielectrics

In this chapter we define electric potential energy and utilize the energy-conservation law. We first find the potential energy between a pair of point-charges and then extend this to a system of interacting point-charges. We define the electric potential as a scalar quantity whose magnitude at each point in space surrounding the charges permits us to solve easily problems involving changes in the energy of a charged particle. We derive the general relations between the electric potential and the electric field and discuss their graphical representation through equipotential surfaces and electric field lines. We consider electric potential as related to conducting objects.

Next we treat capacitors, devices for storing separated charges and electric potential energy. The capacitance for some simple geometries and the rules for combining capacitors in series and in parallel are derived. We define the dielectric constant. Finally, we find the energy of a charged capacitor and the energy density of the electric field.

23-1 ELECTRIC POTENTIAL ENERGY OF TWO POINT-CHARGES

We can, of course, solve all problems involving electric forces simply by knowing the coulomb force and applying Newton's laws, but this is neither necessary in all problems nor desirable. Because the coulomb force is a conservative force (Sec. 10-3), one depending only on the separation distance between interacting charges, we may associate an electric potential energy with the coulomb interaction. In introducing the potential-energy concept, we deal with a *scalar* quantity rather than a vector quantity, and the powerful conservation of energy principle applies. We have already seen the utility of the potential-energy concept in dealing with such conservative forces as the gravitational force and the elastic force of a deformed body.

We first find the electric potential energy U associated with two interacting positive point-charges q_1 and q_2. We imagine q_2 to be fixed in position while the charge q_1 is moved from point 1 to point 2 at separation distances r_1 and r_2, respectively, as shown in Fig. 23-1. (We could, of course, do this in reverse: Keep q_1 fixed and move q_2.)

Recalling the general expression for difference in potential energy $(U_f - U_i)$ [Eq. (10-4)] between some initial and final states, we have for the conservative electric force $\mathbf{F}_e$:

$$U_f - U_i = -W_{i \to f}(\text{by } \mathbf{F}_e) = -\int_i^f \mathbf{F}_e \cdot d\mathbf{s} \qquad (23\text{-}1)$$

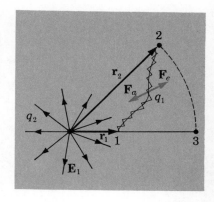

FIG. 23-1. Charge q_1 is moved from point 1 to point 2 so that its separation distance from charge q_2 goes from r_1 to r_2. The series of small radial and circular segments is equivalent to the continuous path, and we can imagine the charge q_1 to go from point 1 radially outward to point 3 and then to point 2 along the circular arc.

The change in potential energy is simply the negative of the work $W_{i \to f}$ done by the interaction force. Since the two charges are imagined to be of like sign, they repel one another, and the electric force $\mathbf{F}_e$ on charge q_1 is radially outward.

Although the path shown in the figure is quite general, evaluating the integral of (23-1) is easy. For one thing, we can imagine the complicated path to be replaced by a series of small segments which closely approximate the actual path, these segments lying either radially outward or along circular arcs centered at charge q_2. No work is done along the circular segments, since the force $\mathbf{F}_e$ is perpendicular to the displacement $d\mathbf{s}$ there. Only the radial displacements contribute. This is equivalent to saying that we can imagine the charge q_1 to be moved radially outward to point 3 at the same distance r_2 as point 2, and then moved, with no change in potential energy, from point 3 to point 2 along the circular arc. Therefore, (23-1) becomes

$$U_2 - U_1 = -\int_{r_1}^{r_2} \frac{kq_1q_2}{r^2} \, dr = kq_1q_2 \left(\frac{1}{r_2} - \frac{1}{r_1} \right) \qquad (23\text{-}2)$$

Note that in taking the outward route from point 1 to point 3, the vectors $\mathbf{F}_e$ and $d\mathbf{r}$ are in the *same* direction.†

Equation (23-2) gives the *difference* in electric potential energy for two different separation distances of the point-charges. This is all it can give, since choice of the zero of potential energy is always arbitrary. It is customary to choose the zero of potential en-

---

†Because of the similarity between the electrostatic and the gravitational interactions, there is a direct parallel between various expressions for the potential energy between electric point-charges and those between gravitational point-masses.

ergy to correspond to that configuration in which the force between the interacting particles is zero. Thus, we shall take U to be zero when particles are infinitely separated, putting

$$U_2 = 0 \quad \text{when } r_2 = \infty$$

Then (23-2) becomes, without the subscript,

$$U = \frac{kq_1q_2}{r} \tag{23-3}$$

We will have to bear in mind that electric potential energy, unless otherwise noted, is always zero for infinite charge separation.

Equation (23-3) is the basis of *all* further considerations in this chapter, and it is important to understand its implications. First, we can speak of electric potential energy only because the coulomb force is conservative. Said differently, the *work done* by this force in taking a charge around any *closed loop in an electrostatic field is zero:*

$$\oint \mathbf{F}_e \cdot d\mathbf{s} = 0 \tag{10-5}$$

In returning a charge to its starting point we restore the system to its initial state, doing no net work. Equivalently, the *work done* in moving one charge relative to another is *independent of the path* and *depends only on the end points*.

Whenever one can associate a potential energy with an interaction, as in the case of the gravitational force, the spring force and, now, the electric force, the *conservation of energy principle* applies. That is, if the system is isolated from external work-producing forces, the total energy E, comprised of the kinetic energies of the particles and the potential energy between the particles, is constant. For two particles a and b, we may write

$$E = K_a + K_b + U_{ab} = \text{constant} \tag{23-4}$$

where U_{ab} is the potential energy *of the system* (not of one or the other particles).

We see from both (23-2) and (23-3) that the potential energy associated with two positive charges is *positive*. This is as it should be, since two such particles repel each other. If the positive charges are brought together from infinity, the potential energy *increases* (see Fig. 23-2a). The same remarks apply when one has two negative charges: U is again positive. On the other hand, if a positive and a negative charge interact, the force is attractive and the potential energy is negative (see Fig. 23-2b). This follows directly from (23-3) (q_1 is $+$ and q_2 is $-$; hence U is $-$). It follows also from a simple physical argument. Suppose the two unlike charges are far separated and initially at rest. The total energy of this isolated system is initially zero and must remain so. If the charges are released and move toward one another, they gain kinetic energy and, since the total energy E remains zero, the potential energy must be-

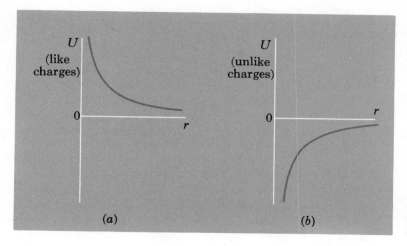

FIG. 23-2. Electric potential energy as a function of the separation distance between (a) two like point-charges and (b) two unlike point-charges.

come *negative*. The matter of signs will arise often in this chapter. One can always deal properly with this tricky matter by scrupulously assigning the right signs to q and to U. Still better, physical arguments will always determine whether the signs have been correctly assigned.

We have found that the electric potential energy of two point-charges separated by distance r is kq_1q_2/r. It can be shown that the total electric potential energy U for a system of three (or more) interacting charges (see Fig. 23-3) is simply

$$U = U_{12} + U_{23} + U_{13} = \frac{kq_1q_2}{r_{12}} + \frac{kq_2q_3}{r_{23}} + \frac{kq_1q_3}{r_{13}} \quad (23\text{-}5)$$

The total potential energy U of the system is independent of the way in which the system is assembled. One may bring charges together in any order or by any route. The potential energy is always the same. It is a property of the system, not of any one charge, and it depends only on the separation distances between the charged particles.

A word on units for electric potential energy. Since we shall use only SI units, U carries the unit joules (newton-meter) with electric charges in coulombs. A more convenient energy unit for systems of atomic particles is the electron volt (eV), where by definition, 1 eV = 1.6×10^{-19} J (we shall shortly see how this definition arises). The electron volt is a particularly convenient energy unit because two electrons (or any other two charged particles with charge magnitude e) when separated by atomic distances (10^{-10} m = 1 Å) have electric potential energies of the order of a few electron volts, as one can easily verify from (23-3).

Example 23-1

Consider two oppositely charged spheres of mass M and radius R; each carries an electric charge of magnitude Q. Initially the spheres

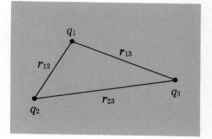

FIG. 23-3. Three interacting point-charges.

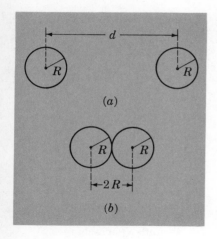

FIG. 23-4

are separated by a distance d, measured between centers. If the spheres are released from rest they move toward each other because of the Coulomb force of attraction. What is the speed of either sphere at the instant the two spheres collide? Fig. 23-4a shows the spheres in their initial positions and Fig. 23-4b indicates their positions at the moment of impact.

Since only conservative forces are involved, it is proper to use the energy-conservation principle: The total energy E is constant, where $E = U + K$. Equating the initial total energy (upon release) to the final total energy (at impact) we have

$$\frac{kQ^2}{d} + 0 = \frac{kQ^2}{2R} + 2(\tfrac{1}{2}Mv^2)$$

where d is the initial distance between centers, $2R$ is the final separation distance, and v is the speed of either sphere at the instant of impact.

After a little algebra we find

$$v = \sqrt{\frac{kQ^2}{M}\left(\frac{1}{d} - \frac{1}{2R}\right)}$$

Example 23-2

In the classical planetary model, an electron in a hydrogen atom moves in a circle of radius r about the proton. (a) What is the total energy of the system? (b) What is the binding energy of the system if the radius of the orbit is 0.53 Å? See Fig. 23-5.

(a) Taking the proton mass to be infinite (relative to the much smaller electron mass), we have for the total energy

$$E = U + K = \frac{kq_1 q_2}{r} + \tfrac{1}{2}mv^2$$

Here q_1 and q_2 are the proton and electron charges, m is the electron mass, and v is the electron speed. We take $q_1 = +e$ and $q_2 = -e$. Then,

$$E = -\frac{ke^2}{r} + \tfrac{1}{2}mv^2 \qquad (23\text{-}6)$$

Note that the potential energy is negative because the two particles attract one another.

We can eliminate the speed v by applying Newton's second law, noting that a radial coulomb force of magnitude ke^2/r^2 causes the electron to travel in a circle at constant speed v:

$$\Sigma F_r = ma_r$$

$$\frac{ke^2}{r^2} = \frac{mv^2}{r}$$

$$\frac{\tfrac{1}{2}ke^2}{r} = \tfrac{1}{2}mv^2$$

For a *circular* orbit, the kinetic energy is just half the *magnitude* of the potential energy. Using this result in (23-6) yields

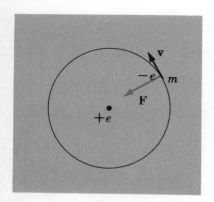

Fig. 23-5. An electron orbiting a proton in the classical planetary model of the hydrogen atom.

$$E = -\frac{ke^2}{r} + \frac{\frac{1}{2}ke^2}{r} = -\frac{\frac{1}{2}ke^2}{r} \qquad (23\text{-}7)$$

The *total* energy E is negative, and the electron is *bound* to the nucleus.

(b) The binding energy E_b is the minimum energy necessary to remove the electron from the proton, both particles then being at rest at infinite separation. From (23-7) it follows that

$$E_b = \frac{\frac{1}{2}ke^2}{r}$$

Taking $r = 0.53$ Å $= 0.53 \times 10^{-10}$ m, we have

$$E_b = \frac{\frac{1}{2}(9.0 \times 10^9 \text{ N-m}^2/\text{C}^2)(1.6 \times 10^{-19} \text{ C})^2 (1 \text{ eV}/1.6 \times 10^{-19} \text{ J})}{(0.53 \times 10^{-10} \text{ m})}$$

$$= 14 \text{ eV}$$

The binding energy of this atomic system is the *ionization energy* of hydrogen, since it is the energy required to change the neutral atom into an electron and positively charged ion (proton).

23-2 ELECTRIC POTENTIAL DEFINED

The concept of electric potential energy leads us naturally to that of the *electric potential*, a quantity which is extraordinarily useful in treating problems involving electric charge distributions.

Suppose that we have a collection of charges *fixed* in position and that still one more charge q_t, a test charge, is brought from infinity at constant speed to point P; see Fig. 23-6. We can readily compute the work required to do this: It is simply the electric potential energy of the system after q_t has been brought to its final position at P less the potential energy of the same charges with q_t infinitely far away. Note that we need *not* specify the route followed by q_t. Suppose that we take the charge q_t away and then bring another charge $2q_t$ to the same position P as that at which we had earlier placed q_t. Again, we wish to know the total work required. There is no need to go through the whole computation again; we know that in the second instance the work is just twice that in the first. Thus, if we know the work required to bring a

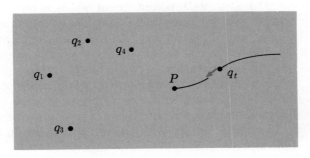

FIG. 23-6. Test charge q_t brought from infinity to point P in the vicinity of other charges held fixed in position.

unit positive charge from infinity to some location, we know at once the work required to bring any other charge there. Note that the assembly of charges into which we introduce the additional charge must be such that these charges remain fixed in position; equivalently, we may imagine the test charge to be so small that it does not affect the charge distribution.

The work W required to bring a unit positive charge from infinity to some position is defined as the *electric potential V* at that point. In symbols,

$$V = \frac{W}{q} \tag{23-8}$$

where W is the work done by the *external* force and V is the corresponding *increase* in potential. We have assumed, of course, that the electric potential energy of any pair of charges is chosen to be zero when the charges are infinitely separated. Given a fixed distribution of charges we can then associate an electric potential, a scalar quantity, with every point in space.

The electric potential can be measured in units of joules per coulomb. A special name, the *volt,* is given to this ratio:

$$1 \text{ volt} = 1 \text{ J/C}$$

The volt is named after A. Volta (1745–1827) who developed the first rudimentary battery. Units related to the volt (abbreviated V) are the microvolt (1 μV = 10^{-6} V), the millivolt (1 mV = 10^{-3} V), and the kilovolt (1 kV = 10^3 V).

If it takes 10 J of work to bring a positive charge of 1 C from infinity to a point P_1, then the potential assigned to P_1 is +10 V. By the same token, we know at once that to bring a charge of −4 C from infinity to P_1 will require, from (23-8), work in the amount

$$W = qV = (-4 \text{ C})(+10 \text{ V}) = -40 \text{ J}$$

Rather than our actually having to do work in bringing the negative charge to P_1, 40 J of work is done on us. Here we see the motivation for introducing the concept of electric potential: If this quantity is known or can be readily calculated for a given charge distribution, we can at once find the work done in moving any charge along any path from one location (with one potential) to a second location (with, possibly, another potential).

Just as with potential energy, it is only the difference in potential that is important in physical problems, and the choice of zero potential is arbitrary. In the interactions among point-charges described above it is convenient to choose the potential to be zero at infinite separation. On the other hand, in circuit problems it is often convenient to arbitrarily assign zero potential to the earth (or ground), which is a reasonably good conductor. This is useful because many conducting objects (including ourselves) are usually in electrical contact with the earth and would then be at zero potential.

As an example of potential difference, we consider the definition of the electron volt (eV). The electron volt is the energy unit chosen so that the work done in moving a particle with elementary charge $e = 1.60 \times 10^{-19}$ C across a potential difference of 1 V is, by definition, 1 eV. That is, with $q = e$ and $V = 1$ V, (23-8) gives

$$W = qV = (1.60 \times 10^{-19} \text{ C})(1 \text{ V}) = 1.60 \times 10^{-19} \text{ J} = 1 \text{ eV}$$

Thus, if an electron is accelerated through a potential difference of $+100$ V, its potential energy changes by $qV = (-e)(+100 \text{ V}) = -100$ eV; the electric potential energy *decreases* by 100 eV. This means that its kinetic energy must *increase* by 100 eV.

A familiar device which maintains an essentially constant potential difference across its terminals is the electrochemical cell, which changes internal chemical potential energy into electrostatic potential energy at its terminals. The common 12-V automobile battery consists of six lead cells, each maintaining a potential difference of approximately 2.0 V across its terminals. That the potential difference across cell terminals is typically a few volts is no accident. It is a direct consequence of the fact that the electric potential energy of ions separated by atomic distances is also a few electron volts.

23-3 ELECTRIC POTENTIAL OF POINT–CHARGES

The electric potential energy U_1 of a charge q_1 and a test charge q_t at a distance r_1 from it is given by

$$U_1 = \frac{kq_1 q_t}{r_1} \tag{23-3}$$

It follows that the electric potential V_1 due to the charge q_1 at the location of the test charge is

$$V_1 = \frac{U_1}{q_t} = \frac{kq_1}{r_1} \tag{23-9}$$

If one has a number of point-charges $q_1, q_2, q_3, \ldots$, at the respective distances $r_1, r_2, r_3, \ldots$ from the test charge q_t, the total potential V is simply the scalar sum of the potentials contributed by each charge separately (electric potential energies add as scalars):

$$V = V_1 + V_2 + \cdots = k \left(\frac{q_1}{r_1} + \frac{q_2}{r_2} + \cdots \right)$$

$$V = k \Sigma \frac{q_i}{r_i} \tag{23-10}$$

Equation (23-10) gives the potential for *any* charge distribution, since we can regard even a continuous distribution of charge as arising from a large number of point-charges. The procedures for

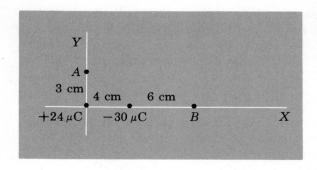

FIG. 23-7

finding the potential from a continuous charge distribution are treated in the next section.

Example 23-3

Charges of +24 and −30 μC are placed at (0, 0) and (4.0 cm, 0), respectively, as shown in Fig. 23-7. (a) Find the electric potential arising from these charges at points A (0, 3.0 cm) and B (10.0 cm, 0), (b) What is the external work required to move an electron from A to B? (c) What is the work required to move a Zn^{2+} ion from A to B?

(a) Using (23-10), we find the potentials at points A and B to be

$$V_A = 9.0 \times 10^9 \text{ N-m}^2/\text{C}^2 \left(\frac{+24 \times 10^{-6} \text{ C}}{3.0 \times 10^{-2} \text{ m}} + \frac{-30 \times 10^{-6} \text{ C}}{5.0 \times 10^{-2} \text{ m}} \right)$$

$$= 1.8 \times 10^6 \text{ V}$$

$$V_B = 9.0 \times 10^9 \text{ N-m}^2/\text{C}^2 \left(\frac{+24 \times 10^{-6} \text{ C}}{10.0 \times 10^{-2} \text{ m}} + \frac{-30 \times 10^{-6} \text{ C}}{6.0 \times 10^{-2} \text{ m}} \right)$$

$$= -2.3 \times 10^6 \text{ V}$$

Therefore, $V_{AB} = V_A - V_B = [(+1.8) - (-2.3)] \times 10^6 \text{ V} = 4.1 \times 10^6 \text{ V}$.

(b) The potential goes "downhill" from point A to point B. Therefore, the change in potential energy when the electron moves from A to B is, by (23-8),

$$q(V_B - V_A) = (-e)(-4.1 \times 10^6 \text{ V}) = +4.1 \text{ MeV}$$

(One MeV is one million electron volts.) To provide this increase in electric potential energy, external work of 4.1 MeV must be done on the electron.

(c) To take a doubly charged positive Zn^{2+} ion from A to B, the work required is

$$W_{A \to B} = q(V_B - V_A) = (+2e)(-4.1 \times 10^6 \text{ V}) = -8.2 \text{ MeV}$$

23-4 RELATIONS BETWEEN V AND $\mathbf{E}$

Given the electric field $\mathbf{E}$ as a function of position for a known charge distribution, how does one find the corresponding electric potential V as a function of position?

The potential energy U is related to the conservative electrical force $\mathbf{F}_e$ between a charge q_t and the charge distribution by

$$U_2 - U_1 = -\int_1^2 \mathbf{F}_e \cdot d\mathbf{s} \qquad (23\text{-}1)$$

Dividing both sides of this equation by the charge q_t gives

$$\frac{U_2}{q_t} - \frac{U_1}{q_t} = -\int_1^2 \frac{\mathbf{F}_e}{q_t} \cdot d\mathbf{s}$$

The left-hand side represents the electric potential difference $V_2 - V_1$, and the electric field $\mathbf{E} = \mathbf{F}_e/q_t$ appears on the right. Therefore, we have

$$V_2 - V_1 = -\int_1^2 \mathbf{E} \cdot d\mathbf{s} \qquad (23\text{-}11)$$

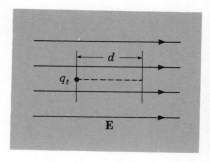

FIG. 23-8. A test charge q_t moved along the field lines of a uniform electric field.

The potential difference is the negative of the line integral of the electric field between the end points. If we are able to choose $V = 0$ for an infinitely distant point (say, point 1), and we designate point 2 as point p, (23-11) becomes

$$V_p = -\int_\infty^p \mathbf{E}_n \cdot d\mathbf{s} \qquad (23\text{-}12)$$

Let us find how the electric potential is related to $\mathbf{E}$ for the simplest possible geometry, that of a constant, or uniform, electric field. We imagine that a charge q_t is moved along the direction of the uniform electric field a distance d, as shown in Fig. 23.8. The potential difference V between two points separated by d is given by (23-11):

$$V = -\int_1^2 \mathbf{E} \cdot d\mathbf{s} \qquad (23\text{-}11)$$

The integration above is simple because $\mathbf{E}$ is constant and parallel with $d\mathbf{s}$, and we obtain

$$V = -Ed \qquad (23\text{-}13)$$

For a uniform electric field the magnitude of $\mathbf{E}$ is simply the potential difference V divided by the displacement d along the direction of field lines. We see, then, that the electric field may be expressed in volts per meter as well as in newtons per coulomb. Note the minus sign in (23-13). It corresponds to the fact that, when we travel *along* the direction of the electric-force lines, we move to points of progressively *lower* electric potential.

Example 23-4

Find the electric potential as a function of the distance r from an infinitely long and uniformly charged wire.

In Sec. 21-3, we found that the electric field is perpendicular to the wire and has magnitude

$$E = \frac{\lambda}{2\pi\varepsilon_0 r} \qquad (21\text{-}5)$$

where λ is the constant linear charge density.

We find the potential difference between two points r_1 and r_2 from (23-11):

$$V_2 - V_1 = -\int_{r_1}^{r_2} \mathbf{E} \cdot d\mathbf{r}$$

Taking the wire to be positively charged, the outward electric field $\mathbf{E}$ is in the same direction as $d\mathbf{r}$, and the above relation becomes

$$V_2 - V_1 = -\int_{r_1}^{r_2} \frac{\lambda}{2\pi\varepsilon_0} \frac{dr}{r} = -\frac{\lambda}{2\pi\varepsilon_0} \ln \frac{r_2}{r_1} \qquad (23\text{-}14)$$

For this configuration we *cannot* take $V = 0$ for $r = \infty$, for then (23-14) would lead to an infinite potential for all finite r (the physical reason for this mathematical difficulty is that an *infinitely* long wire would always look infinite in length, even if viewed from an infinite distance away). This is not troublesome, however, inasmuch as we always deal with potential differences.

23-5 EQUIPOTENTIAL SURFACES

Through the concept of electric potential we may associate an algebraic number with every point in space surrounding a charge distribution. Through the electric field concept we associate a vector (with three scalar components) with every point in space. We can map the electric field with electric field lines. Similarly, we can map the electric potential with *equipotential surfaces*.

Consider the simple case of a single point-charge q. The electric field is radial, and the potential V at any point depends only on the distance r from the charge ($V = kq/r$). Therefore, for a given radial distance r, *all* points on the sphere of radius r have the *same* potential. Indeed, we can imagine the point-charge to be surrounded by a number of concentric spherical surfaces, every point on a given sphere having the same potential and each spherical equipotential surface differing from its neighboring surfaces in potential by a constant amount (of course, in a two-dimensional figure the equipotential surfaces are represented as equipotential lines); see Fig. 23-9. The electric field lines, radially outward for a positive charge, are perpendicular to the equipotential surfaces.

It is easy to show that for *any* charge configuration the equipotential surfaces are always normal to the corresponding electric field lines. For the sake of argument suppose that $\mathbf{E}$ did *not* lie perpendicular to constant V surfaces. Then there would be a component of the electric force acting along the equipotential surface. But, by definition, a charge can be moved over an equipotential surface without doing work on it. Therefore, $\mathbf{E}$ lines must be normal to constant V surfaces. Moreover, the direction of $\mathbf{E}$ is always

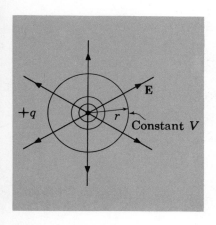

FIG. 23-9. Electric field lines and equipotential surfaces for a point-charge.

toward surfaces of *lower* potential. This is illustrated in Fig. 23-9 where **E** is radially outward, perpendicular to the spherical potential surfaces and directed toward lower potentials.

We have found the general relation which gives V, knowing **E** [Eq. (23-11)]. We wish now to find the inverse relation, **E** in terms of V. Let the vector $d\mathbf{s}$ represent the small displacement made when one moves *along the direction of* **E** between two neighboring equipotential surfaces differing in potential by dV. Since $d\mathbf{s}$ is parallel to **E**, that route has been taken which involves the *largest* potential difference for a given path length ds, and (23-11) gives

$$dV = -E\,ds$$

$$E = -\frac{dV}{ds} \quad (23\text{-}15)$$

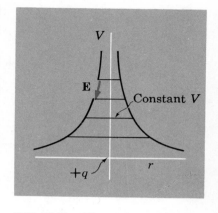

FIG. 23-10. Electric potential V plotted against distance from a positive point-charge; the electric field E at any point is the negative gradient of the potential.

The electric field is the negative of the spatial derivative of the electric potential *taken along that direction for which V changes most rapidly with position.* It is clear that in the case of a constant **E**, we have $E = -V/d$, as found earlier.

When the electric potential is given as a function of the space coordinates $V(x, y, z)$, the rectangular components of the electric field are given by

$$E_x = -\frac{\partial V}{\partial x} \quad E_y = -\frac{\partial V}{\partial y} \quad E_z = -\frac{\partial V}{\partial z} \quad (23\text{-}16)$$

As an illustration of (23-15) we may represent the relationship between V and **E** for a single positive point-charge (see Fig. 23-9) by a diagram in which V is plotted against r, as in Fig. 23-10. By (23-15), the electric field is the negative of the slope of the potential curve. Constant values of the electric potential are similar to constant values of gravitational potential energy for an actual hill; the electric field is called a measure of the *gradient*, that is, the *steepness* of the grade.

A model of this sort is useful for visualizing the interaction between two point-charges. If a small object is rolled up the potential hill, its path, when viewed from above, corresponds to the path of the scattered charge; it is an hyperbola.

Example 23-5

Find the electric potential as a function of r for a uniformly charged thin spherical dielectric shell of radius R.

We found earlier that a uniformly charged spherical shell produces an electric field which is equivalent to that of a point-charge at its center for all points *exterior* to the shell, and that the electric field is *zero* for all points within the shell. It follows that the potential exterior to the shell is just like that of a point-charge, namely $V = kq/r$, where q is the total charge on the shell. On the interior, $\mathbf{E} = 0$. Since $E = -dV/dr$, from (23-15), V must be *constant* at all points with-

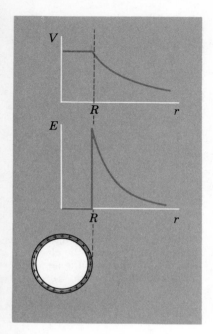

FIG. 23-11. Electric potential and electric field as a function of the distance r from the center of a spherically symmetrical thin shell of charge of radius R.

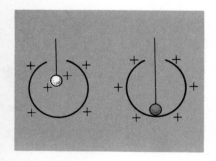

FIG. 23-12. Any charge brought to the interior of a conductor goes to the conductor's outside surface.

in the shell. Therefore, the potential at all interior points must be just the potential at the surface, $V = kq/R$; see Fig. 23-11. Any charge placed within the shell is not subject to an electric force and can be moved about without work being done on it.

23-6 ELECTRIC POTENTIAL AND CONDUCTORS

Recall some important results (Sec. 22-5) for all electric conductors: (1) Even if a conductor of arbitrary shape carries a net charge, *all* of the net charge resides on the outer conductor surface; (2) the *electric field inside* any conductor is exactly *zero;* and (3) at the surface the *electric field lines* are always *perpendicular* to the exterior conductor *surface.*

Since $\mathbf{E} = 0$ inside a conductor, we know at once that the *electric potential* at all points *within* a conductor, charged or uncharged, in electrostatic equilibrium is *constant,* quite apart from the shape of the conductor; thus a conductor has an equipotential volume, the constant electric potential at all locations within a conductor being just the electric potential at the surface.

Suppose that two or more metallic objects are connected by a conducting wire and electrostatic equilibrium has been achieved. Then we no longer have a number of separate conductors, but really just a *single* conductor. This means that any conducting objects connected together with a conducting wire will, when the free charges no longer move, be at the *same* electric potential everywhere on and within the objects.

As an example, consider an initially charged hollow spherical conducting shell into which a second smaller charged conducting sphere is introduced while attached to an insulating thread as in Fig. 23-12. The small sphere is made to touch the interior of the large shell. Where is the small sphere's charge after this? Once the two objects are brought in contact, we have, in effect, a *single* conductor. Therefore, all points on the interior of the shell will achieve a single potential, and all of the net charge will be found on the *outside* of the shell, no matter how much charge was originally on the exterior of the shell.

Example 23-6

How is the charge distributed over the surface of a conductor carrying a net charge?

For a sphere the surface charge density is constant. We wish to show that for a nonspherical surface the charge is concentrated at regions of small radius of curvature, or points, on a conductor (see Fig. 23-13b). Let us first consider the situation shown in Fig. 23-13a, that in which two spheres of radii r_1 and r_2 and carrying charges q_1 and q_2, respectively, are connected by a long conducting wire. We wish to find the charge densities σ_1 and σ_2 on the two spheres.

Since the spheres and connecting wire comprise a single conductor, the potential is the same for both spheres. Therefore,

$$V_1 = \frac{kq_1}{r_1} = \frac{kq_2}{r_2} = V_2 \qquad (23\text{-}17)$$

$$\frac{q_1}{q_2} = \frac{r_1}{r_2}$$

Most of the charge is on the *larger* sphere.

By definition,

$$\sigma_1 = \frac{q_1}{4\pi r_1^2} \quad \text{and} \quad \sigma_2 = \frac{q_2}{4\pi r_2^2}$$

Therefore,

$$\frac{\sigma_1}{\sigma_2} = \frac{q_1}{q_2}\left(\frac{r_2}{r_1}\right)^2$$

Using the relation for q_1/q_2 given above, we have

$$\frac{\sigma_1}{\sigma_2} = \frac{r_2}{r_1}$$

The *charge density* is greatest on the *smaller* sphere. Since the electric field near any conducting surface is proportional to the charge density $E = \sigma/\varepsilon_0$ (Sec. 21-4), the electric field is strong at the surface of the small sphere, compared with the field at the surface of the larger sphere.

The two connected spheres in Fig. 23-13a may be thought of, very roughly, as a single conductor of the shape shown in Fig. 23-13b. Therefore, the charge is concentrated mostly at the pointed end, and the electric field is greatest there. The very strong electric fields near the sharp points of charged conductors can accelerate the few ions always present in air (normally a good insulator) to energies such that when the ions collide with neutral molecules of air, they produce more ions and there is an electric breakdown of the air. For air at standard temperature and pressure, an electric field of about 3×10^6 V/m will produce this electric breakdown; the phenomenon is known as *corona discharge*.

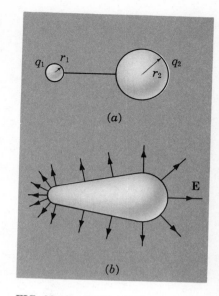

FIG. 23-13. *(a)* Two spheres connected by a long conducting wire. *(b)* A charged conductor. Note that the electric field is strong at points of small radius of curvature.

Example 23-7

In the Van de Graaff electrostatic generator, an important type of particle accelerator, a moving insulated charged belt is brought inside a large spherical conducting shell. The charge is removed from the belt by conducting needles connected to the sphere, as shown in Fig. 23-14. The electric potential of the sphere relative to ground can reach *millions* of volts, and this high potential can be used to accelerate charged particles to relatively high energies (up to 10 MeV per charge e). *(a)* Why a *sphere*? *(b)* Why a *large* sphere? *(c)* Why is the charge introduced through the *inside* of the sphere?

(a) The sphere is that closed shape which, for a given volume, has the smoothest surface. Corona discharge is minimized by using this shape.

(b) Corona discharge eventually limits the charge that can be accumulated on the sphere. Let E_m represent that electric field at the spherical surface for which corona discharge first occurs. The electric field at the surface of a sphere of radius R is $E = kq/R^2$, and the potential there is $V = kq/R$. Therefore, $E_m = V/R$. For a given

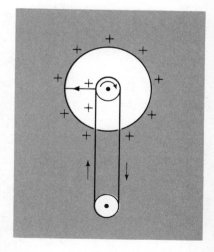

FIG. 23-14. Schematic diagram of a Van de Graaff generator.

Electric Potential and Conductors SEC. 23-6

high potential V, the least electric field at the surface is achieved through the use of a large R, hence a large sphere. For example, for air at STP, $E_m = 3 \times 10^6$ V/m. To obtain a potential V of say, 3×10^6 V would require a sphere of radius $R = V/E_m = (3 \times 10^6 \text{ V})/(3 \times 10^6 \text{ V/m}) = 1$ m. The charge on the 1-m sphere would then be more than 300 μC.

(c) The charge is brought to the hollow sphere through the inside because, once there, the charge is *not* repelled by charge already on the outside of the sphere. All charge coming in contact with the interior surface then goes at once to the outside.

The Van de Graaff generator is used to accelerate charged particles to high energies by having the particles travel across a very high potential difference. Corona discharge is suppressed by surrounding the spherical shell by gas under high pressure, and the sphere may then be raised to a potential of up to 10 million V above ground. Protons or other singly charged particles, accelerated from rest across this potential difference, acquire a kinetic energy of 10 MeV. These highly energetic particles may then be used in nuclear experiments in which the beam of particles is directed at a target; see Fig. 23-15a.

The maximum energy of particles accelerated by a Van de Graaff electrostatic generator can be *doubled* through a remarkably simple procedure. Singly charged *negative* hydrogen ions (H⁻) are first accelerated, for example, from ground potential to the positive potential of the sphere; then the two electrons are stripped from the highly energetic hydrogen ions to produce *positive* protons, which are accelerated a second time, now from the high positive potential to ground. Such a device is called a tandem Van de Graaff accelerator; see Fig. 23-15b.

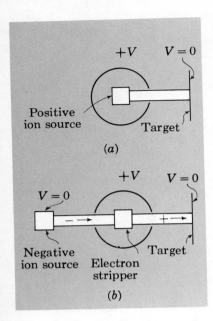

FIG. 23-15. (a) A Van de Graaff generator accelerates positive ions from a high potential to ground potential. (b) A tandem Van de Graaff accelerates particles twice with a single high potential by changing the charge of the particles. Negative ions are first accelerated from ground potential to a high positive potential; their charge is changed in sign at the electron stripper; then the positive ions are accelerated again from the high potential to ground potential.

23-7 CAPACITANCE DEFINED

To charge a conductor requires work. For example, after the first electron has been placed on a spherical conducting shell, other electrons brought to the conductor are repelled. Then, after the conductor has acquired a total charge Q, it can be thought of not only as a device which stores this charge but also as one which establishes an electric field around it; that is, it has an electric potential V representing the storage of electric potential energy.

We have seen (Example 23-5) that, when a spherical conductor of radius R acquires a charge Q, its electric potential V (relative to zero potential at infinity) is given by

$$V = \frac{kQ}{R} \qquad (23\text{-}18)$$

For this simple geometrical configuration the conductor's potential is directly proportional to the charge on it and inversely proportional to the conductor's radius. This behavior is true for conductors of other shapes: The stored charge is always proportional to the potential and also to a characteristic dimension of the conductor.

Of course, if one places a negative charge on one conductor, some other object must acquire a positive charge of equal magnitude. In what follows we shall always be concerned with two conductors (plates) carrying opposite charges of equal magnitude Q and having a potential difference V between them, as shown in Fig. 23-16. Since all points on the surface and interior of any one conductor, charged or uncharged, are at a single potential, V represents the potential difference between *any* point on one conductor and any point on the other conductor. Such a pair of oppositely charged conductors is said to constitute a *capacitor*. The electrical property one ascribes to a capacitor is *capacitance C*, where, by definition,

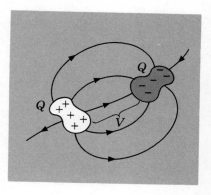

FIG. 23-16. Most general form of capacitor: two conductors carrying equal but opposite charges of magnitude Q and having an electrical potential difference V between them.

$$C = \frac{Q}{V} \qquad (23\text{-}19)$$

the capacitance being the charge magnitude (on *either* of the two oppositely charged conductors) per unit potential difference between the conductors. The capacitance C of a given configuration of conductors is a constant, independent of Q and V. Thus, a capacitor's capacitance is a quantitative measure of its capacity for storing separated charges and thereby for storing electric potential energy. A capacitor is symbolized in electric circuit diagrams by ─┤├─ or by ─┤(─.

From (23-19) we see that capacitance has the units of coulombs per volt. A special name, the *farad* (abbreviated F) — after the famed experimentalist in electromagnetism, Michael Faraday — is given to this ratio:

$$1 \text{ F} = 1 \text{ C/V}$$

As we shall see shortly, a capacitance of one farad is enormous. More commonly used units for capacitance are the microfarad (1 μF = 10^{-6} F) and the micromicrofarad, or picofarad (1 $\mu\mu$F, or pF, = 10^{-12} F).

23–8 CAPACITANCE FOR SOME SIMPLE CONFIGURATIONS

Any two oppositely charged conductors comprise a capacitor. A practical capacitor, however, is small, has its oppositely charged plates easily insulated from one another, and is so arranged that external fields will not disturb the distribution of charge on the plates. These requirements are met by capacitors with parallel plates or with coaxial cylindrical conductors, and the symmetry of these geometries allows the capacitance to be computed easily.

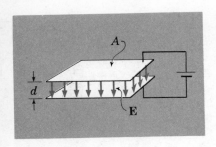

FIG. 23-17. A parallel-plate capacitor with plate area A and plate separation distance d.

Parallel-plate Capacitor. A capacitor with two parallel plates, each of area A and separated by a distance d, is shown in Fig. 23-17. The plate separation d is small compared with the size of the plates, so that the electric field **E** is uniform and confined almost entirely to the region between the plates, fringing of the electric-force lines at the boundaries being negligible. The two plates must be held apart by an insulating material, not only to prevent charges from going from one plate to the other but also to hold apart the oppositely charged plates under the action of the attractive force between them. The insulator may be a dielectric material sandwiched between the plates, in which case the capacitance is enhanced over its value when a vacuum exists between the plates. In what follows here we shall assume the plates to be immersed in a vacuum.

One may charge a capacitor by transferring electrons from one plate to the other against the opposing electric force produced by the field between the plates. Alternatively, one may charge a capacitor very simply by connecting the two capacitor plates with conducting wires to the opposite terminals of a battery. Since the potential difference between the terminals of a battery is constant, the capacitor plates acquire the same potential difference V as that of the battery. The capacitance C of a parallel-plate capacitor is easily found. From the definition of C [Eq. (23-19)], we must know the charge Q in terms of the potential difference V. The link is the electric field **E**. Since the field is uniform over the plate separation distance d,

$$E = \frac{V}{d} \qquad (23\text{-}13)$$

In Example 22-4 Gauss' law was used to show that the electric field near the surface of any conductor is related to the surface charge density by

$$E = \frac{\sigma}{\varepsilon_0} \qquad (23\text{-}20)$$

For a parallel-plate capacitor (see Fig. 23-18) σ is constant over the area A of each plate and is related to the charge Q by

$$\sigma = \frac{Q}{A} \qquad (23\text{-}21)$$

Using Eqs. (23-13), (23-20), and (23-21) in (23-19), we obtain

$$C = \frac{Q}{V} = \frac{Q}{Ed} = \frac{\varepsilon_0 Q}{\sigma d} = \frac{\varepsilon_0 A Q}{Q d} = \frac{\varepsilon_0 A}{d} \qquad (23\text{-}22)$$

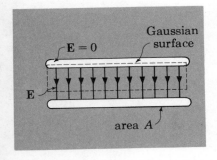

FIG. 23-18. Gauss' law applied to a parallel-plate capacitor. The cross section of the gaussian surface is shown with dashed lines.

As (23-22) shows, the capacitance is (1) proportional to the plate area and (2) inversely proportional to the plate separation. Both make sense on the basis of simple physical arguments: (1) the larger the plate area (for a given V and d) the larger the amount of charge which can be stored on the plates; (2) when Q and A are fixed (and E thereby constant), bringing the plates closer together reduces the distance over which this field exists and hence reduces V.

That the farad represents, in fact, an enormous capacitance follows directly from (23-22). Suppose that we have two large plates, each with an area of 1.0 m², separated by 1.0 mm = 10^{-3} m. Then, $C = \varepsilon_0 A/d = (8.85 \times 10^{-12} \text{ C}^2/\text{N-m}^2)(1.0 \text{ m}^2)/(10^{-3} \text{ m}) = 8.85 \times 10^{-9}$ F, or somewhat less than $\frac{1}{100}$ μF. With this separation distance one would need a plate area of 10^8 m² to have a capacitance of 1 F.

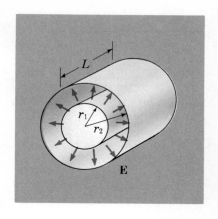

FIG. 23-19. A coaxial cylindrical capacitor.

Coaxial Capacitor. A coaxial capacitor consisting of two concentric cylindrical conductors of radii r_1 and r_2 is shown in Fig. 23-19. The two cylindrical surfaces are equipotential surfaces between which there exists a potential difference, $V_1 - V_2$. We recall another arrangement, in which the equipotential surfaces were coaxial cylinders: that of an infinitely long charged wire having a linear charge density $\lambda = Q/L$, where Q is the magnitude of the charge on the conductor of length L. We found earlier that

$$V_1 - V_2 = \frac{\lambda}{2\pi\varepsilon_0} \ln \frac{r_2}{r_1} = Q \left(\frac{1}{2\pi\varepsilon_0 L} \right) \ln \frac{r_2}{r_1} \qquad (23\text{-}14)$$

This formula applies, now, to the coaxial capacitor. The capacitance is then given by

$$C = \frac{Q}{V_1 - V_2} = \frac{2\pi\varepsilon_0 L}{\ln (r_2/r_1)} \qquad (23\text{-}23)$$

The radius r_2 is always to be taken as that of the outer conductor, to ensure that C in (23-23) is positive.

The capacitance of a coaxial capacitor can also be derived directly through Gauss' law. The reader is urged to carry out such an exercise.

Dielectrics. By definition, the *dielectric constant* κ of an insulating material is the capacitance C_d of a capacitor filled with dielectric divided by the corresponding capacitance C when the capacitor plates are immersed in vacuum:

$$\kappa = \frac{C_d}{C} \qquad (23\text{-}24)$$

Inserting a dielectric into a capacitor increases the capacitance†;

---

†When an insulating material is introduced between the plates of a capacitor with fixed charges on its two plates, the electric field between the plates arising from the charge on the plates causes the electric charges within the insulating material to be redistributed in such a way that electric charges of opposite sign appear on the two faces of the dielectric material. These dielectric charges in turn produce an electric field which is opposite to and partially annuls the

Table 23-1

Material	Dielectric constant κ
Air (1 atm)	1.00059
Air (100 atm)	1.0548
Pyrex glass	5.6
Quartz	3.8
Paraffined paper	2
Mica	5
Barium titanate	1200

clearly, κ is always greater than 1.00. Table 23-1 lists dielectric constants at room temperature for several common materials. By definition, the dielectric constant for a vacuum is exactly 1. Note that κ for air is very close to 1.

The *electrical permittivity of an insulating material*, designated by ε, is related to κ and ε_0 by the equation

$$\varepsilon = \kappa \varepsilon_0 \tag{23-25}$$

When $\kappa = 1$, $\varepsilon = \varepsilon_0$, the permittivity of the vacuum.

It is an easy matter to generalize the formulas giving the capacitances for various configurations, whatever the dielectric material between the plates: Replace ε_0, wherever it appears, by $\kappa \varepsilon_0 = \varepsilon$. Thus, the capacitance of a dielectric-filled parallel-plate capacitor is, from (23-22),

$$C = \frac{\kappa \varepsilon_0 A}{d} = \frac{\varepsilon A}{d} \tag{23-26}$$

Dielectric materials serve useful functions in actual capacitors: They insulate the plates from one another, and they enhance the capacitance by a factor κ. Thus, an ordinary "paper condenser" is formed by sandwiching thin paper between two metallic foils and rolling an essentially parallel-plate capacitor into cylindrical shape.

23-9 CAPACITOR CIRCUITS

Electric circuits comprised entirely of capacitors are neither very interesting nor useful. However, they are worthy of attention because they illustrate some fundamental concepts common to all circuits. Furthermore, one may use the rules for combining capacitors in series and in parallel combinations to advantage in computing the capacitance of capacitors filled entirely or partially with dielectric materials.

Consider first the simple circuit shown in Fig. 23-20. In traversing a simple closed path around the circuit (as indicated by the white arrow), the potential rise across the battery is—of necessity—of the same magnitude as the potential drop across the capacitor.

Capacitors in Series. Suppose that several capacitors C_1, C_2, C_3 are connected in series and attached across battery terminals as

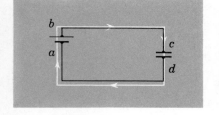

FIG. 23-20. Circuit consisting of a battery and capacitor, with a loop *abcda* around the circuit passing along the conducting wires.

---

electric field produced by the charges on the two conducting plates. Consequently, the net electric field **E** between the two plates is reduced, as is the associated potential difference $V = Ed$, where d is the plate separation distance. Therefore, for fixed capacitor charge Q, the effect of the dielectric is to *decrease* the potential difference V. Since by definition $C = Q/V = Q/Ed$, the overall effect of introducing a dielectric material into a capacitor is, then, that of *enhancing* its capacitance over the corresponding value obtained with a vacuum between the capacitor plates.

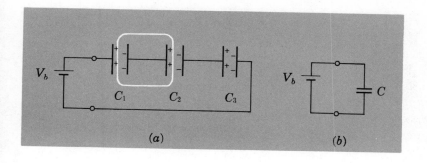

FIG. 23-21. (a) Capacitors in series; the net charge with the white loop is zero. (b) The equivalent single capacitor.

shown in Fig. 23-21a. We wish to find the value C of the single capacitor which is equivalent to this group of capacitors.

We see from Fig. 23-21a that the *net* charge on the single conductor shown within the white loop must remain zero, since this section is isolated electrically from everything else. The battery serves only to separate the charges on this conductor, equal amounts of positive and negative charge appearing on the plates of adjoining conductors. Thus, all capacitors in series have the *same charge:*

$$Q_1 = Q_2 = Q_3$$

Choosing the route along the wires connecting the capacitors, we see that

$$V_b = V_1 + V_2 + V_3 \qquad (23\text{-}27)$$

where V_1, V_2, and V_3 are the respective potential drops across the capacitors. By definition, $V_1 = Q_1/C_1$; similarly for the other two capacitors. Using these relations in (23-27), we have

$$V_b = \frac{Q_1}{C_1} + \frac{Q_2}{C_2} + \frac{Q_3}{C_3} = Q\left(\frac{1}{C_1} + \frac{1}{C_2} + \frac{1}{C_3}\right)$$

where we have used Q to represent $Q_1 = Q_2 = Q_3$. Now, if a single equivalent capacitor C, as shown in Fig. 21b, were connected to the battery terminals, its effect on the rest of the circuit would be the same as that of the capacitors in series. That is, its potential difference, $V = Q/C$, would equal V_b. Therefore,

$$\frac{Q}{C} = Q\left(\frac{1}{C_1} + \frac{1}{C_2} + \frac{1}{C_3}\right)$$

$$\frac{1}{C} = \frac{1}{C_1} + \frac{1}{C_2} + \frac{1}{C_3} \qquad (23\text{-}28)$$

In general, for capacitors in series the reciprocal of the single equivalent capacitance is the sum of the reciprocals of the individual capacitances. As a consequence, the equivalent capacitance is always less than the smallest of the series capacitors. For example, with four capacitors, each 20 μF and connected in series, the equivalent capacitance is 5 μF.

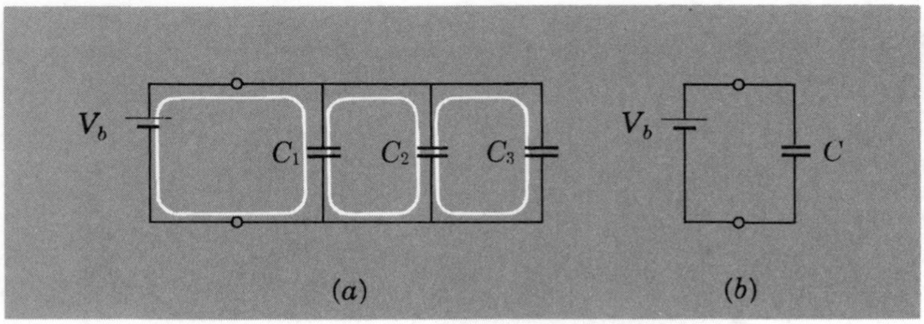

FIG. 23-22. (a) Capacitors in parallel; the net potential difference around any white loop is zero. (b) The equivalent single capacitor.

Capacitors in Parallel. The circuit arrangement for capacitors connected in parallel is shown in Fig. 23-22a. What single capacitor, as in Fig. 23-22b, is equivalent to this group of capacitors? The charges and potential differences for C_1, C_2, and C_3 are again designated, respectively, Q_1, Q_2, Q_3 and V_1, V_2, V_3. By taking potential differences along the closed loops shown by white lines in the figure, we establish that the potential differences across each capacitor is the same as that across the battery terminals. *Parallel capacitors have the same potential difference;* that is,

$$V_b = V_1 = V_2 = V_3$$

Since the circuit contains only two conductors attached to the battery terminals (the upper plates with their connecting wires and the lower plates with theirs), the total charge Q held on a single capacitor equivalent to those in parallel is given by

$$Q = Q_1 + Q_2 + Q_3$$

or $$Q = C_1 V_1 + C_2 V_2 + C_3 V_3 = V_b (C_1 + C_2 + C_3)$$

But $$C = \frac{Q}{V_b}$$

Therefore, $$C = C_1 + C_2 + C_3 \qquad (23\text{-}29)$$

In general, for parallel capacitors the equivalent capacitance is the sum of their several capacitances. The equivalent capacity always exceeds that of the largest capacitor in parallel.

Example 23-8

A 1.0-μF capacitor and a 3.0-μF capacitor are connected in parallel to a 100-V battery. The battery is then removed and the two charged capacitors are connected as shown in Fig. 23-23a. What is the charge on each capacitor (a) before the switch is closed and (b) after the switch is closed?

(a) We have immediately that

$$Q_1 = C_1 V = (1.0 \times 10^{-6} \text{ F})(100 \text{ V}) = 1.0 \times 10^{-4} \text{ C}$$
$$Q_3 = C_3 V = (3.0 \times 10^{-6} \text{ F})(100 \text{ V}) = 3.0 \times 10^{-4} \text{ C}$$
$\qquad (23\text{-}19)$

Note that by taking potential drops around the entire loop of Fig.

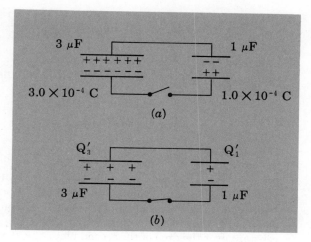

FIG. 23-23

23-23a we can establish that the potential difference across the open switch is 200 V.

(b) After the switch has been closed, as shown in Fig. 23-23b, there is a redistribution of charge on the plates. The *net* charge on the upper two plates must be, by the law of conservation of charge, $(3.0 \times 10^{-4}) + (-1.0 \times 10^{-4}) = +2.0 \times 10^{-4}$ C. After the switch has been closed, the potential differences across the two capacitors are the same:

$$\frac{Q'_1}{C_1} = \frac{Q'_3}{C_3} \quad \text{or} \quad \frac{Q'_1}{1\ \mu\text{F}} = \frac{Q'_3}{3\ \mu\text{F}}$$

where
$$Q'_1 + Q'_3 = +2.0 \times 10^{-4}\ \text{C}$$

Q'_1 and Q'_3 representing the charges on the respective capacitors after the switch has been closed. Solving the two equations above we find

$$Q'_1 = 0.50 \times 10^{-4}\ \text{C} \quad \text{and} \quad Q'_3 = 1.5 \times 10^{-4}\ \text{C}$$

Moreover, the common potential difference is now

$$V'_3 = V'_1 = \frac{Q'_1}{C_1} = \frac{0.50 \times 10^{-4}\ \text{C}}{1.0 \times 10^{-6}\ \text{F}} = 50\ \text{V}$$

23-10 ENERGY OF A CHARGED CAPACITOR AND OF THE ELECTRIC FIELD

Work is required to charge a capacitor because, after the first small charge is placed on either plate, this charge repels other charges of like sign that are subsequently added to the plate. We wish to find the total work U_e required to charge the capacitor to a final potential difference V_f with a final charge of magnitude Q on each plate. The potential difference between plates changes from zero to V_f as the capacitor is being charged; therefore, we must integrate from the initial zero charge to the final charge Q:

$$U_e = \int_0^Q V\,dq = \int_0^Q \frac{q}{C}\,dq = \frac{Q^2}{2C}$$

We may write U_e, which represents the electric potential energy of the charges on the capacitor plates, in equivalent forms by using $C = Q/V_f$:

$$U_e = \frac{Q^2}{2C} = \tfrac{1}{2}CV_f^2 = \tfrac{1}{2}QV_f \qquad (23\text{-}30)$$

Example 23-9

What is the attractive force between the two oppositely charged plates of a charged parallel-plate capacitor?

The force F is easily found by recalling that the electric field of *one* plate at the site of the other is $E = Q/2\varepsilon A$ [Eq. (21-6)]. Since **E** is constant over the area of the second plate, the attractive force on this plate is

$$F = QE = \frac{Q^2}{2\varepsilon A}$$

Substituting $Q = CV = (\varepsilon A/d)V$ into the above equation gives

$$F = \tfrac{1}{2}\varepsilon A \left(\frac{V}{d}\right)^2 \qquad (23\text{-}31)$$

Equation (23-31) provides a method of measuring the potential difference between parallel plates; one simply measures A, d, and the external force necessary to hold the plates at the separation distance d. A device which measures potential difference is called an *electrometer* and may consist simply of an ordinary balance so arranged that a weight on one side is balanced by the electric force on the capacitor's upper plate suspended on the other side of the balance.

A charged capacitor has electric potential energy because work is required to assemble the charges at the plates. We may say the same thing differently: A charged capacitor has energy because an electric field has been established in the region between the plates. That is, instead of attributing the potential energy to the relative positions of positive and negative charges, we may alternatively and equivalently ascribe the energy to the electric field itself. Using (23-26) and (23-13) for a parallel-plate capacitor, we have

$$U_e = \tfrac{1}{2}CV_f^2 = \frac{1}{2}\left(\frac{\varepsilon A}{d}\right)(Ed)^2 = \tfrac{1}{2}\varepsilon E^2 (Ad)$$

Neglecting fringing, the uniform field is confined to the volume Ad between the plates. Therefore, the electric energy per unit volume, the *electric energy density* u_e, is

$$u_e = \frac{U_e}{Ad} = \tfrac{1}{2}\varepsilon E^2 \qquad (23\text{-}32)$$

The energy density is proportional to the *square* of the electric field. Although derived for the special case of a parallel-plate capacitor, (23-32) is altogether general. It gives the electric energy density for *any* electric field. That one may properly associate energy with the electric field itself is justified by the fact that the fields for electromagnetic waves are found to exist in space *without* being attached to electric charges.

SUMMARY

The electric potential energy U of two point-charges separated by the distance r is

$$U = \frac{kq_1q_2}{r} \qquad (23\text{-}3)$$

For more than two point-charges the total electric potential energy of the system, which represents the total work required to assemble the charges in the system, is the sum of the potential energies between pairs of point-charges.

The electric potential V at any point P is equal to the total work W required to move a unit positive charge from infinity to P while all other charges remain fixed in position:

$$V = \frac{W}{q} \qquad (23\text{-}8)$$

Electric potential and electric potential difference are measured in volts. The electron volt (1 eV = 1.60×10^{-19} J) is the energy acquired by a particle of charge e which falls across a potential difference of 1 V.

The electric potential V arising from a collection of point-charges is given by

$$V = k \sum \frac{q_i}{r_i} \qquad (23\text{-}10)$$

Knowing the electric field $\mathbf{E}$ as a function of position, one can derive the potential difference between two points according to the relation

$$V_2 - V_1 = -\int_1^2 \mathbf{E} \cdot d\mathbf{s} \qquad (23\text{-}11)$$

Conversely, knowing the electric potential as a function of position, one can derive the electric field by using the relation

$$E = -\frac{dV}{ds} \qquad (23\text{-}15)$$

where the displacement $d\mathbf{s}$ is taken along that direction in which

V changes most rapidly with position. For a *uniform* electric field,

$$E = -\frac{V}{d} \qquad (23\text{-}13)$$

The capacitance C of a capacitor is given by

$$C = \frac{Q}{V} \qquad (23\text{-}19)$$

where Q is the charge magnitude of either of two oppositely charged conductors and V is the potential difference between them. The unit of capacitance is the farad.

The capacitance of a parallel-plate capacitor of plate area A and plate separation distance d is

$$C = \frac{\varepsilon A}{d} \qquad (23\text{-}26)$$

Capacitors in series all have the same charge; the single equivalent capacitance C is given by $1/C = \Sigma 1/C_i$. Capacitors in parallel have a common potential difference; the equivalent capacitance is $C = \Sigma C_i$.

The dielectric constant κ of a dielectric material is defined as

$$\kappa = \frac{C_d}{C} = \frac{\varepsilon}{\varepsilon_0} \qquad (23\text{-}24), (23\text{-}25)$$

where C_d and C are the respective capacitances of a capacitor filled with, and empty of, the dielectric material.

The energy of a charged capacitor is

$$U_e = \frac{Q^2}{2C} = \tfrac{1}{2}CV^2 = \tfrac{1}{2}QV \qquad (23\text{-}30)$$

The energy density of an electric field is

$$u_e = \tfrac{1}{2}\varepsilon E^2 \qquad (23\text{-}32)$$

PROBLEMS

23-1. Initially two protons are stationary and separated by a distance $d = 1.0 \times 10^{-10}$ m. Subsequently the particles are released from their original positions and are then free to move in reponse to the coulomb force between them. Ignoring the gravitational force, find (a) the speed and (b) the kinetic energy of either of these protons when they have separated to a distance of 10 $d = 1.0 \times 10^{-9}$ m.

23-2. An α particle (helium nucleus with electric charge $+2e$) moving initially with kinetic energy 8.0 MeV approaches a gold nucleus (charge, $+79e$) head-on and momentarily comes to rest before being scattered back along its initial path. (a) What is the electric potential energy (in MeV) of the α particle-gold nucleus system when the α particle is at rest? (The gold nucleus is much more massive than the α particle and can be taken to remain at rest.) (b) What is the minimum separation distance between the α particle and the gold nucleus? (c) With what initial kinetic energy would the α particle have to move in order to approach the gold nucleus within half of the distance of part (b)?

23-3. Three positive point charges each of magnitude Q are to be placed at the corners of an equilateral triangle with edge length L. How much work must be done (a) to place the first charge on a corner, (b) to add the third charge to the third corner (with the first and second charges already fixed in place)? (c) What is the total work required to assemble the system of three charges? (d) What is this system's total electric potential energy?

23-4. (a) At what distance from a 1.0-C point-charge is the potential 1.0 V? (b) What is the magnitude of the electric field at this location?

23-5. What electric potential difference will bring to rest a copper ion Cu^{2+} (with electric charge $+2e$) initially having a kinetic energy of 12 eV?

23-6. Determine the energy released (in MeV) in the fission of the $^{238}_{92}U$ nucleus when the nuclear fragments are $^{146}_{56}Ba$ and $^{90}_{36}Kr$, initially separated by 1.4×10^{-14} m. The pre-subscript to the chemical symbol gives the electric charge of the nucleus in units of the electronic charge e; the pre-superscript gives the total number of protons and neutrons.

23-7. The electric potential at point A is $+10$ V, at point B at $+15$ V, and at point C -12 V. How much work (sign and magnitude in units of eV) must be done to move a He^{2+} ion (a helium nucleus, with charge $+2e$) at constant speed: (a) from A to B to C; (b) from A to C to B; (c) from A to C to B to A; and (d) from C to B to A?

23-8. A charge of 2.0×10^{-2} μC is located at $x = 0$; a charge of 1.0×10^{-2} μC is located at $x = 10$ cm. What is the minimum speed with which an electron must be projected to the right from $x = 2.0$ cm to reach the point $x = 8.0$ cm?

23-9. An electron and a proton are both accelerated from rest by the same magnitude of electric potential difference. The proton mass is 1836 times that of the proton. What is the ratio of the electron-to-proton (a) kinetic energy, (b) momentum, and (c) speed?

23-10. Three particles of equal mass and having charges of $+Q$, $+Q$, and $-2Q$ are located at the corners of an equilateral triangle of edge length L. (a) How much work must be done to completely separate the three particles from one another? (b) If the three particles are held fixed at the corners of the equilateral triangle, how much work must be done in bringing a charge $-q$ from infinity to the center of the triangle?

23-11. (a) Graph the potential energy between a point-charge q and two fixed point-charges, each of charge q, as a function of position along the line joining the two fixed charges. (b) Show that for small displacements from the midpoint the potential energy between the charge q and the two fixed charges is parabolic. (c) Is the charge q in stable or unstable equilibrium with respect to a small displacement along any direction?

23-12. Consider the following electric charge configuration: $+10$ μC at $x = -10$ cm, $y = 0$; -3.0 μC at $x = 0$, $y = +6.0$ cm; and -5.0 μC at $x = +10$ cm, $y = 0$. Sketch the equipotential lines in the XY plane for (a) $V = +3.0$ MV, (b) $V = -3.0$ MV, and (c) $V = 0$.

23-13. The stopping potential (the electric potential difference which brings changed particles to rest) for a beam of electrons is found to be 100 V. What is the initial speed of the electrons?

23-14. In a tandem Van de Graaff accelerator a negative hydrogen ion is accelerated from rest through a potential difference of $+10$ MV. Then the two electrons are stripped from the ion and the remaining proton is accelerated through a potential difference of -10 MV. What are (a) the final kinetic energy (MeV) and (b) speed of the proton? The proton's mass is 1.7×10^{-27} kg.

23-15. What is the electrostatic potential energy (eV) between two protons separated by (a) an atomic distance, 1.0×10^{-10} m? (b) A nuclear distance, 1.0×10^{-15} m?

23-16. A Geiger counter consists of two coaxial cylindrical conductors insulated from one another. Assume the outer conductor is a metal cylinder of radius 1.0 cm and the inner conductor is a thin wire of radius 0.01 cm. A voltage of 800 V is applied across the inner and outer conductors. (a) What is the electric field at the surface of the wire? (b) What is the electric field at the inner surface of the outer metal cylinder? (c) Find the linear charge density along the wire.

23-17. One device used in reducing air pollution is the electrostatic precipitator, which consists of a thin negatively charged wire extending down a large cylindrical positively charged vertical duct. Assume the diameter of the duct and wire to be 30 and 0.20 cm, respectively; a potential difference of 100,000 V is applied across the two. (a) Find the electric field in the region near the wire. This field is strong enough to produce a corona discharge near the wire, electrons being pulled off the wire. Most of these freed electrons then attach themselves to the waste particles moving upward through the duct; these negatively charged particles then move outward and can be collected off the walls of the duct. (b) As the flue gas moves upward through the precipitator and the pollution particles are removed at the outer duct wall, energy must be provided to keep the process going. Why?

23-18. Calculate the work necessary to charge a conducting sphere of radius R to a total charge Q.

23-19. Consider the electrostatic generator in Example 23-7. If the radius of the hollow sphere is 50 cm and the sphere is to be charged to a potential of 1.0 MV relative to ground: *(a)* How much work must be done against the electric field of the sphere in bringing a small charge Δq from the lower end of the belt in Fig. 25-14 to the inside of the sphere, assuming the sphere has a total charge q at this time? *(b)* How much work must be done in charging the sphere to a potential of 1.0 MV? *(c)* What is the total mass of electrons transferred from the sphere to ground in charging the sphere?

23-20. Consider charges q and $-q$ which are fixed in position a distance $2a$ apart; such a charge pair is called an *electric dipole,* with *dipole moment* $p = q(2a)$. See Fig. P23-20. (The electric dipole is also treated in Problems 21-26 and 23-21.) Show that, for distances which are very large compared to the charge separation, the electric potential due to the dipole is given by $V = (kp/r^2)\cos\theta$, where k is the coulomb force constant, p is the dipole moment, r the distance from the midpoint of the dipole, and θ is the angle measured with respect to the dipole axis, which is considered as being directed from $-q$ to $+q$. (*Hint:* In the limit for large distances, the lines from either charge to a field point are essentially parallel to the line from the dipole center to the field point.)

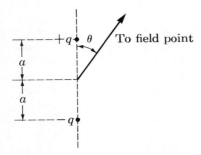

FIG. P23-20

23-21. Charges q and $-q$ are fixed in position on the X axis with coordinates $(a, 0)$ and $(-a, 0)$, respectively. See Fig. P23-21. *(a)* Show that the electric potential V for points on the X axis is given as follows:

$$x > a: \quad V = \frac{kp}{a}\left[\frac{1}{\rho^2 - 1}\right]$$

$$-a < x < a: \quad V = \frac{kp}{a}\left[\frac{\rho}{1 - \rho^2}\right]$$

$$x < -a: \quad V = \frac{kp}{a}\left[\frac{1}{1 - \rho^2}\right]$$

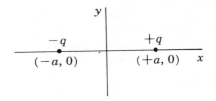

FIG. P23-21

where $\rho = x/a$ and $p = q(2a)$. (The quantity p is called the dipole moment.) *(b)* Sketch V as a function of x for $-\infty < x < +\infty$.

23-22. One simple electrical model of the earth and its atmosphere is that in which the system is considered to be a spherical capacitor: The solid earth is the inner conductor with radius 6.4×10^6 m; the highly conductive atmospheric region approximately 50 km above the earth's surface is the outer conducting shell. *(a)* Find the capacitance of this system. *(b)* If the electric potential of the conducting shell is 400 kV above that of the solid earth, find the surface charge density σ at the earth's surface.

23-23. *(a)* Show that the capacitance of two concentric spherical conductors of radii r and $r + d$ with a vacuum between is $C = r(r + d)/kd$. *(b)* Show that this reduces to the capacitance of a parallel-plate capacitor for $d \ll r$.

23-24. A coaxial cable has inner and outer conductors of 1.0- and 2.0-mm radius, respectively. The region between the conductors is filled with a glass dielectric having a dielectric constant 5.0 and dielectric strength 10 MV/m. *(a)* Find the capacitance per meter of this cable. *(b)* What is the maximum potential difference that can be applied across the two conductors?

23-25. *(a)* Show that the capacitance of an isolated sphere of radius r immersed in a dielectric is given by $C = 4\pi\varepsilon r = \kappa r/k$, where ε, κ, and k are, respectively, the permittivity of the dielectric, the dielectric constant, and the coulomb force constant. (*Hint:* Consider that the other capacitor plate is at infinity.) *(b)* On this basis calculate the capacitance of the earth, given that the earth's radius is 6.4×10^6 m.

23-26. Find the minimum area of a 1000 picofarad parallel-plate capacitor rated at 1.0×10^4 volts if the dielectric is to be paraffin ($\kappa = 2.0$). Paraffin paper will break down permanently if the potential gradient across it exceeds 4.0×10^7 V/m.

23-27. Find the capacitance of a parallel-plate capacitor which is filled with two dielectric slabs of equal thickness, as shown in Fig. P23-27.

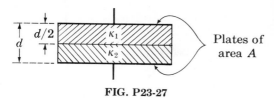

FIG. P23-27

23-28. Show that in a parallel-plate capacitor filled with a dielectric with dielectric constant κ, the area charge density σ_κ on the dielectric surface adjacent to the positive terminal is given by $\sigma_\kappa = -(\kappa - 1)\sigma$, where σ is the area charge density on the positive plate of the capacitor.

23-29. You are given four capacitors each of 12 pF. (a) How should the capacitors be connected to yield the smallest equivalent capacitance? (b) What is this capacitance?

23-30. A 20 μF capacitor is connected across the terminals of a 12-V battery. (a) What is the electric charge on either capacitor plate? (b) What is the energy of the charged capacitor?

23-31. An unlimited number of 5.0-μF capacitors are each rated for breakdown whenever the applied potential difference exceeds 500 V. Using the minimum number of these capacitors, find the arrangement that will provide (a) a 5.0-μF capacitor with applied voltages up to 1,000 V, (b) a 2.0-μF capacitor with applied voltages up to 1,000 V.

23-32. Two capacitors, one of 1.0 μF and the other of 2.0 μF, are each charged initially by being connected to a 10-V battery. Then the two capacitors are connected together. What is the total electric energy stored in each capacitor, if the capacitors are connected such that (a) plates of like charge are brought together and (b) plates of opposite charge are brought together? (c) Account for the lost energy in part (a).

23-33. Given three capacitors each of 10.0 pF and a 1.50 V battery, (a) how should the capacitors be connected in a circuit to yield the largest possible energy stored in the capacitors and (b) what is the magnitude of this largest possible energy?

23-34. A conducting sphere of radius R carries an electric charge Q. (a) What is the electric energy density at a point just outside the conducting sphere? (b) How does the electric energy density vary with distance r from the center of the sphere (for $r > R$)?

23-35. (a) Show that the force of attraction between the plates of a parallel-plate capacitor (in vacuum) is given by $F = Q^2/2\varepsilon_0 A$, where Q is the charge on the capacitor, ε_0 is the permittivity of free space, and A is the plate area. (b) Explain why this force is *not* given by $QE = Q^2/\varepsilon_0 A$, where E is the magnitude of the electric field in the region between the plates. (*Note:* This result is the basis of Kelvin's absolute electrometer, in which a charge Q can be determined by measuring F.)

CHAPTER 24

Electric Current and Resistance

This chapter is concerned with the electric currents arising from electric charges in motion. Electric current is defined and related to the properties of electric charge carriers. The fundamental laws of electric charge conservation and energy conservation are applied to electric circuits, and the general relation giving the electric power delivered by an energy source to a load is derived. Finally, we give a very brief and qualitative treatment of the characteristics which determine the resistivity of any material.

24–1 ELECTRIC CURRENT

Electric charges in motion constitute an *electric current*. By definition, the current i through a chosen surface is the total net charge dQ passing through that surface divided by the elapsed time dt:

$$i = \frac{dQ}{dt} \qquad (24\text{-}1)$$

The direction of the so-called conventional current is taken to be that in which *positive* charges move. Thus, protons (charge, $+e$) traveling to the right make a current to the right, while electrons (charge, $-e$) traveling to the right make a current to the left. To obtain the total current i, we must always count the *net* charge passing through the chosen surface per unit time. For example, if positive ions move to the right through a surface and at the same time negative ions move to the left through the same surface, as in the case of currents in liquids, *both* types of ions contribute to an electric current to the *right;* see Fig. 24-1.

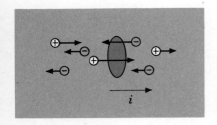

FIG. 24-1. Both the positive charges moving to the right and the negative charges moving to the left contribute to a conventional current i to the right.

In the SI system the unit of current is the *ampere* (abbreviated A) named after André M. Ampère (1775–1836), who made significant contributions to understanding the relation between an electric current and its associated magnetic field. Recall that 1 A corresponds to a net charge of 1 C passing through a chosen surface in the time of 1 s. Whereas the coulomb is an uncommonly large amount of charge by laboratory standards, laboratory currents of a few amperes are typical. Related units are the milliampere (1 mA = 10^{-3} A) and the microampere (1 μA = 10^{-6} A).

A direct current (dc) implies that net positive charges move in one direction only but *not* necessarily at a constant rate. For an alternating current (ac), on the other hand, the current changes direction periodically.

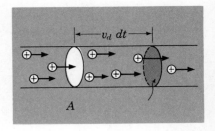

FIG. 24-2. Charges drifting at speed v_d through a cylinder of cross-sectional area A in the time dt.

How is the current i related to the properties of the charge carriers? Suppose that a current consists of n identical charge carriers per unit volume *d*rifting at the speed v_d along a single direction, each carrier having a positive charge q (see Fig. 24-2). Then, in the time dt each charge carrier advances a distance $v_d\,dt$, and all the charge carriers within a cylindrical volume of length $v_d\,dt$ and cross-sectional area A with have passed through one end of the cylinder. By the same token, an equal number of charge carriers will have entered the cylindrical column through the other end. The total charge dQ passing through the area A in time dt is the charge per unit volume nq multiplied by the cylinder's volume $Av_d\,dt$:

$$dQ = nqAv_d\,dt$$

Consequently, the current i contributed by these charge carriers is

$$i = \frac{dQ}{dt} = nqAv_d \qquad (24\text{-}2)$$

If several species of charge carriers, differing in n, q, or v_d, contribute to the current, we may generalize this relation by writing

$$i = A\Sigma n_\alpha q_\alpha v_{d\alpha} \qquad (24\text{-}3)$$

We must emphasize that it need *not* be assumed that each charge carrier moves at a constant velocity. The charged particles may, in fact, have a very complicated motion, as do the free electrons in ordinary conductors. The drift speed v_d is the average displacement per unit time along the direction of the current.

The most common device used for measuring current is the *ammeter* which depends in its operation upon the torque produced by a magnet on a current-carrying coil of wire (Sec. 26-6). A still simpler (but usually not very practical) current-measuring device is based on the electrolysis phenomenon: One measures the current by measuring the mass plated on an electrode in a conducting liquid (an electrolyte).

Example 24-1

A current of 1.0 A exists in a conducting copper wire whose cross-sectional area is 1.0 mm². What is the average drift speed of the conduction electrons under these conditions?

In copper, as in other typical metallic conductors, there is approximately one free (conduction) electron per atom. We can find the density n of charge carriers by computing the number of atoms per unit volume from Avogadro's number N_0 (the number of atoms per kg-mol), the atomic mass w (the number of kilograms per kg-mol), and the mass density ρ_m (the number of kilograms per unit volume). Clearly,

$$n = \frac{\text{charge carriers}}{\text{volume}} = \left(\frac{1 \text{ charge carrier}}{\text{atom}}\right)\left(\frac{\text{atoms}}{\text{volume}}\right)$$

$$= \left(\frac{1 \text{ charge carrier}}{\text{atom}}\right)\left(\frac{\text{atoms}}{\text{kg-mol}}\right)\left(\frac{\text{kg-mol}}{\text{kilogram}}\right)\left(\frac{\text{kilogram}}{\text{volume}}\right)$$

or

$$n = (1)(N_0)\left(\frac{1}{w}\right)(\rho_m)$$

$$= \left(\frac{1 \text{ charge carrier}}{\text{atom}}\right)\left(\frac{6.0 \times 10^{26} \text{ atoms}}{\text{kg-mol}}\right)\left(\frac{1 \text{ kg-mol}}{64 \text{ kg}}\right)(9.0 \times 10^3 \text{ kg/m}^3)$$

$$= 8.4 \times 10^{28} \text{ charge carriers/m}^3$$

From (24-2),

$$v_d = \frac{i}{nqA} = \frac{(1.0 \text{ A})}{(8.4 \times 10^{28} \text{ m}^{-3})(1.6 \times 10^{-19} \text{ C})(1.0 \times 10^{-6} \text{ m}^2)}$$

$$= 7.4 \times 10^{-5} \text{ m/s} = 0.074 \text{ mm/s}$$

The free electrons drift through the conductor with an average speed which is actually less than 0.1 mm/s, a remarkably low speed. It should *not* be inferred, however, that when such a current is established at one end of a conducting copper wire it takes almost 10 s for the signal to travel a mere 1 mm. The speed at which the electric field driving the free electrons is established is close to the speed of light. One must distinguish here between the speed with which the charged particles drift and the speed at which the signal is propagated, just as one must distinguish between the speed (possibly very low) at which a liquid drifts through a pipe and the much higher speed at which a change in pressure is propagated along the pipe. The *drift* speed of conduction electrons is much less than the *random* thermal speeds of electrons at any finite temperature.

24-2 CURRENT AND ENERGY CONSERVATION

Here we set down a fundamental relation concerning the transfer of electric energy to a device through which charges flow. Our arrangement is altogether general: We have some energy source, such as a battery or an electric generator, which maintains a potential difference $V_{ab} = V_a - V_b$ between the terminals of a device, or load, through which a current i passes. The source supplies electric energy to the load; see Fig. 24-3. The load need not be specified in detail (except that it can contain no energy sources). For example, it might be a single conductor, or an electric motor, or a transistor. All that matters is that a current i exists within the device and that a potential difference V_{ab} be somehow produced across its terminals. We imagine V_a to exceed V_b, so that the conventional current is "in" at a and "out" at b (electrons then go "uphill" in potential from b to a).

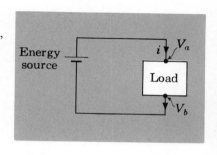

FIG. 24-3. An energy source delivering energy to a load.

If a positive charge dQ moves through the load, the work dW done on this charge by the electric forces associated with the potential difference V_{ab} is

$$dW = (dQ)V_{ab} \qquad (24\text{-}4)$$

Since $i = dQ/dt$,

$$dW = iV_{ab}\,dt \qquad (24\text{-}5)$$

or the instantaneous power P, the rate of doing work, is

$$P = \frac{dW}{dt} = iV_{ab} \qquad (24\text{-}6)$$

The instantaneous power to the load is simply the product of current and potential difference.

Equations (24-4) and (24-5) hold if the current and potential difference are constant or if they vary with time. When i and V_{ab} are constant, P is constant, and the total work W done on the charges transported through the device in time t is

$$W = iV_{ab}t \qquad (24\text{-}7)$$

Otherwise, it is

$$W = \int P\,dt = \int iV\,dt$$

If the current i is in amperes and the potential difference V is in volts, the power P has the units of watts: (A)(V) = (C/s)(J/C) = J/s = W.

Let us be clear on the meaning of (24-6). The delivered power P is simply the rate of energy transfer from the energy source to the load, whatever the details of the source or load. If the source establishes a potential difference across a motor and charges pass through this motor, the power to the motor appears (mostly) as work done on an external system. If the load consists of nothing more than a beam of charged particles accelerated across a potential difference, then the power P is the rate at which these particles gain kinetic energy.

An important case is that in which the load is a conductor or a resistor. Then the energy appears as heat, or thermal energy, which raises the temperature of the conductor or is radiated or conducted away. The power does *not* depend upon the way in which V and i are related. Their relation may, in fact, be quite complicated, as in the case of a vacuum-tube diode or the Vi relation may be simple, as in the case of a homogeneous solid conductor (taken up in the next section).

24–3 RESISTANCE AND OHM'S LAW

It is instructive to plot the relation between the voltage across and the current through various dissipative current elements, that is, any current-carrying device in which electric energy is dissipated and appears as thermal energy. For some elements the V-i plot is complicated, as illustrated in Fig. 24-4a and b.

Figure 24-4a is the voltage-current relation between two electrodes with a vacuum tube or an enclosed tube containing a gas. Another example of a complicated relation between V and i is that of semiconducting devices, such as transistors, shown in Fig. 24-4b.

One important situation in which the relation between V and i of a dissipative circuit element is simple is that of a homogeneous solid electrical conductor (including many poor conductors, or insulators). When a potential difference V is applied between any two points of a conductor of any shape, a current i exists in the conductor, and it is found that a linear relation exists between V and i; see Fig. 24-4c. For a conductor the ratio of V to i, defined as the electrical resistance R, is a *constant* independent of V or of i. This relation, found by Georg S. Ohm (1787–1854) to hold for a large variety of materials and for an enormous range of currents and potential differences, is known as Ohm's law.

Ohm's law: $$R = \frac{V}{i} = \text{a constant} \tag{24-8}$$

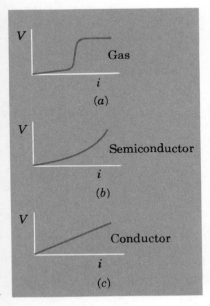

FIG. 24-4. Voltage-current characteristics for *(a)* a gas, *(b)* a semiconducting device, and *(c)* a conductor obeying Ohm's law.

The electrical resistance may be defined as the voltage-current ratio for any dissipative circuit element, for example, for any one of the three situations of Fig. 24-4. Ohm's law is the experimental finding that this ratio is a constant for certain materials, leading to the linear relation shown in Fig. 24-4c.

The SI unit for resistance is the ohm (abbreviated by the Greek omega, Ω). A resistor has a resistance of 1 Ω when a current of 1 A exists with an applied potential difference of 1 V. Related units are the megohm (1 MΩ = 10^6 Ω) or the micro-ohm (1 $\mu\Omega$ = 10^{-6} Ω). In circuit diagrams a resistor is depicted by the symbol ⌇⌇⌇, and a variable resistor, or rheostat, is represented by ⌇⌇⌇, or ⌇⌇⌇. A dissipative circuit element whose resistance is *not* independent of V (or, therefore, of i) is known as a nonohmic, or nonlinear, resistor (its Vi plot is not a straight line through the origin).

It is remarkable that Ohm's law holds precisely over an enormous range of currents and potential differences (over a range of 10^{10} for many materials); it is even more remarkable, at least at first sight, that it holds at all. If a potential difference is maintained between two points on a conductor, a nonzero electric field must exist within the conductor. It is this applied field, superimposed on the fluctuating local electric fields of the charged particles within the conductor, which drives the charge carriers through the conductor and so maintains the electric current. If the applied potential difference is constant, so too is the applied electric field at each point in the conductor. Then a net electric force acts on the free electrons comprising the current. But if an electron is acted on by a constant force, it moves with constant *acceleration,* not constant velocity. Such is the motion of an electron under the influence of an electric field in a vacuum tube. We have seen, however, that the conduction electrons in a conductor have a *constant* drift speed. Actually, there is no contradiction here, when we recognize that the conduction electrons undergo frequent collisions with the atoms of the conductor. In these collisions the kinetic energy acquired by the electrons from the external field is lost to the conductor lattice, and the power *delivered to the conductor is dissipated* as thermal energy.

When a circuit element obeys Ohm's law and $V = iR$, then the power delivered to and dissipated in a resistor can be written

$$P = iV = i^2R = \frac{V^2}{R} \qquad (24\text{-}9)$$

That the thermal energy appearing in a resistor (the so-called i^2R loss) is, in fact, exactly equal to the electric energy delivered to it was first established by the historic experiments of James Joule, and the effect is referred to as Joule heating. The thermal energy (in joules) is given by $W = i^2Rt = (V^2/R)t = iVt$, where t is the elapsed time.

Ohm's law holds for a conductor of any shape and with the leads attached to any two points. Suppose that a potential difference is established between the ends of a cylindrical conductor of length L and cross-sectional area A, as in Fig. 24-5. For a given material, experiment shows, quite reasonably, that the resistance for this simple geometrical configuration is directly proportional to the length and inversely proportional to the cross-sectional area. We may write the resistance as

$$R = \rho \frac{L}{A} \qquad (24\text{-}10)$$

where ρ, called the *resistivity,* is a property of the material of which the conductor is made but does not depend on the conductor's physical shape. Resistivity carries the units ohm-meters (Ω-m) or ohm-

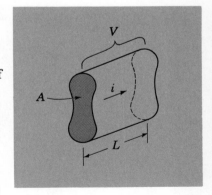

FIG. 24-5. A potential difference V is maintained across the ends of a cylindrical conductor of length L and cross-sectional area A through which a current i passes.

Table 24-1

Material	Resistivity at room temp. (Ω-m)
Silver	1.47×10^{-8}
Copper	1.72×10^{-8}
Aluminum	2.83×10^{-8}
Tungsten	5.51×10^{-8}
Iron	10×10^{-8}
Manganin (Cu 84%, Mn 12%, Ni 4%)	44×10^{-8}
Constantan (Cu 60%, Ni 40%)	44.1×10^{-8}
Nichrome	100×10^{-8}
Graphite	8.0×10^{-6}
Carbon	3.5×10^{-5}
Germanium	0.43
Silicon	2.6×10^{3}
Rock (granite)	$10^{5} - 10^{7}$
Wood (maple)	4×10^{11}
Mica	9×10^{13}
Quartz (fused)	5×10^{16}

centimeters (Ω-cm). Table 24-1 lists the resistivities of a number of common materials, both good and poor as conductors. Note the great range in resistivities: The best conductors are better than the worst conductor by a factor of about 10^{24}. From (24-10) it is seen that the resistance of a cube 1 m along an edge, whose opposite faces are maintained at different potentials, is equal numerically to the material's resistivity in ohm-meters.

The reciprocal of the resistivity ρ is called the *conductivity* σ:

$$\sigma = \frac{1}{\rho} \qquad (24\text{-}11)$$

Similarly, the reciprocal of the resistance (in ohms) is the conductance (in mhos).

Typically the resistivity of a conductor increases as the temperature rises. However, some materials—particularly semiconducting materials such as carbon—show the reverse behavior: their resistivity falls as the temperature rises (they have a negative temperature coefficient). Such materials have larger resistivities than metallic conductors, principally because they have far fewer conduction electrons. When the temperature rises, the number of free electrons in a semiconductor increases and the conductivity thereby increases to such a degree that it overcomes the decrease in conductivity resulting from increased thermal vibration.

Certain elements show a remarkable variation in electrical

resistivity with temperature: Below a certain critical temperature, the resistivity falls to zero—not merely a very small resistivity, but exactly zero! Such materials are appropriately called *superconductors*. For example, the element lead exhibits superconductivity below 7.175 K. One expects that, since a superconductor offers no resistance to the flow of electric charge and dissipates no energy, an electric current once established in a superconducting loop will persist indefinitely. Insofar as experiment will disclose, this is precisely what happens. Currents of several hundred amperes induced in a superconducting lead ring have been observed to persist with no measurable diminution for several years! Although bizarre, the phenomenon of superconductivity is now understood in terms of the quantum theory.

SUMMARY

The electric current i through any surface is given by

$$i = \frac{dQ}{dt} \qquad (24\text{-}1)$$

where dQ is the total net charge passing through the surface in time dt. By convention, the direction of current is that direction in which positive charges move. Current is measured in amperes.

The rate at which an energy source delivers electrical energy to a load is given by

$$P = iV \qquad (24\text{-}6)$$

where V is the potential difference across the load and i is the current through it.

The resistance R of a circuit element is defined by

$$R = \frac{V}{i} \qquad (24\text{-}8)$$

Ohm's law is the experimental finding that for many conducting materials the resistance, so defined, is independent of V or of i.

The resistance R is related to the length L and cross-sectional area A of a cylindrical conductor by

$$R = \rho \frac{L}{A} \qquad (24\text{-}10)$$

where the resistivity ρ is a constant, characteristic of the conducting material. The resistivity depends on the chemical and physical purity of a material and on its temperature.

PROBLEMS

24-1. A 10.0-eV proton beam consisting of 4.0×10^6 protons/cm³ travels to the left, while a 10.0-eV electron beam consisting of 4.0×10^6 electrons/cm³ travels to the right. The area of each beam is 2.0 mm². What are the direction and magnitude of the resultant current?

24-2. What is the current per transverse area in a copper bar carrying 50 A if the cross-sectional area is 20 mm²?

24-3. A 2.0-MeV proton beam consisting of 2.5×10^{11} protons/m³ travels to the right. The area of the beam is 2.0 mm². (a) Find the current and the power of the proton beam. (b) How long (years) would it take this beam to deposit 1 g of hydrogen? The mass of a proton is approximately 1.7×10^{-27} kg.

24-4. A circular dielectric disk has a uniform surface charge density of 1.0 μC/m². If the disk has a radius of 10 cm, at what angular speed must the disk be rotated to produce a current of 5.0×10^{-8} A across a straight line, fixed in space, from the center to the edge of the disk?

24-5. Assume that an electron is a concentrated charge of -1.6×10^{-19} C moving in a circular orbit of radius 0.5 Å about a fixed proton under the coulomb attractive force. (a) What is the average current due to the electron in its orbit? (b) Sketch the instantaneous current in part (a) as a function of time.

24-6. A particular electron synchrotron, a machine for accelerating electrons to very high energies, produces a pulsed beam of electrons, each electron with an energy of 10 GeV (1.0×10^{10} eV). The pulses are 2.5×10^{-6} s in duration and occur at the rate of 60 pulses/s with a beam intensity of 1.0×10^{11} electrons/pulse. Find the average and peak values (during the pulse) of (a) the electron current and (b) the power delivered to the electron target.

24-7. A 50-Ω resistor is rated at 0.50 W. What are the maximum current through and maximum voltage across the resistor?

24-8. An electric heater raises the temperature of 1.0 kg of water 40 C° when turned on for 10 min. If the current in the resistor is 10 A, what is its resistance?

24.9 A certain 12-V battery carries an initial charge of 120 A-h (A-h = ampere-hour, a unit for electric charge). Assuming for simplicity that the potential difference across its terminals remains at 12 V until the battery is fully discharged, for how many hours can such a battery deliver 60 W?

24-10. A coil of wire dissipates energy at the rate of 5.0 W when a potential difference of 200 V is applied to it. A second coil made of the same wire dissipates 15 W when the same potential difference is applied to it. What is the ratio of the length of wire in the second coil to that in the first coil?

24-11. Electrons are emitted from the filament of a vacuum tube at the rate of 5.0×10^{18} electrons/s. All the emitted electrons are collected at a plate, whose potential is 100 V above that of the filament. What are (a) the current at the plate and (b) the power dissipated at the plate? If the plate potential is now raised to 200 V, what are (c) the plate current and (d) the power now dissipated at the plate?

24-12. In a 100-W incandescent bulb designed for use at 120 V, the filament is maintained at about 2820°C. (a) What is the corresponding filament resistance? (b) At room temperature (20°C) the lamp's resistance is 10 Ω. Find $\Delta R/\Delta T$, the average change in filament resistance per K° over this temperature interval. (This dependence of resistance on temperature is typical for metals.)

24-13. What are (a) the voltage drop per unit length (in V/m) and (b) the power loss per unit length (in W/m) in No. 12 gauge aluminum wire when a 1500-W load is connected to a 120-V circuit? At 20°C (68°F) No. 12 wire is 2.503 mm in diameter and the resistivity of aluminum is 2.828×10^{-8} Ω-m.

24-14. According to standard practice No. 14 gauge annealed copper wire is approved for 15 ampere service in wiring for houses. The resistivity of annealed copper is 1.724×10^{-8} Ω-m and No. 14 gauge wire is 1.628 mm in diameter at 20°C (68°F). (a) What is the power dissipation per unit length (in W/m) of this wire at the rated current? (b) What is the power dissipation per unit length if No. 18 gauge (1.024 mm diameter) copper wire is used instead? Assume that the temperature is maintained a constant at 20°C.

24-15. What length of No. 40 gauge Nichrome wire is needed to construct a 10,000 Ω resistor accurate to ± 1.0 percent? At 20°C (68°F) the resistivity of Nichrome is 150×10^{-8} Ω-m; the diameter of No. 40 gauge wire is 0.07987 mm.

24-16. A resistor design which is resistant to corrosion over long periods of time consists of a film of noble metal which is vacuum-deposited on a ceramic core. What is the thickness of a platinum film needed to construct a 100-Ω resistor on a ceramic cylinder 5.00 mm in diameter and 5.00 cm long? At 20°C (68°F) the resistivity of platinum is 10.47×10^{-8} Ω-m.

24-17. In recent years the availability and cost of copper has prompted a reevaluation of the relative merits of copper and aluminum for use in the transmission of electric power. To illuminate some of the factors involved, it is useful to make calculations such as those which follow. *(a)* How many kg of aluminum are required to make a conductor with the same electrical resistance per km as 100-kg of copper? *(b)* What is the ratio of the costs of the two alternative metals if copper sells for 2.5 times as much as aluminum on the basis of weight? The mass densities of copper and aluminum are 8.89 g/cm³ and 2.70 g/cm³, respectively; the resistivities are listed in Table 24-1.

24-18. A thin silver wire of 0.20-mm diameter carries a current of 2.0 A. *(a)* What is the current per unit cross-sectional area in the wire? *(b)* What is the drop in potential along a 1.0-m length of the wire? *(c)* What is the drift speed of electrons in the conductor? Assume that silver has 1 free electron per atom, a mass density of 10 g/cm³, and an atomic mass of 108 kg/kg-mol.

24-19. A solid rectangular parallelepiped has edges of lengths l, l, and $10l$. Find the ratio of each of the quantities *(a)* resistance, *(b)* current, and *(c)* power dissipated, when a constant potential difference is applied across parallel sides separated by $10l$, then across opposite sides separated by l.

24-20. What potential difference must be applied across opposite faces of a cube of maple wood 1 cm in edge length to produce a current of 1.0 μA?

24-21. Two resistors are made of the same material and have identical resistances. Resistor A has a small cross section, resistor B a larger cross section. If both resistors are placed across identical batteries, which resistor is expected to show the smaller temperature rise?

24-22. A solid piece of carbon is in the shape of a right circular cylinder of diameter 2.0 mm and length L. When 10 V is applied across the two ends of the cylinder, there is a current of 2.0 A. What is the length L?

24-23. Find the average thermal speed of electrons within a conductor at room temperature, assuming that the electrons follow the classical laws of kinetic theory, and compare this with the speed of 10^6 m/s, which can be derived using quantum mechanics.

24-24. *(a)* Assuming that Ohm's law is valid for all drift speeds, what is the maximum current density j (current per unit transverse area) for electrons in a copper conductor at room temperature if the speed of electrons cannot exceed the speed of light (3.0×10^8 m/s)? *(b)* Obviously, Ohm's law does not hold for speeds approaching the average thermal speed of 10^6 m/s, since the conductivity σ does not remain constant. Show how the Vi curve deviates from the linear curve of Fig. 24-4c.

24-25. *(a)* Combine Eqs. 24-8), (24-10), and (24-11) to show that Ohm's law can be written as

$$j = \sigma E$$

where $j = i/A$, $\sigma = 1/\rho$, and E are, respectively, the current density, conductivity, and electric field within the conductor. (This particular equation can be generalized as the *vector* formulation of Ohm's law: $\mathbf{j} = \sigma \mathbf{E}$, which is valid at interior points in an ohmic conductor.) *(b)* Find j and E for a copper wire 1.0 mm² in cross section when the current is 1.0 A. The resistivity of copper is 1.72×10^{-8} Ω-m at 20°C.

24-26. Show that P', the power dissipated per unit volume of conductor, can be written in the form $P' = \rho j^2$, where j is the local current density as defined in Problem 24-25 above. (*Hint:* Apply Eqs. (24-9) and (24-10) to a resistor in the form of a cube of edge length L and make appropriate substitutions and identifications.)

24-27. In the *Daniell cell*, a copper rod is immersed in a water solution of copper sulfate (CuSO$_4$) and in an adjacent but separate chamber a zinc rod is immersed in a water solution of zinc sulfate (ZnSO$_4$). The two solutions are separated by a barrier which is impermeable to sulfate ions (SO$_4^{2-}$) but which will allow copper ions (Cu^{2+}) and zinc ions (Zn^{2+}) to pass through it. (The sulfate ions are formed in the two sulfate solutions.) See Fig. P24-27. When used as a source of electrical current, the zinc rod (the *cathode*) and the copper rod (the *anode*) are connected together by a conducting circuit, as by a copper wire. The basic events in the operation of the cell can be summarized as follows: At the cathode metallic zinc atoms go into solution as zinc ions (Zn^{2+}), each of which has a net charge of +2e, and each

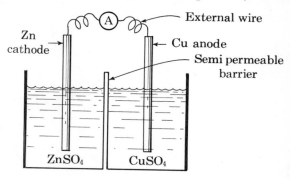

FIG. P24-27

such zinc ion contributes two surplus electrons to the cathode it leaves behind. These two surplus electrons flow through the external electrical circuit to the anode. At the anode Cu^{2+} ions formed from the dissociation of $CuSO_4$ absorb surplus electrons to become neutral atoms of metallic copper, which deposit onto the anode. The net result of this activity is that two electrons flow through the external circuit for each zinc atom which is dissolved from the anode and for each copper atom which plates out onto the anode. *(a)* Find the mass of copper deposited on the anode and *(b)* the mass of zinc dissolved from the cathode when the cell has operated for 10 hours at a current of 10 A. The atomic weights of copper and zinc are 63.5 g/g-mol and 65.7 g/g-mol, respectively.

24-28. The energy dissipated in a resistor of resistance R in the time interval dt in which an instantaneous current i exists is given by $i^2R\, dt$. Suppose that the current is a sinusoidally-varying alternating current given by $i = I_0 \sin \omega t$, where I_0 is the amplitude of the ac current and the frequency ν and period T of the current are, respectively, $\nu = \omega/2\pi$ and $T = 2\pi/\omega$. Then energy dissipated in the resistor, computed over one period of oscillation, is given by

$$\int_0^T i^2 R\, dt = R \int_0^T I_0^2 \sin^2 \omega t\, dt$$

(a) Show that the energy dissipated in the resistor over the time interval of one period T is $\tfrac{1}{2}I_0^2 RT$. *(b)* Show that the time-average rate at which energy is dissipated in the resistor is given by $\tfrac{1}{2}I_0^2 R$. *(c)* Show that the rate of energy dissipation can also be written as $I_{rms}^2 R$, a form analogous to that applicable for dc, where the *root-mean-square*, or rms, value of the current is given by $I_{rms} = I_0/\sqrt{2}$.

CHAPTER 25

DC Circuits

This chapter concerns circuits in which the elements are batteries and resistors. First the emf of a battery is defined in terms of the work done by chemical forces. The rules for handling resistors in series or in parallel are developed next. Then Kirchhoff's rules, the principles of energy conservation and charge conservation as applied to circuits for dealing with multiloop circuits in general, are treated. Finally some dc circuit instruments are discussed: the ammeter, voltmeter, Wheatstone's bridge, and the potentiometer.

25-1 THE EMF OF A BATTERY

In all that follows we shall always speak of so-called conventional electric currents, in which the direction of the current is taken to be that in which positive charges move. Thus, to traverse a resistor in the direction of the current is to go from high to lower electric potential. Similarly, when the battery is discharging, positive charges are taken to be leaving the positive terminal of a battery and entering the negative terminal. We realize, of course, that in ordinary conductors it is the free (conduction) electrons which actually move. Since these charge carriers are negatively charged, they travel "uphill" in potential, from a lower to a higher potential, and leave the negative terminal of a battery.

Consider the simple dc circuit of Fig. 25-1, in which a resistor R is connected across the terminals of a battery. The positive terminal of the battery is labeled a and the negative terminal b. Whether the switch is open or closed, the battery maintains the potential difference $V_a - V_b = V_{ab}$ across its terminals; note that V_{ab} is always positive.

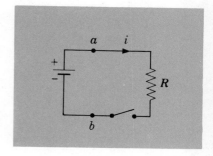

FIG. 25-1. A simple dc circuit.

When the switch is open, the potential difference across the resistor R is zero. Almost immediately after the switch is closed the potential difference V_{ab} appears across R and the current in the circuit is $i = V_{ab}/R$. We assume that the resistance of the connecting wires is negligible compared to R.

We know that if the electrostatic potential difference across the battery terminals is V_{ab}, the electrical power P_{ab} supplied by the battery to the load resistor R is given by

$$P_{ab} = iV_{ab} \tag{24-6}$$

The resistor R continuously converts the electric energy supplied to it into thermal energy at the rate

$$P_R = i^2 R \tag{24-9}$$

When any two *dissimilar* metals are immersed in a conducting medium, it is found that a potential difference exists between them; this is the most rudimentary form of an electrochemical cell. The chemical reactions which are the origin of this potential difference and the details of what goes on within a battery lie in the area of electrochemistry and will not be dealt with here. Suffice it to say, however, that the potential difference has its origin in the differences in the binding energy of electrons to atoms of different types. In this sense the nonelectrical, or chemical, forces are ultimately electrical in origin.

We may characterize any energy source that is capable of driving charges around a circuit against opposing potential differences as an *emf*. This is the abbreviation for *electromotive force*, a term so misleading (the emf is *not* a force) that we shall hereafter refer to it simply as the emf or, symbolically, as $\mathscr{E}$. By the $\mathscr{E}$ of a battery is meant the energy per unit positive charge gained (or the work per unit charge done by the *chemical* forces within a battery) in the transfer of a charge *within* the battery from a negative to the positive terminal:

$$\mathscr{E} = \text{work per unit positive charge done by the } \textit{chemical forces} \tag{25-1}$$

Since energy per unit charge, or joules per coulomb, is equivalent to volts, an emf, like a potential difference, is expressed in volts.

There is always some energy dissipation *within* an actual battery. One can account for this energy loss by ascribing an *internal resistance* r to the battery, where the rate of internal energy loss is i^2r. Thus, a battery with emf $\mathscr{E}$ and internal resistance r, connected across a load resistor R, is represented by the circuit diagram shown in Fig. 25-2. Let us apply energy conservation to this circuit. Since the emf is by definition the energy per unit charge gained from the battery, the time rate at which nonelectric chemical potential energy is transformed into electric energy, or the power associated with the emf, is $\mathscr{E}i$. Some of this electric energy is dissipated into

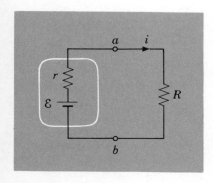

FIG. 25-2. The white lines show what is within the battery: an emf and an internal resistance.

thermal energy within the battery at the rate i^2r; the remaining energy is delivered to the load resistance at the rate $iV_{ab} = i^2R$. Therefore, energy conservation around a closed loop gives

$$\mathscr{E}i = i^2r + i^2R = i(ir + iR)$$

$$\mathscr{E} = i(r + R) \tag{25-2}$$

The current i in every element of the circuit is then

$$i = \frac{\mathscr{E}}{r + R} \tag{25-3}$$

Since iR represents the potential difference V_{ab} across the load resistor and also across the battery terminals, (25-2) can be written as

$$V_{ab} = \mathscr{E} - ir \tag{25-4}$$

Thus, the potential difference across the battery terminals is the battery's emf less the potential drop across its internal resistance. The potential difference V_{ab} across the battery terminals will equal the emf $\mathscr{E}$ if the current i is zero but, as we can see from (25-2), this occurs only when the load resistor R is infinite. Said differently, the potential difference appearing across the battery terminals on open circuit ($R = \infty$) is the battery's emf.

The emf of a particular type of cell depends only on the chemical identity of its parts. As a battery ages or loses its "charge" (strictly, of course, a battery loses only its internal chemical potential energy), the internal resistance increases but the emf remains unchanged. As (25-3) and (25-4) show, when the battery is "discharged," V_{ab} goes to zero as r increases, even for a relatively small load resistance R.

An electrochemical cell, or battery, is but one example of a device characterized by an emf, that is, a device which can convert nonelectric energy into electric energy. There are other arrangements in which an emf exists: (1) A thermocouple consists of two dissimilar conducting wires connected together in a circuit, the two junctions being maintained at different temperatures; an electric current exists in such a circuit, driven by a *thermoelectric* emf. (2) In a photovoltaic cell, exemplified by a photographic exposure meter, visible light (radiant energy) strikes a sensitive material and generates an electric current. (3) In an electric generator, mechanical energy is transformed into electric energy. (4) In electromagnetic induction, an emf is produced by a changing magnetic flux (we shall have much to say about this in Chap. 28).

In every instance a *nonelectrostatic force* is responsible for transporting charges against an opposing electric potential difference. It is appropriate to speak of a battery as the "seat" of an emf: Obviously, the current stops when the battery is removed from the circuit, and we know that the origin of the emf is the chemical re-

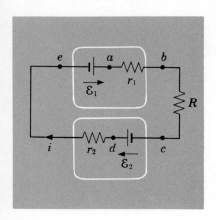

FIG. 25-3

actions taking place within. In other cases, however, one cannot necessarily identify one particular spot, or device, as the "seat" of the emf. It is useful, therefore, to speak of the emf of the circuit as being due to the particular nonelectrostatic force which converts nonelectric energy into electric energy or vice versa.

25–2 SINGLE–LOOP DC CIRCUITS

We wish to find the general relation between the emf's in a circuit and the potential differences. As a specific example, consider the circuit of Fig. 25-3, which contains two batteries with emf's $\mathscr{E}_1$ and $\mathscr{E}_2$ and three resistors r_1, r_2, and R connected in series (r_1 and r_2 are the internal resistances of $\mathscr{E}_1$ and $\mathscr{E}_2$, respectively). The arrows associated with the two batteries, pointing from the negative to the positive terminal in both cases, indicate the directions of the electric field associated with the nonelectrostatic forces; they give, so to speak, the "directions" of the emf. The "direction" of the current i is, in fact, precisely the same at each point in the circuit loop according to the law of electric charge conservation. It implies that electric charge does not pile up or become depleted at any one point in the circuit.

Consider how the conservation of energy principle applies to Fig. 25-3. Proceeding in a clockwise direction around the loop, we simply match the total energy delivered *by* the energy sources with the total energy delivered *to* the circuit elements (here, resistors). The power delivered *by* the batteries is

$$\mathscr{E}_1 i + \mathscr{E}_2 i = i\Sigma\mathscr{E} \tag{25-5}$$

Both terms on the left side are taken as positive, since the "directions" of $\mathscr{E}$ of i are the same for both batteries. The total power delivered *to* the circuit elements is

$$iV_{ab} + iV_{bc} + iV_{de} = i\Sigma V \tag{25-6}$$

Note especially the ΣV includes only the potential differences across circuit elements, *not* the potential differences of the batteries.

Equating (25-5) and (25-6), we have

$$\Sigma\mathscr{E} = \Sigma V \tag{25-7}$$

This is the fundamental relation for solving single-loop dc circuits. One must be careful about signs in applying this relation. It must be emphasized that $\Sigma\mathscr{E}$ is the *algebraic* sum of the emf's in the circuit; that is, each emf is taken as positive if the battery is traversed from the negative to the positive terminal; on the other hand, the emf must be taken as negative if it is traversed in the other direction; see Fig. 25-4a and b. Similarly, ΣV is the *algebraic* sum of potential *drops* across circuit elements. Each V is taken as positive if we traverse the circuit element from one

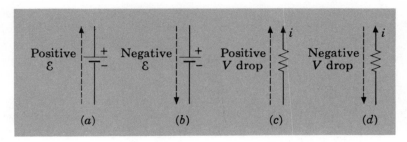

FIG. 25-4. Sign conventions for emf's and potential drops. The dashed arrows indicate the direction in which the circuit element is traversed. (a) The emf is taken as *positive* if the battery is traversed from the *negative* to the *positive* terminal. (b) The emf is *negative* if traversed from the *positive* to the *negative* terminal. (c) The potential *drop* is *positive* if the resistor is traversed in the *same* direction as the current. (d) The potential *drop* is *negative* if the resistor is traversed in the direction *opposite* to that of the current.

potential to a lower potential *in* the direction of the conventional current; each V is taken as negative, if we traverse the circuit element in the direction opposite to that of the current; see Fig. 25-4c and d. We *exclude* potential differences across battery terminals.

The potential difference between any two points a and b of a circuit can be related to the potential drops and emf's (potential increases) between the two points by using (25-7) along any path from a to b:

$$V_{ab} = V_a - V_b = \Sigma V - \Sigma \mathscr{E} \qquad (25\text{-}8)$$

The signs of the potential drops and emf's are determined by Fig. 25-4. If one proceeds around a closed loop, $V_{aa} = 0$ and (25-8) becomes (25-7).

Example 25-1

Two batteries are connected in opposition, as shown in Fig. 25-5; the emf's and internal resistances are 18.0 V and 2.0 Ω, 6.0 V and 1.0 Ω, respectively. (a) What is the current in the circuit? (b) What is the potential difference across the battery terminals? (c) At what rate does the discharging battery charge the charging battery? (d) At what rate is energy dissipated as heat in the 6.0-V battery?

(a) We decide to traverse the circuit of Fig. 25-5 in the *clockwise* sense. Moreover, we take this clockwise sense as the direction of the current i, knowing that the current will, in fact, be clockwise, since its direction will be controlled by the emf of the larger, 18.0-V battery. But it is *not* necessary to do so. We may choose the current direction arbitrarily and then, if it is not correct, the current will appear as *negative* in the solution; that is, the current will be shown actually to exist in the sense opposite to that chosen.

When the circuit traversal and current directions are taken as shown in Fig. 25-5, we have

$$\Sigma \mathscr{E} = \Sigma V \qquad (25\text{-}7)$$

$$18 \text{ V} - 6 \text{ V} = 2i + 1i$$

Note that the emf of the 6-V battery must be assigned a minus sign, inasmuch as we pass through this battery from the positive to the negative terminal (see Fig. 25-4b). The potential drops across the two

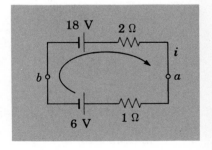

FIG. 25-5

Single-loop DC Circuits   SEC. 25-2

resistors, $2i$ and $1i$, are both positive, inasmuch as we traverse each resistor in the same direction as the current. Solving for the current i in the relation above gives

$$i = \frac{12.0 \text{ V}}{3.0 \text{ }\Omega} = 4.0 \text{ A}$$

(b) Ignoring the internal resistance for the moment, we know that the potential difference across a battery's seat of emf is just equal in magnitude to the emf of the battery. Thus, the *rise* in potential going from the negative to the positive terminal of the 18-V battery is 18 V. Across the internal resistance of 2 Ω is a potential drop

$$V = iR = (4.0 \text{ A})(2.0 \text{ }\Omega) = 8.0 \text{ V}$$

This 8-V potential difference occurs internally within the 18-V battery. Therefore, across the 18-V battery terminals is the potential difference 18.0 V − 8.0 V = 10.0 V. We can arrive at this result also by applying (25-8). If we proceed from a to b along the upper path, we have

$$V_{ab} = -\Sigma \mathscr{E} + \Sigma V = -(-18 \text{ V}) + (-4.0 \text{ A})(2.0 \text{ }\Omega) = 10.0 \text{ V} \quad (25\text{-}8)$$

Note that both $\mathscr{E}$ and iR are negative. If a 10.0-V potential difference exists across the 18-V battery terminals, this same potential difference appears as well across the 6.0-V battery terminals. Let us confirm this result by using (25-8) along the lower path from a to b.

$$V_{ab} = \Sigma V - \Sigma \mathscr{E} = (4.0 \text{ A})(1.0 \text{ }\Omega) - (-6 \text{ V}) = 10 \text{ V}$$

(c) The rate at which any source of emf delivers energy is given by $\mathscr{E}i$. If $\mathscr{E}i$ is positive—that is, if the emf and current are both in the same direction—chemical energy from the battery is delivered *to* other circuit elements. On the other hand, if $\mathscr{E}$ and i are of opposite sign, as is the case with the 6-V battery here, energy is delivered by other sources *to* it. Thus, the 18-V battery delivers energy (that is, loses chemical potential energy) at the rate

$$\mathscr{E}i = (18.0 \text{ V})(4.0 \text{ A}) = 72 \text{ W}$$

The 6-V battery has energy delivered to it (that is, gains chemical potential energy) at the rate $\mathscr{E}i = (6.0 \text{ V})(4.0 \text{ A}) = 24$ W. The battery with the larger emf is being discharged at the rate of 72 W, while the smaller battery is being charged at the rate of 24 W. What happens to the difference, 72 W − 24 W = 48 W? It is dissipated in the two resistors.

(d) The rate at which energy is dissipated within the 6-V battery is

$$P = i^2 R = (4.0 \text{ A})^2 (1.0 \text{ }\Omega) = 16 \text{ W} \quad (24\text{-}9)$$

and the rate at which thermal energy is developed within the 12-V battery is $P = i^2 R = (4.0 \text{ A})^2 (2.0 \text{ }\Omega) = 32$ W. The total power dissipated in the resistors, 16 W + 32 W = 48 W, is just the difference between the power delivered by the discharging battery (72 W) and the power delivered to the charging battery (24 W).

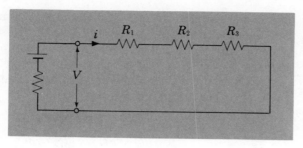

FIG. 25-6. Resistors in series.

25–3 RESISTORS IN SERIES AND IN PARALLEL

What is the equivalent resistance of resistors in series? By the equivalent resistance of a number of resistors in series, such as those shown in Fig. 25-6, is meant the resistance of that single resistor which, when replacing the separate resistors, does *not* change the current drawn from the energy source.

Clearly, the same current i passes through each resistor. Applying Eq. (25-7) to the circuit, we have

$$V = V_1 + V_2 + V_3$$
$$= iR_1 + iR_2 + iR_3 = i(R_1 + R_2 + R_3) \qquad (25\text{-}9)$$

where V_1, V_2, and V_3 are the potential drops across resistors R_1, R_2, and R_3, respectively, and V is the potential difference across the battery terminals. The same current i will exist in the circuit when the single equivalent R replaces the group in series:

$$V = iR \qquad (25\text{-}10)$$

Comparing (25-9) and (25-10), we have

Resistors in series: $\qquad R = R_1 + R_2 + R_3 \qquad (25\text{-}11)$

The equivalent resistance is the sum of the separate resistances in series. The *same current* exists in each series resistor; the potential drop across the entire group is the same as that across the single equivalent resistance.

What is the equivalent resistance of resistors connected in parallel, as shown in Fig. 25-7? We designate the potential drops

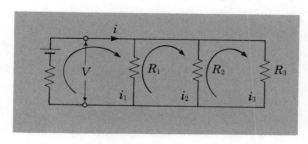

FIG. 25-7. Resistors in parallel.

across the resistors R_1, R_2, and R_3 as V_1, V_2, and V_3, respectively. The current through the battery is i, and the currents through the resistors are i_1, i_2, and i_3. Because we have more than one circuit loop in Fig. 25-7, we do *not* have the same current in all circuit elements. In fact, the current i from the battery is divided into three currents through the resistors:

$$i = i_1 + i_2 + i_3$$

$$\frac{dQ}{dt} = \frac{dQ_1}{dt} + \frac{dQ_2}{dt} + \frac{dQ_3}{dt} \tag{25-12}$$

This merely expresses the law of electric charge conservation; it implies that, since no charge accumulates at any point in the circuit, the charge dQ from the battery in time dt must appear as charges dQ_1, dQ_2, and dQ_3 through the resistors in the same time.

Now suppose that we apply (25-7) in traversing the left-hand circuit loop of Fig. 25-7:

$$\Sigma \mathscr{E} = \Sigma V \tag{25-7}$$

$$V = V_1 = i_1 R_1$$

Applying the same relation to the second loop of Fig. 25-7, we have

$$0 = -V_1 + V_2 \quad \text{or} \quad V_1 = V_2$$

Note that we use a negative sign for the potential "drop" V_1, since we traverse this resistor in the opposite direction to that of the current i_1. In similar fashion we find that $V_2 = V_3$. In summary,

$$V = V_1 = V_2 = V_3 \tag{25-13}$$

The potential difference is the *same* across each of the elements connected in parallel, and we designate this single potential drop by V.

From Ohm's law, we know that

$$V = i_1 R_1 = i_2 R_2 = i_3 R_3 \tag{25-14}$$

Substituting (25-14) in (25-12) gives

$$i = \frac{V}{R_1} + \frac{V}{R_2} + \frac{V}{R_3} = V \left[\frac{1}{R_1} + \frac{1}{R_2} + \frac{1}{R_3} \right]$$

A single resistor R replacing the parallel resistors obeys the relation

$$i = \frac{V}{R}$$

Comparing the last two equations gives, finally,

Resistors in parallel: $\quad \dfrac{1}{R} = \dfrac{1}{R_1} + \dfrac{1}{R_2} + \dfrac{1}{R_3} \tag{25-15}$

The reciprocal of the equivalent resistance is equal to the

sum of the reciprocals of the separate parallel resistances. The equivalent resistance is always less than the smallest of the parallel resistances. The essential feature of parallel connections is this: All elements have the *same potential difference.* Note also that, for any two resistors in parallel, the ratio of the currents is in the inverse ratio of the respective resistances, as (25-14) shows. Thus, with 1.0- and 3.0-Ω resistors in parallel, the 1.0-Ω resistor always has three times the current of the 3.0-Ω resistor, to ensure that the potential difference across both is always the same.

Example 25-2

Consider the circuit of Fig. 25-8. Find *(a)* the current in the battery, *(b)* the current in the 3.0-Ω resistor, *(c)* the potential difference

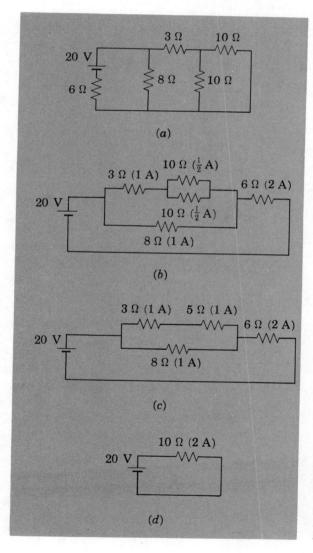

FIG. 25-8. A circuit *(a)* and its evolution into progressively simpler equivalent forms: *(b)* to *(d)*. The currents in the various resistors (shown in parentheses) are found by starting with part *(d)* and working backward to *(a)*.

across the 6.0-Ω resistor, and *(d)* the rate at which thermal energy is dissipated in the 8.0-Ω resistor. Note that in Fig. 25-8, the currents in parentheses are not given, but are calculated.

We first recognize that *any* of the questions that can be asked concerning a circuit like that in Fig. 25-8a (potential differences, power dissipation, and the like) require that we first find the current through each resistor. Other quantities then can be easily computed. To find the current through each resistor, we must first reduce the complex of resistors, through the rules for combining resistors in series and in parallel, until we are left with a single equivalent resistor connected across the battery.

Figure 25-8b to d shows the evolution of the circuit, Fig. 25-8a, into progressively simpler forms. (The numerical values of the resistances have been so chosen here that the computations can, without difficulty, be carried out in one's head.) We see that the current through the single equivalent resistor is 2 A. Now we work backward, through Fig. 25-8c, b, and a, in turn, to find the current in each resistor. Here we use the facts that the current through all series resistors is the same and that the potential difference across all parallel resistors is the same. Of course, as soon as we find the current through a given resistor, we can immediately compute the potential drop iR across it and the power $i^2R = iV$ dissipated in it. In this way we have, finally:

    a. Current through battery = 2 A
    b. Current through 3-Ω resistor = 1 A
    c. Potential difference across 6-Ω resistor = 12 V
    d. Power dissipated in the 8-Ω resistor = 8 W

25-4 MULTILOOP CIRCUITS

Some circuits involving more than one current loop, such as the one shown in Fig. 25-8, can be solved simply by applying the rules for combining resistances in series and in parallel. In general, however, this is *not* possible. As a specific example, consider the relatively simple circuit shown in Fig. 25-9. Here we have a circuit with *three* loops: a left inside loop, a right inside loop, and an outside loop going all the way around the circuit. There is no way to reduce this multiloop circuit into one involving a single battery and resistor.

A general method for solving multiloop circuit problems is to apply *Kirchhoff's rules*. These two rules are simply statements, in the language of electric circuits, of the fundamental *conservation laws* of (1) *electric charge,* and (2) *energy.*

1. Consider a *junction,* a point in the circuit at which three or more conducting wires are joined together, as shown in Fig. 25-10. The currents in the several wires are labeled i_1, i_2, i_3, and i_4, current i_1 being into the junction and the other currents being out of the junction. Now, according to the law of conservation of electric charge, no net charge can accumulate or be depleted at any point in the circuit, and certainly

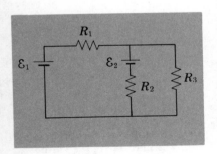

FIG. 25-9. A simple multiloop circuit.

not at a junction. Therefore, the net charge flow per unit time into any junction must equal the net charge flow per unit time out of the junction. That is, the net current into the junction equals the net current out of the junction. For the situation shown in Fig. 25-10 this implies that

$$i_1 = i_2 + i_3 + i_4 \quad \text{or} \quad i_1 - i_2 - i_3 - i_4 = 0$$

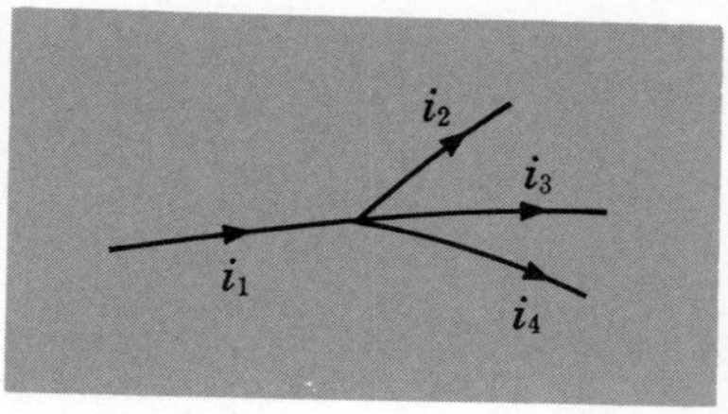

We may interpret the negative terms appearing in the second relation above as representing *negative* currents *into* the junction, which are, of course, equivalent to positive currents *out* of the junction. A general formulation of the junction requirements for currents is, then,

$$\Sigma i = 0 \tag{25-16}$$

where it is understood that currents into and out of a junction are identified, respectively, as positive and negative currents. Equation (25-16) is the *junction theorem*, the first Kirchhoff rule.

FIG. 25-10. A circuit junction. The total current into the junction is zero.

2. Recall the general relation which we derived earlier for a single-loop circuit:

$$\Sigma \mathscr{E} = \Sigma V \tag{25-7}, (25-17)$$

The algebraic sum of the emf's around a loop $\Sigma \mathscr{E}$ equals the algebraic sum of the potential drops ΣV (*including* the potential drops across battery internal resistors) around the loop (we *exclude*, however, the potential differences appearing across a seat of emf). Equation (25-17) is merely an expression of the conservation of energy law: The left-hand side represents the energy per unit charge supplied by energy sources in the circuit loop, and the right-hand side represents the energy per unit charge delivered to circuit elements around the loop. Kirchhoff applied (25-17) to any closed electric circuit loop; thus, this equation is known as the *loop equation*, or *Kirchhoff's second rule*.

Let us see how these rules apply to the circuit of Fig. 25-9, which is shown again in Fig. 25-11; we employ the same conventions as before. The currents in the *three branches* i_1, i_2, and i_3 are assigned directions quite arbitrarily. These currents must be regarded as the unknowns to be solved for. From (25-16) we have

$$i_1 - i_2 - i_3 = 0 \tag{25-18}$$

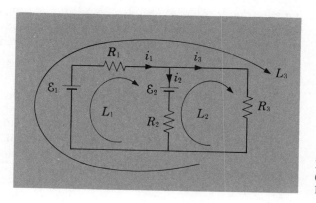

FIG. 25-11. Three circuit loops (L_1, L_2, and L_3) for applying Kirchhoff's rules.

Multiloop Circuits SEC. 25-4

This gives one equation in the three unknown currents. Straightforward application of (25-17) to each of the loops in Fig. 25-11 would give three more equations—but only two are needed. However, one easily finds that one of these equations is not independent of the other two. A simple procedure for removing redundancies of this sort is to choose only loops in which one always goes in the same direction as the current. For the example shown in Fig. 25-11 we must choose loops L_1 and L_3. Applied to loop L_1, (25-17) yields

$$\mathscr{E}_1 - \mathscr{E}_2 = R_1 i_1 + R_2 i_2 \qquad (25\text{-}19)$$

Applied to loop L_3, we have

$$\mathscr{E}_1 = R_1 i_1 + R_3 i_3 \qquad (25\text{-}20)$$

We now have three equations [Eqs. (25-18) to (25-20)] and three unknowns (i_1, i_2, and i_3). Using these three equations to solve for the three currents we find

$$i_1 = \frac{R_3(\mathscr{E}_1 - \mathscr{E}_2) + R_2 \mathscr{E}_1}{R_1 R_2 + R_1 R_3 + R_2 R_3} \qquad (25\text{-}21a)$$

$$i_2 = \frac{R_3(\mathscr{E}_1 - \mathscr{E}_2) - R_1 \mathscr{E}_2}{R_1 R_2 + R_1 R_3 + R_2 R_3} \qquad (25\text{-}21b)$$

$$i_3 = \frac{R_2 \mathscr{E}_1 + R_1 \mathscr{E}_2}{R_1 R_2 + R_1 R_3 + R_2 R_3} \qquad (25\text{-}21c)$$

Whereas current i_3 is always positive, i_1 and i_2 can be either positive or negative, depending on the relative sizes of the emf's and of the resistances.

Solving any multiloop circuit, however complicated, is simply a matter of applying Kirchhoff's rules and solving a number of simultaneous linear equations. One can work this in reverse, using electric circuits to solve linear equations. One arranges the emf's and resistances in a circuit to correspond to the parameters of the linear equations; solving for the unknowns consists merely of measuring the currents in the various branches of the circuit. This is one simple example of an *analog computer* in which one studies the physical behavior of a system obeying a well-known mathematical relationship, to solve for mathematical unknowns.

25-5 DC CIRCUIT INSTRUMENTS

Here we consider the essential features of four instruments commonly used in dc circuits: the ammeter, used for measuring electric current; the voltmeter, used for measuring electric potential difference; the Wheatstone bridge, used for comparing resistances; and the potentiometer, used for comparing emf's.

Ammeter and Voltmeter. Before speaking of any construction details of an ammeter or voltmeter, it is important that we be clear on how these instruments are used to measure current and potential difference. Figure 25-12a shows a simple circuit of a battery and resistor with a current i. When an ammeter (symbolized by —Ⓐ—) is introduced in series with the resistor to measure the current, as in Fig. 25-12b, it changes the circuit so that the measured current i' is less than i, the original current. If the resistance R_a of the ammeter is much less than R, then $i' \approx i$. In Fig. 25-12c a voltmeter (symbolized by —Ⓥ—) is connected in parallel across the resistor. Again the circuit is changed. The current i'' through the battery is now greater than i. Only if the voltmeter resistance is large compared with R will $i'' \approx i$.

Ideally, no current or potential differences are altered by the addition of ammeters or voltmeters in a circuit; an ammeter has such a low resistance that the potential difference across it usually is negligible, and a voltmeter has such a high resistance that the current through it usually is negligible.

Both ammeters and voltmeters of the common variety employ a needle whose angular position registers the current or potential difference. The needle is attached to a coil through which current passes, and this coil is immersed in the magnetic field of a permanent magnet. Such a device is known as a *galvanometer*; it is an instrument whose details we shall discuss in Sec. 26-6. Suffice it to say here that a galvanometer (symbolized by —Ⓖ—) is capable of registering extremely small currents. For example, a galvanometer showing full-scale deflection for a current of 1.0×10^{-6} A = 1.0 μA and having a resistance of 1,000 Ω might be typical (very much more sensitive galvanometers, giving full-scale deflection for currents as small as 10^{-12} A, have been constructed). When placed into a circuit such a galvanometer will certainly register a full-scale deflection when 1.0 μA passes through it. At the same time, there will exist a potential difference $V = iR = (1.0 \times 10^{-6}$ A$)(1.0 \times 10^3$ Ω$) = 1.0 \times 10^{-3}$ V = 1.0 mV across its terminals. If this 1.0-mV potential difference is small compared with potential differences existing across other circuit elements, the galvanometer's influence is negligible.

By itself, such a galvanometer could be used as a microammeter to measure currents up to 1 μA (provided that the galvanometer's resistance was much less than that of circuit elements with which it was in series), or as a millivoltmeter to measure potential differences up to 1 mV (provided that its resistance was much greater than that of circuit elements with which it was in parallel).

Suppose, however, that one wishes to construct an ammeter which registers 1.0 A full-scale, using this galvanometer. Then one places a very small resistance R_p in parallel with the galvanometer, as shown in Fig. 25-13; said differently, one "shunts" the galva-

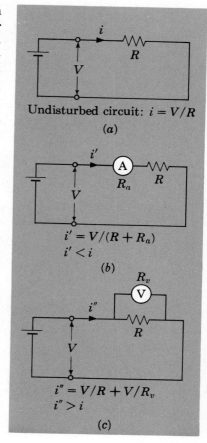

FIG. 25-12. A circuit (a) undisturbed, (b) with an ammeter in series with the resistor, and (c) with a voltmeter in parallel with the resistor.

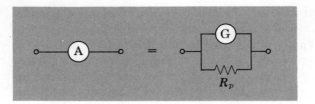

FIG. 25-13. An ammeter is a galvanometer in parallel with a small resistance.

nometer with a small resistance R_p. What is the required shunt resistance? We can assume that essentially all of the current, 1.0 A, through the ammeter will pass through the shunt resistance R_p, since only 1.0 μA is permitted through the galvanometer itself. Moreover, the potential differences across the galvanometer and its shunt must both be 1.0 mV at full-scale deflection. It follows that $R_p = V/i = (1.0 \times 10^{-3} \text{ V})/(1.0 \text{ A}) = 1.0 \times 10^{-3} \text{ } \Omega$.

The galvanometer, with $R = 1{,}000 \text{ } \Omega$ and full-scale deflection for 1.0 μA, or for 1.0 mV across its terminals, is, in effect, a voltmeter registering 1.0 mV full-scale. Suppose, however, that we wish to use this same galvanometer to construct a voltmeter registering 10 V full-scale. We construct a voltmeter from a galvanometer by placing a high resistance R_s in series with it, as shown in Fig. 25-14. Since the galvanometer alone can have a potential difference of only 1.0×10^{-3} V across its terminals, the potential difference V across the resistor R_s must be essentially 10 V, while the current i through it is 1.0×10^{-6} A. Therefore, $R_s = V/i = (10 \text{ V})/(1.0 \times 10^{-6} \text{ A}) = 1.0 \times 10^7 \text{ } \Omega = 10 \text{ M}\Omega$. Such a voltmeter has so large an internal resistance that it produces little perturbation of a circuit unless used across circuit elements whose resistance becomes comparable to 10 MΩ.

It is easy to construct multirange ammeters (or voltmeters) by having several shunt (series) resistors in the meter. Problems 25-17 and 25-18 illustrate multirange meters.

A simple way of measuring resistance is shown in Fig. 25-15a: One measures the current i through the device with an ammeter and the potential difference V across it with a voltmeter, and then applies Ohm's law, $R = V/i$. Strictly, however, the current measurement is not altogether correct, inasmuch as the ammeter of Fig. 25-15a registers the current through both the resistor *and* the voltmeter. One might correct this by placing the ammeter inside the connections to the voltmeter, as in Fig. 25-15b. Then the ammeter reads the current through R alone, but now the voltmeter

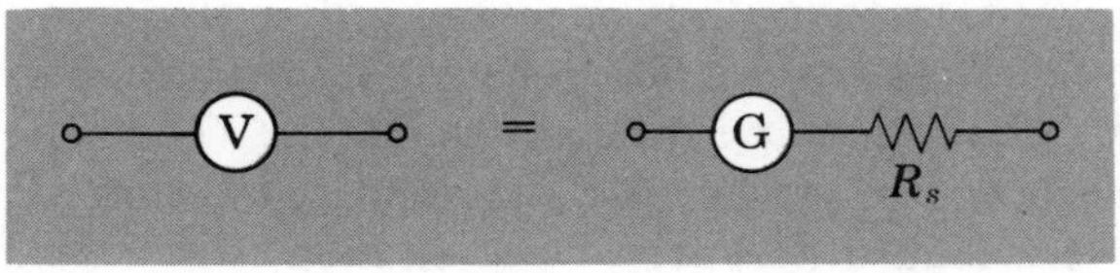

FIG. 25-14. A voltmeter is a galvanometer in series with a large resistance.

reads the potential difference, not across R alone, but across the resistance *and* ammeter.

Wheatstone's Bridge. These difficulties are eliminated when one measures a resistance or, more properly, compares a resistance with a known standard resistance R_s, using the bridge circuit devised by C. Wheatstone (1802–1875); see Fig. 25-16. The bridge, in its simplest form, consists of four resistors, a battery, and a sensitive galvanometer. R_1, R_2, and R_s are all known; R_x is the unknown resistance. Like the ordinary beam balance, which indicates equal masses on its two pans when the needle shows no deflection from the vertical, the Wheatstone bridge is a *null instrument*. With a given unknown resistance R_x, the resistors R_1, R_2, and R_s are so adjusted that the galvanometer registers no current. Since points b and c are then at the same potential, the current i_1 through resistor R_1 is the same as the current through R_2; likewise, R_x and R_s carry the current i_2. The potential differences V_{ab} and V_{ac} are equal, as are V_{bd} and V_{cd}. It follows that

$$V_{ab} = i_1 R_1 = V_{ac} = i_2 R_x$$
$$V_{bd} = i_1 R_2 = V_{cd} = i_2 R_s$$

Eliminating i_1 and i_2 from these relation yields

$$\frac{R_x}{R_s} = \frac{R_1}{R_2} \qquad (25\text{-}22)$$

When the bridge has been balanced, one finds the unknown resistance R_x in terms of the standard resistance R_s and the ratio R_1/R_2, using (25-22).

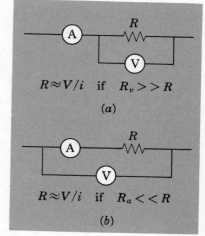

$R \approx V/i$ if $R_v >> R$

(a)

$R \approx V/i$ if $R_a << R$

(b)

FIG. 25-15. Two arrangements for measuring the resistance R with an ammeter and a voltmeter: (a) voltmeter across R; (b) ammeter in series with R.

Potentiometer. Suppose one is to measure the emf of a battery. The *approximate* emf is registered by a voltmeter placed across the battery terminals. Such a voltmeter reading V is always somewhat less than the true emf, since the potential difference appearing across a battery's terminals is the emf $\mathscr{E}$ *less* the potential drop ir across the battery's internal resistance r:

$$V = \mathscr{E} - ir \qquad (25\text{-}4)$$

Since all batteries have some internal resistance, the emf $\mathscr{E}$ and potential difference V are the same only if *no* current passes through the battery.

The potentiometer circuit permits the emf of a battery to be measured under the condition in which the battery current is actually zero. Strictly, the potentiometer permits an unknown emf $\mathscr{E}_x$ to be compared with the precisely known emf $\mathscr{E}_s$ of a *standard cell* by a *null* method. The circuit is shown in Fig. 25-17. A so-called *w*orking battery with emf $\mathscr{E}_w$, which need not be known but must be greater than $\mathscr{E}_x$ (or $\mathscr{E}_s$), maintains a constant current i

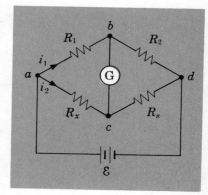

FIG. 25-16. A Wheatstone bridge circuit.

DC Circuit Instruments   SEC. 25-5

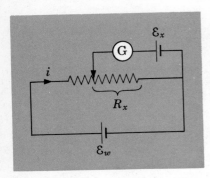

FIG. 25-17. A potentiometer circuit for comparing emf's.

through the resistor. The adjustable tap (see Fig. 25-17 where the left end of the galvanometer wire connects to the resistor) at the resistor is set so that the current through the sensitive galvanometer is zero. Then the potential drop across the galvanometer is zero, as is also the potential drop across the internal resistance of the unknown battery. Thus, the total potential drop across the branch in the circuit containing the unknown battery is equal to the battery's emf $\mathscr{E}_x$. But this is also the potential drop iR_x across the resistor from the tap to the right end. Therefore, $\mathscr{E}_x = iR_x$.

Now, if the unknown battery is replaced by the standard cell and the resistor tap is again adjusted for balance (zero galvanometer current), we have $\mathscr{E}_s = iR_s$, where R_s is the corresponding resistance from the tap to the end of the resistor. Eliminating i from these relations yields

$$\frac{\mathscr{E}_x}{\mathscr{E}_s} = \frac{R_x}{R_s} \tag{25-23}$$

From this equation we see that comparing emf's with a potentiometer circuit consists of comparing resistances (or, if the adjustable resistor is a wire of uniform cross section, of comparing lengths).

The term *potentiometer* is used in another sense in electric circuits—that of a *voltage divider;* see Fig. 25-18. An input voltage V_i is applied across a variable resistor of total resistance R_i with a center tap. The output voltage is V_o, and R_o represents the resistance between the tap and the lower end of the resistor. Clearly, $V_o/V_i = R_o/R_i$.

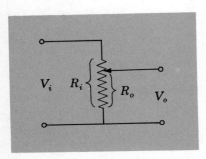

FIG. 25-18. A "potentiometer" used as a voltage divider.

SUMMARY

The emf $\mathscr{E}$ of an energy source is defined as the work done per unit positive charge by the nonelectrostatic forces in the circuit.

The equivalent resistance R of resistors connected in series is given by $R = \Sigma R_i$. The equivalent resistance for resistors connected in parallel is $1/R = \Sigma 1/R_i$. Resistors in series all carry the same current; resistors in parallel all have the same potential difference across their terminals.

One solves for the current in any branch of a multiloop circuit by applying Kirchhoff's two rules:

1. The junction theorem: The total current into any junction is zero:

$$\Sigma i = 0 \tag{25-16}$$

2. The loop theorem: The algebraic sum of the emf's around a loop equals the algebraic sum of the potential drops across resistors:

$$\Sigma \mathscr{E} = \Sigma V \tag{25-17}$$

Kirchhoff's first and second rules stem from the conservation principles of charge and of energy applied to electric circuits.

PROBLEMS

25-1. A battery is connected to a resistor, and a current of 1.0 A exists in the circuit. When an additional 12-Ω resistor is added to the circuit in series, the current drop to 0.33 A. What is the emf of the battery?

25-2. Two batteries of different emf's and internal resistances are connected in series with one another and with an external load resistor. The current is 4.0 A. Then the polarity of one battery is reversed, and the current becomes 1.0 A. What is the ratio of the emf's of the two batteries?

25-3. Suppose that a 10.0-Ω resistor is connected in series with the terminals of a 9.0-V battery (six dry cells in series) which has an emf of 9.24 V and an internal resistance of 0.50 Ω. (a) At what rate is chemical energy being converted into electric energy? Find (b) the voltage across the output terminals, (c) the power delivered to the load resistor, and (d) the rate at which heat is generated within the battery. (When fresh, a dry cell has an emf of about 1.55 V and an internal resistance in the range from 0.02 Ω to 0.3 Ω, depending on the intended use.)

25-4. The lead-acid storage battery is capable of delivering the high currents required to operate the type of electric motor used to start gasoline engines. A standard measure of this property is the "cold-cranking rating" devised by the *Society of Automotive Engineers* (SAE). This rating quotes the current which a given "12-volt" battery will deliver for 30 s at 0°F (−17.8°C) while maintaining a minimum terminal voltage of 7.20 volts. (Such a battery consists of six lead-acid cells in series; when freshly charged the emf of a lead-acid cell is 2.13 V.) Consider a particular battery whose SAE cold-crank rating — as described above — is 450 A. (a) Find the internal resistance of this battery and (b) the energy drain (in J and kcal) from it during this test. (c) What fraction of this energy is dissipated *in the battery*?

25-5. An electric generator dissipates energy internally at the rate of 20 W when the difference in potential across its output terminals is 120 V and it has a current of 2.0 A through it. What is the generator's emf?

25-6. An energy source, or generator, of emf $\mathscr{E}$ and internal resistance R_G is connected to a load of resistance R_L. (a) Show that the power P_L that the generator delivers to the load is given by $\mathscr{E}^2 R_L/(R_G + R_L)^2$. (b) Show that the power delivered from the generator to the load is a maximum when the load resistance is matched to the generator resistance, that is, show that P_L is a maximum when $R_L = R_G$.

25-7. In the circuit shown in Fig. P25-7 the internal resistance of the battery is negligible compared to the other resistances in the circuit. (a) Find that single resistor which would cause the same current to flow through the battery. Calculate (b) the current in each resistor and (c) the total power delivered to the circuit by the battery.

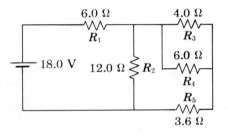

FIG. P25-7

25-8. Three resistors of 3.0, 4.0, and 6.0 Ω, respectively, are connected in parallel with a 60-V battery. (a) Draw a schematic diagram of the circuit. Find (b) the current through each resistor and (c) the total current through the battery. (d) What single resistance is equivalent to these three in parallel? (Assume that the internal resistance of the battery can be neglected.)

25-9. Twelve 1-Ω resistors are aligned along the edges of a cube and connected together at the corners. (a) What is the equivalent resistance measured between two diagonally opposite corners of the cube? (*Hint:* Use the symmetry of the arrangement to find how the current divides at the junctions.) (b) Find the fraction of the total current through each resistor.

25-10. Consider the circuit shown in Fig. P25-10, where the internal resistance of the battery is 0.50 Ω. (a) What is the equivalent resistance of the circuit ex-

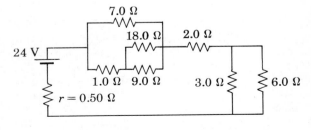

FIG. P25-10

409

ternal to the battery? Find *(b)* the current drawn from the battery and *(c)* the potential difference across its terminals. *(d)* What is the current through the 9.0-Ω resistor? *(e)* At what rate is heat developed in the 3.0-Ω resistor?

25-11. Solve for the currents in Eqs. (25-18) to (25-20), and show that they are given by (25-21).

25-12. Assume the following values for the parameters in Fig. 25-11: $\mathcal{E}_1 = 6$ V, $\mathcal{E}_2 = 12$ V, and $R_3 = 2$ Ω. For what values of R_1 and R_2 will *(a)* the current i_1 be negative, *(b)* the current i_2 be negative?

25-13. Consider the circuit shown in Fig. P25-13. *(a)* Find the magnitude and direction of the currents through the two batteries and the 1-Ω resistor. *(b)* Find $V_{ab} = V_a - V_b$ by proceeding along three different paths.

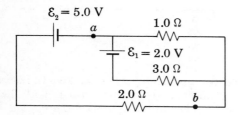

FIG. P25-13

25-14. Consider the circuit shown in Fig. P25-14. *(a)* Find the direction and magnitude of the currents through each of the three resistors. *(b)* What is the potential difference between points A and B?

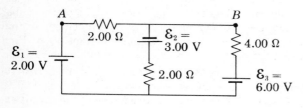

FIG. P25-14

25-15. Consider the two arrangements of Fig. 25-15 used for estimating an unknown resistance by the ratio of the voltmeter-to-ammeter readings. If the resistances of the ammeter and voltmeter are 1.0 and 20 Ω, respectively, and the resistance $R = 10$ Ω, what estimates are obtained for the voltmeter-ammeter arrangements in *(a)* and *(b)*?

25-16. An ammeter having a resistance of 0.10 Ω and a voltmeter having an internal resistance of 10,000 Ω are to be used in combination to measure an unknown resistance R. For what values of R will the resistance be given as V/i to within 1 percent, irrespective of the order of connection of the ammeter and voltmeter to the resistance R (Fig. 25-15)?

25-17. The internal wiring of a multirange voltmeter is shown in Fig. P25-17. The galvonometer alone gives full-scale deflection when a current of 1.0 mA exists in it. The galvanometer's resistance is 1,000 Ω. Find the values of the resistors *(a)* R_a, *(b)* R_b, and *(c)* R_c.

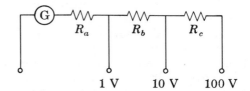

FIG. P25-17

25-18. The internal wiring of a multirange ammeter is shown in Fig. P25-18. When each of the three terminals is used, the full-scale reading of the ammeter is 1.0 A, 10 A, or 100 A. The galvanometer gives full-scale deflection when a current of 1.0 mA exists in it; its resistance is 1,000 Ω. Find the values of the resistors *(a)* R_a, *(b)* R_b, and *(c)* R_c.

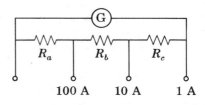

FIG. P25-18

25-19. *(a)* Using a galvanometer of resistance 1,000 Ω which gives a full-scale deflection when a current of 1.0 mA exists in it, find the shunt resistance necessary to construct a 100-A ammeter. *(b)* Find the length of copper wire (diameter, 1.0 mm; resistivity, 1.7×10^{-8} Ω-m) that would provide this shunt resistance.

25-20. An ammeter is a low-resistance instrument connected in a series, whereas a voltmeter is a high-resistance instrument connected in parallel. Which instrument would be damaged when improperly connected, and why?

25-21. Although precision instruments of resistance can be done using the Wheatstone bridge (Sec. 25-5), it is very useful to have a less elaborate instrument when less precision is needed. The circuit shown in Fig. P25-21 illustrates some of the basic ideas involved in the design of such an instrument which is known as an *ohmmeter*. An unknown resistor R_x is connected across the terminals A-B, completing the circuit so that the battery emf $\mathscr{E}$ will produce a current. A part of this current passes through the galvanometer G, and the value of R_x is indicated by the deflection of the galvanometer, measured against an appropriate scale. When R_x is infinite (open circuit) the meter does not deflect; this point on the scale is marked $R = \infty$. By choice, the meter shows a full-scale deflection when the terminals A-B are short-circuited; this point on the scale is marked $R = 0$. A convenient procedure for the detailed design of an ohmmeter intended for a particular range of resistance values is as follows: Choose that value of R_x which is to produce a meter deflection of one-*half* full scale (*hs*); call this particular resistance value R_{hs}. If the terminals A-B were short-circuited ($R_x = 0$), the meter current would be doubled (to full scale), which means that the resistance in series with the emf $\mathscr{E}$ has been halved. Therefore, $R_e = R_{hs}$, where R_e is the equivalent series resistance of the resistance network inside the instrument. Knowing R_e we can then calculate the current supplied by the emf $\mathscr{E}$; given the specifications of the galvanometer movement (its series resistance R_m and i_{fs}, the current required for *full-*scale deflection), we can calculate the values of R_{sh} and R_{se}. Having selected these particular circuit parameters, the remainder of the resistance scale on the ohmmeter dial can be calculated. As the battery ages its internal resistance increases. To compensate for this R_{sh} is made variable (increasing R_{sh} will force a greater fraction of the battery current to go through the meter). (a) With this outline as a guide, design an ohmmeter that will indicate half-scale for $R_x = 1{,}000\ \Omega$ if $\mathscr{E} = 3.0$ V, galvanometer resistance $r_m = 10{,}000\ \Omega$, and galvanometer current for *f*ull-scale

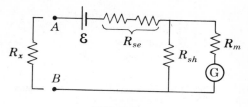

FIG. P25-21

deflection $i_{fs} = 100\ \mu$A. *(b)* Calculate the value of R_x corresponding to a deflection of three-fourths of full scale.

25-22. Equation (25-23), giving the ratio of emf's in terms of the ratio of resistances, assumes that the current i through the lower loop of Fig. 25-17 remains constant. If the temperature of the resistor is higher when the standard cell is balanced than when the unknown cell is balanced, will the value $\mathscr{E}_x$ determined by Eq. (25-23) be greater or less than the true value? Assume that the resistance increases with temperature.

25-23. The circuit for the potentiometer (Fig. 25-17) is basically the same as the multiloop circuit of Fig. 25-11. Derive Eq. (25-23) for the potentiometer from Eqs. (25-21).

25-24. Show that if the potentiometer circuit shown in Fig. 25-17 is unbalanced, the magnitude of the current through the galvanometer is given by

$$i_g = \frac{\mathscr{E}_w - (1 + R_1/R_3)\mathscr{E}_x}{R_1 + (1 + R_1/R_3)R_2}$$

where R_1 is the sum of the internal resistance of the working battery and that part of the slide wire resistance to the left of the adjustable tap, R_2 is the sum of the internal resistances of $\mathscr{E}_x$ and the galvanometer, and $R_3 = R_x$, that part of the slide wire resistance which is to the right of the adjustable tap.

CHAPTER 26

The Magnetic Force

One usually associates magnetism with the behavior of magnets and of magnetic materials. To be sure, magnets are one important aspect of magnetism, but the *fundamental* magnetic effect is this: An electric charge in motion may produce a force on a second moving electric charge in addition to the coulomb (electric) force. This velocity-dependent force between charges is the *magnetic force*.

Just as the electric force between charged particles may be thought to act via the electric field—charge q_a creates an electric field **E** at the site of charge q_b, and q_b, finding itself in this field, is acted upon by an electric force—so too the magnetic interaction between charged particles may be described as taking place via a magnetic field. In this view, moving charge q_a creates a *magnetic field* **B** at the site of moving charge q_b, and q_b, finding itself in this field, is acted upon by the so-called magnetic force. The magnetic force is more complicated than the electric force in that it depends on the velocities of the two interacting charges as well as the magnitude and sign of the charges and their separation distance. For this reason it is useful to discuss the magnetic interaction in two parts: (1) the creation of the magnetic field by one moving charge and (2) the effect of the magnetic field on a second moving charge. This chapter will deal with the second part.

26–1 MAGNETIC INDUCTION FIELD DEFINED

The term magnetic field immediately conjures up in the imagination pictures of one magnet attracting or repelling a second magnet or causing the alignment of iron filings or compasses. This conception of the magnetic field, although familiar and indeed correct, is not fundamental. To define properly the magnetic field, one must first investigate the motion of charged particles in the field, established, for example, by permanent magnets or, as we shall see, by electric currents. How the magnetic field is produced is not our concern here; this will be discussed in Chap. 27. We shall merely assume that a magnetic field can be maintained in a region where we place a beam of moving charged particles. We further assume that no electric or gravitational forces act on the particles.

We may explore the properties of the magnetic force and therefore, also, the magnetic field, by directing a beam of charged particles into a region where a magnetic field exists. For example, one might use a small oscilloscope, in which a beam of electrons falls on a phosphorescent screen and produces a bright spot at its center. The speed of the particles is assumed to be variable and known. By investigating the paths of the charged particles, we can deduce the magnetic force acting on them. Whenever the magnetic force $\mathbf{F}_m$ acts, the experimental facts are these:

1. The magnetic force is always perpendicular to the particle's velocity $\mathbf{v}$:

$$\mathbf{F}_m \perp \mathbf{v}$$

2. There exists one particular orientation of the velocity $\mathbf{v}$ for which the force $\mathbf{F}_m$ is zero; see Fig. 26-1a. That is, there is a particular line in space along which charged particles can move without experiencing a magnetic force. *By definition*, the line of the magnetic field is this line.†

$$F_m = 0 \quad \text{when} \quad \mathbf{v} \parallel \mathbf{B}$$

There remains the matter of assigning a *direction* to the magnetic field $\mathbf{B}$.

3. The magnetic force is proportional to the particle's charge q, both in magnitude and sign:

$$F_m \propto q$$

If one replaces negative charges by positive charges moving in the same direction, the direction of $\mathbf{F}_m$ is reversed (see Fig. 26-1b).

4. For all directions of the velocity $\mathbf{v}$, the magnitude of $\mathbf{F}_m$ is directly proportional to the component $v_\perp = v \sin \theta$ of the velocity at right angles to the line of the magnetic field:

$$F_m \propto v_\perp = v \sin \theta \tag{26-1}$$

where θ is the angle between the direction of $\mathbf{v}$ and the direction of $\mathbf{B}$. Thus, the force is zero when $\theta = 0$ (or 180°), the vectors $\mathbf{v}$ and $\mathbf{B}$ then being parallel (or antiparallel). When the particles move at right angles to the magnetic field ($\theta = 90°$), the magnetic force is a maximum. See Fig. 26-1b.

The above experimental facts can be simply expressed by defining the magnitude of the magnetic field $\mathbf{B}$ through the relation

$$B = \frac{F_m}{qv_\perp} = \frac{F_m}{qv \sin \theta} \tag{26-2}$$

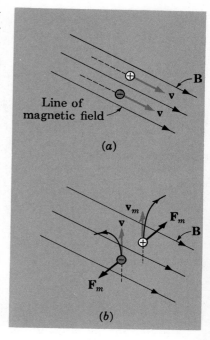

Fig. 26-1. (a) When a charged particle moves through a magnetic field and is *not* subject to a magnetic force, the line of its velocity is, by definition, the line of the magnetic field. (b) When a charged particle moves at right angles to a magnetic field, it is deflected by a magnetic force at right angles to both v and B.

---

†The direction of a magnetic field so defined is, in fact, also the line along which a small compass, or permanent magnet, will align itself.

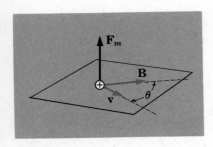

FIG. 26-2. The magnetic force F_m is perpendicular to the plane of **v** and **B**.

where θ is the angle between **v** and **B**. Since all quantities on the right are measurable, the magnitude of **B** can be determined.

5. The magnetic force is always at right angles to the plane containing two lines, the line of the velocity and the line of the magnetic field (that is, the line along which a particle travels when no magnetic force is acting on it). The direction and magnitude of $\mathbf{F}_m$ observed in experiment are then correctly related to those of **v** and of **B** through the vector relation

$$\mathbf{F}_m = q\mathbf{v} \times \mathbf{B} \tag{26-3}$$

Here we have arbitrarily *chosen* a direction for the magnetic field through this vector relation and the convention of the *right*-hand rule for cross products; see Fig. 26-2.

The electric force and the electric field *differ greatly* from the magnetic force and the magnetic field: The electric force acts whatever the state of motion of the charged particle, and the direction of the electric force on a positive charge is the *same* as the direction of the electric field.

The magnetic force, on the other hand, depends on the charge's velocity, and the force is perpendicular to both **v** and **B**.

Equation (26-3) implies that the magnetic field **B** is a vector. Certainly, **B** has magnitude, and we have now established a convention for its direction.

The magnetic field is indeed a vector in the following sense: Two or more magnetic fields acting simultaneously on a charged particle produce a magnetic force, given by (26-3), if by **B** is meant the *vector sum* of the separate fields. This is a result of experiment.

We see from (26-2) that appropriate units for the magnetic induction field are newtons per ampere-meter (N/A-m), as follows:

$$\text{N/C-(m/s)} = \text{N/(C/s)-m} = \text{N/A-m}$$

Thus, a charge of 1 C moving at 1 m/s at right angles to a magnetic field of 1 N/A-m is subject to a magnetic force of 1 N.

In the SI system of units the magnetic field is also assigned units of *webers per square meter,* or simply *tesla* (T). By definition,

$$1 \text{ T} = 1 \text{ Wb/m}^2 = 1 \text{ N/A-m}$$

A magnetic field of 1 Wb/m² is, by laboratory standards, a field of relatively large magnitude. It is therefore often useful to specify magnetic fields in units of the *gauss,* where

$$1 \text{ T} = 1 \text{ Wb/m}^2 = 10^4 \text{ G}$$

The gauss is, in fact, the unit for magnetic field in the cgs gaussian system of units; see Appendix III. The magnetic field of the earth is about 0.5 G. A typical small permanent magnet might produce a field of 100 G. Large electromagnets can produce magnetic fields up to 20,000 G, or 2 Wb/m². Still larger magnetic fields are produced only by rather special procedures.

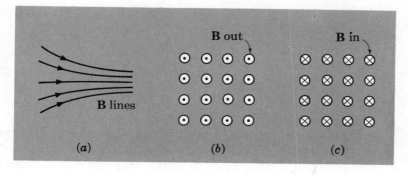

FIG. 26-3. (a) Magnetic field lines. Field lines (b) out of the paper and (c) into the paper.

26-2 MAGNETIC FLUX

Just as one may represent an electric field by electric field lines whose direction and density give the direction and magnitude of the electric field, so too one may represent the direction and magnitude of a magnetic field **B** by magnetic field lines. One should speak of magnetic *field* lines rather than magnetic lines *of force*, since the magnetic force is never parallel to the magnetic field lines. The magnetic field is strong where the magnetic lines are crowded and weak where they are far apart. A constant, or uniform, magnetic field is represented by uniformly spaced parallel straight lines.

We shall have occasion to portray magnetic field lines going into or out of the plane of the paper. In such cases we use the symbol · to show a magnetic field *out* of the paper and the symbol × to show a magnetic field *into* the paper; see Fig. 26-3. These symbols remind us, respectively, of an arrow point emerging from the paper, and the feathers on the tail of an arrow going into the paper.

The electric flux $d\phi_E$ over the surface element dS was defined as

$$d\phi_E = \mathbf{E} \cdot d\mathbf{S} \qquad (22\text{-}1)$$

Similarly, we define the *magnetic flux* $d\phi_B$ through the surface element dS, where the magnetic field has the value **B**, as

$$d\phi_B = \mathbf{B} \cdot d\mathbf{S} = (B \cos \theta)\, dS$$

That is, we take the component $B \cos \theta$ of the magnetic field lying along the normal to the surface in computing the flux through this surface element; see Fig. 26-4.

The total magnetic flux ϕ_B through a finite surface area is then given by

$$\phi_B = \int \mathbf{B} \cdot d\mathbf{S} \qquad (26\text{-}4)$$

where the integration is carried out over the surface area through which we wish to find the total magnetic flux.

For the special case in which a magnetic field **B** is uniform

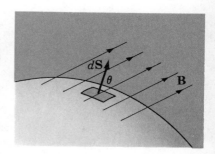

FIG. 26-4. Magnetic flux through a surface element dS.

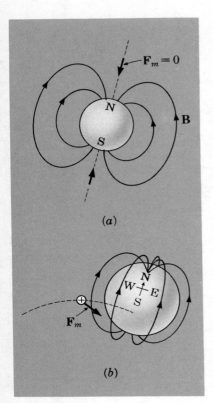

FIG. 26-5. (a) Magnetic field of the earth; charged particles approaching along the magnetic axis are undeflected. (b) Positively charged particles approaching the earth at the equator are deflected toward the east.

over a flat area A oriented at right angles to the magnetic field lines, (26-4) becomes

$$\phi_B = BA$$

or

$$B = \frac{\phi_B}{A} \qquad (26\text{-}5)$$

The magnetic field B is, from (26-5), the magnetic flux density, that is, the magnetic flux divided by the transverse area through which it penetrates. Thus, **B** can be thought of as the number of magnetic lines per unit transverse area, corresponding to the numerical value of **B**. In the SI system the unit for magnetic flux, which is a measure of the *total* number of the magnetic lines crossing a chosen transverse area, is the weber. Clearly, then, the corresponding unit for magnetic field **B** (magnetic flux density) is the Wb/m².

Example 26-1

The cosmic radiation consists of highly energetic, positively charged particles (mostly protons) which rain upon the earth in all directions from outer space. How does the magnetic field of the earth affect protons approaching it *(a)* toward the north or south poles and *(b)* at the equator?

The earth's magnetic field lines are shown in Fig. 26-5. Protons approaching the earth at the poles travel along magnetic field lines and are, consequently, undeflected (Fig. 26-5a). On the other hand, protons approaching the earth at the equator cross magnetic field lines going from geographic south to north at right angles. Using (26-3), we see that the particles are acted upon by a magnetic force which deflects them toward the east (Fig. 26-5b).

The incoming cosmic-ray particles have a large range of energies. Some are highly energetic (energies up to 10^{15} MeV), and the earth's magnetic field is feeble (of the order of 1 G near the surface). Yet, because the earth's magnetic field extends far out into space, it is able to influence the motion of these particles. For low-energy particles arriving at the equator, the earth's magnetic field deflects them so strongly that they even miss hitting the earth's atmosphere. As a consequence, the intensity of the cosmic radiation is found to be greater at the north and south poles than near the equator (the so-called latitude effect). This shows that the incoming particles are electrically charged. Moreover, experiment shows that those particles arriving at the equator come preferentially from the west (the so-called east-west effect). This shows that the primary particles carry a *positive* charge.

26-3 MOTION OF CHARGED PARTICLES IN A UNIFORM MAGNETIC FIELD

The fundamental relation for the magnetic force on a moving charge is

$$\mathbf{F}_m = q\mathbf{v} \times \mathbf{B} \qquad (26\text{-}3)$$

This force is unlike any of the other fundamental forces we have encountered heretofore, such as the gravitational force or the electric force. For one thing, the magnetic force is *velocity-dependent*. This means that it is not possible to associate a scalar potential energy with the magnetic interaction. Furthermore, the magnetic force can do *no work*. A charged particle does *not* have its kinetic energy changed as it moves through a constant magnetic field. This is easy to prove. The displacement $d\mathbf{r}$ of a moving charge over any small time interval dt is $d\mathbf{r} = \mathbf{v}\,dt$. Since the force $\mathbf{F}_m$ always acts at right angles to the particle's velocity $\mathbf{v}$, there is no component of the force along the direction of the particle's displacement $\mathbf{v}\,dt$ and, hence, no work is done by $\mathbf{F}_m$. Thus, a particle subject to a magnetic force will have its velocity changed but not its speed; it will be deflected, but it will not gain or lose energy.

What is the most general path of a charged particle in a uniform magnetic field? First consider two special cases, (1) $\mathbf{v}$ parallel (or antiparallel) to $\mathbf{B}$ and (2) $\mathbf{v}$ perpendicular to $\mathbf{B}$.

1. As (26-3) shows—and as our definition of the direction of the magnetic field requires—a charged particle moving initially in (or opposite to) the direction of a magnetic field coasts in this direction with unchanged speed; see Fig. 26-6a.
2. When the particle's *velocity* is at *right angles* to the *magnetic field*, we have the condition for *uniform circular motion*. The magnetic force points at each instant toward the center of the circular path in which the charged particle moves with the constant speed v. The charged particle encircles the magnetic field lines in a plane at right angles to $\mathbf{B}$; see Fig. 26-6b.

Taking the charge's speed to be $v_\perp$ when the velocity is perpendicular to $\mathbf{B}$, we may write Newton's second law as

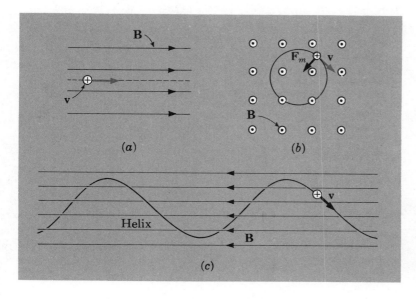

FIG. 26-6. A charged particle moving in a uniform field: *(a)* with v parallel to B, the path is a straight line; *(b)* with v perpendicular to B, the path is a circle; *(c)* for other angles between v and B, the path is a helix.

$$\mathbf{F}_r = m\mathbf{a}_r$$

where the resultant radial force $\mathbf{F}_r$ is the magnetic force $F_m = qv_\perp B$, and the radial (or centripetal) acceleration is $a_r = v_\perp^2/r = \omega^2 r$. Therefore,

$$qv_\perp B = \frac{mv_\perp^2}{r} = m\omega^2 r \qquad (26\text{-}6)$$

In this equation r is the radius of the circle and ω is the angular speed of the orbiting charged particle.

What is the path of a charged particle whose velocity $\mathbf{v}$ is *not* at right angles to $\mathbf{B}$? There is, then, a velocity component $v_\parallel = v\cos\theta$ along $\mathbf{B}$, this component being unchanged in direction and magnitude, and there is a component $v_\perp = v\sin\theta$ perpendicular to $\mathbf{B}$, this component being constant in magnitude but continuously changing direction. The resulting motion is the superposition of drift at constant speed along a straight line and uniform circular motion in a plane perpendicular to the straight line. That is, the charged particle moves in a *helix* whose axis of symmetry is the direction of the magnetic field; see Fig. 26-6c.

Thus, whenever a charged particle is projected into a uniform magnetic field, it travels at constant speed in a helix wrapped around the magnetic field lines. Suppose, now, that we have a number of charged particles of one type, such as electrons, differing in initial velocity and energy and all injected into the same constant magnetic field. Each particle moves in a helical path (or, in special cases, a circle and a straight line). The electrons differ in speed from one another, but each one maintains a constant speed. Now, it is an extraordinary fact that *all* such electrons, despite differences in their energies, speeds, and paths, will complete one loop in precisely the *same time!* All cycling particles will have the same frequency. Let us prove it.

We may write the perpendicular velocity component $v_\perp$ as $v_\perp = \omega r$, where ω is the angular velocity and r is the radius of the helix. Then (26-6) may be written

$$q\omega r B = m\omega^2 r$$

$$\omega = \frac{q}{m}B$$

the frequency $f = \omega/2\pi$ is, then,

$$f = \frac{q}{2\pi m}B \qquad (26\text{-}7)$$

The characteristic frequency given in (26-7) is called the *cyclotron frequency* (for reasons to be evident in Sec. 39-7). The cyclotron frequency f is proportional to the magnetic field and to the charge-to-mass ratio q/m of the particles, but f does *not* depend on the particle speed v nor on the radius of the orbit r.

When a charged particle moves in a nonuniform magnetic field, its path is, in general, quite complicated. Nevertheless, one can still recognize the essentially helical path if the magnetic field does not depart too greatly from a constant field. For example, charged particles are trapped in the Van Allen belts surrounding the earth, as shown in Fig. 26-7. The particles spiral about the magnetic-field lines and are actually "reflected" at regions where the magnetic field becomes strong near the earth's north and south poles. This is one illustration of the curious possibilities of an inhomogeneous magnetic field trapping charged particles in what can appropriately be called a "magnetic bottle."†

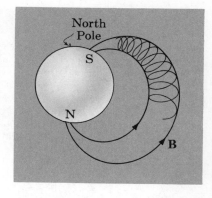

FIG. 26-7. Charged particles trapped in the earth's Van Allen belt. Note that the south magnetic pole is near the north geographic pole.

Example 26-2
What is the cyclotron frequency for electrons in a magnetic field of 3,000 G = 0.30 Wb/m²?
From Eq. (26-7) we have

$$f = \frac{qB}{2\pi m} = \frac{(1.6 \times 10^{-19} \text{ C})(0.30 \text{ N/A-m})}{(2\pi)(9.1 \times 10^{-31} \text{ kg})} = 8.4 \times 10^9 \text{ Hz}$$

This cyclotron frequency of 8.4 GHz lies in the microwave region of the electromagnetic spectrum. When the free electrons in materials immersed in an external magnetic field of 3,000 G are irradiated with microwaves of this frequency (with wavelengths of a few centimeters), the electrons absorb energy in an effect known as *cyclotron resonance*.

26-4 CHARGED PARTICLES IN UNIFORM B AND E FIELDS

Neglecting the gravitational force, the resultant force **F** on a charged particle is merely the vector sum of the electron force $\mathbf{F}_e$ and the magnetic force $\mathbf{F}_m$:

$$\mathbf{F} = \mathbf{F}_e + \mathbf{F}_m = q\mathbf{E} + q\mathbf{v} \times \mathbf{B} = q(\mathbf{E} + \mathbf{v} \times \mathbf{B}) \quad (26\text{-}8)$$

Equation (26-8) is often referred to as the *Lorentz force* relation, after H. A. Lorentz (1853–1928), who made many important contributions to the theory of electromagnetism. If one knows **E** and **B** at each point in space and the particle's charge and initial velocity, one may use the Lorentz formula to predict in complete detail the future motion of a charged particle.

Here we consider only the special cases in which both **E** and **B** are constant. We know that a charged particle moving in a uniform **E** field alone follows a parabolic path whose symmetry axis lies along the direction of **E**. When **v** is initially parallel or antiparallel to **E**, the parabola reduces into a straight line; see

---

†See S. C. Brown, Plasma Physics, *The Physics Teacher* **2**, 103 (1964), on magnetic reflection of charged particles, and W. G. V. Rosser, The Van Allen Radiation Zone, *Contemporary Physics* **5**, 198 (1963) and **6**, 255 (1964).

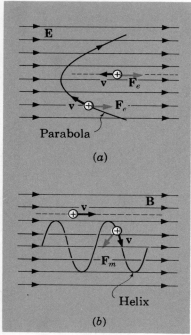

FIG. 26-8. (a) In a uniform electric field the path of a charged particle is a parabola. (b) In a uniform magnetic field the path of a charged particle is a helix.

Fig. 26-8a. As we have seen, when a charged particle moves in a constant **B** field alone, it follows a helical path whose symmetry axis lies along the direction of **B** (Fig. 26-8b). When **v** is initially parallel or antiparallel to **B**, the helix reduces into a straight line.

Suppose that we so arrange a uniform **E** field and a uniform **B** field such that the resultant force on a charged particle is always zero. This can be accomplished by having the particle move at right angles to both **E** and to **B**, which are mutually perpendicular, as shown in Fig. 26-9. The electric force F_e and magnetic forces F_m are then both at right angles to the particle's velocity and opposite to one another. F_e and F_m are of equal magnitude when

$$F_e = F_m$$

$$qE = qvB$$

$$v = \frac{E}{B} \tag{26-9}$$

If we so adjust the relative magnitudes of **E** and **B** that one charged particle will pass undeflected through the region of the two crossed fields, we see from (26-9) that *any other charged particle with the same initial velocity* will also pass through undeflected, quite apart from differences in the magnitude or sign of the charge, or of the particle's mass. Such a device is known as a *velocity selector*, since it will pass only those particles whose speed satisfies (26-9). When a beam of polyenergetic particles is projected into a velocity selector, all particles save those of the proper speed given by (26-9) will fail to pass undeviated through the selector.

Next consider a *momentum selector*. It is easy to show that a uniform *magnetic field* provides a direct method of measuring the magnitude of a particle's *linear momentum* $\mathbf{p} = m\mathbf{v}$. When a charged particle moves in a circular path of radius r in a uniform magnetic field **B**, we have

$$F_r = ma_r$$

$$qvB = \frac{mv^2}{r}$$

$$p = mv = qrB \tag{26-10}$$

The linear momentum is directly proportional to the path radius and to the magnitude of the magnetic field. If B and q are known and r is measured, the momentum p can be computed. Thus, one can easily measure the momentum of a charged particle whose wake of small bubbles (or droplets) is photographed in a bubble chamber (or cloud chamber) by measuring r and B (the product rB is sometimes referred to as the "magnetic rigidity"); see Fig. 36-14.

Finally, we consider a *kinetic-energy selector*. A uniform electric field provides a direct method of measuring a particle's kinetic energy $K = \tfrac{1}{2}mv^2 = p^2/2m$. Consider the simple arrangement shown in Fig. 26-10a, where a uniform field **E** is established

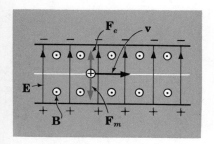

FIG. 26-9. A velocity selector: A charged particle moves through crossed E and B fields.

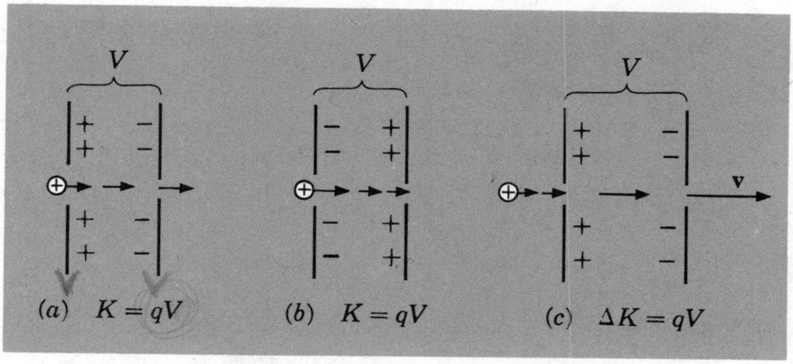

FIG. 26-10. A charged particle moving across *(a)* an accelerating electric potential from rest, *(b)* a retarding potential to rest, *(c)* a potential which changes the kinetic energy.

by a potential difference V across two parallel oppositely charged conducting plates separated by the distance d. From Eq. (23-13) we have $E = V/d$, so that measuring the potential difference, which is easily done, immediately gives the magnitude of the uniform field. A particle accelerated from rest by a potential difference V acquires a kinetic energy K, where

$$K = qV = qEd \qquad (26\text{-}11)$$

Similarly, a charged particle with initial kinetic energy K traveling in the opposite direction and parallel to the electric field lines is brought to rest just in front of the second conducting plate (and therefore fails to register an electric current there) when $K = qV$, where V is now the retarding potential; see Fig. 26-10b. For example, electrons with an initial kinetic energy of 50 eV are brought to rest by a retarding potential of 50 V. Even more generally, whenever a charged particle moves through a potential difference V, as in Fig. 26-10c, its kinetic energy changes by

$$\Delta K = qV$$

The potential difference V may either increase or decrease the particle's kinetic energy, depending on whether the electric field is an accelerating or retarding field.

In summary, knowing **B** we find the momentum $\mathbf{p} = m\mathbf{v}$, and knowing the potential difference V (or **E**) we may find the kinetic energy $K = \tfrac{1}{2}mv^2$. We have also seen that crossed electric and magnetic fields provide a means of measuring the charged particle's speed v [Eq. (26-9)]. It is clear, then, that there are a variety of ways in which one might construct a *mass selector:*

1. Measure v and $p = mv$; that is, use a velocity selector and a momentum selector as shown in Fig. 26-11a.
2. Measure p and $K = \tfrac{1}{2}mv^2 = p^2/2m$; that is, use a kinetic-energy selector starting with particles at rest and a momentum selector, as shown in Fig. 26-11b.
3. Measure v and $K = \tfrac{1}{2}mv^2$; that is, use a velocity selector to produce an undeflected beam, then turn off the magnetic field, leaving a kinetic-energy selector, as shown in Fig. 26-11c.

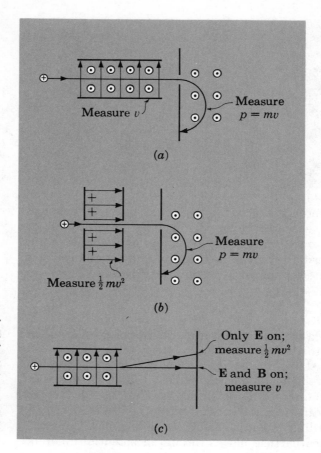

FIG. 26-11. Arrangements for measuring the mass of a charged particle: *(a)* a velocity selector followed by a momentum selector, *(b)* an accelerating electric field followed by a momentum selector, and *(c)* a velocity selector followed by a deflecting electric field (the Thomson arrangement for measuring the electron q/m ratio).

Still other combinations can, of course, be devised. The arrangement shown in Fig. 26-11c was used by J. J. Thomson in 1897, to measure the mass of the electron. Strictly, any of these procedures will yield only the charge-to-mass ratio q/m, not the mass alone. Therefore, one must have an independent measurement of a particle's charge, as through the Millikan oil-drop experiment, to permit the mass to be computed. Although Thomson's experiment gave q/m for an electron, the electron mass was known only after $q = e$ was measured in the Millikan experiment (Sec. 21-5). The experimental arrangements described above are used in mass spectrometers for measuring atomic masses with high precision. It is a relatively easy matter to distinguish among the isotopes of an element; indeed, one can measure atomic masses to a few parts in 10^5.

Example 26-3

A bubble-chamber photograph shows a proton moving in a circular arc 20 cm in radius at right angles to a magnetic field of 0.30 T. What is the proton's kinetic energy?

Using Eq. 26-10), we have

$$K = \tfrac{1}{2}mv^2 = \frac{p^2}{2m} = \frac{(qrB)^2}{2m}$$

$$= \frac{(1.6 \times 10^{-19}\text{ C} \times 0.20\text{ m} \times 0.30\text{ N/A-m})^2}{2(1.67 \times 10^{-27}\text{ kg})}$$

$$= 1.0 \times 10^{-12}\text{ J} = 17\text{ MeV}$$

Although the classical relations for the kinetic energy and linear momentum are applicable to proton kinetic energies of several million electron volts, for much higher energies one must use relativistic relations for K and p. The formulas of classical mechanics apply only when the particle speed is much less than the speed of light, 3×10^8 m/s. (See Chap. 35.) In the case of electrons, the relativistic relations are required at much lower energies; that is, the classical relations are seriously in error for electron kinetic energies much above a few times 10^4 eV.

26-5 MAGNETIC FORCE ON A CURRENT–CARRYING CONDUCTOR

Since a magnetic field exerts a force on a moving charge, it must also exert a force on the moving charges in a current-carrying conductor. Consider Fig. 26-12, where the free electrons move to the left at the drift speed v_d through a conductor of small cross-sectional area A. The conventional current i is then to the right. The conductor is immersed in a constant magnetic field **B**, whose direction is into the paper. Applying Eq. (26-3), we would expect that each free electron, subject to the magnetic force $\mathbf{F}_m$, would move clockwise in a circular orbit within the conductor. However, those electrons near the upper surface of the conductor cannot leave the surface because of the attractive force exerted by the conductor's fixed lattice ions. Thus, electrons reaching the upper surface experience an electric force downward which balances the magnetic force and the electrons continue to move to the left at the drift speed v_d. The upper side quickly becomes negatively charged and the lower side positively charged, thereby producing an electric field upward within the conductor—the so-called Hall effect. Free electrons within the conductor then experience no net force, the upward magnetic force counterbalanced by the downward electric force, and the electrons drift to the left. In this manner the magnetic force is transmitted to the lattice of the conductor, and the conductor as a whole is subject to an upward magnetic force $\mathbf{F}_m$ whose direction is perpendicular to both the magnetic field lines and the direction of the current.

Note that with a current-carrying conductor we need *not* be concerned with a possible electric force on the conducting wire from an external electric field in addition to the magnetic force. Any section of the conductor is electrically neutral. One might

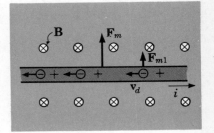

FIG. 26-12. Magnetic force on a conduction electron in a current-carrying conductor.

then say that the magnetic force on a current-carrying conductor represents a pure magnetic effect, since there is no possible complicating additional electric force.

Let us compute the magnitude of the magnetic force $\mathbf{F}_m$ acting on a straight current-carrying conductor of length L, oriented at right angles to an external magnetic field $\mathbf{B}$, when a current i exists in it. The total force $\mathbf{F}_m$ acting over the length L is simply the magnetic force $\mathbf{F}_{m1}$ on *one* electron multiplied by the total number of free electrons within this length. If the number of conduction electrons per unit volume is n, then the total number within the length L and cross-sectional area A is nAL. We have, then,

$$F_m = (nAL)F_{m1}$$

where, by (26-3),

$$F_{m1} = qv_d B$$

Therefore,

$$F_m = nALqv_d B$$

Now, the current i is given by

$$i = nqv_d A \qquad (24\text{-}2)$$

Substituting this relation in the equation for F_m above, we have

$$F_m = iLB \qquad (26\text{-}12)$$

Thus, the magnetic force on a conductor 1 m long, carrying a current of 1 A, and immersed in a transverse field of 1 Wb/m² is 1 N. One may, in fact, use this relation to measure a magnetic field: One measures the magnetic force on a conductor of known length carrying a known current.

It is easy to generalize (26-12) by noting that, if the current-carrying conductor makes an angle θ with respect to the magnetic field, the force F_{m1} on one electron becomes $qv_d B \sin \theta$, and (26-12) contains an additional factor, $\sin \theta$. Then we may write (26-12) in vector form as follows:

$$\mathbf{F}_m = i\mathbf{L} \times \mathbf{B} \qquad (26\text{-}13)$$

The vector $\mathbf{L}$ represents, in magnitude, the length of the conductor; its direction is chosen to correspond to the direction of positive charge flow, or the direction of conventional current.

Equation (26-13) holds only for a *straight* conductor. We can, of course, never have an infinitely long straight conductor, inasmuch as there must always be a return loop of some sort to complete the electric circuit. Therefore, it is useful to have a relation giving the magnetic force $d\mathbf{F}_m$ on only a short segment of conductor of length $d\mathbf{L}$. From (26-13) we have

$$d\mathbf{F}_m = i\, d\mathbf{L} \times \mathbf{B} \qquad (26\text{-}14)$$

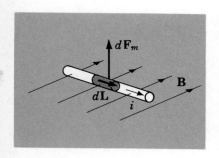

FIG. 26-13. Magnetic force $d\mathbf{F}_m$ on an element of length $d\mathbf{L}$ of a current-carrying conductor.

Once again the direction of $d\mathbf{L}$ is taken to be the direction of the current; see Fig. 26-13. One may find the resultant force on a conductor of arbitrary shape by integrating the magnetic force on the length elements that comprise it, taking into account, as well, that the magnetic field **B** may not have the same magnitude and direction at all points in the current loop. Such an integration is, in general, quite complicated, although it may be simple in certain special cases, as the following example illustrates.

Example 26-4

A conductor of arbitrary shape carrying current i runs from point A to point B in two dimensions, as shown in Fig. 26-14a. A constant magnetic field of magnitude B acts into the plane of the paper. What is the resultant force on this segment of conductor?

We may approximate the continuously varying conductor shape by a series of straight-line segments lying, respectively, parallel and perpendicular to the straight line L_{AB} from point A to B. The magnetic forces on these segments are, from Eq. (26-14), in the directions shown in Fig. 26-14b. We see that the total force component parallel to the line L_{AB} is zero, each force to the right being matched by an equal force to the left. This leaves only the magnetic forces perpendicular to L_{AB}. The resultant force perpendicular to the line L_{AB} is exactly the same as the magnetic force on a straight conductor running from A to B. Therefore, the net magnetic force has a magnitude $F_m = iL_{AB}B$, and its direction is perpendicular to the line joining the end points. The arbitrarily shaped conductor is equivalent to a straight wire between its end points carrying the same current.

Now suppose that we form a current loop by imagining points A and B to be brought together. From the arguments above, the resultant magnetic force on the loop must be *zero*. A little thought will show that this conclusion applies, whatever the shape of the loop and whether the loop lies entirely in a plane or not, and that it also is independent of the loop's orientation relative to the magnetic field lines. *The magnetic field must, however, be uniform.* Although a closed current loop is subject to no resultant magnetic force in a uniform magnetic field, it may be subject to a resultant torque, as we shall see in the next section.

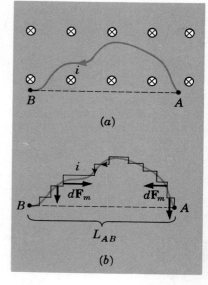

FIG. 26-14. *(a)* A current-carrying conductor of arbitrary two-dimensional shape in an external magnetic field. *(b)* Equivalent conductor comprised of segments parallel and perpendicular to the line joining the end points.

26-6 MAGNETIC TORQUE ON A CURRENT LOOP

Consider the situation shown in Fig. 26-15a: A rectangular loop, with sides W and L and carrying a current i, is placed in a uniform magnetic field. The angle between the magnetic field lines and the normal to the plane of the loop is θ. We wish to find the magnetic forces on the four sides of the loop. (One cannot, of course, have a current loop of this sort unconnected to a source of current. There must be two lead wires, one to bring the current into the loop and another to take the current out of it. If these two lead wires are placed next to each other, as by wrapping or twisting them together, then the resultant magnetic force on the lead wires is zero, since there are equal currents in opposite directions.)

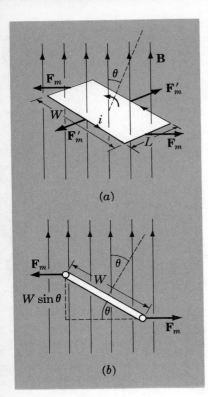

FIG. 26-15. (a) Magnetic forces on a rectangular current-carrying conducting loop in a magnetic field. (b) A side view of part (a).

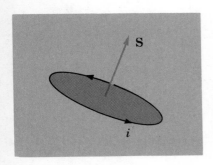

FIG. 26-16. A current-carrying loop enclosing an area S. The vector S is normal to the surface and related to the current sense through the right-hand rule.

We see that there are two pairs of equal but opposite magnetic forces acting on the four sides of the current loop, and the resultant *force* is *zero*. Although the forces on the sides of length W are along the same line, the two forces on the sides of length L are not. Therefore, these equal and opposite forces apply a torque to the loop, tending to align its normal with the direction of the magnetic field. Let us find the torque.

Equation (26-12) gives the magnitude of each force on the sides of length L:

$$F_m = iLB$$

We choose as axis for computing the torques the lower of the two sides of length L; see Fig. 26-15b. Then, in magnitude, we have

$$\tau = r_\perp F = (W \sin \theta)(iLB)$$

We may write this in simpler form by recognizing LW as the area A of the current loop:

$$\tau = iAB \sin \theta \qquad (26\text{-}15)$$

The direction of the torque is out of the paper; that is, the magnetic torque on the loop turns it in the counterclockwise sense. We may write (26-15) in vector form as follows:

$$\boldsymbol{\tau} = i\mathbf{S} \times \mathbf{B} \qquad (26\text{-}16)$$

Here the magnitude of the vector $\mathbf{S}$ represents the loop's area. The direction of $\mathbf{S}$ is perpendicular to the plane of the loop and is so chosen that we may use the right-hand rule to relate the sense of the current to the direction of the normal.† See Fig. 26-16.

Although we have derived (26-16) for a rectangular current loop, this relation gives the magnetic torque on a plane current loop of *any* shape. To see this, we imagine the conducting wire to be approximated by small straight segments at right angles to one another. Of course, if a loop consists of N turns (in the *same* sense) rather than a single turn, the torque of (26-16) is increased by the factor N.

The magnetic torque on a current-carrying conducting loop is the basic physical principle used in the galvanometer. The current to be measured passes through a coil of many turns placed in the magnetic field of a permanent magnet; see Fig. 26-17. The coil is attached to an elastic helical spring which applies a restoring torque to the coil whenever the coil is turned from its equilibrium position. According to Hooke's law (Sec. 9-4), the restoring torque τ_r is proportional to the angular displacement, which here we shall call ϕ,

---

†The quantity $i\mathbf{S}$, where i is the current and $\mathbf{S}$ is the vector representing the area, is called the *magnetic moment* of the conducting loop. It can be shown that the torque $\boldsymbol{\tau}$ exerted on the magnetic moment by the magnetic field can be written $\boldsymbol{\tau} = \boldsymbol{\mu} \times \mathbf{B}$, where $\boldsymbol{\mu} = i\mathbf{S}$, and that the potential energy U_B of the magnetic moment is given by $U_B = -\boldsymbol{\mu} \cdot \mathbf{B}$.

$$\tau_r = \kappa\phi$$

where κ, the torque constant of the spring, is typically very small so that the galvanometer coil is extremely sensitive to angular displacements. When a current exists in a coil of N turns, the magnitude of the *m*agnetic torque acting on it is, by (26-15),

$$\tau_m = NiAB \sin\theta$$

The angle θ is set equal to 90° when the coil is in the zero-current equilibrium position, and $\phi = 0$ when $\theta = 90°$. Since $\theta + \phi = 90°$, we have

$$\tau_m = NiAB \cos\phi$$

Equating the two torques gives

$$\kappa\phi = NiAB \cos\phi$$

If the deflection angle ϕ is not too large, $\cos\phi \approx 1$, and

$$i \propto \phi \tag{26-17}$$

The angular deflection is proportional to the current.

In actuality, the pole faces of galvanometers (see Fig. 26-17) are shaped so that the magnetic field **B** is always perpendicular to the loop, $\tau_m = NiAB$, and (26-17) holds for any angle ϕ.

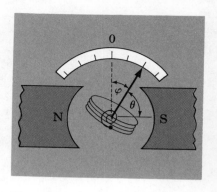

FIG. 26-17. Elements of a galvanometer.

SUMMARY

The fundamental magnetic effect is that a moving electric charge creates, in addition to the electric field, a magnetic field which acts to produce a magnetic force on a second moving charge.

The magnetic force on a particle of charge q and velocity **v**, moving in magnetic induction field **B**, is

$$\mathbf{F}_m = q\mathbf{v} \times \mathbf{B} \tag{26-3}$$

In the SI system of units the magnetic induction field is measured in newtons per ampere-meter, or webers per square meter, or tesla (N/A-m = Wb/m² = T). The magnetic flux ϕ_B is defined as

$$\phi_B = \int \mathbf{B} \cdot d\mathbf{S} \tag{26-4}$$

The magnetic force, always acting at right angles to the moving charge's velocity, does *no* work. A particle of charge q and mass m travels in a helix in a uniform magnetic field at the cyclotron frequency

$$f = \frac{q}{2\pi m} B \tag{26-7}$$

The linear momentum p of a charged particle traveling in a circle of radius r at right angles to the field lines of the magnetic field B is given by

$$p = qrB \qquad (26\text{-}10)$$

The magnetic force $d\mathbf{F}_m$ on an element of length $d\mathbf{L}$, in which a current i flows (along the direction of $d\mathbf{L}$) and which is immersed in a magnetic field $\mathbf{B}$, is

$$d\mathbf{F}_m = i\, d\mathbf{L} \times \mathbf{B} \qquad (26\text{-}14)$$

The magnitude of the magnetic torque on a conducting loop of area A, carrying current i and immersed in a magnetic field B, is

$$\tau = iAB \sin\theta \qquad (26\text{-}15)$$

where θ is the angle between the normal to the plane of the loop and the magnetic field.

PROBLEMS

26-1. Express the unit for magnetic field, the tesla, in terms of fundamental units (kg, m, s, and C).

26-2. (a) Find the magnitude of the magnetic field required to confine a 10-keV electron to an orbit 1.0 m in diameter. (b) What is the electron's orbital period? (The electron mass is 9.1×10^{-31} kg.)

26-3. A charged particle having a charge-to-mass ratio equal to that of the electron is found to be moving in a circle 20 cm in radius in the clockwise sense, when a constant magnetic field of 5.0 G acts upward, normal to the plane of path. (a) Is the particle an electron or a positron (a positively charged electron)? (b) What is its speed? (c) What is its kinetic energy?

26-4. At the equator the earth's magnetic field is horizontal, toward the north, and has the magnitude 0.30 G. (a) In what direction (east or west) should a proton be fired to encircle the earth at constant speed? (b) What is the proton's momentum? (c) What is the proton's kinetic energy? (d) Is the gravitational force significant here? (Assume—incorrectly—that Newton's formulation of mechanics is valid.)

26-5. A beam of electrons in an oscilloscope has an energy of 10 keV/electron. If the beam is oriented such that the electrons travel from east to west, and the earth's magnetic field has a horizontal component of 0.20 G north and a vertical component of 0.55 G down: (a) What is the path of the electron beam? (b) What is the beam's radius of curvature?

26-6. An electron and a proton are injected simultaneously into a uniform magnetic field of 0.50 Wb/m². The injection velocity is transverse to the magnetic field. Assuming that the proton mass is 1,836 times the electron mass, how many revolutions will the electron make before it again meets the proton at the injection point?

26-7. At some location in the northern hemisphere on earth, the magnitude of the earth's magnetic field is 0.40 G and the dip angle, the angle between the direction of the field lines and the horizontal (pointing north), is 70°. A magnetic compass points to geographic north at this location. What is the magnetic flux through a circular loop of 10-cm² area when (a) the loop is in the horizontal plane; (b) it is in the vertical east-west plane?

26-8. Find the angle with respect to the direction of the uniform magnetic field $\mathbf{B}$ at which one must project a charged particle into this field, such that the distance it moves along the field line is just equal to the distance it moves along a circular arc perpendicular to the field.

26-9. A beam of charged particles is directed along the field lines of a constant magnetic field. If the particle velocities coincide exactly with the field lines, the particles are, of course, undeflected. Now suppose that the particles are very slightly misaimed. Show that the magnetic field acts as a sort of lens, in the sense that the particles are returned to the desired direction.

26-10. An electron and a proton in the same uniform magnetic field trace out circles of the same radius. (a) What is the ratio of the electron to proton momentum? (b) What is the ratio of the electron to proton kinetic energy? (The proton is 1836 times more massive than the electron.)

26-11. A certain charged particle traces out a circular orbit of radius R when it is projected into a uniform magnetic field at right angles to the direction of the magnetic field lines. What is the radius of the particle's circular path if it is injected with twice the kinetic energy?

26-12. Protons injected into an evacuated chamber perpendicular to a uniform magnetic field describe circular orbits with a period of 1.0 μs and a radius of 1.0 m. Find (a) the magnitude of the magnetic field and (b) the kinetic energy of the protons.

26-13. Suppose that we wish to support a straight section of No. 10 copper wire by means of the magnetic interaction between a current in the wire and the earth's magnetic field at a point where $B = 0.32$ G (**B** is horizontal and points north). The linear mass density of the wire is 0.468 g/cm. (a) What should be the orientation of this piece of wire if the magnetic force on it is to be the maximum possible? What are the (b) direction and (c) magnitude of the *electron* current (in amperes) which will cause the wire to be suspended in mid-air, that is, with an upward magnetic force just balancing the weight?

26-14. Protons are observed moving in a circle 10 cm in radius in a uniform magnetic field $B = 1.0$ T. What are the (a) momentum, (b) speed, and (c) kinetic energy of the protons?

26-15. A 10-eV electron enters a uniform magnetic field of 0.20 T, the angle between the electron's initial velocity and the magnetic field direction being 45°. What is the electron's kinetic energy after it has traveled a total distance of 50 cm through the magnetic field?

26-16. A charged particle moving through a strictly uniform magnetic field traces out a helix whose symmetry axis coincides with the direction of the magnetic field. Suppose that a positively charged particle moves through a magnetic field whose magnitude increases gradually with distance, such that the particle moves toward regions of increasingly stronger magnetic field. Will the particle's path—an approximate helix—expand or contract in radius as the particle progresses into the regions of stronger magnetic field?

26-17. Consider a particle of charge q and mass m which is moving at constant speed v_0 in a helical path in a uniform magnetic field **B**. The constant angle between the particle velocity $\mathbf{v}_0$ and **B** is θ. Show that the pitch of the helix (defined as the distance, measured along the axis, between corresponding points on successive turns) is given by $2\,(mv/qB)\cos\theta$.

26-18. A trapped electron in one of the Van Allen belts (see Fig. 26-7) is momentarily in a region where the earth's magnetic field is 1.0×10^{-2} G. (a) How long does it take for the electron to make one revolution in the circular orbit perpendicular to the magnetic field?

(b) What is the radius of the electron orbit if it has an orbital speed of 10^5 m/s?

26-19. When protons traveling north enter a uniform magnetic field of 0.80 Wb/m² in the downward direction, they are bent into horizontal circles of a 25-cm radius. What are the magnitude and direction of a uniform electric field, applied over the same region as the magnetic field, which will allow the protons to pass through undeflected?

26-20. Charged particles of mass m and charge q are accelerated from rest through a potential difference V. Then these particles pass at right angles to a magnetic field of flux density B and move in circular arcs of radius r. Derive an expression for q/m in terms of the measured quantities $V, B,$ and r.

26-21. Moving transverse to a uniform magnetic field of magnitude B, a beam of ions, each of charge q and mass m, travels in a circular arc of radius R. Find the magnitude of the uniform electric field which, when applied perpendicular to the plane of the particle velocity and the field **B**, will cause the ions to move in a straight line.

26-22. Consider the mass-measuring device shown in Fig. 26-11a, in which the magnetic field in the velocity selector is B_1 and in the momentum selector is B_2. Show that the particle mass is given by $m = qB_1B_2R/E$ where q is the particle charge, E is the velocity-selector electric field, and R is the radius of the trajectory in the momentum selector.

26-23. Show that the kinetic energy of a charged particle entering the selector of Fig. 26-11c is inversely proportional to the transverse displacement of the particle.

26-24. A beam of singly ionized chlorine atoms, composed of a mixture of the two isotopes of masses 35 and 37 u (1 u = 1.66×1.0^{-27} kg) enters perpendicularly a uniform magnetic field of 0.50 T. All ions have the same speed, 2.0×10^5 m/s. After bending through 180°, the ions strike a photographic plate. (a) What is the separation distance between the two regions on the film where the ions strike? (b) If the beam also contained some doubly ionized chlorine atoms, where would they fall on the film?

26-25. A conducting rod of mass 100 g slides without friction on two parallel fixed conducting bars separated by 50 cm and attached to an inclined plane of angle 10°. A constant magnetic field of 0.10 Wb/m² acts upward. See Fig. P26-25. What current must exist in the movable conductor to have it ascend the inclined plane at constant speed?

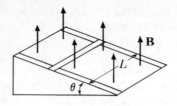

FIG. P26-25

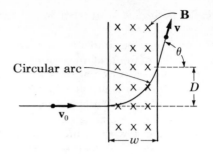

FIG. P26-28

26-26. (a) Find the internal force between adjoining differential segments of a plane circular loop, 5.0 cm in radius and carrying a current of 100 A, which is immersed in a uniform magnetic field of 2.0 T perpendicular to the plane of the loop. (b) Is the force tensile or compressive?

26-27. A coil 2.0 cm in diameter has 300 turns. What is the maximum torque on this coil when it carries a current of 10 mA while immersed in a constant magnetic field of 500 G?

26-28. A collimated beam of particles of charge q, mass m, and velocity $\mathbf{v}_0$, is incident on a region of width W (measured along $\mathbf{v}_0$) in which there is a uniform magnetic field $\mathbf{B}$ at right angles to $\mathbf{v}_0$. See Fig. P26-28. (a) Show that the beam emerges from the magnetic field at an angle with respect to $\mathbf{v}_0$ of $\theta = \sin^{-1}(qBW/mv)$, having been deflected laterally by a distance $D = (mv/qB)(1 - \sqrt{1 - (qBW/mv)^2})$. Note that, if the entering beam has a finite transverse diameter, the emerging beam would *not* have a *circular* cross section. (b) Would the transverse dimension of the emerging beam in the plane of Fig. P26-28 be contracted or expanded relative to that of the entering beam?

CHAPTER 27

The Sources of the Magnetic Field

In the last chapter one part of the magnetic interaction was treated: the magnetic force on a charged particle moving in a magnetic field. Now we complete our discussion of the magnetic force by discussing how the magnetic field is created by moving electric charges, or electric currents.

27–1 THE OERSTED EXPERIMENT

As we have seen, the presence of a magnetic field can be determined by observing whether a charged particle in motion is acted upon by a magnetic force. In a similar way, the existence of a magnetic torque on a compass needle or a current-carrying loop would indicate the presence of a magnetic field.

Electric charges in motion create a magnetic field. This was first demonstrated in the fundamental experiments of H. C. Oersted (1777–1851) in 1820. Oersted showed that a current-carrying conductor could produce torques on small magnets surrounding it; see Fig. 27-1. The magnetic field lines consist of circular loops surrounding the conductor and lying in planes perpendicular to that of the conductor. It is found that the sense of the magnetic field is related to the sense of the electric current through a right-hand rule: If the thumb of the right hand points in the direction of i, the right-hand fingers give the sense of the **B** lines. Reversing the current then reverses the direction of the magnetic field at each point. By observing the frequency at which a small magnet oscillates when displaced from alignment with the magnetic field (high frequency for high B, low frequency for low B), one finds that the magnitude of the magnetic field drops off as one goes away from the conductor.

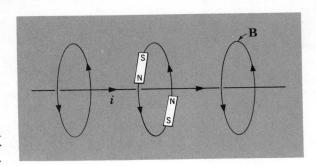

FIG. 27-1. The Oersted effect: A current-carrying conductor generates magnetic field loops.

One also finds that the magnitude of **B** at each point in space is directly proportional to the current i.

27–2 THE BIOT-SAVART LAW

Here we set down the fundamental relation giving the magnetic field produced by a moving electric charge or, equivalently, by a small element of a current-carrying conductor. It is called the *Biot-Savart law*. Since the magnetic field from a single moving charge or from a short element of current is very feeble, one tests the correctness of the Biot-Savart law indirectly by noting that all predictions of the magnetic field made from it are in accord with observation.

Consider Fig. 27-2, where we see a positive point-charge q in motion with the velocity **v**. We wish to find the magnetic field at some point in space, say P. The radius vector from the moving charge to the point P is **r**, and the angle between the vectors **v** and **r** is ϕ. The direction of the magnetic induction field **B** at point P, perpendicular to the plane containing the vectors **v** and **r**, is given by a right-hand rule: For a positive charge, if the vector **v** is imagined to be turned through the smaller angle into alignment with **r** by the right-hand fingers, then the right-hand thumb gives the direction of **B**.

The magnitude of **B** at point P is given by

$$B = \frac{\mu_0}{4\pi} \frac{qv \sin \phi}{r^2} \tag{27-1}$$

The magnetic field is proportional to the charge and its speed and inversely proportional to the square of the distance. Since B also varies as $\sin \phi$, the field is zero along the line containing the velocity vector.

The quantity μ_0, called the *permeability of free space*, is a constant and, in the SI system of units, is *assigned* the numerical value

$$\mu_0 = 4\pi \times 10^{-7} \text{ Wb/A-m} \tag{27-2}$$

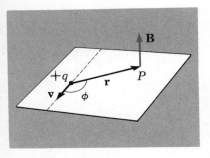

FIG. 27-2. A moving charged particle and its magnetic field B at point P.

The reader should verify that, when μ_0 is given the units webers per ampere-meter, the magnetic field B in (27-1) has the units webers per square meter or their equivalent, newtons per ampere-meter. The factor 4π appears in the Biot-Savart relation *so that* it will not appear later in the mathematical formulation of Ampère's law [Eq. (27-12)]. (This parallels the situation for the electric force: In the rationalized unit system, 4π appears in Coulomb's law but then does not appear in Gauss' law.)

From (27-1) we see that the magnetic induction field at a perpendicular distance of 1 m from a 1-C charge moving at 1 m/s is $B = 10^{-7}$ Wb/m², or 10^{-3} G.

We can incorporate the relative directions of **v**, **r**, and **B** in Biot-Savart's law by writing (27-1) in vector form with the cross product:

$$\mathbf{B} = \frac{\mu_0}{4\pi} \frac{q\mathbf{v} \times \mathbf{r}}{r^3} \qquad (27\text{-}3)$$

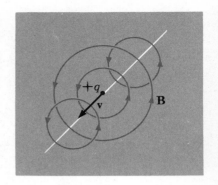

FIG. 27-3. Magnetic field loops from a moving charged particle.

Note that the distance r now appears to the third power in the denominator, to compensate for the additional factor r in the numerator.

It is easy to see that the magnetic field lines associated with a moving charge consist of circular loops centered about the moving charge and lying in planes perpendicular to the particle's velocity, as shown in Fig. 27-3. Imagine all quantities on the right side of (27-3), except the direction of **r**, to be fixed. Then turn the vector **r** about the vector **v**, while keeping the angle ϕ between them fixed. The vector **B**, unchanged in magnitude, then turns through a circle in a transverse plane.

Although the Biot-Savart relation may be expressed fundamentally in terms of the magnetic field produced by a moving charge, we shall deal primarily with current-carrying conductors in applying this relation. Therefore, we wish to have an expression for the magnetic field $d\mathbf{B}$ produced by an infinitesimal current element dl carrying the current i. The direction of the length element $d\mathbf{l}$ is chosen to correspond to the direction of conventional current through it; that is, $d\mathbf{l}$ corresponds to the direction of moving positive charges. We found earlier that

$$q\mathbf{v} = i\,d\mathbf{l} \qquad (26\text{-}3),\ (26\text{-}14)$$

Using this identity in (27-3), we have the Biot-Savart law for a current element:

$$d\mathbf{B} = \frac{\mu_0}{4\pi} \frac{i\,d\mathbf{l} \times \mathbf{r}}{r^3} \qquad (27\text{-}4)$$

The field $d\mathbf{B}$ is now written in differential form. This is necessary, since all elements of a current loop contribute their respective differential fields at a given point. The resultant field **B** is the vector sum of the infinitesimal fields; see Fig. 27-4.

The Biot-Savart Law   SEC. 27-2

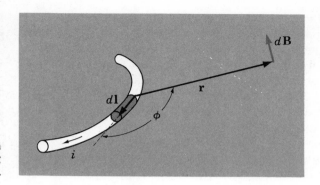

FIG. 27-4. Magnetic field from an element $d\mathbf{L}$ of a current-carrying conductor.

Now that we know the magnetic field created by one electric charge in motion [Eq. (27-3)] and also the magnetic force it exerts on a second moving charge [Eq. (26-3)], we can write down the entire magnetic interaction. Taking charge q_a as creating the magnetic field $\mathbf{B}_a$, which then acts on moving charge q_b, (27-3) and (26-3) give

Magnetic interaction between two charges:

$$\mathbf{B}_a = \frac{\mu_0}{4\pi} \frac{q_a \mathbf{v}_a \times \mathbf{r}}{r^3} \quad \text{and} \quad \mathbf{F}_b = q_b \mathbf{v}_b \times \mathbf{B}_a \qquad (27\text{-}5)$$

The radius vector $\mathbf{r}$ extends from q_a to q_b.

The corresponding relations for the electric force between the two charges are the following.

Electric interaction between charges:

$$\mathbf{E}_a = \frac{1}{4\pi\varepsilon_0} \frac{q_a \mathbf{r}}{r^3} \quad \text{and} \quad \mathbf{F}_b = q_b \mathbf{E}_a \qquad (27\text{-}6)$$

We now know the part of the force between charged particles that does not depend on their speeds (the electric force), and we also know the velocity-dependent part of the interaction between charged particles (the magnetic force). We might be inclined to think that, insofar as basic electromagnetism is concerned, this is the whole story. It is not! Not only can electric and magnetic fields be created by electric charges at rest or in motion, but an electric field can be created by a changing magnetic field (Chap. 28) and a magnetic field can be created by a changing electric field (Sec. 30-1). Furthermore, we shall see (Chap. 30) that electric and magnetic fields can become detached from accelerated electric charges to exist independently as electromagnetic waves.

27-3 THE MAGNETIC FIELD FOR SOME SIMPLE GEOMETRIES

The law of Biot-Savart, Eq. (27-4), gives the magnetic field produced by an *infinitesimal* current element. No such isolated current element can exist by itself, however. There is always a complete loop of current in any circuit. Therefore, to find the magnetic field arising from a current-carrying conductor at any point, we must sum up, or integrate, the contributions from all significant current elements.

Center of Circular Loop. First consider the magnetic field at the center of a plane circular loop of radius r carrying current i; see Fig. 27-5. We must, of course, have current leads to and from the loop, but if these two lead wires are placed together, their magnetic effects cancel exactly because they have equal currents in opposite directions. At the center of the loop the direction of the magnetic field $d\mathbf{B}$ produced by the element $d\mathbf{l}$ along the circumference is perpendicular to the plane of the paper and is out of the paper. Indeed, all elements of the loop contribute to the field at the center in the same direction, so that we may sum these contributions algebraically. The angle ϕ between $\mathbf{r}$ and $d\mathbf{l}$ is 90°; therefore, (27-4) gives for the magnitude of $\mathbf{B}$:

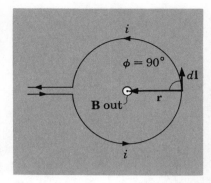

FIG. 27-5. Magnetic field at the center of a circular conducting loop.

$$B = \int dB = \frac{\mu_0}{4\pi} i \frac{\int dl}{r^2} = \frac{\mu_0}{4\pi} \frac{i 2\pi r}{r^2}$$

The radius vector $\mathbf{r}$ here is equal in magnitude to the radius of the loop. In summing the length element around the circle above we get simply the circumference $2\pi r$, so that the field B at the center of the loop is

$$B = \frac{\mu_0}{4\pi} \frac{2\pi i}{r} \tag{27-7}$$

Figure 27-6 shows the magnetic field lines in a plane transverse to the plane of the loop through the axis. Near the conducting wires the field lines are almost circular, since at such locations the more distant current elements are unimportant. The field lines are symmetrical about the loop's axis of symmetry. We may, of course, find the direction of the magnetic field at the loop's center by applying the right-hand rule to an element of the loop, taking the right-hand thumb to give the current direction and the right-hand fingers to give the magnetic field loops. But we can also relate the current and field directions as follows. Let the fingers of the right hand give the sense of the *current* around the loop; then the right-hand thumb points in the direction of the magnetic *field* at the loop's center.

To get an idea of the magnitude of the field from a circular loop, suppose that a current of 10 A flows through a loop having a radius of 2π cm. Then, from (27-7), the field at the center is

$$B = \frac{(4\pi) \times 10^{-7} \text{ Wb/A-m})(10 \text{ A})}{2(2\pi \times 10^{-2} \text{ m})} = 10^{-4} \text{ Wb/m}^2 = 1 \text{ G}$$

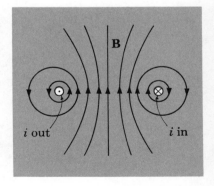

FIG. 27-6. Magnetic field lines in a plane transverse to a circular conducting loop.

The field in this situation is just slightly greater than the earth's magnetic field. Of course, if there were N nearly coincident circular loops, the field at all points would be enhanced by the factor N.

Long Straight Conductor. Now we find the magnetic field produced by a current i through an infinitely long straight conducting wire. Of course, no such conductor can be constructed, since there must always be return leads; therefore, the results we shall find will apply whenever the distance from a straight wire is small compared with the distance out to where the straight wire bends.

We first concentrate on the field $d\mathbf{B}$ at the point P produced by a straight current element $i\,d\mathbf{x}$, Fig. 27-7. For the upward current direction shown, the field at P, a perpendicular distance R from the conductor, is into the paper. The angle between the current element and the radius vector $\mathbf{r}$ is ϕ. From (27-4) we have

$$dB = \frac{\mu_0}{4\pi}\frac{i\,dx\sin\phi}{r^2} \tag{27-8}$$

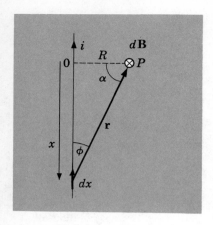

FIG. 27-7. Magnetic field from a long straight current-carrying conductor.

Every current element of the conductor produces a field into the paper at P, so that we merely integrate this equation over the entire length of the conductor, to find the total field. The quantities x, r, and ϕ are not independent; in fact, we can write (27-8) in terms of a single variable. The integration is most easily carried out if we choose the angle α, the complement of ϕ, as the variable. From the geometry of Fig. 27-7 we see that

$$x = R\tan\alpha$$

Hence, $$dx = R\sec^2\alpha\,d\alpha$$
Also, $$r = R\sec\alpha$$
and $$\sin\phi = \cos\alpha$$

Making these substitutions in (27-8), we have

$$dB = \frac{\mu_0}{4\pi}\frac{i(R\sec^2\alpha\,d\alpha)(\cos\alpha)}{(R\sec\alpha)^2} = \frac{\mu_0}{4\pi}\frac{i}{R}\cos\alpha\,d\alpha$$

To account for the entire infinite length of the conductor, we integrate from $\alpha = -90°$ to $90°$. Therefore,

$$\int dB = \frac{\mu_0 i}{4\pi R}\int_{-\pi/2}^{\pi/2}\cos\alpha\,d\alpha$$

$$B = \frac{\mu_0}{4\pi}\frac{2i}{R} \tag{27-9}$$

The magnetic field consists of circular loops circling the conductor in planes transverse to the wire, the magnitude of the field falling off inversely as the distance from the conductor, as was first tested in 1820 by J. B. Biot (1774–1862) and F. Savart (1791–1841).

[Using (27-9) it is easy to show that, for $i = 10$ A and $R = 2$ cm, the field B is just 1 G.]

It is interesting to compare (27-9), which gives the magnetic field from a long straight current-carrying conductor, with (21-5), which gives the electric field of a uniformly charged infinite wire: $E = (1/4\pi\varepsilon_0)(2\lambda/r)$. Both are of similar form, and this is hardly surprising. The electric and magnetic fields both fall off inversely as r, a consequence of the fact that the electric and magnetic interactions between *point*-charges are both inverse-*square*. The current i is a measure of the source of the magnetic field, and the linear charge density λ is a measure of the source of the electric field. The quantities ε_0 and μ_0 are proportionality constants for the electric and magnetic interactions, respectively.

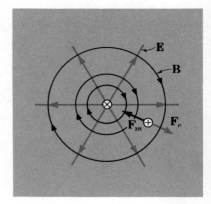

FIG. 27-8

Example 27-1

A straight conducting wire with current i into the paper carries a uniform positive charge of linear charge density λ. At what speed can a charged particle be fired parallel to the straight conductor and in the direction of the current so as to maintain its constant speed in a straight line?

The radial electric field and the circular concentric magnetic field loops are shown in Fig. 27-8. A positively charged particle traveling at speed v into the paper is subject to an electric force $F_e = qE$ radially outward and a magnetic force $F_m = qvB$ radially inward. (A *negatively* charged particle also traveling *into* the paper is subject to electric and magnetic forces in the reverse directions. On the other hand, for a charged particle of either sign traveling *out* of the paper, both the electric and magnetic forces are in the *same* direction.)

The charged particle is undeflected when

$$F_e = F_m \quad \text{or} \quad qE = qvB$$

$$v = \frac{E}{B}$$

Note that the relation is the same as for the velocity selector, Eq. (26-9). The magnetic and electric fields are given by

$$B = \frac{\mu_0 i}{2\pi R} \tag{27-9}$$

$$E = \frac{\lambda}{2\pi\varepsilon_0 R} \tag{21-5}$$

Therefore, substituting for E and B above, we have

$$v = \frac{\lambda}{i(\varepsilon_0 \mu_0)}$$

Note that the charged particle's distance R from the wire cancels out.

It is easy to show that the magnetic force is typically much less than the electric force by assuming that a charged particle moves at 1.0 m/s parallel to a wire with a current of 1.0 A. Then the relation for v, given above, shows that the required linear charge density is a mere 10^{-17} C/m = 10^{-5} pC/m.

The Magnetic Field for Some Simple Geometries   SEC. 27-3

27–4 THE MAGNETIC FORCE BETWEEN CURRENT-CARRYING CONDUCTORS

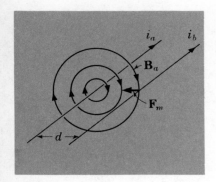

FIG. 27-9. The magnetic force of conductor a on the parallel conductor b.

Let us find the magnetic force between two long straight parallel wires separated by a distance d. We first suppose that the currents i_a and i_b are in the same direction; see Fig. 27-9. The current i_a produces magnetic field loops with a magnetic field at the site of wire b given by (27-9),

$$B_a = \frac{\mu_0}{4\pi} \frac{2i_a}{d} \qquad (27\text{-}9)$$

The magnetic force $\mathbf{F}_b$ on current i_b in the magnetic field $\mathbf{B}_a$ is, from (26-13), to the left and of magnitude

$$F_b = i_b L B_a \qquad (26\text{-}13)$$

where L is the length of conductor b. Combining these equations we have

$$F_b = \frac{\mu_0}{4\pi} \frac{2 i_a i_b L}{d} \qquad (27\text{-}10)$$

It is easy to show that the magnetic force on conductor a arising from the magnetic field of i_b is of the same magnitude as F_b and that this force is to the *right;* that is, two straight conductors carrying currents in the *same* direction magnetically *attract* each other. On the other hand, if the currents are in *opposite* directions, the two conductors *repel* one another, as can be confirmed by applying the vector form of (26-13). Thus, parallel currents attract and antiparallel currents repel.

One may take the magnetic attraction between two parallel current-carrying conductors as *the fundamental magnetic effect*. Since the conductors are electrically neutral, there can be no complicating electric forces. Moreover, there is no need to employ magnetic materials as such. The magnetic interaction between straight conductors was first studied in 1822 by A. M. Ampère (1775–1836), who found that the force varied inversely as the separation distance, in agreement with (27-10).

As an example, we find the magnitude of the magnetic force from (27-10), assuming two conductors each carrying a current of exactly 1 A and separated by exactly 1 m. The magnetic force per unit length, F/L, on either conductor is

$$\frac{F}{L} = \frac{\mu_0}{4\pi} \frac{2 i_a i_b}{d} = \frac{(10^{-7} \text{ Wb/A-m})(2)(1 \text{ A})^2}{1 \text{ m}}$$

$$= 2 \times 10^{-7} \text{ N/m}$$

Note that 1 Wb = 1 N-m/A. The force per meter is *exactly* 2×10^{-7} N.

Recall that the permeability constant μ_0 is *assigned* the value $4\pi \times 10^{-7}$ Wb/A-m. Actually, the relation giving the magnetic force between two parallel straight conductors serves to *define the ampere* as the unit of electric current in the SI system of units. At long last we have the *definition* of the ampere (and therefore also of the coulomb, which is one ampere-second): The equal currents in two parallel straight conductors separated by 1 m in a vacuum are each, *by definition,* exactly 1 A when the magnetic force per unit length between them is exactly 2×10^{-7} N/m. The quantities can be measured with precision: the magnetic force compared through a balance with a standard weight, and the lengths calibrated against the standard meter. In practice, the so-called current balance used in measuring the magnetic force employs two parallel concentric plane circular *loops* of current, rather than two straight conductors, because complications arising from noninfinite straight conductors are thereby avoided.

27-5 GAUSS' LAW FOR MAGNETISM

First recall Gauss' law for electricity:

$$\phi_E = \oint \mathbf{E} \cdot d\mathbf{S} = \frac{q}{\varepsilon_0} \tag{22-3}$$

The total electric flux ϕ_E through any closed surface is proportional to q, the net charge enclosed by the surface. Gauss' law implies that the coulomb force between point-charges is inverse-square, that electric field lines terminate on electric charges, and that electric fields may be superposed as vectors.

A similar law can be written for the magnetic flux ϕ_B:

$$\phi_B = \int \mathbf{B} \cdot d\mathbf{S} \tag{26-4}$$

Since the magnetic force between moving point-charges is also inverse-square, one may properly represent the flux density **B** of a magnetic induction field by magnetic field lines. Such lines indicate the direction of the field at any point in space by their tangent, and their "density" through a transverse area is proportional to the magnitude of the field **B**.

We find the total magnetic flux $\oint \mathbf{B} \cdot d\mathbf{S}$ through any closed surface by adding the contributions $\mathbf{B} \cdot d\mathbf{S}$ from all small surface elements. By convention, the vector $d\mathbf{S}$ is taken as positive when it points toward the outside of a closed surface; see Fig. 27-10. Then, the magnetic flux is positive when the magnetic lines pass outward through a surface and is negative when the lines pass inward. If the total magnetic flux over a closed surface is zero, it is implied that an equal number of magnetic lines enter and leave the surface. This is precisely what is found in every case. All experimental observations are consistent with the statement:

$$\phi_B = \oint \mathbf{B} \cdot d\mathbf{S} = 0 \tag{27-11}$$

This is *Gauss' law for magnetism.* In effect, it says that magnetic

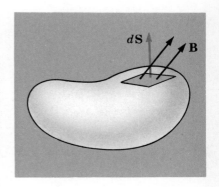

FIG. 27-10. Magnetic flux through the outward surface element $d\mathbf{S}$.

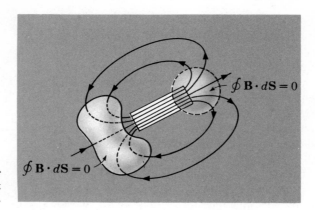

FIG. 27-11. Gauss' law for magnetism: The net magnetic flux through any closed surface is zero.

field lines always form closed loops. If this were not the case, one would find situations in which the magnetic flux out of some closed surface would not match exactly the magnetic flux into the same surface. That is, if the magnetic flux were not always zero through a closed surface, the magnetic field lines would terminate on magnetic "charges," or single magnetic poles, in the same way that electric field lines terminate on electric charges.

What about permanent magnets, then, with which one clearly sees (in magnetized iron filings) field lines emanating from what appear to be magnetic poles at the ends of the magnet, as in Fig. 27-11? Here the field lines *pass through* the interior of the magnet, again forming closing loops.

Gauss' law for magnetism asserts that no isolated magnetic poles, or magnetic monopoles, exist in nature. The most subtle experiments confirm this.

27–6 AMPÈRE'S LAW

Ampère's law is an alternative to the Biot-Savart relation for electric currents and the magnetic field they generate. In some ways it is analogous to Gauss' law for electricity, which is an alternative formulation of Coulomb's law. Ampère's law is particularly useful for finding the magnetic field in situations of high geometrical symmetry.

First consider again the relation for the magnetic field produced by an infinitely long straight conductor:

$$B = \frac{\mu_0}{4\pi} \frac{2i}{r} \qquad (27\text{-}9)$$

Rearranging, we have

$$B(2\pi r) = \mu_0 i$$

We recognize the left-hand side of this equation as the field B, at all points a distance r from the conductor, multiplied by $2\pi r$, the circumference of a circular loop at this distance; see Fig. 27-12. We may write this as

$$\oint \mathbf{B} \cdot d\mathbf{l} = \mu_0 i \qquad (27\text{-}12)$$

where the line integral on the left-hand side of the equation implies that we take the component of **B** along the path element $d\mathbf{l}$ around a closed loop. Of course, with a circular loop about a straight conductor, the direction of **B** always coincides with the direction of $d\mathbf{l}$, provided that we traverse the loop in the same direction as the magnetic field line. The current i is the current enclosed by the loop about the conductor.

Now, the remarkable thing about (27-12), which is the formal statement of *Ampère's law*, is this: It holds for *any closed* path about *any* configuration of electric conductors when the current i is taken to be the net current passing through any area enclosed by the loop. Let us prove it.

We first restrict ourselves to the field from a single infinitely long straight conductor and to path loops in a plane perpendicular to the conductor. Consider the loop of Fig. 27-13a. Here we have a path consisting of circular arcs coinciding with the field lines, and of radial lines at right angles to the field lines. The field has magnitude B_1 for radius r_1 and B_2 for radius r_2. From (27-9), $B_1 r_1 = B_2 r_2$; that is, the field falls off inversely as the distance from the conductor. Along the radial lines there is no contribution to the line integral, since $\mathbf{B} \perp d\mathbf{l}$ along these segments. Moreover, if the path is traversed in the same sense as the magnetic field lines, we see that, for the circular arcs, the decrease from B_1 to B_2 in going from r_1 to r_2 is exactly matched by the increase in the arc lengths. For this loop, then, we again have

$$\oint \mathbf{B} \cdot d\mathbf{l} = \mu_0 i \qquad (27\text{-}12)$$

Now suppose the path loop does *not* enclose the conductor, as in Fig. 27-13b. Again there is no contribution to the line integral along the radial lines. Once more $B_1 r_1 = B_2 r_2$, so that the contributions along the two circular arcs are of the same magnitude. But the signs now differ. If the path is traversed in the same direction as the field along the arc of radius r_2, it is traversed in the opposite sense for r_1, and conversely. Consequently, the *total* line integral is now zero:

$$\oint \mathbf{B} \cdot d\mathbf{l} = 0$$

This corresponds to zero current threading the chosen path loop.

It is easy to deal with more complicated paths. One may regard a path of arbitrary shape as replaced by a number of circular arcs and radial segments; see Fig. 27-14.

As we found earlier (see Fig. 27-13) the radial sections do not contribute to the line integral, and all circular sections subtending

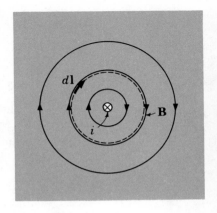

FIG. 27-12. Magnetic field for a straight conductor.

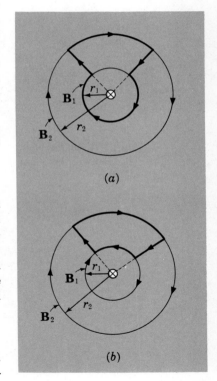

FIG. 27-13. *(a)* When the loop encloses the current, the line integral $\oint \mathbf{B} \cdot d\mathbf{l}$ is $\mu_0 i$; *(b)* when it does not, the line integral is zero.

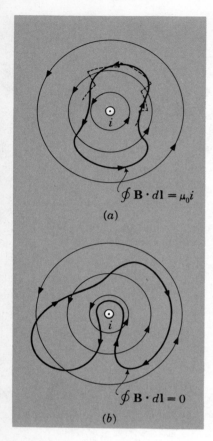

FIG. 27-14. Ampère's law applied to a general loop shape (a) enclosing the conductor and (b) not enclosing the conductor.

the same angle give the same contribution. In every case, then, the line integral of the magnetic induction field around a closed path is equal to the current threading the loop multiplied by μ_0. Notice, further, that if we reverse the sense in which the path is traversed, a factor -1 appears on the left side of (27-12). This corresponds to the fact that when the current direction is reversed, the sense of the magnetic field lines is also reversed. A path that does *not* lie in a plane tranverse to the conductor introduces no complications, since the magnetic field from a long straight conductor is confined entirely to transverse planes; traversing a path element parallel to the conductor will not contribute to the line integral.

Now suppose that we have two or more infinitely long conductors, not necessarily parallel to one another nor carrying the same current. This introduces nothing new. Magnetic fields are added by using the rules of vector addition; therefore, if Ampère's law applies to any one conductor, it also applies to a collection of conductors. We must, however, take the current i to be the *net* current enclosed by the chosen path loop, using *different* signs for currents into and out of one side of the path loop. To show that Ampère's law holds for conductors of any shape, not merely long straight conductors, one proves that *any* current distribution and its associated magnetic field can be replaced by equivalent straight conductors (the proof is complicated and will be omitted here†).

In summary, if we know the line integral of the magnetic field around a closed loop, we know at once the net current threading the loop. If the line integral is zero, there can be no net current within. Currents *outside* the loop make no contribution. Although Ampère's law is altogether general, it is easy to apply in practice only when the magnetic field has a geometry that can be deduced in advance on the basis of symmetry. Then we know what sort of path to choose. (This is reminiscent of Gauss' law for electricity: If the geometry of the electric field is known in advance on the basis of symmetry, we know how best to choose the gaussian surface.)

One final remark concerning Ampère's law must be made at this point: As written in (27-12), it is *not* complete. In Sec. 30-1 we shall see that a changing electric flux, as well as an enclosed current, can contribute to the line integral of the magnetic field.

Example 27-2

A coaxial conductor consists of a solid inner cylinder, of radius r_1 and carrying a current I, and a concentric outer conductor, of inner and outer radii r_2 and r_3 and carrying a current I in the opposite direction; see Fig. 27-15a. What is the magnetic field at all points (a) between the two conductors, (b) outside the outer conductor, and (c) within the inner conductor?

---

†For a simple proof, see the short article by F. Reines in *Am. J. Phys.* **39,** 838 (1971).

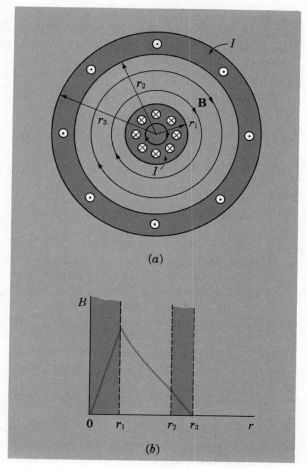

FIG. 27-15. (a) A pair of coaxial conductors carrying equal but opposite currents. (b) Magnitude of B as a function of r for part (a).

(a) On the basis of symmetry alone one can assert that the magnetic field, wherever it is not zero, must consist of circular loops concentric with the conductors. For the region between the two conductors, we choose a circular path of radius r, to apply Ampère's law. Only the current I *within this loop* contributes to the field; the current in the outer conductor, whatever its magnitude or direction, has *no* effect on the region inside. From Ampère's law we have, then,

$$\oint \mathbf{B} \cdot d\mathbf{l} = \mu_0 i$$

$$B(2\pi r) = \mu_0 I$$

$$B = \frac{\mu_0 I}{2\pi r} \quad \text{for} \quad r_1 < r < r_2$$

(b) For a circular path around the *outside* of both conductors, the *net* enclosed current $I - I$ is zero. Consequently, the field is zero at all exterior points:

$$B = 0 \quad \text{for} \quad r > r_3$$

Ampère's Law   SEC. 27-6   443

(c) We assume the current I to be uniformly distributed over the cross section of the inner conductor. For a circular path of radius r inside the inner conductor, the fraction of the current within the path loop is just in the ratio of the circular areas, $\pi r^2/\pi r_1^2$; that is, within radius r the current is $(r/r_1)^2 I$. Then Ampère's law gives

$$B(2\pi r) = \mu_0 \left(\frac{r}{r_1}\right)^2 I$$

$$B = \frac{\mu_0 r}{2\pi r_1^2} I \quad \text{for} \quad 0 < r < r_1$$

The variation of B with r is shown in Fig. 27-15b.

27-7 THE SOLENOID

A solenoid consists of a tightly wound helix of conducting wire. It is, in effect, a series of circular conducting loops arranged into a cylindrical shell. The magnetic field within a long solenoid is essentially uniform near its center, as shown in Fig. 27-16. Altogether, the external magnetic field configuration for a long solenoid is quite similar to that of a long permanent magnet.

Let us find the magnitude of the magnetic field at the center of an infinitely long solenoid. Strictly of course, this implies a solenoid whose length is much greater than its diameter. Each loop carries a current i_1 and the *number of turns per unit length* is designated n. Consider, then, the closed path shown in Fig. 27-17, to which we apply Ampère's law. The magnetic field just outside the solenoid at A must be essentially zero since the same number of lines outside are spread over an infinite area, and there is no contribution to the line integral along portion A. The field inside at point B must be parallel to the solenoid axis; if it were not, this would imply a noninfinite solenoid length. Consequently, the field is perpendicular to the dashed path along the transverse portions C and D, and these parts of the closed path do not contri-

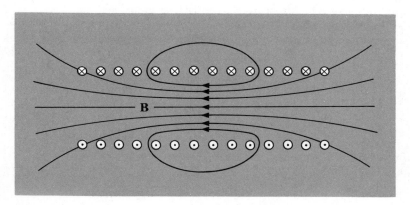

FIG. 27-16. Magnetic field lines for a long solenoid.

bute to the line integral of **B**. The only contribution, then, comes from the inside segment of length L. Within the length L the *total* current enclosed is nLi_1. Therefore, Ampère's law gives

$$\oint \mathbf{B} \cdot d\mathbf{l} = \mu_0 i$$
$$BL = \mu_0 nLi_1$$
$$B = \mu_0 ni_1 \qquad (27\text{-}13)$$

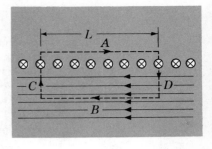

FIG. 27-17. Ampère's law applied to a solenoid.

The magnetic field at any point near the center of a long solenoid is uniform, independent of the transverse distance from the axis, and dependent only on the number of turns per unit length and the current i_1 through each.

SUMMARY

The magnetic field **B** generated by a charge q moving at velocity **v** is given by the Biot-Savart law,

$$\mathbf{B} = \frac{\mu_0}{4\pi} \frac{q\mathbf{v} \times \mathbf{r}}{r^3} \qquad (27\text{-}3)$$

where **r** is the radius vector from the charge to the point at which **B** is measured. The velocity-dependent magnetic force between two moving point-charges varies inversely as the square of the distance between them. The permeability of the vacuum, μ_0, is assigned the magnitude $4\pi \times 10^{-7}$ Wb/A-m.

Another form of the Biot-Savart law, applicable for the field $d\mathbf{B}$ produced by a current element $d\mathbf{l}$, is

$$d\mathbf{B} = \frac{\mu_0}{4\pi} \frac{i\, d\mathbf{l} \times \mathbf{r}}{r^3} \qquad (27\text{-}4)$$

Applying the Biot-Savart relation to an infinitely long straight conductor, one finds that the magnetic field falls off inversely as the perpendicular distance from the conductor. Likewise, the magnetic force between two parallel current-carrying conductors is inversely proportional to their separation distance. The ampere is defined in terms of the magnetic force in such an arrangement.

The net magnetic flux through any closed surface is always zero.

$$\oint \mathbf{B} \cdot d\mathbf{S} = 0 \qquad (27\text{-}11)$$

According to Ampère's law, the line integral of the magnetic induction field around any closed path is proportional to the net current through any area enclosed by the path:

$$\oint \mathbf{B} \cdot d\mathbf{l} = \mu_0 i \qquad (27\text{-}12)$$

PROBLEMS

27-1. The beam of a particle accelerator consists of 20-MeV protons with a current of 1.0 μA. (a) What is the magnitude of the magnetic field at a distance of 1.0 mm from the beam, assumed to be of infinitesimal cross section? (b) What is the magnitude of the electric field at the same distance from this beam?

27-2. A 5.0-MeV proton moves along the positive X axis at constant speed. Find the directions and magnitudes of the electric and magnetic fields at the point $x = 1.0 \times 10^{-10}$ m, $y = 1.0 \times 10^{-10}$ m at the instant the proton passes through the origin.

27-3. An electron and a proton travel side by side in the same direction at the speed of 1.0×10^6 m/s. They are a distance of 10^3 Å apart. (a) What is the electric force between the two particles? (b) What is the magnetic force between the two particles?

27-4. At the equator the earth's magnetic field is 0.30 G north. This field is to be annulled by the magnetic field at the center of a circular conducting loop 4.0 cm in radius. What is (a) the required orientation of the loop and (b) the direction and magnitude of the loop current?

27-5. The electric force is a *central* force which lies along the line connecting two electrically charged particles. Show that the magnetic force between two electrically charged particles is *not*, in general, a central force. (*Hint:* Consider, for example, one particle traveling north, another east.)

27-6. Two identically charged particles travel side by side with equal speeds, each v, in the same direction. (a) Show that the particles repel through an electric force but attract one another through a magnetic force. (b) Show that the ratio of the magnitudes of the magnetic to electric forces is given by $(\varepsilon_0\mu_0)v^2$. (In Sec. 30-3 it is shown that the speed of light c is equal to $1/(\varepsilon_0\mu_0)^{1/2}$; therefore the ratio of magnetic to electric force above is equal to $(v/c)^2$.

27-7. Two long straight wires separated by L each carry a current i. Find the magnitude of the magnetic field at point P in Fig. P27-7, when the currents are (a) in the same direction and (b) in opposite directions.

27-8. A conductor consists of a circular arc of 60° and two lead wires to this arc, which are radial from the center of the arc and effectively infinite in length; see Fig. P27-8. Find the magnetic field at the center of the arc. (*Hint:* Consider separately the fields from the straight wires and the circular arc.)

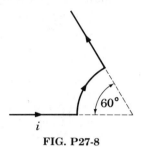

FIG. P27-8

27-9. A square loop L on a side is placed a distance R from an infinitely long straight wire carrying a current i. The plane of the loop contains the conducting wire, and two sides of the square are parallel to this wire. See Fig. P27-9. Compute the total magnetic flux through the square.

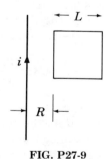

FIG. P27-9

27-10. Consider a circular coil of N turns and radius a which is carrying a current i. Show that the magnitude of the magnetic field at a point along the axis of the loop a distance x from the center is given by

$$B = \frac{(\mu_0/4\pi)(2NiS/x^3)}{(1 + a^2/x^2)^{3/2}}$$

where $S = \pi a^2$ is the area of the coil. Note that at large distances $a/x \to 0$ so that when B varies inversely as the *cube* of the distance from the center of the coil: $B = (\mu_0/4\pi)(2NiS/x^3)$. This result is very similar to

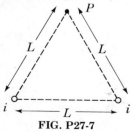

FIG. P27-7

that found for the electric dipole (see Problem 21-26). For this reason, the quantity $NiS = \mu_d$, the product of the current in the loop and its area, is called the *magnetic dipole moment*.

27-11. The *geomagnetic* field—the magnetic field of the earth—is essentially that of a magnetic dipole (see Problem 27-10). At the earth's surface and beyond it, it is proper to use the relations which describe the field under conditions where $a/x \to 0$. At the north magnetic pole the magnitude of the earth's *axial* field is 0.62 G. Find the magnetic moment which represents the geomagnetic field. (Because of regional and local anomalies on and within the earth's surface, the principal usefulness of the magnetic dipole model of the geomagnetic field is in its quantitative description of the field at points in space within a few earth-radii of the surface. The *geomagnetic* poles do *not* coincide with the *geographic* poles: The earth's magnetic axis is inclined about 11.5° to the axis of rotation.)

27-12. Use the equation for the magnetic field along the axis of a current loop (see Problem 27-10) to show that the magnitude of the magnetic field at a point on the axis of a solenoid of *finite* length is given by $B = (\mu_0/4\pi)(2\pi ni)(\cos\theta_2 - \cos\theta_1)$, where n is the number of turns per unit length along the axis and the angles θ_1 and θ_2 are shown in Fig. P27-12. (*Hint:* In performing the integration required, it is convenient to change variables from x to θ, using the relation $a/x = \tan\theta$.)

27-13. Use Eq. (27-13) for the field on the axis of an infinitely long solenoid to find the magnetic field *at one end* of a very long solenoid. (*Hint:* Conceptually, an infinitely long solenoid can be considered at any point along its length to be divided into two semi-infinite solenoids which are joined together at that point.)

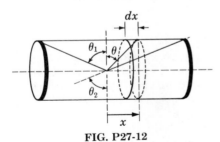

FIG. P27-12

27-14. Two long conducting wires, each carrying a current i, are parallel and in a vertical plane near the earth's surface. The upper wire is firmly supported mechanically, and both wires are assumed to remain essentially straight. Find the current i for which a section of the lower conductor is to be supported against its weight by means of an upward magnetic force. The linear mass density of the wires is 1.0 g/m and they are 10 cm apart. Take $g = 9.8$ m/s^2.

27-15. A charged particle is moving at 1.0×10^6 m/s parallel to a long straight wire carrying a current of 1.0 A. If the wire is also electrically charged, the particle can continue moving at constant speed parallel to the wire at *any* distance from the wire. What is the required linear charge density?

27-16. A square conducting loop of length 10 cm on a side carries current of 2.0 A and is a distance of 4.0 cm from an infinitely long straight wire carrying a current of 1.0 A, as shown in Fig. P27-16. (*a*) Find the resultant magnetic force on the loop. (*b*) Find the resultant magnetic torque on the loop.

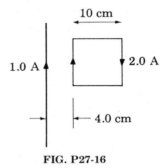

FIG. P27-16

27-17. Two parallel straight conductors 50 cm long carry currents each of 10 A in opposite directions. At what separation distance is the repulsive force between them 1.0×10^{-4} N?

27-18. A conducting wire of radius 1.00 mm carries a current of 100 A. What is the magnetic field (*a*) at the surface of the wire, (*b*) at a point 0.50 mm from the center of the wire, and (*c*) 2.00 mm from the center of the wire?

27-19. A large number N of long straight conducting wires are arranged symmetrically on the curved surface of a right circular cylinder of radius R, the wires being parallel to the cylinder axis. Each conductor carries the same current i. Using Ampère's law, describe the magnetic field inside and outside the cylinder.

27-20. Refer to Example 27-2. Show that within the region $r_2 < r < r_3$ the magnetic field is given by

$$B = \frac{\mu_0 I}{2\pi r} \frac{r_3^2 - r^2}{r_3^2 - r_2^2}$$

27-21. The two basic relations for the magnetic interaction between electrically charged particles

$$\mathbf{F} = q\mathbf{v} \times \mathbf{B} \quad \text{and} \quad \mathbf{B} = \frac{(\mu_0/4\pi)(q\mathbf{v} \times \mathbf{r})}{r^3}$$

involve a cross product and, therefore, the right-hand rule for cross products. Suppose that some person consistently uses a *left*-hand rule for cross products, rather than the conventional right-hand rule. Would she get right or wrong the actual *motions* of particles predicted by the relations above?

27-22. The magnitude of the electric field for three simple arrangements of electric charge, a point charge, an infinite straight line of charge, and a uniformly charged infinite plane, varies with the distance from the source, respectively, as $1/r^2$, $1/r^1$, and $1/r^0$ (that is, inverse-square, inverse, and constant). The magnetic force between *point* electric charges varies as $1/r^2$. Arguing by analogy, (a) what is the expected variation in magnetic field magnitude with distance from an infinitely long straight conductor? (b) What is the expected variation in magnetic field magnitude from an infinite sheet of current, that is, an infinite collection of infinitely long conductors placed side-by-side in a plane?

27-23. Two parallel straight wires each carry a current of the same magnitude. Find the relation for the magnitude of the resultant magnetic field at the point P which is in the plane of the two wires and between them. Specify the location of P by its distance x from the midline, everywhere a distance a from the two wires. See Fig. P27-23. (a) Show that $B = 2\rho/(1 - \rho^2)B_1$ when the currents are parallel and (b) $B = 2/(1 - \rho^2)B_1$ when the currents are antiparallel. The dimensionless ratio ρ is defined as $\rho = x/a$ and $B_1 = (\mu_0/4\pi)2i/a$ is the magnitude of the magnetic field produced by a current in *one wire only* when P is on the centerline and $x = 0$. (c) When the currents are antiparallel, what is the width of the strip on either side of the midline where B is within 3.0 percent of its magnitude on the centerline?

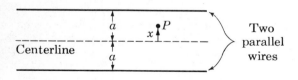

FIG. P27-23

27-24. A solid conducting cylinder of radius R carries a current i. How must the current per unit area vary with distance r from the center of the conductor, if the magnitude of the magnetic field within the conductor is to have the same value at all points?

27-25. A "conducting sheet" is produced by laying together on a flat surface many long straight wires, each carrying current i in the same direction. Find the magnetic field at any point near the conducting sheet.

27-26. Use symmetry arguments and Ampère's law to find the magnetic field between two infinite sheets conducting equal currents of magnitude i in opposite directions.

27-27. A solenoid consists of 200 turns of wire wound around a cylindrical shell 2.0 cm in radius and 50 cm in length. (a) What is the magnetic field at the solenoid's center when a current of 2.0 A is sent through the solenoid? (b) What is the average magnetic field at one end?

27-28. A toroid is a solenoid bent into the form of a doughnut. Show that the magnetic field at any point within the current loops varies inversely with the distance from the center of the toroid and that the field is zero at all other points.

27-29. Show that the magnetic field contributed by a current i in a straight section of wire of *finite* length is given by $B = (\mu_0/4\pi)(i/R)(\sin \alpha_2 - \sin \alpha_1)$, where the angles are shown in Fig. P27-29. [*Hint:* Reexamine the analysis which leads to Eq. (27-9).]

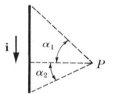

FIG. P27-29

27-30. Use the relation developed in Problem 27-29 to show that the magnetic field at the center of a square coil of side L and of N turns each carrying a current i is given by $B = (\mu_0/4\pi)8\sqrt{2}\,Ni/L$.

27-31. Two identical plane circular coils of N turns each, with axes coinciding, are placed parallel to each other on their common axis a distance R apart, where R is the radius of either coil. See Fig. P27-31. A current i exists in each coil in the same sense, so that the resultant magnetic field in the central region between the coils is enhanced. (a) Use the relation for the axial field of a circular current (see Problem 27-10) to show that the magnitude of the axial field at the midpoint m on the axis, a distance $R/2$ from each coil, is given by

$$B_m = \left(\frac{16\sqrt{5}}{25}\right)\beta = 1.431\beta$$

where $\beta = (\mu_0/4\pi)(2NiS/a^3) = \tfrac{1}{2}\mu_0 Ni/a$; $S = \pi a^2$ is the

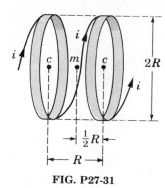

FIG. P27-31

area of either coil. (b) Show that the combined field at the center c of either coil is

$$B_c = \left(1 + \frac{\sqrt{2}}{4}\right)\beta = 1.354\beta$$

Finally, note that $B_c/B_m = 1.354/1.431 = 94.58$ percent. (This arrangement was first exploited by Helmholtz, and such a pair are called *Helmholtz coils*. They are of great practical importance for achieving a uniform magnetic field over an appreciable region of space.)

27-32. In the Bohr planetary model of the hydrogen atom an electron is imagined to move at 2.2×10^6 m/s in a circular orbit of radius 5.3×10^{-11} m, with a proton at the center. Find (a) the equivalent electron current, and (b) the magnitude and direction of the magnetic field at the proton produced by the electron current.

CHAPTER 28

Electromagnetic Induction and Inductance

Up to this point, whenever we considered an electric field, this field had its origin directly in electric charges. In this chapter we shall deal with electric fields not originating from electric charges directly, but having their origin in a changing magnetic flux. The phenomenon is known as electromagnetic induction, and its quantitative statement is, after its originator, Faraday's law.

We begin by considering one simple example, the emf produced in a conducting loop moved through a magnetic field. Next the relationship between the induced emf and the magnetic flux change is found. Other examples of electromagnetic induction are given, and the general form of Faraday's law, which relates the rate of magnetic flux change to the nonconservative electric field that it generates, is obtained.

We define the self-inductance of a conductor and compute it for a simple geometry. Finally, we derive the energy of a current-carrying inductor and the energy density of the magnetic field.

28-1 MOTIONAL EMF

When an electric current passes through a conductor immersed in a magnetic field, a magnetic force acts on the conductor to move it. Does the reverse happen? That is, if one applies a force to a conductor to move it through a magnetic field, is an electric current generated in the conductor? The answer is *yes*, as we shall see. Figure 28-1 shows a conductor of length l in a uniform magnetic flux density **B** directed into the paper. Suppose that the conductor

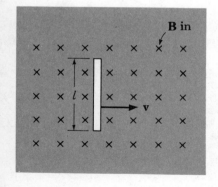

Fig. 28-1. A conductor moved through a magnetic field.

is moved to the right with the velocity **v**, this velocity being at right angles to both the conductor and the magnetic field lines. As the conductor is pulled to the right, so are the free electrons within it. Now, if a negatively charged particle moves to the right through a magnetic field into the paper, it is subject to a magnetic force $\mathbf{F}_m$ acting downward, as shown in Fig. 28-2.

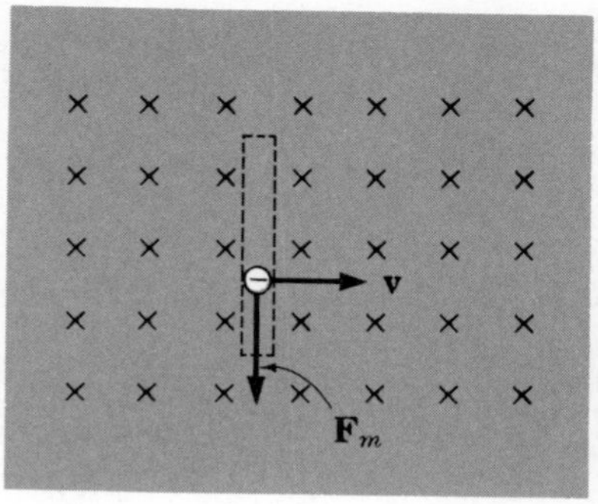

Thus, moving the conductor to the right causes the free electrons within it to be forced toward the lower end of the conductor and to leave the upper end. (Of course, comparable magnetic forces also act on the positive charges and on the bound electrons, but these charges are locked to the conductor lattice and do not move up or down.) The lower end of the conductor then acquires a net negative charge while the upper end acquires a positive charge of the same magnitude. Electrons will not, however, continue to move toward the lower end indefinitely. The first electrons arriving at the lower end repel the remaining free electrons. That is, an electric field is created within the rod by the charge separation. As more charge accumulates at the ends, the electric field increases, until the electric force on a free electron within the rod balances the magnetic force on the electron and no further charge separation takes place.

The electric field within the rod gives rise to a potential difference across its ends. This potential difference exists only as long as the conductor is in motion through the magnetic field. When the conductor is brought to rest, there is no charge separation and the potential difference drops to zero. When the conductor is moved toward the left, the potential difference is reversed in polarity.

Suppose, now, that our conductor is made the right end of a rectangular conducting loop, as shown in Fig. 28-3. We imagine (unrealistically) that the external magnetic field drops abruptly to zero at left and right boundaries and that the left end of the conducting loop lies outside the magnetic field initially. Then, as the loop is moved to the right, there is no charge separation in the left conductor end, but there is one in the right end. Moreover, no magnetic force acts on electrons within the two horizontal segments of the loop.† As the loop is moved to the right, the excess electrons at the lower end of the conductor can travel clockwise around the loop and enter at the upper end. That is, as we move the conductor through the magnetic field (the left end, however, remaining outside the magnetic field), we generate a counterclockwise current i in the conducting loop. The current is driven, of course, by the charge separation maintained across the ends of the right-hand conductor, and this charge separation has its origin in the work done by an external agent to maintain the constant velocity **v**. The induced current exists as long as the right end passes through the magnetic field. It ceases when the conductor is

FIG. 28-2. Magnetic force $\mathbf{F}_m$ on a free electron within a moving conductor.

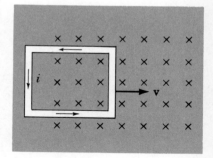

FIG. 28-3. A conducting loop one end of which moves through a transverse magnetic field B.

---

†Strictly, a magnetic force acts to produce a potential difference across the wire diameter (the so-called *Hall* effect).

Motional EMF   SEC. 28-1   451

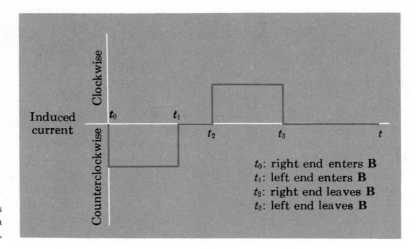

FIG. 28-4. The induced current as a function of time for the situation shown in Fig. 28-3.

at rest, and the current circulates about the loop in the opposite sense when the direction of motion is reversed.

When the entire conducting loop is traveling within the region of the constant magnetic field, a charge separation occurs at *both* the right and left conducting ends. Then there is no induced current. Moreover, if we continue moving the conductor until the right end emerges from the magnetic field while the left end is still within the field, the charge separation occurs in the left conducting end only, and the direction of the induced current is now reversed; see Fig. 28-4. We may summarize these effects by saying that an *induced current is generated* in the loop *when the magnetic flux through* the loop *changes* and that this changing magnetic flux produces a so-called motional emf in the loop. This is but one example of the general phenomenon of *electromagnetic induction*.

Let us now analyze the situation of Fig. 28-3 in more detail, to find how the induced emf is related to the changing magnetic flux; see Fig. 28-5. We must first show that a constant external force **F** must be applied to the conductor, to move it through the magnetic field at the constant velocity **v**.

As described above, the excess electrons originally at the lower end of the conducting rod in Fig. 28-2 are free to move clockwise around the closed conducting loop in Fig. 28-5. This decreases the upward electric force on the remaining free electrons in the right segment of Fig. 28-5. However, the downward magnetic force on the free electrons remains the same, assuming that the conducting loop moves to the right at constant velocity **v**. Therefore, there is again a net downward force on each electron, resulting in a downward component to its velocity and an induced current i.

With a downward velocity component, the electrons will be deflected to the left by the magnetic field until they reach the back edge of the conducting segment. There, they experience an electric force to the right from the rod, just balancing the magnetic force to

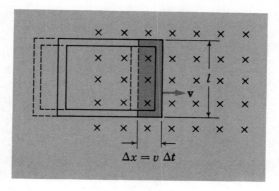

FIG. 28-5. The loop area through which the magnetic field lines pass increases with time.

the left so that the electrons remain in the rod. The electrons then move at a constant oblique velocity, with a component v to the right and a component v_d downward.

The charge separation between the back and front edges of the conductor continues until the electric force (due to the charge separation) acting to the right on each of the remaining conduction electrons just balances the magnetic force acting to the left. After this, all the conduction electrons within the conductor move with the same oblique velocity, v to the right and v_d downward.

The magnetic force to the left on each electron is balanced by an electric force to the right due to the unbalanced charge in the conductor. By Newton's third law, there is an electric force on the conductor to the left. The total force on the conductor due to all the moving electrons can be expressed in terms of the current i by Eq. (26-12):

$$F = Bil \qquad (26\text{-}12), (28\text{-}1)$$

If the conductor is to continue moving to the right at constant velocity **v**, a constant external force **F** to the right and of magnitude Bil must be applied to the conductor. Since the conductor in Fig. 28-5 is displaced a distance $\Delta x = v\,\Delta t$ in the time Δt, the work ΔW, done by the external force in this time is, from (28-1),

$$\Delta W = F\,\Delta x = (Bil)(v\,\Delta t) \qquad (28\text{-}2)$$

The current i is $\Delta q/\Delta t$, where Δq is the net charge transferred in the time interval Δt. Therefore, with $\Delta q = i\,\Delta t$, (28-2) can be written

$$\Delta W = Blv\,\Delta q \qquad (28\text{-}3)$$

The work ΔW done by the external force can be interpreted also as work done by an emf in circulating charges around the conducting loop. Since the emf $\mathscr{E}$ is, in general, the work done per unit charge by an energy source in a circuit,

$$\mathscr{E} = \frac{\Delta W}{\Delta q} \qquad (25\text{-}1)$$

We may write (28-3) in the form

$$\mathscr{E} = Blv \tag{28-4}$$

The emf $\mathscr{E}$ exists as long as the conductor is in motion and *only one* of its ends is moving through the magnetic field. Equation (28-4) applies, however, only when **v**, **B**, and the direction of the conducting wire are at right angles. If a 1-m conductor is moved at 1 m/s through a constant field of 1 Wb/m², the induced emf is 1 V.

To relate the emf to the change in the magnetic flux enclosed by the conducting loop, we recall that the magnetic flux ϕ_B is given, in general, by

$$\phi_B = \int \mathbf{B} \cdot d\mathbf{S} \tag{26-4}$$

As Fig. 28-5 shows, when the loop advances a distance Δx, the loop area which is penetrated by that part of the magnetic field lines increases by $\Delta A = l \, \Delta x$. Thus, the change in magnetic flux through the loop in time Δt is

$$\Delta \phi_B = B \, \Delta A = Bl \, \Delta x$$

We may then write (28-4) as

$$\mathscr{E} = Bl \frac{\Delta x}{\Delta t} = \frac{B(l \, \Delta x)}{\Delta t}$$

$$\mathscr{E} = -\frac{\Delta \phi_B}{\Delta t} = -\frac{d\phi_B}{dt} \tag{28-5}$$

The emf induced in the conducting loop equals the rate at which the magnetic flux through the loop changes. [The meaning of the minus sign in (28-5) will be discussed in Sec. 28-3; it relates to the direction of the emf.] Note that, when the entire conducting loop is immersed in the magnetic field and the flux through the loop is unchanged, (28-5) predicts a zero emf, in accord with our earlier arguments.

Although we have arrived at (28-5) by considering a rather special situation, this relation is altogether general and is the basic relation of electromagnetic induction. It gives the emf induced by a changing magnetic flux for all possible situations.

An external agent does work on the conducting loop at a constant rate as it is moved through the magnetic field [see (28-3)]; yet the speed of the conducting loop is constant. Where does the energy go? It is easy to show that the power into the loop generated by the external agent is exactly equal to the power dissipated as thermal energy in the loop by the induced current.

The power P_a delivered to the loop by the *a*gent is

$$P_a = Fv \tag{9-13}$$

Using (28-1) in (9-13), we have

$$P_a = (Bil)v \tag{28-6}$$

In terms of the emf $\mathscr{E}$ and the resistance R of the loop the induced current i is given by

$$i = \frac{\mathcal{E}}{R} = \frac{Blv}{R} \qquad (28\text{-}7)$$

from (28-4). Therefore, (28-6) can be written

$$P_a = i^2 R \qquad (28\text{-}8)$$

But $i^2 R$ is the power P_d dissipated as thermal energy in the conductor loop [see Eq. (24-9)]. Therefore, the power P_a delivered to the loop is equal to the power P_d dissipated as thermal energy.

28-2 EXAMPLES OF ELECTROMAGNETIC INDUCTION

We have seen one simple example of electromagnetic induction. There are many others. Most were investigated and discovered by Michael Faraday starting in 1831 (and independently by Joseph Henry at the same time), and Eq. (28-5) is referred to as *Faraday's law*.

We see a number of examples in Fig. 28-6. In each example the induced emf arises because the magnetic flux through a con-

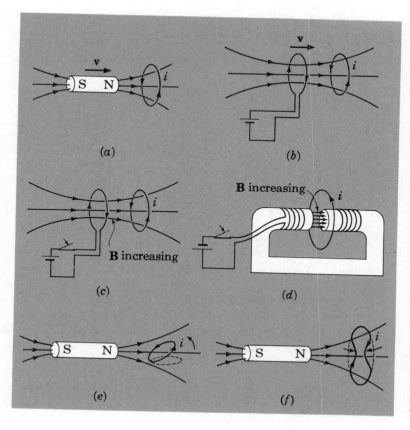

FIG. 28-6. Examples of electromagnetic induction, with each induced current produced in the conducting loop by a changing magnetic flux through the loop. When one object moves and another is at rest, one finds exactly the same induced current if the second object is moved in the opposite direction, with the first object at rest. (a) A magnet approaches a loop. (b) A current-carrying conducting loop approaches a second loop at rest. (c) The current through the left-hand (primary) coil changes as the switch is closed, and a current is induced in the secondary coil (this is the fundamental transformer effect). (d) Closing the switch on an electromagnet changes the field between the poles; note that a current is induced in the loop even though the magnetic field at the *loop* is essentially *zero*. (e) A loop is turned in the region of a magnetic field. (f) A loop is deformed; here the magnetic flux changes because the area through which it passes is reduced.

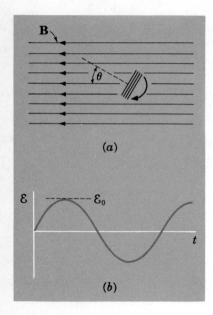

FIG. 28-7. *(a)* A coil rotating at constant angular speed in a uniform magnetic field. *(b)* The resulting sinusoidally varying alternating emf across the leads to the rotating coil.

ducting loop changes, and in every case we can compute the induced emf through Faraday's law:

$$\mathscr{E} = -\frac{d\phi_B}{dt} \tag{28-5}$$

Example 28-1

A coil of N turns, each of area A, is rotated at constant angular speed ω in a uniform magnetic field of magnitude B. What is the emf induced in the coil?

The arrangement here is that of a simple electric generator. The magnetic flux through each turn of the coil is given by

$$\phi_B = AB \cos \theta$$

where θ is the angle between the normal to the plane of the coil and the direction of **B**; see Fig. 28-7a. If the coil is rotated at the constant angular speed ω, then $\theta = \omega t$, and we may write

$$\phi_B = AB \cos \omega t$$

The emf induced in *one* turn is given by

$$\mathscr{E} = -\frac{d\phi_B}{dt} \tag{28-5}$$

Here we have N conductor turns connected in series, so that the total emf across the terminals of the coil is enhanced by a factor N, and

$$\mathscr{E} = -\frac{N d\phi_B}{dt} = \omega N A B \sin \omega t = \mathscr{E}_0 \sin \omega t$$

The output of the coil is a sinusoidally varying alternating emf, as shown in Fig. 28-7b.

In the electric generator described here mechanical power supplied by an external source is transformed into electric power. Conversely, one can operate the arrangement in Fig. 28-7 as an electric motor by supplying a sinusoidal electric current to the coil. Then there will be a magnetic torque on the coil resulting in electric power being converted into mechanical power. Thus, an electric motor is an electric generator operated in reverse.

28–3 LENZ' LAW

There are always *two* possible directions for the electric current in any conducting loop, but when a current is induced by a changing magnetic flux, it goes, of course, in one direction only. Which is it? The basis for finding the direction of the induced current, or of the induced emf, is Lenz' law, named for H. F. E. Lenz (1804–1865).

Let us return to the situation shown in Fig. 28-3. Here a current is induced in a conducting loop pulled by an agent through a magnetic field. We found, by analyzing the magnetic force on the electrons within the conductor, that the current was counterclockwise when the loop was moved to the right and with only the right end of the loop immersed in the field. As a consequence, the magnetic force on the right-end conductor segment was to the left; that is, an external agent had to apply a force to the right to overcome the retarding magnetic force. Then the work done by the agent appeared as thermal energy dissipated through the conducting loop's resistance.

Suppose it were otherwise; that is, suppose that the current were clockwise in Fig. 28-3. Then the magnetic force on the right-end conductor would be to the right, and an agent would not be required to pull the conductor in that direction. Once given a little push to the right, the conductor would be pushed further toward the right by a magnetic force, it would accelerate, the current would grow, and the conductor would go still faster, gaining kinetic energy, and all the while there would be increasing amounts of thermal energy dissipated in the conductor. Clearly, this is impossible, because it violates the principle of energy conservation. In this example, as in all other examples of induced currents, *the direction of the current is always such as to preclude a violation of energy conservation.* This is one way of stating Lenz' law.

There are other equivalent ways of stating Lenz' law. It is often stated as: The direction of *the induced current is always such as to oppose the change (in magnetic flux) that produces it.* Thus, in Fig. 28-3, the magnetic force to the left tends to keep the conductor from being moved to the right; that is, the magnetic force "tries" to keep the magnetic flux through the loop from changing. If the conductor were moved to the left rather than to the right, the direction of the induced current and of the retarding magnetic force would be reversed.

Another situation illustrating Lenz' law is shown in Fig. 28-8a. Here one current-carrying loop is being moved toward a second fixed loop, in which a current is then induced. The magnetic flux through the fixed loop on the right increases with time to the *right*. Consequently, the situation in the fixed loop must be as follows: The magnetic field produced by the induced current must be in such a direction that it annuls (or tends to annul) the magnetic field producing the increasing flux to the right through the fixed loop. That is, the induced current must produce a magnetic field to the *left*. Clearly then, from the right-hand rule relating the current to the magnetic field, the current in the fixed loop must be in the sense shown in Fig. 28-8a.

There is another way of looking at this. We may imagine any current-carrying loop to be equivalent to a magnet, with magnetic field lines emanating from the north pole and converging into the south pole. Then, we see from Fig. 28-8b that the magnets associated with the two loops *repel* each other. Actually, of course, it

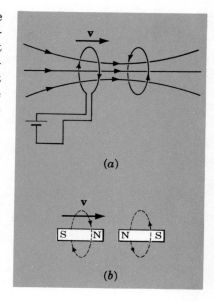

FIG. 28-8. An example of Lenz' law. (a) The left-hand current-carrying conductor is moved to the right, inducing a current in the second coil in the opposite sense. (b) Magnets equivalent to the coils of part (a).

is the magnetic force between the two current-carrying conducting loops which is responsible for the repulsion between them. We have already found that the two currents are in opposite senses. This is in accord with the general result that two parallel current-carrying conductors *repel* one another when the currents are in *opposite* directions (Sec. 27-4).

For practice in applying Lenz' law, the reader should confirm that all the directions of induced current shown in Fig. 28-6 are correct.

The minus sign appearing in the mathematical statement of Faraday's law, $\mathscr{E} = -d\phi_B/dt$, is a symbolic representation of Lenz' law. We shall treat the relative "directions" of the induced emf and magnetic flux change in the next section.

28-4 FARADAY'S LAW AND THE ELECTRIC FIELD

We know that a changing magnetic flux produces an induced electric current in a conducting loop; the changing magnetic flux creates an electric field which drives electric charges around the conducting loop. Does a changing magnetic flux create an electric field in space, even when no electric charges are present? The answer is *yes*. Indeed, the fundamental electromagnetic induction effect is this: *A changing magnetic flux generates an electric field.*

First we relate an emf $\mathscr{E}$ to an associated electric field **E**. Since the emf is just the work done on a unit charge Q carried through all elements of a closed loop, the most general definition of an emf can be written

$$\frac{W}{Q} = \mathscr{E} = \oint \mathbf{E} \cdot d\mathbf{l} \tag{28-9}$$

If the emf is produced by a changing magnetic field, we can combine Faraday's law, Eq. (28-5), with (28-9) to write the law of electromagnetic induction in the following form:

$$\oint \mathbf{E} \cdot d\mathbf{l} = -\frac{d\phi_B}{dt} \tag{28-10}$$

A changing magnetic flux does generate an electric field.

We must distinguish carefully between (1) an electric field originating from electric charges directly and (2) an electric field arising from a changing magnetic flux and given by (28-10).

1. The electric field originating from electric charges is a *conservative* electric field. The line integral $\oint \mathbf{E} \cdot d\mathbf{l}$ is *zero*; that is, if one transports an electric charge around a closed loop in a *conservative* electric field, there is *no net work* done on the charge. Consequently, one can associate an electric potential with a conservative electric field. The electric field lines originate from positive charges, terminate on negative charges, and are continuous in between.

2. The electric field originating from a changing magnetic flux is quite different. Of course, by definition, it is still the electric force per unit positive point-charge, but the electric field is now *nonconservative*, and the line integral $\oint \mathbf{E} \cdot d\mathbf{l}$ is *not* zero. Rather, this line integral (the emf) depends upon the rate at which the magnetic flux ϕ_B changes through the closed path. One *cannot* associate a scalar electric potential with the nonconservative electric field; the work done on a unit positive charge taken around a closed path is *not* zero but is equal, in fact, to the emf around that loop. Moreover, since the electric field lines do *not* originate from and terminate upon electric charges, but must nevertheless be continuous, the *electric field lines* from a changing magnetic flux *form closed loops*. The following example shows what the electric field lines generated by a changing magnetic flux look like for one situation.

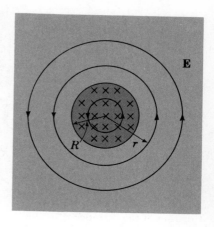

FIG. 28-9. A uniform magnetic field is into the paper over a region of radius R. The field increases with time. The induced electric field lines consist of counterclockwise circular loops.

Example 28-2

A uniform magnetic field directed into the paper is confined to a circular region, as shown in Fig. 28-9. The magnitude of the field increases with time. Find the electric field **E** induced by this changing magnetic field.

 By symmetry, we know that the closed electric field loops must consist of circles concentrically surrounding the magnetic field. To find the sense of the electric lines, we imagine a circular conducting loop around the changing magnetic field, and we determine the direction of the induced current (hence, the direction of **E**) by Lenz' law. In the center of the loop a counterclockwise induced current will produce a magnetic field out of the paper. This field opposes the increasing magnetic flux of the external field, as required by Lenz' law. Therefore, the induced-current field is counterclockwise, as shown.

 Now let us see how the magnitude of **E** depends upon the distance r from the center of the magnetic field. We first consider the region of space *outside* the magnetic field ($r > R$). Evaluating the line integral of (28-10) around the circular loop of radius r, we have

$$E(2\pi r) = -\left(\frac{d\phi_B}{dt}\right)_r = -\left(\frac{d\phi_B}{dt}\right)_R$$

or
$$E = -\left(\frac{1}{2\pi r}\right)\left(\frac{d\phi_B}{dt}\right)_R \qquad \text{for } r > R \qquad (28\text{-}11)$$

where $(d\phi_B/dt)_R$ represents the *total* time rate of flux change inside the region of radius R (and also inside r, since $B = 0$ for $r > R$). We shall presently remark on the significance of the minus sign in (28-11).

 The electric field at any point *outside* the changing magnetic field falls off inversely as the distance r. Note a curious circumstance: We have a *nonzero* electric field in a region of space in which there is *no* magnetic field. What matters is not whether the magnetic *field* is changing *at* the location of the loop, but whether the total magnetic *flux* is changing at any region *within* the loop. Put an electric charge in such an electric field and it is accelerated.

Faraday's Law and the Electric Field SEC. 28-4

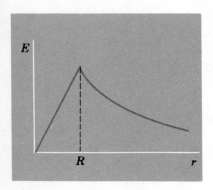

FIG. 28-10. Electric field E as a function of the distance r for the situation shown in Fig. 28-9.

In finding the electric field for $r < R$, we must recognize that only a fraction of the *total* time rate of flux change within R, $(d\phi_B/dt)_R$, penetrates through a transverse area of radius r. Indeed, that fraction is $\pi r^2/\pi R^2$, assuming the magnetic field to be uniform. Evaluating the line integral of (28-10), we then have

$$E(2\pi r) = -\left(\frac{r}{R}\right)^2 \left(\frac{d\phi_B}{dt}\right)_R$$

or

$$E = -\frac{r}{2\pi R^2}\left(\frac{d\phi_B}{dt}\right)_R \quad \text{for } r < R \quad (28\text{-}12)$$

Inside R, we find that the electric field is directly proportional to r.

The magnitude of the circumferential electric field is shown in Fig. 28-10 as a function of r. This plot is similar to that for the electric field from a uniformly charged sphere, Figure 22-10, but the situations are altogether different. Here the electric field is tangential; there it is radial. For $r > R$ the field here varies as $1/r$; there it varies as $1/r^2$. More importantly, here the electric field is nonconservative and originates from a changing magnetic flux; there the electric field is conservative and originates from charges at rest.

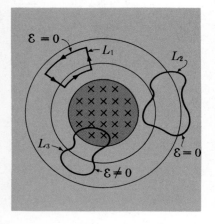

FIG. 28-11. Some loops for computing the induced emf in a region where the magnetic flux is changing.

It is instructive to consider closed loops other than circles to represent the situation shown in Fig. 28-9. For the loop L_1 of Fig. 28-11, consisting of two circular arcs and two radial segments, the line integral $\oint \mathbf{E} \cdot d\mathbf{l}$ is zero, since $\mathbf{E}$ varies inversely with r. This does not mean that $\mathbf{E}$ is zero at all points around L_1. In more general terms, the emf around this loop is zero because the magnetic flux through it does not change. Similarly, there is no emf around the loop L_2. There *is*, however, an emf around the loop L_3 since $d\phi_B/dt$ changes through the area enclosed by this loop.

Knowing how the magnetic flux changes, we can always find the direction of the induced electric field by means of Lenz' law. We can also use the minus sign appearing in Faraday's law [Eq. (28-5)] for this purpose. First recall that we use the right-hand rule for relating the directions of an electric current and the magnetic field loops that surround it; see Fig. 28-12a. In mathematical

FIG. 28-12. (a) Relation between the electric current and the surrounding magnetic field loops, following the right-hand rule. (b) Relation between the magnetic flux change and the surrounding electric field loops.

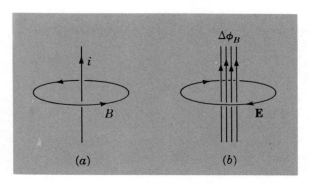

terms, the relation between B and i is given by Ampère's law [Eq. (27-12)], $\oint \mathbf{B} \cdot d\mathbf{l} = \mu_0 i$; *no* minus sign appears here. The relative "directions" of the electric field $\mathbf{E}$ and the magnetic flux change $\Delta \phi_B$ are shown in Fig. 28-12b. The right-hand rule does *not* apply here; that is, if the right-hand thumb gives the direction in which the magnetic flux is *increasing,* the right-hand fingers are in a direction *opposite* to that of the induced electric field lines. This is indicated formally by the minus sign in Faraday's law.

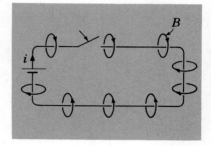

FIG. 28-13. The buildup of current and the associated magnetic field upon the closing of a switch in a conducting loop connected to a battery.

28–5 SELF–INDUCTANCE DEFINED

When the current through one conductor varies, an emf is induced in a nearby conductor, but in addition an emf is induced in the first conductor. This is the phenomenon of self-induction, and a conductor's inductance is a quantitative measure of the effect.

Consider the simple circuit of Fig. 28-13. When a battery is connected to a conducting loop, the current in the circuit increases from zero to the final value determined by the battery's emf and the circuit's resistance. At the same time, the magnetic flux through the conducting loop changes. Consequently, from Faraday's law, there must be an emf induced in the loop itself whose direction, from Lenz' law, is such as to oppose the increase in the magnetic flux through the loop; thus, the induced emf is opposite to the battery's emf. Similarly, if the current has been established in the circuit and the switch is now opened, the current drops to zero. As the current decreases, so does the magnetic flux through the circuit, and again there is an induced emf, whose direction now is such as to tend to maintain the magnetic flux through the loop; the induced emf is in the same direction as the battery emf.

In general, then, whenever the current and, therefore, the magnetic flux change in a circuit, an induced emf (or "back" emf) is generated in that circuit, its direction always being in the direction which opposes the magnetic flux *change*. This self-induction effect is of small consequence when a circuit consists simply of a single current loop. It becomes significant, however, when the magnetic flux is concentrated in a relatively small region of space, as is the case when a conducting wire is wound into the shape of a coil or a solenoid. For a coil of N turns, the magnetic flux through each loop is enhanced by a factor N over the flux for a single turn. Furthermore, the induced emf is enhanced by another factor N, inasmuch as the emf is induced in each of the N turns. Thus, winding a conductor into a coil with N turns increases the self-induction effects by a factor of N^2.

Any device, such as a coil, showing the effect of self-induction is called an *inductor;* see Fig. 28-14. It is represented in a circuit diagram by the symbol ⊸ℓℓ⊸. An inductor can be characterized by its self-inductance or, simply, *inductance,* which we now relate to the time rate of change of the current.

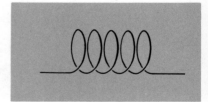

FIG. 28-14. An inductor.

According to Ampère's law [Eq. (27-12)] the magnetic field produced by any current element at any point in space is directly proportional to the current i. Therefore, the magnetic flux, which is proportional to the magnetic field, is also proportional to i. For a coil of N turns we may write

$$N\phi_B = Li \tag{28-13}$$

where $N\phi_B$ is the total flux and the proportionality constant L, the coil's inductance, depends only on the dimensions of (and the material within) the conductor.

Now, from Faraday's law the induced emf $\mathscr{E}$ is given by

$$\mathscr{E} = -\frac{d(N\phi_B)}{dt} \tag{28-5}$$

Substituting (28-13) into (28-5), we then have

$$\mathscr{E} = -L\frac{di}{dt} \tag{28-14}$$

or

$$L = -\frac{\mathscr{E}}{di/dt} \tag{28-15}$$

The inductance is defined either through (28-13) or the equivalent (28-15). We use (28-13) for computing the inductance of a particular conductor arrangement, and we use (28-15) to describe the behavior of an inductor in an electric circuit.

From (28-15) we see that inductance has the units of volts per ampere per second, or V/(A/s) = V-s/A. A special name, the *henry* (in honor of J. Henry), is given to this combination of units:

$$1\text{ H} = 1\text{ V-s/A}$$

Thus, if the current through an inductor changes at the rate of 1 A/s and an induced emf of 1 V is produced, the inductor's inductance is 1 H. Related units are the millihenry (1 mH = 10^{-3} H) and the microhenry (1 μH = 10^{-6} H). Air-filled laboratory coils of moderate size may have inductances on the order of several millihenries. With cores of magnetic materials the inductance may rise to several henries.†

To compute a conductor's inductance is, in general, quite complicated, because one must know the magnetic field at each point in space around the conductor. As in the case when computing a capacitor's capacitance, it is only with geometrically simple situations that the inductance is readily found. Of these, the simplest is that of a toroid, a finite but long solenoid (the length

---

†See R. T. Weidner and R. L. Sells, *Elementary Classical Physics*, 2d ed., sec. 33-3, Allyn and Bacon, 1973.

being much longer than the diameter) bent into a circle whose ends are joined together to form a doughnut shape, as shown in Fig. 28-15. The magnetic field is then confined within the turns of the toroid. If the radius of the toroid is large compared with the radius of any one turn, the magnetic field within the coils is uniform.

We found that the magnetic field B at any point near the center of a long solenoid (and therefore at any interior point for the toroid) is given by

$$B = \mu_0 n i \qquad (27\text{-}13), (28\text{-}16)$$

Here i is the current and n is the number of turns per unit length; that is, $n = N/l$, where the length l is the circumference of the toroid.

From the definition of inductance,

$$L = \frac{N\phi_B}{i} = \frac{NBA}{i} \qquad (28\text{-}13), (28\text{-}17)$$

where A is the cross-sectional area of any one turn and B is the magnetic field anywhere within the turns.

Substituting (28-16) into (28-17), we have

$$L = \mu_0 N n A$$
$$= \mu_0 \frac{N}{l} n(Al) = \mu_0 n^2 (Al)$$
$$L = \mu_0 n^2 V \qquad (28\text{-}18)$$

We have used V to represent the volume Al of a toroidal section of length l.

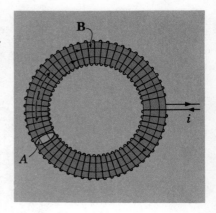

FIG. 28-15. A toroid produces a magnetic field confined entirely within the turns of the toroid.

Example 28-3

What is the inductance of a 2,000-turn toroid with a mean radius of 10.0 cm and having a cross-sectional area of 2.0 cm²?

The total volume V over which the magnetic field exists is the cross-sectional area A multiplied by the toroid's mean circumference $2\pi r$; that is, $V = 2\pi r A$. If N is the total number of turns, then $N = 2\pi r n$. Using (28-18), we have

$$L = \mu_0 n^2 V = \mu_0 \left(\frac{N}{2\pi r}\right)^2 (2\pi r A) = \frac{\mu_0 N^2 A}{2\pi r}$$

$$= \frac{(4\pi \times 10^{-7} \text{ Wb/A-m})(2000)^2(2.0 \times 10^{-4} \text{ m}^2)}{2\pi (0.10 \text{ m})} = 1.6 \text{ mH}$$

The inductance of this moderate-sized air-filled inductor is a mere 1.6 mH. The inductance can be increased more than a thousandfold by winding the toroidal turns about a strongly magnetic material, such as iron.

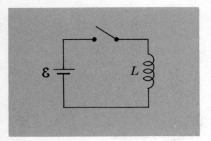

FIG. 28-16

28-6 ENERGY OF AN INDUCTOR AND OF THE MAGNETIC FIELD

Because a back emf exists in any closed circuit when the current is changing, it is necessary to do work in order to establish a current in an inductor. Since energy is not dissipated by a pure inductor, we may regard any work done to establish a current as energy stored in the inductor. Let us find it.

Consider the circuit of Fig. 28-16, in which a battery is connected to a circuit containing an inductor. The loop theorem yields

$$\mathscr{E} = L \frac{di}{dt} \qquad (28\text{-}19)$$

Multiplying this equation by the instantaneous current i, we have

$$\mathscr{E}i = Li \frac{di}{dt} \qquad (28\text{-}20)$$

The term $\mathscr{E}i$ represents the rate at which energy is supplied to the inductor. Since no energy is dissipated, $Li(di/dt)$ must be the rate at which energy is supplied to and stored in the inductor. Labeling the instantaneous power into the inductor P_L and the energy associated with the inductor U_L, we may then write

$$P_L = Li \frac{di}{dt}$$

and

$$U_L = \int_0^t P_L \, dt = \int_0^t Li \frac{di}{dt} \, dt = \int_0^i Li \, di = \tfrac{1}{2} Li^2 \qquad (28\text{-}21)$$

The energy stored in an inductor carrying a current i is proportional to the inductance and to the square of the current. The magnetic energy stored in a 1-H inductor carrying 1 A is, from (28-21), 0.5 J. As inductors go, such an inductor is relatively large.

A current-carrying inductor stores energy in its magnetic field. We wish to compute the energy density u_B, or magnetic energy per unit volume, of a magnetic field of flux density B.

The energy U stored in the magnetic field of any inductor carrying current i is

$$U = \tfrac{1}{2} Li^2$$

We imagine the inductor to be a toroid, because only in such a case is the magnetic field confined entirely to the region within the windings. The inductance L is given by (28-18), and the current i by (28-16). Then (28-21) becomes

$$U = \tfrac{1}{2}(\mu_0 n^2 V)\left(\frac{B}{\mu_0 n}\right)^2 = \frac{1}{2} \frac{B^2}{\mu_0} V$$

Recall that V is the volume of the inductor. Therefore, the magnetic energy density $u_B = U/V$ is

$$u_B = \frac{B^2}{2\mu_0} \tag{28-22}$$

The energy density of the magnetic field is proportional to the *square* of the magnetic flux density.

Although derived for the special case of a toroid, (28-22) can be shown to hold in general. The relation for the magnetic energy density is analogous to that for the electric energy density with B replaced by E and $1/\mu_0$ replaced by ε_0 [Eq. (23-32)].

SUMMARY

A changing magnetic flux induces an emf where, by Faraday's law,

$$\mathscr{E} = -\frac{d\phi_B}{dt} \tag{28-5}$$

According to Lenz' law, the direction of induced emf is always such as to preclude a violation of energy conservation; alternatively, the direction of the induced current is always such as to oppose the change which produces it.

The line integral of the *nonconservative* electric field **E** induced by a changing magnetic flux is given by

$$\oint \mathbf{E} \cdot d\mathbf{l} = -\frac{d\phi_B}{dt} \tag{28-10}$$

The electric field lines induced by a changing magnetic flux form closed loops in space.

By Faraday's law, changing the current through a circuit loop induces an emf in the loop itself. By Lenz' law, this self-induced emf $\mathscr{E}$ is always in a direction opposing the current and magnetic flux change. The self-inductance L (in henries) is defined by

$$L = \frac{N\phi_B}{i} \tag{28-13}$$

or

$$\mathscr{E} = -L\frac{di}{dt} \tag{28-14}$$

When current i passes through an inductor L, the energy stored in the magnetic field is

$$U_L = \tfrac{1}{2}Li^2 \tag{28-21}$$

The magnetic energy per unit volume, or the magnetic energy density, is

$$u_B = \frac{B^2}{2\mu_0} \tag{28-22}$$

PROBLEMS

28-1. A 1.0-m conducting rod moves in a uniform magnetic field of 500 G directed into the paper. The rod lies in the plane of the paper and moves at a constant velocity of 5.0 m/s to the right, with the rod at an angle of 30° with respect to its velocity. (a) What is the potential difference across the ends of the rod? (b) Assuming the induced electric field within the conductor to be uniform and that the conductor has 10^{29} free electrons/m^3, what fraction of the electrons will be located at the end of the rod?

28-2. A circular coil of flexible wire with initial radius 10 cm and 200 turns is placed in a uniform magnetic field of 0.50 Wb/m^2 and normal to the plane of the coil. Find the direction and magnitude of the average emf when (a) the coil shrinks to an area of 100 cm^2 in 0.10 s; (b) the coil is rotated 90° about an axis perpendicular to the magnetic field in a time of 0.10 s.

28-3. Confirm that the induced emf is given in units of volts when the magnetic flux is in units of webers.

28-4. A coil lies on a horizontal plane at the equator. It is turned over about an east-west axis at a constant angular speed. (a) What is the direction of the normal to the plane of the coil when the induced emf is a maximum? Suppose that the coil is turned over about a north-south axis. (b) What is now the orientation of the normal to the plane when the emf is a maximum?

28-5. As one end of a permanent magnet is moved downward toward a circular conducting loop lying in the horizontal plane, there is a clockwise induced current in the loop. Is the end of the magnet a north pole or south pole?

28-6. At the equator the earth's magnetic field is essentially horizontal and north and has a magnitude of approximately 0.30 G. A satellite circles the earth (near the surface) at the equator and has a 1.0-m antenna oriented perpendicular to the earth's surface. What is the potential difference between the ends of the antenna measured by (a) an observer on earth and (b) an astronaut riding in the satellite?

28-7. A solenoid consisting of 800 turns wound uniformly on a cylindrical shell is 50 cm in length and 1.0 cm in radius. A secondary coil of 10 turns is placed around the center of the solenoid. (a) When the solenoid is connected to a battery, the current through it initially changes at the rate 5.0 A/s. What is the induced emf across the terminals of the secondary coil at this time? (b) If instead the battery were connected to the coil and the current through the coil initially changed at the rate of 5.0 A/s, describe how one would proceed in order to find the induced emf in the solenoid.

28-8. A coil of wire has a radius of 5.0 cm, 100 turns, and a total resistance of 50 Ω. At what rate must the transverse magnetic field through the coil change to produce joule heating in the coil at the rate of 1.0 mW?

28-9. An electric generator produces an alternating sinusoidal emf having an amplitude of 150 V and a frequency of 60 Hz. The generator coil has an area of 3.0×10^{-3} m^2 and it rotates in a constant magnetic field of 10,000 G. How many turns are in the generator coil?

28-10. A magnetic field of 1.0 Wb/m^2 acts vertically downward. A conducting rod 0.50 m in length rotates clockwise (looking down) in a horizontal plane about a vertical axis at one end and at an angular speed of 200 turns/s. (a) What is the potential difference between the two ends of the rod? (b) Which ends of the rod, the fixed one or the moving one, carries a positive charge?

28-11. A rectangular conducting loop, parallel to the vertical plane, of width 1.0 m and height 0.50 m, moves downward into a uniform, horizontal field of 0.50 Wb/m^2. The mass of the loop is 250 g, and its resistance is 2.0 Ω. When the lower end is in the field but the upper end is not, there is one constant speed at which it can move. Find this speed.

28-12. A rectangular conducting loop lies in a plane containing an infinitely long conducting wire, as shown in Fig. P28-12. The loop has dimensions of w and d, and at its closest point is a distance r from the long straight wire. The current through the long straight wire changes at the rate di/dt. What is the emf induced in the conducting loop?

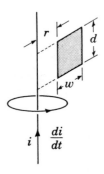

FIG. P28-12

28-13. A conducting bar of length L moves at a constant velocity $\mathbf{v}$ in a direction transverse to its length through a field-free region of space. The rod then enters a uniform magnetic field which is perpendicular to the velocity and length of the rod. While traveling through the uniform magnetic field, the rod is observed to move at only one-half its initial speed. Explain what velocity you might expect the rod to have upon emerging from the magnetic field, and discuss any energy transformations taking place.

28-14. A magnetic field is uniform over a cylindrical region of space of radius 0.20 m, and it changes magnitude at the rate of 50 G/s. What is the electric field in a plane perpendicular to the magnetic field lines and at a distance from the center of the magnetic field of (a) 0.10 m, (b) 0.25 m, and (c) 0.50 m?

28-15. A conducting rod slides across the U-shaped fixed conductor of Fig. P28-15 with an initial velocity $\mathbf{v}_0$ to the right. A constant magnetic field $\mathbf{B}$ exists normal to the paper. The resistance R in the U-shaped wire can be assumed to be constant, and the mass of the sliding rod is m. (a) Find the magnetic retarding force $\mathbf{F}_m$ on the rod in terms of $\mathbf{B}, L, R,$ and $\mathbf{v}$. (b) Show that the velocity decreases as a function of time according to $v = v_0 e^{-t/\tau}$, where $\tau = mR/B^2L^2$. (c) Show that the total distance traveled by the rod before coming to rest is τv_0. (d) Show that the total thermal energy dissipated is just equal to the rod's original kinetic energy, $\frac{1}{2}mv_0^2$.

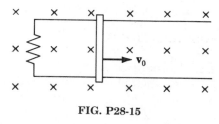

FIG. P28-15

28-16. Show that the henry is equivalent to kg-m²/C².

28-17. When the current through a certain coil is 1.0 A and the current is changing at the rate 0.50 A/s, the potential difference across the coil terminals is 6.0 V. When the current through the coil is again 1.0 A but changing at the rate of 0.50 A/s in the opposite direction, the potential difference across the coil terminals is 4.0 V. What are (a) the inductance and (b) the internal resistance of the coil?

28-18. A single circular loop of wire has a radius of 1.0 cm. (a) Estimate the inductance of the wire by assuming the magnetic field to be uniform over the area of the loop. (b) What is the inductance of a coil of the same radius with 100 turns?

28-19. Show that the inductance per unit length of two concentric cylindrical conductors of radii r_1 and r_2 ($r_2 > r_1$) having equal currents in opposite directions is $L = (\mu_0/2\pi) \ln (r_2/r_1)$. Assume that the inner conductor is so small in radius that we may properly assume that the magnetic field is entirely between the two conductors, a negligible field being found within the surface of the inner conductor.

28-20. Show that the equivalent self-inductance for inductors connected in series is given by $L = \Sigma L_i$ and for inductors in parallel by $1/L = \Sigma 1/L_i$. Assume that there is no magnetic coupling between separate inductors, that is, that the magnetic field from one inductor does not affect a second inductor. In formal terms, this amounts to saying that there is no *mutual inductance* between any two inductors.

28-21. A solenoid 2.0 cm in radius and 100 cm long is wound closely and uniformly with 3,000 turns of wire. Ignore the end effects. (a) What is the inductance of the solenoid? (b) What is the total magnetic energy stored in the inductor if a current of 2.0 A moves through the wire?

28-22. A long straight conductor carries a current i. What is the magnetic energy density of the magnetic field at a distance r from the conductor?

28-23. A flat circular coil of N turns, each of radius r, carries a current i. What is the energy density of the magnetic field at the center of the coil?

28-24. For a long straight solenoid having 30 turns/cm and a current of 2.0 A, find the magnetic energy density (a) near the center of the solenoid and (b) at the end of the solenoid (recall Sec. 27-7).

CHAPTER
29

Electric Oscillations. *RC* and *RL* Circuits

First, the simplest form of electric oscillator, an inductor connected across an initially charged capacitor, is discussed. The frequency of free oscillations is derived, and some analogies between electrical and mechanical oscillations are given.

We then derive the time constant characterizing the decay (or growth) of charge on a capacitor in an *RC* circuit. Finally, we compute the characteristic time constant for the decay (or growth) of current in an *RL* circuit.

29-1 ELECTRIC FREE OSCILLATIONS

Consider the circuit of Fig. 29-1, where a capacitor C, initially carrying a charge of magnitude Q_m on each plate, is connected across an inductor L by the closing of a switch. For simplicity we assume that the circuit contains no resistance. If the connecting wire across the capacitor plates had no self-inductance, the charges on the capacitor would immediately be neutralized. With the inductor present, however, any change in the current through the inductor, and therefore also any change in the magnetic flux through the coil, will give rise to a self-induced emf whose direction is always such as to tend to maintain a constant current.

At the instant the switch closes, the capacitor starts to discharge. The inductor responds by setting up a back emf, opposing the discharge. When the charge q on either plate has reached zero, a current still exists in the circuit, and this current continues to exist under the influence of the induced emf, which now opposes a decrease in this current. But as the current continues and charges again accumulate on the capacitor plates in the reverse sense, the

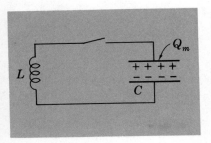

FIG. 29-1. An electric oscillator.

current decreases in magnitude as the first charges arriving on the capacitor plates repel other charges arriving later. The current falls to zero, and now the capacitor is again fully charged, but with opposite polarity. At this point the electric oscillator has completed exactly one-half of an oscillation cycle. After this the process is repeated, but in the opposite sense. The capacitor again loses its charge, a current is created, now in the opposite direction, until the capacitor has again reached its initial charge state. Then one oscillation has been completed.

The oscillations in the electric charge and electric current continue. No energy is dissipated, since we have assumed no resistance. Indeed, the electric oscillations consist of a continuous alternation of energy stored in the electric field of the charged capacitor and energy stored in the magnetic field of the current-carrying inductor. The two circuit elements play different roles: (1) The capacitor C stores energy in its electric field when charged, but it tends to lose its charge and be restored to its equilibrium state of electric neutrality, and (2) the inductor L stores magnetic energy when carrying a current, and it displays an electrical inertia in that its self-induced emf tends to maintain the charges in motion. These results are all qualitative. Now let us consider the electric oscillator analytically.

Applying Kirchhoff's second rule (energy conservation) to the circuit loop of Fig. 29-1, we have

$$\Sigma \mathscr{E} = \Sigma V$$

$$-L \frac{di}{dt} = \frac{q}{C} \tag{29-1}$$

Here $-L(di/dt)$ is the inductor's emf [Eq. (28-14)], q/C the potential difference across the capacitor [Eq. (23-19)], and q the charge on one capacitor plate at any instant. By definition, $i = dq/dt$; substituting this into (29-1), we have

$$-L \frac{d^2q}{dt^2} = \frac{q}{C}$$

$$\frac{d^2q}{dt^2} + \frac{1}{LC} q = 0 \tag{29-2a}$$

$$\frac{d^2q}{dt^2} + \omega_0^2 q = 0 \tag{29-2b}$$

where

$$\omega_0^2 = \frac{1}{LC} \tag{29-3}$$

by definition. The subscript zero signifies that there is no energy lost in this oscillator (because there is zero resistance). Equation (29-2b) is a linear second-order differential equation, whose solution is

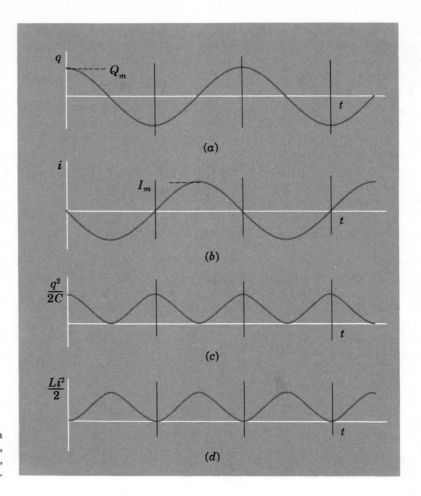

FIG. 29-2. Time variation for an electric oscillator of *(a)* charge, *(b)* current, *(c)* capacitor's energy, and *(d)* inductor's energy.

$$q = Q_m \cos \omega_0 t \qquad (29\text{-}4)$$

as can easily be verified by substituting (29-4) into (29-2b). Q_m is the maximum charge on the capacitor; it is also, in this problem, the initial charge, since $q = Q_m$ at $t = 0$. (See Sec. 14-1 for the treatment of an analogous problem in simple harmonic motion.)

The charge varies sinusoidally with time, as shown in Fig. 29-2, and the angular frequency ω_0 of the free oscillations is given by

$$\omega_0 = \frac{1}{\sqrt{LC}} \qquad (29\text{-}5)$$

The frequency $f = \omega_0/2\pi$ and its reciprocal the period T is given by

$$f = \frac{1}{T} = \frac{1}{2\pi\sqrt{LC}} \qquad (29\text{-}6)$$

The instantaneous current $i = dq/dt$ also oscillates sinusoidally, as we confirm by taking the time derivative of (29-4),

$$i = -\omega_0 Q_m \sin \omega t = -I_m \sin \omega_0 t \qquad (29\text{-}7)$$

where $I_m = \omega_0 Q_m$ is the maximum value of i.

Comparing (29-7) and (29-4) (and Fig. 29-2a and b), we see that the charge (here varying as the cosine) and the current (here varying as the sine) are 90° out of phase. That is, when the capacitor is fully charged and the energy resides entirely in the capacitor's electric field, the current through the inductor and the magnetic field associated with it are zero, and conversely.

The circuit's total energy U remains constant; it consists of the capacitor's energy $U_C = q^2/2C$ and the inductor's energy $U_L = Li^2/2$, as given by (23-30) and (28-21), respectively.

$$U = U_C + U_L = \frac{q^2}{2C} + \frac{Li^2}{2} = \frac{Q_m^2 \cos \omega_0 t^2}{2C} + \frac{L(I_m \sin \omega_0 t)^2}{2} \qquad (29\text{-}8)$$

From (29-7), $I_m = \omega_0 Q_m$, and from (29-5), $\omega_0^2 = 1/LC$; therefore,

$$LI_m^2 = L\omega_0^2 Q_m^2 = \frac{Q_m^2}{C}$$

and (29-8) becomes

$$U = \frac{Q_m^2}{2C}(\cos^2 \omega_0 t + \sin^2 \omega_0 t) = \frac{Q_m^2}{2C} \qquad (29\text{-}9)$$

Although the energies of the capacitor and inductor vary sinusoidally with time, as Fig. 29-2c and d show, their sum is constant with time.

As (29-6) shows, the frequency of free oscillations of an LC circuit depends on L and C. For example, with $C = 1.0$ μF and $L = 1.0$ mH, we find from (29-6) that $f \approx 5.0 \times 10^3$ Hz $= 5.0$ kHz. The frequency rises as the magnitudes of L and C decrease. Thus, if one is to construct an electric oscillator of very high frequency, the capacitor and inductor must be small, not only in the magnitudes of C and L, but also in the actual dimensions of these circuit elements.

Figure 29-3 shows the evolution of an ordinary LC circuit, with obvious capacitance and inductance elements, into two varieties of high-frequency oscillators, for which these circuit parameters are less easily recognized. In Fig. 29-3a first the inductance is reduced considerably by replacing the coil with a single conducting wire; then the capacitance also is reduced considerably by reducing the area of the capacitor plates. There is left a single conducting loop broken by a gap at one point. This is indeed an electric oscillator, and if the loop's size is of the order of 1 m with a gap of perhaps 1 cm, it oscillates at a frequency of tens of megahertz, a frequency lying in the radiofrequency region of the electromagnetic spectrum.

Oscillators of just this type were, in fact, used in the historic experiments of Heinrich Hertz (1857–1894), who first demonstrated

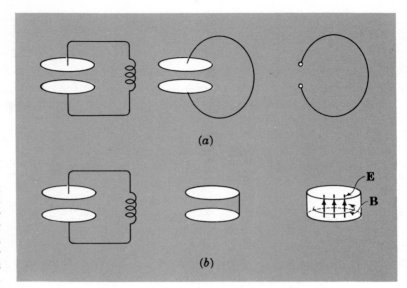

FIG. 29-3. Evolution of an ordinary electric oscillator into (a) a high-frequency oscillator consisting of a single conducting loop with a gap and (b) a microwave cavity oscillator consisting of a hollow right circular cylindrical conductor.

the existence of electromagnetic radiation in 1887. Hertz used two such oscillators, both resonant at the same frequency. The oscillations were observed through a spark at the gap. The electromagnetic radiation was detected by observing that, when the second oscillator was moved relatively far from the first oscillator, the sparks across its gap persisted, although this effect could *not* be attributed to the direct action of the electric and magnetic fields of the first oscillator on the second.

Figure 29-3b shows the evolution of a simple LC circuit into a different type of high-frequency oscillator. Here the inductance is reduced first by connecting the capacitor plates with a single straight conducting wire. Then the inductance is reduced still further by connecting additional wires in parallel with the first. Indeed, one constructs an entire cylindrical surface between the two capacitor plates, and so forms a hollow, closed right circular cylinder of conducting material. At least superficially this certainly does not look like an LC oscillator of the ordinary variety. It is, in fact, one simple form of a *microwave* oscillator. For dimensions of the order of a few centimeters, the free oscillations occur at microwave frequencies of the order of tens of kilomegahertz (or electromagnetic waves having wavelengths of a few centimeters). The oscillating electric field is confined entirely within the closed cylinder, and so is the oscillating magnetic field. Here it becomes more useful to describe the electric oscillations, not in terms of the current through the circuit or the potential difference across various pairs of the points — although this is still possible and proper — but in terms of the electric and magnetic fields and their configurations within the closed cylinder. These best characterize the microwave oscillator.

Table 29–1

Mechanical quantity	Symbol	Electrical quantity	Symbol
Displacement	x	Charge	q
Velocity	$v = dx/dt$	Current	$i = dq/dt$
Acceleration	$a = dv/dt = d^2x/dt^2$	Time rate of change of current	$di/dt = d^2q/dt^2$
Mass	m	Inductance	L
Spring constant	k	Reciprocal of capacitance	$1/C$

29–2 ELECTRICAL–MECHANICAL ANALOGS

Recall the differential equation describing the free mechanical oscillations of a system consisting of a mass m attached to a spring of force constant k:

$$\frac{d^2x}{dt^2} = -\frac{k}{m} x \qquad (14\text{-}4), (29\text{-}10)$$

This equation is of exactly the same form as (29-2) for the electric oscillations. Indeed, one may transform (29-2) into (29-10) by making replacements as indicated in Table 29-1.

The correspondence is more than formal. Like the mass m, which corresponds to the inertial tendency of a particle to maintain constant velocity v, the inductance shows such a tendency to maintain a constant current i. Likewise, the spring's restoring constant k corresponds to what might be called the capacitor's restoring constant $1/C = V/q$. (Although both an inductance coil and a helical spring are typically represented in a diagram by the same symbol, it must be emphasized that the capacitor is analogous to the elastic spring and the inductance coil to the mass.)

Furthermore, the electric potential difference q/C corresponds to the elastic restoring force kx, and the induced emf $L\,(di/dt)$ to the force $m\,(dv/dt)$. There is also a parallel to the mechanical dissipative force $F = rv$, proportional to the particle's velocity. It is the electrical resistance R, across which the potential difference is Ri. Similarly, the potential energy $\frac{1}{2}kx^2$ of the stretched spring and the kinetic energy $\frac{1}{2}mv^2$ of the moving particle correspond, respectively, to the electric energy $q^2/2C$ of the charged capacitor and the magnetic energy $Li^2/2$ of the inductor. See the compilation of still more electrical-mechanical analogs given in Table 29-2.

There is a complete analogy, both in mathematical terms and in physical behavior, between mechanical and electrical "circuit" elements. Indeed, it is often useful to analyze a complicated mechanical system by constructing its electrical analog with ordinary electric circuit elements. Then the measured charges, currents, and

Table 29-2

Mechanical quantity	Symbol	Electrical quantity	Symbol
Elastic restoring force	kx	Potential difference	q/C
Force	$m(dv/dt)$	Induced emf	$L(di/dt)$
Dissipative force	rv	Potential drop across a resistance	Ri
Potential energy of a spring	$\frac{1}{2}kx^2$	Electrical energy of a charged capacitor	$\frac{1}{2}(1/C)q^2$
Kinetic energy	$\frac{1}{2}mv^2$	Magnetic energy of an inductor	$\frac{1}{2}Li^2$

potential differences give the corresponding particle displacements, velocities, and forces.

In any actual circuit it is inevitable that some resistance will be present. As a result the total energy of the electric oscillator decreases exponentially with time, and the characteristic frequency is slightly lower than ω_0, its value in the zero-resistance case. This behavior is exactly analogous to that of a mechanical oscillator.

In an *LCR* circuit with these three elements in series, resonance occurs when the frequency of the driving emf is equal to the natural frequency. At resonance there is an enhanced response—the charge on the capacitor and the current in the circuit reach their maximum values.

29-3 THE *RC* CIRCUIT

We are now able to analyze a simple two-component electric circuit comprised of a resistance and a charged capacitor in series, as shown in Fig. 29-4. A capacitor C, having an initial charge of magnitude Q_0 on each plate and a potential difference $V_0 = Q_0/C$ between the plates, is discharged across a resistor R. It is easy to describe qualitatively what happens after the instant the switch is closed. The free electrons move clockwise around the circuit until the capacitor is finally electrically neutral. A current, initially of magnitude $i_0 = V_0/R$, exists in the circuit, this current falling to zero as the charge Q and potential V of the capacitor also reach zero. At the same time, the electric potential energy associated with the opposite charges initially residing on the capacitor plates (or, if you will, associated with the electric field between the capacitor plates) is dissipated as thermal energy while the charges move through the resistor.

Let us see how the charge Q and current i vary with time. We know that the net change in potential around a closed path is zero. That is, going counterclockwise around the circuit of Fig. 29-4, the potential drop iR across the resistor plus the potential

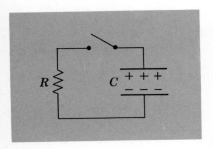

FIG. 29-4. A circuit consisting of a charged capacitor and a resistor.

drop $(-Q/C)$ across the capacitor must equal zero at each instant of time:

$$iR - \frac{Q}{C} = 0$$

The negative sign for the potential drop across the capacitor indicates that there is an increase in crossing the capacitor from the negative to the positive plate. If we substitute $i = -dQ/dt$, this relation becomes

$$R\frac{dQ}{dt} = -\frac{Q}{C}$$

This equation holds at any instant of time. Rearranging terms and recalling that R and C are independent of time, we have

$$\int_{Q_0}^{Q} \frac{dQ}{Q} = -\frac{1}{RC}\int_0^t dt$$

$$\ln\frac{Q}{Q_0} = -\frac{t}{RC}$$

$$Q = Q_0 e^{-t/RC} \qquad (29\text{-}11)$$

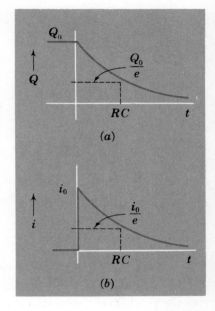

FIG. 29-5. Variation with time of (a) the capacitor charge and (b) the current for an RC circuit.

The charge Q on each capacitor plate decreases exponentially with time, as shown in Fig. 29-5a. The decay rate is controlled by the characteristic time constant τ of the RC circuit

$$\tau = RC \qquad (29\text{-}12)$$

As (29-11) shows, when $t = RC$, then $Q = Q_0/e$; thus, the constant RC is the time elapsing until the capacitor's charge or potential is $(1/e)$th, or 37 percent, of its initial charge. For example, in a circuit containing a 1.0-MΩ resistor and a 1.0-μF capacitor, the time constant is $RC = (1.0 \times 10^6 \text{ V/A})(1.0 \times 10^{-6} \text{ C/V}) = 1.0$ s.

How does the current i vary with time? Using (29-11) for Q, we have

$$i = -\frac{dQ}{dt} = \frac{Q_0}{RC} e^{-t/RC}$$

Since $Q_0/RC = V_0/R = i_0$, we may write this relation as

$$i = i_0 e^{-t/RC} \qquad (29\text{-}13)$$

The current, which is zero before the switch is first closed, rises abruptly to i_0 at $t = 0$ and then decays exponentially in time with the characteristic decay time RC, as shown in Fig. 29-5b.

29-4 THE LR CIRCUIT

Now consider a two-component circuit comprised of a resistance R and an inductance L as shown in Fig. 29-6. We assume that a con-

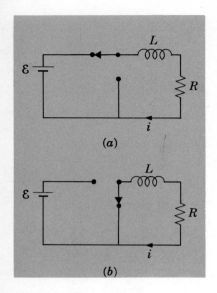

FIG. 29-6. An LR circuit (a) in series with battery $\mathscr{E}$ with steady current $i = \mathscr{E}/R$ and (b) with battery $\mathscr{E}$ short-circuited so that it does not drive current through the LR branch of the circuit.

stant current $i = \mathscr{E}/R$ has been established by the battery. Now suppose that the switch in Fig. 29-6a is suddenly closed. This disconnects the battery so that it no longer supplies its emf $\mathscr{E}$ to drive current through the LR branch of the circuit. The current in the closed circuit does not, however, drop to zero instantaneously. Although the battery's emf no longer drives charges around the circuit, the changing current through the inductor creates an emf which attempts to replace the battery's emf. That is, the inductor's emf is in the same sense as the battery's emf was formerly, inasmuch as the inductor attempts to sustain the current.

Applying the loop theorem (Kirchhoff's second rule, energy conservation) to the closed circuit of Fig. 29-6b, we have

$$\Sigma \mathscr{E} = \Sigma V \qquad (25\text{-}7)$$

$$-L \frac{di}{dt} = iR \qquad (29\text{-}14)$$

To find how the current i varies with time t, we rearrange (29-14) and integrate the current from its initial value i_0 at $t = 0$ to the value i at time $t = t$:

$$\int_{i_0}^{i} \frac{di}{i} = -\frac{1}{L/R} \int_0^t dt$$

$$\ln \frac{i}{i_0} = -\frac{t}{L/R}$$

$$i = i_0 e^{-t/(L/R)} \qquad (29\text{-}15)$$

The current decays exponentially in time, as shown in Fig. 29-7. The characteristic time constant for the LR circuit is

$$\tau = \frac{L}{R} \qquad (29\text{-}16)$$

As (29-15) shows, after the time $t = \tau = L/R$ has elapsed, the current is $i = i_0/e$. That is, the time constant L/R is the time elapsing as the current falls to $1/e$, or 37 percent, of its initial value. For example, if an LR circuit consists of an inductor of 1.0 H and a resistor of 1.0 Ω, the current drops to $1/e$ of its initial value in 1.0 s.

Any real inductor will have some finite resistance, simply because the conducting wire of which it is made has a nonzero resistance. Therefore, a pure inductance cannot exist isolated from resistance, and any inductor will have its own characteristic decay time.

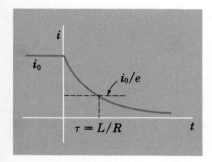

FIG. 29-7. The exponential time decay of the current in the circuit of Fig. 29-6b.

The current through a short-circuited current-carrying inductor decays exponentially in time. It is easy to show that immediately after a constant emf has been applied to an inductor, the current grows exponentially. Consider the circuit of Fig. 29-8. We wish to find how the current i varies in time from the moment the switch has been closed until the current reaches its final maximum value $i_m = \mathscr{E}/R$. Clearly, as the current first passes through the inductor and the magnetic flux through the inductor changes, an emf is induced which, by Lenz' law, opposes the battery's emf. We apply the loop theorem as follows:

$$\Sigma \mathscr{E} = \Sigma V$$

$$\mathscr{E} - L\frac{di}{dt} = iR \tag{29-17}$$

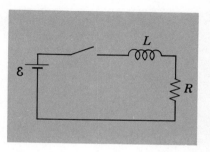

FIG. P29-8

where $\mathscr{E}$ is the constant emf of the battery. The term $-L(di/dt)$ again appears on the left side as an emf. Rewriting, we have

$$\mathscr{E} = L\frac{di}{dt} + iR \tag{29-18}$$

It is not difficult to solve this differential equation for i as a function of t. The solution is

$$i = \frac{\mathscr{E}}{R}(1 - e^{-t/(L/R)}) \tag{29-19}$$

This may easily be verified by substituting i and di/dt from (29-19) into (29-18).

The current grows exponentially in time with the same characteristic time constant $\tau = L/R$ as the circuit of Fig. 29-6. After the time L/R has elapsed, the current i differs from its final value $\mathscr{E}/R$ by $1/e$, or 37 percent.

Example 29-1

Both the inductance and the internal resistance of a certain coil are unknown. Experiment shows that when short-circuited the inductor's current decays to $1/e$ of its initial value of 0.67 ms. When a 4.0-Ω resistor is placed in series with the inductor, the inductor's initial current falls to $1/e$ of its initial value in 0.50 ms. What are the inductor's inductance L and internal resistance R_i?

From (29-16),

$$\tau = \frac{L}{R_i} = 0.67 \times 10^{-3} \text{ s}$$

and

$$\frac{L}{R_i + 4.0 \text{ }\Omega} = 0.50 \times 10^{-3} \text{ s}$$

Solving the equations, we get

$$L = 8.0 \text{ mH}$$
$$R_i = 12 \text{ }\Omega$$

SUMMARY

When an initially charged capacitor is connected across a resistance-less inductor, the charge and current in the circuit oscillate sinusoidally at the angular frequency ω_0:

$$\omega_0 = \frac{1}{\sqrt{LC}} \qquad (29\text{-}5)$$

The inductor shows inertia to a change in the current through it, and the capacitor tends to restore its charge state to electrical neutrality.

When a charged capacitor C discharges through a resistor R, the charge on the capacitor decreases exponentially with time, the characteristic time constant of the decay being RC.

The time constant for the exponential growth or decay of current through an inductor in a series LR circuit is L/R.

PROBLEMS

29-1. What capacitance must be used with an inductance coil of 4.0 H to cause the circuit to resonate at 60 Hz?

29-2. A resonant circuit in a radio receiver has a fixed inductor in series with a variable capacitor. By what factor does the capacitance change when the radio is tuned over the entire broadcast band from 500 kHz to 1.5 MHz?

29-3. The physical dimensions of all circuit components in a resonant electric circuit are reduced by a factor of exactly 5. By what factor is the resonant frequency increased?

29-4. An electric oscillator consisting of a parallel-plate capacitor and a solenoid resonates at the frequency f. The capacitor plate separation distance is reduced by a factor 2, and the number of turns in the solenoid is increased by a factor 2. What is the new resonance frequency of the circuit?

29-5. An electric oscillator consisting of a 10-μF capacitor in series with a 1.0-mH resistanceless inductor has a maximum potential difference across the capacitor plates of 50 V. (a) Find the maximum value of the current in the circuit. (b) On a v-t diagram sketch the separate voltage v_C across the capacitor and v_L across the inductor, and show that the resultant voltage across the combination is always zero. (c) What is the total energy of the oscillator at any instant of time?

29-6. A capacitor of 0.10 μF initially has a potential difference of 100 V across its terminals. The capacitor is then connected across a resistanceless inductor of 3.0 mH. (a) What is the total energy of the electric oscillator at any instant of time? (b) What is the maximum instantaneous current through the inductor?

29-7. The potential difference between the plates of a slightly leaky capacitor of 0.10 μF is found to drop to half its initial value in 1.5 s. What is the equivalent resistance between the capacitor plates?

29-8. Show that, when a charged capacitor is discharged through a resistor, the total energy initially stored in the capacitor, $Q^2/2C$, is equal to the total energy dissipated in the resistor, $\int_0^\infty i^2 R\, dt$.

29-9. A parallel-plate capacitor is filled with quartz between the plates. The dielectric constant and electrical resistivity of quartz are 3.8 and 5×10^{16} Ω-m, respectively. If this capacitor is charged, how long will it take for it to lose half the charge?

29-10. Show that the time constant for a leaky parallel-plate capacitor (one filled with a slightly conducting dielectric) is independent of both the area of the plates and their separation distance.

29-11. An LR circuit is connected to a battery, as shown in Fig. 29-8. (a) Show that the rate at which the current initially changes with time is $i_f/(L/R)$, where i_f is the final steady-state current. (b) Find the ratio of the voltage drop across the inductor to that across the resistor when $i = i_f/2$.

29-12. Consider the circuit shown in Fig. P29-12. Assume that the switch S has been open for a long time. (a) What are the currents through R_1, R_2, and L? (b) If the switch is now closed at time $t = 0$, what are the

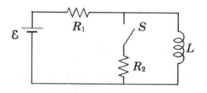

FIG. P29-12

expression for the currents through R_1, R_2, and L as functions of time, from Kirchhoff's rules? Check your answer by finding the values of the currents as t approaches infinity.

29-13. It is found that the time constant for the decay of current through a certain coil is halved when a 10-ohm resistor is added in series to the coil. Furthermore, when a pure inductance of 30 mH is added in series to the original coil and series resistor, the time constant is the same as that for the coil alone. What are the coil's (a) inductance and (b) internal resistance?

29-14. The initial current in a series LR circuit is 1.0 A, where $L = 0.10$ mH and $R = 10$ Ω. Estimate the time for the current to decay to 1 electron/s.

29-15. A current of 2.0 A passes through an inductor of 30 mH. What is the total energy dissipated as thermal energy when the inductor is connected across a resistance and disconnected from everything else?

29-16. In the simple series circuit shown in Fig. 29-8, assume that $\mathscr{E} = 10$ V, $R = 2.0$ Ω, and $L = 4.0$ mH. Four milliseconds after the switch is closed, what is (a) the power being supplied by the battery, (b) the power being dissipated as thermal energy, (c) the power being supplied to the magnetic field of the inductor?

29-17. A coil's internal resistance is measured to be 40 Ω, and when short-circuited the current through the coil is found to decay to half its initial value in 0.80 ms. (a) What is the coil's inductance? (b) How long does it take for the energy stored in the coil to fall to one-half its value?

29-18. A coil having an inductance of 4.0 mH and a resistance of 10 Ω is connected to a battery with an emf of 12 V and internal resistance of 2.0 Ω. How long must one wait after the switch is closed until (a) the current is 99.9 percent of its steady-state value and (b) the energy stored in the inductor is 99.9 percent of its steady-state value?

CHAPTER 30

Maxwell's Equations and Electromagnetic Waves

All of classical electromagnetism can be summarized in four fundamental equations relating electric and magnetic fields to one another and to electric charge and current. These equations, known as Maxwell's equations, have been discussed earlier: Gauss' law for electricity, Gauss' law for magnetism, Faraday's law, and Ampère's law. In this chapter we assemble in one place the fundamental laws of electromagnetism and then use these equations to derive the principal features of electromagnetic waves.

First, we generalize Ampère's law by considering the so-called displacement current, or *changing* electric flux, which can generate a magnetic field in the same fashion as do moving charges. Next, after summarizing Maxwell's four equations for electromagnetism, we derive expressions for the following properties of electromagnetic waves traveling in free space: the speed of electromagnetic waves, which is found to be a universal constant; the energy density and intensity of the radiation; the radiation force and pressure produced by electromagnetic waves impinging upon a material; the linear momentum of the radiation.

We discuss the generation of electromagnetic waves by the sinusoidal oscillations of electric charge. The electromagnetic spectrum is shown and the various regions briefly discussed. Methods of measuring the speed of light are also given.

30-1 AMPÈRE'S LAW AND MAXWELL'S DISPLACEMENT CURRENT

According to Ampère's law a magnetic field is created by the electric current i, the line integral of the magnetic field around any closed loop being merely the current (multiplied by μ_0) crossing *any surface* bounded by this loop

$$\oint \mathbf{B} \cdot d\mathbf{l} = \mu_0 i \qquad (27\text{-}12)$$

As given above, Ampère's law recognizes only one source of a magnetic field: moving electric charges. There is, however, a second origin of a magnetic field: a changing electric field (or changing electric flux), also called a displacement current.

To see how the so-called displacement current arises, consider the situation shown in Fig. 30-1. Suppose that the switch in the circuit has just been closed. Initially the current jumps from zero to some maximum value; very soon thereafter it falls to zero again, after the capacitor has been fully charged. We are interested in the time during which the current through the circuit is changing. Actually, to modify this statement slightly, we are interested in the time the current through the *conducting wire* is changing, since a real current does *not* exist at any time in the region between the two capacitor plates, inasmuch as no charged particles ever pass through this region.

Let us now apply (27-12) to a circular loop around the lower section of Fig. 30-1. Figure 30-2 illustrates this for one loop with two possible surfaces bounded by this loop. If we choose the flat area whose boundary is the circular path Fig. 30-2a, then clearly there is a real current penetrating this surface and producing a magnetic field around the closed path. Although simple, this flat surface is not the only one we may choose. Ampère's law relates to the current through the loop, *without* specifying the nature of the surface bounded by the loop; that is, we may choose a surface of *any* shape in applying the law. In every case the total current through the surface determines the magnetic field around the loop. If we choose the hemispherical surface which passes through the region between the charging capacitor plates in Fig. 30-2b, there is no real current through the hemispherical surface. This would imply, by (27-12), that there is *no* magnetic field around the loop, in contradiction to the field predicted by the flat surface. Realizing this inconsistency, in 1865 J. C. Maxwell (1831–1879) proposed a more general form for Ampère's law by assuming that in addition to the real current there was a displacement current within the capacitor. His theoretical conjecture, based on symmetry and a self-consistent generalized Ampère law, was later verified experimentally.

To allow for contributions from a fictitious displacement current i_d as well as a real current i, Ampère's law must be written as

$$\oint \mathbf{B} \cdot d\mathbf{l} = \mu_0 (i + i_d) \qquad (30\text{-}1)$$

So that the magnetic field around the loop will be the same, irrespective of the surface chosen, in the simple situation in Fig. 30-2 we must have

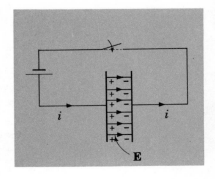

FIG. 30-1. A parallel-plate capacitor being charged just after the switch in the circuit is closed. The current i is increasing with time.

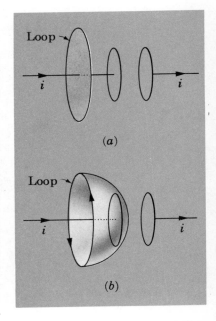

FIG. 30-2. The charging capacitor of Fig. 30-1. The loop for applying Ampère's law surrounds the conducting wire. In (a) this loop bounds a flat surface and in (b) the loop bounds a hemispherical surface.

$$i \text{ (in conductor)} = i_d \text{ (in capacitor)} \qquad (30\text{-}2)$$

What is it that is happening between the capacitor plates to which we can relate i_d? If there are no moving charges, what is it that generates the magnetic field? We shall see that i_d is related to the electric flux ϕ_E between the capacitor plates. By definition,

$$\phi_E = \oint \mathbf{E} \cdot d\mathbf{S} \qquad (22\text{-}2), (30\text{-}3)$$

The instantaneous current $i = dq/dt$ gives the rate at which charges pass any point in the conductor; dq/dt is also the rate at which charges accumulate on each of the capacitor plates. For the uniform electric field $\mathbf{E}$ between the plates of the parallel-plate capacitor, we know that

$$E = \frac{\sigma}{\varepsilon_0} = \frac{q}{\varepsilon_0 A} \qquad (22\text{-}9)$$

or

$$q = \varepsilon_0 E A = \varepsilon_0 \phi_E \qquad (30\text{-}4)$$

Here we have identified EA, using (30-3), as the electric flux ϕ_E through a transverse area parallel to the capacitor plates. From (30-4) and (30-2), we have

$$i_d = i = \frac{dq}{dt}$$

$$i_d = \varepsilon_0 \frac{d\phi_E}{dt} \qquad (30\text{-}5)$$

The displacement current i_d through any chosen area is the time rate of change of the electric flux through that area multiplied by ε_0. Using (30-5) in (30-1) gives

$$\oint \mathbf{B} \cdot d\mathbf{l} = \mu_0 \left(i + \varepsilon_0 \frac{d\phi_E}{dt} \right) = \mu_0 i + \varepsilon_0 \mu_0 \frac{d\phi_E}{dt} \qquad (30\text{-}6)$$

This is the generalized form of Ampère's law. It implies that a magnetic field may be produced by a changing electric flux, even in the absence of electric charges, that is, in the absence of a real current. In short, $\mathbf{B}$ may be generated not only by a steady *or* changing real current, but also by a fictitious displacement current which exists whenever the electric flux is *changing*. Although we have derived the generalized form of the Ampère law from the rather special case of a charging parallel-plate capacitor, this relation is altogether general, holding for *any* changing electric flux.

When no real current exists and a magnetic field is generated entirely by a changing electric flux in space, (30-6) becomes

$$\oint \mathbf{B} \cdot d\mathbf{l} = \varepsilon_0 \mu_0 \frac{d\phi_E}{dt} \qquad (30\text{-}7)$$

This relation is similar in form to Faraday's law:

$$\oint \mathbf{E} \cdot d\mathbf{l} = -\frac{d\phi_B}{dt} \qquad \text{(28-10), (30-8)}$$

In the first instance a changing electric flux produces a magnetic field; in the second, a changing magnetic flux produces an electric field. This symmetry between the electric and magnetic fields was one reason why Maxwell hypothesized the displacement current.

A comparison of the magnetic and electric fields produced by changing electric and magnetic fluxes is illustrated in Fig. 30-3.

How was the correctness of Maxwell's hypothesis concerning the displacement current tested experimentally? The magnetic field induced by a changing electric flux is relative small and, therefore, not easily detected directly. The crucial test of the generalized Ampère law came from Maxwell's theoretical prediction that an electromagnetic disturbance, once excited in empty space, can be self-generating. That is, an electromagnetic disturbance can become detached from electric charges and propagate through space as an *electromagnetic wave*. A simple form of the theoretical argument is given in Sec. 30-3. The direct observation of electromagnetic waves (short-wave radio waves) by Heinrich Hertz in 1887 verified Maxwell's prediction.

30-2 MAXWELL'S EQUATIONS

We have discussed Maxwell's contributions to electromagnetic theory through his hypothesis of the displacement current. Maxwell played an even more significant role in bringing together into one unified electromagnetic field theory the hitherto disparate facts of electric and magnetic phenomena. It was Faraday who invented the field concept as a useful and picturesque means of visualizing electric and magnetic effects. Maxwell took the electric and magnetic fields seriously and developed the mathematical expressions for their properties and interrelations. These fundamental relations, which say everything there is to say about classical electromagnetism, are called *Maxwell's equations*.

Before writing the four Maxwell equations for free space, we must be clear on the definitions of the electric field **E** and the magnetic field **B**. From the so-called Lorentz equation, which gives the total force on an electric charge arising from electric and magnetic fields, we have

$$\mathbf{F} = q(\mathbf{E} + \mathbf{v} \times \mathbf{B}) \qquad \text{(26-8)}$$

This equation defines **E** and **B**. That part of the force, $q\mathbf{v} \times \mathbf{B}$, which depends on the charge's velocity is the magnetic force; the remaining part, $q\mathbf{E}$, is the electric force.

Table 30-1 lists the four Maxwell equations for vacuum, and gives the common name and the primary experimental evidence for each.

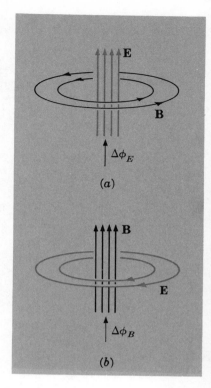

FIG. 30-3. *(a)* Magnetic field loops surrounding a region in which the electric flux is increasing. *(b)* Electric field loops surrounding a region in which the magnetic flux is increasing. Notice the different senses of the field loops.

Table 30-1

Name	Equation		Experimental evidence
Gauss' law for electricity (Coulomb's law)	$\varepsilon_0 \oint \mathbf{E} \cdot d\mathbf{S} = q$	(22-3)	Electric force is inverse-square; no net charge on interior of hollow charged conductor under steady-state conditions
Gauss' law for magnetism	$\oint \mathbf{B} \cdot d\mathbf{S} = 0$	(27-11)	No isolated magnetic poles
Faraday's law	$\oint \mathbf{E} \cdot d\mathbf{l} = -\dfrac{d\phi_B}{dt}$	(28-10)	Electromagnetic induction effects
Ampère's law	$\oint \mathbf{B} \cdot d\mathbf{l} = \mu_0 i + \varepsilon_0 \mu_0 \dfrac{d\phi_E}{dt}$	(30-6)	Magnetic force between current-carrying conductors; electromagnetic waves

30-3 ELECTROMAGNETIC WAVES FROM MAXWELL'S EQUATIONS

As one important application of Maxwell's equations we shall derive the fundamental properties of electromagnetic waves in empty space, far removed from any electric charges or currents. With both q and i equal to zero, Maxwell's equations (Table 30-1) reduce to the following:

1. The surface integrals of **E** and **B** over any closed surface are both equal to zero.

$$\oint \mathbf{E} \cdot d\mathbf{S} = 0 \qquad (22\text{-}3)$$

$$\oint \mathbf{B} \cdot d\mathbf{S} = 0 \qquad (27\text{-}11)$$

2. A changing magnetic field (strictly, a changing magnetic flux) generates an electric field following *Faraday's law:*

$$\oint \mathbf{E} \cdot d\mathbf{l} = -\frac{d\phi_B}{dt} \qquad (30\text{-}8)$$

3. A changing electric field (strictly, a changing electric flux) generates a magnetic field following the *generalized Ampère law:*

$$\oint \mathbf{B} \cdot d\mathbf{l} = \varepsilon_0 \mu_0 \frac{d\phi_E}{dt} \qquad (30\text{-}7)$$

We assume that a rectangular magnetic pulse with magnetic field **B** is moving along the X axis at some speed c, as shown in Fig. 30-4. This field is imagined to move through empty space, far from

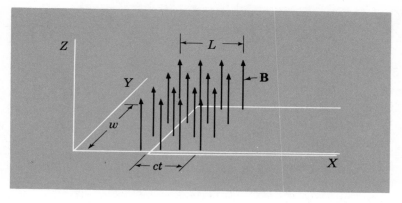

FIG. 30-4. Traveling magnetic pulse.

any electric charges or currents. How such a magnetic field can be separated from charges and currents will be discussed later.

The width of the pulse along the X axis is L; along the Y and Z axes it is infinite in extent. At all points within the pulse the magnetic field is in the $+Z$ direction and of magnitude B. Outside this interval B is zero.

To understand the implications of Maxwell's equations in empty space, we consider a fixed imaginary rectangular loop parallel to the XY plane. The width of the loop along the Y direction is w, and the length in the X direction is indefinitely long; the dashed lines in Fig. 30-4 indicate one such loop. If the leading edge of the magnetic pulse passes the left end of the loop at the time $t = 0$, it will, after a time t has elapsed, have progressed a distance ct into the loop. Since the magnetic flux through the loop is changing, there is an induced electric field, and we can evaluate this electric field by using Faraday's law, Eq. (30-8).

The flux through the loop at time t is

$$\phi_B = BA = B(wct) \tag{30-9}$$

When $t = L/c$, the trailing edge of the pulse will have entered the loop; thereafter ϕ_B will be constant. The rate of flux change while the pulse passes the left end of the loop is

$$\frac{d\phi_B}{dt} = Bwc \tag{30-10}$$

Assuming the induced electric field to be in the Y direction, the only contribution to the integral $\oint \mathbf{E} \cdot d\mathbf{l}$ taken around the loop in Fig. 30-4 is along the left side, where the integral yields $-Ew$. Thus,

$$\oint \mathbf{E} \cdot d\mathbf{l} = -Ew \tag{30-11}$$

Substituting (30-10) and (30-11) into (30-8) gives

$$Ew = Bwc$$

$$E = Bc \tag{30-12}$$

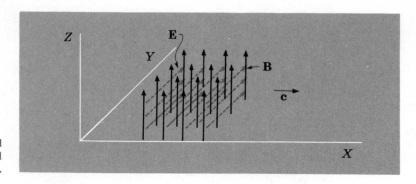

FIG. 30-5. Electric field accompanying the magnetic field of Fig. 30-4.

Equation (30-12) gives the magnitude of the electric field induced in an imaginary loop in space, which is to say, it gives the magnitude of the electric field produced in empty space.

What about the direction of **E**? We find this from Lenz' law. As the magnetic pulse passes into the loop, the magnetic flux through the loop increases in the $+Z$ direction. Consequently, an induced emf and an induced electric field are set up to oppose the increase in magnetic flux. Along the left side of the loop this requires that **E** be in the $+Y$ direction (into the paper in Fig. 30-4). As long as the magnetic pulse passes the left end of the loop, there will exist an electric field **E**, in the $+Y$ direction, of magnitude Bc. At any time that the pulse is *not* passing the left end of the loop, there is *no* electric field there. Thus, the original transverse magnetic pulse is accompanied by a transverse electric pulse, the two fields being perpendicular to each other and to the direction of propagation of both fields; see Fig. 30-5.

Let us review what has been done so far. We began with a moving magnetic pulse and showed, using Faraday's law, that an electric pulse must accompany it. The magnitudes of **E** and **B** are related by (30-12). Now we do just the reverse. We begin with an electric pulse and show that a magnetic pulse must accompany it.

In a fashion similar to that employed for the magnetic pulse in Fig. 30-4, we consider a rectangular pulse of electric field **E,** as shown in Fig. 30-6. The field is in the direction of the $+Y$ axis and travels at speed c along the $+X$ axis. We now consider an imaginary rectangular loop lying in the XZ plane, whose width in the Z direction is l. We evaluate the magnetic field produced in this loop by the passing electric pulse by using Ampère's law:

$$\oint \mathbf{B} \cdot d\mathbf{l} = \varepsilon_0 \mu_0 \frac{d\phi_E}{dt} \qquad (30\text{-}7), (30\text{-}13)$$

The only contribution to the integral $\oint \mathbf{B} \cdot d\mathbf{l}$ is along the left end of the loop, where the direction of **B** is taken to be along the Z axis:

$$\oint \mathbf{B} \cdot d\mathbf{l} = Bl \qquad (30\text{-}14)$$

The electric flux $\phi_E = EA$ through the loop at time t is

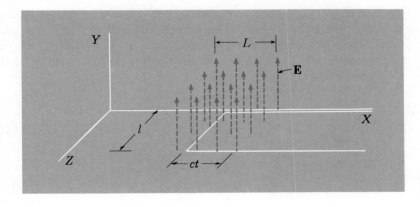

FIG. 30-6. Traveling electric pulse.

$$\phi_E = EA = E(lct)$$

After $t = L/c$, the trailing edge of the moving pulse will have passed the left end of the loop, and thereafter the flux within the loop will be constant. While the flux changes,

$$\frac{d\phi_E}{dt} = Elc \qquad (30\text{-}15)$$

Substituting (30-14) and (30-15) into (30-13) yields

$$Bl = \varepsilon_0\mu_0 Elc$$
$$B = \varepsilon_0\mu_0 Ec \qquad (30\text{-}16)$$

The direction of **B** at the left end of the loop is out of the paper (in the $+Z$ direction). This must be so because the displacement current (electric flux change) is in the $+Y$ direction; the accompanying magnetic field surrounding this current follows from the right-hand rule. Thus, a transverse magnetic pulse accompanies the electric pulse. When the electric field is zero, so is the magnetic field; see Fig. 30-7. The combined electric and magnetic fields are the same as shown in Fig. 30-5. The relative directions of **E**, **B**, and **c** are exactly the same as we found before!

Beginning with a magnetic pulse one finds an accompanying

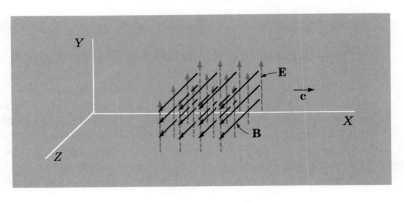

FIG. 30-7. Magnetic field accompanying the electric field of Fig. 30-6.

Electromagnetic Waves from Maxwell's Equations   SEC. 30-3

electric pulse; beginning with an electric pulse one finds an accompanying magnetic pulse. The directions of the fields are self-consistent. In addition, the magnitudes of **E** and **B**, given by (30-12) and (30-16) must be self-consistent. Substituting (30-12) into (30-16) gives

$$B = \varepsilon_0 \mu_0 B c^2$$

$$c = \frac{1}{\sqrt{\varepsilon_0 \mu_0}} \quad (30\text{-}17)$$

By arbitrary choice (Sec. 27-2),

$$\mu_0 = 4\pi \times 10^{-7} \text{ Wb/A-m} = 12.57 \times 10^{-7} \text{ N/A}^2 \quad (27\text{-}2)$$

As a consequence of this choice, it is found experimentally that

$$\varepsilon_0 = 8.85 \times 10^{-12} \text{ C}^2/\text{N-m}^2 \quad (20\text{-}6)$$

Substituting the values for the electric and magnetic constants, ε_0 and μ_0, into (30-17) gives

$$c = \frac{1}{\sqrt{(8.85 \times 10^{-12} \text{ C}^2/\text{N-m}^2)(12.57 \times 10^{-7} \text{ N/A}^2)}}$$

$$c = 3.00 \times 10^8 \text{ m/s}$$

Thus, an electromagnetic disturbance can travel through empty space, but it must travel at the unique speed $c = 3.00 \times 10^8$ m/s — which is exactly the measured speed at which light travels through empty space. In fact, light is just one form of electromagnetic radiation. This discovery was the greatest triumph of classical electromagnetic theory. The measured speed of light agrees with the predicted speed based on electric and magnetic parameters ε_0 and μ_0. Hereafter the symbol c shall denote the speed of light in vacuum.

Figures 30-5 and 30-7 show that the directions of the **E** and **B** fields must be perpendicular to each other and to the direction of propagation. What are the relative magnitudes of **E** and **B**? We find this by using the results of (30-17) in (30-16):

$$B = \frac{E}{c} \quad (30\text{-}18)$$

In our derivation the electric and magnetic fields change with time, but in a simple way: The fields are "turned on and off" abruptly at the leading and trailing edges of the pulse. We shall, of course, wish to consider waves in which the electric and magnetic fields change continuously with time, particularly sinusoidally. Nothing different need be done, however, in treating such waves, inasmuch as it has been shown above that whenever the electric field changes magnitude in space so does the magnetic field accompanying it. Moreover, through the superposition principle, which applies without restriction to electromagnetic waves, we may add, or superpose, any number of pulses to approximate a sinusoidal wave.

30-4 ELECTROMAGNETIC ENERGY DENSITY, INTENSITY, AND THE POYNTING VECTOR

First recall that the energy density u_E of an electric field of magnitude E is given by

$$u_E = \tfrac{1}{2}\varepsilon_0 E^2 \qquad (23\text{-}32), (30\text{-}19)$$

and that the magnetic energy density u_B is

$$u_B = \frac{1}{2\mu_0} B^2 \qquad (28\text{-}22), (30\text{-}20)$$

Of course, if the **E** and **B** fields vary with time, so do the energy densities u_E and u_B. These relations give the *instantaneous* energy densities at any point in space. Using (30-18) in (30-20), we have

$$u_B = \frac{1}{2\mu_0}\left(\frac{E}{c}\right)^2 = \frac{\varepsilon_0 \mu_0}{2\mu_0} E^2 = \tfrac{1}{2}\varepsilon_0 E^2$$

Comparing this with (30-19), we find that

$$u_B = u_E \qquad (30\text{-}21)$$

The energy densities of the electric and the magnetic fields of an electromagnetic wave are *equal;* that is, the energy carried by an electromagnetic disturbance is shared equally by the electric and magnetic fields.

The total energy density u, or energy per unit volume, of an electromagnetic wave is

$$u = u_E + u_B = 2u_E = \varepsilon_0 E^2 \qquad (30\text{-}22)$$

The energy density is proportional to the *square* of the electric field (equivalently, we may write $u = 2u_B = B^2/\mu_0$).

The intensity I of any wave is defined as the energy flow per unit time (the power P) across a unit area oriented at right angles to the direction of wave propagation; that is,

$$I = \frac{P}{A} \qquad (30\text{-}23)$$

where A is the transverse area through which the energy flows. We wish to find an expression for the intensity I in terms of the energy density u of a wave traveling with wavespeed c.

Consider a cylinder of cross-sectional area A and length L through which electromagnetic radiation is propagated along the direction of the cylinder's axis, as shown in Fig. 30-8 (the cylinder's volume is assumed to be so small that the energy density u is

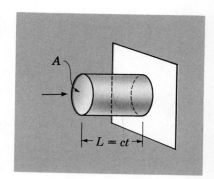

FIG. 30-8. Electromagnetic energy flux through the area A.

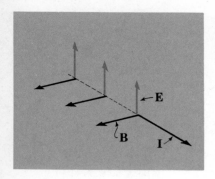

FIG. 30-9. The relative directions E, B, and I for an electromagnetic wave.

constant throughout). The length can be written as $L = ct$, where c is the speed of light and t is the time required for a wave to travel from one end of the cylinder to the other. Therefore, in the time t, all the energy originally contained in the cylinder's volume $AL = Act$ will have passed through the area A. We can write

$$\text{Intensity} = \frac{\text{energy}}{\text{area} \times \text{time}} = \frac{(\text{energy/volume}) \times (\text{volume})}{\text{area} \times \text{time}}$$

$$I = \frac{u \times Act}{At}$$

$$I = uc \tag{30-24}$$

We have not used the explicit form for the energy density u; therefore, (30-24) is a general relationship applying to *any* kind of wave. For an electromagnetic wave, we have, using (30-22),

$$I = \varepsilon_0 E^2 c \tag{30-25}$$

We may write the relation for the intensity differently by using (30-17) and (30-18) in (30-25):

$$I = \varepsilon_0 E^2 c = \varepsilon_0 E(Bc)c = \frac{\varepsilon_0 EB}{\varepsilon_0 \mu_0} = \frac{EB}{\mu_0}$$

Recall that for an electromagnetic wave the electric field **E** and the magnetic field **B** are perpendicular to each other and to the direction of wave, or energy, propagation, as shown in Fig. 30-9. In fact, the direction of the cross product **E** × **B** is the direction of wave propagation (the thumb of the right hand points in the direction in which the electromagnetic energy flows when we turn the **E** vector into the **B** vector through 90° with the right-hand fingers). Thus, we can assign a *direction* to the intensity **I** and regard it as a *vector* pointing in the direction of the electromagnetic energy flux. We write the last equation in the vector form

$$\mathbf{I} = \mathbf{E} \times \frac{\mathbf{B}}{\mu_0} \tag{30-26}$$

The vector intensity **I** is referred to as the *Poynting vector*, named for J. H. Poynting (1852–1914). If we wish to find the electromagnetic energy flow per unit time, or the power P, through a surface having an element $d\mathbf{S}$ we merely integrate $\mathbf{I} \cdot d\mathbf{S}$ over the entire surface:

$$P = \int \mathbf{I} \cdot d\mathbf{S} \tag{30-27}$$

Example 30-1

The intensity of the sun's radiation at the earth is approximately 1,000 W/m². (*a*) What is the electromagnetic energy density of this radiation at the earth? (*b*) What are the magnitudes of the electric and magnetic fields for this light?

(a) From (30-24),

$$u = \frac{I}{c} = \frac{10^3 \text{ W/m}^2}{3 \times 10^8 \text{ m/s}} = 3.3 \times 10^{-6} \text{ J/m}^3$$

(b) Using (30-22), we have

$$E = \sqrt{\frac{u}{\varepsilon_0}} = \sqrt{\frac{3.3 \times 10^{-6} \text{ J/m}^3}{8.85 \times 10^{-12} \text{ C}^2/\text{N-m}^2}}$$

$$= 6.1 \times 10^2 \text{ N/C} = 610 \text{ V/m}$$

and from (30-18),

$$B = \frac{E}{c} = \frac{6.1 \times 10^2 \text{ N/C}}{3.0 \times 10^8 \text{ m/s}}$$

$$= 2.1 \times 10^{-6} \text{ Wb/m}^2 = 0.021 \text{ G}$$

The magnitude of the magnetic field of an electromagnetic wave is *much* less (by the factor c) than that of the electric field. So small a magnetic field cannot easily be detected; an electromagnetic wave is detected in absorption by the much larger effects of the electric field.

It should be noted that all the relations derived in this exercise apply for a point in space and an instant of time. If the electric and magnetic fields were to vary with time, as would be the case of visible light, then the values of **E** and **B** computed above are strictly the root-mean-square values.† It is significant, however, that *all* the above formulas apply without regard to the time variation, or frequency, of the electromagnetic wave; that is, the magnitudes of **E** and **B** for a 1.0 kW/m² radio beam are the same as for a 1.0 kW/m² light beam.

30-5 RADIATION FORCE AND PRESSURE: THE LINEAR MOMENTUM OF AN ELECTROMAGNETIC WAVE

An electromagnetic wave carries energy. It is easy to show that an electromagnetic wave also has *linear momentum* in the direction of propagation and that such a wave can exert a force, or *radiation pressure*, on a material upon which it impinges. The first measurement of the pressure of light was made by P. N. Lebedev (1866–1912) in 1901.‡

We first assume that an electromagnetic wave is incident on a material which absorbs all the energy striking it, reflecting and transmitting none. This implies that the electric field **E** must do work on a charged particle within the material, the energy removed per unit time from the electromagnetic wave being exactly the power absorbed by the material. Note that we say the work done

---

†See R. T. Weidner and R. L. Sells, *Elementary Classical Physics*, 2d ed., sec. 35-7, Allyn and Bacon, 1973.
‡For an interesting review of experiments on the pressure of light, including laser beams, see A. Ashkin, *Scientific American*, February, 1972, p. 62.

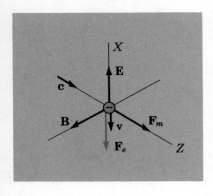

FIG. 30-10. The electric force F_e and the magnetic force F_m exerted on an electron by an electromagnetic wave.

by the *electric* field. The magnetic force, since it always acts at right angles to a charged particle's velocity, does *no work*.

Consider the situation shown in Fig. 30-10. Here an electromagnetic wave travels along the positive Z axis, the electric field **E** being along the positive X axis and the magnetic field **B** being along the positive Y axis. We are interested in the forces produced by **E** and **B** on an electron within the material. The electric force $\mathbf{F}_e = q\mathbf{E}$ on the electron has a magnitude $F_e = eE$. This force acts in the direction opposite to that of **E** and accelerates the electron in a direction transverse to that of the wave propagation.

What is the effect of the magnetic field? We assume the electron to be moving with speed v along the negative X axis. The field **B** is along the positive Y axis. In general, the direction and magnitude of the magnetic force is given by $\mathbf{F}_m = q\mathbf{v} \times \mathbf{B}$. In this case, the magnitude of $\mathbf{F}_m$ is evB, and its direction is along the positive Z axis. The electromagnetic wave produces a *force* on the electron (and, therefore, also on the material to which the electron is bound) *along the direction of wave motion!* In short, when an electromagnetic wave impinges on an electric charge, the **E** field accelerates the charge in the transverse direction and does work on it, while the **B** field, acting on the moving charge, produces a longitudinal force.

We wish to find relations for the radiation force and pressure, and the linear momentum of the electromagnetic field, in terms of such quantities as the intensity I and power P of the wave. We found that the radiation force F_r is given by

$$F_r = evB$$

Since the magnitudes of **E** and **B** are related by $B = E/c$, this equation may be rewritten as

$$F_r = \frac{v}{c} eE \tag{30-28}$$

But eE is just the magnitude of the electric force, $\mathbf{F}_e$, the force that does work. In general, the rate of doing work, or the power P, is given by

$$P = Fv \tag{9-13}$$

where F is the force doing work and v is the speed of the particle acted on. Then (30-28) may be written

$$F_r = \frac{v}{c} F_e$$

Total absorption:
$$F_r = \frac{P}{c} \tag{30-29}$$

The radiation force of an electromagnetic wave on a material that *absorbs it completely* is simply the power of the wave divided by the speed of light.

Of course, the electrons in an absorbing material are in constant motion, although bound to the material; thus, they are *not* at rest until acted upon by the impinging electromagnetic wave. Nevertheless, the simple derivation for the radiation force given above still applies. The forces arising from the electric and magnetic fields of the electromagnetic wave are in addition to any other forces present. Even if the electric and magnetic fields vary with time, the relations hold at each instant.

We have found that a longitudinal force P/c is exerted by an electromagnetic wave on a material absorbing it completely. It follows that when a material *emits* radiation of power P in one direction, the emitter must recoil under the action of a recoil radiation force of magnitude P/c. We can see this most easily by noting that emission is, so to speak, absorption run backward in time. That is, in emission accelerating charges within the material *lose* energy and create an outgoing electromagnetic wave. With time reversal the directions of the electric field **E**, the electric force $\mathbf{F}_e$, and the radiation force $\mathbf{F}_r$ remain *the same;* but the direction of the *velocity* **v** is reversed, as is also the direction of the *magnetic field* **B**. Thus, under time reversal, that is, with emission rather than absorption, the direction of the Poynting vector is reversed, and energy now flows away from the material rather than toward it.

What is the radiation force on a material which reflects all the radiation striking it, absorbing none? We imagine the reflection process to take place in two stages: absorption of the incident radiation followed by re-emission in the reverse direction. Since a radiation force of magnitude P/c acts on the material both in absorption and in emission, the radiation force for complete reflection is

Complete reflection: $$F_r = \frac{2P}{c} \tag{30-30}$$

The fact that the radiation force for complete reflection is twice that for complete absorption has an exact analog in mechanics. When a particle with initial momentum $+mv$ strikes and sticks to an object, the linear momentum transferred to the struck object is $+mv$, but when a particle with initial momentum $+mv$ is "reflected" from the struck object, rebounding with the same speed, the particle's final momentum is $-mv$, its momentum having been changed by $\Delta(mv) = -mv - (+mv) = -2mv$. Thus, for reflection the struck object acquires a momentum $+2mv$, just *twice* the momentum acquired in absorption. Equivalently, the (average) force on the struck object is twice as great for reflection as for absorption.

The radiation force given in (30-29) and (30-30) applies for radiation which is incident in a direction *perpendicular* to the absorbing or reflecting surface. For oblique incidence, with an angle θ between the direction of wave propagation and the normal to the plane of the absorber or reflector, the radiation force normal to the surface is obtained by multiplying F_r by the factor $\cos\theta$. For total reflection when $\theta \neq 0$, there is no tangential force.

It is useful to write relations giving the *radiation pressure* p_r, the radiation force F_r per unit transverse area A, for absorption and reflection. Since the pressure p is, by definition, F/A, we have for complete absorption,

$$p_r = \frac{F_r}{A} = \frac{P}{cA}$$

The intensity I is given by $I = P/A$. Therefore,

Complete absorption: $\qquad p_r = \dfrac{I}{c}$

(30-31)

Complete reflection: $\qquad p_r = \dfrac{2I}{c}$

Example 30-2

A 3-W beam of electromagnetic radiation shines on, and is completely absorbed by, a black object. *(a)* What is the radiation force on the absorber? *(b)* What is the recoil force on the source emitting the beam?

(a) For complete absorption,

$$F_r = \frac{P}{c} = \frac{3\text{ W}}{3 \times 10^8 \text{ m/s}} = 10^{-8}\text{ N}$$

which is a very small force indeed.

(b) The source emitting the 3-W beam, whether it be a source of light or a radio transmitter, will, as long as the emitted waves travel outward in a single direction, recoil under the action of a force of 10^{-8} N. Thus, a flashlight emitting light constitutes a very elementary form of a rocket.

For any sources of moderate intensity or power the radiation force is extraordinarily small. Yet it can be measured with a torsion pendulum, the same instrument used to measure the extremely small gravitational force between laboratory-size objects in the Cavendish experiment (Sec. 13-2). The results are found to be in exact agreement with the predictions of electromagnetic theory. Moreover, in stellar phenomena the radiation force may equal or exceed the gravitational force, as evidenced by an exploding star, or supernova.

Clearly, if electromagnetic radiation can exert a force and transfer linear momentum to an object upon which it impinges, linear momentum must be associated with the electromagnetic field itself. It is easy to derive the expression for the momentum M of an electromagnetic wave (we use the symbol M for linear momentum, rather than the conventional symbol p, to avoid confusion with the pressure p and the power P). By definition, the force F is related to the momentum M by

$$F = \frac{dM}{dt}$$

Similarly, the power P is related to the energy by

$$P = \frac{d}{dt} \text{ (energy)}$$

From (30-29),

$$F_r = \frac{P}{c}$$

$$\frac{dM}{dt} = \frac{1}{c} \times \frac{d}{dt} \text{ (energy)}$$

We use the relation for the radiation force in *absorption* since we wish to count the energy transfer only *once*. Integrating yields

$$M = \frac{\text{electromagnetic energy}}{c} \tag{30-32}$$

The magnitude of the linear momentum of an electromagnetic wave is the energy of the wave divided by c; the direction of the momentum is along the direction of energy propagation, that is, along the direction of the Poynting vector **I**.

We have seen that one can and must attribute energy and linear momentum to an electromagnetic wave. One can also attribute *angular* momentum (actually, *spin* angular momentum) to circularly polarized electromagnetic waves. It can be shown that the magnitude of the electromagnetic angular momentum of a circularly polarized wave of frequency f is the energy of the wave divided by $2\pi f$ and that the direction of the angular momentum vector is parallel or antiparallel to the direction of wave propagation.[†]

30-6 ACCELERATING CHARGES AND ELECTROMAGNETIC WAVES

In our derivation of the speed ($1/\sqrt{\varepsilon_0 \mu_0}$) of an electromagnetic wave in Sec. 30-3 we assumed the electric and magnetic fields to exist in space unattached to any charges. We then showed that such an electromagnetic disturbance in space was self-generating, that is, it was an electromagnetic *wave*. But how is such a wave launched? How can the electric and magnetic fields become detached from the electric charges which produce them? The subject of the generation of electromagnetic waves can, of course, be treated by applying Maxwell's equations, a rather complex analysis lying beyond the

---

[†]See R. T. Weidner and R. L. Sells, *Elementary Classical Physics*, 2d ed., sec. 40-6, Allyn and Bacon, 1973.

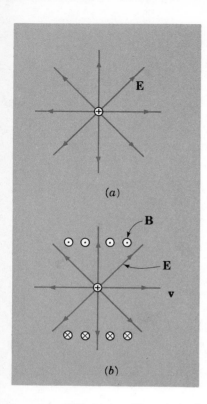

FIG. 30-11. Electric and magnetic fields surrounding an electric charge (a) at rest and (b) in motion at constant velocity.

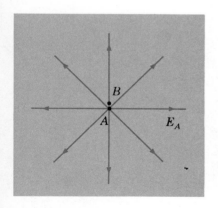

FIG. 30-12a. Electric field lines surrounding and electric charge before being displaced from point A.

scope of this book. What shall be done here is to give qualitative and plausible arguments to show how an accelerating electric charge generates an electromagnetic wave.

Note that we say an *accelerating* electric charge. If one views an electric charge from an inertial frame in which it is at rest, one sees only a static electric field radiating outward from the charge. On the other hand, if one views the charge from an inertial frame in which the charge moves with constant velocity, the electric charge has, in addition to the electric field, an accompanying magnetic field which encircles it; see Fig. 30-11. As long as the electric charge has a *constant* velocity, zero or nonzero, the electric and magnetic fields remain attached to the charge and move along with it.

Now let us consider the behavior of the electric field lines from a single positive charge which was accelerated for a short time. Figure 30-12 shows the charge q which has been at rest for a long time at point A. The electric field lines extend radially outward from A, as shown in Fig. 30-12a. One such line is E_A. At some time, say $t = 0$, the charge is accelerated upward for a short time δt, after which it then continues at a constant speed (we assume $v \ll c$) upward. At time $t \gg \delta t$ it passes point B in the figure. As the charge moves at constant v, the electric lines *near the charge q* extend outward from the moving charge and move upward with it. The field lines are indicated in Fig. 30-12b at the instant of time when the charge is at point B. Near the charge, the field lines emanate radially outward from B. One such line E_B is shown in Fig. 30-12b. But far away, the field lines still emanate from point A, as shown by line E_A. This is so because the electromagnetic effects produced by the accelerated charge travel outward from the charge at the *finite* speed c. Note that the far electric line E_A and the near electric line E_B are the *same* field line. Thus, at the instant shown in Fig. 30-12b, a small charge q_C at point C will not have "known" that q has accelerated from A to B, because q_C is still experiencing the field E_A. The kink in the electric line of force, that is, the transverse component, is traveling toward q_C with the speed of light and reaches it at some later time. On the other hand, a charge q_D at point D has already experienced a transverse electric force arising from the acceleration of q, and at the instant shown in Fig. 30-12b it experiences the field E_B. A charge q_R at R will experience a smaller transverse electric force than q_C, because the kink in the electric line is less pronounced than at C. Finally, a charge q_S at S, which lies along the direction of the acceleration (from A to B) of q, experiences *no* transverse force.

At a point such as R, both before and after the passage of the transverse kink, a small charge is acted upon only by a *radial* electric field, whose strength varies inversely as the *square* of the distance AR (or BR), in accordance with Coulomb's law. The radial electric field is the *static* electric field, always attached to the accelerated charge; the transverse electric field is the *radiated* electric field, which becomes detached from the charge and moves outward from it. The magnitude of the transverse electric field can be shown to vary inversely as the distance AR. Therefore, at large distances from the charge q, the radial component (varying as $1/r^2$) is negligible compared with the transverse component (varying as $1/r$) during the time when the latter component passes. Although both the static and radiation fields fall off with distance, the radiation field survives over the static field at great distances from the charge.

It can be shown that the transverse electric field E at a distance r from an electric charge q having an acceleration a is given by

$$E \propto \frac{qa \sin \theta}{r} \qquad (30\text{-}33)$$

where θ is the azimuthal angle, the angle between the direction of the charge's acceleration and that of the radius vector **r**; see Fig. 35-13a.

Since the magnitude of the radiated electric field varies as $\sin \theta$, the radiated energy is a maximum in the equatorial plane (perpendicular to the direction of the charge's acceleration) with no radiation emitted along the polar axis. According to (30-25) the intensity I of an electromagnetic field is proportional to E^2. Therefore, from (30-33),

$$I \propto \frac{q^2 a^2 \sin^2 \theta}{r^2} \qquad (30\text{-}34)$$

The radiated intensity falls off inversely as the square of the distance r from the accelerating charge and varies with direction according to $\sin^2 \theta$.

The angular variation is shown in Fig. 30-13b, which is a polar diagram showing the intensity I as a function of the azimuthal angle θ. The curve is so drawn that the length of the vector **I** for a given angle θ is proportional to the intensity of the radiation in that direction (following $\sin^2 \theta$). The pattern is symmetrical with respect to an axis along the direction of the charge's acceleration.

We have focused our attention on the electric field radiated by an accelerated electric charge. Because the charge is in motion, there is, in addition, an associated magnetic field, transverse to the electric field and also traveling outward from the accelerated charge at the speed of light. We know, of course, that any electromagnetic disturbance carries energy. The radiated energy arises from the fact that an external agent does work on the charge in accelerating it. This acquired energy does *not*, however, appear as energy residing in the *static* fields. It leaves the charge and travels through space. When the radiation is absorbed, the electric field of the electromagnetic wave does work on a charge in the absorber.

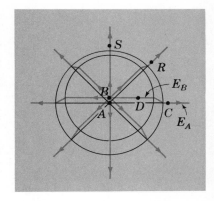

FIG. 30-12b. Electric field lines surrounding an electric charge after being displaced from A to B.

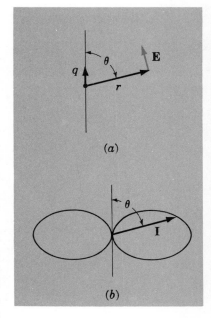

FIG. 30-13. (a) Electric field radiated from an accelerated charge. (b) Radiation pattern from an accelerated charge giving the variation in intensity I with the azimuthal angle θ.

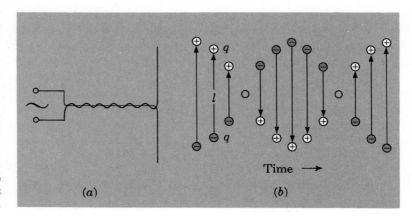

FIG. 30-14. (a) Electric dipole oscillator. (b) Equivalent oscillating electric charges.

30-7 SINUSOIDAL ELECTROMAGNETIC WAVES

We have discussed a pulse of electromagnetic radiation arising from a single accelerated charge. Let us now turn to the more interesting and useful situation in which a continuous electromagnetic wave having a well-defined frequency and wavelength is produced. A common source of sinusoidal electromagnetic radiation is an electric-dipole oscillator, represented schematically in Fig. 30-14a. An applied voltage varying sinusoidally produces an alternating current in the two straight-line conductors (an antenna). At one end of the wire the charge is alternatively positive and negative; at the other end, the charge similarly alternates, but is 180° out of phase. We can think of this behavior in terms of the motion of two equal but opposite electric charges undergoing simple harmonic motion along the axis of the antenna at the same frequency as that of the alternating-current source, as in Fig. 30-14b. Such a pair of oscillating charges constitutes an oscillating electric dipole. The intensity pattern from each charge is that of Fig. 30-13. Therefore, the radiation pattern of an oscillating electric dipole is also that shown in Fig. 30-13.

Figure 30-15 is a representation of the electric and magnetic fields produced by such an electric dipole oscillator at one instant of time. We can see several features in this figure.

1. At short distances from the dipole the electric lines extend from the positive to the negative charge of the dipole. At large distances the electric lines have become detached from the oscillating dipole and form closed loops.
2. The electric field **E** and the magnetic induction **B** are perpendicular to one another and also to the direction (outward) of propagation of the radiated electromagnetic waves.
3. At great distances the wavefronts, surfaces of constant phase, are spheres centered about the dipole.
4. The wave is *linearly* polarized; that is, the transverse electric field at

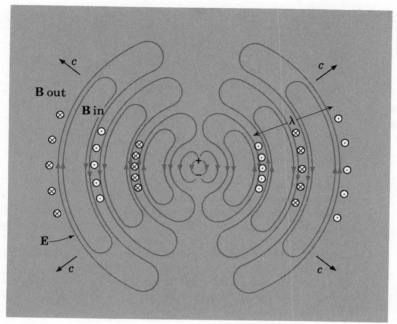

FIG. 30-15. Electric and magnetic fields generated by a sinusoidally varying electric dipole.

any point in space oscillates along a fixed straight line, this direction being called the direction of polarization of the wave (we shall have more to say about the polarization properties of electromagnetic waves in Chap. 33).

5. The radiation pattern is symmetrical about the axis of the dipole, in accord with Fig. 30-13b. (The radiation pattern of Fig. 30-13b applies only when the distance r is large compared with the wavelength.)
6. The wavelength λ of the electromagnetic wave is the distance between two adjacent wavefronts in the same phase. As the oscillator completes one oscillation cycle, a wavefront travels outward a distance λ. The frequency f of the oscillator is precisely the frequency of the electromagnetic wave. Therefore, as for other types of waves,

$$c = f\lambda \qquad (30\text{-}35)$$

Consider now the sinusoidally varying electromagnetic wave along any radial line at a large distance from the dipole. Locally, the wave will appear as a plane wave, that is, a wave in which points in the same phase form a plane in space, or a plane *wavefront*. The variation in space of the electric and magnetic fields of such a sinusoidal electromagnetic wave at one instant of time is shown in Fig. 30-16. The *length* of the lines representing the electric and magnetic fields indicates the strength of the fields.

Example 30-3
A radio wave having an intensity of 1.5×10^{-10} μW/m² is incident upon a 1.0-m antenna oriented along the direction of the electric field

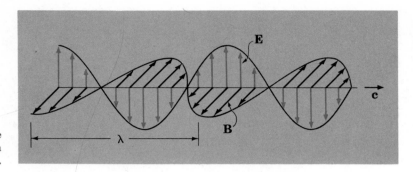

FIG. 30-16. Representation of the electric and magnetic fields of a sinusoidal electromagnetic wave.

of the electromagnetic wave. What is the maximum instantaneous potential difference across the ends of the antenna?

Eq. (30-25) gives the *instantaneous* intensity I of an electromagnetic wave in terms of E^2, the square of the *instantaneous* magnitude of the electric field. We assume that the radio wave varies sinusoidally with time so that $E = E_0 \sin \omega t$, where E_0 is the electric field amplitude. We must interpret E^2 to mean the *time average*, or $\overline{E^2}$. For such a sinusoidal wave it can be shown† that $\overline{E^2} = \tfrac{1}{2}E_0^2$. Combining this with (30-25) we find

$$I = \varepsilon_0 c \overline{E^2} = \tfrac{1}{2}\varepsilon_0 c E_0^2 \tag{30-36}$$

$$E_0 = \sqrt{\frac{2I}{\varepsilon_0 c}}$$

$$= \sqrt{\frac{2(1.5 \times 10^{-16}\ \text{W/m}^2)}{(8.85 \times 10^{-12}\ \text{C}^2/\text{N-m}^2)(3.0 \times 10^8\ \text{m/s})}}$$

$$= 3.4 \times 10^{-7}\ \text{V/m}$$

Across the 1-m conductor the maximum potential difference is a mere $0.34\ \mu\text{V}$.

30–8 THE ELECTROMAGNETIC SPECTRUM

All electromagnetic waves, whatever the frequency and wavelength, travel at the speed $1/\sqrt{\varepsilon_0 \mu_0}$ through vacuum and have the fundamental properties we have already discussed. The complete spectrum of electromagnetic radiation is shown in Fig. 30-17, where the frequency and wavelength are plotted on a logarithmic scale. At the time of Hertz' experiments in 1887, which confirmed Maxwell's electromagnetic theory of light, only two types of radiation were recognized, visible light and radio waves. Now it is known that other types of radiation exist; they differ in wavelength and frequency as well as in the origin of the radiation and their effects on substances upon which they impinge.

The waves of lowest frequency (and longest wavelength) are

---

†See R. T. Weidner and R. L. Sells, *Elementary Classical Physics,* 2d ed., sec. 34-2, Allyn and Bacon, 1973.

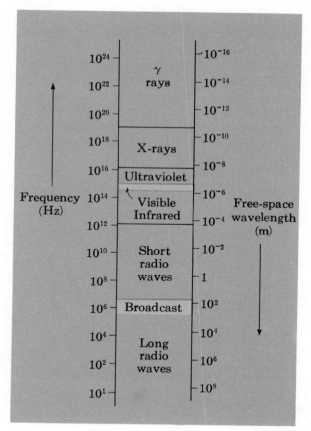

FIG. 30-17. The electromagnetic spectrum.

radio waves; such waves are generated by oscillating electric currents. The short-wavelength radio waves, or microwaves, have wavelengths comparable to those of audible sound through air. Infrared radiation is produced by heated solids or the molecular vibrations and rotations in gases and liquids. Visible light is produced by rearrangements of the outer electrons in atoms. The very narrow range of wavelengths, from 4000 Å (1 Å = 10^{-10} m) to 7000 Å from violet to red light) to which the human eye is sensitive corresponds, in musical terminology, to slightly less than one octave (a factor of 2 in frequency) and is to be compared with the enormous frequency range (20 to 20,000 Hz) to which the human ear is sensitive. Ultraviolet radiation immediately adjoins the visible spectrum. X-rays have wavelengths of the approximate size of atoms, and they originate in the rearrangement of innermost electrons of atoms. γ rays are the electromagnetic waves of the highest frequency and shortest wavelength; they originate in rearrangements among the particles within the atomic nucleus.

The boundaries between the adjoining regions are not sharply defined. For example, one cannot distinguish between a short-wavelength x-ray and a long-wavelength γ ray. The various types of electromagnetic radiation differ in their effects on materials. For example, materials which are opaque to visible light are transparent to x-rays.

In Chap. 32 we shall discuss procedures for measuring the wavelengths of electromagnetic radiation, particularly for relatively short-wavelength radiation, such as visible light. Suffice it to say here that a wavelength measurement typically involves an arrangement in which some characteristic dimension is of the same order of magnitude as the wavelength.

30-9 MEASUREMENTS OF THE SPEED OF LIGHT

Experiment indicates that nothing travels faster than light or any other form of electromagnetic radiation. In fact, light's speed is so great that it appears to the casual observer to move at an infinite speed. It takes considerable insight to raise the question whether the propagation of light is *not* instantaneous and considerable experimental ingenuity and finesse to measure light's speed of propagation with precision. A variety of methods have been developed over the past 300 years to measure the fundamental constant c. Although differing in many respects, all methods have this in common: One must either measure the time for light to travel a very large distance, so as to make the time interval reasonably large (or, with smaller distances, one must measure the very short time interval with great precision), or else one must somehow combine the velocity of light with another high velocity in an effect depending upon their combination.

Roemer Method. The first rough measurement of c was made in 1666 by Ole Roemer (1644–1701), a Danish astronomer. Roemer's measurement was based upon observations of the orbit period of Io, one of Jupiter's moons. (Recall that the orbit period of any satellite is essentially constant.)

Roemer observed a nonperiodicity in the orbit of Io and realized that the apparent discrepancies were due to changes in the time required for light to travel (at constant speed) the ever-changing distance between earth and the satellite. The changes in distance were due primarily to the orbital motion of the earth and—to a lesser extent—to the orbital motion of Jupiter. If we combine Roemer's measurements with modern values for the dimensions of the earth's orbit, we get a value of c which is within a few percentage points of the accepted value.

Stellar Aberration. In this method, first employed by the astronomer James Bradley (1692–1762) in 1725, the velocity of light is combined vectorially with the orbital velocity (3×10^4 m/s) of the earth around the sun.

Suppose first that one is to catch raindrops as they fall vertically to the bottom of a long tube. Of course, the tube must be held vertical. However, if the tube is moving horizontally at the same time, it must be tilted at an angle θ relative to the vertical to prevent the raindrops from striking the sides of the tube. This is similar to the situation arising when an astronomer views a distant star. His tube, a telescope, is to catch light from a distant "fixed" star. However, the telescope is attached to the earth, which, by virtue of its orbital motion around the sun, is moving at a speed of 3×10^4 m/s relative to the sun and the fixed stars; see Fig. 30-18.

When a star at the zenith is being viewed, the telescope must be tilted through the angle θ when the earth is in position a; see Fig. 30-18. Six months later, when the earth is traveling in the opposite direction relative to the fixed stars, the telescope must be tilted in the opposite sense. The change in telescope orientation 2θ can be measured, and the speed of light c can be computed from the relation $\tan \theta = v/c$. The effect is known as *stellar aberration*. The angle 2θ is a mere 41 s of arc.

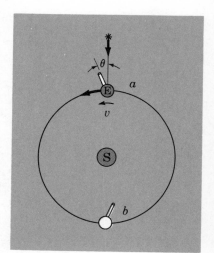

FIG. 30-18. Inclination of a telescope arising from stellar aberration.

Fizeau Method. The most direct way of measuring the speed of light is to time its travel in a round trip from some starting point to a more or less distant mirror and back. This method requires *pulses* of light rather than a continuous waves; one can use the same device which chops the continuous light beam into pulses to measure their total travel time t, where $c = 2d/t$, d being the distance from the light source to the mirror. Figure 30-19 shows the simple arrangement used by A. H. L. Fizeau (1819–1896) in 1849. A rotatable wheel has regularly spaced openings along its rim. When the wheel is at rest, the light can pass through an opening, travel to the distant mirror, and return to pass through the *same* opening. When the wheel is in rotation, however, the reflected light will not enter the telescope unless the wheel has turned so that the *adjoin-*

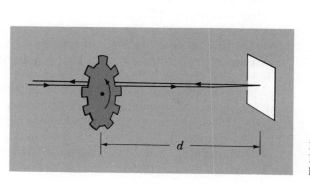

FIG. 30-19. Arrangement used by Fizeau in measuring the speed of light.

ing opening is in the same position as was the opening through which the light first left the light source. It is then an easy matter to compute the time for the round trip from the measured speed of rotation and the angular separation between adjacent openings. In Fizeau's experiment, $d \approx 9$ km, and thus $t \approx 6 \times 10^{-5}$ s. Present-day techniques allow the measurement of time intervals to a fraction of a nanosecond (10^{-9} s).

Laser Resonance Method. The most precise determination of the speed of light to date was carried out in 1972 by a group at the National Bureau of Standards in Boulder, Colorado. They measured the frequency f and the wavelength λ of a particular resonance radiation from a helium-neon laser from which they calculated the speed of light in vacuum to be $c = f\lambda$. (The laser was stabilized to a specific absorption line in methane gas.)

The measurement of f was related to the cesium atomic clock and λ was measured in terms of the wavelength of the orange-red light from krypton 86. (The cesium clock and the krypton light are, respectively, the international standards for the measurement of frequency and length.)

The value of c obtained by this technique is

$$c = 2.997924562 \times 10^8 \text{ m/s}$$

with an uncertainty of only 1.1 m/s or 4 parts in 10^9.

Why is it desirable to know the speed of light with such high precision? First, of course, we thereby confirm the electromagnetic theory of light in finding that the computed value of $c = 1/\sqrt{\varepsilon_0 \mu_0}$ based on electrical measurements of ε_0 is in agreement with the measured value. There are other compelling reasons, however. If one knows c with high precision, then one can reverse the procedure; that is, by timing a light pulse in a round trip, one can measure astronomical and large terrestrial distances with precision. Thus, the distance to the moon or to Jupiter can be measured precisely by radar timing of radio pulses. Furthermore, the speed of light plays a dominant role in the theory of relativity and in the quantum theory of atomic and nuclear structure. One can make precise comparisons between theory and experiment only if c is known precisely.

As will be shown in relativity theory (Chap. 34) the speed of light in vacuum is the same for all observers and circumstances. Although the speed is the same, the wavelength and the frequency may differ according to the state of motion of source and observer, in much the same fashion as for sound. In fact, this is the Doppler effect as it applies to electromagnetic radiation.

SUMMARY

Maxwell's four equations describe all of classical electromagnetism. Table 30-1 lists these equations.

Maxwell's equations correctly predict the following properties of electromagnetic waves:

Radiation is produced by accelerated charges; for nonrelativistic velocities the radiated intensity varies as

$$I \propto \frac{q^2 a^2}{r^2} \sin^2 \theta \qquad (30\text{-}34)$$

Electromagnetic waves travel through a vacuum at the speed of light.

$$c = \frac{1}{\sqrt{\varepsilon_0 \mu_0}} = 3.00 \times 10^8 \text{ m/s} \qquad (30\text{-}17)$$

The **E** and **B** fields are perpendicular to each other and also perpendicular to the direction of propagation of the electromagnetic energy. The relative magnitudes of the **E** and **B** fields are related by

$$B = \frac{E}{c} \qquad (30\text{-}12)$$

The energy density u and linear momentum density M/V of an electromagnetic wave are given by

$$u = \varepsilon_0 E^2 \qquad (30\text{-}22)$$

$$\frac{M}{V} = \frac{u}{c} \qquad (30\text{-}32)$$

where E is the instantaneous electric field and V is the volume.

Both the direction and magnitude of the instantaneous intensity, the electromagnetic power per unit transverse area, is given by

$$\mathbf{I} = \mathbf{E} \times \frac{\mathbf{B}}{\mu_0} \qquad (30\text{-}26)$$

A completely absorbed electromagnetic wave of power P and intensity I exerts a radiation force F_r and radiation pressure p_r on its absorber, given by

$$F_r = \frac{P}{c} \qquad (30\text{-}29)$$

$$p_r = \frac{I}{c} \qquad (30\text{-}31)$$

For complete reflection, the force and pressure are doubled.

PROBLEMS

30-1. At a particular instant the electric field in a beam of electromagnetic radiation is 3.0×10^{-3} V/m, directed east, and the magnetic field is in the vertical plane and points up. Find (a) the magnitude of the magnetic field and (b) the magnitude and direction of the Poynting vector at this instant.

30-2. A charged parallel-plate capacitor has a uniform electric field of 1.0×10^6 N/C between the plates. The capacitor, with the plates horizontal and the upper plate positively charged, is moved at a constant speed of 10^3 m/s to the right. (a) Find the magnitude and direction of the induced magnetic field **B** that moves along with the electric field. (b) Find the magnitude and direction of the induced electric field **E′** due to the moving magnetic field. (c) Why do these fields not constitute an electromagnetic wave?

30-3. Consider a parallel-plate capacitor with circular plates 1.0 cm in radius and 1.0 mm apart in vacuum. (a) Find the displacement current through the space between the plates which will induce a magnetic field of 1.0 G at a distance of 1.0 cm from the axis of the circular plates. Find the corresponding time rates of change of (b) the electric flux and (c) the electric field between the plates. (d) What is the time rate of change of the electric potential across the capacitor? (e) Finally, find the time rate of change of charge on the capacitor.

30-4. The rms value of the electric field in a beam of electromagnetic radiation is 1.0×10^6 V/m. Find (a) the energy density, (b) the radiation intensity, (c) the pressure which this beam exerts on a perfect absorber, and (d) the rms value of the magnetic field.

30-5. A laser produces a beam of light with a circular cross section 2.0 mm in diameter. The laser emits monochromatic light of wavelength 6328 Å at a power of 2.0 mW. What are the rms values of (a) the electric field and (b) the magnetic field of this beam?

30-6. A 100-W point-source radiates light isotropically in all directions. At a distance of 2.0 m from the light source, find the amplitude of (a) the electric field E and (b) the magnetic field B.

30-7. The intensity of electromagnetic radiation from the sun is 1,400 W/m² at the earth's surface. The earth is 1.5×10^{11} m from the sun. Find the amount of electromagnetic energy within a sphere of radius 1.5×10^{11} m, with the sun at the center of the sphere.

30-8. What is the inward-directed pressure on the sun's surface resulting from total radiated power of 3.8×10^{26} W?

30-9. At what distance from a point source of radiation with the *luminosity* of the sun (3.8×10^{26} W) would the radiation pressure be 1.0×10^{10} atm? (The luminosity of a star is simply its total radiated power.)

30-10. (a) If the intensity of all electromagnetic radiation from the sun at the earth's surface is 1,400 W/m², what is the force of the sun's radiation on the earth, assuming complete absorption? (b) Compare this radiation force with the gravitational force between the sun and earth.

30-11. A beam of electromagnetic radiation strikes a block of translucent material which reflects 25 percent of the incident radiation and absorbs the rest. (a) What is the radiation pressure on the block if the rms value of the magnetic field is $B = 0.10$ T? (b) What is the volume density of momentum in the incident beam?

30-12. At the Lawrence Livermore Laboratory of the University of California initial investigations of the feasability of laser-induced fusion are underway. These studies employ high-power pulsed lasers made of neodymium-doped optical glass. Such a laser compresses a total energy of 1.0×10^3 J into a pulse 1.0×10^{-10} s in duration. Find (a) the pulse length in m, (b) the volume density of linear momentum, (c) the volume density of energy, and (d) the radiation pressure if the beam is focused on a target 50 microns in diameter which absorbs the pulse without reflection.

30-13. Theoretical calculations suggest that it may be feasible to use radiation pressure from an array of pulsed laser oscillators to create conditions of temperature and pressure suitable for the practical generation of electrical power based on nuclear fusion of heavy hydrogen isotopes (deuterium and tritium). The pulse duration must be greater than 10^{-11} s to ensure that the energy has the time to propagate throughout the target (an initially frozen pellet) and less than 10^{-9} s so that the now-vaporized target does not have time to disperse. For these reasons the pulse duration must be of the order of 10^{-10} s. (a) Find the power during the pulse needed to create a pressure of 1.0×10^{10} atm on the surface of a spherical fuel pellet 100 microns in diameter. (Assume that the power is completely absorbed and that there is no reflection.) What are (b) the intensity of the focused beam and (c) the corresponding volume density of electromagnetic energy? Find the rms values of the electric and magnetic fields in the focused beam. (According to the U.S. Federal Power Commission, in 1972 the total installed capacity for the generation of electrical power in the United States was 400 million kW.)

30-14. A spaceship of mass 2.0×10^3 kg, initially at rest and far from any other bodies, navigates by the radiation force produced by the ship's "photon rocket", which emits an electromagnetic beam. *(a)* What must be the power output of the photon rocket if the ship is to be moving at 3.1×10^5 m/s after a constant acceleration of 1 yr? *(b)* Is such an acceleration possible with the present sources of laser beams?

30-15. An evacuated spherical shell is black on its inside and has an isotropic point source of electromagnetic radiation at its center. The source emits radiation at the rate of 10^7 W. If the sphere is subject to an air pressure of 1.0×10^{-1} N/m^2 (one-millionth of atmospheric pressure) on its outside surface, for what radius will the shell have equal interior and exterior pressures?

30-16. A uniform beam of light of intensity 1.0×10^2 W/m^2 is incident normally on a thin sheet of material for 1.0×10^3 s. The material absorbs all the light and has a mass and area of 0.50 g and 1.0 cm^2, respectively. Assume that no other forces act on the sheet and that it is initially at rest. *(a)* Find the translational kinetic energy of the sheet; *(b)* find the increase in thermal energy of the sheet. *(c)* If the specific heat capacity of the sheet is 0.40 J/g-C°, what will be the rise in temperature of the sheet?

30-17. A horizontal electromagnetic beam having a constant intensity of 10 MW/m^2 is incident upon a perfectly absorbing rectangular sheet of area 1.0 m^2 which hangs freely and rotates about a horizontal axis at its top edge. The sheet's mass is 50 g. At what angle with respect to the vertical does the sheet remain in equilibrium?

30-18. Show that the electromagnetic power radiated by a particle of charge q and acceleration a is proportional to $q^2 a^2$.

30-19. *(a)* Show that the electromagnetic power radiated from a sinusoidal electric-dipole oscillator varies as the fourth power of the oscillation frequency and as the square of the oscillating charges' amplitude. *(b)* An electric dipole emits radio waves of 10 MHz with an amplitude of the oscillating charges of 0.10 m. What must be the oscillation amplitude of an atomic radiator emitting light of 5000-Å wavelength and having the same power output?

30-20. The current best value for the speed of light is in error by no more than 4 parts in 10^9. Suppose that the radar method is used to determine the distance to the moon (3.8×10^8 m). *(a)* To what fraction of a second must one measure the time for a pulse of electromagnetic radiation to complete a round trip to the moon, if one is to exploit fully the precision in the known value of the speed of light? *(b)* What would be the uncertainty (in meters) of this measurement of the earth-moon distance?

30-21. Using a modification of the Fizeau method, in 1927 A. A. Michelson measured the speed of light by timing the flight of a light beam on a round trip between Mount Wilson and Mount San Antonio, 22 mi away. This distance was measured to within $\frac{1}{8}$ in. What is the maximum allowable error in the time interval for one round trip, if the errors in the distance and time measurements are to be alike?

30-22. In a certain microwave measurement of the speed of "light," accurate to 1 part in 3×10^7, radiation having a frequency of 10^{10} Hz was used. What error, in terms of the number of wavelengths of 6000-Å light, is permissible in the measurement of the wavelength of the microwaves?

CHAPTER 31

Ray Optics: Reflection, Refraction, and Lenses

In this chapter we deal with waves,† primarily electromagnetic waves of light, traveling in two and three dimensions and encountering the boundaries between media under those conditions in which the wavelength is small compared with the size of obstacles or apertures. Then the only phenomena occurring at the interfaces, reflection and refraction, can be understood and the progress of a wave can be charted by a simple geometrical procedure, ray-tracing. In this chapter we shall be concerned with ray optics, or geometrical optics, and we shall exclude such distinctive wave effects as interference and diffraction (to be treated in Chap. 32).

31-1 RAY OPTICS AND WAVE OPTICS

An opaque object with a sharp boundary casts a sharp shadow when illuminated with visible light from a small source. Light travels, or appears to travel, strictly along a straight line. It is not obvious, certainly not to the casual observer, that light is, in fact, an electromagnetic *wave* phenomenon. Indeed, the rectilinear propagation of light through a uniform medium suggested to early physicists that light consists of particles, or corpuscles. Let us first consider under what conditions light, or any other wave disturbance, can be considered to follow the "paths" given by the rays associated with wavefronts (Sec. 16-5), that is, let us note under what conditions one may ignore the distinctive wave effects.

---

†The fundamentals of mechanical waves appear as Chaps. 15 and 16.

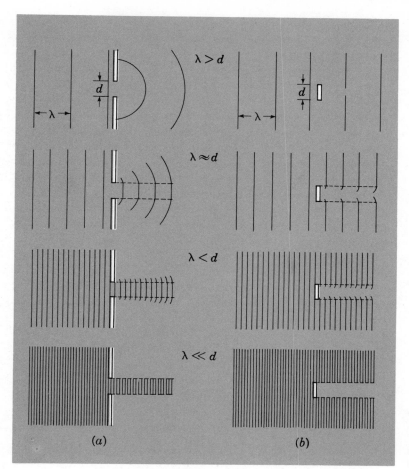

FIG. 31-1. Waves of decreasing wavelength λ encountering (a) an aperture and (b) an obstacle of size d.

Figure 31-1a shows waves of various wavelengths λ impinging on an opaque object with a circular aperture of width d. When λ > d, the waves spread outward from the aperture in all directions, and the wavefronts are circular. For shorter wavelengths, the spreading, or diffraction, of the waves beyond the limits of the geometrical shadow is less pronounced. Finally, when λ ≪ d, the wavefronts remain straight lines as they pass through the aperture, and the wave disturbance lies strictly within the limits of the "shadow" of the opening. Said differently, any ray which enters the opening continues through the opening without a change in direction.

A similar behavior is seen in waves encountering an isolated opaque object, as shown in Fig. 31-1b. When the wavelength is relatively large compared with d, the rays are bent and the wavefronts are curved as a result of diffraction by the object. When λ ≪ d, however, the object casts a sharp shadow. Diffraction effects are negligibly small, and we may draw any ray as undeviated.

In the remainder of this chapter we shall assume that the condition $\lambda \ll d$ is always met; the diffraction effects which occur when $\lambda \gtrsim d$ will be dealt with in Chap. 32.

Under what conditions, then, may we treat visible light by the procedures of ray, or geometrical, optics rather than wave, or physical, optics? The wavelengths of visible light are somewhat less than 10^{-3} mm. Therefore, ray optics suffices so long as we deal with objects or openings of ordinary size, that is, much larger than 10^{-3} mm (10^4 Å). The requirement for ray "optics" applies equally well to other wave types. Thus, an audible sound wave through air with a frequency of 1,000 Hz and a wavelength of 30 cm, or an electromagnetic microwave with a wavelength of several centimeters, cannot be traced by geometrical optics unless such waves strike objects whose characteristic dimensions are much larger than a few centimeters.

Although we shall deal with wavelengths much shorter than the width of apertures and obstacles, we shall always assume that the wavelength is large compared with the microscopic objects responsible for the reflection and refraction effects. For example, it is the individual atoms and their associated electrons in a solid or liquid that cause visible light to be reflected and refracted; but the spacing between adjacent atoms, a few angstroms typically, is much less than the wavelength of light. By the same token, a sheet of ordinary chicken wire, porous to visible light, acts as an opaque reflector for radio waves several meters in wavelength.

31-2 THE RECIPROCITY PRINCIPLE

Suppose that we view a motion picture of a wave phenomenon, perhaps a wave traveling along the surface of a liquid, as in a ripple tank, assuming no energy dissipation. Now, if the motion picture is run backward—if we view the wave motion with time reversed—we see the wavefronts moving in the opposite directions, the directions of rays associated with these wavefronts having been reversed. Both wave motions, the first one with time running forward and the second with time reversed, are *possible* motions. Both are consistent with the laws governing wave motion. Thus, if we know the "path" of a wave from one point to a second point by knowing the configuration of the ray (or rays) connecting the two points, then the reverse path, from the second to the first point, is found simply by reversing the arrows on the rays. In short, if a ray goes from A to B, a ray also goes from B to A by the same route. This is the *reciprocity principle,* which asserts that any two points connected by a ray are reciprocal in the sense that the directions of wave propagation may be interchanged with no alteration in the pattern of the wavefronts. Also termed the principle of *optical reversibility,* this principle applies to *all* nondissipative wave motion, not merely visible light.

Consider Fig. 31-2. Here we see a portion of the wavefronts and their corresponding rays, which radiate from the point source and reflect off a parabolic reflector (the remaining waves emitted by the point source continue to expand outward as spherical waves). The beam consists of plane waves. Thus, a parabolic reflector changes diverging rays into parallel rays; as it were, it changes a point source into a plane wave source. Reversing the ray directions (imagining time to run backward), we see that a beam of plane wavefronts incident upon the parabolic reflector is brought to a focus, all the rays intersecting at a single point. Thus, a parabolic reflector serves equally well as a transmitter or receiver of a parallel beam. More generally, if we know the radiation pattern of a transmitting antenna, which gives the intensity radiated as a function of direction, we thereby know as well the behavior of the same antenna used as a receiver of radiation incident upon it.

Another example of reciprocity is shown in Fig. 31-3a. Here light from a point source passes through a lens to form an image of the source. Upon ray (or time) reversal, the image becomes the source, and the source becomes the image. Strictly, the arrows on the rays are *not* necessary, except to indicate for convenience which of the two possibilities is under consideration.

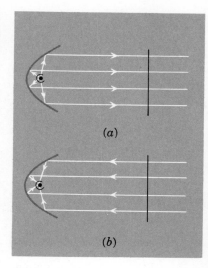

FIG. 31-2. Example of "optical" reversibility: a parabolic reflector as *(a)* a transmitter and *(b)* a receiver.

31-3 RULES OF REFLECTION AND REFRACTION

When a wave encounters a boundary between two media, the incident wave is partially transmitted into the second medium and partially reflected back into the first medium. Since in two or

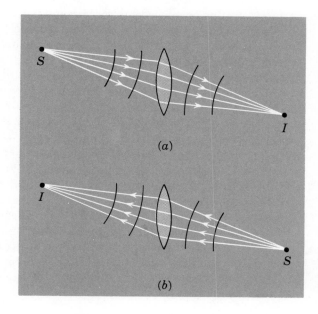

FIG. 31-3. *(a)* A source *S* forms an image *I* when rays pass through a lens. *(b)* Reversing ray directions interchanges source and image.

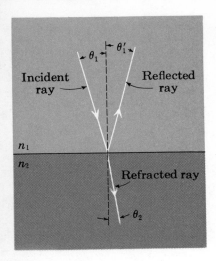

FIG. 31-4. Reflection and refraction at an interface.

three dimensions the transmitted wave is usually bent, or *refracted*, at the interface, we shall call it the *refracted* ray.

We first state the rules for reflection and refraction and later show how they follow simply from fundamental principles. Figure 31-4 shows a ray incident at the interface between two media; for example, a narrow pencil of light is entering water from air. By convention the directions of the incident ray θ_1, of the reflected ray θ_1', and of the refracted ray θ_2 are measured relative to the normal to the interface.

1. Both the reflected ray and the refracted ray lie in the plane defined by the incident ray and the normal.
2. The angle of incidence equals the angle of reflection.

$$\theta_1 = \theta_1' \tag{31-1}$$

3. For a given frequency, the angles of incidence θ_1, and of refraction, θ_2, are related by

$$n_1 \sin \theta_1 = n_2 \sin \theta_2 \tag{31-2}$$

where n_1 and n_2 are constants characteristic of media 1 and 2, respectively. This relation is known as Snell's law, after its discoverer W. Snell (1591–1626). The constants, called *indices of refraction*, will be related to the wavespeeds and wave-lengths in the respective media in Sec. 31-5. According to the principle of reciprocity, we may reverse the ray directions; thus, the designation of one angle as that of "incidence" and the other as that of "refraction" is actually arbitrary.

This is the long and the short of ray optics. Given a table listing the values of n for various materials, one may, in principle, trace the rays in complete detail as they pass through a whole succession of surfaces. In *specular* reflection, as from the polished surface of a flat or curved mirror, irregularities in the surface are not large. Even in *diffuse* reflection, which occurs when the surface is not smooth, the law of reflection holds exactly at any portion of the surface which is small enough to be regarded as smooth. Although Fig. 31-4 shows the interface between the media as straight, these rules apply also to curved surfaces; one merely chooses an interface small enough to be regarded as a plane. Although the program of ray optics is simple in principle, involving nothing more than geometrical constructions following the rules given above, the actual design of optical systems, such as lenses, or collections of lenses, or radar antennas, is, except for certain special situations, rendered extraordinarily difficult by the fact that one must trace an extremely large number of rays. The simpler elements of the theory of refraction from curved surfaces and of lens design are discussed in Sec. 31-8.

One aspect of reflection and refraction *cannot* be treated by the methods of ray optics: the relative intensities of the reflected and transmitted beams. The reflection-transmission ratio can, in fact, be computed from the n_1/n_2 ratio by applying Maxwell's

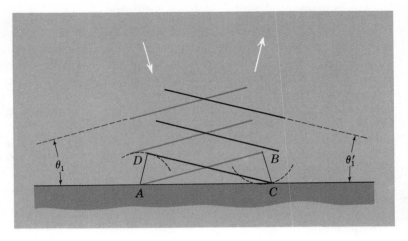

FIG. 31-5. Plane wavefronts incident upon a plane reflecting surface.

equations to electromagnetic waves, or by applying energy and momentum conservation to mechanical waves. We do consider one important case: that in which the incident wave is *totally* reflected from the interface (Sec. 31-7).

31–4 REFLECTION

In ray optics the law of reflection can be understood by considering the propagation of plane waves toward a plane surface. Figure 31-5 shows a succession of plane wavefronts incident upon a plane surface, the angle of each wavefront *relative to the surface* being the angle of incidence θ_1. We find the reflected wavefronts by using Huygen's principle (Sec. 16-7), that is, by taking the envelope of the Huygens wavelets generated along the wavefront to find a future wavefront. At the interface Huygens wavelets are generated in both the forward and the *backward* direction. There is a physical basis for this construction: When an electromagnetic wave travels through a medium, its electric field sets electrons in forced oscillation. These electric oscillators generate electromagnetic waves which propagate both forward *and* backward; but within the refracting medium the net backward radiation is *zero*. At the boundary, however, the symmetry required for cancellation of the backward wave no longer obtains, and both reflected and refracted waves are generated.

As Fig. 31-5 shows, the left end A of the incident wavefront touches the surface when the right end B is a distance BC away from the surface. After a time $t = BC/c$ has elapsed (c is the wavespeed in the incident medium), the right end of the wavefront reaches point C. At this same instant the left end is at D, where $AD = ct$. Therefore, $BC = AD$, and $\theta_1 = \theta_1'$.

Now consider spherical wavefronts from a point source S which are incident on a plane surface, as shown in Fig. 31-6. We

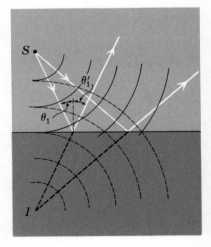

FIG. 31-6. Spherical wavefronts striking a plane reflecting surface and forming an image I.

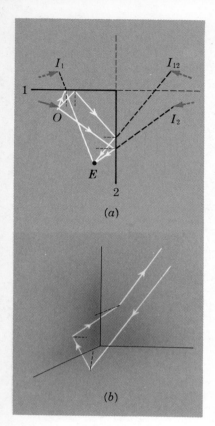

FIG. 31-7. (a) Images formed by reflections from two mirrors at right angles. (b) A corner mirror.

find the reflected rays and the reflected wavefronts by applying the rule $\theta_1 = \theta_1'$ to each ray. The reflected rays appear to diverge from a single point I. To an observer viewing the reflected rays only, these rays and their associated spherical wavefronts seem to come from the point *image I* rather than from their true origin, the point source S. The human eye is naïve in that it interprets all rays reaching it as having always traveled in unbroken straight lines. Thus, to the eye (or a camera) it would appear that the source is located at the position of the image, which, from the geometry of Fig. 31-6, is symmetrically located with the source relative to the reflecting boundary midway between them. This image is said to be *virtual,* inasmuch as the rays appear to originate from location I, although they do not actually do so.

Example 31-1

An object is placed near two plane mirrors at right angles to one another. What images of the object are seen in the mirrors?

See Fig. 31-7a, where the object is represented by O and the eye by E. Applying the rule $\theta_1 = \theta_1'$ at each reflection, we find that there are *three* virtual images: image I_1 formed by reflection from mirror 1, image I_2 formed by reflection from mirror 2, and image I_{12} formed by reflections from both mirrors. Image I_{12} can be said to be the image in mirror 2 of the image I_1. The object and its three images are located symmetrically with respect to the lines representing the mirrors.

Note that the ray reaching the eye from image I_{12} is *parallel* to the same ray leaving the object O; that is, a ray undergoing two reflections at a corner mirror always emerges parallel to the ray entering the object. This same behavior is seen when billiard balls make two "reflections" upon colliding with a corner of a billiard table. When *three* mutually perpendicular mirrors form a corner reflector, any ray undergoing a reflection from each of the three mirrors will emerge parallel to the incident ray, as shown in Fig. 31-7b. Whatever its direction, *any* ray encountering a corner mirror is reflected back undeviated in direction (but displaced laterally). For example, a number of corner reflectors, usually made of red glass, are used on the rear fenders of bicycles; any light shining on them is returned toward the light source. Large-scale corner reflectors are in common use as "targets" for radar signals.

31–5 INDEX OF REFRACTION

We first recognize that when a wave travels from medium 1, in which its wavespeed is v_1, to a second medium 2, in which the wavespeed is v_2, the *frequency f* of the wave is *unchanged.* If the wavelengths in the two media are λ_1 and λ_2, respectively, we may write

$$v_1 = f\lambda_1 \quad \text{and} \quad v_2 = f\lambda_2 \qquad (15\text{-}8), (31\text{-}3)$$

The wavelength is greater in the medium with higher wavespeed.

We denote the speed of electromagnetic waves *in vacuum* by c. One can then specify the speed in any medium by an *index of refraction* n, the ratio of c to that of the wavespeed in the medium. Therefore,

$$v_1 = \frac{c}{n_1} \quad \text{and} \quad v_2 = \frac{c}{n_2} \qquad (31\text{-}4)$$

where n_1 and n_2 are called the indices of refraction for media 1 and 2. By definition, the index of refraction of a vacuum, or empty space, is exactly 1. Light travels through media at a *lower* speed than c. Therefore, the indices of refraction always exceed 1. For example, the speed of light through water is found to be 2.25×10^8 m/s, so that water's index of refraction for visible light is $n = (3.00 \times 10^8 \text{ m/s})/(2.25 \times 10^8 \text{ m/s}) = 1.33$. For air near the earth's surface n is 1.00029, nearly the same as in a vacuum.

Table 31-1 lists indices of refraction to three significant figures for some common transparent materials.

The *relative* index of refraction of medium 2 to medium 1, represented by n_{21}, is given by

$$n_{21} = \frac{n_2}{n_1} \qquad (31\text{-}5)$$

which, from (31-4), is equivalent to

$$n_{21} = \frac{v_1}{v_2}$$

It follows, of course, that

$$n_{21} = \frac{1}{n_{12}}$$

In the next section we shall see that the index of refraction, here defined as the ratio of wavespeeds in two media, is the same as the refractive index defined by Snell's law, Eq. (31-2). Thus, one can measure the value of the relative refractive index quite directly by observing the refraction of waves at an interface. The wavelengths in two media can be related to the respective indices of refraction by means of (31-3) and (31-4);

$$\frac{\lambda_1}{\lambda_2} = \frac{n_2}{n_1} \qquad (31\text{-}6)$$

If the index is large, the wavelength is small. For example, when blue light of wavelength 4000 Å (in free space) enters glass having an index of refraction of 1.500, the wavelength of this *blue* light in glass is *less*, namely 4000 Å/1.500 = 2667 Å. It is customary, however, to characterize a particular color of visible light by its *wavelength in free space*, rather than by its frequency, simply because the wavelength can be measured directly, whereas the frequency is computed from a knowledge of the wavespeed.

Table 31–1

Material	Refractive index
Diamond	2.42
Ethyl alcohol	1.36
Glass (crown)	1.52
Glass (light flint)	1.60
Ice	1.31
Quartz (fused)	1.46
Sodium chloride	1.54
Stibnite (Sb_2S_3)	4.46
Water	1.33

Table 31–2

Wavelength, in air (Å)	Refractive index, quartz
4000	1.470
5000	1.463
6000	1.458
7000	1.455

FIG. 31-8. Change in wavelength arising from a change in wavespeed in a ripple tank. (From PSSC *Physics*, D. C. Heath and Company.)

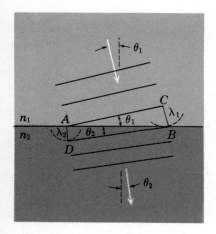

FIG. 31-9. Refraction of plane wavefronts at an interface.

In the case of mechanical waves one can give only relative indices of refraction for a pair of media; there is no unique "medium" corresponding to free space for electromagnetic waves, in which the index of refraction is 1.

Measurements show that the relative index of refraction for a given medium usually depends on the frequency. The index of refraction of a material transparent to visible light rays usually *increases* with frequency. Thus, violet light travels through glass at a lower speed (approximately 1 percent) than does red light, of longer wavelength and lower frequency. Whereas all the component frequencies of white light travel through vacuum at the same speed c, the speeds of the various frequencies differ as the light enters a refracting medium. Table 31-2 shows the refractive indices of quartz for light of several wavelengths (in air).

The surface waves on a liquid of varying depth show a behavior similar to that of light waves in that the wavespeed decreases as the depth of the water is reduced. Thus, in the ripple tank one sees the wavefronts compressed, corresponding to a decrease in wavelength, as the waves enter a region in which the depth is reduced; see Fig. 31-8.

31–6 REFRACTION

Figure 31-9 shows wavefronts incident from medium 1, in which the wavespeed is v_1, the wavelength λ_1, and the refractive index n_1 into medium 2, where the corresponding quantities are v_2, λ_2, and n_2. The angle of incidence θ_1 is also the angle between the incident wavefronts and the interface; similarly, the angle of refraction θ_2 can be measured between the interface and the wavefronts in medium 2. We see from the figure that, in the time interval in which the right end of the wavefront advances one wavelength λ_1 in medium 1, the left end of the same wavefront has advanced one wavelength λ_2 in medium 2. From the geometry of Fig. 31-9, we have for triangle *ABC*

$$\sin \theta_1 = \frac{CB}{AB} = \frac{\lambda_1}{AB}$$

and for triangle *ABD*

$$\sin \theta_2 = \frac{AD}{AB} = \frac{\lambda_2}{AB}$$

Eliminating *AB* gives

$$\frac{\sin \theta_1}{\sin \theta_2} = \frac{\lambda_1}{\lambda_2}$$

By using (31-6), this can be written

$$n_1 \sin \theta_1 = n_2 \sin \theta_2 \qquad (31\text{-}2), (31\text{-}7)$$

This is Snell's law.

An alternative form of Snell's law, employing the relative refractive index, is

$$\frac{\sin \theta_1}{\sin \theta_2} = n_{21} = \frac{1}{n_{12}} \qquad (31\text{-}8)$$

Since the refraction of a ray at an interface arises from a change in the wavespeed, we may compute the relative wavespeeds in two media simply by applying Snell's law to find the relative refractive index. In 1862 J. B. L. Foucault performed an experiment highly significant in the history of the theory of light when he showed that the speed of visible light through water is *less* than through air.†

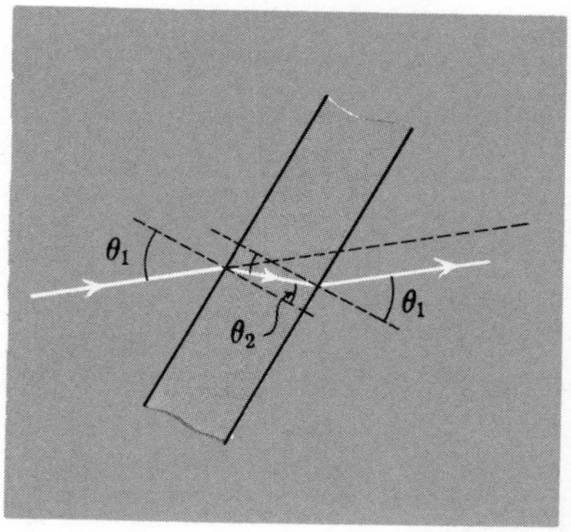

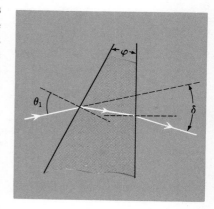

FIG. 31-10. Refraction of a ray through a plate with parallel faces.

Consider Fig. 31-10, which shows a ray incident on a slab of refracting material with parallel faces. Since Snell's law governs the refraction at both interfaces and the interior angles θ_2 are the same at both, the emerging ray is exactly parallel to the incident ray. The ray is displaced laterally by an amount which depends on the thickness of the slab and its refraction index, but without deviation from its initial direction. Thus, objects viewed through a window of glass with parallel surfaces are not distorted but merely displaced laterally by the refraction of light rays.

A ray is deviated from its original direction when it passes through a slab of refracting material with *nonparallel* faces. Such a device is known as a *prism;* see Fig. 31-11. The angle of deviation δ between the incident and emergent rays depends, for a given incident angle θ_1, upon the refractive index of the prism, and its angle ϕ. Thus, if white light is sent through a glass prism, the various frequency components are deviated by different angles, or dispersed, since the refractive index varies with frequency. The index n is greatest for high frequencies (violet) and least for low frequencies (red); consequently, violet light is deviated most and red least, all intermediate colors of the visible spectrum lying between.

FIG. 31-11. Refraction of a ray through a prism. For a given refractive index, the deviation angle δ increases as the prism angle ϕ increases.

Isaac Newton first observed the dispersion of the visible spectrum by a prism and noted that one single color, such as green, *cannot* be further resolved into component colors. He also noted that two prisms can be used first to disperse and then to reunite the various components into white light; see Fig. 31-12.

A prism is commonly used in a device, known as a prism spectrometer, for analyzing visible light into its component wavelengths; see Fig. 31-13. Light from the source goes through a nar-

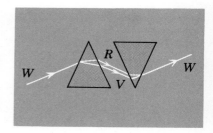

FIG. 31-12. Dispersion of white light by a prism and its (approximate) recombination in a second prism.

---

†Foucault's experiment refuted the particle theory of light in which the refraction of a light ray is attributed to a change in the particle's velocity at the interface. Snell's law can be derived from the particle model, but this model predicts a *higher* speed in a refracting medium than in vacuum. See Prob. 31-8.

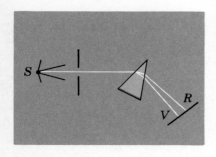

FIG. 31-13 Simple elements of a prism spectrometer.

row slit and then through the prism. If the emitted light is not continuous over a range of frequencies but rather consists of certain discrete frequencies, as does light emitted from excited atoms, the eye sees, or a photographic film records, a succession of "lines," each line being an image of the slit at a particular frequency. One cannot make absolute determinations of wavelength with a prism spectrometer, but only relative measurements. An independent determination of wavelength, as is possible with a diffraction-grating spectrometer (Sec. 32-10), is used to calibrate the prism spectrometer.

From an atomic point of view, any refracting material consists simply of a number of electrically charged particles immersed in empty space. When light travels through a refracting medium it travels, strictly, at the speed of light through a vacuum. A refractive index greater than 1 indicates that, from an atomic point of view, the electromagnetic radiation is reradiated and scattered by many charged particles within the medium.

31-7 TOTAL INTERNAL REFLECTION

Typically, a beam incident upon the interface between refracting media is partially transmitted and partially reflected. In one situation, however, the incident beam *cannot* be transmitted, and consequently it is *totally reflected*.

Consider Fig. 31-14, which shows a series of rays traveling from an optically less dense to a more dense medium (into a medium with larger n). Clearly, the largest possible angle θ_1 is 90°. From Snell's law the corresponding angle for θ_2 is

$$\frac{\sin \theta_1}{\sin \theta_2} = \frac{1}{\sin \theta_2} = \frac{n_2}{n_1}$$

Under these conditions θ_2 is the *critical angle* θ_c, where

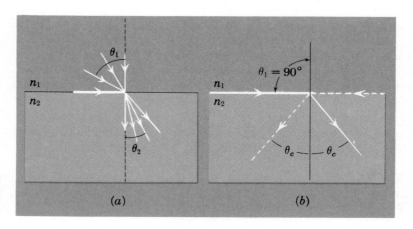

FIG. 31-14. (a) The range of permitted refraction angles; (b) the critical angle θ_c.

$$\sin \theta_c = \frac{n_1}{n_2} \qquad (31\text{-}9)$$

and θ_c is the largest possible angle of refraction in medium 2.

Now consider all the directions of the rays in Fig. 31-14 to be reversed, as in Fig. 31-15. A ray incident upon the interface from medium 2 at the angle θ_c will emerge into medium 1 to travel along the interface. All rays incident at lesser angles will emerge in medium 1. But it is impossible for a ray to be refracted into medium 1 if its angle in medium 2 is greater than θ_c. Consequently, all rays incident at angles greater than the critical angle will be *totally reflected* back into the optically more dense medium. For example, at an air-water interface, $\theta_c = \sin^{-1}(1/1.33) = 49°$.

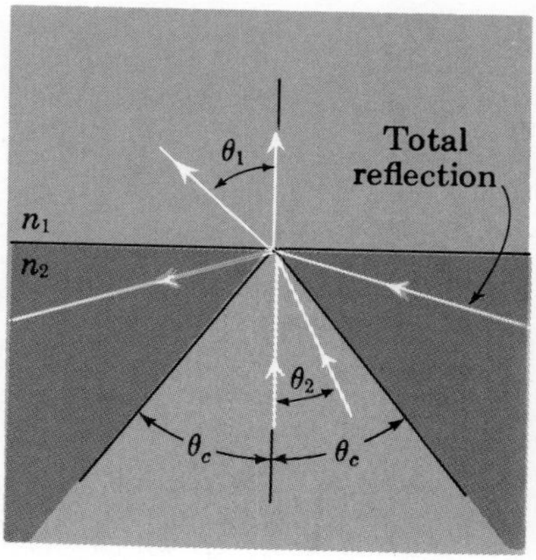

Thus, an observer located within an optically dense material such as water will see, when looking toward the surface, a transparent circular hole, through which all rays above the surface enter, surrounded by a mirror. The total internal reflection effect has important applications. When light is incident upon a glass prism, as in Fig. 31-16, at an angle greater than θ_c, it is reflected from the interior face as from a perfectly reflecting mirror. Likewise, the particular brilliance of a gem, such as a diamond, arises from the very large refractive index of crystalline carbon, high dispersion in this material, and multiply reflected rays. A *light pipe* is shown in Fig. 31-17; light striking the sides of the pipe is totally reflected, and thereby trapped within.

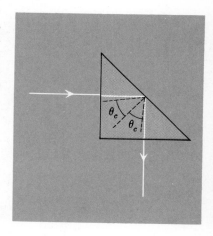

FIG. 31-15. Total internal reflection for $\theta_2 > \theta_c$.

FIG. 31-16. Total internal reflection from a 45°-90°-45° prism for normal incidence.

31-8 THIN LENSES

When a ray passes through a plate of glass with parallel surfaces, the ray is refracted at each interface, but the emerging ray is parallel to the incident ray, as shown in Fig. 31-18a. The ray is displaced laterally, but *not* deviated in direction. On the other hand, when a ray enters a slab of glass with nonparallel faces (a prism),

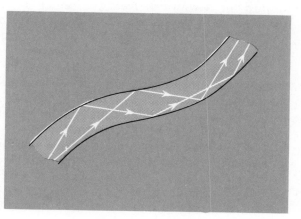

FIG. 31-17. A light pipe.

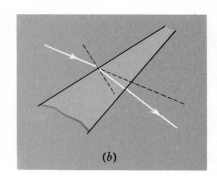

FIG. 31-18. *(a)* Refraction through a slab with parallel faces results in a lateral displacement of the rays, but no deviation. *(b)* A ray is deviated by a prism.

the emerging ray is deviated relative to the incident ray, as shown in Fig. 31-18*b*.

Now consider the structure shown in Fig. 31-19, two prisms and a plate with parallel faces. A number of horizontal rays inci-

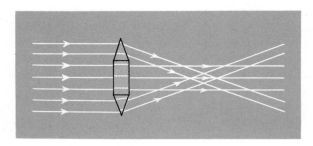

FIG. 31-19. Focusing by a crude lens.

dent from the left pass through this rather crude lens and intersect in a relatively small region to the right. Rays through the center are undeviated. Those rays which pass through the top and bottom prisms are deviated down and up, respectively, both emerging at a single deviation angle. We wish to devise a structure in which *all* of the parallel incident rays intersect at a *single point*. It is clear what is required: The rays' deviation must increase gradually upward (or downward) from the center of the lens, rather than abruptly, as in the figure. That is, the external surfaces must be smoothly curved rather than flat. In practice, this is done by making lenses with *spherical* surfaces; they are relatively easy to grind. Such a lens refracts parallel rays so that they intersect (almost) at a single point. Here we shall consider thin lenses only; that is, we assume that the radii of curvature of the surfaces are large compared with the thickness of the lens at its center, as shown in Fig. 31-20.

Figure 31-21*a* shows a thin lens with *convex* spherical surfaces. After refracting through the lens, incident rays initially parallel to the axis now intersect at a single point, or at least in a very much smaller region than shown in Fig. 31-19. This point is known as a *principal focal point*, or *principal focus*, of the lens.

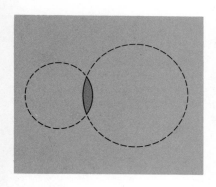

FIG. 31-20. A thin lens with spherical surfaces.

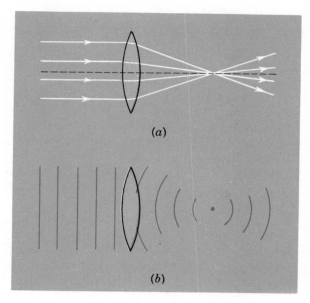

FIG. 31-21. (a) Parallel incident rays focused by a lens. (b) Plane wavefronts focused by a lens.

Thus, if a source is placed at an *infinite* (or very large) distance from the lens, so that the rays incident upon the lens are effectively parallel over a small solid angle, these parallel rays intersect at the principal focus after traversing the lens. The focusing arises from the fact that, in going upward (or downward) from the center, each ray is deviated slightly more than the one below (or above) it. The corresponding behavior for incident plane wavefronts is shown in Fig. 31-21b. The wavefronts undergo a change in curvature at each of the two faces, so that the emerging wavefronts are converging spherical wavefronts which collapse into a point at the principal focus (and then expand outward as diverging spherical waves). This arises from the fact that the relatively thicker central portions of the lens slow a wavefront more than do the thinner portions near the edges.

Any focus may be defined in several equivalent ways: (1) the point at which *all rays intersect,* (2) the point at which *wavefronts collapse into a point,* thereby changing from converging into diverging spherical wavefronts, (3) the point at which the *intensity* of the beam is a *maximum.*

The principle of reciprocity, or time reversal (Sec. 31-2), asserts that we can reverse the directions of rays (or imagine time to run backward) without changing the light paths. Thus, if a point source is placed at a principal focus, as in Fig. 31-22, rays emerge from the lens as a beam of parallel rays. Equivalently, spherical wavefronts diverging from the principal focus and passing through the lens emerge as plane wavefronts.

Suppose that we reverse the thin lens, interchanging the two faces, to allow rays to pass through in the opposite direction.

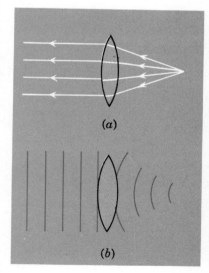

FIG. 31-22. (a) After refraction, rays which originate at the focal point emerge as parallel rays. (b) Diverging spherical wavefronts become plane wavefronts after refraction through the lens.

Thin Lenses   SEC. 31-8

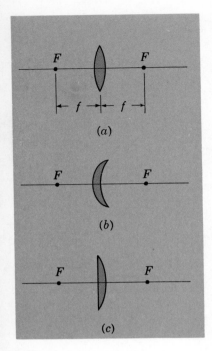

FIG. 31-23. Three examples of converging lenses: *(a)* a double convex lens, *(b)* a concave-convex lens, and *(c)* a plano-convex lens.

Incident parallel rays again converge to a focus at the *same* distance from the lens' center. Thus, every lens has *two* principal focal points, one on each side. Both principal foci, denoted by F, are at the same distance f, the *focal length*, from the center of a thin lens; see Fig. 31-23a.

This result holds not only for a thin lens with two convex surfaces, but also for a lens, such as that shown in Fig. 31-23b, with a concave and a convex surface, or a planoconvex lens with one surface of infinite radius of curvature, as in Fig. 31-23c. The essential requirement is that the lens be thicker at its center than at its edges. (Clearly, the focal length depends on the radii of curvature of the two surfaces and on the refractive index of the lens material.)

Imagine now that the lens is tilted a little, as in Fig. 31-24a, so that its symmetry axis (shown dashed) does not lie along the direction of incident parallel rays. With a small tilt angle, there is no change: The rays are focused at essentially the same point as before, a distance f from the center of the lens. Now look at the very same situation but with the lens axis horizontal, as in Fig. 31-24b. For obliquely incident parallel rays, the focal point is displaced transversely from the lens axis. Clearly, the ray through the lens' center is undeviated; it passes, in effect, through two parallel surfaces. Thus, rays from an infinitely distant source are brought to focus in a *plane*, the *focal plane*, a distance f from the lens. This result holds, however, only if the angle between the rays and the lens axis is small. Such rays, nearly parallel to the

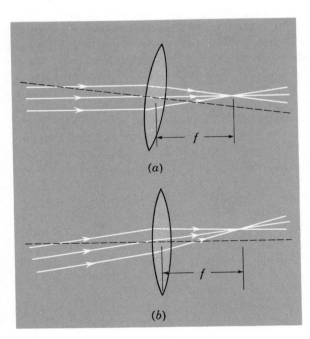

FIG. 31-24. *(a)* Tilting the lens through a small angle does *not* change the focal point. *(b)* Equivalently, oblique rays also come to a focus a distance f from the lens.

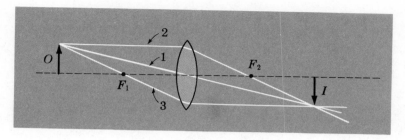

FIG. 31-25. Formation of an image *I* by an object *O*.

lens axis, are called *paraxial rays*. We shall assume hereafter that the lens is very thin and the rays are paraxial.

The type of lens illustrated thus far is known as a *converging lens*. (Strictly, a lens thicker at its center than at its edges is a converging lens only if its material is optically more dense than its surroundings, as for a glass lens for visible light immersed in air.) Any converging lens increases the degree of convergence of rays and wavefronts passing through it, or decreases the degree of their divergence.

We can illustrate the convergence property of a converging lens for a source located at *any* location along the lens axis. Suppose that some luminous source, or object represented by *O*, is located near a thin converging lens, as shown in Fig. 31-25. We wish to find the place where the light from the upper tip of *O* comes to a focus after passing through the lens. Three rays are easy to trace:

1. A ray passing through the center of the lens is *unchanged*, inasmuch as it is neither deviated (the lens faces are parallel here) nor displaced laterally (the lens is very thin).
2. A ray incident *parallel* to the lens axis is deviated by the lens through the principal focus F_2 on the *far* side.
3. A ray which passes through the *near principal focus* F_1 is deviated to emerge *parallel* to the lens axis.

These three rays intersect, or focus, at a single point; indeed, *all* other paraxial rays from the upper tip of *O* intersect to form a point image of *O*. We have traced rays from the uppermost point of object *O*. If we choose some other point on an extended object lying in the same transverse plane, we can again find the corresponding image point. Thus, for paraxial rays through a thin lens, we obtain the image *I* of the object *O*.

The method of ray construction is simple. We need only a prior knowledge of the lens' focal length *f*; then we may locate the image by drawing the three rays as in Fig. 31-25. Ray-tracing, strictly a geometrical procedure, is not always practicable, and we wish to find the mathematical relation between the focal length *f*, the distance *s* of the object from the lens, and the distance *s'* of the image from the lens. Consider Fig. 31-26, which is merely Fig. 31-25 redrawn with identifying letters. The object distance *s* is

Thin Lenses SEC. 31-8

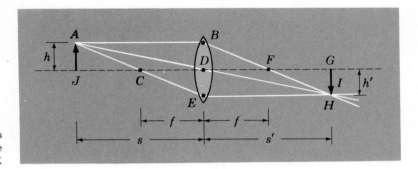

FIG. 31-26. Geometrical relations among object distance s, image distance s', and focal length f.

AB, the image distance s' is $EH = DG$, and the focal length f is $CD = DF$. The object height h is $AJ = BD$; the image height h' is $GH = DE$. Triangles AJD and DGH are similar. Therefore,

$$\frac{AJ}{JD} = \frac{GH}{DG}$$

$$\frac{h}{s} = \frac{h'}{s'} \tag{31-10}$$

The ratio of image-object *distances*, s'/s, is equal to the ratio of image-object *sizes*, h'/h. This ratio h'/h is known as the *lateral magnification*.

Triangles BDF and FGH are also similar. Therefore,

$$\frac{DF}{BD} = \frac{FG}{GH}$$

$$\frac{f}{h} = \frac{s' - f}{h'}$$

Using (31-10), this becomes

$$\frac{f}{s} = \frac{s' - f}{s'} = 1 - \frac{f}{s'}$$

and dividing by f and rearranging yields

$$\frac{1}{f} = \frac{1}{s} + \frac{1}{s'} \tag{31-11}$$

This is the formula† for locating images formed by a thin lens. Everything we can do with it in computing image or object distances can be done equally well by ray construction [that is how we derived (31-11)]. For example, if $f = 12$ cm and $s = 15$ cm, we find, either by ray construction or by (31-11), that $s' = 60$ cm. Moreover, the image size is 4 times that of the object.

---

†When object and image distances are measured from the lens, the lens formula is said to be in *gaussian form*. In the *newtonian form* object and image distances are measured from the principal foci; see Prob. 31-23.

If we place $s = \infty$ in (31-11), then we find $s' = f$; that is, the image of a very distant object is found at the principal focus, in accord with its definition. Conversely, if $s = f$, then $s' = \infty$.

The image shown in Figs. 31-25 and 31-26 is said to be a *real* image; rays actually pass through this location. If a sheet of paper were placed at the distance s' from the lens, one would see an actual image in focus on it. The image would be inverted in the transverse focal plane: Up and down would be interchanged relative to the object, as would left and right. Equation (31-11) shows that as s increases, s' decreases, and conversely, so that when the object is displaced along the lens axis, its image is displaced in the *same* direction. Thus, a three-dimensional object, although inverted in the transverse plane, is *not* inverted along the lens axis.

When we apply the reciprocity principle to Figs. 31-25 and 31-26, the rays are then reversed and our original image becomes object, and vice versa. The two locations, one for s and one for s', are said to *conjugate* points. When $s = 2f$, then $s' = 2f$. For this special case, the lens merely inverts the object without changing its size.

A number of optical devices are based on a converging lens forming a real image. The most familiar is the eye, which forms a real image on the retina. Here the image distance s' from the eye lens to the retina is fixed; objects at various distances from the eye are brought into focus on the retina when the eye muscles change the focal length f of the eye lens by changing the radii of curvature of the lens surfaces. A camera forms a real image in the focal plane of the photographic film, and a projector forms a much enlarged and inverted image of a slide or film on the screen.

Now consider what happens when an object is placed closer to a converging lens than the principal focal point, as in Fig. 31-27. We locate the image geometrically by drawing exactly the same three rays as before. Ray 1 through the lens center is undeviated. Ray 2, initially parallel to the lens axis, passes through the far focal point. Ray 3 is so drawn that its direction after leaving the object is the same as that of a ray starting at the near focal point and passing through the top of the object; therefore, this third ray emerges from the lens parallel to the axis. These rays do *not*

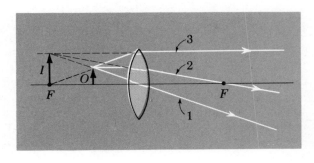

FIG. 31-27. Formation of a virtual image.

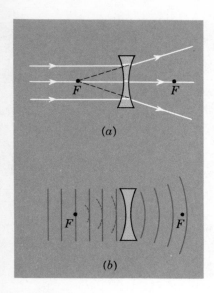

FIG. 31-28. Divergence of *(a)* the rays and *(b)* the wavefronts by a diverging lens.

intersect. What sort of image is now formed? What does an eye see when looking from the right toward the lens?

The eye (or a camera) is naïve in that it interprets any ray reaching it as always having traveled strictly along a straight line. Said differently, the eye recognizes only the *final* direction of rays entering it. Thus, the three rays appear to have originated from that point on the left of the lens where their backward extensions intersect. They appear to come from what is called a virtual image *I*. If a sheet of paper were placed at the location of *I*, no image would be seen on it. But a person viewing the object through the lens would see an *erect, enlarged,* and *virtual* image. Used in this fashion, with an object closer to the lens than the focal length, a converging lens is a *magnifying glass*, or *simple magnifier*.

The lens equation (31-11) can be used to compute the image distance s' for a virtual image if s' is taken to be negative, as may be proved in detail by analyzing the geometry of Fig. 31-27 in the fashion of Fig. 31-26. For example, if $f = 40$ cm and an object is placed at $s = 30$ cm, we find from (31-11) that $s' = -120$ cm. Moreover, from (31-10) we find the virtual image's size to be $\frac{120}{30} = 4$ times that of the object.

Thus far we have considered converging lenses; such lenses are thicker in the center than at the edges (when the refractive index of their material exceeds that of the surrounding medium). Now consider the reverse case, a thin lens with spherical surfaces thinner at its center than at its edges, as shown in Fig. 31-28. Such a lens is a *diverging* lens because it increases the degree of divergence of rays and wavefronts passing through it. Incident parallel rays diverge after passing through the lens and appear to have originated from a point source at the principal focus on the same side of the lens (Fig. 31-28a). Equivalently, incident plane waves emerge from a diverging lens as diverging spherical wavefronts (Fig. 31-28b). We may again use (31-11) to relate image distance, object distance, and focal length, provided that the focal length of a diverging lens is taken to be *negative*. The object distance *s* is, as before, taken as positive. Again there are two principal foci, one on each side of the lens, and the focusing of paraxial rays is independent of the tilt angle of the lens or of the face of the lens exposed to the object.

We locate an image produced by a diverging lens by using the same ray-construction procedure as before. See Fig. 31-29. Ray 1 passes undeviated through the lens' center. Ray 2, parallel to the lens axis, emerges as if originating from the near principal focus *F*. Ray 3 is "aimed" to go through the far principal focus and is, therefore, deviated to emerge parallel to the axis. Upon emerging from the lens, the rays appear to diverge from image *I*. It is reduced, erect, and virtual. Indeed, a diverging lens *always* forms a *virtual* image of a real object. That *I* is virtual is indicated by the fact that the image distance s', as computed from (31-11), is nega-

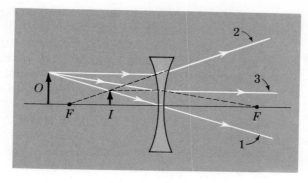

FIG. 31-29. Image formation by a diverging lens.

tive. For example, with a diverging lens having $f = -10$ cm and with an object at the location $s = 15$ cm, we find $s' = -6$ cm.

We have concentrated upon lenses of the usual variety, ones made of optically dense materials, such as glass, and intended for use with visible light. Lens devices may be made of optically light materials. For example, Fig. 31-30 shows converging and diverging lenses formed by cavities within an optically dense medium. Here the converging lens has a pair of concave surfaces, and the diverging lens has a pair of convex surfaces.

The term *lens* need not be restricted to devices which focus visible light. There are also acoustic lenses and microwave lenses. Indeed, magnetic and electric field arrangements which focus a beam of charged particles are referred to respectively as magnetic and electrostatic lenses.

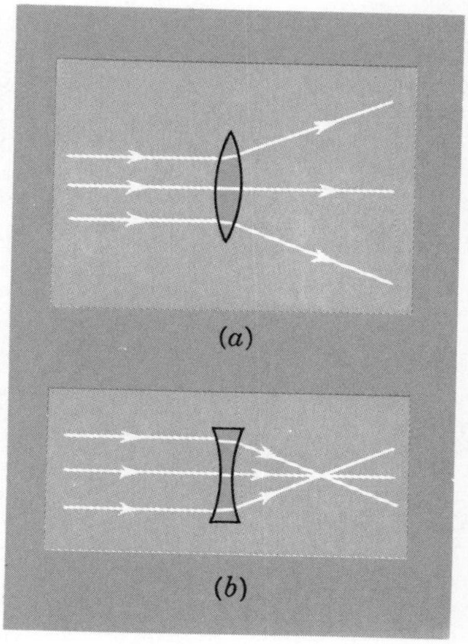

FIG. 31-30. Lenses immersed in an optically more dense medium: *(a)* a diverging lens and *(b)* a converging lens.

31–9 OPTICAL INSTRUMENTS

To trace rays through two or more lenses in sequence, one uses the following procedure: Treat the image for the first lens as the object for the second lens, treat the image for the second as the object for the third lens, and so forth.

The lens combination shown in Fig. 31-31 illustrates the

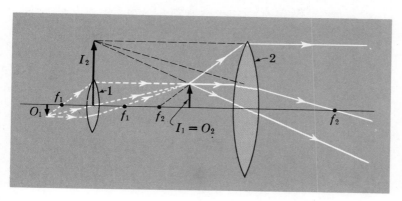

FIG. 31-31. A compound microscope.

compound microscope. Here the object O_1 is placed close to the principal focus f_1 of the objective lens; the image I_1 is enlarged and real. The eyepiece is used to form a still further enlarged but virtual image I_2. Without the aid of an optical instrument, we can see objects more clearly and have retinal images of greatest size by bringing the objects as close to the eye as the focusing properties of the eye will permit, typically 25 cm from the eye. Therefore, a microscope is most effective if the final image I_2 is located at 25 cm from the eye.

Figure 31-32 shows a combination of two converging lenses. Lens 1 forms a real inverted image I_1 of the object O_1. The image I_1 becomes the object O_2 for lens 2. In the particular arrangement shown here, the object O_2 falls just inside the focal point of the second lens. Consequently, the final image I_2 formed by lens 2 is virtual, being to the left of lens 2. Note that the rays chosen to find the image I_1 are *not* continued through the second lens; instead, new rays are chosen, whose deviations through lens 2 are found in the fashion shown in Fig. 31-27. Also note that, for the purposes of ray construction, the lenses are assumed to be of infinite transverse size.

The *astronomical telescope* corresponds closely to the lens arrangement in Fig. 31-32. The object O_1 is, of course, far from lens 1; its real image I_1 is examined by lens 2. This lens acts as a magnifying glass, giving a greatly magnified image formed at a great distance from the lens. Lens 1 (closer to the object) is known

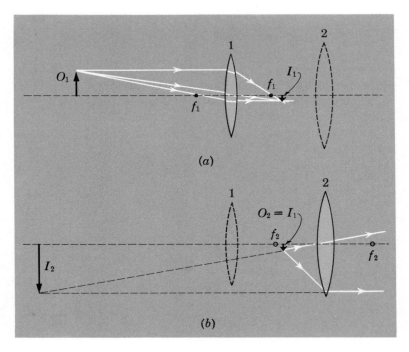

FIG. 31-32. Image formation (a) by lens 1 followed (b) by lens 2 in an astronomical telescope.

as the *objective lens;* lens 2 (closer to the eye) is known as the *eyepiece,* or *ocular.* For an object O_1 at a great distance from a telescope and a final image I_2 also at a great distance, it is clear that the total distance between the objective lens and the eyepiece is close to $f_1 + f_2$. This is the minimum length of the telescope.

SUMMARY

In ray optics one ignores diffraction effects and is concerned only with the change in direction of rays at an interface arising from reflection and refraction. In reflection,

$$\theta_1 = \theta_1' \qquad (31\text{-}1)$$

and in refraction (Snell's law),

$$n_1 \sin \theta_1 = n_2 \sin \theta_2 \qquad (31\text{-}2)$$

where the indices of refraction n_1 and n_2 are given by

$$v_1 = \frac{c}{n_1} \quad \text{and} \quad v_2 = \frac{c}{n_2} \qquad (31\text{-}4)$$

$$n_{21} = \frac{n_2}{n_1} = \frac{1}{n_{12}} = \frac{\lambda_1}{\lambda_2} \qquad (31\text{-}5), (31\text{-}6)$$

Refraction arises from the slowing down of waves in an optically dense medium.

A ray is totally reflected internally when refraction cannot take place, the critical angle θ_c in medium 2 being given by

$$\sin \theta_c = \frac{n_1}{n_2} \qquad (31\text{-}9)$$

The equation for a thin lens is

$$\frac{1}{s} + \frac{1}{s'} = \frac{1}{f} \qquad (31\text{-}11)$$

where the object and image distances, s and s', and the focal length f are measured from the lens center. The sign conventions are as follows:

$f > 0$ Converging lens
$f < 0$ Diverging lens
$s > 0$ Real object
$s < 0$ Virtual object
$s' > 0$ Real image
$s' < 0$ Virtual image

PROBLEMS

31-1. Many sensitive instruments involving the measurement of a small angular displacement, such as the torsion pendulum in the Cavendish experiment, have a mirror mounted on the object and undergoing an angular displacement. Show that, if the mirror turns through an angle θ, a light ray reflected from the mirror from a fixed light source is turned through an angle 2θ.

31-2. Two plane mirrors make an angle θ, as in Fig. P31-2. Find images of the object O for θ equal to (a) 90°, (b) 60°, and (c) 30°.

FIG. P31-2

31-3. What is the minimum height of a mirror placed in a vertical plane that will permit a person 6.0 ft tall to see himself from head to toe?

31-4. At what angle relative to the normal must a ray strike the interface between a medium with refractive index n_1 and a medium with refractive index n_2 for the reflected and transmitted rays to be at right angles?

31-5. An underwater swimmer looks toward a water surface to see light reaching his eye at an angle of 45° with respect to the surface. What is the angle between the ray in air and the water surface? (For water, $n = 1.33$.)

31-6. Light of 4500-Å wavelength goes from one medium in which the angle relative to the normal is 60° into a second medium in which the ray's angle relative to the normal is 30°. What is the relative refractive index of the second relative to the first medium?

31-7. A man standing to the side of an empty pool cannot see its bottom. See Fig. P31-7. What depth of water ($n = \frac{4}{3}$) must be added to the pool if the man, remaining at the position, is to see the bottom edge?

31-8. Snell's law ($\sin \theta_1 / \sin \theta_2 =$ a constant) may be derived from a particle model of light in which it is assumed that the particles of light travel at constant velocity within any uniform medium and that the component of a particle's velocity parallel to an interface between two refracting media is unchanged. See Fig. P31-8. Note that $v_{t1} = v_{t2}$. (a) Derive Snell's law from the particle model and show that $\sin \theta_1 / \sin \theta_2 = v_2/v_1$, where v_1 and v_2 are the respective particle speeds in media 1 and 2. (b) Show that, according to this particle model, the speed of light propagation through a refracting medium exceeds that through vacuum, in contradiction to the experimental findings.

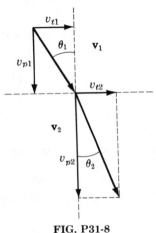

FIG. P31-8

31-9. In the first century A.D. Ptolemy knew Snell's law in the approximate form $\theta_1/\theta_2 =$ constant. Assuming the relative refractive index to be 1.50, what percentage error will there be in the refracted angle if one used the approximate form of Snell's law for angles of incidence of (a) 20° and (b) 40°?

31-10. A light ray is incident upon a 45°-45°-90° prism with refractive index 1.50 as shown in Fig. P31-10. What is the angle ϕ?

31-11. A ray of light passes through several parallel slabs of material with different refractive indices. Suppose that the order of the materials is changed; for example, 2, 1, 4, 3, rather than 1, 2, 3, 4. Assume, further, that total internal reflection does *not* occur. Show that the angle of the emerging ray is *not* changed.

31-12. The "apparent depth" of an object immersed in an optically dense refracting medium is less than the

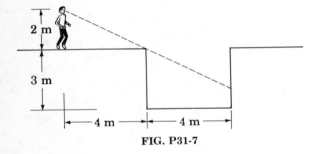

FIG. P31-7

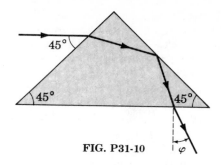

FIG. P31-10

true depth when viewed from directly above. Show that the apparent depth d' is related to the true depth d by $d' = d/n$, where n is the relative refractive index of the medium in which the object is immersed. See Fig. P31-12. One may assume the angles to be so small that the sine of an angle can be replaced by the angle itself.

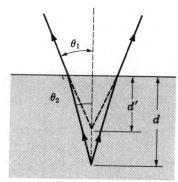

FIG. P31-12

31-13. The apparent depth of an object within a refracting medium is smaller than its true depth by the factor n, where n is the refractive index in the optically dense medium (Prob. 31-12). A microscope is focused on a scratch made on the upper surface at the bottom of a small container. Water is added to the container to a depth of 3.00 mm. Through what vertical displacement must the microscope lens be raised to bring the scratch into focus again?

31-14. Figure P31-14 shows two parallel rays, 1 and 2, incident upon a prism with angles 45°, 45°, and 90°.

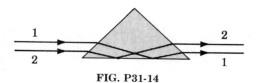

FIG. P31-14

The index of refraction for the glass is 1.5. Trace the rays to show that they emerge from the prism inverted. Of what practical use is a prism of this type?

31-15. A fish looking upward toward the water-air interface sees a circular transparent hole surrounded by a mirror. What is the radius of this hole when the fish's eye is a distance of 1.00 m from the water surface? (n for water is 1.33.)

31-16. The angle through which a ray passing through a prism is deviated depends, of course, on the angle of incidence of the ray with respect to the first surface, the angle of the prism, and its refractive index. There is, however, one angle of incidence for which the deviation angle δ is a minimum. See Fig. 31-11. Show that for minimum deviation the angles of incident and emergent rays with respect to the front and back surfaces of the prism are *equal*. (*Hint:* Apply the reciprocity principle.)

31-17. When a ray is incident upon a prism having an angle ϕ in such a direction as to produce minimum deviation, the incident and emergent rays make the same angle with respect to the prism surfaces (Prob. 31-16). Show that if the emergent ray is deviated through an angle δ under the conditions of minimum deviation, the refractive index of the prism is given by $n = \sin[(\delta + \phi)/2] \sin(\phi/2)$; see Fig. 31-11. This provides a direct means of measuring the prism's index of refraction.

31-18. Laws of reflection and refraction can be regarded as particular examples of a remarkable general principal called the principle of least time or *Fermat's principle,* first propounded by P. Fermat in 1650. According to this principle the actual path followed by a light beam differs from other alternate conceivable nearby paths in that the actual path is one of *least time.* (a) Show that the law of reflection can be deduced from Fermat's principle. (b) Show that Snell's law (the law of refraction) can be deduced from Fermat's principle by using Fig. P31-18. (This figure shows a ray from point A in medium 1 incident upon the interface with medium 2 at point B and proceeding to point C. Points A and C, at the respective perpendicular distances h_1 and h_2 from the interface, may be regarded as fixed whereas point B is regarded as a variable distance x from point D.) (c) Show that the total travel time from A to B to C is

$$t = \frac{(h_1{}^2 + x^2)^{1/2}}{c/n_1} + \frac{(h_2{}^2 + (L-x)^2)^{1/2}}{c/n_2}$$

(d) Show that $n_1 \sin \theta_1 = n_2 \sin \theta_2$ by taking the derivative (to find the minimum travel time).

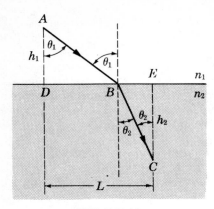

FIG. P31-18

31-19. A simple thin converging lens has a focal length of 40 cm. Find the image and describe its nature (real or virtual) for an object placed at the following distances from the lens: (a) 120 cm, (b) 80 cm, (c) 40 cm, and (d) 20 cm.

31-20. A simple thin diverging lens has a focal length of 40 cm. Find the image and describe its nature (real or virtual) for an object placed at the following distances from the lens: (a) 120 cm, (b) 80 cm, (c) 40 cm, and (d) 20 cm.

31-21. Show that the minimum distance between an object and its real image formed by a thin converging lens of focal length f is $4f$.

31-22. A point isotropic radiator of light is located at $x = 0$ along the axis of a converging lens having a focal length of 20 cm and located at $x = 40$ cm. The lens' aperture has a radius of 5.0 cm. Plot the intensity of the radiation (in arbitrary units) as a function of x over the range $x = 0$ to $x = 80$ cm.

31-23. Equation (31-11) is the *gaussian* form of the general lens equation; object and image distances, s and s', are measured relative to the center of the thin lens. Show that if object and image distances, S and S', are measured *from the two principal foci,* the lens equation can be written: $SS' = f^2$, which is the *newtonian* form of the lens equation.

31-24. An object and its real image are to be separated by 50 cm. (a) What are the two locations of a thin lens relative to the object position, one of which will produce an image twice the size of the object? (b) What is the focal length of the lens?

31-25. A certain person can see most clearly objects placed 25 cm from his eyes when he wears converging spectacle lenses having a focal length of 1.00 m (+1 diopter). Where does he see objects most clearly without his glasses?

31-26. A photograph film having a width of 8 mm is to be projected onto a screen 4.0 m distant, the width of the projected image being 1.0 m. What must be the focal length of the projection lens?

31-27. Two identical thin converging lenses, each 40 cm in focal length, are separated by 20 cm. An object is placed 80 cm from the first lens. Where is the final image relative to the position of the object?

31-28. Relative to a converging lens of focal length f at what distance must an object be placed if the image is to be twice the object size, and (a) real and (b) virtual?

31-29. Show that the equivalent focal length f of a lens combination consisting of two thin lenses in contact is given by $1/f = 1/f_1 + 1/f_2$, where f_1 and f_2 are the focal lengths of the two lenses.

31-30. Through what distance must the lens of a camera having a fixed focal length of 50 mm be displaced, when its focus is changed from that for an infinitely distant object to one 2.0 m away?

31-31. Show that, in general, if a lens, or combination of lenses, forms a virtual image of a real object, then a virtual object at the location of this image will form a real image.

31-32. An astronomical telescope is to be constructed. Its eyepiece is to have a focal length of 1.5 cm. The magnifying power of the telescope is to be 100. (a) What is the required focal length of the objective lens? (b) What will be the overall length of the telescope?

31-33. (a) Will the focal length of a simple converging lens for green light be greater than or less than the lens' focal length for blue light? (b) Which color has the greater focal length for a simple diverging lens?

31-34. Assume that a magnifying glass of focal length f (in centimeters) is placed immediately adjacent to the eye when it is viewing an object located close to the principal focal point. The image formed by the magnifying glass is at a distance of 25 cm from the eye, the so-called near point for most distinct vision. Show that the angular magnification of this simple magnifier is given by $25/(f-1)$.

CHAPTER 32

Wave Optics: Interference and Diffraction

The phenomena of interference and diffraction arise from the superposition of waves from two or more coherent sources. We shall investigate the geometrical relations governing the phase difference between the waves from two (or more) sources and consequently obtain the intensity of the resultant wave.

32-1 SUPERPOSITION AND INTERFERENCE OF WAVES

The resultant wave disturbance produced by two or more waves is governed by the *superposition principle* (Sec. 15-2): To find the net effect of separate waves, one adds, or superposes, the component disturbances at every instant of time and every point in space. Thus, the resultant electric field arising from two electromagnetic waves is given by $\mathbf{E} = \mathbf{E}_1 + \mathbf{E}_2$, where the electric fields $\mathbf{E}_1$ and $\mathbf{E}_2$ of the individual waves are now superposed as vectors.† The superposition principle implies, then, that to find the resultant wave disturbance from two sources, we first find the wave disturbance from each source separately and then add these disturbances.

An important consequence of the superposition principle is this: If two waves "collide," both passing through the same region in space at the same time, each wave emerges from the "collision" as if the second wave had not been present at all. The progress of one wave is completely independent of the presence of other waves;

---

†We could equally well choose to find the resultant magnetic field $\mathbf{B}$. However, it is conventional to give the electric rather than the magnetic field for an electromagnetic wave.

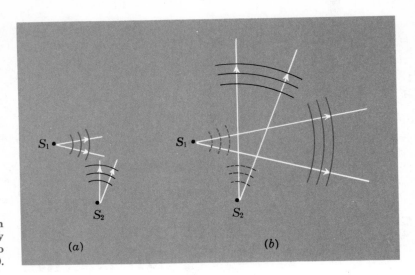

FIG. 32-1. When superposition occurs, the waves generated by two independent sources (a) do not interact (b).

see Fig. 32-1. Indeed, the validity of the superposition principle is established by the fact that "interfering" waves do not act on one another. For example, what an observer sees in looking at some object emitting or reflecting light is unaffected in shape, color, and brightness by any electromagnetic radiation, light or otherwise, passing through the line connecting the observer with the sighted object. The superposition principle holds (almost) exactly for electromagnetic radiation, however great the intensities of the individual waves.

Whenever two or more waves merge, they are said to *interfere*. If the resultant wave disturbance from two separate sources at some point is greater than that from either one alone, the waves are said to interfere *constructively*. If, on the other hand, the resultant wave disturbance is less than that of either of the individual disturbances, the waves are said to exhibit *destructive interference*. A point in space at which component waves always show complete destructive interference, with a *zero* resultant wave disturbance at *all* times, is called a *node*.

Hereafter in this chapter we shall be concerned chiefly with the interference from two point sources radiating sinusoidal electromagnetic waves of the same frequency. Although we shall concentrate on the electric field and intensity of radio waves and visible light, it must be understood that the analysis holds equally well for all other types of waves propagating in two and three dimensions. For example, to apply the theory to sound waves, we merely replace the (vector) electric field by the (scalar) pressure difference.

Recall that the time average of the intensity of a continuous sinusoidal electromagnetic wave is given by

$$I = \tfrac{1}{2}\varepsilon_0 c E_0^2 \qquad (30\text{-}36), (32\text{-}1)$$

where E_0 is now the amplitude of the *resultant* electric field. The intensity is proportional to the *square* of the resultant electric field. Thus, to find the intensity pattern of two radiators we *first* superpose the individual waves to find E_0 and then square E_0 to find I. When dealing with two identical sources oscillating continuously, we *cannot* combine the intensities of the two sources; that is, in general I does *not* equal $I_1 + I_2$. In short, one superposes the wave functions, not the intensities.

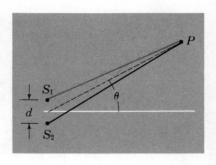

FIG. 32-2. Two point sources S_1 and S_2 whose resultant radiation is observed at point P.

32–2 INTERFERENCE FROM TWO POINT SOURCES

We first consider the radiation pattern from two identical radio antennas, S_1 and S_2, oscillating in phase at the same frequency f and thereby radiating waves of wavelength $\lambda = c/f$. For example, we might have two electric dipole oscillators (Sec. 30-7) oriented at right angles to the plane of the paper, as shown in Fig. 32-2. Then, considering radiation emitted in the plane of the paper, each source *separately* radiates uniformly in all directions; the intensity of each source is dependent only on the distance from the source, and the electric field is perpendicular to the plane of the paper. We wish to find the resultant intensity I at any point P (due to the *two* sources) as a function of the angle θ measured from the perpendicular bisector of the line joining S_1 and S_2. The point P is a much larger distance from the sources than their separation distance d. Our task is first to find the resultant amplitude E_0 at any point P.

What controls E_0? We can easily see the significant factors by examining a simple example. Consider Fig. 32-3a. Here two sources oscillate *in phase* with one another and are separated by one half-wavelength ($d = \lambda/2$). At point P_1 (for $\theta = 0$) the path lengths from S_1 and S_2 are the *same;* consequently, the electric field E_0 at that point is simply the sum of E_1 and E_2, the electric field amplitudes at P_1 arising from the two individual sources. Since E_1 and E_2 are in phase at P_1, there is constructive interference at this point. It will be useful to indicate the phase differences of the separate oscillating fields by vectors. We see, then, that at P_1 the resultant electric field is *twice* that arising from one source alone and, since the average intensity is proportional to E_0^2, the intensity at P_1 is *four* times that for a single source.

Now consider point P_2 ($\theta = 90°$) in Fig. 32-3a. Here the path lengths differ by $\frac{1}{2}\lambda$, the wave from S_2 traveling $\frac{1}{2}\lambda$ farther than that from S_1. Although the sources oscillate in phase, the electric fields from S_1 and S_2 are *out of phase* by 180° at P_2. If P_2 is far from the two sources, the separate electric fields are of essentially the same magnitude, and there is complete destructive interference at this point. At P_2 the intensity is zero.

The situation at point P_3 with $\theta = 30°$ is a little more complicated. The path difference is $d \sin \theta = (\frac{1}{2}\lambda)(\sin 30°) = \frac{1}{4}\lambda$. The

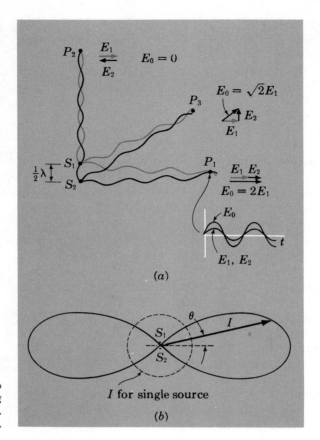

FIG. 32-3. (a) Radiation from two identical point sources oscillating in phase and separated by $\tfrac{1}{2}\lambda$. (b) The radiation pattern.

electric fields *at point P_3* are 90° out of phase. We may symbolize this by drawing E_1 and E_2 as vectors 90° out of phase. The resultant amplitude of the electric field is $\sqrt{2}E_1$ and the intensity is proportional to $(\sqrt{2}E_1)^2 = 2E_1^2$. Thus, for $\theta = 30°$ the average intensity is twice what it would be from a single radiator.

We could readily compute the intensity at still other points at the same distances from the oscillators, to find I as a function of θ. The results can be plotted on a polar diagram, the intensity I as a function of θ, as in Fig. 32-3b. Electromagnetic energy is radiated primarily in the directions $\theta = 0°$ and $\theta = 180°$, with none at right angles. The radiation pattern of the two radio antennas is said to consist of two *lobes*. If the two antennas had been exactly at the same place, rather than separated, the radiation pattern would have been a circle. Interference between the two sources, although it does not change the total energy radiated, redistributes this energy in space.

In general, the time average of the intensity I is proportional to the square of the resultant field amplitude E_0, and E_0 depends on the *phase difference ϕ* between the two arriving waves at the *observation point P*. The phase difference ϕ depends, in turn, on

(1) the *intrinsic phase difference* Φ between the sources, that is, the difference in phase of the two oscillating point sources, and (2) the *difference in path length* Δr from the two points to P.

This is the long and the short of interference. It is simply a matter of geometry to find the path differences from two or any number of oscillators to the point at which one wishes to find the intensity.

Suppose that the two point sources S_1 and S_2 of Fig. 32-2 are separated by an arbitrary distance d and that these sources oscillate *in phase* at the same frequency $f = c/\lambda$. The intrinsic phase difference Φ is then zero; ϕ, the phase difference at the observation point, is related to the path difference Δr by

$$\frac{\text{Phase difference (at observation point)}}{2\pi} = \frac{\text{difference in path length}}{\lambda}$$

which we may write as

$$\frac{\phi}{2\pi} = \frac{\Delta r}{\lambda} \qquad (32\text{-}2)$$

We assume point P to be so far from S_1 and S_2, compared with d, that the two rays to P make the same angle θ with respect to the perpendicular bisector of the line joining S_1 and S_2. From the geometry of Fig. 32-4 the path difference is then

$$\Delta r = d \sin \theta \qquad (32\text{-}3)$$

As (32-2) shows, the phase difference at P is zero (or an integral multiple of 2π), and there is constructive interference whenever the path difference is an integral multiple of the wavelength; that is, constructive interference corresponds to $\Delta r = m\lambda$, where $m = 0, 1, 2, \ldots$. Thus, from (32-3), the directions θ for maximum radiated intensity are given by

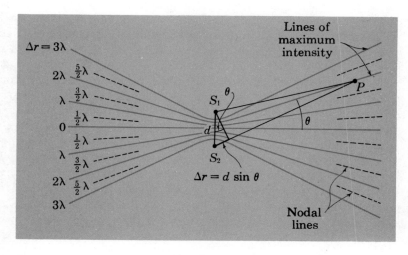

FIG. 32-4. Lines of maximum intensity and nodal lines for two oscillating point sources oscillating in phase.

Interference from Two Point Sources   SEC. 32-2

FIG. 32-5. Interference from two point sources in a ripple tank. (From PSSC *Physics*, D. C. Heath and Company.)

Maximum I: $\quad \Delta r = m\lambda = d \sin \theta \quad$ with $m = 0, 1, 2, \ldots \quad$ (32-4)

On the other hand, when the path difference is an odd multiple of $\lambda/2$, there is destructive interference, the directions of the nodal points in the intensity being given by

Zero I: $\quad \Delta r = m \dfrac{\lambda}{2} = d \sin \theta \quad$ with $m = 1, 3, 5, \ldots \quad$ (32-5)

In Fig. 32-4 point P corresponds to a path difference of 2λ, and the locus of points, all with this same path difference, defines a curve of maximum intensity. The curve is an hyperbola, which follows from the definition of an hyperbola as the locus of points whose distances from two fixed points (here S_1 and S_2) differ by a constant (here 2λ). The asymptotes of the hyperbola make the angle θ relative to the bisector of the line connecting S_1 and S_2. A series of such hyperbolas gives the locations of maximum intensity. A second set of hyperbolas represents the lines of zero intensity, or *nodal lines*: The path differences for these curves is an odd multiple of $\lambda/2$.

The interference pattern from two sinusoidally oscillating point sources can be demonstrated strikingly with water surface waves generated by two objects oscillating transversely on the water surface. A photograph of such a ripple-tank interference pattern is shown in Fig. 32-5.

The variation in intensity with angle may also be portrayed in the polar diagram of Fig. 32-6. Here the sources are separated by $\tfrac{7}{2}\lambda$, and the number of lobes in the radiation pattern is increased over the two lobes shown in Fig. 32-3b.

32–3 REFLECTION AND CHANGE OF PHASE

By an ingenious arrangement known as *Lloyd's mirror* (devised by H. Lloyd in 1834) one can produce interference effects by using a *single* wave source rather than two sources. (As we shall see, it is essential to have a single source to observe interference effects with visible light.) One places the single source S_1 close to a reflecting sheet, or mirror, as in Fig. 32-7. Then waves reach point P both through a direct path from S_1 and through a path involving a reflection from the mirror. We may regard the reflected ray as originating from a source S_2 located below the mirror surface at the position of S_1's mirror image. We assume that P is far from S_1 and that all radiation striking the mirror is reflected. We might expect, then, that the intensity pattern resulting from interference between the waves from the real source and those from its mirror image (a virtual source) would be just that found earlier for two real sources oscillating in phase. Certainly, the path differences and geometrical relations are unchanged. This is not, however, what is observed. The locations of maximum and minimum intensity are

FIG. 32-6. Intensity pattern from two sources separated by $\tfrac{7}{2}\lambda$.

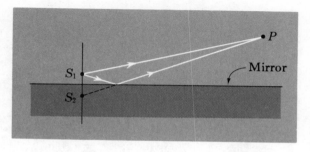

FIG. 32-7. Rays from point source S_1 and its mirror image S_2.

found to be just reversed: Nodal lines are found where lines of maximum intensity were found before, and conversely. Apparently, an additional phase change of 180°, not attributable to a path difference, arises here. Indeed, an electromagnetic wave undergoes a *180° shift in phase* whenever it is *reflected from an optically more dense medium or from a conducting surface*.

To make this plausible, we recall the results we have found for a transverse wavepulse on a stretched string, reflected from the interface between two media (two strings of different density); see Fig. 32-8. A wave reflected from an interface leading to a medium in which the wavespeed is less undergoes a 180° change in phase. No phase change occurs, however, when a wave in a "slow" medium is reflected from an interface leading to a "fast" medium. Exactly the same behavior arises with electromagnetic waves: A 180° phase change occurs when the wave impinges upon and is reflected from an interface leading to a more dense medium. This important result can, of course, be deduced from Maxwell's equations. The phase relations for reflection are summarized in Fig. 32-9. Thus, when visible light in air strikes an optically dense material, such as glass, the reflected waves are reversed in phase; but when visible light impinges upon an interface from glass into air, or into a medium with a smaller index of refraction, there is no phase

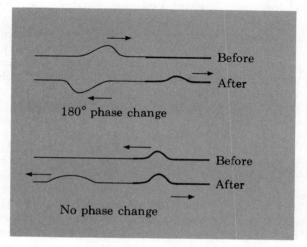

FIG. 32-8. Phase changes for reflection at a boundary of a transverse wavepulse on a stretched string. The wavespeed is greater in the light string than in the dense string.

Reflection and Change of Phase SEC. 32-3

539

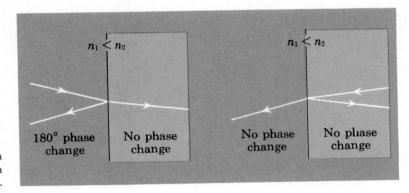

FIG. 32-9. Phase changes for an electromagnetic wave at an interface.

change. A 180° phase change also occurs when an electromagnetic wave is reflected from a conducting surface, for example, radio waves incident upon a metal surface or light upon a mirror. The experimental evidence for the phase change comes from interference experiments, such as Lloyd's mirror.

32–4 COHERENT AND INCOHERENT SOURCES

We have assumed thus far that two interfering wave sources generate sinusoidal waves continuously and that any intrinsic phase difference between the sources, is maintained indefinitely. This would surely be the case for two electric-dipole oscillators wired to a single radio oscillator operating continuously. Any two such sources which maintain a constant phase relation are said to be *coherent* sources, and the waves they generate are said to show *coherence* both in space and in time.

Now suppose that one source oscillates continuously while a second, generating waves of the same frequency and wavelength, is turned on and off *at random*. The separate electric fields E_1 and E_2 from the two sources, and the resultant field, are shown as a function of time in Fig. 32-10. Since E_2 is not a *continuous* monochromatic wave, neither is the resultant $E_1 + E_2$. It has the same frequency as the two sources, but both the phase and amplitude of the resultant wave change abruptly and randomly in time. Two such sources are said to be *incoherent*. Their interference in time at any point in space leads to an intensity, proportional to $(E_1 + E_2)^2$, which fluctuates erratically in time. This is to be contrasted with the intensity variation in time from two coherent sources, as shown in Fig. 32-10e. The time average of the resultant intensity over many cycles is proportional to $(E_1 + E_2)^2$ for coherent sources but proportional to $E_1^2 + E_2^2$ for incoherent sources.

The positions of the nodal lines and the lines of maximum intensity depend, as we have found, on the intrinsic phase difference between the oscillators. If one oscillator changes phase by 180°, nodal lines become lines of maximum intensity, and conversely. That is, a shift in phase causes an angular shift in the radiation pattern. Thus, if two oscillators do not run coherently, because either one oscillator or both oscillators are turned on and off at random, we do not find a fixed interference pattern in

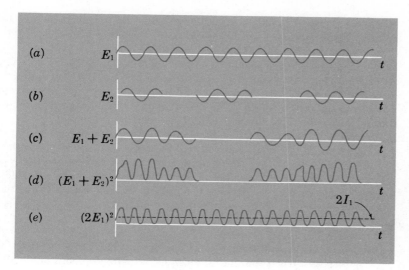

FIG. 32-10. Electric field as a function of time for (a) a continuous oscillator, (b) an oscillator turned on and off at random, (c) the resultant wave. Square of the resultant field for (d) two incoherent and (e) two coherent sources.

space but, rather, one which fluctuates rapidly and whose time average is independent of the angle θ. Another way of putting it is this: The interference is washed out.

It is a simple matter to devise a pair of coherent oscillators for sound waves (connect two loudspeakers to a single audio oscillator) or for long-wavelength electromagnetic radiation (connect two radio antennas to a single oscillating electric circuit), but it is *not* possible, without some care, to devise a pair of coherent oscillators for visible light. Visible light is emitted when atoms in certain excited energy states make transitions discontinuously to lower energy states, the loss of energy by the atom being the energy of the light emitted. A typical atom emitting visible light is "on," that is, radiating, for only 10^{-8} s. Although very short, this time is long compared with the oscillation period for typical visible-light frequencies, $\approx 10^{-15}$ s. Most light sources depend upon the heating of the atoms to produce the atomic transitions which raise the atoms to higher energy states, the precondition for radiation. Excitation of atoms arises from random thermal motion. Consequently, the radiation by a large number of atoms, each over a time of about 10^{-8} s, is also random. That is to say, the radiation of visible light from thermally excited light sources is *incoherent*. Of course, the radiation from any two emitting atoms *is* coherent over a period of about 10^{-8} s, but since we register the intensity of visible light in the eye or on photographic film over a time much greater than 10^{-8} s, the light appears essentially incoherent.

Coherent sources of visible light have been made available in a device known as a *laser,* the acronym for *l*ight *a*mplification by the *s*timulated *e*mission of *r*adiation. A thoroughgoing explanation of laser operation is possible only through the quantum theory, and what follows here are merely some general features. In a laser the atoms are brought to excited states in preparation for radiation, not by thermal excitation, but by *optical pumping,* in which light of higher frequency than that emitted by the atoms is *absorbed* by the active material (for example, chromium atoms within ruby). These atoms remain in the upper energy states for much longer than 10^{-8} s. When light of the frequency to be amplified by the laser enters the material, it stimulates the atoms in upper energy states to make transitions to lower energy states, thereby emitting light of the *same* frequency as the "stimulating" radiation.

The waves of light emitted by stimulated atoms must be in phase with the waves stimulating the emission. If the emitted waves were out of phase, the resultant intensity would be reduced, a result which would violate energy conservation. Therefore, light is emitted in synchronism — that is, is in coherence — with the light stimulating the emission. There is light amplification. In practice, the active material which has been optically pumped to the excited states is held between two perfectly parallel reflecting boundaries, the emitted light being made to traverse the region between the boundaries repeatedly through multiple reflections. The intensity of the wave grows because the light first present causes additional atoms to emit light in coherence. The useful light — highly monochromatic, unidirectional, intense, and coherent — leaves the laser by passing through a *partially* reflecting end mirror. The technological applications of lasers derive from the fact that they produce electromagnetic radiation in the visible region that has coherence properties heretofore available only in radio waves.

32–5 YOUNG'S DOUBLE–SLIT EXPERIMENT

Clearly, one *cannot* observe interference effects with visible light simply by using two separate ordinary light sources. Two such sources would be *incoherent*. If one is to observe interference with visible light, one must, apart from using two laser beams (locked in phase), divide the waves from a single point source into two beams and then recombine the beams. This can be accomplished by illuminating two slits with light from a single source, as shown in Fig. 32-11.

Here light from source S passes through a very narrow single slit to fall upon the double slits. As noted in Sec. 31-1, waves do

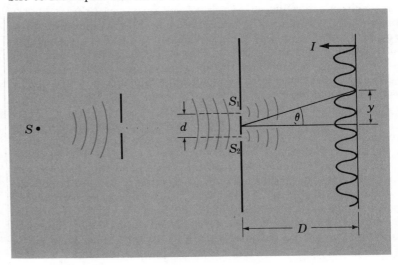

FIG. 32-11. Young's double-slit experiment. (Schematic diagram, not drawn to scale.)

not emerge within the geometrical shadow of a very narrow slit. The light is diffracted; that is, the wavefronts spread beyond the geometrical boundaries corresponding to ray optics. We consider diffraction later in this chapter. Suffice it to say here that a single wavefront illuminates both slits. The slits are, in effect, coherent wave sources, S_1 and S_2, which produce the characteristic interference pattern on the distant screen.

An arrangement analogous to that shown in Fig. 32-11 was used by Thomas Young (1773–1829) in 1801 in an experiment of great historical significance. Young's observation of the interference pattern of light demonstrated unambiguously that light does, in fact, consist of waves. It was not known at the time of Young's experiment that light consists of *electromagnetic* waves, but the interference effect showed that the particle model for light was simply impotent to account for these experimental results. If light did indeed consist of particles traveling in straight lines, then one surely could not find, as Fig. 32-11 shows, an intensity *maximum* at the center of the screen ($\theta = 0$), exactly that point in the "shadow" between the two slits where one would expect to have zero intensity.

Figure 32-12 is a photograph of a double-slit interference pattern. Note that the intensity of the maxima in the pattern is *not* constant, but falls off away from the center of the pattern (at $\theta = 0$). There would be essentially uniform intensity maxima if the two slits were of infinitesimal width. But with slits of finite width, additional *diffraction* effects appear. Of course, since a laser is a coherent light source, one may use a simpler experimental arrangement with it to produce interference effects. One need merely place the double slits in the path of a laser beam (and omit the first single slit).

FIG. 32-12. Interference fringes for Young's double-slit experiment. (Courtesy of Klinger Scientific Apparatus Corporation.)

The alternating regions of brightness and darkness on the screen are known as interference fringes. The locations of the bright interference fringes are given by the relation derived earlier,

$$m\lambda = d \sin \theta \quad \text{where } m = 0, 1, 2, \ldots \quad (32\text{-}4), (32\text{-}6)$$

where d is now the separation distance between the two slits. Angle θ may be measured quite simply by noting that when adjacent interference fringes are very closely spaced (say, a few millimeters apart), as is typically the case when the screen distance from the slits is tens of centimeters, the angle θ is very small, and we may write

$$\sin \theta \approx \theta \approx \frac{y}{D}$$

so that (32-6) may be written as

$$m\lambda = \frac{d}{D} y \quad (32\text{-}7)$$

All quantities in (32-7), save λ, are directly measurable, and we may use this result to compute the wavelength of visible light (a quantity which ranges from 0.0004 to 0.0007 mm).

For a given m and d, the distance y in (32-7) is proportional to the wavelength. Consequently, the various wavelengths of white light produce different interference patterns. The central interference fringe is white (waves of any wavelength constructively interfere for $m = 0$); for all other fringes, the locations of the bright regions depend directly on the wavelength, according to (32-7). Thus, for a given fringe (a given m) the long wavelengths (for example, red) will be farther from the center than the shorter wavelengths (violet). Each fringe, except the central one, then consists of a dispersed spectrum of white light.

32-6 INTERFERENCE IN THIN FILMS

Familiar examples of interference involve thin films. The variegated colors observed from a thin film of oil on water have their origin in interference between the light waves reflected from the two surfaces of the oil film.

Consider the arrangement shown in Fig. 32-13. Here a microwave transmitter sends a monochromatic beam of radiation toward two parallel planes A and B, each containing a grid of wires and acting as a partial mirror. A fraction of the incident waves is reflected from each plane. A receiver at a relatively large distance from the reflectors detects the resultant electric field of the waves reflected from mirrors A and B, which are separated by a distance d. The two reflected beams are, of course, coherent, since they originate from the single wave generator. The reflected waves will interfere constructively if the difference in path length (here approximately $2d$ for nearly normal incidence and reflection) is an integral multiple of the wavelength. That is, a strong signal is registered in the receiver when

$$2d = m\lambda \qquad \text{where} \quad m = 0, 1, 2, \ldots$$

Both reflected waves undergo a 180° phase change.

The device shown in Fig. 32-13 is a simple type of interferometer. Such an instrument may be used to measure wavelengths; conversely, knowing the wavelength, one can measure the displacement of one of the mirrors. Suppose, for example, that the mirrors A and B are initially separated by such a distance as to given constructive interference. Then, if mirror B is displaced by $\frac{1}{4}\lambda$, the overall path length of the ray reflected from B is changed by $\frac{1}{2}\lambda$, and its interference with the wave from A at the receiver is now destructive. In Sec. 32-7 we discuss an interferometer with which one can measure displacements to a fraction of the wavelength of visible light, that is, to a few thousand angstroms.

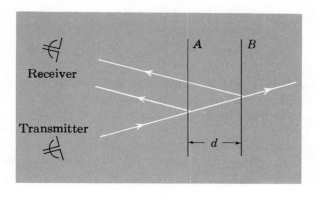

FIG. 32-13. Interference from partially reflected radio waves at interfaces A and B.

Now consider the situation involving visible light, shown in Fig. 32-14. Here light rays are incident nearly along the normal to two parallel interfaces separating media with refractive indices n_1, n_2, and n_3. In determining whether the reflected rays interfere constructively or destructively, we must take into account the following facts: (1) A reflected wave undergoes a 180° phase change relative to the incident (and transmitted) waves, if the wave is incident from a medium of lower to one of higher refractive index (Sec. 32-3); (2) the wavelength λ_m within any medium of refractive index n is

$$\lambda_m = \frac{\lambda}{n} \qquad (31\text{-}6), (32\text{-}8)$$

where λ is the free-space wavelength.

The shrinking of wavelengths in an optically dense medium gives rise to the idea of *optical path length*, defined as the geometrical path length multiplied by the refractive index of the medium and equivalent to the path length in vacuum. For example, the wavelength of red light in vacuum is 6000 Å, whereas the wavelength of the same radiation in glass, with $n = 1.50$, is 6000 Å/1.5 = 4000 Å. A path length of 6.0 mm in vacuum contains 10,000 such wavelengths, whereas these same 10,000 wavelengths extend over only 4.0 mm in glass. Thus, the geometrical path *in glass* is 4.0 mm; the equivalent optical path is (4.0 mm)(1.5) = 6.0 mm, since this is the geometrical path length in vacuum containing the same number of wavelengths. Clearly, then, in computing phase differences one can take into account the change in wavelength in a refracting medium by replacing the geometrical path length by the optical path length.

Suppose that in Fig. 32-14 we have a thin film of oil on water, with $n_1 = 1.00$, $n_2 = 1.46$, and $n_3 = 1.33$, illuminated at normal incidence with monochromatic light of wavelength λ (in vacuum). Since $n_1 < n_2$, but $n_2 > n_3$, a 180° phase change occurs at the reflections from surface A but not from surface B. Then the condition for maximum reflected intensity is

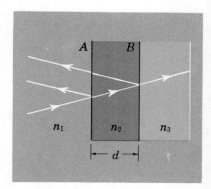

FIG. 32-14. Interference arising from reflections from two interfaces.

Interference in Thin Films SEC. 32-6

$$2d = (m + \tfrac{1}{2})\lambda_{m_2} \quad \text{where} \quad m = 0, 1, 2, \ldots$$

or, equivalently,
$$2d = (m + \tfrac{1}{2})\frac{\lambda_{m1}}{n_{21}}$$

where λ_{m1} is the free-space wavelength. Light shining on an oil film reaches the eye from various sources, and the path differences vary according to the angle of incidence on the film. In addition, the film may not be of constant thickness. Finally, the various colors have different wavelengths, so that if there is constructive interference for one wavelength, there may be destructive interference for another. Consequently, one sees a variety of colors in the film.

A simpler situation obtains when two perfectly flat plates of glass are slightly inclined to produce a wedge-shaped film of air between them, as in Fig. 32-15a. Because the film thickness changes uniformly, in viewing the nearly parallel plates one sees a succession of parallel interference fringes, as in Fig. 32-15b. This arrangement provides a direct test for optical flatness (flatness to within a fraction of the wavelength of light) of a surface. Departure from a plane surface is manifest in the interference pattern by non-parallel or unequally spaced interference fringes.

A modification of the arrangement with two flat plates is shown in Fig. 32-16a. Here a plano-convex lens is in contact with a flat plate of glass. The air wedge has cylindrical symmetry, and the interference fringes appear as alternating bright and dark circles, as in Fig. 32-16b. This phenomenon is known as *Newton's rings*. Note particularly, that the central region is *dark*, not bright. At the point of contact there is *zero* path difference between the rays reflected from the two adjoining interfaces, but the reflected rays are out of phase: One interface is between glass and air, whereas the other is between air and glass.

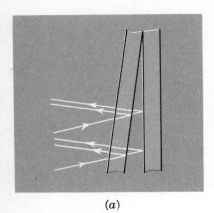

(a)

(b)

FIG. 32-15. *(a)* Interference from an air wedge. *(b)* The observed fringes. (Courtesy of Bausch & Lomb, Rochester, N.Y.)

32–7 THE MICHELSON INTERFEROMETER

This remarkable optical instrument was invented by A. A. Michelson (1852–1932), the first American Nobel laureate in physics. Its essential parts are shown in Fig. 32-17. An incident light beam is split by a partially silvered mirror M into two beams, which travel separate perpendicular paths to mirrors M_1 and M_2 and are then recombined to interfere as they enter the observer's eye. M is inclined at 45° relative to the incident ray. The compensator plate P, parallel to M and equal to it in thickness, is placed as shown in the figure, to ensure that the optical path lengths up-down and right-left are at least approximately equal.

With mirrors M_1 and M_2 exactly perpendicular to rays incident at their centers, the observed interference pattern is like that in Fig. 32-18, a series of circular interference fringes, reminiscent of Newton's rings. (Noncentral rays incident on the mirrors travel slightly different path lengths from those traveled by rays striking the mirrors at their centers.)

Suppose that a dark circle first appears at the center of the interference pattern. Then, the central rays taking the two routes have interfered destructively. Now imagine the mirror M_2 to be displaced $\frac{1}{4}\lambda$. The path length of the up-down ray has thereby been changed by $\frac{1}{2}\lambda$, and the combining central rays reaching the eye now interfere constructively. Consequently, the central spot has been changed from dark to bright (and white and black have been interchanged in all the surrounding interference fringes). A further displacement of $\frac{1}{4}\lambda$ restores a dark central spot. As the mirror M_2 is moved, one sees a succession of interference rings collapsing into the center of the pattern, every alternation of black and white corresponding to a mirror displacement of $\frac{1}{4}\lambda$. Thus, one may measure the wavelength of light by counting the total number of fringe shifts for a known displacement of the mirror. Conversely, knowing the wavelength, one measures displacements of the mirror to within a fraction of the wavelength of light.

Since the meter is defined as 1,650,763.73 wavelengths of krypton-86 light, one may measure a distance in meters in terms of the wavelengths of this light.

An optical Michelson interferometer is easily capable of detecting a difference in path length as small as one optical wavelength. For example, with $\lambda = 6000$ Å, a mirror displacement of $\frac{1}{2}\lambda$ corresponds to a difference in round-trip travel times of only $(\frac{1}{2} \times 6000$ Å$)/(3.0 \times 10^8$ m/s$) = 10^{-15}$ s. The possibility of measuring

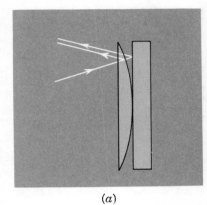

(a)

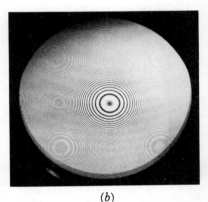

(b)

FIG. 32-16. (a) Interference leading to Newton's rings. (b) The observed rings. (Courtesy of Bausch & Lomb, Rochester, N.Y.)

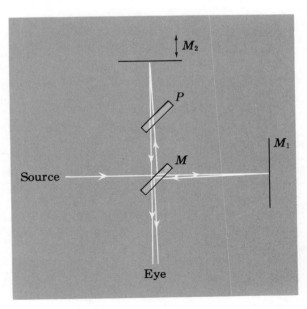

FIG. 32-17. The elements of a Michelson interferometer; M is a partially silvered mirror, P a compensator plate, M_1 a fixed mirror, and M_2 a moveable mirror.

FIG. 32-18. Interference pattern seen in a Michelson interferometer when mirrors M_1 and M_2 of Fig. 32-17 are exactly at right angles.

indirectly such very small differences in travel time by light is the basis of the use of the Michelson interferometer in the Michelson-Morley experiment, a fundamental test of special relativity theory. See Chap. 34.

32–8 RADIATION FROM A ROW OF POINT SOURCES

The term *diffraction* is used to describe the distinctive wave phenomena resulting from the interference of *many* (even an infinite number of) wave point sources oscillating coherently. Thus, diffraction is nothing more than the interference of waves from very many sources. The physics is the same as that for two sources; the geometry may be more complicated.

Consider Fig. 32-19, in which a large number (here, 12) of identical point oscillators are arranged in a row of total width w, each source being separated from neighboring sources by the distance d. For example, one might have 12 equally spaced electric-dipole radio antennas. All point sources oscillate at the same frequency and in phase; there is *no intrinsic phase difference* among them. The sources generate waves of length λ, this wavelength being *large* compared with the distance d.

We wish to find the relative intensity, observed at some very distant point P, as a function of the angle θ between the normal to the line of oscillators and the line joining any of the oscillators to the point P; that is, we wish to find the radiation pattern of this array of equally spaced oscillators. With no intrinsic phase difference between any two oscillators, the phase difference ϕ at a distant point arises solely from the difference Δr in path length, where

$$\phi = \frac{2\pi \Delta r}{\lambda} \tag{32-9}$$

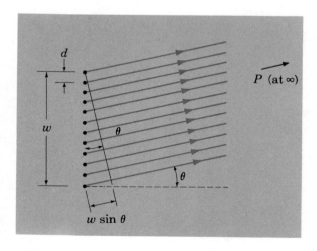

FIG. 32-19. A row of equally-spaced point oscillators ($d \ll \lambda$).

Clearly, when $\theta = 0$, all rays drawn horizontally to an infinitely distant point have the same length, and the sources interfere constructively. Thus, the resultant electric field at the angle $\theta = 0$ is N (here, 12) times the electric field of any one single oscillator.

At which angle is the intensity zero? Our procedure is this: We choose pairs of oscillators such that the resultant field at P from any such pair is zero. Thus, we must so choose the oscillators and the angle θ that the difference in path length between each pair is $\frac{1}{2}\lambda$ (or an odd multiple of $\frac{1}{2}\lambda$); then the pair of oscillators will interfere destructively. Suppose, then, that the angle θ is such that the difference in path length between oscillator 1 and oscillator 7 is $\frac{1}{2}\lambda$. At a distant point, the resultant electric field from this pair is zero. But, by the geometry of Fig. 32-19, we see that oscillator 2 and oscillator 8 then also differ in path length by $\frac{1}{2}\lambda$. Indeed, we can match up all of the oscillators in pairs—1 and 7, 2 and 8, 3 and 9, etc.—so that the resultant electric field at point P from all oscillators is zero. The first zero in intensity occurs when

$$\frac{N}{2} d \sin \theta = \frac{\lambda}{2}$$

First intensity zero: $\qquad Nd \sin \theta = \lambda$

We may write this alternatively in terms of the total width w of the array. From Fig. 32-19, $w = (N-1)d \approx Nd$ for large N. Thus, for a large number N of oscillators,

First intensity zero: $\qquad w \sin \theta = \lambda$

where $w \sin \theta$ is the path difference between oscillator 1 and oscillator 12, those at the extreme ends of the array.

The second intensity zero is found in similar fashion. We now divide the array into four zones, or groups of oscillators: oscillators 1 through 3, 4 through 6, 7 through 9, and 10 through 12. Angle θ must now be larger, so large, in fact, that the path difference between oscillator 1 and oscillator 4 is $\frac{1}{2}\lambda$. This pair of sources then interfere destructively at an infinite distance. In like fashion we match up oscillators 2 and 5, 3 and 6, etc., so that again the resultant field of the array is zero. For large N, the path difference $w \sin \theta$ between the sources at the extremes of the array is now 2λ, and we have

Second intensity zero: $\qquad w \sin \theta = 2\lambda$

It is then apparent that the angles for *zero* intensity are given, in general, by

$$w \sin \theta = n\lambda \qquad (32.10)$$

where $n = 1, 2, 3, \ldots$ (but *not* zero) and the number of sources N is large.

Figure 32-20 shows the radiated intensity I as a function of θ. This *diffraction pattern* consists of an intense maximum at $\theta = 0$,

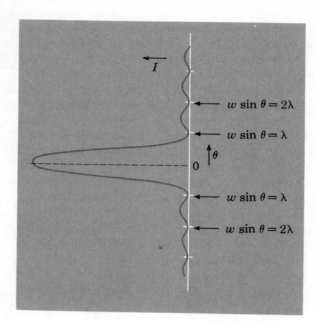

FIG. 32-20. Intensity as a function of angle θ.

which is twice the width of the relatively weaker secondary maxima, which appear symmetrically to its sides. The zeroes are given by (32-10). The same information is portrayed in a different graphical form in Fig. 32-21, which shows the radiation pattern of the array of oscillators; it consists of two strong narrow central lobes, into which most of the radiated energy is directed, together with small side lobes. Thus, an array of equally spaced antennas, with $d \ll \lambda$, will radiate strongly only in the directions perpendicular to the line of antennas.

32–9 DIFFRACTION BY A SINGLE SLIT

Given a monochromatic light source illuminating a single narrow slit at a great distance from the source, what is the intensity pattern of the light falling on a far distant screen? We have already solved this problem! We merely recognize that when a plane wavefront of light impinges upon an opening in an otherwise opaque plane, we may, through the Huygens construction, imagine each point on the wavefront as a new point source of radiation. These

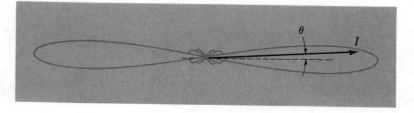

FIG. 32-21. Radiation pattern for a row of equally spaced point oscillators.

550 CHAP. 32 Wave Optics: Interference and Diffraction

continuously distributed coherent point sources oscillate in phase. Effectively an infinite array of point sources is spread over the finite width of the slit, and the intensity pattern radiated from the plane of the slit must be precisely what we have already derived.

An arrangement for single-slit diffraction is shown in Fig. 32-22. For small displacements y, the angle θ is given by $\sin\theta \approx \theta \approx y/D$. The light source S is at so great a distance from the slit that the spherical or cylindrical wavefronts impinging on the slit are effectively plane waves; additionally, the observation screen, at a distance D from the slit of width w, is far from the slit.

Under these conditions one observes *Fraunhofer diffraction*. Fraunhofer diffraction applies when both the source and screen are *infinitely* distant from the slit or, equivalently, when the wavefront at the slit is *plane*. These conditions can be met, even when the source and screen are physically close to the slit, by placing the light source at the near principal focus of a converging lens. Then the spherical or cylindrical wavefronts diverging from the source become *plane* wavefronts upon emerging from the lens and striking the slit. Similarly, one may place the observation screen at the far principal focal plane of a second converging lens; then parallel rays leaving the slit are brought to a focus in the focal plane of the screen; see Fig. 32-23. The general case, with no restrictions on distances or wavefronts, is known as *Fresnel* diffraction. Fraunhofer diffraction is a special case of Fresnel diffraction.

Figure 32-24, a photograph of a single-slit Fraunhofer diffraction pattern, corresponds to Fig. 39-20. The weak secondary diffraction fringes have the same angular width, half that of the central fringe. Note that, following (32-10), for a given wavelength the diffraction pattern expands as the slit width decreases, and light is diffracted far beyond the geometrical shadows of the slit edges. Indeed, the diffraction pattern bears no resemblance to the pattern of constant intensity with sharp edges that is observed when light passes through a wide slit.

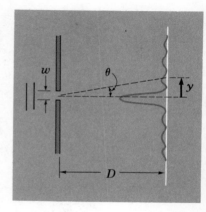

FIG. 32-22. Diffraction pattern from a single slit.

Example 32-1

The central dark fringes in the diffraction pattern from a single slit are separated by 2.0 mm when the slit is illuminated with light of 6000-Å wavelength and the screen is placed 1.0 m from the slit. What is the slit width?

From Eq. (32-10),

$$w \sin\theta \approx w\frac{y}{D} = n\lambda$$

where D is the slit-screen distance and y is the distance from the central maximum in the diffraction pattern to a zero. Here $2y = 2.0$ mm, with $n = 1$. Thus,

$$w = \frac{n\lambda D}{y} = \frac{(1)(6.0 \times 10^{-7} \text{ m})(1.0 \text{ m})}{1.0 \times 10^{-3} \text{ m}} = 0.60 \text{ mm}$$

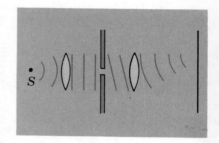

FIG. 32-23. Conditions for Fraunhofer diffraction, achieved with source and screen at *finite* distances from the slit by use of converging lenses.

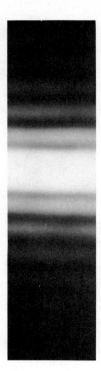

FIG. 32-24. A single-slit Fraunhofer diffraction pattern.

Suppose now that we have *two* parallel slits, each of width w, their centers being separated by a distance d. The two slits are illuminated by a monochromatic plane wave. If we cover one slit and leave the second exposed, the diffraction pattern is, of course, that of a single slit (Fig. 32-24); if we cover the second slit and expose the first, we see the *same* diffraction pattern on the screen (shifted by the distance d). With both slits exposed to a beam with plane wavefronts, the pattern on a distant screen is as shown in Fig. 32-25. There is a diffraction pattern, as before, but now with additional rapid variations in the intensity arising from double-slit *interference* (Sec. 32-5). The intensity variations in Fig. 32-25 are controlled by two influences. The first is the *interference* between the two slits, which is responsible for the narrow interference fringes, all of equal width determined by the *slit separation distance d*.

The second influence is the diffraction from each of the two slits, which is responsible for the slow variations with distance in the intensity of the interference fringes. Dark lines in the diffraction pattern are determined by the slit width w. One might describe this curve by saying that the interference pattern is "modulated" by the diffraction pattern.

32–10 THE DIFFRACTION GRATING

The diffraction grating, invented by J. Fraunhofer (1787–1826), is a high-resolution instrument for analyzing polychromatic radiation into its component wavelengths. A so-called transmission grating consists of a series of N parallel slits separated from one another by a distance d. For example, closely spaced lines (perhaps 10,000 slits in 1 cm) may be scratched onto a plate of glass. As another example, a Venetian blind could be used as a diffraction grating for sound waves.

The essential elements of a diffraction grating are illustrated in Fig. 32-26. This arrangement is, in fact, just like the array of equally spaced sources shown in Fig. 32-19. Here we assume, how-

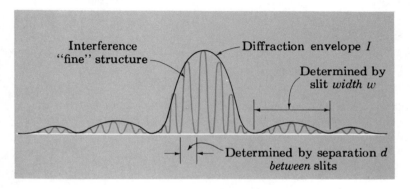

FIG. 32-25. Intensity variation for a double slit.

CHAP. 32   *Wave Optics: Interference and Diffraction*

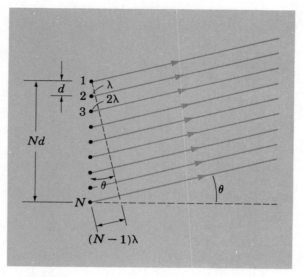

FIG. 32-26. Arrangement of sources in a diffraction grating ($d > \lambda$).

ever, that *the separation distance d between each of the N adjoining slits is greater than one wavelength.*

We wish to find the angular locations of the principal intensity maxima on a distant screen. Suppose that the path difference between slits 1 and 2 to a distant point is one wavelength. The waves from these two slits constructively interfere, just as for the first noncentral interference maximum in the Young double-slit experiment (Sec. 32-5). The path difference $d \sin \theta$ is, then, one wavelength between any two adjoining slits, and all slits send waves that add constructively in the direction θ given by $d \sin \theta = \lambda$. The intensity of this peak is far stronger than that in the double slit: With N slits, not two, the resultant electric field is multiplied by a factor N and the intensity by a factor N^2. Since the total energy is increased by a factor N, not N^2, the width of the strong peak must, by energy conservation, be *reduced* by a factor N.

There are still other strong interference peaks. All slits will interfere constructively at a distant point when the path difference, $d \sin \theta$, between adjacent slits is an integral multiple m of the wavelength. Thus, strong interference peaks occur when

$$m\lambda = d \sin \theta \qquad (32\text{-}11)$$

where $m = 0, 1, 2, \ldots$. The integer m denotes the order of the interference peak.

Note that the angle θ depends on the wavelength. If the incoming radiation consists of many wavelengths, each component wavelength will be deviated differently, the red end of the visible spectrum more than the violet end. Of course, when $m = 0$, then $\theta = 0$ for all wavelengths; the zero-order "spectrum" consists of white light. But the first-order spectrum ($m = 1$), second-order spectrum ($m = 2$), etc., show the light increasingly dispersed. The wavelength is readily calculated if one knows the slit separation

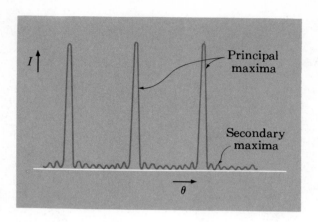

FIG. 32-27. Intensity variation for a diffraction grating.

(or its reciprocal, the number of lines per centimeter on the grating) and measures θ.

Polychromatic light can be dispersed into a spectrum both with a diffraction grating (through interference and diffraction effects) and with a prism (through frequency-dependent refraction). Although a prism forms a more intense spectrum than a grating (it directs *all* the light into *one* spectrum, rather than diluting it into several spectra), it does not permit a direct measurement of wavelengths since one must know precisely how the prism's index of refraction varies with wavelength. A distinct advantage of the grating is its resolution; with a large enough number of slits a grating can resolve two radiations of very nearly the same wavelength.

Figure 32-27 shows the intensity pattern from a diffraction grating with only a few slits.

Example 32-2

A certain diffraction grating has 5,000 rulings over 1.0 cm. What is the angular separation in the first order of the visible spectrum (from 4000- to 7000-Å wavelength)?

The grating spacing d is $\frac{1}{5000}$ cm. From (32-11),

$$\theta = \sin^{-1}\frac{m\lambda}{d}$$

For $\lambda = 4000$ Å,

$$\theta = \sin^{-1}(4.0 \times 10^{-7} \text{ m})(5.0 \times 10^5 \text{ m}^{-1}) = 11.5°$$

Similarly, for $\lambda = 7000$ Å, we find $\theta = 20.5°$. The entire visible spectrum is found in the 9° between the violet and red limits.

32-11 OTHER EXAMPLES OF DIFFRACTION

Diffraction was first observed in 1655 by F. M. Grimaldi (1618–1663). He noticed that the shadow of a sharp straight edge is *not* perfectly sharp, as predicted by ray optics, but shows characteristic diffraction variations in the intensity near the edge. When a plane wave is incident upon an opaque straight edge, the intensity varia-

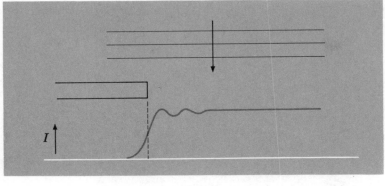

(a)

(b)

FIG. 32-28. Diffraction of plane waves at an opaque straight edge: (a) intensity as a function of distance and (b) the diffraction pattern.

tions along a *nearby* screen are as shown in Fig. 32-28. The intensity decreases gradually to zero *beyond* the geometrical edge of the shadow, shows variations near the edge, but becomes constant for points far from the edge.

Now suppose that we have a wide slit with opaque boundaries illuminated by a plane wave. The intensity pattern we see close to the slit will be that of the two straight edges, as shown in Fig. 32-29a. The intensity is essentially constant at the center of the wide slit, and diffraction effects appear only near the boundaries. Indeed, for small wavelengths (or large slit width), the intensity is essentially that predicted by ray optics: darkness in the geometrical shadow of the slit and constant intensity between. This result is hardly surprising. If ray optics successfully accounts for light's properties when the wave effects are negligible—that is, when the wavelength is negligibly small compared with the dimensions of any apertures or objects—then the more comprehensive wave theory *must* yield the results of ray optics in the limit, $\lambda/L \to 0$, where L is a characteristic length. Thus, we may write symbolically

$$\underset{\lambda/L \to 0}{\text{Limit}} \text{ (wave optics)} = \text{ray optics}$$

Figure 32-29 shows the transition from ray to wave optics for single-slit diffraction. In Fig. 32-29a the screen is very close to the slit; in Fig. 32-29b the screen has been moved very far from

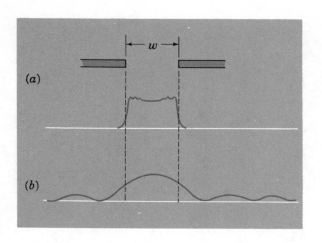

FIG. 32-29. Diffraction from a single slit: *(a)* screen close to slit, *(b)* screen very far from slit.

the slit, so that the conditions of Fig. 32-22 obtain, and we have the characteristic single-slit diffraction pattern of Fig. 32-20. Note that the "shape" of the slit is clearly evident only in Fig. 32-29a.

We can make the transition from wave optics to ray optics in a different way. Suppose that the screen is sufficiently far from the slit to produce the diffraction pattern of Fig. 32-29b. If the wavelength is sufficiently decreased, the pattern of Fig. 32-29a appears on the screen.

32–12 DIFFRACTION AND RESOLUTION

Suppose that a point source of light is far from an opaque surface with a circular aperture of diameter d. Effectively plane wavefronts arrive at the hole, and the (Fraunhofer) diffraction pattern observed on a screen far distant from the aperture is as shown in Fig. 32-30, a bright central spot surrounded by diffraction rings. The pattern is similar to that observed for a single slit (Fig. 32-24), but with circular rather than straight diffraction fringes.

FIG. 32-30. Fraunhofer diffraction pattern through a circular aperture. (From M. Cagnet, M. Francon, J. C. Thrierr, *Atlas of Optical Phenomena*, Berlin-Heidelberg-New York: Springer, 1962. Courtesy of Springer-Verlag, Heidelberg.)

Through an analysis similar to that carried out in Sec. 32-8, it can be shown that the angular separation θ between the center of the bright spot and the first zero in the diffraction pattern is given by†

$$\sin \theta \approx \theta = 1.22 \frac{\lambda}{d} \qquad (32\text{-}12)$$

where λ is the wavelength. See Fig. 32-31.

The inexact, nonpoint character of the image from a point object is not in itself a serious difficulty if one looks at a *single* point object. But if the two images, each a blur, from two point objects overlap too closely, they cannot be distinguished from the

---

†The number 1.22 is the smallest root of the first-order Bessel function, which enters into situations having cylindrical symmetry in a fashion analogous to the appearance of sinusoidal functions in situations having rectangular symmetry.

image of a single point object. Thus, there is a fundamental limitation, having its origin in diffraction and in the wave character of light, in the resolution of *two* point objects. Of course, any extended object is merely a collection of point objects.

Suppose, for example, that two stars have an angular separation ϕ when viewed directly. When an aperture is interposed,† such as the objective lens of an astronomical telescope, one sees two disks each of "radius" θ, as given by (32-12), separated by ϕ, as shown in Fig. 32-32a. The question is, then, what is the minimum ϕ that will permit two definite images to be resolved? Figure 32-32b and c shows two diffraction images from two point sources. The images have an angular separation ϕ just equal to the angular radius θ of either image separately. The center of one image falls exactly at the first zero in the intensity of the second image. One can tell that there are, in fact, *two* images. This convenient situation corresponds to *Rayleigh's criterion* for resolution, the minimum angle ϕ for resolution then being taken as

$$\phi_{min} = \theta = 1.22 \frac{\lambda}{d}$$

from (32-12).

FIG. 32-31. Diffraction through a circular aperture; angle θ gives the angular radius of the first dark fringe.

All optical devices — mirrors, lenses, apertures, and the eye itself — are ultimately limited in their resolving power by diffraction effects. The images are intrinsically fuzzy, and their resolution can be improved only by using waves of shorter length or apertures of larger size. Thus, the motivation for constructing telescopes, whether for visible or radio electromagnetic radiation, with large-diameter lenses or mirrors is not only to capture more radiation and thereby increase the intensity of the image, but also to improve the resolution of the images; that is, big telescopes produce bright, sharp images.

†One always views through an aperture, if only that of the eye's pupil.

FIG. 32-32. (a) Two point objects of angular separation ϕ clearly resolved as two distinct images, each of angular radius θ. (b) The Rayleigh criterion for resolution, $\phi_{min} = \theta$. (c) Intensity pattern corresponding to part (b). (From M. Cagnet, M. Francon. J. C. Thrierr, *Atlas of Optical Phenomena*, Berlin-Heidelberg-New York: Springer, 1962. Courtesy of Springer-Verlag, Heidelberg.)

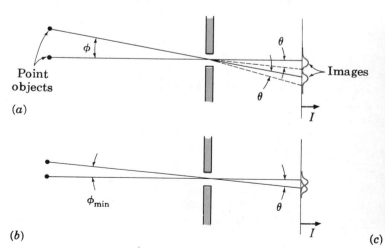

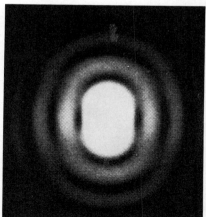

Diffraction and Resolution   SEC. 32-12

SUMMARY

To find the resultant intensity from two or more wave sources, we apply the superposition principle, adding the separate wave functions at any instant of time and any point in space to find the resultant wave amplitude whose square is proportional to the intensity.

For two coherent sources radiating waves of the same wavelength λ, the resultant amplitude is determined by the phase difference ϕ at any observation point,

$$\frac{\phi}{2\pi} = \frac{\Delta r}{\lambda} \qquad (32\text{-}2)$$

where Δr is the path difference and the *intrinsic* phase difference between the sources is zero. If the two sources oscillate in phase and are separated by a distance d, the intensity at a distant point is maximum when

Maximum I: $\qquad \Delta r = m\lambda = d \sin \theta$

$$\text{with } m = 0, 1, 2, \ldots \qquad (32\text{-}4)$$

where θ is the angle measured from the perpendicular bisector of the line joining the sources. In Young's double-slit experiment the point sources are replaced by two parallel slits illuminated through a single slit.

Electromagnetic waves undergo a 180° phase shift when reflected from an interface leading to an optically more dense medium or from a conducting surface.

Diffraction is interference from a large number of coherent sources. The zeros in the radiated intensity of a row of many sources, which are oscillating in phase and separated by equal distances much less than a wavelength λ, are given by

$$w \sin \theta = n\lambda \qquad (32\text{-}10)$$

where w is the total width of the row and $n = 1, 2, 3, \ldots$. This relation then gives as well the zeros in the intensity pattern for diffraction from a single slit under the conditions that both the source and observation point are far from the slit.

The strong and sharply resolved interference peaks of a diffraction grating occur when

$$m\lambda = d \sin \theta \qquad (32\text{-}11)$$

where d is the distance between adjacent slits and $m = 0, 1, 2, \ldots$.

Diffraction effects due to any aperture limit the resolution of two point-objects.

PROBLEMS

32-1. Two separated isotropic oscillators generate waves of slightly different wavelengths. Discuss qualitatively (a) how the radiated intensity varies with time at a fixed location and (b) how the nodal lines in the interference pattern vary with time.

32-2. Two identical point sources oscillate in phase. Locate the points of minimum oscillation along the line joining the two sources when (a) the sources are separated by a distance of 5 wavelengths and (b) the sources are separated by a distance of $5\frac{1}{2}$ wavelengths. (c) What term describes the interference in this region?

32-3. The two speakers of a stereo system are wired in phase to the same 165-Hz source, which produces sound waves of 2.0-m wavelength. The speakers are separated by 2.0 m. How far back from one speaker must a listener move until he hears a minimum in sound intensity? Assume he moves perpendicular to the line joining the two speakers.

32-4. What is the ratio of the area within the radiation pattern for the two sources in Fig. 32-3b to the area within the pattern for a single source?

32-5. Two microwave generators separated by 6.0 cm oscillate in phase at the frequency of 10 kMHz. What is the angular separation between the lobes in their radiation pattern?

32-6. The radiation pattern of two identical isotropic radio oscillators consists of six lobes. What is the separation distance between the oscillators if the oscillators are radiating at a frequency of 10 MHz?

32-7. A simple test to determine whether the two speakers of a stereo system are properly wired is this: Compare the bass response with the speakers wired in one sense with that in which the wires to one speakers are reversed in polarity; the correct wiring corresponds to the larger bass response. Verify this procedure in terms of interference effects.

32-8. Two oscillators are similar to those in Fig. 32-3, except that the angle between the symmetry axis of the lobes and the line joining the oscillators is 30°. (a) Find the intrinsic phase difference between the two oscillators. (b) Sketch the radiation pattern.

32-9. A plane electromagnetic wave of 30.0-cm wavelength traveling in a horizontal plane is incident upon two vertical electric dipole receiving antennas separated by 10.0 cm along a north-south line. Measurements show that the signal received in the southern antenna lags 60° in phase behind the signal received in the northern antenna. What are the possible directions of propagation of the incident wave?

32-10. A point generator of waves on a water surface is 6.0 in from a plane reflecting boundary. The water waves have a wavelength of 3.0 in. At what angles, measured relative to the normal to the boundary, will the intensity be zero at great distances from the generator?

32-11. A microwave oscillator generating 1.0-cm microwaves is placed 3.0 cm from a large conducting sheet. A microwave receiver is a distance of 30.0 cm (measured along the sheet) from the oscillator. What minimum (nonzero) distance of the receiver from the sheet will result in zero received intensity?

32-12. The radar antenna on a boat transmits and receives waves of 10-cm wavelength. The antenna is 10 m from the water surface. At what minimum height (nonzero) above the water surface must an airplane be located, when it is 5.0 km from the boat, to elude detection by the radar set?

32-13. The phenomenon of "fading" in radio reception arises in large measure from the interference between two waves from the transmitter to the receiver, one following a direct route and another involving reflection from a conducting layer of charged particles (the ionosphere), lying above the earth. Suppose that a transmitter and receiver are separated by 10 km. What is the minimum height of an ionospheric layer which will produce a minimum in the received signal for radio waves of 710 kHz?

32-14. Because a laser beam is coherent, one can modulate it with a signal in the fashion that a radio beam is modulated with an audio signal. Taking the bandwidth required for intelligent speech to be 2 kHz, how many nonoverlapping voice bands can be accommodated in the green region of the visible spectrum (from 5000 to 5700 Å)?

32-15. In a Young double-slit experiment light of 6000-Å wavelength illuminates two slits separated by 0.40 mm. The interference pattern is observed on a screen 1.0 m from the slits. What is the separation between adjacent fringes in the interference pattern?

32-16. Two identical parallel slits, illuminated by light of 7000-Å wavelength, produce an interference pattern on a screen 0.80 m from them in which adjacent fringes are separated by 0.70 mm. (a) What is the separation distance between the two slits? (b) What is the distance between adjacent fringes if one illuminates the slits with light of 4000-Å wavelength?

32-17. A thin wedge of air is formed by two glass plates in contact along one edge and separated by a sheet of paper at the other. The distance between the two ends of the plates is 6.0 cm. One sees 10 interference fringes per centimeter between these ends when the plates are illuminated from above with light of 6000-Å wavelength. What is the thickness of the sheet of paper?

32-18. A plano-convex lens touches a plane plate of glass, and one sees Newton's rings with a dark central spot when viewing from above. Show that the radii of the dark interference fringes are given approximately by $\sqrt{m\lambda R}$, where λ is the wavelength, R is the radius of curvature of the spherical interference, and $m = 0, 1, 2, \ldots$.

32-19. Reflection from a camera lens may be appreciably reduced if the lens is coated with a thin film of transparent material. Assuming that the film thickness is so chosen as to prevent the reflection of yellow light of 5800-Å wavelength (at the center of the visible spectrum), what is the minimum thickness, if the refractive index of the coating is 1.30 and that of the lens glass is 1.60? (Such a coated lens appears purple, a mixture of red and violet light, because light from the extremes of the visible spectrum, having wavelengths widely different from those of light at the center of the spectrum, is not discriminated against by such a coating.)

32-20. A thin film of transparent material having a refractive index of 1.50 is placed in one of the beams of a Michelson interferometer (Fig. 32-17). When light of wavelength 6000 Å is used, it is observed that the insertion of the film causes a displacement of 40 fringes. What is the film thickness?

32-21. Consider a straight line of 20 radio antennas with a separation of 40 cm between adjacent antennas. Find the angular width of the strong central band in the reception pattern for incoming radio waves of wavelength (a) 5.0 m and (b) 2.5 m.

32-22. Show that the largest order of zero intensity in Eq. (32-10) is less than or equal to w/λ.

32-23. The diffraction pattern on a screen 1.0 m from a single slit shows the central maximum bounded by dark fringes separated by 2.0 mm. The wavelength of the light is 6000 Å. What is the slit width?

32-24. Plane waves of sound having a frequency of 375 Hz are incident upon a very tall window 8.0 ft wide, leading into a very large room. It is found that at the far wall the intensity of the sound falls continuously to zero at an angle of 30° relative to the normal to the window surface. What is the speed of the sound waves?

32-25. A single-slit diffraction pattern is seen on a screen. What changes will occur in the pattern if the entire apparatus, from light source to screen, is immersed in a medium of refractive index n?

32-26. Light from two identical line sources passes through a single slit of width w and thence to a distant screen. The sources are a distance D from the slit and produce light of wavelength λ. What is the minimum separation distance between the line sources that will permit them to be resolved into two distinct diffraction patterns on the screen?

32-27. What is the closest two point objects can be together to be resolvable by the eye (diameter of pupil, 2.0 mm) when the objects are 25 cm from the eye? Assume the wavelength of light to be 5000 Å. Use Rayleigh's criterion.

32-28. The Mt. Palomar telescope has a mirror 200 in in diameter. What is the diameter of a reflecting mirror for a radio telescope operating with 21-cm wavelengths which will have the same resolving power as the Palomar telescope?

32-29. A magnifying glass of focal length 5 cm and diameter 1.0 cm is used to aid the eye (diameter of pupil, 2.0 mm) in looking at two point objects close together. How close can the two objects be so that they will be resolvable as two distinct objects? Assume wavelength of light to be 5000 Å and the near point for the eye to be 25 cm.

32-30. A Venetian blind having a "grating spacing" of 1.75 in is to be used as a diffraction grating for sound waves of 0.50 in. At what angle relative to the normal to the "grating" will one find the first-order intensity maximum?

CHAPTER
33

Polarization

All types of waves show interference and diffraction effects, but only *transverse* waves, those in which the propagated disturbance is a *vector* quantity, show the phenomenon of *polarization*. Polarization effects may be illustrated, for example, by transverse waves traveling along a string under tension. Here we are concerned with the polarization properties of electromagnetic waves. We first consider how the superposition of simple harmonic motions gives rise to linear, circular, and elliptical polarization. Then we treat the polarization properties of radiation from an electric dipole oscillator. Next we turn to the meaning of polarization for visible light and the means by which polarized light is achieved.

33-1 SUPERPOSITION OF SIMPLE HARMONIC MOTIONS

Figure 33-1 shows two simple harmonic motions at the same frequency and at right angles to one another, but having varying relative phases, together with their resultant. The oscillation along the X axis is represented by $\mathbf{E}_x$; that along Y, by $\mathbf{E}_y$. The resultant of $\mathbf{E}_x$ and $\mathbf{E}_y$ is $\mathbf{E}$. These oscillations may be identified with the electric field of electromagnetic waves propagating at right angles to the plane of the paper.†

In Fig. 33-1a the two oscillations are *in phase*, and their resultant is also simple harmonic motion along a single line. The corresponding wave is said to be *linearly* polarized, the plane of polarization being that containing the resultant electric field vectors and the direction of wave propagation. The angle of the

---

†The magnetic field $\mathbf{B}$ is, of course, always at right angles to the electric field $\mathbf{E}$, the directions of $\mathbf{E}$, $\mathbf{B}$, and $\mathbf{I}$ (the vector intensity of the electromagnetic wave) being related by the Poynting formula $\mathbf{I} = \mathbf{E} \times \mathbf{B}/\mu_0$, Eq. (30-26).

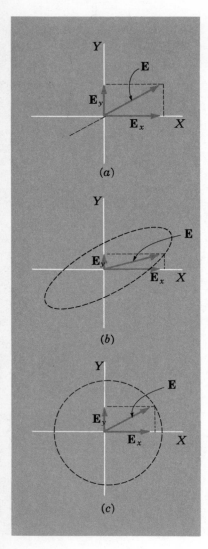

FIG. 33-1. Simple harmonic motions of the same frequency at right angles, with (a) no phase difference, (b) a finite phase difference, and (c) a 90° phase difference with equal amplitudes.

resultant field relative to the X axis depends on the relative amplitudes of $\mathbf{E}_x$ and $\mathbf{E}_y$. Conversely, a single linearly polarized wave along some oblique direction relative to the X axis is equivalent to two linearly polarized waves in phase along X and Y.

In Fig. 33-1b the two component oscillations are *not* in phase, and the tip of the resultant electric field vector $\mathbf{E}$ traces out an ellipse in the plane transverse to the direction of wave propagation. The orientation of the ellipse's major axis depends on both the amplitude and the phase difference between the two oscillations. For example, if the amplitudes are equal and the phase difference is 45°, the major axis of the ellipse makes an angle of 45° relative to the X axis. A resultant wave produced by two linearly polarized waves oscillating in mutually perpendicular directions, but differing in phase, is said to be *elliptically polarized*.

Now suppose that the two component oscillators are 90° out of phase and of equal amplitude, as in Fig. 33-1c. The resultant $\mathbf{E}$ is now of constant magnitude, and it rotates at the frequency of the component waves, the tip of the $\mathbf{E}$ vector tracing out a circle. This is a *circularly* polarized wave. The rotation sense depends on the relative phases of $\mathbf{E}_x$ and $\mathbf{E}_y$. If $\mathbf{E}_y$ lags behind $\mathbf{E}_x$ by 90°, the vector $\mathbf{E}$ rotates in the counterclockwise sense, and the corresponding wave assumed to be emerging out of the paper is said to be left circularly polarized. On the other hand, if $\mathbf{E}_x$ lags behind $\mathbf{E}_y$ by 90°, the rotation sense of $\mathbf{E}$ is clockwise, and the wave is right circularly polarized.

The three examples shown in Fig. 33-1 and all others may be demonstrated by the motion of a pendulum bob in the horizontal plane. With a simple pendulum of small amplitude, the bob moves in simple harmonic motion along a single line; as a conical pendulum the bob moves in a horizontal circle. Its most general path is an ellipse. These same patterns may be seen on an oscilloscope screen by applying sinusoidal variations to the horizontal and vertical deflecting plates.

33-2 POLARIZATION PROPERTIES OF WAVES FROM AN ELECTRIC DIPOLE OSCILLATOR

Recall some results given in Sec. 30-6 concerning the electromagnetic waves radiated by an electric dipole oscillator. The waves are generated by electric charges of opposite sign oscillating sinusoidally along a line, the dipole axis. At large distances from the dipole, the wavefronts are spheres. Moreover, the radiated waves are *linearly polarized*, the electric field lying in the plane containing the electric dipole and the direction of wave propagation. The magnitude of the electric field is *not* the same in all directions; $\mathbf{E}$ is maximum in the plane perpendicular to the dipole axis and zero along the dipole axis, as shown in Fig. 33-2. Since the radiation intensity varies as E^2, no energy is radiated along the dipole axis.

Figure 33-3a shows an electric dipole oscillator transmitting waves to a nearby electric dipole receiver. The relative orientations of the two dipole antennas must be proper if the maximum energy is to be transferred from transmitter to receiver. Thus, if the receiving dipole is parallel to the electric field of the arriving waves, this electric field will drive charges along the axis of the dipole and thereby produce a large signal at the receiver. But if the receiving antenna were turned through 90°, as in Fig. 33-3b, no signal would be received.

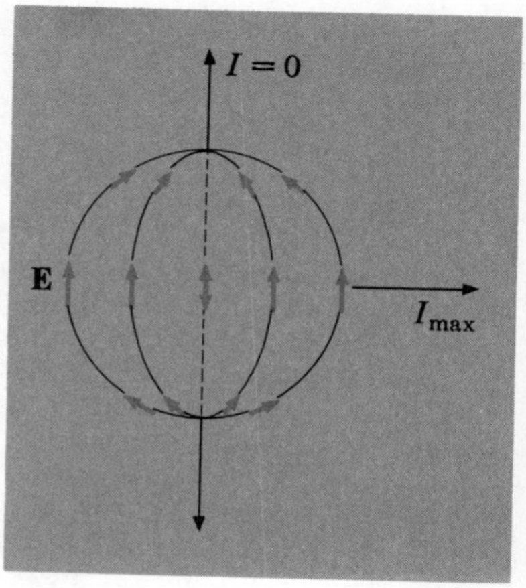

Now suppose that we place a screen consisting of a series of parallel wires or rods between the transmitter and receiver, the conducting wires first being oriented parallel to the direction of polarization of the electromagnetic waves, as shown in Fig. 33-4a. Since the electric field oscillates along the long direction of the wires, it causes the charges within the wires to oscillate, and the incident wave is strongly absorbed by the wire grid. This is confirmed by a reduced signal at the receiver. On the other hand, if the wire grid is rotated through 90°, so that the incident electric field is perpendicular to the direction of "easy absorption" by the wires, electromagnetic waves pass through the grid with little absorption, as indicated by the stronger signal at the receiver.

These fundamental effects, which have an exact counterpart in the polarization properties of ordinary light waves, can be readily demonstrated with microwaves a few centimeters in wavelength.

FIG. 33-2. Radiated electric field for an electric dipole oscillator.

33-3 POLARIZATION OF VISIBLE LIGHT

When an electric dipole radio oscillator is driven continuously at a constant frequency, the emitted waves are coherent and linearly polarized. This is not the case for visible light. Except

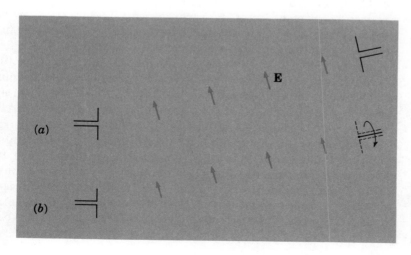

FIG. 33-3. (a) A dipole transmitter radiates to a dipole receiver. (b) No signal is received when the axes of the two dipoles are at right angles.

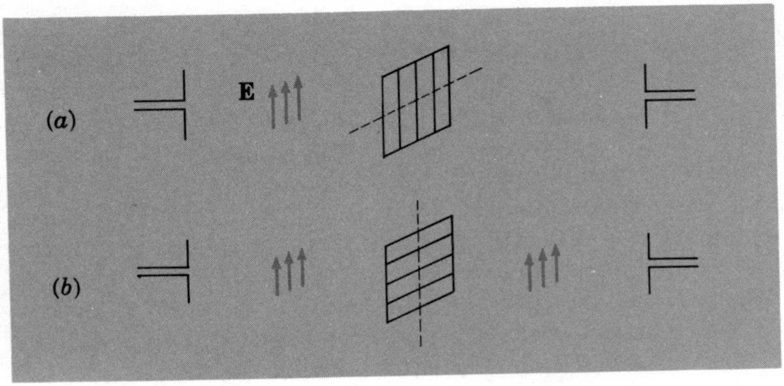

FIG. 33-4. (a) No signal reaches the receiver when the wires in the grid are parallel to the polarization direction of the radiation. (b) When the wires are turned 90°, the signal reaches the receiver.

for lasers (Sec. 32-4), visible light from ordinary sources arises from the random radiation of many individual atoms, each atom radiating for a time of approximately 10^{-8} s. Not only is the emitted light incoherent in the sense that the phases of successive individual light trains are not the same, the polarization direction of the light from individual atoms also varies over all possible orientations, so that the electric field in the plane transverse to the direction of propagation ranges over all possible directions. Thus, if one were somehow able to measure the polarization of light from a source containing very many radiating atoms, one would first find one direction of polarization for the light from one atom, then another polarization direction for the light from the next atom, and so on. The individual wave trains arrive, however, at such very small time intervals (much less than 10^{-8} s) that it would be impracticable to specify the polarization direction of the light at each instant. When the polarization direction changes randomly and rapidly, so that one cannot follow it in time, the light is said to be *unpolarized*.

Three ways of representing unpolarized light are shown in Fig. 33-5. The various electric field vectors, viewed from along the direction of propagation of the separate wave trains, are shown as randomly oriented in Fig. 33-5a. Alternatively, each electric field vector may be replaced by its rectangular components. Therefore, the unpolarized light is equivalent to two mutually perpendicular and linearly polarized electric field oscillations of equal magnitude, which for convenience we have taken to be horizontal and vertical. See Fig. 33-5b. (It must be emphasized, however, that the X and Y components are *not* individually continuous and coherent. Since the separate wave trains in an unpolarized light beam are random in phase as well as in polarization direction, the two rectangular components fluctuate randomly in phase.) Finally, we may also represent unpolarized light as in Fig. 33-5c: Here mutually perpendicular lines, transverse to the propagation direction, represent the two equivalent linearly polarized waves of random phase.

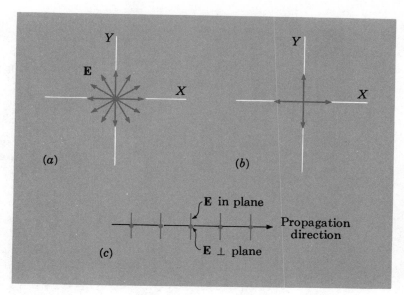

FIG. 33-5. Various representations of unpolarized waves: *(a)* electric field vectors for an unpolarized wave traveling in or out of the plane of the paper; *(b)* the equivalent fields (of random phase); and *(c)* an unpolarized wave with two mutually perpendicular electric fields.

A simple way to produce linearly polarized visible light is to pass the unpolarized light beam through a sheet of the commercial material Polaroid. This material consists of needlelike molecules (herapathite) aligned mostly along one direction, this alignment being achieved by stretching the flexible transparent sheet as the material solidifies. In its polarization properties for visible light, a sheet of Polaroid is similar to the grid of parallel wires for microwaves: With the electric field parallel to the long axis of the molecules, the wave is absorbed, and with the electric field perpendicular to the long axis, the wave is (mostly) transmitted.

It is customary to designate the direction of the electric field for easy transmission within the Polaroid sheet as the *polarization direction*. Then, if an unpolarized beam is incident upon one Polaroid sheet, only the electric field oscillations parallel to the polarization direction are transmitted. We take the two equivalent and mutually perpendicular polarization directions in the unpolarized beam to be parallel and perpendicular to the polarization direction of the Polaroid; see Fig. 33-6.

Now suppose that the emerging linearly polarized light is incident upon a second Polaroid sheet, whose polarization direction makes an angle θ relative to that of the light incident upon it. We may resolve the electric field **E** into rectangular components, $E \cos \theta$ parallel to the second Polaroid's polarization direction and $E \sin \theta$ perpendicular to it. Only the component $E \cos \theta$ is transmitted; the other is absorbed.

What, then, is the intensity of the light after passing through the second Polaroid? Recalling that the intensity I varies as E^2, where **E** is the resultant electric field of an electromagnetic wave, it follows that the transmitted intensity is given by

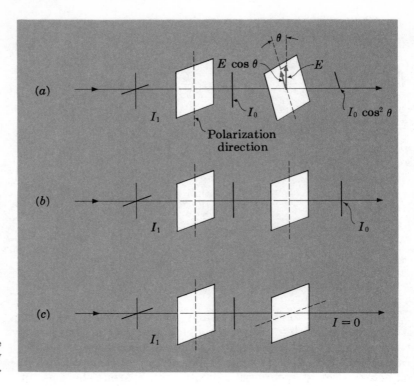

FIG. 33-6. Polaroid sheets whose polarization directions differ by (a) θ, (b) $0°$, and (c) $90°$.

$$I = I_0 \cos^2 \theta \qquad (33\text{-}1)$$

where I_0 is the maximum intensity, corresponding to $\theta = 0$.

When $\theta = 90°$, that is, when the polarization directions of the two Polaroid sheets are at right angles, the transmitted intensity is zero. Said differently, when the *polarizer* and *analyzer* are at right angles, there is *extinction*. Equation (33-1), which governs the intensity through *any* two devices, each of which transmits only one polarization direction, is known as *Malus' law*, after its discoverer, E. L. Malus (1775–1812).

The polarization direction of such a polarizing device as a Polaroid sheet cannot be established uniquely unless there is an independent means of determining the polarization direction of a wave alone. This follows from the fact that experiments with two Polaroid sheets give the *relative* polarization directions of two sheets as differing by $90°$ when no light is transmitted through them, but such experiments alone do not give the polarization direction of any one sheet. Polaroid is one well-known example of a number of materials exhibiting *dichroism,* the attenuation of one polarization component and the transmission of the mutually perpendicular component. The naturally occurring crystal, tourmaline, is dichroic.

Example 33-1

An unpolarized beam of light passes through a *single* sheet of Polaroid. What is the intensity of the transmitted beam as compared with that of the incident beam?

If the incident light is truly unpolarized, then the two equivalent perpendicular oscillations into which the beam may be resolved, shown in Fig. 33-5b, must be of equal magnitude. We may choose the orientation of one oscillation to be parallel to the polarization direction of the Polaroid sheet. This component is completely transmitted; the other is completely absorbed. Therefore, the intensity of the transmitted beam is *half* that of the incident beam.

Now, if there were no absorption whatsoever for oscillations parallel to the polarization direction of a Polaroid sheet, adding a second Polaroid sheet with its polarization direction parallel to that of the first would produce *no* further attenuation in the light intensity. This ideal behavior is not found, however, because partial absorption occurs even in the preferred orientation.

Example 33-2

An unpolarized light beam falls on two Polaroid sheets so oriented that no light is transmitted through the second sheet. A third Polaroid sheet is then introduced between the first two sheets. How does the intensity of transmitted light vary with the orientation of the third sheet? (The intensity is *not* zero for all orientations!)

The polarization directions of sheets 1 and 2 are mutually perpendicular. We take the angle between the polarization directions of sheets 1 and 3 to be θ. Then, as Fig. 33-7a shows, if **E** is the magnitude of the electric field transmitted through the first sheet, the component emerging through the third (the one placed *between* the first and second sheets) is $E \cos \theta$. From Fig. 33-7b we see that the component emerging through the last sheet has the magnitude $E \cos \theta \cos (90° - \theta) = E \cos \theta \sin \theta$. Thus, the intensity varies according to

$$I = I_0 \cos^2 \theta \sin^2 \theta = \tfrac{1}{4} I \sin^2 2\theta$$

where I_0 is the intensity transmitted through the first sheet. Note that there are *four* positions at which the intensity is maximum: 45°, 135°, 225°, and 315°. Note also that interchanging sheets 1 and 3 does not change the dependence of I on θ.

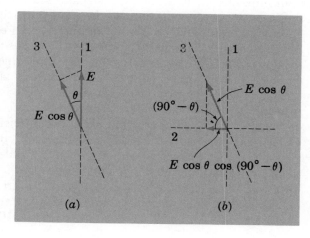

FIG. 33-7

33-4 POLARIZATION IN SCATTERING

Imagine linearly polarized light to be incident upon an object much smaller than the wavelength of light, for example, a molecule. The electric field of the incident wave acts on the charges within the molecule and causes electrons to oscillate along the polarization direction. But charges oscillating along a line constitute an electric dipole oscillator, and the molecule becomes, in effect, a dipole oscillator and radiates accordingly. It absorbs radiation from an incident wave and reradiates this energy in various directions. The incident wave is said to be *scattered* by the molecule.

Let us examine the polarization properties of the scattered radiation. Figure 33-8a shows a linearly polarized wave incident upon a scattering center. Recalling the polarization and intensity properties of the radiation emitted by an oscillating electric dipole (Fig. 33-2), we see that if the incident wave is linearly polarized in the plane of the paper and the axis of the induced oscillator is therefore parallel to the electric field of the incident wave, *no* radiation is scattered outward along the dipole axis, that is, scattered through 90° in the plane of the paper. The scattered radiation is also linearly polarized in the plane determined by the dipole axis and the propagation direction of the scattered wave. The intensity is a maximum in the plane perpendicular to the oscillating dipole but falls to zero as the scattering direction approaches 90°.

Figure 33-8b shows the more general case: unpolarized light incident upon a scattering center. The dipole oscillates both in and perpendicular to the plane of the paper. Now radiation is scattered in *all* directions. The radiation scattered through 90° in the plane of the diagram is completely linearly polarized, and partially polarized in all other directions (except that along which the incident wave travels). This is also shown in Fig. 33-8c.

Electrons in a typical atom or molecule have natural oscillation frequencies corresponding to ultraviolet light. When visible light impinges on an atomic or molecular scattering center, it drives the electrons into oscillation at a frequency which is lower than the natural, or resonant, oscillation frequency. Of course, violet light comes closer to the natural frequency than does red light, and it is to be expected that, other things being equal, blue light is scattered more effectively than red. Detailed analysis shows that the intensity of the scattered radiation varies as the fourth power of the frequency. Since the ends of the visible spectrum differ in frequency by a factor of approximately 2, the intensity of the scattered violet light is roughly 2^4, or 16, times that of red light.

Consider the scattering of the sun's visible radiation by particles and molecules in the earth's atmosphere. The sky appears blue overhead, because blue is scattered to a greater degree by air molecules and dust particles than is red light (the same effect produces the bluish cast of a smoke-filled room). On the other hand,

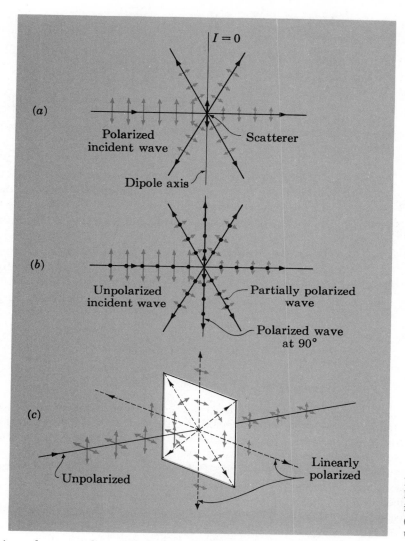

FIG. 33-8. *(a)* Scattering of a linearly polarized wave by an induced electric dipole oscillator. *(b)* and *(c):* Scattering of an unpolarized wave.

viewing the sun directly, especially through a thick dust-filled atmosphere at sunset, shows predominantly colors other than blue, inasmuch as the blue light has been scattered from the forward direction. As one would expect, the scattered blue light in the sky is at least partially polarized, and one may use the polarization properties of scattered light to navigate by. Indeed bees use this navigation scheme; each of the many eyes on a bee's head can detect the state of polarization of the light scattered from the sun.

Polarization effects also enter in the *double* scattering of electromagnetic waves. Suppose that unpolarized light is incident upon a first scatterer, and the radiation scattered 90° then excites a second scatterer. We are interested in the intensity of the doubly scattered light. As Fig. 33-9 shows, the intensity of waves scattered by the second scattering center is zero along a direction perpendicu-

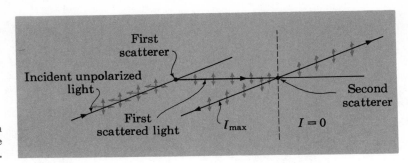

FIG. 33-9. Polarization characteristics of double scattering.

lar to the plane containing the incident beam and the two scattering centers. This effect was observed for x-rays by C. G. Barkla (1877–1944) in 1906, who established thereby that x-rays are transverse waves.

In addition to polarized waves arising from the scattering of electromagnetic radiation, polarized waves may also be produced by light undergoing reflection and refraction at the interface between two media or in transmission through an anisotropic refracting medium.†

---

†See R. T. Weidner and R. L. Sells, *Elementary Classical Physics*, 2d ed., secs. 40-5 and 40-7, Allyn and Bacon, 1973.

PROBLEMS

33-1. The simple harmonic oscillations along the Y axis lead the simple harmonic oscillations of the same frequency and amplitude along the X axis by 45°. What is the orientation of the major axis of the ellipse produced by the superposition of the two oscillations?

33-2. Simple harmonic motions along two mutually perpendicular directions are described by $X = A \sin \omega t$ and $Y = B \sin(\omega t + \delta)$, where A, B, ω, and δ are constants. (a) Show that when $\delta = 0$ but $A \neq B$, the path traced out in the XY plane is a straight line inclined relative to the X and Y axes. (b) Show that in the general case, when $\delta \neq 0$, the path is an ellipse, but with symmetry axes not coincident with the X and Y axes.

33-3. What must be the angle between the polarization axes of two ideal polarizing sheets, if the transmitted light intensity is to be 25.0 percent of the intensity incident on the first sheet?

33-4. The polarization direction of a beam of linearly polarized light is to be rotated 90°. Find the intensity of the resultant beam if one uses (a) two polarizing sheets, each rotating the polarization direction by 45°, and (b) 90 polarizing sheets, each rotating the polarization direction by 1.0°.

33-5. Some sunglasses are made of Polaroid. The polarizing sheets serve the purpose of minimizing the intensity of light reflected from surfaces in a horizontal plane. Should the polarization axis of the Polaroid be horizontal or vertical?

33-6. The sun is in the east. A man looks into the overhead sky through a polarizing sheet. For what orientation of the polarization axis of the sheet, east-west or north-south, will the intensity of the light be a minimum?

33-7. Unpolarized light travels vertically upward and is scattered. Some of this scattered light travels north and is then scattered a second time. In what direction will the intensity of the double scattered light be zero?

33-8. Show that any linearly polarized electromagnetic wave can be regarded as being the superposition of two circularly polarized waves, one clockwise and the other counterclockwise.

33-9. Show that when a circularly polarized electromagnetic wave impinges upon an absorber, a torque acts on the absorber and that, therefore, such a wave carries angular momentum.

33-10. Circularly polarized light passes through an ideal polarizing sheet. What fraction of the incident intensity is absorbed in the sheet?

CHAPTER 34

Relativistic Kinematics: Space and Time

The theory of relativity, primarily the creation in 1905 of Albert Einstein (1879–1955), ranks as one of the two great advances in twentieth-century physics. Whereas quantum theory extends physics to the domain of the very small, relativity theory extends physics to the domain of the very fast.

Although often thought to be esoteric and recondite, the principal features of relativity theory can be set forth with minimal mathematical sophistication on the basis of two fundamental postulates. The theory of relativity can no longer be regarded as conjectural; even its most bizarre predictions have been amply confirmed by experimental test. Many of these predictions conflict with our common-sense conceptions; indeed, relativity theory shows classical physics to be downright wrong when applied to high-speed phenomena. Therefore, we must be prepared to re-examine radically the most obviously apparent presuppositions about the nature of space and time and find them drastically revised.

In this chapter we shall be concerned exclusively with relativistic kinematics—space, time, velocities, and their interrelationship—as arising from the postulated constancy of the speed of light. In the next chapter we explore the implications of the relativity theory in dynamics—momentum, energy, mass, and other mechanical quantities—as arising from the postulated invariance of the laws of physics.

34–1 THE CONSTANCY OF THE SPEED OF LIGHT

All observers measure the speed of light through vacuum to be the same constant c. This is the basic postulate upon which all of relativistic kinematics hinges. The assumption was made by Einstein in his formulation of the theory, probably without direct knowledge of the very experimental evidence which supported this seemingly curious postulate.

The postulate is curious in that it asserts *all* observers will measure the speed of electromagnetic waves through empty space to have the same constant value, quite apart from the state of motion of observer or of the source of electromagnetic radiation. In this respect the propagation of electromagnetic waves differs drastically from all other wave propagation. Consider, for example, sound waves through air, whose speed at room temperature is 340 m/s *relative to the medium*—air—in which the sound waves propagate. Thus, if an observer is at rest in air he measures the pressure disturbance of a sound pulse to advance a distance of 340 m in 1 s. But if the observer is in motion relative to the air at, say 40 m/s, in the same direction as that in which the pulse of sound moves, he finds that the speed of the sound pulse is, relative to him, 340 − 40 = 300 m/s. By the same token, if this observer travels toward the source at the same speed, the sound pulse advances at a rate, again relative to him, of 340 + 40 = 380 m/s. Only if the observer is at rest relative to the medium propagating sound waves does he measure its speed to be 340 m/s.

Now, if the analogous situation prevailed for the propagation of electromagnetic waves, then there would, first, be the necessity of assuming a medium for the propagation of light. This conjectured medium—in the past referred to as the *ether*—would then constitute the only reference frame in which the speed of light would be c. An observer in motion relative to the ether—for example, an observer attached to the earth as it circles the sun—would necessarily find the speed of light to be greater than or less than c by the magnitude of his speed relative to the ether. The speed of light would be found by him to be exactly c only if he happened to be at rest relative to the ether. In short, if electromagnetic waves were like other waves, if their speed were c only when measured by an observer at rest in the medium (the ether) in which these waves propagate, then the measured speed of light would differ from c when the observer is in motion relative to the unique reference frame in which the ether is at rest. On the other hand, if speed of light is always c, quite apart from the motion of the observer or source of light, then electromagnetic wave propagation is fundamentally different from other types of waves, there is no need to assume the existence of an ether, and the basic postulate of relativity on the constancy of the speed of light is indeed a fact of nature.

An experimental demonstration that the speed of light is indeed independent of any circumstances is rendered particularly difficult by virtue of the large magnitude of c, 3.00×10^8 m/s. Measuring the speed of light is difficult; measuring a small change in c is extraordinarily difficult and requires subtle experimental procedures. Historically, the first significant test was made by A. A.

Michelson (1852–1931)† and E. W. Morley (1838–1923) in the famed Michelson-Morley ether-drift experiment of 1887. The basis of this and later equivalent experiments is the following: If an ether exists and is not rigidly fixed to the earth, then surely at some time during a year the earth will, because of its orbital motion about the sun at a speed of 3×10^4 m/s, be drifting through the ether at this speed, and the speed of light along or against the direction of drift will differ from c by 3×10^4 m/s, or 1 part in 10^4, while the speed of light directed at that time at right angles to the drift direction will be unaffected. As a consequence, the round-trip travel time for a pulse of light going "upstream" and then "downstream" (or the converse) will differ from the round-trip travel time for a pulse of light over the same distance but at right angles to the drift direction. (See Prob. 34-1.) The very minute difference in travel time for the two routes was measured indirectly by Michelson and Morley by examining the interference effects between two beams of light sent outward at right angles in a Michelson interferometer (Sec. 32-7) and then recombined, or more specifically, by looking for a shift in the interference pattern arising from a rotation of the instrument and the concomitant interchange of the upstream-and-downstream and right-angle routes.

The Michelson-Morley experiment showed a null effect—no effect attributable to the ether; therefore, no necessity for assuming the existence of an ether; therefore, a demonstration of constancy of the speed of light for all observers—and so too have all subsequent experiments of an analogous type, whether conducted with visible light from ordinary sources or from lasers or with radio waves.

Not only does experiment emphatically confirm the correctness of the postulated constancy of the speed of light, electromagnetic theory suggests it as well. Recall that the speed of light is given in classical electromagnetism as $c = 1/(\varepsilon_0 \mu_0)^{1/2}$ (see Sec. 30-3). Here ε_0 and μ_0 are fundamental constants which measure, respectively, the strengths of the electric and magnetic interactions between electrically charged particles in empty space. These constants can be identified with the properties of an otherwise empty space (an ether) only through convoluted arguments.

To appreciate the profound, not to say startling, implications of the assertion that the speed of light is a unique constant of nature consider the following: Suppose that a light source is somehow set in motion at the extraordinarily high speed of $0.9c$ toward an observer. Then, if the fundamental postulate of relativity theory is valid, this observer finds a pulse of electromagnetic radiation advancing toward him, not at the speed $1.9c$, but at the unique speed c. Or, if the observer retreats from the light source at the speed

---

†Michelson was the first American to be awarded the Nobel prize in physics.

0.9c, he again measures the speed of light relative to him to be c, not $c - 0.9c = 0.1c$.

34-2 TIME DILATION

By the speed of a particle or of a signal is meant, of course, the spatial interval it traverses divided by the corresponding time interval, both measurements being made by the same observer. But if light has a unique speed for *all* observers, then the constancy of c implies that space intervals and time intervals, whose ratio yields c, may not have unique values for all observers. Said differently, if c is absolute, space intervals and time intervals cannot be absolute, but may have values that somehow depend on the state of motion of an observer.

In prerelativity physics the absolute character of space and time was taken as axiomatic and self-evident: If one observer measures the distance between two separated points as, say, 1 m, then other observers will agree that their separation distance is exactly 1 m. Similarly, if one observer clocks a time interval between two events as, say, 1 second, other observers will likewise measure the time interval to be precisely 1 second. But these seemingly obvious claims, certainly in accord with experience for all ordinary speeds, must be reexamined in the light of the constancy of c, and will indeed be found to be fundamentally incorrect.

In what follows we shall be concerned with two prototype observers in relative motion at constant velocity; their separation distance will increase (or decrease) at a constant rate. In fact, all that we can say is that they are in *relative* motion, not whether one or the other is in motion or at rest. As the law of inertia (Newton's first law) reminds us, even in classical physics there is no fundamental distinction between the states of rest and motion.

One observer in an inertial frame of reference will be denoted S; the symbol S will also denote the reference frame itself. The three spatial and one temporal coordinates required to specify an event in this coordinate frame will be given as (x, y, z, t). More specifically, we may think of S as an observer with a meter stick that measures the location (x, y, z) of points in his reference relative to his origin and with a clock that records the time t of any event in this reference frame. (Strictly, S must be conceived to consist of an infinite number of observers distributed through all points in space in the S reference frame; each such observer carries a clock to register the time of an event at his particular location, all such clocks having been synchronized and calibrated to run in identical fashion.)

The second observer is denoted S'; spatial and temporal coordinates registered in this reference frame are denoted (x', y', z', t'). We can make *no* a priori assumptions about the relationship be-

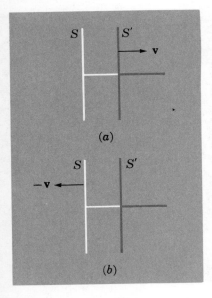

FIG. 34-1. (a) Reference frame S' has velocity v relative to S; (b) reference frame S has velocity $-v$ relative to S'.

tween x and x', t and t', etc. Reference frame S' moves with constant velocity **v** relative S, as shown in Figure 34-1a. (For simplicity we imagine that the X and Y axes of the two reference frames are aligned.) But if S' has a constant velocity to the right (along the X axis) relative to S of magnitude v, then S has a constant velocity to the left relative to S' also of magnitude v. See Fig. 34-1b.

We wish to find the relationship for time intervals between the same two events as measured in S and S'. Consider the following hypothetical experiment. Observer S has a tube of length H which is at rest and aligned along the Y axis (at right angles to the direction of relative motion between S and S'). Observer S sends a pulse of light from the base of the tube along the tube axis until it reaches a mirror at the top end and is then reflected back to the base. See Fig. 34-2. Thus, the departure of the light pulse and

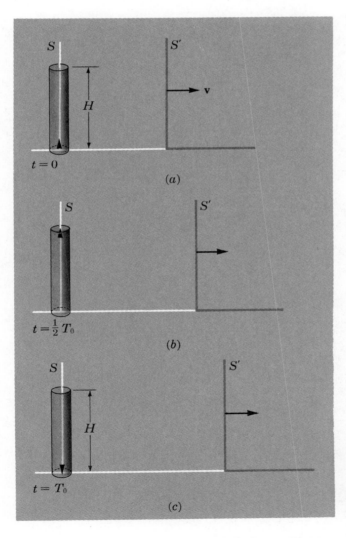

FIG. 34-2. A light pulse travels vertically up and then down through tube of height H at rest in reference frame S. The pulse (a) departs the base at $t = 0$, (b) arrives at the top at $t = \frac{1}{2}T_0$, and (c) returns to the base at $t = T_0$.

its later return to the base take place at the *same* location in reference frame S. To emphasize that the time interval elapsing from the departure to the return of the light pulse is measured by an observer who is at *rest* with respect to the tube, we indicate the time interval S measures as T_0. Since the pulse of light traverses the total round-trip distance of $2H$ in the time interval T_0 at the speed c, we have

$$c = \frac{2H}{T_0} \tag{34-1}$$

Now consider the same events as seen by observer S'. Relative to S', reference frame S and the tube travel left at speed v. The path followed by the light pulse now consists of two oblique lines. See Figure 34-3. The pulse does not, as S' observes it, re-

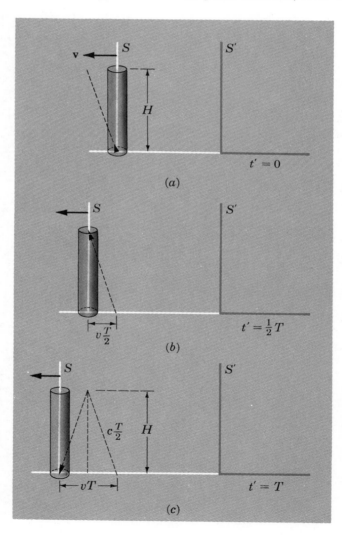

FIG. 34-3. The same events as in Fig. 34-2, but now as observed in reference frame S'. The pulse *(a)* departs the base at $t' = 0$, *(b)* arrives at the top at $t' = \frac{1}{2}T$, and *(c)* returns to the base at $t' = T$, all as registered on the clock of observer S'. Over the round trip the tube advances to the left, relative to S', through a distance vT. The speed of the light pulse over the two oblique path segments is also c relative to S'.

turn to the same location in *his* reference frame. In fact, the overall pathlength observed by S' *exceeds* that $(2H)$ which S observes. But the speed of the light pulse must be the *same* for both observers. So it follows at once that the time interval T recorded by S' between the departure and return of the pulse to the base of the tube must *exceed* T_0. As Fig. 34-3 shows, the tube moves left a total distance vT in the time interval T, or a distance $vT/2$ in time $T/2$. During the half-time $T/2$, the light pulse covers the oblique distance $cT/2$. Therefore, from the geometry of the triangle in Fig. 34-3, the distances are related by

$$H^2 + \left(\frac{vT}{2}\right)^2 = \left(\frac{cT}{2}\right)^2$$

Using (34-1) to eliminate H† from the above equation, we have

$$T = \frac{T_0}{\sqrt{1-(v/c)^2}} \qquad (34\text{-}2)$$

This is the fundamental *time-dilation* equation. It is crucial that the meaning of the terms appearing in it be clearly understood. T_0 is the time interval elapsing between two events occurring at the *same location* and measured on the clock of an observer at rest at this location; T_0 is termed the *proper time* (or rest time). On the other hand, T is the time interval between the *same two events* as registered on the clock of an observer traveling at speed v relative to the location at which the two events take place (and who, therefore, sees the two events taking place at *different* locations relative to his reference frame). The clocks of the two observers when compared *at rest* with respect to one another give *identical* readings. Equation (34-2) shows that in general $T > T_0$; relative to a moving observer time intervals are increased, or dilated. The effect is a result of the constancy of the speed of light; it is not attributable to a physical cause, but reflects the relativistic properties of time (and space). Therefore, the phenomenon is termed *time* dilation.

Time-dilation effects are significant only at speeds approaching that of light. For example, if $v = 0.1c$ (a speed about 4,000 times greater than that of a satellite orbiting the earth), then $T = T_0/[1 - (0.1)^2]^{1/2} = T_0/\sqrt{0.99} = 1.005T_0$, and T exceeds T_0 by only one-half of 1 percent. At speeds close to c the effects are dramatic, however. Suppose that $v = 0.98c$. Then $T = T_0/[1 - (0.98)^2]^{1/2} = T_0/\sqrt{0.04} = 5T_0$, and T exceeds T_0 by a factor of five. Thus, if the timer for a bomb is set to make the bomb explode in one hour (when the bomb is at rest), but the bomb is set in motion at $0.98c$, then an observer seeing it move by at this speed would find that, as read

---

†The vertical distance H, at right angles to the direction of relative motion of the two reference frames, is assumed to be the same for both observers. Although we shall see that spatial intervals along the direction of relative motion are not the same for all observers (Sec. 34-3), the transverse distances remain unchanged.

on his clock (and on the clocks of his associates stationed throughout his reference frame), it takes 5 hours for the bomb to explode.

No one has observed time-dilation effects with bombs moving at high speeds. But an exactly analogous effect *has* been observed, and the time-dilation effect thereby confirmed, in experiments with high-speed unstable subatomic particles, such as muons. A muon is created (born) when another unstable particle (a pion) decays; a muon decays (dies) in turn into an electron (together with two uncharged, massless particles called neutrinos). One cannot specify when any one muon will be born or die, but the *average* lifetime of a *large* number of muons can be given quite precisely. The half-life of a muon is 1.52×10^{-6} s. This means that if one has a large number of muons at some initial time and these muons remain at rest, exactly one-half will have survived after 1.52×10^{-6} s has elapsed, while the other half will have decayed. On the other hand, if a large number, say 10,000, of muons are in motion at the speed $v = 0.98c$, then the number of surviving muons is 5,000 only after the elapsed time is $T = 5T_0 = 5(1.52 \times 10^{-6} \text{ s}) = 7.60 \times 10^{-6}$ s. Muons in flight at high speed live significantly longer than muons at rest. The increased lifetime arising from time dilation may be measured directly by noting that, if high-speed muons live longer than muons at rest, they travel correspondingly greater distances before decaying.†

Since the increased lifetime of an unstable particle in motion at high speed reflects the properties of time itself and not of any physical mechanism, time dilation applies to any process taking place in time, including those in biochemical systems. Consider the famous twin paradox, first introduced by Einstein. Suppose that there are two identical twins. One stays home while the other goes on a round trip in a spaceship to and back from some distant point at so high a speed that time-dilation effects are significant, and is finally reunited with her stay-at-home sister. Time is dilated for the traveling twin, and when the sisters compare their ages upon being reunited they agree that the stay-at-home twin is older than the traveling twin, for whom the time-dilation effect has introduced a shorter period of elapsed time. (The effect is *not* reciprocal, however; whereas the stay-at-home twin has an uneventful history, the traveling twin experiences three profound shocks — the first shock as the spaceship takes off suddenly, the second as it comes suddenly to rest upon arriving at its far destination and immediately reaccelerates to start the homeward portion of the trip, and the third shock as the spaceship arrives home and is suddenly brought to rest.)

---

†Indeed, significant numbers of high-speed muons, created in the upper portions of the earth's atmosphere when energetic particles from outer space (cosmic radiation) strike the earth, can be observed at sea level because of the time-dilation phenomenon. Without this effect, most muons would not have survived decay long enough, and therefore traveled far enough, to reach the bottom of the atmosphere at sea level.

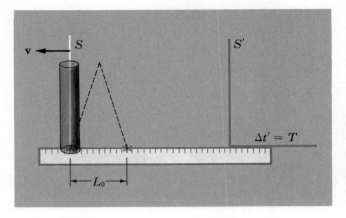

FIG. 34-4. The light pulse of Fig. 34-2 makes burn marks on a meter stick attached to reference frame S' as it departs from and returns to the base of the tube. The separation distance is L_0 and the elapsed time interval is T, both relative to S'.

34–3 SPACE CONTRACTION

Just as a time interval between two events depends of the state of motion of the observer, so too a spatial interval, or length, may, because of the constancy of c, depend on the state of motion of the observer.

We use the same arrangement as before (Figs. 34-2 and 34-3) to analyze the space-contraction effect. A tube through which a light pulse travels up and down is at rest in reference frame S. As observed by S', the tube and observer S travel left at speed v, as shown in Fig. 34-4. To make the events more specific we suppose that observer S' places his meter stick along his X' axis and that the light pulse makes one burn mark on this meter stick when it leaves the base of the tube and a second burn mark when it returns to the base. The distance between the two burn marks on the meter stick of S' is designated L_0. The subscript zero is to emphasize that this is the length measured by observer S', who is *at rest* relative to his own meter stick.

Now, as indicated earlier, the time interval elapsing between the markings of the two spots, as measured by S', is the dilated time T. Since reference frame S moves left at speed v over a distance L_0 in time interval T, all as measured by S', we have

$$v = \frac{L_0}{T}$$

Now consider the distance between the two burn marks on the meter stick of S', but as measured by S. Observer S sees S' and his meter stick in motion to the right at speed v, as shown in Fig. 34-5. The distance between the two burn marks on the moving meter stick, as measured by observer S, is designated L. (Observer S, in measuring the length of an object in motion with his own meter stick, must be sure to mark the locations of the two burn marks *simultaneously*.) Relative to S, reference frame S' ad-

FIG. 34-5. The situation of Fig. 34-4, but now as observed by observer S, who uses *his* meter stick to measure the distance L between the burn marks on the meter stick of S'. Reference frame S' advances a distance L in the time interval T_0, as observed by S.

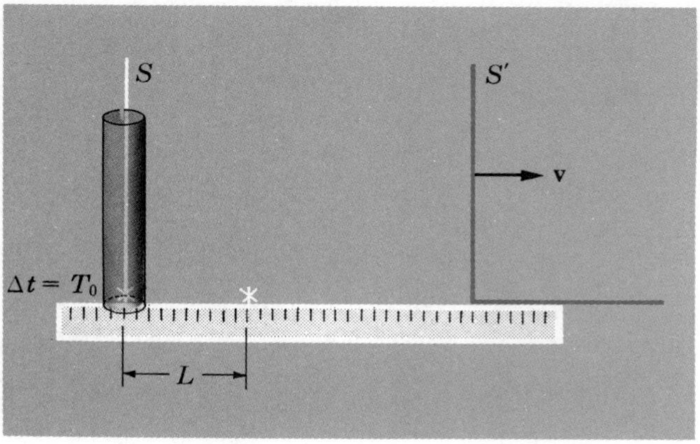

vances to the right at speed v over a distance L; moreover, this occurs in the *undilated* time interval T_0 between the marking of the two burn marks. Therefore, observer S may write

$$v = \frac{L}{T_0}$$

Eliminating v from the two equations above we have

$$L = L_0 \frac{T_0}{T}$$

and using the time-dilation relation (34-2) we have finally

$$L = L_0 \frac{T_0 \sqrt{1 - (v/c)^2}}{T_0}$$

or
$$L = L_0 \sqrt{1 - \left(\frac{v}{c}\right)^2} \quad (34\text{-}3)$$

Equation (34-3) is the basic *space-contraction* relation. The meaning of the terms is this: L_0 is the spatial separation, or *proper length*, between two points (lying along the line of relative motion of S and S') and measured by an observer at rest with respect to these points. The contracted length L is the distance between the same two points as measured by an observer traveling at speed v relative to these points. It must be emphasized again that if one is properly to measure the length of an object in motion, the two ends must be marked simultaneously. Although lengths are contracted along the direction of relative motion, there is *no* space contraction at *right angles* to the direction of relative motion.

As (34-3) shows, in general $L < L_0$. Length contraction is significant only at high speeds; for example, with $v = 0.1c$, $L = 0.995L_0$, whereas for $v = 0.98c$, $L = \frac{1}{5}L_0$. At the limit of low speeds ($v/c \to 0$), the relative spatial intervals and time intervals of relativity physics become effectively the absolute space and time intervals of classical physics: $L = L_0$ and $T = T_0$.

The term *space* contraction is used to emphasize that the effect is not due to a physical cause, that is, shrinking of an object because of external pressure or a drop in temperature, but rather reflects the changed character of space and time intervals imposed by the fundamental requirement that all observers measure the same speed for light through a vacuum.

34-4 THE LORENTZ COORDINATE TRANSFORMATIONS

According to observer S the four space-time coordinates of an event are (x, y, z, t); the four space-time coordinates of the *same event* for an observer S' in motion relative to S at speed v are (x', y', z', t'). How are (x, y, z, t) related to (x', y', z', t')?

The appropriate equations are the relativistic Lorentz coordinate transformation relations (named for H. A. Lorentz, who first formulated them in 1903). To simplify their form we suppose: (1) that XYZ and $X'Y'Z'$ are mutually parallel; (2) that the velocity **v** of S' relative to S is along the X (or X') axis; and (3) that when the two origins coincide, both clocks read zero (that is, $x = x' = 0$, when $t = t' = 0$).

A detailed analysis, based on the assumed constancy of c, yields for the Lorentz coordinate transformations†

$$x' = \frac{x - vt}{\sqrt{1 - (v/c)^2}}$$
$$y' = y$$
$$z' = z \quad (34\text{-}4)$$
$$t' = \frac{t - (v/c^2)x}{\sqrt{1 - (v/c)^2}}$$

These equations give the four primed coordinates on the left in terms of the unprimed coordinates on the right. To find the inverse equations one could, of course, solve the equations above for x, y, z, and t. But another consideration will yield these equations immediately, with no algebraic manipulation: If the velocity of S' relative to S is v, the velocity of S relative to S' is $-v$. Therefore, the inverse equations are arrived at simply by interchanging primed and unprimed coordinates in (34-4) and, at the same time, replacing v wherever it appears by $-v$. On this basis, the first of the four relations in (34-4) becomes $x = (x' + vt')/[1 - (v/c)^2]^{1/2}$.

We can easily see that the above equations yield, as they must, the familiar relations applicable at low speeds—the so-called galilean transformation relations (Sec. 6-3)—by taking the low-speed limit, $v/c \to 0$ (or, equivalently, by imagining the speed of

---

†For a derivation of the Lorentz transformations from fundamental principles see R. T. Weidner and R. L. Sells, *Elementary Modern Physics*, alternate 2d ed., Sec. 2-5, Allyn and Bacon, 1973.

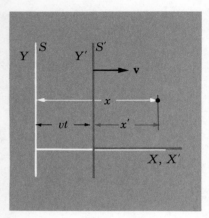

FIG. 34-6. In the low-speed limit of the Galilean coordinate transformations the coordinates x and x' in reference frames S and S', respectively, are related by $x' = x - vt$.

light to be effectively infinite). Equations (34-4) then become

$$x' = x - vt \qquad y' = y \qquad z' = z \qquad \text{and} \qquad t' = t \qquad (6\text{-}7), (34\text{-}4a)$$

The only nontrivial equation is the first, and its meaning is clear from Fig. 34-6. The galilean transformations express in formal, mathematical terms, the assumed absolute character of space and time in classical physics.

In general, however, the Lorentz transformations are characterized by the appearance of the factor $\sqrt{1 - (v/c)^2}$, which remains real, rather than imaginary, only if $v < c$. Since space and time coordinates are meaningful only if real, this implies that the speed of light c is not only the unique speed for the propagation of electromagnetic radiation, but also the upper speed limit of the universe. No particle or signal can exceed this speed.†

The time-dilation and space-contraction relations can, of course, be derived directly from (34-4); see Probs. 34-13 and 34-12, respectively.

Note the striking implication contained in the fourth of Eqs. (34-4), that for time. We see that $t' \neq t$, not merely because of the denominator (which relates to the time-dilation effect), but also because of the presence in the numerator of the term $-(v/c^2)x$. This says, in effect, that the time t' depends not only on the time t and on the speed ratio v/c, but also on the *space coordinate x*. Time and space are intermingled in relativity into what can rightly be called a four-dimensional space-time continuum. A fundamental consequence—one which Einstein considered to be the most profound—is the *relativity of simultaneous events*.

It goes like this: Suppose that one observer S finds that two spatially separated events, say two explosions, are simultaneous when recorded on his clock. Then these same two events *cannot* be simultaneous when recorded on the clock of another observer S' in motion; the two explosions, relative to this second observer, occur in sequence. More specifically, suppose that events A and B are simultaneous for observer S, both taking place at $t = 0$ and at the respective locations $x = 0$ and $x = x$. Then the times t' for events A and B as observed by S' may be computed from (34-4) and are given below:

Event	Relative to S		Relative to S'
	x	t	t'
A	0	0	0
B	x	0	$-\dfrac{(v/c^2)x}{\sqrt{1-(v/c)^2}}$

---

†The theory of relativity does not preclude the existence of particles, called *tachyons*, whose speed *always exceeds* c. Although the possible existence of such particles has been conjectured, no experimental evidence of their existence has yet been produced.

Event B occurs at a negative t'; event A occurs at $t' = 0$, which is to say, relative to S' event B comes first, then A.

34–5 RELATIVISTIC VELOCITY RELATIONS

In nonrelativistic, classical physics the rule for relating velocities of an object observed in two different reference frames having a constant relative velocity is simple: The velocity vectors add as vectors (Sec. 6-3). Even more simply, when all velocities lie along a single line, one merely treats the velocities as algebraic quantities. Suppose, for example, that a particle is moving east at 24 m/s relative to the earth. The same particle is observed from a train traveling west at 15 m/s. Then, relative to an observer in the train, the particle's velocity is $24 + 15 = 39$ m/s east. In more formal terms, the rules for combining velocities much less than c are based on an assumed absolute character for space and time, and on the galilean transformation relations which express these assumptions in mathematical language.

But, as we have seen, space and time intervals are not absolute, and the space and time coordinates for events recorded by observers in two different reference frames are related by the Lorentz transformation relations. Therefore, the velocity transformation relations applicable at all speeds and compatible with the requirements of relativity theory can be derived from the Lorentz coordinate transformation equations.

We denote the X component of the velocity observed in reference frame S by $\dot{x}$. By definition $\dot{x}$ is the ratio, displacement Δx divided by the corresponding time interval Δt, where Δx is, so to speak, measured on the meter stick of S and Δt is registered on his clock. The X component of the velocity observed in reference frame S' is denoted $\dot{x}'$, where by definition $\dot{x}' \equiv \Delta x'/\Delta t'$. Here $\Delta x'$ is the displacement measured by the meter stick of observer S' and $\Delta t'$ is the time interval registered on his clock.

In taking the differential, here indicated by Δ, of both sides of Eqs. (34-4) we have

$$\Delta x' = \frac{\Delta x - v\, \Delta t}{\sqrt{1 - (v/c)^2}}$$

$$\Delta y' = \Delta y$$

$$\Delta z' = \Delta z$$

$$\Delta t' = \frac{\Delta t - (v/c^2)\, \Delta x}{\sqrt{1 - (v/c)^2}}$$

We now divide each of the first three equations above by the fourth equation to arrive at the velocity relations:

$$\frac{\Delta x'}{\Delta t'} = \frac{(\Delta x/\Delta t) - v}{1 - (v/c^2)(\Delta x/\Delta t)}$$

$$\frac{\Delta y'}{\Delta t'} = \frac{(\Delta y/\Delta t)\sqrt{1 - (v/c)^2}}{1 - (v/c^2)(\Delta x/\Delta t)}$$

$$\frac{\Delta z'}{\Delta t'} = \frac{(\Delta z/\Delta t)\sqrt{1 - (v/c)^2}}{1 - (v/c^2)(\Delta x/\Delta t)}$$

These equations become simpler when we use the definitions $\dot{x}' \equiv \Delta x'/\Delta t'$, $\dot{x} \equiv \Delta x/\Delta t$; $\dot{y}' \equiv (\Delta y'/\Delta t')$, etc.:

$$\dot{x}' = \frac{\dot{x} - v}{1 - (v/c^2)\dot{x}}$$

$$\dot{y}' = \frac{\dot{y}\sqrt{1 - (v/c)^2}}{1 - (v/c^2)\dot{x}} \qquad (34\text{-}5)$$

$$\dot{z}' = \frac{\dot{z}\sqrt{1 - (v/c)^2}}{1 - (v/c^2)\dot{x}}$$

Equations (34-5) are the Lorentz transformation equations for velocity components.† One surprising result implicit in (34-5) is that the Y and Z components of the velocity of a particle as measured in S' depend on the X component of the velocity as measured in S! Equations 34-5 give the primed quantities (left side) in terms of the unprimed quantities (right side). As in the case of the coordinate transformation equations, we can arrive at the inverse transformation equations (giving unprimed quantities on the left in terms of primed on the right) simply by interchanging primed and unprimed quantities and, at the same time, replacing v wherever it appears by $-v$.

Example 34-1

Show that the Lorentz velocity transformation relations yield, as they must, the result that all observers measure the speed of light to be c.

For simplicity we suppose that a beam of light is directed along the X axis and that in reference frame S the velocity components are, therefore,

$$\dot{x} = c \qquad \dot{y} = 0 \qquad \dot{z} = 0$$

From Eqs. (34-5) we then have

$$\dot{x}' = \frac{\dot{x} - v}{1 - (v/c^2)\dot{x}} = \frac{c - v}{1 - (v/c^2)c} = \frac{c - v}{c - v} c = c$$

$$\dot{y}' = 0$$

$$\dot{z}' = 0$$

---

†Note that for speeds sufficiently small compared to c we can imagine $v/c \rightarrow 0$, and the Lorentz velocity transformation equations reduce, as they must, to the familiar galilean transformation equations for velocity components:

$$\dot{x}' = \dot{x} - v \qquad \dot{y}' = \dot{y} \quad \text{and} \quad \dot{z}' = \dot{z} \qquad (6\text{-}9)$$

Observer S', traveling at *any* velocity v (less than c) also measures the speed of the light to be c.

Example 34-2
A particle moves east at the speed 24×10^7 m/s ($= 0.8c$) relative to the earth. What is this particle's speed as measured by an observer in a spaceship traveling west relative to the earth at the speed 15×10^7 m/s ($= 0.5c$). (The circumstances here are like those given in the introductory paragraph of this section, but with the speeds increased by a factor 10^7.)

Let S be a reference frame attached to the earth, and S' a reference frame attached to the spaceship. We are then given that

$$\dot{x} = 0.8c$$
$$v = -0.5c$$

Therefore from (34-5) we have

$$\dot{x}' = \frac{\dot{x} - v}{1 - (v/c^2)\dot{x}} = \frac{(0.8 + 0.5)c}{1 + (0.5)(0.8)} = \frac{1.3c}{1.4} = 0.93c$$

or
$$\dot{x}' = 28 \times 10^7 \text{ m/s}$$

Note that whereas the (inapplicable) classical velocity combination rule would have yielded the observed particle speed as $0.8c + 0.5c = 1.3c$, a speed *in excess* of the speed of light, the relativistic relation ensures that the particle speed not exceed c (here, 93 percent of c).

SUMMARY

The fundamental postulate of the theory of relativity, that the speed of light through vacuum is the same for all observers, quite apart from the motion of the light source or of the observer, is a fact of nature.

The dilated time-interval T is related to the proper-time interval T_0 by

$$T = \frac{T_0}{\sqrt{1 - (v/c)^2}} \qquad (34\text{-}2)$$

where v is the speed of an observer measuring T relative to the single location where the two events marking the beginning and ending of T_0 take place.

The space-contracted length L is related to the proper length L_0 by

$$L = L_0 \sqrt{1 - \left(\frac{v}{c}\right)^2} \qquad (34\text{-}3)$$

where L is measured by an observer traveling at speed v relative to a reference frame in which the length is L_0 along the line of relative motion.

The velocity components measured in S ($\dot{x}$, $\dot{y}$, $\dot{z}$), which travels along the X direction at speed v relative to $S(\dot{x}', \dot{y}', \dot{z}')$, are given by the Lorentz (relativistic) velocity transformation relations:

$$\dot{x}' = \frac{\dot{x} - v}{1 - (v/c^2)\dot{x}}$$

$$\dot{y}' = \frac{\dot{y}\sqrt{1 - (v/c)^2}}{1 - (v/c^2)\dot{x}} \qquad (34\text{-}5)$$

$$\dot{z}' = \frac{\dot{z}\sqrt{1 - (v/c)^2}}{1 - (v/c^2)\dot{x}}$$

PROBLEMS

34-1. To see how the Michelson-Morley experiment would reveal the existence of a unique reference frame, or ether, for the propagation of light, *(a)* suppose that an observer in motion relative to the ether sends a light pulse along a tube, as in Fig. 34-3, oriented at right angles to the direction of motion of the tube relative to the ether, and show that the round-trip travel time for a tube of length H in this orientation is $(2H/c)/[1 - (v/c)^2]^{1/2}$. *(b)* Suppose that the tube is then reoriented so that it lies along the observer's X axis, that is, along the line of relative motion relative to the ether. Then the speed of the light pulse along the tube is $c + v$ in the "downstream" direction relative the ether and $c - v$ in the "upstream" direction. Show that the round-trip travel time is now given by $(2H/c)/[1 - (v/c)^2]$. *(c)* Show that the difference in round-trip travel time intervals for parts *(a)* and *(b)* is approximately Hv^2/c^3. *(d)* Show that if speed v is 3.0×10^4, the orbital speed of the earth about the sun, the *fractional* difference in round-trip travel times is equal to 5 parts in 10^9 (a detectable change in the original Michelson-Morley experiment).

34-2. Assume that in the Michelson-Morley experiment (see Prob. 34-1) the length of the cylinders is 10.0 m and the wavelength of the light 5000 Å. *(a)* If the speed of the earth through the ether is 3.0×10^4 m/s, what fraction of a wavelength shift will there be between the two beams when they recombine? *(b)* At what speed must the apparatus move through the ether to have the two beams recombine 180° out of phase with one another?

34-3. Unstable muons, having a half-life of 1.52×10^{-6} s in the rest frame of the muons, are passing an observer at the speed $0.98c$. How far do the muons travel relative to this observer before their number drops to *(a)* one-half and *(b)* one-quarter of their initial number?

34-4. At what speed must a clock be in motion relative to an observer if the ticks of the moving clock are to be observed by him (and other observers at rest in his reference frame) to come every second rather than every half second?

34-5. *(a)* At what speed must a clock travel, relative to an earth observer, to run slow by 0.50 s in 1 yr (3.16×10^7 s)? *(b)* Compare this speed to that of a satellite orbiting earth close to its surface at 7,900 m/s.

34-6. A direct experimental test has been made of the twin paradox through the use of unstable muons sent at high speed around an accelerator ring and returned to their starting point. It was found that the traveling muons lived 12 times longer than muons at rest. What was the speed of the traveling muons?

34-7. A spacecraft must travel 6×10^{12} m to reach home. The life-support systems within the spacecraft will last an additional 20 hours. What is the minimum speed of the spacecraft relative to the home planet if the crew is to survive the trip?

34-8. Twin A stays at home, while identical twin B goes on a round trip always at speed $0.98c$ to a point 10 light-years distant (one light-year is the distance traversed by light in a period of one year). Twins A and B are both 20 years old when B departs. *(a)* What is A's age when B arrives back home? *(b)* What is B's age at this time?

34-9. In a recent experimental test of the time-dilation phenomenon an atomic clock carried on a jet plane flying on a coast-to-coast round trip was compared with the time interval registered on an atomic clock which remained at rest on earth. Find the approximate number of seconds by which the traveling clock is behind the stationary clock for a one-way trip distance of 4,000 km for a plane traveling at a constant speed of 1,000 km/s. (In actuality the computation is more complex by virtue of the fact that the earth is in orbital motion about the sun and in rotational motion about its axis.)

34-10. Relative to a high-speed observer moving toward it a sphere appears squashed because of the space contraction phenomenon. At what speed must an observer approach the earth if its "thickness," measured along the direction of travel of the observer, is one-fifth of its diameter?

34-11. A rigid rod is oriented at an angle of θ_1 with respect to the X_1 axis. What is the rod's orientation θ_2 as measured by an observer traveling at a speed v along the positive X_1 axis?

34-12. Derive the space-contraction relation (34-3) from the Lorentz coordinate transformation relations (34-4). (*Hint:* Consider the first of Eqs. (34-4), which can be written in differential form as $\Delta x' = (\Delta x - v\,\Delta t)/[1 - (v/c)^2]^{1/2}$. Recall that the length of a moving object must be measured by marking both ends of the object *simultaneously*. If Δt is taken to be zero, which of $\Delta x'$ and Δx is L_0 and which is L?)

34-13. Derive the time-dilation relation (34-2) from the Lorentz coordinate transformation relations (34-4). (*Hint:* Consider the fourth of Eqs. (34-4), which can be written in differential form as $\Delta t' = [\Delta t - (v/c^2)\,\Delta x]/[1 - (v/c)^2]^{1/2}$. Recall that the proper time interval T_0 is measured in a reference frame in which the events marking the beginning and ending of the interval take place at the *same location*. If Δx is taken to be zero, which of Δt and $\Delta t'$ is T_0 and which is T?)

34-14. A luminous cube has an edge length L_0 when at rest. A camera takes a photograph of the cube when the cube is traveling at a high speed v relative to it. It is far from the cube, which moves at right angles to the line joining cube and camera; see Fig. P34-14a. We concentrate on the light that comes from cube edges A, B, and C and enters the camera during the very short time the shutter is open for the photographic image. Light pulses that leave points A, B, and C simultaneously (in the reference frame of the camera) do *not* reach the camera simultaneously; light has a finite speed, and it travels farther from A than from B and C and, consequently, during the brief moment that the camera shutter is open light from the farther edge A enters the camera at the same time as light that has been emitted from edges B and C at a

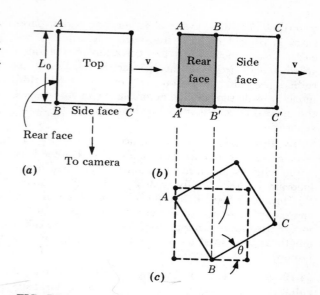

FIG. P34-14. A moving cube being photographed at right angles to its direction of motion: *(a)* view from above; *(b)* the photograph of the side of the cube; *(c)* an interpretation of *(b)* in which the cube has been rotated.

later time. The developed photograph appears as shown in Fig. P34-14b; the rear face AB (rear with respect to the motion of the cube) is visible, while the side face BC is shortened by space contraction. Show that a person examining such a photograph might infer that the cube was not moving with high speed at right angles to the camera but, rather, was at rest after being rotated through an angle θ, as shown in Fig. P34-14c, where $\sin\theta = v/c$. This illustrates a general effect; a high-speed object moving transverse to the observation point is seen to be *rotated*.

34-15. What is wrong with the following argument: Meter stick B is in motion relative to meter stick A, so that the length of B is contracted relative to A. But meter stick A is also in motion relative to B, so that A is contracted relative to B. How can meter stick A be both longer and shorter than meter stick B? (*Hint:* What requirement must be met in properly measuring the length of a moving object? Are two spatially separated events simultaneous for all observers?)

34-16. Solve algebraically for x, y, z, and t in terms of x', y', z', and t' in Eqs. (34-4). This proves the statement that the inverse Lorentz transformations result from changing v to $-v$ and interchanging subscripts.

34-17. Two spaceships moving toward one another on a head-on collision course are separated by a distance of 4.0×10^{10} m, according to an observer on earth. At this time, according to the observer, one of the spaceships is coincident with the earth. Relative to the earth system both ships are traveling at the same speed, 0.98c. (a) How much later do the two ships collide, according to the observer on earth? (b) How much later do they collide according to an observer on the spaceship coincident with the earth at the initial separation?

34-18. Protons leave a particle accelerator at a speed of 0.8c to enter and travel through an evacuated tube 1.0 m long, as measured by an observer at rest in the laboratory, and finally reach a detecting device. (a) How long does it take for a proton to travel from one end of the tube to the other according to an observer traveling with a proton? (b) What is the tube's length as measured by an observer traveling with a proton?

34-19. A spacecraft is moving relative to a nearby galaxy at a speed of 0.98c. Observers within the spacecraft find that it takes 6×10^{-8} s for the spacecraft to pass a small space marker at rest relative to the galaxy. (a) As measured by observers riding with it, what is the overall length of the spacecraft? (b) What is the length of the spacecraft as measured by observers at rest with respect to the space marker?

34-20. A fast train traveling at a speed of 0.8c passes two posts fixed to the ground and adjoining the train tracks. The posts are separated by a distance of 125 m, measured by an observer on the ground. Observers on the ground find that the front and back ends of the train coincide simultaneously with the two posts as the train passes. (a) What is the length of the train according to an observer traveling in the train? (b) How long does it take, according to an observer on the ground, for the train to pass one post? (c) What is the time interval, measured by an observer standing at one point in the train, between the passing of the first and second posts? (d) What is the distance between the posts according to observers riding in the train? (e) The coincidence of the ends of the train with the two posts was simultaneous from the viewpoint of an observer fixed on the ground. Which end of the train, front or back, first coincided with a post, as according to an observer on the train?

34-21. A long, fast spaceship passes an observer stationed at a post fixed to the earth. The spaceship travels at the speed 0.80c relative to the observer at the post, who notes that the back end of the spaceship passes the post 4.0×10^{-5} s after the front end. (a) What is the time interval, relative to observers at rest within the spaceship, elapsing between the instant when the front end of the spaceship coincides with the post and the later instant when the back end of the spaceship coincides with the post? (b) How long is the spaceship as measured by observers traveling with it? (c) What is the length of the spaceship as measured by observers fixed to earth?

34-22. Monochromatic light passes through a single slit and produces a diffraction pattern on a distant screen. The light source, slit, and screen are all at rest with respect to a first observer, who marks the locations of intensity zeros on the screen. Suppose that a second observer is in motion at a very high speed along the line in which the light travels from the source to the slit. (a) Will this observer find the intensity zeros to come at the marks made by the first observer? (b) Is the distance between the slit and observation screen changed? Now a third observer travels at a very high speed along the sheet containing the slit. (c) Does this observer find the intensity zeros at the locations marked by the first observer? (d) Is the slit width changed for the third observer?

34-23. A muon is an unstable particle with an average lifetime of 2.20×10^{-6} s (from the moment of its creation until it decays) as measured by an observer at rest with respect to it. (a) If an average muon travels a distance of 600 m during one lifetime, according to an observer in the laboratory, what is the muon's speed? (b) How long did this muon live according to an observer in the laboratory? (c) How far did the muon travel between birth and death, according to an observer traveling with the muon?

34-24. Train A travels east at 0.80c relative to a station; train B travels north at 0.80c relative to the same station. Find the velocity of train A relative to train B.

34-25. A fast train moves at the speed 0.8c relative to earth. Inside the train a fast runner moves at a speed 0.8c relative to the train. Assume that the directions of the two velocities are the same. What is the speed of the runner relative to earth?

34-26. In Example 34-1 it was shown that if observer S shines a beam of light along his X axis, any observer S' also measures the speed of light to be c along the X' axis. Suppose, instead, that the beam of light shines along the positive Y axis in the reference frame of S. (a) Show that the speed measured by any other observer S' is also c. (b) In what direction does the light beam travel relative to the positive Y' axis of observer S'?

34-27. The linear density (number per unit length) of electrons in a certain wire conductor is λ_0 when no electric current exists in the wire and the electrons can be imagined to be at rest. When a current is established in the wire, the electrons drift along the wire at the drift speed v. Find the linear density λ of the drifting electrons in the wire in terms of λ_0, v, and c, as measured by an observer at rest with respect to the wire.

34-28. A rod 1.25 m long is to pass through a window 1.00 m wide. More precisely, the rod of rest length 1.25 m, in motion as shown in Fig. P34-28, passes through an opening that is 1.00 m between edges when measured by an observer at rest with respect to the window. The rod's speed is such that the two ends of the rod coincide simultaneously with the two window edges, according to an observer in the reference frame in which the window is at rest. (a) What is the speed of the rod? (b) Now consider events from the point of view of an observer traveling with the rod. What is the width of the window opening relative to this observer? (c) From the point of view of an observer traveling with the rod, do the two rod ends coincide simultaneously with the two window edges and, if not, which rod end, the leading or the trailing one, first coincides with a window edge? (d) What is the time interval between the two coincidences as measured by an observer traveling with the rod?

34-29. (a) As seen by observer S, one explosion occurs at the location $x = 0$ at the time $t = 0$. A second explosion occurs at a nearby location a little later, at $x = 1.0$ km and $t = 1.0 \times 10^{-6}$ s. What must be the velocity of a second observer S' relative to S who observes both explosions to occur simultaneously? (b) A second set of explosions is seen by observer S' as follows: one explosion at $x' = 0$ and $t' = 0$ and a second at $x' = 1.0$ km and $t' = 1.0 \times 10^{-6}$ s. Where and when are these events as observed by S?

34-30. There are some physical quantities which, within an isolated region of space, are conserved; that is, when viewed from some inertial system, the amount of the physical quantity within that region always remains the same. A familiar example of a conserved quantity is electric charge. Consider an isolated volume in which two equal, but opposite, charges are separated by a distance d and are at rest. Now, in this inertial system there is no violation of the conservation of the total charge if the two charges disappear, are annihilated, *simultaneously*. At any instant of time the total charge within the volume is always the same: namely, zero. Show that, if total charge is to be conserved *in all inertial systems,* the two charges must be together ($d = 0$) at the time of annihilation. This is an example of a general principle concerning any conserved quantity; if a quantity is to be conserved in all inertial systems, then it must be locally conserved (conserved within any arbitrarily small volume).

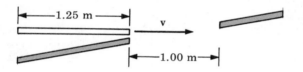

FIG. P34-28. Rod passing through a window.

CHAPTER 35

Relativistic Dynamics: Momentum and Energy

In the last chapter we were concerned with the strictly kinematical aspects of special relativity—how the constancy of the speed of light implies radical changes in the very concepts of space and time and in the associated kinematical relations. In this chapter we extend our treatment to relativistic dynamics; in particular, we wish to find the appropriate forms for such basic dynamical quantities as momentum, mass, force, and energy, which, like the kinematical quantities, also differ in significant ways from their classical counterparts.

35–1 THE INVARIANCE OF THE LAWS OF PHYSICS

The constancy of c—a postulate of special relativity theory and a firmly established experimental fact—implies that the rules for combining velocities are altered in relativistic physics. But such basic mechanical quantities as the momentum and kinetic energy of a particle involve the particle's velocity, so that these quantities are also altered for high speeds. How does one find their proper relativistic forms?

The starting point is a second postulate of special relativity theory. This postulate, which is also fundamental to classical mechanics, is often unstated or ignored, not because it is so subtle, but because it is so transparently obvious. It is this: *The laws of physics are the same in all inertial frames of reference.* Stated more formally, the laws of physics are invariant† in all inertial frames.

---

†Sometimes it is said that the speed of light is *invariant*, that is, it has the *same magnitude* for all observers, while the laws of physics are *covariant*, that is, they have the same *mathematical form* for all inertial observers.

An inertial frame is, of course, a frame of reference in which the law of inertia, or Newton's first law of motion, holds: In an inertial frame an undisturbed object has a constant velocity.† Now the invariance of the laws of physics for all inertial frames means simply this: If such fundamental propositions as Newton's laws of motion, the conservation of momentum principle, and the conservation of energy principle are to be truly *laws* of physics—propositions that are universally valid—then these propositions cannot depend on the particular inertial frame in which they are applied. One inertial frame must be just as good as any other.

Let us see what this means when expressed in more specific terms. Suppose that the collisions at low speeds between elastic spheres in a billiard game are studied. One observer finds that the momentum conservation principle and the energy conservation principle describe such collisions. Then *all* observers in inertial frames must agree that momentum and energy are conserved, quite apart from the observers' states of relative motion. So, if the billiard table is in a train moving in a straight line at constant speed relative to the earth, an observer in the train will say that momentum and energy are each conserved. And so will an observer at rest on earth who studies the very same collisions, even though, for this second observer, the actual speeds of the billiard balls before and after the collisions will differ from the corresponding values measured by the observer in the train.

It is a simple matter to demonstrate that the momentum-conservation principle of classical mechanics applies equally well in all inertial frames when we use the transformation relations (the galilean transformations, Sec. 6-3) that describe adequately low-speed phenomena. Suppose that a head-on collision occurs between particles m and M. In inertial frame S, the velocities of particle m before and after the collision are, respectively, $\dot{x}_b$ and $\dot{x}_a$, while the corresponding velocities for particle M are $\dot{X}_b$ and $\dot{X}_a$.

Then the momentum conservation principle, according to which the total momentum of an isolated system is the same before and after any collision, when applied to this collision by observer S, yields

$$m\dot{x}_b + M\dot{X}_b = m\dot{x}_a + M\dot{X}_a \qquad (35\text{-}1)$$

Relative to observer S', who is in motion at the constant velocity v along the X direction relative to S, the particle velocities are

$$\dot{x}'_b = \dot{x}_b - v \qquad \dot{X}'_b = \dot{X}_b - v$$
$$\dot{x}'_a = \dot{x}_a - v \qquad \dot{X}'_a = \dot{X}_a - v$$

where we have used the simple algebraic galilean velocity transformations [Eq. (6-9)] which are applicable at low speeds. See Fig. 35-1.

---

†The *special* theory of relativity is restricted to *inertial* frames of reference; the *general* theory of relativity includes *accelerated* reference frames as well.

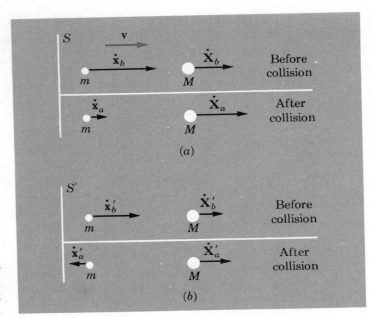

FIG. 35-1. Head-on collision of two particles as viewed by two observers moving with constant relative velocity v.

Substituting the equations above into (35-1) we have then

$$m(\dot{x}'_b + v) + M(\dot{X}'_b + v) = m(\dot{x}'_a + v) + M(\dot{X}'_a + v)$$

or
$$m\dot{x}'_b + M\dot{X}'_b = m\dot{x}'_a + M\dot{X}'_a \qquad (35\text{-}2)$$

The terms involving v cancel, and we find that (35-2) is identical in mathematical form to (35-1); one equation has primed velocities (for observer S'), the other does not (for observer S). That is to say, observer S' and observer S agree that the total momentum before collision equals total momentum after collision.

What has been demonstrated, then, is that the classical-momentum conservation principle is an acceptably good law of physics at low speeds: It applies equally to all inertial observers. Exactly analogous arguments show that Newton's second law and the classical energy conservation principle are also invariant under a galilean velocity transformation. In short, all of classical mechanics meets the basic requirement of invariance of the laws of mechanics. But we know that only the Lorentz transformation equations properly describe the interrelationships between space intervals, time intervals, and velocities observed in inertial frames in relative motion at *any* speed. Surely the *classical* forms of the basic mechanical laws of physics cannot be invariant when the universally correct *relativistic* transformation relations are used.

Our general strategy in what follows will be this: To find the appropriate form of relativistic dynamical quantities we apply the two basic postulates of relativity theory, which are (1) *The speed of light is the same constant for all observers* (with the associated consequences for space, time, velocity, and transformation relations) and (2) *the laws of physics are the same for all inertial observers.*

35-2 RELATIVISTIC MASS AND MOMENTUM

We first consider the relativistic momentum **p** of a particle which, by analogy with the classical relation for momentum, we take to be given by

$$\mathbf{p} = m\mathbf{v} \tag{35-3}$$

where m is the so-called relativistic mass. *By definition,* then, a particle's relativistic mass m is taken to be that physical quantity by which the particle's velocity **v** must be multiplied to yield the vector quantity **p** such that the total momentum $\Sigma \mathbf{p}$ of an isolated system is conserved. Although the relativistic form for momentum can be expected to differ from the classical relation, we know that for low speeds the relativistic momentum *must* reduce to the familiar form $m_0 v$, where m_0 represents the particle's mass at zero speed or, at least, a speed v much less than c, the speed of light. More generally, relativistic dynamics *must,* for particle speeds much less than c, reduce to classical Newtonian dynamics.

Our specific strategy is as follows: We first consider an elastic collision between two identical objects which is so completely symmetrical that we may be assured that the law of momentum conservation is valid. We then examine the same collision from the point of view of another inertial observer and require that the momentum-conservation law also describe the collision from this second observer's point of view.

Suppose that we have two particles A and B with the same mass m_0 when compared at rest. The particles are fired at one another in opposite directions at equal speed and they collide elastically. See Fig. 35-2a. The velocities of the particles are equal in magnitude but opposite in direction both before and after the collision. Because of the symmetry of the collision, we can be sure that the collision is observed in the center-of-mass reference frame, the inertial frame in which the system's total momentum is zero at all times (Sec. 6-5). The collision appears even more symmetrical if we rotate the coordinate axes, as shown in Fig. 35-2b, so that the velocities of both particles make the *same* angles with respect to the X_{CM} and Y_{CM} axes. Take the velocity component along X_{CM} of either particle to have the magnitude v_x. Then we can imagine particle A as having been thrown and then caught by an observer, whom we shall also call A, who travels to the right at the speed v_x and at a distance $\frac{1}{2}y$ below the X_{CM} axis. Similarly, we can imagine particle B to have been thrown and then caught by an observer B, who travels to the left at the speed v_x and at a distance $\frac{1}{2}y$ above the X_{CM} axis.

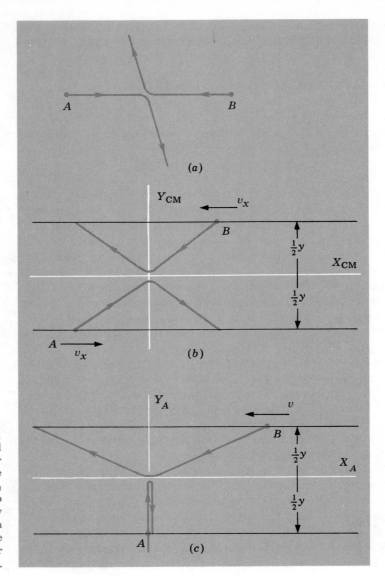

FIG. 35-2. Identical particles A and B colliding elastically at equal speeds: *(a)* viewed by an observer in the center-of-mass reference frame; *(b)* same as *(a)* but with X_{CM} and Y_{CM} axes so oriented as to make the collision completely symmetrical for the observer in the center-of-mass reference frame; *(c)* viewed by an observer moving along the X axis with A.

A third observer, one fixed with respect to the $X_{CM}Y_{CM}$ inertial frame and therefore different from observers A or B, can then assert, simply on the basis of symmetry, that both particles A and B were thrown *simultaneously*, collided at the origin, and then were caught *simultaneously* by the respective observers. In other words, the *time interval T_{CM}, as observed in the center-of-mass reference frame*, between the throwing and the catching of particle A is the *same* as that between the throwing and the catching of particle B.

Moreover, an observer in this center-of-mass reference frame can be certain that momentum is conserved in the collision; indeed, the total vector momentum of the two-particle system along any one direction must be zero. This implies that the change Δp_{yA} in the momentum component along the Y_{CM} direction for particle A before and after collision must equal in magnitude the change Δp_{yB} in the momentum component along Y_{CM} for particle B. Thus, the law of momentum conservation requires that

$$\Delta p_{yA} = \Delta p_{yB} \tag{35-4}$$

Particle A travels a distance $\tfrac{1}{2}y$ along Y_{CM} before, and another $\tfrac{1}{2}y$ after, collision; all told, it travels a distance y in the time interval T_{CM}, so that the Y component of A's velocity is y/T_{CM}. The momentum *change* along Y_{CM} is twice the magnitude of the momentum component along Y_{CM} before or after collision. Therefore,

$$\Delta p_{yA} = \frac{2m_A y}{T_{\text{CM}}}$$

where m_A is the relativistic mass of particle A. In similar fashion

$$\Delta p_{yB} = \frac{2m_B y}{T_{\text{CM}}}$$

where m_B is the relativistic mass of particle B. Equation (35-4) then becomes, for an observer in the center-of-mass inertial frame,

$$\frac{m_A y}{T_{\text{CM}}} = \frac{m_B y}{T_{\text{CM}}}$$

Because of the symmetry of the collision we know that the two particles have the same relativistic mass: $m_A = m_B$.

Now if relativistic momentum conservation is to hold for *all* inertial frames, then (35-4) must be satisfied for any other inertial frame moving at constant velocity relative to that of the system's center of mass. For convenience we now examine the collision from the inertial frame in which observer A is at rest; see Fig. 35-2c. The differences are as follows. Observer A now sees his particle A traveling back and forth along the Y_A axis while particle B travels obliquely before and after the collision. We denote the velocity of observer B relative to observer A by **v**. Observer A, stationed at a point $\tfrac{1}{2}y$ below the X_A axis, sees observer B traveling to the left at speed v and at a distance $\tfrac{1}{2}y$ above the X_A axis. Note that the transverse distance y between the two observers A and B is the same in the new $X_A Y_A$ reference frame as in the earlier $X_{\text{CM}} Y_{\text{CM}}$ reference frame, since there is *no space contraction* at right angles to the direction of relative motion.

As measured by observer A, however, the time intervals between the tossings and the catchings of the two particles are *not* the same: particle A is thrown and caught at the *same* location in reference frame $X_A Y_A$, but particle B is thrown from a location on the right and later caught at another location on the left. Recall this general relativistic feature of time intervals: If two events occur at the same location in one observer's reference frame, then the time interval between them is the proper interval T_0; but if these same two events are observed from a reference frame traveling at a speed v, then the time interval between them is longer (dilated) and is given by

$$T = \frac{T_0}{\sqrt{1-(v/c)^2}} \qquad (34\text{-}2), (35\text{-}5)$$

Thus, if we take the time interval between the tossing and the catching of particle A as observed by A to be T_0, then the time interval between the tossing and the catching of particle B *as observed by* A is T, a *longer* interval.

Seen by observer A, the sequence of events is as follows. First B throws particle B, then A throws particle A, then the particles collide, then A catches particle A, and finally B catches particle B. We have an example here of the general rule that events that are simultaneous in one reference frame (the tossing and catching of particles A and B in the $X_{CM}Y_{CM}$ reference frame) are not simultaneous in a second reference frame in motion with respect to the first.

In the $X_A Y_A$ reference frame shown in Fig. 35-2c the momentum changes along Y_A are, then,

$$\Delta p_{yA} = \frac{2m_A y}{T_0} \quad \text{and} \quad \Delta p_{yB} = \frac{2m_B y}{T}$$

Substituting these relations in (35-4), which guarantees that momentum conservation still applies, yields

$$\frac{m_A y}{T_0} = \frac{m_B y}{T}$$

which, using (35-5), gives

$$m_B = \frac{m_A}{\sqrt{1-(v/c)^2}} \qquad (35\text{-}6)$$

where m_A and m_B are the respective masses of the particles observed in the $X_A Y_A$ reference frame, and v is the speed of observer B relative to observer A. Clearly, this equation cannot be satisfied unless m_A and m_B are different. Since the two masses were taken to be equal when measured at rest with respect to an observer, it is apparent, as we anticipated, that the relativistic mass of a particle must depend in some way on its speed.

In the collision shown in Fig. 35-2c *both* particles are in motion. To find out how the mass of a particle depends upon its motion with respect to an observer, we now consider the special collision occurring when particle A is at rest in the $X_A Y_A$ reference frame. That is, we suppose that the Y components of the speeds of both particles A and B approach zero. Particle A is then at rest with respect

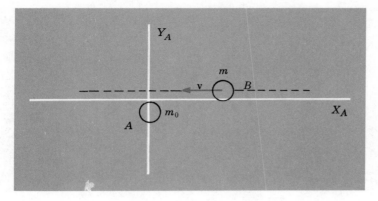

FIG. 35-3. Collision of moving particle B with particle A at rest, as viewed by observer A; the collision is derived from that shown in Fig. 35-1c in the limit of zero transverse velocity component.

to the reference frame $X_A Y_A$, and we label its mass m_0, the *rest mass*. An observer at rest in $X_A Y_A$ now sees particle B just making a grazing collision with particle A while particle B approaches and then recedes effectively along a single straight line. See Fig. 35-3. The speed of particle B is then just the speed v of observer B (with respect to which it is now at rest). When at rest, particle B's mass was also m_0, but now, when it is in motion at speed v relative to an observer in $X_A Y_A$, its mass is different; we label it m. For the grazing collision with $m_A = m_0$ and $m_B = m$, (35-6) becomes

$$m = \frac{m_0}{\sqrt{1 - (v/c)^2}} \qquad (35\text{-}7)$$

By the same token observer B sees his own particle B at rest with mass m_0 and particle A in motion at speed v with mass m. Any observer then finds that, if a particle has a rest mass m_0 when measured at rest with respect to him, its mass when in motion at speed v is $m_0[1 - (v/c)^2]^{-1/2}$.

Figure 35-4 shows a particle's relativistic mass plotted as a function of its speed, as given by (35-7). Clearly, m deviates markedly from m_0 only when a particle's speed is comparable to that of light; for example, with $v/c = \frac{1}{10}$ it exceeds m_0 by only 0.5 percent.

Now recall our basic definition of relativistic mass m. When it is said that a particle's mass depends on its speed, what is implied basically is that the relativistic momentum $\mathbf{p}$ of a particle with velocity $\mathbf{v}$ is given by the relation

$$\mathbf{p} = m\mathbf{v} = \frac{m_0 \mathbf{v}}{\sqrt{1 - (v/c)^2}} \qquad (35\text{-}8)$$

where m_0 is a constant, independent of speed. Experiments with particles traveling at speeds approaching that of light demonstrate conclusively that (35-8) correctly gives a particle's momentum at all speeds up to c. Of course, for low-speed particles, with $v/c \ll 1$, the relativistic momentum reduces to the classical form, $\mathbf{p} = m_0 \mathbf{v}$. Whereas in prerelativity physics space intervals, time intervals,

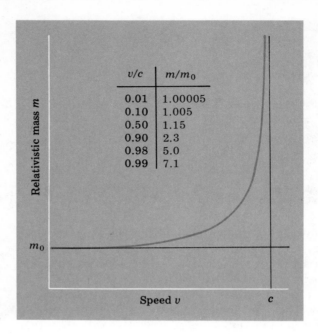

FIG. 35-4. Variation of relativistic mass with speed.

and mass were regarded as absolute and the speed of light was regarded as relative to some unique inertial frame (the ether), the relativity physics of Einstein requires that, because the speed of light is absolute and the laws of physics are invariant, space intervals, time intervals, and mass (defined as the momentum-velocity ratio) be relative and depend on the relative motion between the object and the observer.

In relativistic dynamics we take the force **F** on an object to be the time rate of change of its relativistic momentum. Therefore, Newton's second law is written as

$$\mathbf{F} = \frac{d\mathbf{p}}{dt} = \frac{d}{dt}(m\mathbf{v}) = m\frac{d\mathbf{v}}{dt} + \mathbf{v}\frac{dm}{dt} \qquad (35\text{-}9)$$

In classical mechanics, with the mass of any object constant, $dm/dt = 0$; then the two forms of Newton's second law, $\mathbf{F} = (d/dt)\,m\mathbf{v}$ and $\mathbf{F} = m\,d\mathbf{v}/dt = m\mathbf{a}$, are equivalent. In relativistic dynamics, however, the two forms usually are not equivalent; the mass m varies with the particle's speed, so that if the *speed* changes with time, so does the mass and $dm/dt \neq 0$. The general form, $\mathbf{F} = d\mathbf{p}/dt$, is always correct.

One very important situation, in which a particle's speed is constant even though its velocity changes, is that of a particle traveling in a circular arc. The particle must be under the influence of a force toward the center of the arc. Because the force is radial and at right angles to the particle's motion, the particle travels at constant speed but with a continuously changing velocity direction.

Consider a particle of relativistic mass m and electric charge Q moving with velocity **v** at right angles to a uniform magnetic field **B**. The magnetic force, which is perpendicular to the velocity and therefore causes the particle to move in a circle, has the magnitude

$$F = QvB \qquad (26\text{-}3)$$

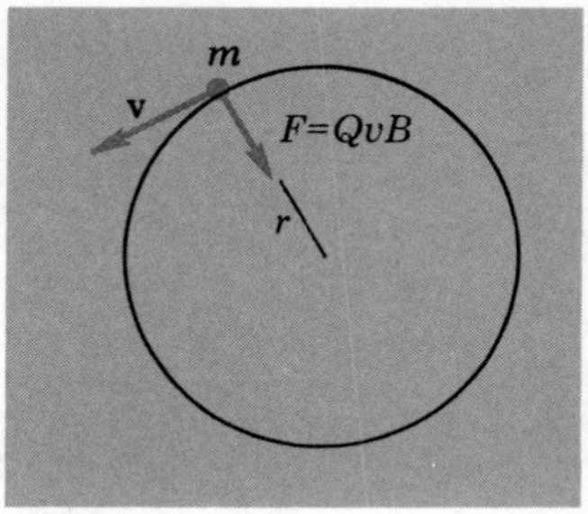

Because the particle moves with a constant speed, its relativistic mass is constant, and hence $dm/dt = 0$. Equation (35-9) then becomes $\mathbf{F} = m\, d\mathbf{v}/dt = m\mathbf{a}$, where **a** is the centripetal acceleration with a magnitude v^2/r, and where r is the radius of the circular path. See Fig. 35-5. Using the relation for the magnetic force, we have

FIG. 35-5. Motion of charged particle in a uniform transverse magnetic field.

$$QvB = \frac{mv^2}{r}$$

$$p = mv = QBr \qquad (35\text{-}10)$$

It must be especially noted that, although this equation is of exactly the same form as that obtained by the methods of classical mechanics, the mass m appearing in it is the *relativistic* mass, not the rest mass.

Equation (35-10) is the basis of a simple method of determining the relativistic momentum of a charged particle. For if Q is known, and B and r are measured, then p may be computed from the equation. Furthermore, a charged particle's velocity v can be measured by using crossed electric and magnetic fields (Sec. 26-4), so that its relativistic mass m can be computed from its measured momentum $p = mv$ and its measured velocity v. This method was first used by A. H. Bucherer in 1909 to verify the variation of relativistic mass with speed.

Example 35-1

A particle of rest mass m_0 initially moves at speed $0.40c$. If the particle's speed is doubled, how will its new momentum compare with its initial momentum?

The initial momentum is given by Eq. (35-8):

$$p_i = \frac{m_0 v_i}{\sqrt{1-(v_i/c)^2}} = \frac{0.40 m_0 c}{\sqrt{1-(0.40)^2}} = 0.44 m_0 c$$

When the speed is doubled, the particle moves at speed $v = 2(0.40c) = 0.80c$, and the momentum becomes

$$p = \frac{m_0(2v)}{\sqrt{1-(0.80)^2}} = 1.33 m_0 c$$

Although the ratio of the final speed to the initial speed is 2.0, the ratio of the final momentum to the initial momentum is 3.0.

35-3 RELATIVISTIC ENERGY

We now have expressions for relativistic mass, Eq. (35-7), and relativistic momentum, Eq. (35-8), and the correct form of Newton's second law of motion, Eq. (35-9). We next ask, "What is the relativistic kinetic energy E_k?" We may take the kinetic energy to be, as in classical physics, the total work done in bringing a particle from rest to the final speed v under a constant force F:

$$E_k = \int_0^s F\,ds = \int_0^s \frac{d}{dt}(mv)\,ds = \int_0^t \frac{d}{dt}(mv)\,v\,dt$$

$$= \int v\,d(mv) = \int (v^2\,dm + mv\,dv)$$

To integrate, we must recognize that both m and v are variables, the dependence of one on the other being given by (35-7). We shall find it simpler to express v in terms of m and then integrate with respect to the variable m. The expressions for v and dv can be obtained by rewriting (35-7) in the form

$$1 - \frac{v^2}{c^2} = \frac{m_0^2}{m^2}$$

Differentiating, we have

$$-2v\,\frac{dv}{c^2} = -2m_0^2\,\frac{dm}{m^3}$$

Combining the two equations, we have

$$mv\,dv = (c^2 - v^2)\,dm$$

Substituting for $mv\,dv$ in the equation for the kinetic energy E_k, we then have

$$E_k = \int_{m_0}^m [v^2\,dm + (c^2 - v^2)\,dm] = c^2 \int_{m_0}^m dm = mc^2 - m_0 c^2$$

$$E_k = (m - m_0)c^2 \qquad (35\text{-}11)$$

Thus, the relativistic kinetic energy is the increase in mass, arising from the particle's motion, multiplied by c^2. At high speeds the relativistic kinetic energy is markedly different from the classical kinetic energy. As (35-11) shows, in relativity physics to say that a particle has kinetic energy is to say that its mass exceeds its rest mass. Of course, the relativistic kinetic energy must reduce to the familiar classical kinetic energy, $\tfrac{1}{2}m_0 v^2$ for $v/c \ll 1$, and to show that it does, we expand (35-11) by the binomial theorem:

$$E_k = m_0 c^2 \left\{ \left[1 - \left(\frac{v}{c}\right)^2 \right]^{-1/2} - 1 \right\}$$

$$= m_0 c^2 \left[1 + \frac{1}{2}\left(\frac{v}{c}\right)^2 + \frac{3}{8}\left(\frac{v}{c}\right)^4 + \cdots - 1 \right]$$

$$= \tfrac{1}{2}m_0v^2\left[1 + \frac{3}{4}\left(\frac{v}{c}\right)^2 + \cdots\right]$$

or
$$\lim_{v/c \to 0} E_k = \lim_{v/c \to 0} (m - m_0)c^2 = \tfrac{1}{2}m_0v^2$$

Note that the relativistic kinetic energy is *not* given by $\tfrac{1}{2}mv^2$, where m is the relativistic mass.

We have seen that an increase in a particle's kinetic energy corresponds to an increase in its mass. This is true of energy in general: A change in a system's total energy corresponds to a change in the system's mass. Equation (35-11) may be written more generally as

$$E_k = E - E_0 = mc^2 - m_0c^2 \tag{35-12}$$

where E represents the particle's *total energy*,

$$E = mc^2 \tag{35-13}$$

ind E_0 is the particle's *rest energy*

$$E_0 = m_0c^2 \tag{35-14}$$

For a system of particles the rest energy E_0 and the rest mass m_0 are the *system's* total energy and total mass when its center of mass is at rest.

Equation (35-13) is the famous Einstein relation. It implies the equivalence of energy and mass. Mass and energy can be interpreted as different manifestations of the same physical entity. A particle at rest with respect to an observer has a rest mass m_0 and a rest energy m_0c^2; in motion the mass and energy are m and mc^2. Because mass and energy may be regarded as equivalent and interchangeable, we no longer have the separate laws of energy conservation and mass conservation; rather, relativity physics combines these into a single, simple law: The conservation of mass-energy, which holds in any inertial system.

In physics a particle's momentum is usually a more significant quantity than its velocity (we have a momentum conservation law but not a velocity conservation law). Therefore, it is often convenient to express energy E in terms of p rather than v. We can eliminate v as follows: Squaring (35-7) and multiplying both sides by $c^4[1 - (v/c)^2]$ gives

$$m^2c^4 - m^2v^2c^2 = m_0^2c^4$$

Substituting (35-8), (35-13), and (35-14) for mv, mc^2, and m_0c^2 in this equation immediately gives the desired relation between E and p:

$$E^2 = (pc)^2 + E_0^2 \tag{35-15}$$

Let us examine the relativistic equations under two limiting conditions, very low speeds and very high speeds.

$v \ll c$. This is the classical region. Newtonian mechanics is adequate, and the relativistic quantities reduce to their familiar classical forms, namely,

$$m \approx m_0$$
$$p \approx m_0 v$$
$$E_k \approx \tfrac{1}{2} m_0 v^2$$

In the low-speed domain a particle's kinetic energy is much less than its rest energy; that is,

$$E_k \ll E_0$$

because
$$\frac{E_k}{E_0} = \frac{\tfrac{1}{2} m_0 v^2}{m_0 c^2} = \frac{1}{2}\left(\frac{v}{c}\right)^2 \ll 1$$

$v \approx c$. This is the extreme relativistic region. Therefore (35-7), (35-12), and (35-15) become

$$m \gg m_0$$
$$E \gg E_0$$
$$p \approx \frac{E}{c}$$
$$E_k \approx E$$

If a particle has energy and momentum but has zero *rest* mass—a possibility that makes no sense from a classical viewpoint but is admissible in relativity theory—then the equations above are *exactly true,* and a zero-rest-mass particle must necessarily travel with the speed of light c. That is, for a zero-rest-mass particle:

$$m_0 = 0 \qquad E = pc \qquad E_k = E \qquad v = c \qquad (35\text{-}16)$$

Conversely, if a particle with a nonzero energy travels with the speed of light, it *must* have a zero rest mass. Such a particle's relativistic mass m is given by $m = E/c^2$, which may be nonzero even though m_0 is zero. The variations of energy (and therefore of mass, because $E = mc^2$) with speed for a proton, an electron, and a particle of zero rest mass are shown in Fig. 35-6. As the figure indicates, for a zero-rest-mass particle the relativistic energy E has a nonzero value only if the particle moves at the *single* speed c, but at this speed its energy (and mass) can have any value from zero to infinity. Since a zero-rest-mass particle has zero rest energy, its total energy is purely kinetic.

The relativistic total energy E is shown as a function of the relativistic momentum p in Fig. 35-7, following the equation

$$E^2 = (pc)^2 + E_0^2 \qquad (35\text{-}15)$$

For low speeds, or $v \ll c$, the relations $m \approx m_0$ and $p \approx p_0 \approx m_0 v$ all hold. Therefore, in this, the classical region, the total energy is

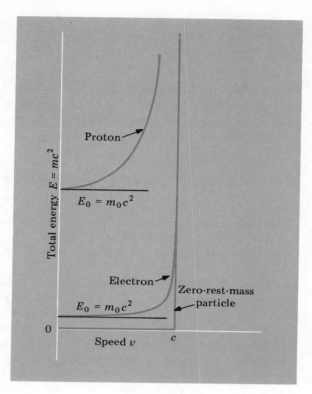

FIG. 35-6. Total relativistic energy vs. speed for three particles having different rest masses. (The rest energies are *not* to scale relative to one another.)

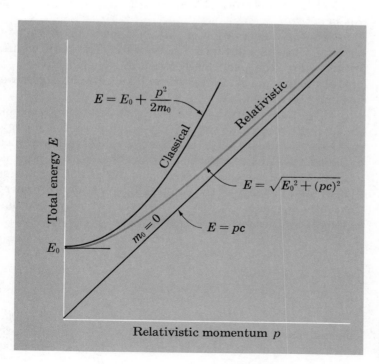

FIG. 35-7. Total relativistic energy vs. relativistic momentum for a particle.

Relativistic Energy   SEC. 35-3

given approximately by

$$E \approx \frac{p_0^2}{2m} + E_0 = \tfrac{1}{2}m_0 v^2 + E_0$$

whereas for very high speeds the relation $E \approx pc$ holds. In the classical region the kinetic energy $E - E_0$ varies as the *square* of the momentum; in the extreme relativistic region the kinetic energy $\approx E$ varies *linearly* as the momentum.

35-4 MASS-ENERGY EQUIVALENCE AND BOUND SYSTEMS

To illustrate the significance of the equivalence of mass and energy and the conservation law of mass-energy, we consider two situations: *unbound systems* and *bound systems*.

Unbound Systems. Consider a collision between two particles, each with a rest mass m_0. They are projected toward one another, each with speed v relative to an observer who is at rest in the center-of-mass reference frame. We assume the collision to be perfectly inelastic; thus, the two particles stick together to form a single particle, whose rest mass is designated M_0. We ask, "How is the final rest mass M_0 of the composite particle related to the rest mass m_0 of each of the separate incident particles?" From the standpoint of classical physics we have M_0 equal to $2m_0$; relativistically, however, this is not so.

By the conservation of linear momentum we know that the system's total momentum must be zero. The total momentum is initially zero, the particles before the collision having equal but opposite momenta; and after the collision the amalgamated particles must be at rest.

The total energy of the two particles before the collision is $2mc^2$, where mc^2 is the rest energy plus the kinetic energy of each particle. To accord with the energy conservation principle, the total energy of the two particles before the collision must equal the total energy $M_0 c^2$ of the composite particle after the collision. Note that the total energy of the composite is entirely rest energy, since the center of mass is at rest. Thus, what we may now characterize as mass-energy conservation gives

$$M_0 c^2 = 2mc^2 = \frac{2m_0 c^2}{\sqrt{1 - (v/c)^2}}$$

or
$$M_0 = \frac{2m_0}{\sqrt{1 - v^2/c^2}} \qquad (35\text{-}17)$$

The rest mass M_0 of the composite object after collision exceeds the rest mass $2m_0$ of the incident particles. This is so because the energy that we observe as the incoming kinetic energy of the two particles becomes a part of the rest energy of the combined particles after the collision.

Although the collision in this example was observed from a reference frame in which the center of mass was at rest, an observer in any inertial system will find that an isolated system's total relativistic momentum is conserved and that the total mass-energy is also conserved.

Example 35-2

Two satellites, each of rest mass 4,000 kg and traveling in orbit in opposite directions at a speed of 8.0 km/s with respect to an earth observer, collide and stick together. Find the increase in the total rest mass of the system.

Because the statellites have equal but opposite momenta, the total momentum is zero; therefore, after the collision the composite object is at rest. Since total energy is conserved, the kinetic energy of the incident satellites is converted into rest mass, and

$$\text{Increase in rest mass} = \Delta m = \frac{2E_k}{c^2}$$

where E_k is the initial kinetic energy of each satellite. The speed of each satellite, 8.0 km/s, is much less than the speed of light, and we can use the classical expression for the kinetic energy, $E_k = \frac{1}{2}m_0 v^2$. Therefore,

$$\Delta m = \frac{2E_k}{c^2} = \frac{2(\frac{1}{2}m_0 v^2)}{c^2} = m_0 \left(\frac{v}{c}\right)^2$$

$$= (4{,}000 \text{ kg})\left(\frac{8.0 \times 10^3 \text{ m/s}}{3.0 \times 10^8 \text{ m/s}}\right)^2 = 2.8 \times 10^{-6} \text{ kg} = 2.8 \text{ mg}$$

Bound Systems. One of the most important consequences of Einstein's relativity theory arises in the binding together by some attractive force of two particles A and B to form a single system. To break the system into its separate components requires work, that is, the adding of energy to the system. Let the rest mass of the composite system be M_0 and the rest masses of the individual particles be m_{0A} and m_{0B}. We observe the bound system and the ensuing separation into A and B from the center-of-mass reference frame.

The breaking up of the bound system is shown symbolically in Fig. 35-8, where E_b is the energy that must be added to the system in order to separate the particles completely. Because the energy needed to break the system is exactly equal to the energy binding the system, E_b itself is called the *binding energy*.

Applying mass-energy conservation to this situation, we have

FIG. 35-8. Symbolic representation of the splitting of two bound particles.

$$M_0 + \frac{E_b}{c^2} = m_{0A} + m_{0B} \qquad (35\text{-}18)$$

Clearly if the system is bound and $E_b > 0$, then from this equation it follows that $M_0 < m_{0A} + m_{0B}$. That is, the rest mass of the bound system must be *less* than the sum of the rest masses of the individual particles when separated. (This is in contrast with the situation of unbound systems, in which the rest mass of the composite *exceeds* the rest masses of the separated particles.) In principle it is then possible to calculate the binding energy E_b if we know merely the rest mass of the system as a whole and the rest masses of its constituents. However, only in the case of the very strong nuclear forces is the binding energy between particles sufficiently great to produce a measurable mass difference.

We are free to choose the zero of *total kinetic and potential energy* of a system of particles—what might be called the total mechanical energy E_m—at any convenient value. The most convenient choice for particles that attract one another is zero when the particles are all infinitely separated and at rest.

When particles are *bound* to one another, the energy E_m of the bound system is *negative:* Energy must be added to the bound system to separate the particles completely and to bring the energy of the system up to zero. Figure 35-9 shows the relationships

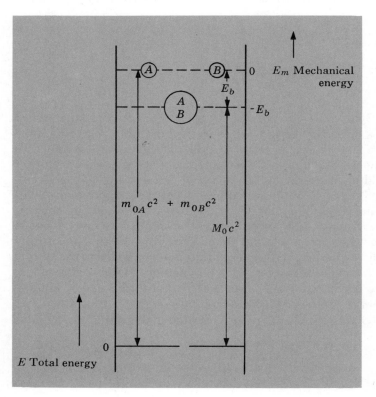

FIG. 35-9. Energy-level diagram showing the energy of two particles when bound together and when completely separated from one another and at rest. The total relativistic energy is plotted on the left; the mechanical energy on the right.

among rest masses, total relativistic energy, binding energy, and mechanical energy.

Example 35-3

The binding energy of an electron and a proton together, forming a stable hydrogen atom, is known experimentally to be 13.6 eV. It is called the ionization energy, since this much energy must be added to a hydrogen atom to separate it into two oppositely charged particles. Saying it differently, the total mechanical energy of a hydrogen atom is -13.6 eV. By using Eq. (35-18) it is possible to compute the difference between the mass of the hydrogen atom, $M_{0_H} = 1.67 \times 10^{-27}$ kg, and the sum of the rest masses of the separated electron, m_{0_e}, and proton, m_{0_p}:

$$m_{0_e} + m_{0_p} - M_{0_H} = \frac{E_b}{c^2} = \frac{13.6 \text{ eV}}{c^2}$$

The fractional change in the mass is

$$\frac{E_b/c^2}{M_0} = \frac{(13.6 \text{ eV})(1.60 \times 10^{-19} \text{J/eV})}{(1.67 \times 10^{-27} \text{ kg})(3.00 \times 10^8 \text{ m/s})^2}$$

$$\frac{E_b}{M_0 c^2} = 1.53 \times 10^{-8}$$

This fractional mass difference, slightly more than one part in 100 million, is much smaller than the fractional error in the measurement of the masses of the hydrogen atom, proton, and electron. Therefore, in a reaction in which the binding energy is *several electron volts* — and all *chemical* reactions are of this order of magnitude — it is impossible to detect directly a change in the total mass of the system. A mass change can, however, be measured in *nuclear* reactions, in which the binding energy typically is *several million electron volts*.

35-5 COMPUTATIONS AND UNITS IN RELATIVISTIC MECHANICS

The classical equations for the momentum and kinetic energy of a particle can be used only when the speed of the particle is much less than the speed of light; for high speeds the relativistic relations must be invoked. It is useful to have a rule of thumb for determining whether a computation in a problem can safely be treated relativistically or classically. Table 35-1 shows the conditions which, if fulfilled, lead to errors no greater than 1 percent in the computed momentum or energy. If the kinetic energy of a particle is a very small fraction of its rest energy, classical mechanics applies; on the other hand, if the total energy or the kinetic energy greatly exceeds a particle's rest energy, then the extreme relativistic relation $E = pc$ (which holds strictly only for $m_0 = 0$) can be applied.

For atomic and subatomic particles a convenient unit of energy is the electron volt or multiples of it:

Table 35-1

	For the condition	Error in relation below is no greater than 1 percent
Classical region	$E_k/E_0 < 0.01$ or $v/c < 0.1$	$E_k \approx \tfrac{1}{2}m_0v^2$ and $p \approx m_0v$
Extreme relativistic region	$E/E_0 > 7$ or $E_k/E_0 > 6$ or $v/c > 0.99$	$E \approx pc$

$$\text{Kilo electron volt} = 1 \text{ keV} = 10^3 \text{ eV}$$

$$\text{Mega electron volt} = 1 \text{ MeV} = 10^6 \text{ eV}$$

$$\text{Giga electron volt†} = 1 \text{ GeV} = 10^9 \text{ eV}$$

From Eq. (35-15) the corresponding unit for momentum is the electron volt divided by the speed of light, eV/c. When momentum is expressed in these units, the unit for pc is just the electron volt, eV. The speed of an atomic particle is most conveniently given in units of the speed of light, that is, as v/c. In these units the speed of any particle must lie somewhere between 0 and 1. These particular units (eV for energy, eV/c for momentum, and v/c for speed) simplify calculations in classical as well as in relativistic problems.

The classical kinetic-energy and momentum relations can be written in terms of a particle's rest energy $E_0 = m_0c^2$, its speed v/c, and the constant c, as follows:

$$E_k \text{ (classical)} = \tfrac{1}{2}m_0v^2 = \tfrac{1}{2}(m_0c^2)\left(\frac{v}{c}\right)^2 = \tfrac{1}{2}E_0\left(\frac{v}{c}\right)^2 \quad (35\text{-}19)$$

$$p \text{ (classical)} = m_0v = \frac{(m_0c^2)(v/c)}{c} = \frac{E_0(v/c)}{c} \quad (35\text{-}20)$$

For example, the kinetic energy and momentum of an electron (rest energy, $E_0 = 0.51$ MeV) moving with a speed of $\tfrac{1}{100}c$, for which the classical relations apply, are easily found to be:

$$E_k = \tfrac{1}{2}E_0\left(\frac{v}{c}\right)^2 = \tfrac{1}{2}(0.51 \text{ MeV})(10^{-2})^2 = 0.26 \times 10^{-4} \text{ MeV} = 26 \text{ eV}$$

$$p = \frac{E_0(v/c)}{c} = \frac{(0.51 \text{ MeV})(10^{-2})}{c} = \frac{0.51 \times 10^{-2} \text{ MeV}}{c} = 5.1 \text{ keV}/c$$

---

†The energy unit BeV (billion electron volts) is also used to denote 10^9 eV. The GeV is preferred, however, because in European usage a billion designates a million million (10^{12}), not a thousand million (10^9).

The masses of particles in atomic physics are most often given in units of the *atomic mass unit* (unified), or "u." *One atomic mass unit is defined as one-twelfth the mass of a neutral carbon atom (isotope 12).* Avogadro's number, 6.02217×10^{23}, gives the number of atoms in 12 g of atomic carbon. Therefore,

$$1 \text{ u} = \frac{1}{12} \left(\frac{12 \text{ g}}{6.02217 \times 10^{23}} \right) = 1.661 \times 10^{-27} \text{ kg}$$

The relation between the atomic mass unit, u, and the energy unit MeV is particularly useful. We find this from the general mass-energy relation, $E = mc^2$,

$$E = mc^2 = (1 \text{ u})c^2$$
$$= \frac{(1.661 \times 10^{-27} \text{ kg})(2.998 \times 10^8 \text{ m/s})^2}{(1.602 \times 10^{-19} \text{ J/eV})(10^6 \text{ eV/MeV})}$$
$$= 931.5 \text{ MeV}$$

Therefore,

$$1 \text{ unified atomic mass unit} = 1 \text{ u} = 931.5 \text{ MeV}/c^2$$

This relation may be regarded as giving the basic conversion factor between mass and energy units. The rest energies of the electron and proton are:

$$\text{Electron rest energy} = 0.51100 \text{ MeV} = 0.00055 \text{ u-}c^2$$
$$\text{Proton rest energy} = 938.26 \text{ MeV} = 1.00728 \text{ u-}c^2$$

It is worth memorizing that the rest energy of an electron is approximately one-half a mega electron volt, $\frac{1}{2}$ MeV, and that the rest energy of a proton is approximately one giga electron volt, 1 GeV. By convention, when a particle is described as, say, a 3.0-MeV particle, this means that the *kinetic* energy, not the total energy E, is three mega electron volts.

Example 35-4

What is the speed of a 2.0-MeV electron? The kinetic energy E_k is 2.0 MeV and the rest energy E_0 is 0.51 MeV. Because $E_k > E_0/100$, a relativistic calculation must be made. Equation (35-7) shows that

$$mc^2 = \frac{m_0 c^2}{\sqrt{1 - (v/c)^2}}$$

$$E = \frac{E_0}{\sqrt{1 - (v/c)^2}} \quad (35\text{-}21)$$

This equation is often useful in relativistic computations. Solving for v/c gives

$$\frac{v}{c} = \sqrt{1 - \frac{E_0^2}{E^2}} = \sqrt{1 - \frac{E_0^2}{(E_k + E_0)^2}} = \sqrt{1 - \left(\frac{0.51}{2.51}\right)^2} = 0.98$$

Example 35-5

What is the momentum of a 20.0-GeV electron? Table 35-1 shows that, because $E_k/E_0 = 20{,}000/0.51 \approx 40{,}000$, we can use $E = pc$ with an error of much less than 1 percent; therefore,

$$p = \frac{E}{c} = \frac{E_k + E_0}{c} = \frac{(20.0 + 0.0005)\text{ GeV}}{c} \approx 20\text{ GeV}/c$$

SUMMARY

The relativistic momentum of a particle is

$$\mathbf{p} = m\mathbf{v} \qquad \text{where} \qquad m = \frac{m_0}{\sqrt{1 - (v/c)^2}} \qquad (35\text{-}8)$$

In relativistic dynamics the total energy and the relativistic mass of a particle are related by the Einstein equation

$$E = mc^2 \qquad (35\text{-}13)$$

A particle's kinetic energy is given by

$$E_k = E - E_0 \qquad (35\text{-}11)$$

where $E_0 = m_0 c^2$ is the rest energy of the particle.

The relativistic energy and momentum are related as follows:

$$E^2 = E_0^2 + (pc)^2 \qquad (35\text{-}15)$$

The rest mass of a bound system of particles is less than the total mass of the separated parts by E_b/c^2, where E_b is the total binding energy.

PROBLEMS

35-1. Show, in a fashion analogous to Eqs. (35-1) and (35-2) for the invariance of classical momentum conservation under galilean velocity transformations, that the classical law of energy conservation is invariant under the galilean velocity transformations. Assume the collision to be head-on and perfectly elastic.

35-2. *(a)* Show by computing the accelerations measured in two inertial frames having a constant relative velocity that Newton's second law of motion is invariant under the galilean transformation relations. *(b)* What form does Newton's second law take if one reference frame is *accelerated* relative to an inertial reference frame?

35-3. A particle moving initially at speed $0.4c$ increases its speed until its momentum has increased by a factor of 10. *(a)* By what factor did the speed increase? *(b)* What is the final speed?

35-4. A polyenergetic beam of protons enters a uniform magnetic field of 1.50 Wb/m² perpendicularly to it. The beam is found to break up as a spectrum whose radii range from 10.0 to 1.00 m. *(a)* What is the range of momenta values (in units of GeV/c) of the proton beam? *(b)* What is the range of kinetic-energy values (in units of GeV)?

35-5. *(a)* What is the radius of curvature of a beam of 20-GeV electrons injected perpendicularly into a uniform magnetic field of 2.0 Wb/m²? *(b)* What would nonrelativistic physics predict for the radius of curvature? *(c)* What would be the radius of a beam of 20-GeV protons injected into the same field?

35-6. A beam of particles each with electric charge q passes undeflected through crossed electric and magnetic fields with magnitudes $\mathscr{E}$ and B, respectively. When the electric field alone is turned off, the beam is deflected into a circular arc of radius r. Show that each particle's relativistic mass is given by $qB^2r/\mathscr{E}$.

35-7. In the high-energy accelerator at Batavia, Illinois, protons may be accelerated to energies as high as 500 GeV. (a) Find the speed of such protons. (b) If the final intensity of the proton beam is 10^{14} protons/s, what is the minimum power (watts) necessary to accelerate the protons?

35-8. It is now possible to accelerate electrons to energies of 20 GeV. (a) By what factor does such an electron's relativistic mass increase? (b) What is the percentage difference between the speed of light and the speed of a 20-GeV electron?

35-9. An electron and a proton, both initially at rest, are accelerated across a potential difference of 500 kV. What is the ratio of the final electron to proton (a) kinetic energy, (b) momentum, and (c) speed?

35-10. Because the electron and proton have the same magnitude of electric charge, the kinetic energies of an electron and a proton that are both accelerated from rest across the *same* electric potential difference are always exactly the same, irrespective of differences in the speeds of electron and proton. Across what minimum potential difference must an electron and proton be accelerated from rest so that their momenta are the same to within 1 part in 10?

35-11. What are (a) the kinetic energy (in eV) and (b) the momentum (in eV/c) of a particle of rest energy 100 MeV moving with a speed of $v/c = \frac{1}{100}$?

35-12. The relation between a particle's total energy E, rest energy E_0, and momentum p may be represented by a right triangle with sides pc, E_0, and E. Draw such a triangle and mark the particle's kinetic energy for (a) $v/c \ll 1$ and (b) $v/c \approx 1$.

35-13. A particle initially moving with speed $0.6c$ has its speed increased by one-third. By what factor does the particle's (a) momentum and (b) kinetic energy increase?

35-14. Show that the classical and extreme relativistic domains may be characterized as follows: In the low-speed limit, when a particle's momentum increases by the factor F, the particle's kinetic energy increases by the factor F^2; in the high-speed limit, when a particle's momentum increases by the factor F, its kinetic energy also increases by the factor F.

35-15. A proton has a speed of $\frac{1}{20}c$. What is (a) its kinetic energy (in MeV) and (b) its momentum (in MeV/c)?

35-16. An electron has a momentum of 10 MeV/c: What are (a) its kinetic energy and (b) its speed? A proton has a momentum of 10 MeV/c: What are (c) its kinetic energy and (d) its speed?

35-17. What is the (a) kinetic energy and (b) speed of an electron having a momentum of 200 keV/c?

35-18. Show that in general the derivative dE/dp is equal to a particle's velocity, where E and p are the particle's energy and momentum.

35-19. What are the momenta of (a) a 1.0-GeV (kinetic energy) electron and (b) a 1.0-GeV (kinetic energy) carbon atom (rest energy $\approx$ 12 GeV)?

35-20. A particle's kinetic energy is 10 times its rest energy E_0. What are its (a) momentum and (b) speed, both in terms of E_0 and c?

35-21. A high-energy accelerator accelerates a beam of electrons through a total potential difference of 2×10^{10} V. The average number of electrons striking the target is 15×10^{13} per second. (a) What is the average electron current? (b) What is the average force exerted on the target by the electron beam as the electrons are brought to rest?

35-22. (a) What is the maximum speed possible of a particle whose kinetic energy may be written $\frac{1}{2}m_0v^2$ with an error in the kinetic energy of no greater than 1 percent? (b) What is the kinetic energy of an electron moving at this speed? (c) What is the kinetic energy of a proton moving at this speed?

35-23. (a) What is the minimum speed of a particle whose kinetic energy may be written as its total energy E and, therefore, as pc with an error in the total energy of no greater than 1 percent? (b) Under such conditions what is the kinetic energy of an electron? (c) Of a proton?

35-24. The total intensity of radiation from the sun at the earth's surface is 8.0 J/cm²-min. Calculate the loss in the sun's mass per second and the fractional loss in the sun's mass in 10^9 years (approximately one-tenth of the age of the universe) resulting from its radiation. The distance from the sun to the earth is 1.49×10^{11} m, and the sun's mass at present is 2.0×10^{30} kg.

35-25. A rocket ship having a final payload rest mass of 6×10^4 kg is accelerated to a speed of $0.95c$. (a) What minimum energy is required to accelerate the rocket ship to this speed? (b) How much equivalent mass does this represent? (c) What amount of nuclear fuel (assume 1 percent conversion of mass to energy) would be needed to achieve this?

35-26. (a) Show that the momentum of a particle may be written as $p = (1/c)(E_k^2 + 2E_0E_k)^{1/2}$. (b) Show that this reduces to m_0v in the classical limit, and to E/c in the extreme relativistic region.

35-27. A 10.2-MeV electron makes a head-on elastic collision with a proton initially at rest. Show that the proton recoils with a speed approximately equal to $(2E_e/E_p)c$ and the fractional energy transferred from the electron to the proton is $2E_e/E_p$, where E_e is the *total* energy of the electron and E_p is the *rest* energy of the proton. (*Hint: (a)* Since the electron's energy is much greater than its rest energy, it may be treated as an extreme relativistic particle, and *(b)* since the rest energy of the proton is much greater than the *total* energy of the electron, the proton may be treated classically.)

35-28. To separate a carbon monoxide molecule, CO, into carbon and oxygen atoms requires 11.0 eV. *(a)* What is the fractional change in mass of a CO molecule when it is broken into the atoms C and O? *(b)* What is the binding energy (in eV) per molecule?

35-29. Indicate whether the mass of the system increases or decreases and by what amount for the following circumstances: *(a)* a charged dry cell rated at 1.25 V and 450 mAh is completely discharged; *(b)* the temperature of 5.0 kg of water is raised by 20 C°; *(c)* a 1.0-kg block of ice at 0° C melts (latent heat of fusion for water, 80 cal/g).

35-30. In inertial system S a particle of rest mass m_0 is observed to be moving at constant speed $0.80c$ along the positive Y axis. *(a)* Find the particle's momentum and total energy as observed in a second inertial system S' that is moving along the positive X axis at speed $0.50c$. *(b)* In what inertial system is the particle's total energy a minimum?

CHAPTER 36

Quantum Effects: The Particle Aspects of Electromagnetic Radiation

36-1 QUANTIZATION IN CLASSICAL PHYSICS

The theory of relativity and the quantum theory constitute the two great theoretical foundations of twentieth-century physics. Just as the theory of relativity leads to new insights into the nature of space and time and to profound consequences in mechanics and electromagnetism, so too the quantum theory leads to drastically new modes of thought that are the basis of an understanding of atomic and nuclear structure.

In the study of the physical world we find two general kinds of physical quantities: quantities that have a continuum of values and quantities that are *quantized*. The latter are quantities that are restricted to certain discrete values; they are sometimes described as having "atomicity" or "granularity."

Examples of classical continuous, or nonquantized, physical quantities are (1) the magnitude of a particle's momentum, angular momentum, and energy, which can take on any value from zero to infinity; (2) the energy of a system of two particles, which can assume any negative value when the particles are bound together; and (3) the angle between the direction of a magnet's dipole moment and an external magnetic field.

Figure 36-1 shows several examples of quantized physical quantities:

a. The observed rest masses of atoms, which do not occur in a continuous range, but which are *nearly* in the ratio of integers.

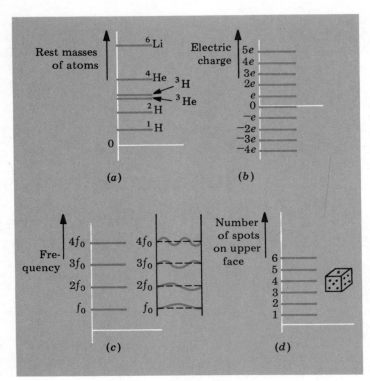

FIG. 36-1. Some examples of classical physical quantities having quantized values.

b. Electric charge, which is quantized in that the total net charge of any object is precisely an integral multiple, either positive or negative, of fundamental electronic charge e.
c. Standing waves and resonance, particularly striking manifestations of quantization in classical physics. The frequency of oscillation of a resonating vibrating string, fixed at both ends, can be only an integral multiple of the lowest, or fundamental frequency of oscillation f_0. The wave on the string may be thought to be repeatedly reflected from the boundaries, or fixed ends, and it constructively interferes with itself, so to speak, to produce standing waves. Resonance occurs when the distance between end points is precisely an integral multiple of half-wavelengths.
d. A rolled die, which can show on its upper face only 1, 2, 3, 4, 5, or 6 spots. Other everyday examples of quantized quantities are the face of a coin, people, and number of pennies.

The quantum theory is in large measure based on the discovery that certain quantities that in classical physics had been regarded as continuous are, in fact, quantized. Historically it had its origins in the theoretical interpretation of electromagnetic radiation from a blackbody (a perfect absorber and radiator). Near the end of the nineteenth century it was found that the experimentally observed variation with wavelength of the intensity of electromagnetic radiation from a blackbody was in disagreement with the theoretical expectations of classical electromagnetism.

Max Planck, formulator of the quantum theory, showed in 1900 that a revision of classical ideas, through the concept of energy quantization, led to satisfactory agreement between experiment and theory. Because a detailed analysis of blackbody radiation involves rather sophisticated arguments, we shall introduce the quantum concepts through the much simpler and in many ways more compelling considerations that arise in the phenomenon of the photoelectric effect.

36-2 THE PHOTOELECTRIC EFFECT

It is an irony of history that the photoelectric effect was discovered by Heinrich Hertz in 1887 during the course of the very experiments which confirmed Maxwell's theoretical prediction (1864) of the existence of classical electromagnetic waves produced by oscillating electric currents.

The photoelectric effect is one of several processes by which electrons may be removed from the surface of a substance. It occurs when electromagnetic radiation shines on a clean metal surface and electrons are released from the surface. The valence electrons in a metal are relatively free to move about through its interior but are bound to the metal as a whole. Such relatively free electrons may become photoelectrons. In the photoelectric effect a beam of radiation supplies an electron with energy that equals or exceeds the energy which binds it to the surface and thereby allows that electron to escape. A more detailed description of the photoelectric effect requires a knowledge, based on experiment, of the relationship among several variables involved in photoelectric emission. They are the frequency ν of the light, the intensity I of the light beam, the photoelectric current i, the kinetic-energy $\frac{1}{2}m_0 v^2$ (we shall see shortly that the classical kinetic energy formula is appropriate), and the chemical identity of the surface from which the electrons emerge.

Figure 36-2 shows a schematic diagram of an experimental arrangement for studying important aspects of the photoelectric effect. Monochromatic light shines on a metal surface, the anode, enclosed in a vacuum tube (thereby eliminating collisions between photoelectrons and gas molecules). When photoelectrons are emitted, some travel toward the cathode and upon reaching it constitute the current flowing in the circuit. Photoelectrons leave the anode with a variety of kinetic energies. The negatively charged cathode tends to repel them. When the work done on a photoelectron by the retarding electrostatic field, of potential difference V, equals the initial kinetic energy of the photoelectron, the latter is brought to rest just in front of the cathode. Thus, $eV = \frac{1}{2}m_0 v^2$, where v is the speed of the photoelectron as it leaves the anode surface and V is the potential difference that stops the photoelectron of rest mass

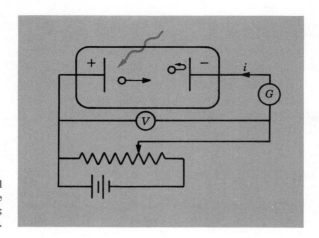

FIG. 36-2. Schematic experimental arrangement for studying the photoelectric effect. V, voltmeter; G, galvanometer.

m_0 and charge e. When the most energetic photoelectrons, of speed $v_{\max}$, are brought to rest in front of the cathode by a sufficiently large potential difference V_0, all the other photoelectrons are, of course, stopped too, and no photocurrent exists: $i = 0$. Then

$$eV_0 = \tfrac{1}{2}m_0 v_{\max}^2 \tag{36-1}$$

At still higher retarding potential differences all photoelectrons are turned *back* before reaching the cathode.

We shall first list below the results of experiment, and then give the results that might be expected on the basis of the classical electromagnetic theory. It will be seen that the experimental results strongly disagree with the classical expectations. Finally, we shall see how the photoelectric effect can be understood on the basis of a quantum interpretation.

Experimental Results of the Photoelectric Effect. The results of experiments on the photoelectric effect are summarized in Fig. 36-3. We shall take them up in the order in which they are given in the figure.

a. The photocurrent begins *almost instantaneously,* even when the light beam has an intensity as small as 10^{-10} W/m². The delay in time, from the instant that the light beam first shines on the surface until photoelectrons are first emitted, is no greater than 10^{-9} s.
b. For any fixed frequency and retarding potential the *photocurrent i is* directly *proportional to the intensity I* of the light beam. Inasmuch as the photocurrent is a measure of the number of photoelectrons released per unit time at the anode and collected at the cathode, the relation signifies that the number of photoelectrons emerging per unit time is proportional to the light intensity (the variation in photocurrent with intensity is utilized in practical photoelectric devices).
c. For a constant frequency ν and light intensity I *the photocurrent decreases with increasing retarding potential V* and finally *reaches zero when V is equal to V_0* [see Eq. (36-1)]. With a small retarding potential the low-speed low-energy photoelectrons are brought to rest; when the

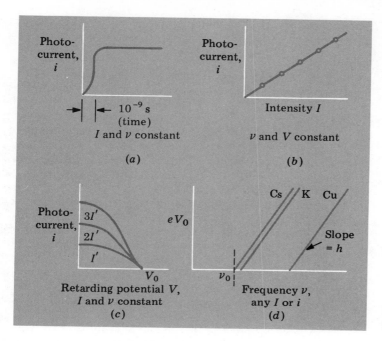

FIG. 36-3. Experimental results of photoelectric emission.

retarding potential is equal to V_0, even the most energetic photoelectrons are brought to rest, and $i = 0$.

d. For any particular surface the value of the *stopping potential V_0 depends on the frequency of the light but is independent of the light intensity* and therefore, from *(b)*, independent also of the photocurrent. Figure 36-3d shows experimental results for the three metals cesium, potassium, and copper. For each there is a well-defined frequency ν_0, the *threshold frequency,* which must be exceeded for photoemission to occur at all; that is, no photoelectrons are produced, however great the light intensity, unless $\nu > \nu_0$. For most metals the threshold frequency lies in the region of ultraviolet light. A typical stopping potential is several volts. The emitted photoelectrons have energies of several electron volts; therefore, we are justified in using the classical kinetic-energy formula for the photoelectrons.

With respect to any one type of metal the experimental results of Fig. 36-3d may be represented by the straight-line equation

$$eV_0 = h\nu - h\nu_0$$

where h, representing the slope of the straight line, is found to be the *same* for *all* metals, and ν_0 is the threshold frequency for the particular metal. Rearranging terms and using (36-1) gives

$$h\nu = \tfrac{1}{2}m_0 v_{\max}^2 + h\nu_0 \qquad (36\text{-}2)$$

Inasmuch as $\tfrac{1}{2}m_0 v_{\max}^2$ has the dimensions of energy, the terms $h\nu$ and $h\nu_0$ must also have the dimensions of energy.

Classical Interpretations of the Photoelectric Effect. We now consider what effects may be expected on the basis of the classical properties of electromagnetic waves for each of the four experimental results on the photoelectric effect given in the preceding paragraphs. As before, we shall discuss them with reference to Fig. 36-3 and in the same order.

a. Because of the apparently continuous nature of light waves we expect the energy absorbed on the photoelectric surface to be proportional to the intensity of the light beam (the power per unit area), the area illuminated, and the time of illumination. All electrons bound to the metal surface with the same energy must be regarded as equivalent, and any one electron will be free to leave the surface only after the light beam has been on long enough to supply the electron's binding energy. Moreover, since any one electron is equivalent to any other electron bound with the same energy, we expect that when one electron has accumulated sufficient energy to be freed, a number of other electrons will have, too. Independent experiments show that in a typical metal the least energy with which an electron is bound to the surface is a few electron volts. A conservative calculation (see Prob. 36-7) shows that in the case of an intensity as low as 10^{-10} W/m², for which delay times no longer than 10^{-9} have been observed, no photoemission can be expected until at least several hundred hours have elapsed! Clearly, the classical theory is unable to account for the essential instantaneous photoelectric emission.

b. Classical theory predicts that as the light intensity is increased, so is the energy absorbed by electrons at the surface. Hence, the number of photoelectrons emitted, or the photocurrent, is expected to increase proportionately with the light intensity. Here classical theory agrees with the experimental result.

c. The results of these observations show that there is a distribution in the speeds, or energies, of the emitted photoelectrons; the distribution is in itself not incompatible with classical theory, because it may be attributed to the varying degrees of binding of electrons at the surface. The fact, however, that there is a very well-defined stopping potential V_0 for a given frequency, independent of the intensity, indicates that the maximum energy of released electrons is in no way dependent on the total amount of energy reaching the surface per unit time. Classical theory predicts no such effect.

d. The existence of a threshold frequency for a given metal, a frequency below which no photoemission occurs, however great the light intensity, is completely inexplicable in classical terms. From the classical point of view the primary circumstance that determines whether or not photoemission will occur is the energy reaching the surface per unit time (or the intensity), but *not* the frequency. Further, the appearance of a single constant h that relates, through (36-2), the maximum energy of photoelectrons to the frequency for any material cannot be understood in terms of any constants of classical electromagnetism.

In short, *classical electromagnetism cannot give a reasonable basis for understanding the experimental results illustrated in Fig. 36-3a, c, and d.*

Quantum Interpretation of the Photoelectric Effect. The photoelectric effect is understood only through the quantum theory. Albert Einstein first applied quantum theory to the nature of electromagnetic radiation in 1905, and this led to a satisfactory explanation of the photoelectric effect.

According to the quantum theory, the apparently continuous electromagnetic waves are quantized and consist of discrete *quanta*, called *photons*. Each photon has an energy E that depends only on the frequency (or on the wavelength) and is given by

$$E = h\nu = h\frac{c}{\lambda} \qquad (36\text{-}3)$$

The constant h is, in fact, the very same h that appears in (36-2). This fundamental constant of the quantum theory is called *Planck's constant* because its value was first determined and its significance first appreciated by Planck in 1900 in his interpretation of blackbody radiation. The value of Planck's constant is found by experiment to be

$$h = 6.626 \times 10^{-34} \text{ J-s}$$

According to the quantum theory, a beam of light of frequency ν consists of particlelike photons, each of energy $h\nu$. A single photon can interact only with a single electron at the metal surface of a photoemitter; it cannot share its energy among several electrons. Inasmuch as photons travel with the speed of light, they must, on the basis of relativity theory, have zero rest mass and an energy that is then entirely kinetic. When a particle with a zero rest mass ceases to move with a speed c, it ceases to exist; as long as it exists, it moves at the speed of light. Thus, when a photon strikes an electron bound in a metal and no longer moves at the unique speed c, it relinquishes its *entire* energy $h\nu$ to the single electron it strikes. If the energy the bound electron gains from the photon exceeds the energy binding it to the metal surface, the excess energy becomes the kinetic energy of the photoelectron.

We are now prepared to interpret on the basis of the quantum theory the experimental results of the photoelectric effect, which we now take in reverse order for convenience, referring to Fig. 36-3.

d. The terms in Eq. (36-2) now give a simple meaning to the energies of the photon and photoelectron:

$$h\nu = \tfrac{1}{2}m_0 v_{\max}^2 + h\nu_0 \qquad (36\text{-}2)$$

The left side of this equation gives the energy carried by a photon and supplied to a bound electron. Those electrons which are least tightly bound leave the surface with maximum kinetic energy. The right side of (36-2) gives the energy gained by the electron from the photon, namely, the kinetic energy and the binding energy. The binding energy of the electrons least tightly bound to the metal surface is called the metal's *work function* ϕ; it represents the work that must be done to remove the least tightly bound electron. Therefore,

$$\phi = h\nu_0 \qquad (36\text{-}4)$$

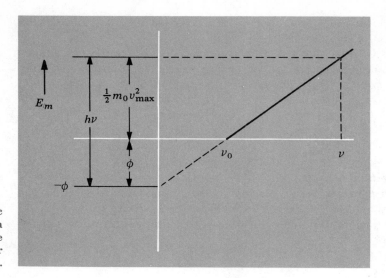

FIG. 36-4. Maximum kinetic energy of photoelectrons as a function of the frequency of the incident photons for a particular material.

and (36-2) may be written in the form

$$h\nu = \tfrac{1}{2}m_0 v_{\max}^2 + \phi \qquad (36\text{-}5)$$

The value of ϕ for a particular material, determined from the photoelectric effect, agrees with the value obtained through independent experiments based on different physical principles. An electron bound with an energy ϕ can be released only if a single photon supplies at least this much energy, that is, if $h\nu > \phi = h\nu_0$, or if $\nu > \nu_0$. Figure 36-3d then takes on new meaning: The ordinate is now identified with the photon energy, as shown in Fig. 36-4.

c. A well-defined maximum kinetic energy of photoelectrons exists for any given frequency, because the frequency of the electromagnetic radiation determines precisely the photon energy ($E = h\nu$).

b. The intensity of a monochromatic electromagnetic wave takes on a new meaning. It is, from the quantum point of view, the energy of each photon multiplied by the number of photons crossing a unit transverse area per unit time. An increase in the intensity of a light beam means, therefore, a proportionate increase in the number of photons striking the metal surface. It is then expected that the number of photoelectrons or the photocurrent i will be proportional to I.

a. Photoemission occurs with no appreciable delay, because whether an electron is released depends, even at the smallest intensity, not upon its accumulating energy, but simply upon the fact of its being hit by a photon that on stopping relinquishes all of its energy to it.

Table 36-1 summarizes the results of experiment, the classical interpretation, and the quantum interpretation for each of the four effects shown in Fig. 36-3.

The photoelectric effect can occur whenever a photon strikes a *bound* electron with enough energy to exceed the binding energy of the electron, for example, a photon freeing a bound electron from a single atom. The kinetic energy of the released electron must,

Table 36-1

Effect Fig. 36-3	Experiment	Classical electromagnetism	Quantum theory
a	Essentially instantaneous photoemission (10^{-9} s)	Emission only after several hundred *hours* (10^6 s) for low intensities	A single photon gives its energy to a single electron essentially instantaneously
b	$I \propto i$	Energy/area-time $\propto i$	$I \propto$ number of photons $\propto i$
c	A well-defined $\frac{1}{2}m_0 v_{max}^2$, dependent only on ν	Inexplicable	A photon gives all its energy to a single electron
d	A threshold for photoemission, independent of I and i: $h\nu = \frac{1}{2}m_0 v_{max}^2 + h\nu_0$	Inexplicable	Photon energy $= h\nu$; work function $= \phi = h\nu_0$

in general, be written in the relativistic form $E - E_0$, and so the more general form of (36-2) becomes

$$h\nu = (E - E_0) + E_b \qquad (36\text{-}6)$$

The photoelectric effect thus provides an indirect method of measuring the energy of a photon. Suppose that the photoelectron's kinetic energy $E - E_0$ is measured and the binding energy E_b is known on some other basis; then $h\nu$ may be computed from (36-6). Conversely, if $h\nu$ and $E - E_0$ are measured, E_b can be determined.

The fundamentally new insight into the nature of electromagnetic radiation that the photoelectric effect provides is the quantization of electromagnetic waves, or the existence of photons. We may properly speak of the quantization of electromagnetic waves because the radiation may be regarded as a collection of particlelike photons, each of energy $h\nu$. When the frequency of the radiation is specified as ν, the photon can have but one energy, $h\nu$. The total energy of a beam of monochromatic electromagnetic radiation is always precisely an integral multiple of the energy $h\nu$ of a single photon. See Fig. 36-5.

The granularity of electromagnetic radiation is not conspicuous in ordinary observations, because the energy of any one photon is very small, and because the number of photons in a light beam of moderate intensity is enormous. The situation is rather like that found in the molecular theory: The molecules are so small and their numbers so great that the molecular structure of all matter is disclosed only in observations of considerable subtlety.

The electromagnetic spectrum, often scaled in units of fre-

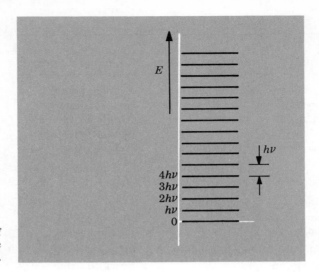

FIG. 36-5. Allowed energies E of a beam of monochromatic electromagnetic radiation.

quency, can be scaled from the point of view of the quantum theory, in units of energy per photon; see Fig. 36-6.

The energy per photon is smallest for radio-wave photons ($\approx 10^{-12}$ eV) and largest for γ-ray photons (≈ 1 GeV). The electromagnetic frequency spectrum corresponds exactly to the energy spectrum of a zero-rest-mass particle, or photon, whose energy can extend from zero to infinity, as shown in Fig. 36-6. The rest energies of the electron and proton are also shown for comparison.

The ideas of wave and particle are apparently mutually incompatible, even contradictory. An ideal particle has vanishing dimensions and is completely localizable, while an ideal wave, one with a perfectly defined wavelength and frequency, must have an infinite extension through space. The fact that in the photoelectric effect light behaves as if it consisted of particles or photons does not mean that we can dismiss the incontrovertible experimental evidence of the wave properties of light; both descriptions must be accepted. An account of the way in which this dilemma is resolved will be postponed (Sec. 37-4) until after we have explored more fully the quantum attributes of light.

36–3 X-RAY PRODUCTION AND BREMSSTRAHLUNG

In the photoelectric effect a photon transfers all its energy to a bound electron. The inverse effect is that in which an electron loses kinetic energy and in so doing creates one or more photons. The process is most clearly illustrated in the production of x-rays.

First consider what happens when a fast-moving electron comes close to and is deflected by the positively charged nucleus of an atom. The electron is accelerated. Classical electromagnetic

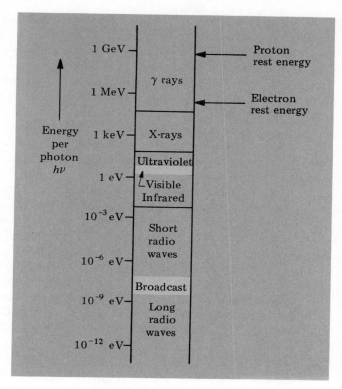

FIG. 36-6. Spectrum of electromagnetic radiation on scale of photon energy.

theory predicts that any accelerated electric charge will radiate electromagnetic energy. But quantum theory requires that any radiated electromagnetic energy consist of discrete photons. It is expected, then, that a deflected and therefore accelerated electron will radiate one or more photons and that the electron will leave the site of the collision with reduced kinetic energy.

The radiation produced in such a collision is often referred to as *bremsstrahlung* ("braking radiation" in German). A *bremsstrahlung* collision is shown schematically in Fig. 36-7, in which an electron approaches the deflecting atom with a kinetic energy E_{k1} and recedes with a kinetic energy E_{k2} after having produced a single photon of energy $h\nu$. The conservation of energy law requires that

$$E_{k1} - E_{k2} = h\nu \qquad (36\text{-}7)$$

Because the atom's mass is at least 2,000 times greater than the electron's, we may ignore the very small energy of the recoiling atom. Whereas classical electromagnetic theory predicts continuous radiation throughout the time that the electron is accelerated, quantum theory requires the radiation of single, discrete photons. That this occurs in the *bremsstrahlung* process is clearly illustrated in the production of x-ray photons.

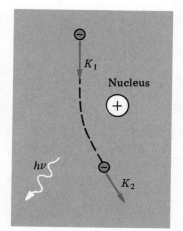

FIG. 36-7. *Bremsstrahlung* collision between an electron and a positively charged nucleus with the emission of a single photon.

X-rays were discovered and first investigated in 1895 by Wilhelm Roentgen, who assigned this name because the true nature of the radiation was at first unknown. X-rays are now known to consist of electromagnetic waves, or photons, having wavelengths of about 10^{-10} m = 1 Å and exhibiting the wave phenomena of interference, diffraction, and polarization. X-rays pass readily through many materials that are opaque to visible light. Because a typical x-ray wavelength is far shorter than the wavelengths of visible light, the experiments require considerable ingenuity. Our chief concern here, however, will be with the energy characteristics of x-ray production.

The essential parts of a simple x-ray tube are shown in Fig. 36-8. Electric current through the filament F heats the cathode C, and the electrons in the cathode are supplied with enough kinetic energy to overcome their binding to the cathode surface and be released in thermionic emission. The electrons are then accelerated through a vacuum by a large electrostatic potential difference V, typically several thousand volts, and strike the target T, which is the anode. While going from the cathode to the anode and before striking the target each electron attains a kinetic energy E_k which is given by

$$E_k = eV$$

where e is the electron charge. We have ignored the electron's kinetic energy as it left the cathode, typically much less than Ve. When the electron strikes the target, it acquires an additional energy, the energy that binds it to the target surface; but this binding energy is also only a *few* electron volts and it too may be ignored.

Upon striking the target, the electrons are decelerated and brought essentially to rest in collisions. Each electron loses its kinetic energy $E_k = eV$ because of its impact with the target. Most of this energy appears as thermal energy in the target, but there is, in addition, the production of electromagnetic radiation through

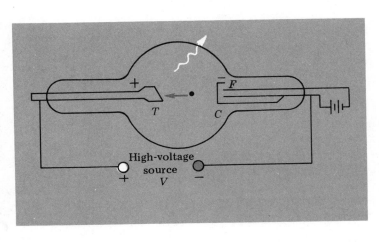

FIG. 36-8. Essential parts of an x-ray tube.

the *bremsstrahlung* process. Any electron striking the target may make a number of *bremsstrahlung* collisions with atoms in the target, thereby producing a number of photons. The *most* energetic photon is produced, however, by an electron whose *entire* kinetic energy is converted into the energy of a *single* photon when the electron is brought to rest in a single collision. Thus $E_{k1} = eV$ and $E_{k2} = 0$, and (36-7) becomes

$$eV = E_k = h\nu_{max}$$

where ν_{max} is the maximum frequency of the x-ray photons produced. More often electrons lose their energy at the target by heating it or by producing two or more photons, the sum of whose frequencies will then be less than ν_{max}. We expect, then, a distribution in photon energies with a well-defined maximal frequency ν_{max} or minimum wavelength $\lambda_{min} = c/\nu_{max}$ given by

$$E_k = h\nu_{max} = \frac{hc}{\lambda_{min}} = eV \tag{36-8}$$

The observed variation in the intensity of emitted x-rays as a function of frequency indicates an abrupt cutoff at the limit ν_{max} of the *continuous x-ray spectrum*, this limit being determined only by the accelerating potential V of the x-ray tube, not the chemical identity of the target material. The value of hc/e can be determined with considerable precision by using (36-8) and simultaneous measurements of λ_{min} and V. The value obtained for Planck's constant h agrees completely with values deduced from experiments on the photoelectric effect and other experiments.

Superimposed on the continuous spectrum are sharp increases in the intensity, or peaks, whose wavelengths are characteristic of the target material; the explanation of these characteristic x-ray lines is to be found in the quantum description of the atomic structure of the target atoms.

36-4 THE COMPTON EFFECT

In the photoelectric effect a photon gives (nearly) all of its energy to a bound electron; it is also possible for a photon to give only part of its energy to a charged particle. This type of interaction is the *scattering* of electromagnetic waves by charged particles. The quantum theory of the scattering of electromagnetic waves is known as the *Compton effect*. We shall first review briefly the classical theory of the scattering of electromagnetic waves by charged particles.

When a monochromatic electromagnetic wave impinges upon a charged particle whose size is much less than the wavelength of the radiation, the charged particle is acted upon principally by the wave's sinusoidally varying electric field. Under the influence

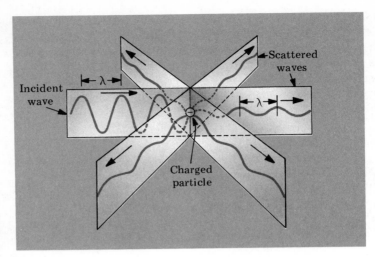

FIG. 36-9. Classical scattering of electromagnetic radiation by a charged particle.

of this changing electric force the particle oscillates in simple harmonic motion at the same frequency as that of the incident radiation (see Fig. 36-9). Since the charged particle is accelerated continuously, it radiates electromagnetic radiation of the *same* frequency in all directions (Sec. 30-6), the intensity being greatest in the plane perpendicular to the line of oscillation and zero along this line. Classical theory predicts, then, that the scattered radiation will have the *same* frequency as that of the incident radiation. The charged particle plays the role of transfer agent, absorbing energy from the incident beam and reradiating at the same frequency (or wavelength) but scattering it in all directions. The scattering particle neither gains nor loses energy, since it reradiates at the same rate as it absorbs. Classical scattering theory agrees with experiment for visible light and other, longer-wavelength radiation. A simple example of the unchanged frequency of coherent scattered radiation is this: Light reflected from a mirror (a collection of scatterers) undergoes *no* apparent change in frequency.

The magnetic field of an incident electromagnetic wave also affects a charged particle. A charge moving in the transverse magnetic field of the electromagnetic wave is acted upon by a magnetic force *along* the direction of wave propagation (Sec. 30-5). When absorption is complete, this results in a radiation force F_r on the charged particle, given by $F_r = P/c$, where P is the power of the incident wave. Moreover, since an electromagnetic wave can exert a force on a scattering center, we attribute momentum p to the wave,

$$p = \frac{E}{c} \qquad (30\text{-}32)$$

where E represents the electromagnetic energy of the incident wave.

Now we consider scattering from the point of view of the quantum theory. Utilizing Einstein's successful photon interpretation of the photoelectric effect, Arthur H. Compton in 1922 used the particlelike, quantum nature of electromagnetic radiation to explain the scattering of x-rays. In the quantum theory electromagnetic radiation consists of photons, each with an energy given by $E = h\nu$. Because a photon may be regarded as a zero-rest-mass particle moving at speed c, Eq. (35-16) shows that the magnitude of its linear momentum $\mathbf{p}$ is given by E/c, in agreement with the classical result. Writing the photon's momentum in terms of its energy, frequency, and wavelength, we have

$$p = \frac{E}{c} = \frac{h\nu}{c} = \frac{h}{\lambda} \qquad (36\text{-}9)$$

Each photon in a beam of monochromatic electromagnetic radiation of wavelength λ has a momentum equal to h/λ. Equation (36-9) shows that the photon's momentum is precisely specified when its wavelength, frequency, or energy is known. The direction of $\mathbf{p}$ is along the direction of wave propagation.

Just as a photon's energy increases with its frequency, so too its momentum increases with frequency. Therefore, the momentum of a high-frequency (or high-energy) photon, such as a γ ray, will exceed by far the momentum of a low-frequency (and low-energy) photon, such as a radio photon.

The distinctive feature introduced by the quantum theory is this: For monochromatic waves the electromagnetic momentum occurs, not in arbitrary amounts, but only in *integral multiples* of the momentum h/λ carried by a single photon.

When we consider a monochromatic electromagnetic beam to consist of a collection of particlelike photons, each with a precisely defined energy and momentum, the scattering of electromagnetic radiation becomes, in effect, a problem involving the collision of a photon with a charged particle. This problem may be solved merely by applying the laws of energy and momentum conservation. Figure 36-10a and b shows the photon and free particle before and after collision. Of course, in applying the conservation laws we need to be concerned, not with the details of the interaction between the photon and particle during the collision, but merely with the total energy and momentum going into and coming out of the collision.

Unlike the classical scattering of electromagnetic waves, in which the particle after collision is assumed to have gained essentially no energy, the quantum treatment requires that it gain at least some momentum and energy. Because these quantities may be great, we must treat the collision relativistically.

We take the particle, of rest mass m_0 and rest energy $E_0 = m_0 c^2$, to be free and initially at rest. Applying energy conservation to the collision of Fig. 36-10 we get

$$h\nu + E_0 = h\nu' + E \qquad (36\text{-}10)$$

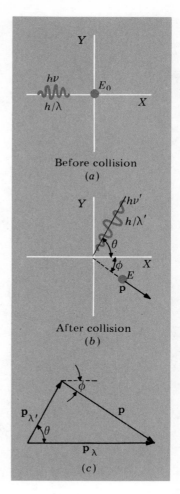

FIG. 36-10. Collision of a photon and a particle initially at rest.

Here E is the energy of the recoiling *particle* after collision, and $h\nu$ and $h\nu'$ are the energies of the incident and scattered photons, respectively. Since the final energy E (rest energy plus kinetic energy) of the recoil particle must exceed its initial energy E_0, we immediately see from (36-10) that $h\nu' < h\nu$. The scattered photon has *less* energy, a lower frequency, and a longer wavelength than the incident photon. This disagrees, of course, with the classical prediction of no frequency change upon scattering. Because the incident and scattered photons have different frequencies, the latter is *not* to be thought of as merely the incident photon moving in a different direction with less energy; rather, the incident photon is annihilated, and the scattered photon is created.

The conservation of linear momentum is implied by the vector triangle of Fig. 36-10c, where $\mathbf{p} = m\mathbf{v}$ is the relativistic momentum of the recoiling particle and the magnitudes of the momenta of the incident and scattered photons are, respectively, $p_\lambda = h\nu/c = h/\lambda$ and $p_{\lambda'} = h\nu'/c = h/\lambda'$. The scattering angle θ is the angle between the directions of $\mathbf{p}_\lambda$ and $\mathbf{p}_{\lambda'}$, the directions of the incident and scattered photons.

We wish to solve for the change in wavelength $\lambda' - \lambda = \Delta\lambda$ in terms of θ. Applying the law of cosines to the triangle in Fig. 36-10c we have

$$p_\lambda^2 + p_{\lambda'}^2 - 2p_\lambda p_{\lambda'} \cos\theta = p^2 \tag{36-11}$$

Multiplying both sides of this equation by c^2 and recalling that for a photon $pc = h\nu$, we have

$$h^2\nu^2 + h^2\nu'^2 - 2h^2\nu\nu' \cos\theta = p^2 c^2 \tag{36-12}$$

We can arrive at a similar expression by using (36-10). We place $h\nu$ and $h\nu'$ on one side of (36-10) and E and E_0 on the other; then squaring the equation, we get

$$h^2\nu^2 + h^2\nu'^2 - 2h^2\nu\nu' = E^2 + E_0^2 - 2EE_0 = 2E_0^2 + p^2c^2 - 2EE_0 \tag{36-13}$$

where we have replaced E^2 with $E_0^2 + p^2c^2$, using Eq. (35-15). Subtracting (36-12) from (36-13), we have

$$-2h^2\nu\nu'(1 - \cos\theta) = 2E_0^2 - 2EE_0$$

$$h^2\nu\nu'(1 - \cos\theta) = E_0(E - E_0) = m_0 c^2 (h\nu - h\nu')$$

$$\frac{h}{m_0 c}(1 - \cos\theta) = c\frac{\nu - \nu'}{\nu\nu'} = \frac{c}{\nu'} - \frac{c}{\nu} = \lambda' - \lambda$$

The increase in wavelength $\Delta\lambda$ is, then,

$$\Delta\lambda = \lambda' - \lambda = \frac{h}{m_0 c}(1 - \cos\theta) \tag{36-14}$$

This is the basic equation of the Compton effect. It gives the increase $\Delta\lambda$ in the wavelength of the scattered photon over that of the incident photon. We see that $\Delta\lambda$ depends only on the rest mass m_0 of the recoiling particle, Planck's constant h, the speed c of light, and the angle θ of scattering. It is perhaps surprising to find that $\Delta\lambda$ is *independent* of the incident photon's wavelength λ. The quantity h/m_0c, appearing on the right-hand side of (36-14) and having the dimensions of length, is known as the *Compton wavelength*. Although the scattering angle θ determines the wavelength increase $\Delta\lambda$ unambiguously, we cannot predict in advance the angle at which any one photon will emerge.

If the recoiling particle is a free electron within the scattering material, then $m_0 = 9.11 \times 10^{-31}$ kg and $h/m_0c = 0.02426$ Å. When a photon emerges at, for example, $\theta = 90°$, the wavelength change, by (36-14), is 0.024 Å. When θ is 180°, with the scattered photon traveling in the backward direction, and the recoil electron forward, the collision being effectively "head-on", the wavelength change is a maximum and equal to 0.049 Å. For such a collision the electron's kinetic energy is also a maximum.

For a 90° scattering, by a free electron, of incident radiation in the visible region, say of 4000 Å, the *fractional* increase in wavelength, $\Delta\lambda/\lambda$, is only 0.0006 percent. Such a wavelength shift is completely masked in visible light by the fact that the electrons in an ordinary scattering material are not at rest but are in thermal motion. An observable shift of, say, $\Delta\lambda/\lambda = 2$ percent can be obtained by using incident radiation of wavelength $\lambda = 1$ Å; then $\Delta\lambda = 0.024$ Å. Thus, the shift is easily observed for x-ray photons and photons of shorter wavelength. For long-wavelength photons the fractional change in wavelength is very small, and the scattered radiation has nearly the same wavelength and frequency as the incident radiation. Classically, the wavelengths of the incident and scattered radiations are essentially equal; hence, Compton scattering agrees with classical scattering in the region of $\Delta\lambda/\lambda \ll 1$.

That the scattering of x-rays agrees with the photon model was shown first by A. H. Compton in 1922. Figure 36-11 gives schematically the experimental arrangement for x-rays incident on a target of carbon, a substance having many free electrons (effectively free, that is). For any fixed angle θ the x-ray detector measures the scattered radiation's intensity as a function of wavelength. Figure 36-12 shows the intensity of the scattered radiation vs. the scattered wavelength for several angles.

At any particular scattering angle θ *two* predominant wavelengths are present in the scattered radiation: one of the same wavelength λ as the incident beam, the *unmodified wave,* and a second, longer wavelength λ', the *modified wave,* given by the Compton equation (36-14). The unmodified wavelength results from the coherent scattering of the incident radiation by the inner electrons of atoms; these electrons are so tightly bound to the atoms that a photon cannot strike one of them without at the same time moving the entire atom. The mass m_0 of a tightly bound electron is, then,

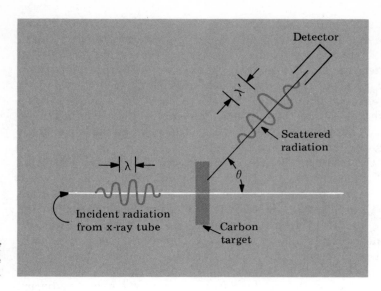

FIG. 36-11. Schematic of experimental arrangement for the Compton effect.

effectively the atom's mass M_0. Therefore, in a Compton collision between a photon and a tightly bound electron the wavelength change $\Delta\lambda$ is $(h/M_0 c)(1 - \cos\theta) \approx 0$, because M_0 is always thousands of times greater than m_0.

The Compton effect provides a simple method of determining the energy of a photon. From (36-10) we have

$$E_k = E - E_0 = h\nu - h\nu'$$

Because $\nu = c/\lambda$ and $\nu' = c/\lambda' = c/(\lambda + \Delta\lambda)$, we may write this as

$$E_k = h\nu \frac{\Delta\lambda}{\lambda + \Delta\lambda} \qquad (36\text{-}15)$$

where $\Delta\lambda$ depends on the scattering angle θ and is given by (36-14). The kinetic energy of the recoil electron is a maximum, $E_{k,\,\text{max}}$, when a head-on collision occurs; then the electron recoils in the forward direction and the scattered photon travels backward. In such a collision $\theta = 180°$ and $\Delta\lambda = 2h/m_0 c$, and (36-15) becomes

$$E_{k,\,\text{max}} = h\nu \frac{2h\nu/m_0 c^2}{1 + 2h\nu/m_0 c^2} \qquad (36\text{-}16)$$

Therefore, if the energy of the most energetic recoil electrons are measured, one can compute the energy of the incident photon from (36-16), and conversely.

The Compton effect clearly illustrates the particlelike aspects of electromagnetic radiation: Not only can a precise energy $h\nu$ be assigned to a photon but also a precise momentum h/λ. Along any direction the total momentum of a monochromatic electromagnetic beam can then assume, *not* any value, but only an exactly integral multiple of the linear momentum of a single photon along that di-

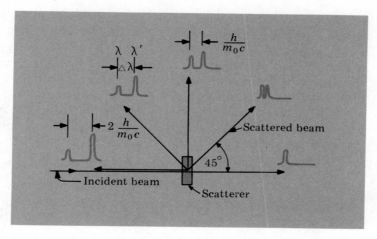

FIG. 36-12. Intensity of scattered radiation vs. wavelength of scattered radiation for several different angles.

rection. In this sense the momentum, as well as the energy, of electromagnetic radiation is quantized.

36–5 PAIR PRODUCTION AND ANNIHILATION

The photoelectric effect, the *bremsstrahlung* process, and the Compton effect are all examples of the conversion of the electromagnetic energy of photons into the kinetic energy and potential energy of material particles, and vice versa. It is natural to ask whether it is possible to convert a photon's energy into *rest* mass—that is, to create pure matter from pure energy—or, on the other hand, to convert rest energy into electromagnetic energy. The answer is yes, provided such conversions do not violate the conservation laws of energy, momentum, and electric charge.

Pair Production. Consider first the minimum energy required to create a single material particle. Since the electron has the smallest nonzero rest mass of all known particles, it requires the least energy for its creation. But a photon has zero electric charge. So the law of electric charge conservation precludes the creation of a *single* electron from a photon. The creation of an electron pair, however, consisting of two particles with opposite electric charges, is possible. The positively charged particle, called a *positron,* is said to be the *antiparticle* of the electron. The electron and positron are similar in all ways except in the signs of their charges, $-e$ and $+e$ (and the effects of this difference). The minimum energy $h\nu_{min}$ needed to create an electron-positron pair is, by energy conservation,

$$h\nu_{min} = 2m_0c^2$$

Since the rest energy m_0c^2 of an electron or positron is 0.51 MeV, the threshold energy $2m_0c^2$ for pair production is 1.02 MeV. The

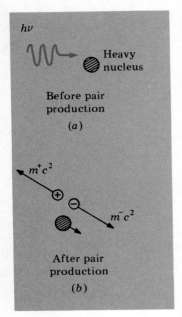

FIG. 36-13. Schematic diagram of pair production.

photon wavelength corresponding to this threshold is 0.012 Å; hence, electron pairs can be produced only by γ-ray photons or by x-ray photons of very short wavelength. The phenomenon in which a particle and its antiparticle are created from electromagnetic radiation is called *pair production*. This is a most emphatic demonstration of the interconvertibility of mass and energy.

If a photon's energy exceeds the threshold energy $2m_0c^2$, the excess appears as kinetic energy of the created pair. Applying the conservation of energy to pair production gives

$$h\nu = m^+c^2 + m^-c^2 = (m_0c^2 + E_k^+) + (m_0c^2 + E_k^-)$$
$$h\nu = 2m_0c^2 + (E_k^+ + E_k^-) \qquad (36\text{-}17)$$

where ν is the frequency of the incident photon, and E_k^+ and E_k^- are the kinetic energies of the created particles. The minimum energy $h\nu_{\min}$, just enough to produce the pair, is obtained, of course, by setting the kinetic energies of the created particles equal to zero: $E_k^+ + E_k^- = 0$.

Pair production cannot occur in empty space. It is easy to prove that energy and momentum cannot simultaneously be conserved in particle-antiparticle production unless the photon is near some massive particle, such as an atomic nucleus. Suppose, for the sake of argument, that a pair has been created in empty space and that we, the observers, are at rest with respect to the center of mass of this two-particle system. Then the total momentum of the pair is zero. But the photon creating the pair would have had some nonzero momentum in this reference frame, since a photon always moves at speed c, whatever the reference frame. Under these imagined circumstances of pair production in empty space, we would have the momentum of the photon before collision but zero momentum after, a violation of the momentum conservation law. In short, a photon cannot decay spontaneously into an electron-positron pair in free space; the process can take place only if the photon encounters a massive particle which acquires the requisite (but negligible) momentum ($E_k = p^2/2M \approx 0$ for large M).

Figure 36-13 is a schematic drawing of pair production, and Fig. 36-14 is a bubble-chamber photograph showing the creation of electron-positron pairs. The paths of the charged particles are visible because of the bubble-producing ionization effects they have as they travel through the liquid; the trajectories of the oppositely charged particles (with approximately equal kinetic energies) show opposite curvatures, the particles having been deflected into oppositely directed circular arcs by a uniform magnetic field.

The energy of a photon producing an electron-positron pair can be computed by means of (36-17), if the kinetic energies of the electron and positron are measured. These energies can be determined from a photograph such as Fig. 36-14, if the magnetic field B and the radius r of curvature of the trajectories are measured. The relativistic momentum p of each particle is given by

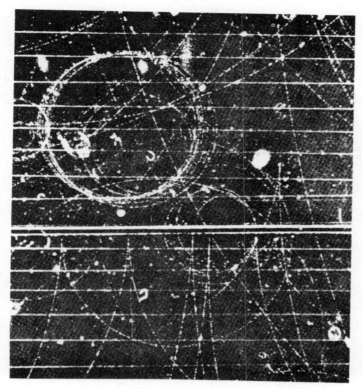

FIG. 36-14. Cloud-chamber photographs showing the creation of electron-positron pairs. Photons of approximately 200 MeV, producing no tracks, appear from the top. Some photons are annihilated, and electron pairs are created in the thin, horizontal lead foil; above the foil and to the right is a pair produced by the collision of a photon and a gas molecule. The external magnetic field, of flux density 1 Wb/m², bends the paths of the electrons and positrons into opposite curvatures. (From *Cloud Chamber Photographs of the Cosmic Radiation*, G. D. Rochester and J. G. Wilson, Pergamon Press, Ltd., 1952. Courtesy of Pergamon Press, Ltd.)

$$p = mv = QBr \qquad (35\text{-}10)$$

and the total energy E or kinetic energy $E - E_0$ of the particle can then be computed by means of using Eq. (35-15), $E^2 = E_0^2 + (pc)^2$.

The existence of positrons was predicted on theoretical grounds by P. A. M. Dirac in 1928. Four years later C. D. Anderson observed and identified a positron during his studies of cosmic radiation. Electron-positron pairs are now a commonly observed phenomenon in the interaction of high-energy photons and matter. Proton-antiproton and neutron-antineutron pairs were first created in the laboratory in 1955. Their threshold energies are several GeV (the proton and neutron masses are approximately equal to 1 GeV and therefore require accelerating machines of very high energy).

Pair Annihilation. The mutual annihilation of particle-antiparticle pairs and the concomitant creation of photons is the pair annihilation process. It is the inverse of pair production. Consider the annihilation of matter and the creation of electromagnetic energy that may occur when an electron and positron are close together and essentially at rest. The total linear momentum of the two particles is initially zero; therefore, a *single* photon cannot be created when the two particles unite and are annihilated, because

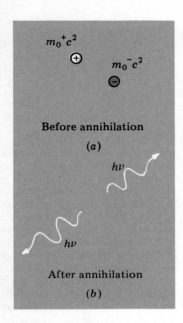

FIG. 36-15. Pair annihilation and the creation of two photons.

that would violate momentum conservation. Momentum can, however, be conserved when *two* photons, moving in opposite directions with equal momenta, are created. Such a pair of photons would have equal frequencies and energies. See Fig. 36-15. (Actually, three or more photons can be created, but with a *much* smaller probability than that of two photons. Similarly, when many electron-positron pairs are annihilated near a heavy nucleus, a small number of the annihilations may produce a single photon.)

Energy conservation implies

$$m_0^+ c^2 + m_0^- c^2 = h\nu_1 + h\nu_2$$

in which the electron and positron are assumed to be at rest initially. But $m_0^+ = m_0^-$, and by momentum conservation, $\nu_1 = \nu_2 = \nu_{min}$; therefore,

$$2h\nu_{min} = 2m_0 c^2$$
$$h\nu_{min} = m_0 c^2$$

Since the minimum energy needed for creating an electron, $h\nu = m_0 c^2$, is 0.51 MeV, this is also the minimum energy of the photon created.

Annihilation is the ultimate fate of positrons. When a high-energy positron appears, as in pair production, it loses its kinetic energy in collisions as it passes through matter, finally moving at low speed. It then combines with an electron, forming a bound system, called a positronium atom, which decays very quickly (10^{-10} s) into two photons of equal energy. Thus, the death of a positron is signaled by the appearance of two annihilation quanta, or photons, of $\frac{1}{2}$ MeV each. The transitoriness of positrons is due, not to an intrinsic instability, but to the high risk of their collision and subsequent annihilation with electrons.

In our part of the universe there is a preponderance of electrons, protons, and neutrons; their antiparticles, when created, quickly combine with them in annihilation processes. It is conceivable, although at present purely conjectural, that there exists a part of the universe in which positrons, antiprotons, and antineutrons predominate.

Pair production and annihilation are particularly striking examples of mass-energy equivalence. They provide irrefutable confirmation of the theory of relativity.

36–6 PHOTON–ELECTRON INTERACTIONS

Figure 36-16 summarizes the important photon-electron interactions, or collisions, discussed in this chapter. In each instance a photon, electron, or positron approaches a slab of material, a collision occurs, and one or more particles emerge. We summarize briefly the salient features of each of these interactions, taking them in the order in which they appear in the figure:

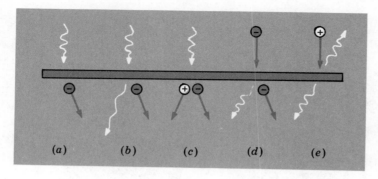

FIG. 36-16. Photon-electron interactions: (*a*) photoelectric effect, (*b*) Compton effect, (*c*) pair production, (*d*) *bremsstrahlung*, (*e*) pair annihilation.

a. The photoelectric effect: A photon strikes a bound electron and disappears, and the electron is dislodged.
b. The Compton effect: A photon collides with a free electron, thereby effecting the creation of a second photon of lower energy and the recoil of the electron.
c. Pair production: A photon is annihilated in the vicinity of a heavy particle, and an electron-positron pair is created.
d. *Bremsstrahlung:* An electron is deflected in the vicinity of a heavy particle, and a photon is created.
e. Pair annihilation: A positron combines with an electron, and a pair of photons is produced.

As we shall see in Sec. 40-1, *all* these photon-electron interactions may be considered to be illustrations of just *one* basic interaction between the particle of the electromagnetic field (the photon) and a particle that can create an electromagnetic field (an electron or any other electrically charged particle). Even the ordinary electric, or coulomb, force between electrically charged particles and, indeed, all other electromagnetic effects, may be ascribed to an interchange of (virtual) photons between charged particles.

The principal features of the photon-electron collisions were derived simply by applying the conservation laws of energy, momentum, and electric charge and by assuming the existence of photons of energy $h\nu$ and momentum h/λ. In no case did we concern ourselves with the details of the interaction, or calculate the probability of the occurrence of any of these processes. For example, in the Compton effect, we are able to predict the wavelength of a photon scattered in any particular direction, but this analysis did not enable us to predict the direction of any particular photon. The probabilities of the occurrence of a photon-electron interaction, however, can be calculated with high precision by the methods of *quantum electrodynamics*.

Highly energetic (at least 10^{19} eV) charged particles from the cosmic radiation enter the earth's atmosphere. They may produce a whole succession of electron-photon interactions, as follows. A collision between a cosmic-ray particle and a nucleus may produce a high-energy γ ray through *bremsstrahlung*. The γ ray may be annihilated on passing near a nucleus and so produce an electron-

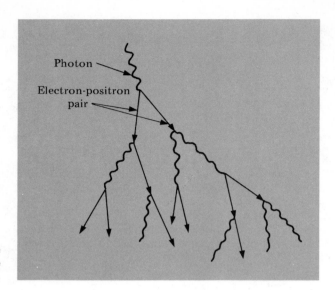

FIG. 36-17. Schematic representation of a cascade shower.

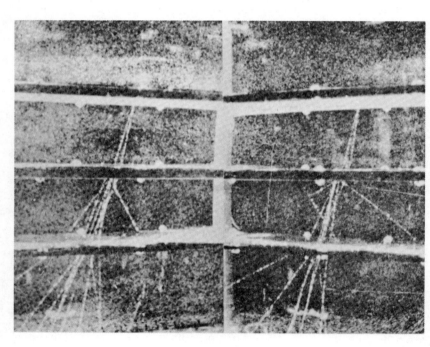

FIG. 36-18. Cloud-chamber photographs of a photon-initiated cascade shower taken simultaneously with two cameras to permit three-dimensional analysis of the tracks. A 700-MeV photon (producing no track) enters from the top, and a positron-electron pair is created at the uppermost horizontal thin lead plate. Photons are created at the lower plates by *bremsstrahlung* collisions, and these photons create more pairs, leading to a shower of electrons, positrons, and photons. (Courtesy of J. C. Street, Harvard University.)

positron pair. The created charged particles, having large kinetic energies, may collide with, and be deflected by, nuclei they encounter on their way to the earth's surface; by virtue of their acceleration after the collisions they radiate high-energy photons by the *bremsstrahlung* process. The positron may combine with the electron, both be annihilated, and two photons be created. The secondary photons may have energies exceeding 1.02 MeV and so may produce more electron pairs. Thus, by the repeated occurrence of pair production, pair annihilation, *bremsstrahlung*, and, to a lesser extent, Compton and photoelectric collisions, a *cascade shower* of electrons, positrons, and photons is produced, the energy of the original photon having been degraded and spread among many particles. The shower is effectively extinguished when pair production becomes energetically impossible. A diagram of photon-electron interactions is shown in Fig. 36-17; spectacular cloud-chamber photographs, such as Fig. 36-18, have confirmed the principal features of cascade showers.

SUMMARY

The quantum theory, formulated by Max Planck in 1900, has shown that many quantities that appear superficially to have continuous values actually have discrete values only.

Electromagnetic radiation, must be considered to consist of photons, each photon having a discrete energy and momentum:

$$E = h\nu \quad \text{and} \quad p = \frac{h}{\lambda} \qquad (36\text{-}3), (36\text{-}9)$$

The following are the basic photon-particle interactions.

Photoelectric effect: The *complete* transfer of *electromagnetic energy* to a *bound electron*:

$$h\nu = E_b + (E - E_0) \qquad (36\text{-}6)$$

where E_b is the binding energy, work function, of an electron.

Bremsstrahlung: The *partial* or *complete* transfer of a particle's *kinetic energy* to *electromagnetic energy*. Photons of maximum frequency (minimum wavelength) are produced when an electron is brought to rest in a single collision:

$$eV = E_k = h\nu_{max} = \frac{hc}{\lambda_{min}} \qquad (36\text{-}8)$$

Compton effect: The *partial* transfer of *electromagnetic energy* to the *kinetic energy* of a particle. When a photon of wavelength λ interacts with a (nearly) free particle essentially at rest, a photon emerges at an angle θ (scattering), and the particle recoils with kinetic energy E_k:

$$\Delta\lambda = \lambda' - \lambda = \frac{h}{m_0 c}(1 - \cos\theta) \qquad (36\text{-}14)$$

$$E_k = h\nu \frac{\Delta\lambda}{\lambda + \Delta\lambda} \qquad (36\text{-}15)$$

where m_0 is the rest mass of the recoil particle.

Pair production and annihilation: The *complete* conversion

of *electromagnetic energy* into the *rest energy* and *kinetic energy* of the created particles, and the reverse:

Pair production: $h\nu = 2m_0 c^2 + (E_k^+ + E_k^-)$ (36-17)

Pair annihilation: $(m^+ + m^-)c^2 = 2h\nu$

PROBLEMS

36-1. The threshold wavelength for the photoemission of electrons from a calcium surface is 3840 Å. *(a)* Calculate the binding energy, or work function, ϕ (in electron volts) of an electron at the surface of calcium. *(b)* What is the maximum kinetic energy (in electron volts) of a photoelectron emitted from the surface when light of 2000 Å strikes this surface?

36-2. *(a)* What is the maximum speed of photoelectrons emitted from a zinc surface ($\phi = 4.23$ eV) when ultraviolet light of 1550 Å is used? *(b)* Does this justify our assumption that $E_{k,\,max} = \tfrac{1}{2} m v_{max}^2$?

36-3. A 13.0-eV photon dislodges an electron from an atom so that the released electron has a kinetic energy of 4.0 eV. What was the binding energy of the electron within the atom?

36-4. A monochromatic beam of light shines on a photo-emitter. When photons of energy E_1 illuminate the photoemitting surface, the maximum energy of the emitted photoelectrons is K_1. When the wavelength of the incident radiation is changed, the maximum energy of the emitted photoelectrons is doubled. What is the energy of the latter photons (in terms of E_1 and K_1)?

36-5. How many photons are radiated per second from a 1.0-mW source of *(a)* 4,000 Å visible light, *(b)* 500 kHz radio waves, *(c)* 1.0 Å x-rays, and *(d)* 1.0 GeV γ rays?

36-6. When monochromatic light of wavelength 4046 Å shines on a certain metal surface, the most energetic photoelectrons are stopped by a retarding potential of 1.60 V; when the wavelength is 5769 Å, the stopping potential is 0.45 V. Assuming h and e unknown, what are *(a)* the work function (in eV) of this photoemitter and *(b)* the value of h/e computed from these data?

36-7. Light of intensity 1.0×10^{-10} W/m² falls normally upon a silver surface, where there is one free electron per atom. The atoms are approximately 2.6 Å apart. Treat the incident radiation classically (as waves), and assume the energy to be uniformly distributed over the surface and all the light to be absorbed by the surface electrons. *(a)* How much energy does each free electron gain per second? *(b)* The binding energy of an electron at a surface is 4.8 eV; how long must one wait, after the beam is switched on, before any one electron gains enough energy to overcome its binding energy and be released as a photoelectron? Compare this with the experimental results.

36-8. A photon enters a so-called lead "radiator" (which radiates photoelectrons) and interacts with an inner electron bound to a lead atom with a binding energy of 89.1 keV. The photoelectron released then enters a uniform magnetic field, and Br (where r is the radius of curvature of the electron in the magnetic field B) is found to be 2.0×10^{-3} Wb/m. *(a)* What is the momentum of the photoelectron (in MeV/c)? *(b)* What is the kinetic energy of the photoelectron? *(c)* What is the energy of the incident photon?

36-9. Photoelectric emission can take place only if the electron is initially *bound,* not if the electron is initially free. Said differently, a photon cannot be absorbed by a single free electron. Prove this by showing that the absorption of a photon by a free particle would violate the conservation laws of energy and momentum.

36-10. A 3.10-Å photon strikes a hydrogen atom at rest, thereby releasing the bound electron (binding energy, 13.6 eV). Suppose that the electron moves in the same direction as that of the incident photon. What are the electron's *(a)* kinetic energy and *(b)* momentum? What are the recoiling ion's *(c)* momentum and *(d)* kinetic energy?

36-11. *(a)* Planck's constant h has units of energy multiplied by time: Show that h has the units of angular momentum. *(b)* A circularly polarized electromagnetic wave of energy E has an angular momentum $L = E/\omega$, where ω is the angular frequency of the wave: Show that from the point of view of the quantum theory such a beam consists of photons each with an angular momentum of $h/2\pi$. An alternative formulization to $E = h\nu$ of the basic quantum condition is this: Every photon, quite apart from its frequency, has the same angular momentum of magnitude $h/2\pi$.

36-12. A well-accommodated human eye is capable of detecting single photons of visible light. At what distance from an eye having a pupil 4 mm in diameter would an isotropic point source radiating 5000 Å light equally in all directions at a rate of 1 W have to be placed so that the number of photons reaching the eye's retina is one per second on the average?

36-13. A monochromatic point source of light radiates continuously. How does the density of photons vary with the distance r from the source in any one direction?

36-14. (a) Show that the energy of a 1.0 Å-wavelength x-ray photon is 12.40 keV. (b) Show that the general relation between the energy E of a photon in MeV and its wavelength λ in Å is $E = (0.01240 \text{ MeV-Å})/\lambda$.

36-15. Electron 1 is accelerated from rest through a potential difference V. Electron 2, also accelerated from rest through the same potential difference, collides with a target, comes to rest, and thereby creates a single photon. Which has the larger momentum, electron 1 or the photon produced by electron 2?

36-16. A beam of electromagnetic radiation comes from a laser source radiating 2.0 MW in a pulse of 1.0×10^{-6} s duration. What is the total momentum of the photons in the beam?

36-17. Visible light of 4,000 Å passes through a camera lens as a photograph is being taken. The light comes from a uniformly radiating (isotropic) point source with a power output of 1.0 W at a distance of 1.0 km from the camera lens. The camera shutter is set for an exposure of $\frac{1}{100}$ s at an aperture of $f/2$ on a 50 mm focal-length camera lens (the effective aperture diameter is 50 mm/2 = 25 mm). What is (a) the total number of photons reaching the film and (b) the total momentum transferred to the film, assuming complete absorption of the incident light?

36-18. (a) Show that a rocket emitting zero-rest-mass particles, or photons, is more efficient than a rocket emitting particles with finite rest mass for a given total energy of emitted particles. (Hint: Write the general relationship for the relativistic momentum of a particle in terms of its total energy E and rest energy E_0.) (b) Suppose that a rocket initially at rest radiates energy (some of its initial mass) in one direction until the fraction of the mass remaining is F and at that point the rocket is traveling at the relative speed v/c. Write down equations indicating the conservation of energy and momentum, and show that $F = (1 - v/c)^{1/2}/(1 + v/c)^{1/2}$. (c) If the final rocket speed is to be such that time-dilation effects involve a factor 5, show that $v/c = 0.98$ and that F is then $\frac{1}{10}$.

36-19. What is the wavelength of a photon having the same (a) energy and (b) momentum as a 2-eV electron?

36-20. (a) Across what minimum potential difference must electrons be accelerated from rest so that upon striking a target they produce photons with a momentum of 1.0 keV/c? (b) Across what potential difference must electrons be accelerated from rest to acquire a momentum of 1.0 keV/c?

36-21. A beam of monochromatic photons each of energy $h\nu$ has a photon flux N; that is, N photons per unit time traverse a unit area at right angles to the beam direction. The intensity of the beam is I, where I is the electromagnetic energy crossing a unit area at right angles to the direction of the beam per unit time. (a) Show that $I = Nh\nu$. (b) Suppose that the beam is completely absorbed. Show that the radiation pressure of the beam on the absorber is given by $I/c = Nh\nu/c$.

36-22. A radar transmitter produces pulses of microwave radiation; the power of each pulse is 10 MW and the duration 1.0 μs. Take the wavelength of the emitted radiation to be 1.0 cm. (a) What is the total momentum of the emitted radar pulse? (b) How many photons are emitted per pulse?

36-23. A 1.0-kW isotropic point source of electromagnetic radiation is at the center of a perfectly absorbing spherical shell. For what radius is the radiation pressure on the inner wall of the sphere equal to that of standard atmospheric pressure (1.0×10^5 N/m²)?

36-24. The intensity of solar radiation at the earth's surface is 1,400 W/m². (a) What is the pressure of this radiation at the earth's surface? (b) Compute the total force on the earth due to this radiation from the sun (earth's radius, 6.4×10^6 m), assuming complete absorption.

36-25. Photons of wavelength 0.0620 Å are scattered by free electrons. What are the wavelengths of those scattered photons whose angle of scattering is (a) 90° and (b) 180°? What are the energies transferred to the free electrons (c) in part (a) and (d) in part (b)?

36-26. An incident photon is scattered by a free electron initially at rest. The *scattered* photon may later produce an electron-positron pair. Show that if the angle between the scattered photon and the incident photon is greater than 60°, the scattered photon cannot create an electron-positron pair, no matter how large the energy of the incident photon.

36-27. A photon of 1.00×10^{-5} Å wavelength undergoes a Compton collision in which the scattered photon is backscattered and has a wavelength of 3.62×10^5 Å. What is the mass of the particle producing the scattering?

36-28. A 5.0-keV electron collides head-on with a second 5.0-keV electron traveling in the opposite direction, and both electrons are brought to rest in the collision. *(a)* What is the minimum number of photons that can be created in this collision? *(b)* What is the maximum photon energy?

36-29. A free *proton* originally at rest is struck by a photon in a Compton collision, and the proton thereby acquires a kinetic energy of 5.7 MeV. What is the minimum photon energy?

36-30. A monochromatic beam of photons strikes a block of metal, and a detector registers the photons emerging from the block at 90° with respect to the incident beam; see Fig. P36-30a. Figure P36-30b shows the energy spectrum of the photons observed at 90°. There are three distinct peaks: at 0.36, 0.51, and 1.24 MeV. *(a)* What is energy per photon in the incident beam? *(b)* Give the physical basis of each of the three peaks.

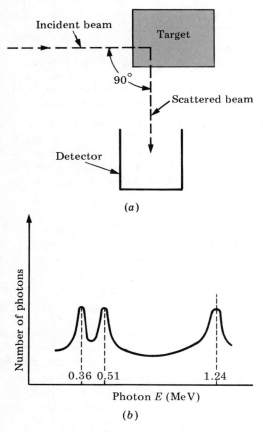

FIG. P36-30

36-31. If the radii of curvature of both the electron and positron created when a photon interacts with a heavy nucleus are 1.0 cm when the particles are bent in a uniform magnetic field of 0.50 Wb/m^2, what is the wavelength of the incident photon?

36-32. An electron and positron are moving together as a system (positronium) at a velocity of $c/2$. If these two particles are annihilated and two photons are created, *(a)* what are the energy and momentum of each photon, and *(b)* what is the angle between the direction of motion of these two photons? (An experiment involving positronium atoms moving at $\frac{1}{2}c$ was performed by D. Sadeh in 1963; the angle between the annihilation quanta was found, as expected, to be less than 180°.)

36-33. A cascade shower is initiated by a photon with an energy of 200 MeV. What is the greatest number of positrons it can produce?

36-34. Photons can be absorbed by a material, that is, removed from the forward direction of the incident beam by three processes: the photoelectric effect (requiring bound electrons), the Compton effect (requiring free electrons), and pair production (requiring massive atomic nuclei). Indicate how one expects the effectiveness of photon absorption to depend on the photon energy for the photoelectric effect and pair production.

36-35. *(a)* A particle of mass m moving at nonrelativistic speed is to strike another particle of mass m, initially at rest, in a completely inelastic collision and thereby dissipate energy E. Show that the minimum kinetic energy of the particle initially in motion is $2E$, not E. (*Note:* Both energy *and* momentum must be conserved in every collision.)

(b) Antiprotons were produced and first identified in the laboratory in 1955 (by Chamberlain, Segre, Wiegand, and Ypsilantis, using the Bevatron accelerator at the University of California Radiation Laboratory). The reation was $p^+ + p^+ \rightarrow p^+ + p^+ + (p^+ + p^-)$; that is, an incident proton struck a proton at rest, and an additional proton and antiproton were created in the collision. Show that the minimum kinetic energy of the incident proton needed for this reaction is 5.6 GeV or *six* times the proton rest energy. Superficially, one might imagine that only *two* times the proton rest energy would be required to create an additional proton and antiproton. However, since both energy *and momentum* must be conserved in every collision, the four particles emerging from the collision must have a total momentum equal to that of the incident proton. Consequently, the particles must emerge from the collision with some kinetic energy, and only a fraction of

the incident proton's kinetic energy is available to create the additional two particles. Moreover, the computation requires relativistic dynamics.

To solve for the threshold energy, we might use the following argument. Viewing from a reference frame at rest with respect to the system's center of mass, we see the incident proton and target proton approach in opposite directions with equal energies and equal momentum magnitudes, collide and, for the threshold energy, create four particles at rest. Thus, as seen from the laboratory reference frame, the four particles after the collision have the *same* momentum, and this momentum is, by momentum conservation, just one-fourth that of the incident proton. On this basis it can be shown that the incident particle must have a momentum of $4(3^{1/2})Mc$, where M is the proton rest mass. The corresponding kinetic energy of the incident proton is $6Mc^2$.

CHAPTER 37

Quantum Effects: The Wave Aspects of Material Particles

37-1 DE BROGLIE WAVES

Electromagnetic radiation has two aspects: a wave aspect and a particle aspect. Interference and diffraction of electromagnetic radiation are explained by assuming electromagnetic radiation consists of waves. Quantum effects, such as the photoelectric and Compton effects, are explained by assuming electromagnetic radiation consists of particlelike photons, each photon with energy E and momentum p, specified precisely by the frequency ν and wavelength λ of the radiation as follows:

$$\nu = \frac{E}{h} \tag{37-1}$$

$$\lambda = \frac{h}{p} \tag{37-2}$$

Note that in these equations two quantities that have clear meanings only when a wave is being described appear on the left, and on the right appear two quantities usually associated with a particle. Thus, the wave-particle duality of electromagnetic radiation is implied in these fundamental relations, and it is the fundamental constant of the quantum theory, Planck's constant h, that relates the wave to the particle characteristics. We may say that electromagnetic waves under some circumstances behave as particles and that photons (zero-rest-mass particles) under some circumstances behave as waves.

Do the two equations, which ascribe both a wave and a particle nature to electromagnetic radiation, have an even greater generality—do they apply to *all* particles, that is, to *finite* as well as to zero-rest-mass particles? This question was first posed by Louis de Broglie in 1924. He conjectured that a material particle might well exhibit wave properties and that the equations above apply also to material particles, such as electrons. Experiments have emphatically confirmed the correctness of de Broglie's hypothesis, and the wave character of material particles is now well established. Because the wavelength enters directly in interference or diffraction effects, we shall concentrate our attention on the second relation, Eq. (37-2).

The wavelength λ of a particle with momentum $p = mv$, where m is the relativistic mass of the particle and v is its velocity, is given by the *de Broglie relation*

$$\lambda = \frac{h}{p} = \frac{h}{mv} \qquad (37\text{-}3)$$

where h is Planck's constant.

To test the deBroglie hypothesis is to determine on the basis of experiment whether particles show interference and diffraction effects. Of course, the question of the physical nature of a particle's wave aspect is a crucial one (if an electron is a wave, what is waving?); we shall, however, postpone it until after we have discussed experiments that confirm (37-3).

The wave nature of electrons was discovered in 1927, when the electron-diffraction experiments of C. Davisson and L. H. Germer confirmed the de Broglie relation, and 29 years after the existence of electrons had been established. Why were the wave characteristics of electrons discovered only many years after their particle nature had been established? The origin of the difficulties is a very small wavelength. As (37-3) shows, a 1.0-kg particle moving at 1.0 m/s has a wavelength of only 6.6×10^{-34} m = 6.6×10^{-24} Å. Clearly, we should not expect, in throwing baseballs through an open window, to find a discernible diffraction pattern of hits on a distant wall any more than we should expect to see one when visible light passes through such a wide "slit." The wavelength of an ordinary particle is very small compared with the dimensions of ordinary objects, so that interference and diffraction effects are subtle. Therefore, if an object's wavelength is to be large enough to produce observable wave effects, its mass and velocity, as we see from (37-3), must be small (clearly, for a large wavelength and low speed nonrelativistic relations may be used).

The diffraction grating having the smallest distance between "lines" is a crystal, a solid in which the atoms are located in a three-dimensional geometrical array. A typical distance between adjacent atoms is of the order of 10^{-10} m, or 1 Å. To observe the diffraction of particles, the particle wavelength must be of comparable size. Since $\lambda = h/mv$, to have large λ we must choose a particle with small m, namely the electron.

Let us compute the kinetic energy of an electron with a wavelength λ of 1.00 Å. An electron with electric charge e, accelerated from rest by an electrostatic potential difference V, acquires a final kinetic energy $\tfrac{1}{2}mv^2$ when

$$eV = \tfrac{1}{2}mv^2 = \frac{1}{2m}p^2 = \frac{1}{2m}\left(\frac{h}{\lambda}\right)^2$$

$$V = \frac{h^2}{2me\lambda^2} = \frac{(6.62 \times 10^{-34}\text{ J-s})^2}{2(9.11 \times 10^{-31}\text{ kg})(1.60 \times 10^{-19}\text{ C})(1.00 \times 10^{-10}\text{ m})^2}$$

$$= 150\text{ V}$$

An electron of 150 eV has a wavelength of 1 Å. Since this wavelength is comparable to that of a typical x-ray photon, we may expect that both electrons and x-rays will show similar diffraction effects when passing through a crystal.

37–2 THE BRAGG LAW

Max von Laue was the first to suggest in 1912 that crystalline solids, in which the arrangement of atoms follows a regular pattern, and in which the distance between atoms is approximately 1 Å, might be used as diffraction gratings for measuring x-ray wavelengths.

Consider a crystal of sodium chloride. Sodium and chlorine atoms are located at alternate corners of identical elementary cubes, each with a distance d along an edge.

The lattice spacing of sodium and chlorine atoms in a crystal of rock salt is 2.820 Å. See Fig. 37-1. This distance, typical of the interatomic spacing of atoms in any solid, is comparable to the wavelength of x-rays or of electrons of 150 eV.

When a wave impinges upon a collection of scattering centers,

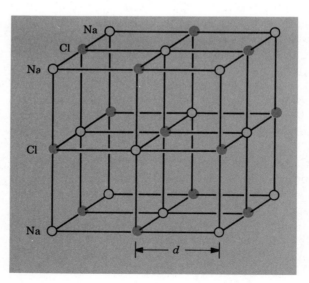

FIG. 37-1. Crystal structure of rock salt (NaCl).

such as the atoms in a crystalline solid, each scattering center generates waves, which radiate outward from it in all directions. The resultant wave from all scattering centers, measured in any one direction, depends, of course, on the interference between all the separate centers. Atoms lying on any one plane within the crystal act with respect to the incident wave as a partially silvered mirror with respect to visible light; that is, they reflect a portion of the wave while allowing the remainder to pass through. These *Bragg planes* and *Bragg reflections* are named after W. H. Bragg, who with his son, W. L. Bragg, developed the fundamental theory of x-ray diffraction by crystals in 1913. Given this fact, we can deal, not with the interference between the waves generated by all of the scattering centers individually but, more simply, with the interference between the waves reflected from parallel Bragg planes.

Consider the reflection of waves from two adjacent and parallel Bragg planes as shown in Fig. 37-2. The directions of the incident and reflected waves, both denoted by θ, are specified by the angle between the direction of propagation of the waves and the Bragg plane (*not* the normal to the reflecting planes). At each plane we regard the incident wave as partially transmitted undeviated and partially reflected.

The incident ray is partially reflected at the first Bragg plane, the reflected ray AB making an angle θ with plane 1. That part, AC, of the incident ray which is transmitted through the first plane is partially reflected from the second plane, also in the direction θ. We concentrate on the wavefront BD perpendicular to the two reflected rays. The reflected rays will constructively interfere at some distant point only if they have the same phase at the points

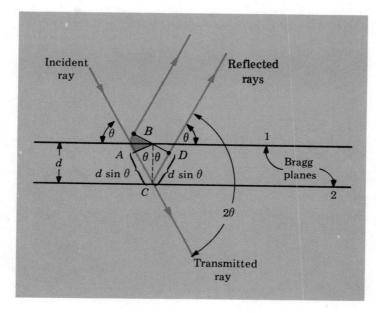

FIG. 37-2. Reflection of waves from two adjacent Bragg planes.

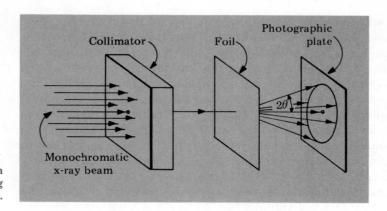

FIG. 37-3. Scattering of monochromatic x-rays by a thin metallic foil from one set of Bragg planes.

B and D. The points B and D are in phase when the path difference $ACD - AB = 2d \sin \theta$ is an integral multiple n of the wavelength λ. The condition, then, for constructive interference of waves reflected from adjacent parallel Bragg planes is

$$n\lambda = 2d \sin \theta \qquad (37\text{-}4)$$

where n, the *order of the reflection,* can have the values $1, 2, 3, \ldots$. This equation,† known as *Bragg's law,* is the basis of all coherent x-ray and electron diffraction effects in crystals. Rays reflected at any angles except those satisfying the equation destructively interfere, and the incident beam is completely transmitted. The Bragg law is the means of measuring wavelengths comparable to interatomic distances, for clearly, if n, d, and θ are known, λ can be computed. Note in Fig. 37-2 that the angle between the transmitted and the reflected rays is 2θ and that the Bragg planes bisect this angle.

---

†Equation (37-4), Bragg's law, bears a resemblance to the equation that applies to an ordinary ruled diffraction grating. The two equations, however, are *not* the same.

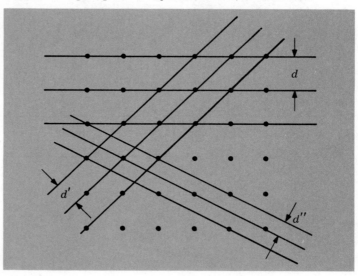

FIG. 37-4. Three sets of parallel Bragg planes with different lattice spacings.

37–3 X-RAY AND ELECTRON DIFFRACTION

Consider a thin metallic foil through which a monochromatic x-ray beam is sent. The foil consists of a very large number of simple, perfect crystals, randomly oriented with respect to one another within the foil. Only those particular microcrystals which are so oriented that the Bragg condition is fulfilled will produce a strongly diffracted beam. Therefore, the emerging beam will consist of two parts: an intense, central, undeviated beam and a beam scattered in a conical shape that makes an angle 2θ with respect to the incident beam; see Fig. 37-3. The angle θ is uniquely determined, for a given order, by the Bragg relation.

When the scattered beam strikes a flat photographic plate, the intensity pattern consists of a strong central spot surrounded by a circle. The radius of the circle is easily measured, and the distance from the scattering foil to the photographic plate is known, and so the angle 2θ can be found; finally, the wavelength of the x-rays can be computed from the Bragg relation.

Any given crystal has, not only one set of parallel Bragg planes, but also many sets of planes in any single crystal. A Bragg plane is any plane that contains atoms, and there are many such planes in a cubic crystal, as shown by Fig. 37-4. They differ in the value of the grating space d; consequently, there will be a number of Bragg angles θ, each satisfying the Bragg relation for a particular set of Bragg planes. The x-ray diffraction pattern will therefore be somewhat more complicated than that indicated in Fig. 37-3; it consists of a number of concentric circles, each corresponding to diffraction from a particular set of Bragg planes. Figure 37-5 is an x-ray diffraction pattern of a sample of polycrystalline aluminum.

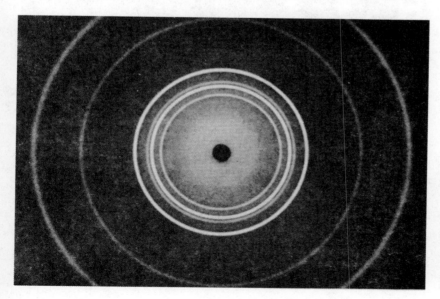

FIG. 37-5. X-ray diffraction pattern of polycrystalline aluminum. The center is dark because a hole was cut in the photographic plate to allow the strong central beam to pass through it. (Courtesy of Mrs. M. H. Read, Bell Telephone Laboratories, Murray Hill, New Jersey.)

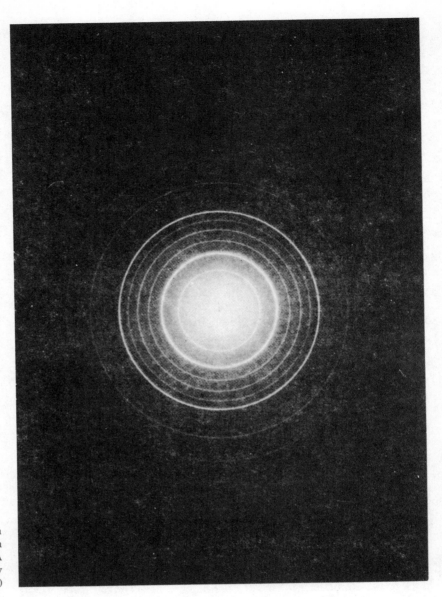

FIG. 37-6. Electron diffraction pattern of polycrystalline tellurium chloride. (Courtesy of RCA Laboratories, Princeton, New Jersey.)

Electrons with a kinetic energy of 150 eV have the same wavelength as x-rays, 1 Å, and a monoenergetic† electron beam shows essentially the same diffraction effects as do x-rays. Figure 37-6 is an electron diffraction pattern from a metallic foil. Such patterns are in complete accord with the Bragg and de Broglie relations. In short, *electron diffraction experiments confirm the relation* $\lambda = h/mv$.

---

†A monoenergetic beam of particles not only has a single energy but also has, of course, a single wavelength and is therefore usually spoken of, with reference to its wave aspect, as "monochromatic."

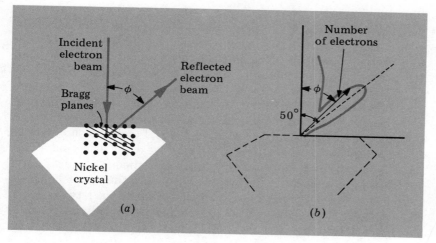

FIG. 37-7. *(a)* Reflection of electron waves by one set of Bragg planes in a crystal of nickel; *(b)* number of electrons reflected from a nickel crystal as a function of angle ϕ is shown by a polar plot.

Electron and x-ray diffraction patterns can be used for measuring the wavelengths of electrons and x-rays when the crystalline structure is known; conversely, when the wavelengths of the x-rays and electrons are known, the diffraction patterns can be used to deduce the geometry of the crystalline structure and the interatomic spacings of the solids.

The experiment of Davisson and Germer in 1927 first confirmed the wave properties of electrons. A beam of electrons incident on a single crystal of nickel was diffracted by reflection from the Bragg planes within the crystal, producing a pronounced peak, which could be attributed to electron diffraction, as shown in Fig. 37-7.

The fact that an object of mass m and velocity v has a wavelength h/mv has been established, not only for electrons, but also for atoms, molecules, and the uncharged nuclear particle, the neutron. Indeed, all of the interference and diffraction effects observed for electromagnetic radiation have been duplicated with particles.

37–4 THE PRINCIPLE OF COMPLEMENTARITY

We attribute both wave and particle characteristics to electromagnetic radiation and to material particles. This wave-particle duality makes us uneasy at first sight. Here we shall examine the origin of our uneasiness and see how this dilemma is resolved by the principle of complementarity.

The concepts of particle and wave are basic in physics, because they represent the only two possible modes of energy transport. In describing any ordinary large-scale phenomenon of energy transport in classical physics, we are always successful in applying one of the descriptions. For example, a disturbance that travels on the surface of a pond is certainly a wave, and a thrown baseball illustrates energy transport by a "particle." There is never any doubt about which description we should apply in such instances.

Now let us turn to somewhat less direct illustrations of wave and particle behavior. The propagation of sound through an elastic medium can be understood as a wave disturbance. We do not "see" the waves, as we did the water waves; nevertheless, we apply the wave description to sound propagation with confidence, because it alone describes diffraction and interference. The propagation of sound can be explained by a *wave model,* because it agrees with *all* experimental observations of sound. Next let us consider particle behavior as it appears in the kinetic theory of gases. We never see the molecules, but we are quite sure that their behavior is like that of very small, hard spheres, because a variety of experiments shows it to be so. A *particle model* is the only appropriate means of describing its behavior. When we describe phenomena remote from our ordinary experience, we still apply one or the other of the two modes of description, because one of them is always successful in accounting for the experimental facts.

The wave and particle descriptions are mutually incompatible and contradictory. If a wave is to have its frequency or its wavelength given with infinite precision, then it must have an infinite extension in space. Conversely, if it is confined to some limited region of space, it *resembles* a particle by virtue of its localizability, but it cannot be characterized by a single frequency and wavelength. An ideal wave, one whose frequency and wavelength are known with certainty, is altogether incompatible with an ideal particle, which has a zero extension in space.

Any energy-transport phenomenon must be described in terms of waves or of particles, but we cannot *simultaneously* apply a particle description and a wave description. We can and must use one or the other, never both at the same time.

Now, what is disturbing about the descriptions of electromagnetic radiation and of material particles is the fact that we apply *both* the wave and particle models. Yet, if we review our interpretation of the experiments discussed thus far, we find that we have *never* applied the descriptions *simultaneously,* which is, as we have seen, logically impossible.

Consider electromagnetic radiation. The wave model describes experiments in interference and diffraction, where we are confronted with alternate light and dark bands that are predicted and accounted for by wave theory. We never apply a particle description to interference and diffraction. Of course, our confidence in the wave model for *propagation* of light is strengthened by the fact that Maxwell's classical electromagnetic theory predicts all the *wave* phenomena, but it would be rash, in view of the open-ended, tentative, and incomplete nature of all physical theory, to conclude that Maxwell's equations are the last word on electromagnetic theory. In fact, classical electromagnetism is incomplete to the degree that it cannot account for quantum effects. In summary, we use the wave model to describe the *propagation* of light; we do not, need not, and cannot apply the particle model to interference and diffraction effects.

Now consider phenomena which call for a particle model of electromagnetic radiation. They are the quantum effects of Chap. 36 with electromagnetic radiation in *interaction* with particles. The radiation consists of photons, each with a specific energy and momentum, with the electromagnetic *particle* localized at a particular point in space, namely at the site of the interaction. Interactions between radiation and matter require the particle description; such interactions are best described as collisions. We can make sense of the photon-interaction experiments only when electromagnetic radiation is assumed to consist of particles making collisions. In short, when we visualize photon-electron interactions, we employ a *particle* model; we cannot, and need not, apply the wave model.

Electromagnetic radiation shows both wave and particle aspects but *not* in the same experiments. Interference or diffraction experiments require a wave interpretation, and it is impossible to apply simultaneously a particle interpretation; a photon-interaction experiment requires a particle interpretation, and it is impossible to apply simultaneously a wave interpretation. Both the wave and the particle aspects are essential features of electromagnetic radiation, and we must accept both. According to the *principle of complementarity,* enunciated by Niels Bohr in 1928, *the wave and particle aspects* of electromagnetic radiation *are complementary.* To interpret the behavior of electromagnetic radiation in any one experiment we must choose either the particle or wave description. The wave and particle aspects are *complementary* in that our knowledge of the properties of electromagnetic radiation is partial unless both the wave and particle aspects are known; but the choice of one description, which is imposed by the nature of the experiment, precludes the simultaneous choice of the other. We make a choice of the description by the experimental arrangement. Electromagnetic radiation is a more complicated entity than can be comprehended in the simple and extreme notions of wave and particle, notions that are borrowed from our direct, ordinary experience with large-scale phenomena.

The complementarity principle applies also to the wave-particle duality of particles, such as electrons. For example, electrons in a cathode-ray tube follow well-defined paths and indicate their collisions with a fluorescent screen by very small, bright flashes. Electrons are particles (or, more properly, a particle model can be used to describe their behavior) in cathode-ray experiments because all of the electron energy, momentum, and electric charge is assigned at any one time to a small region of space. When they interact with other objects, electrons behave *as if* they were particles. The particle nature of electrons is revealed in the cathode-ray experiments and, therefore, by the principle of complementarity, the wave nature of electrons *must* be suppressed.

The wave nature of electrons appears in the experiments showing electron diffraction. Electrons are propagated as waves with an indefinite extension in space, and it is, of course, impossible to specify the location of any one electron. In short, the electron diffraction experiments show the wave nature of electrons and, by the principle of complementarity, the particle nature is necessarily suppressed in these experiments.

37-5 THE PROBABILITY INTERPRETATION OF DE BROGLIE WAVES

The wave nature of electromagnetic radiation is illustrated by oscillating electric and magnetic fields in space; therefore, the *wave associated with a photon is the electromagnetic field*. What about waves associated with a material particle? What is it that is waving when we say that an electron or any other material particle shows wave properties?

Consider a screen illuminated by a monochromatic beam of electromagnetic radiation. When the intensity is great, the screen appears to the eye to be uniformly illuminated; equivalently, a photographic plate placed at the screen will, after being exposed and developed, show a uniform darkening. When the light intensity is great, the number of photons arriving at the screen is so great that the essentially granular and discrete nature of the electromagnetic radiation is obscured by the great number of photons, and the distinct and randomly arranged bright flashes merge into a seemingly continuous and constant illumination.

The intensity I of an electromagnetic beam, the energy per unit transverse area per unit time, is given by

$$I = \varepsilon_0 \mathscr{E}^2 c$$

where $\mathscr{E}$ is the magnitude of the instantaneous electric field at any point on the screen and ε_0 is the electric permittivity of free space. (Sec. 30-4.) Suppose that the intensity of the beam is made extremely weak. Instead of a uniformly illuminated area we have a collection of distinct, bright flashes randomly arranged over the plate, each bright flash corresponding to the arrival of a single photon.† Neither the position nor the time at which a single photon will strike the screen can be predicted. The distribution of photons is completely random. But the *average* number of photons arriving per unit area per unit time can be predicted; this number is the photon flux N. Since $h\nu$ is the energy per photon, the intensity of

---

†Actually, sophisticated instruments, rather than the eye or a photographic plate, can record the arrival of photons one by one.

a monochromatic beam may be given in terms of the photon flux as follows:
$$I = (h\nu)N$$

Instead of increasing the intensity to get a uniformly illuminated screen, we could use a very weak beam, and record on the screen the position of each flash as it comes along. Then we would find that, after a long time has elapsed, the screen is again uniformly covered.

The situation here is rather like that encountered in the kinetic theory of gases: One attributes the apparently continuous pressure of a gas to the combined effects of many individual molecular impacts on the container walls. The arrivals of the molecules are random and discrete, but because of their enormous number, the net effect is a continuous pressure.

Consider a beam of monochromatic light of very low intensity, 1.00×10^{-13} W/m² (approximately one hundred millionth the intensity of starlight at the earth's surface). Suppose that it consists of ultraviolet photons, each with an energy of 5.00 eV = 8.00×10^{-19} J. Then the photon flux is

$$N = \frac{I}{h\nu} = \frac{1.00 \times 10^{-13} \text{ W/m}^2}{8.00 \times 10^{-19} \text{ J/photon}} = 1.25 \times 10^5 \text{ photon/m}^2\text{-s}$$

$$= 12.5 \text{ photon/cm}^2\text{-s}$$

This means that 12.5 bright flashes will be observed over an area of 1 cm² during a period of 1 s. It is, of course, impossible to observe a fraction of a photon, so we shall never see 12.5 photons; but in one interval 11 flashes might be seen, in another 13, and so on, so that the average will be 12.5. Furthermore, the spatial distribution of photons over the 1-cm² area will *not* be the same for all 1-s intervals: The flashes will be distributed randomly and will approach a uniform distribution only over a long period of time. *The photon flux does not give precisely the time and location of any one photon but gives only the probability of observing a photon:*

$$N \propto \text{probability of observing a photon}$$

The intensity of monochromatic electromagnetic radiation can be expressed by either the wave description $I = \varepsilon_0 \mathscr{E}^2 c$ or the particle description $I = h\nu N$. We have in the intensity a quantity that has a precise meaning to both descriptions. The intensity bridges the gap between the two disparate models.

New significance can be assigned to the square of the electric field, $\mathscr{E}^2$, through the photon description of light. Equating the two expressions for the intensity, we have

$$I = h\nu N = \varepsilon_0 \mathscr{E}^2 c$$

Therefore,
$$N \propto \mathscr{E}^2$$

and
$$\mathscr{E}^2 \propto \text{probability of observing a photon}$$

The Probability Interpretation of De Broglie Waves   SEC. 37-5

The probability of observing a photon at any point in space is proportional to the square of the electric field at that point.

From the point of view of the quantum theory the electric field is that quantity, or function, whose square gives the probability of observing a photon at any given place.

We are able to give meaning to the wave nature of a particle, such as an electron, in the following way. We assume the relation between the probability of observing a particle and the square of the amplitude of its wave corresponds exactly to the relation between the probability of observing a photon and the square of the amplitude of its wave (the electric field). The amplitude of the wave associated with a particle is represented by ψ, called simply the *wave function*.

The wave function ψ is that quantity whose square, ψ^2, is proportional to the probability of observing a material particle.

Thus, if ψ represents the wave function at the location x, the probability of observing the particle's being between x and $x + dx$ is given by $\psi(x)^2 \, dx$:

Probability of observing a particle in the interval $dx \propto \psi^2 \, dx$

The wave function of a particle, then, is analogous to the electric field of a photon. Just as $\mathscr{E}$ will, in general, be a function of both position and time, so too, in general, will the wave function ψ.

It is impossible to specify with certainty the particular location of a photon at a particular time, but it is possible to specify by $\mathscr{E}^2$ the probability of observing it. Similarly it is impossible to specify with certainty the particular location of a particle at a particular time, but it is possible to specify by ψ^2 the probability of observing the particle. In short, a particle's wave function gives rise to a *probability interpretation* of the location of a particle.

The interpretation of waves associated with particles through probabilities was first given in 1926 by Max Born. That branch of quantum physics which deals with the problem of finding the values of ψ is known as *wave mechanics,* or *quantum mechanics.* The two principal originators of the wave mechanics of particles were Erwin Schrödinger (in 1926) and Werner Heisenberg (in 1925), who independently formulated quantum mechanics in different but equivalent mathematical forms.

Just as the electromagnetic theory of Maxwell is summarized in the Maxwell equations, which are the basis for computing values of $\mathscr{E}$, the wave mechanics of matter is governed by the *Schrödinger equation,* which is the basis for computing values of ψ in any problem in quantum physics. Here the parallel stops, however. Whereas the electric field, which has its origin in electric charges, gives not only the probability of observing a photon but also the electric force on a unit positive electric charge, the wave function of the

Schrödinger equation has a physical meaning *only* in terms of the probability interpretation: It does *not* indicate any sort of force. The wave function is not directly measurable or observable; it does, however, give the most information one can extract concerning any system of objects, and *all* measurable quantities, such as the energy and momentum, as well as the probability of location, can be derived from it.

Consider the interference effects that arise in the passage of waves through two parallel slits. When either one of the two slits is closed, the pattern is the typical single-slit diffraction pattern: a broad, central maximum flanked by weaker, secondary maxima, as in Fig. 32-22. When both slits are open, the pattern is as shown in Fig. 37-8: interference fine structure within a diffraction envelope. The pattern is *not* merely two single-slit diffraction patterns superposed; the interference between waves traveling through *both* of the slits is responsible for the rapid variations in intensity. In short, in a case in which waves (particles) can take two or more routes from a source to an observation point we solve the problem by first superposing the wave function (or electric fields) from the two separate routes to find the resultant wave function (or electric field) and then squaring to find the probability (or intensity). That is to say, if ψ_1 and ψ_2 represent the wave functions for passage through slits 1 and 2 separately, then $(\psi_1 + \psi_2)^2$, not $\psi_1^2 + \psi_2^2$, gives the probability of observing a particle on the screen. If, then, a single electron or photon is directed toward a pair of slits, we cannot say which of the two slits it will pass through; we must speak in the language of waves and say, in effect, that it passes through *both* slits.

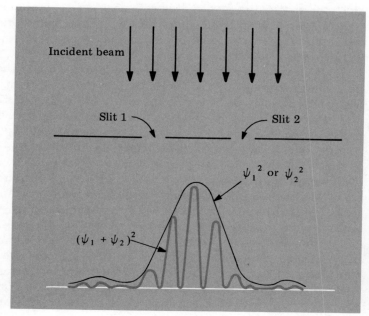

FIG. 37-8. Double-slit diffraction of particles. The wave functions ψ_1 and ψ_2 give the diffraction pattern when either slit 1 or slit 2 is open; the superposed wave function $\psi_1 + \psi_2$ gives the pattern when both slits are open. (The distance between the slits is grossly exaggerated in the figure.)

37-6 THE UNCERTAINTY PRINCIPLE

It is impossible, as indicated in the principle of complementarity, to apply simultaneously the wave and particle descriptions to such an object as an electron, or to electromagnetic radiation. For example, if an object is localized within some small region of space — that is, if the wave function is large within the restricted region of space, but small elsewhere — then the object may be said to resemble a particle. But under these circumstances a precise meaning cannot be assigned to the object's frequency or wavelength, for to do so would require a strictly sinusoidal wave, a wave of *infinite* extent. Conversely, if an object's frequency and wavelength are precisely specified, and the wave function is thereby of infinite extent in space and duration in time, then one cannot speak meaningfully of the object's location at an instant of time. In short, one cannot then speak of the object as a particle.

One *can* specify the location of an object with infinite precision, so that the uncertainty in its position along, say, the X axis is $\Delta x = 0$. But one cannot *simultaneously* give the object's wavelength with precision; since the object's momentum is related to its wavelength by $p = h/\lambda$, uncertainty in wavelength implies a corresponding uncertainty in the object's momentum component Δp_x along the X axis. With $\Delta x = 0$, $\Delta p_x = \infty$. Similarly, uncertainty in the time of observation Δt implies uncertainty in frequency $\Delta \nu$; since $E = h\nu$, uncertainty in frequency implies uncertainty in energy ΔE. With $\Delta t = 0$, $\Delta E = \infty$.

These extreme examples of the intrinsic uncertainties between simultaneous specifications of momentum and position, and of energy and time, are illustrations of the general *uncertainty principle*, or principle of indeterminacy. This characteristic and fundamental feature of the quantum theory was first introduced in 1927 by Werner Heisenberg.

We can arrive at an approximate formulation of the uncertainty principle by first considering how the uncertainty in the frequency $\Delta \nu$ of a wave is related to the finite period of time Δt over which the wave is observed and its frequency measured. It must first be recognized that a strictly sinusoidal wave is one with *infinite* extent in space. A wave restricted to some finite region of space, on the other hand, cannot be characterized by a single well-defined frequency or wavelength. Rather, a number of sinusoidal waves differing from one another in frequency and wavelength must be superposed to yield a resultant confined wave (or what is often referred to as a "wave packet").

Suppose that an observer wishes to measure the frequency of an incoming wave, but he observes the wave only over a *finite* time interval Δt. The observer can be reasonably certain of the frequency ν_1 of the incoming wave, or its period $T_1 = 1/\nu_1$, only if the observation period Δt is at least equal to the time T_1 for one full cycle. That is, $\Delta t \geq T_1 = 1/\nu_1$. For waves with a period greater than T_1, or a frequency less than ν_1, the observer does not observe even one full cycle, and the uncertainties in the frequencies of all such low-frequency components is great. Therefore, we may take the uncertainty in frequency $\Delta \nu$ to equal ν_1, at least approximately. We have then that

$$\Delta t \geq T_1 = \frac{1}{\nu_1} = \frac{1}{\Delta \nu}$$

or
$$\Delta t \, \Delta \nu \geq 1 \tag{37-5}$$

which indicates that observation over a finite time Δt corresponds to an uncertainty in the measured frequency of at least $1/\Delta t$.

An analogous relationship between the uncertainty $\Delta \lambda$ in a measured wavelength and the finite spatial extent Δx in which the observation of wavelength is made is easily arrived at. Since $\lambda = v/\nu$, where v is the speed of the wave, taking the differential of this relation yields

$$\Delta \lambda = \frac{v}{\nu^2} \Delta \nu \tag{37-6}$$

(We ignore the negative sign in the relation above because we are concerned only with the magnitude of the uncertainties.) The uncertainties in spatial location and time are related by $\Delta x = v \, \Delta t$. Using this relation, together with (37-5) and (37-6), we have finally

$$\Delta \lambda \, \Delta x \geq \lambda^2 \tag{37-7}$$

When we take the differential of the basic quantum relations $E = h\nu$ and $p = h/\lambda$, we have (again ignoring negative signs)

$$\Delta E = h \, \Delta \nu$$

and
$$\Delta p_x = \frac{h \, \Delta \lambda}{\lambda^2}$$

When these relations are substituted into (37-5) and (37-7) we obtain

$$\Delta E \, \Delta t \geq h \tag{37-8}$$

$$\Delta p_x \, \Delta x \geq h \tag{37-9}$$

which are two basic formulations of the uncertainty principle. The product of the uncertainties in momentum component and in location along the same direction, $\Delta p_x \, \Delta x$, cannot be infinitesimally small, but must equal in magnitude at least Planck's constant h.†

---

†The value of the constant on the right side of Eqs. (37-8) and (37-9) depends on the precise definition of uncertainty. For the rather conservative and approximate convention used here the constant is h, but if Δp_x and Δx represent the root-mean-square values of a number of independent measurements, then the constant becomes $h/4\pi$.

Likewise, the product of the uncertainties in the energy of a particle, or of a system of particles, and the observation time, $\Delta E \, \Delta t$, must be at least as large as h.

The fundamental limitation on the certainty of measurements of energy and time or of position and momentum is in harmony with the principle of complementarity. If the particle nature of, say, an electron is to be perfectly displayed, then both Δx and Δt must be zero. Therefore, when the particle aspect is chosen, the wave aspect is necessarily suppressed. All the quantities ν, E, λ, and p are then completely uncertain. On the other hand, if the wave characteristics of a material particle or of electromagnetic radiation are to be defined, perfectly, that is, if $\Delta \nu = 0$ and $\Delta \lambda = 0$ (also $\Delta E = 0$ and $\Delta p = 0$), then by the principle of complementarity or by the uncertainty principle we are prevented from giving simultaneously the distinctively particle characteristics of precise location in space and in time, and x and t are completely uncertain.

Suppose that we wish to represent an electron by its wave properties and yet localize it in space to some degree. We cannot use a single, sinusoidal wave: Such a wave extends to infinity and is certainly not localized. We can, however, superpose a number of sinusoidal waves differing in frequency over a range of frequencies $\Delta \nu$ and so have a *wave packet*. The component waves constructively interfere over a limited region of space Δx, identified as the somewhat uncertain location of the "particle," and so yield a resultant wave function ψ of the sort shown in Fig. 37-9. Because there is a range in frequency and a range in wavelength, $\Delta \nu$ and $\Delta \lambda$, the associated momentum and energy are necessarily uncertain, and it is impossible to predict precisely where or when the wave packet will reach another point and what the momentum and energy will then be.

Since the uncertainty relation implies an uncertainty in energy, of magnitude $h/\Delta t$ at least over a time interval Δt, it also implies that the law of energy conservation may actually be violated — by that amount, $\Delta E = h/\Delta t$, but only for the time interval Δt. The greater the amount of energy "borrowed" or "discarded," the shorter

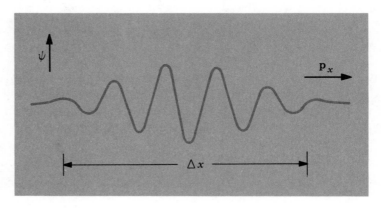

FIG. 37-9. The wave function of a wave packet.

the time interval over which the nonconservation of energy may take place. Similarly, the minimum uncertainty in a particle's momentum, $h/\Delta x$, implies that the law of momentum conservation may be violated, but only by that amount, $\Delta p = h/\Delta x$, over a region of space of Δx.

To see the uncertainty principle illustrated in a specific situation, we consider the diffracting of waves by a single, parallel-edged slit. A monochromatic plane wave is incident on a slit of width w, and the diffraction pattern is formed on a distant screen, as shown in Fig. 37-10. The location of the points of zero intensity is given by the equation (Sec. 32-9)

$$\sin \theta = \frac{n\lambda}{w}$$

where λ is the wavelength and n is 1, 2, 3. . . . Approximately three-fourths of the energy passing through the slit falls within the central region. Its limits are given by the equation above with $n = 1$:

$$\sin \theta = \pm \frac{\lambda}{w} \tag{37-10}$$

We have not yet specified what sort of wave passes through the slit. If it is electromagnetic radiation, the intensity of the diffraction pattern is proportional to $\mathscr{E}^2$, the square of the electric field at the screen. If, on the other hand, the wave consists of a beam of monoenergetic electrons, the intensity is proportional to ψ^2, which is the square of the electron wave function at the screen and gives the probability of finding an electron at any point along the screen. Whatever the type of wave, diffraction effects are pronounced only when the wavelength is comparable to the slit width; at the limit of vanishing wavelength, the intensity pattern on the screen corresponds to a geometrical shadow cast by the edges of the slit.

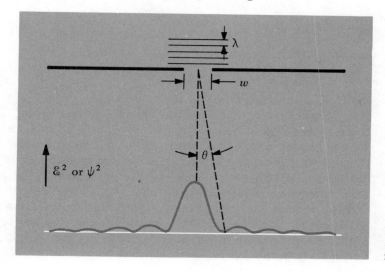

FIG. 37-10. Diffraction pattern of a monochromatic plane wave incident on a slit of width w.

Suppose now that we reduce drastically the amount of incident radiation or the number of electrons, as the case may be. Then, on the screen we no longer see smooth variations but, instead, photons or electrons arriving one by one. Since the intensity is given by $\mathscr{E}^2$ for photons and by ψ^2 for electrons, the quantity plotted in Fig. 37-10 represents the *probability* that a particle will strike a certain spot on the screen. The probability that a particle will fall somewhere within the central region in the diffraction pattern is 75 percent. At very low illumination, bright flashes appear over a large area of the screen. As time passes, more and more particles accumulate on the screen, and the distinct bright flashes merge and form the smoothly varying intensity pattern predicted by wave theory.

There is *no* way of predicting in advance where any one electron or photon will fall on the screen. All that wave mechanics permits us to know is the probability of a particle's striking any one point. Before the particles pass through the slit, their momentum is known with complete precision both in magnitude (monochromatic waves) and in direction (vertically down in this case). When they pass through the slit, their location along a line in the X direction, completely uncertain before they reached the slit, is now known with an uncertainty $\Delta x = w$, the slit width. What is not known, however, is precisely where any one particle will strike the screen. Any particle has approximately a 75 percent chance of falling within the central region, whose boundaries are given by (37-10). There will be an uncertainty in the X component of the momentum p that is *at least* as great as $p \sin \theta$, as may easily be seen in Fig. 37-11. Therefore we write

$$\Delta p_x \geq p \sin \theta$$

Using (37-10) we have

$$\Delta p_x \geq \frac{p\lambda}{\Delta x}$$

and since $p = h/\lambda$, we have

$$\Delta p_x \, \Delta x \geq h$$

the *Heisenberg uncertainty relation*.

Now suppose that Δx in our example is very large. The slit becomes very wide, and the uncertainty in position is increased. We are less certain as to where an electron is located along X. But the uncertainty in the momentum is reduced correspondingly: The diffraction pattern shrinks and, in the limit, essentially all electrons fall within the geometrical shadow. Conversely, if the slit width is reduced and Δx becomes very small, the diffraction pattern is expanded along the screen. For the increase in our certainty of the electron's position we must pay by a correspondingly greater uncertainty in the electron's momentum.

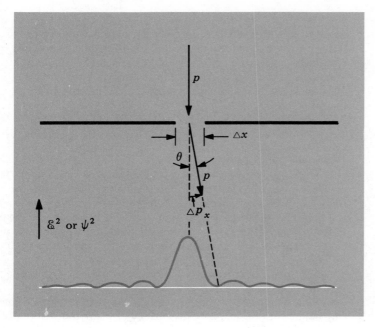

FIG. 37-11. An illustration of the uncertainty in the momentum of a particle that has passed through a single slit.

We see that when the slit width is much greater than the wavelength, particles pass through the slit undeviated to fall within the geometrical shadow. This is in agreement with classical mechanics, where the wave aspect of material particles is ignored. Thus, there is a close parallel in the relationship of wave optics to ray optics, and of wave mechanics to classical mechanics. Ray optics is a good approximation of wave optics whenever the wavelength is much less than the dimensions of obstacles or apertures that the light encounters; similarly, classical mechanics is a good approximation of wave mechanics whenever a particle's wavelength is much less than the dimensions of obstacles or apertures encountered by material particles. Symbolically, we can write

$$\underset{\lambda/w \to 0}{\text{Limit}}\ (\text{wave optics}) = \text{ray optics}$$

$$\underset{\lambda/w \to 0}{\text{Limit}}\ (\text{wave mechanics}) = \text{classical mechanics}$$

No ingenious subtlety in the design of the diffraction experiment will remove the basic uncertainty. We do *not* have here, as in the large-scale phenomena encountered in classical physics, a situation in which the disturbances on the measured object can be made indefinitely small by ingenuity and care. The limitation here is rooted in the fundamental quantum nature of electrons and photons; it is intrinsic in their complementary wave and particle aspects.

The Uncertainty Principle   SEC. 37-6

Example 37-1

As an illustration of the uncertainty principle, we compute the uncertainty in the momentum of a 1,000-eV electron whose position is uncertain by no more than 1 Å = 1.0×10^{-10} m, the approximate size of atoms. From $\Delta p_x \geq h/\Delta x$ it follows that $\Delta p_x = 6.6 \times 10^{-24}$ kg-m/s. Now let us compare this uncertainty in the momentum with the momentum itself, $p_x = (2mE_k)^{1/2} = 17 \times 10^{-24}$ kg-m/s. Therefore, the fractional uncertainty in the momentum is $\Delta p_x/p_x = 6.6/17$, about 40 percent! Because of the uncertainty principle, it is impossible to specify the momentum of an electron confined to atomic dimensions with even moderate precision.

Consider now the uncertainty involved when a 10.0-g body moves at a speed of 10.0 cm/s; that is, an ordinary-sized object is moving at an ordinary speed. Let us further assume that the position of the object is uncertain by no more than 1.0×10^{-3} mm. We wish to find the uncertainty in the momentum and, more especially, the fractional uncertainty in the momentum. We find $\Delta p_x = 6.6 \times 10^{-23}$ kg-m/s and $p_x = 1.0 \times 10^{-3}$ kg-m/s; therefore, $\Delta p_x/p_x = 6.6 \times 10^{-25}$! The fractional uncertainty in the momentum of a macroscopic body is so extraordinarily small as to be negligible compared with all possible experimental limitations. The uncertainty principle imposes an important limitation on the certainty of measurements only in the microscopic domain. In the macroscopic domain the uncertainties are, in effect, trivial.

Figure 37-12 shows the momentum p_x of the electron in our example above, plotted against its position x. The uncertainty principle requires that the shaded area in this figure, which gives the product of the uncertainties in the momentum and the position, be equal in magnitude to Planck's constant h. If the position is known with

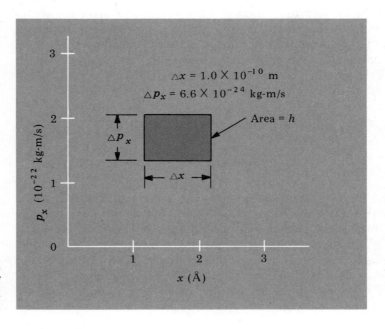

FIG. 37-12. Uncertainties in the simultaneous measurements of position and momentum of an electron.

high precision, the momentum is rendered highly uncertain; if the momentum is specified with high certainty, the position must necessarily be highly indefinite. It is therefore impossible to predict and follow in detail the future path of an electron confined to essentially atomic dimensions. Newton's laws of motion, which are completely satisfactory for giving the paths of large-scale particles, cannot be applied here. To predict the future course of any particle, it is necessary to know not only the forces that act on the particle but also its initial position and momentum. Because *both* position and momentum cannot be known simultaneously without uncertainty, it is not possible to predict the future path of the particle in detail. Instead, *wave mechanics* must be used to find the probability of locating the particle at any future time.

Consider again the 10.0-g body moving at 10.0 cm/s. Figure 37-13 shows its momentum and position. The area h, representing the product of the uncertainties in momentum and position, is so extraordinarily tiny in such macroscopic circumstances that it appears as an infinitesimal point on the figure. In this case the classical laws of mechanics may be applied without entailing appreciable uncertainty.

The *finite size* of Planck's constant is responsible for quantum effects. Quantum effects are subtle because Planck's constant is very small—but not zero. Recall that the relativity effects are subtle because the speed of light is very large—but not infinite. If somehow Planck's constant were zero, the quantum effects would disappear. Thus classical physics may be thought of as the limit of quantum physics as h is imagined to approach zero. Symbolically,

$$\lim_{h \to 0} (\text{quantum physics}) = \text{classical physics}$$

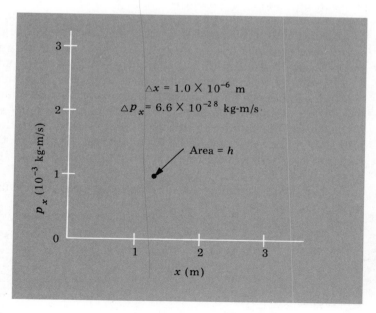

FIG. 37-13. Uncertainties in the simultaneous measurements of position and momentum of a 10-g particle. (On the scale of this drawing the uncertainty, area h, has been exaggerated by a factor 10^{26}.)

37-7 THE QUANTUM DESCRIPTION OF A CONFINED PARTICLE

A particle completely free of external influence will, by Newton's first law, move in a straight line with a constant momentum. In the language of wave mechanics, such a particle, having a constant, well-defined momentum, must be represented by a monochromatic sinusoidal wave with a well-defined wavelength. If the wavelength is to be precisely defined, the wave must have an infinite extension in space. In accordance with the uncertainty principle, when the wavelength and, therefore, the momentum of the particle are specified precisely, the position of the particle is altogether uncertain and indeterminate.

In Fig. 36-1c we saw an example from classical physics of waves that had perfectly defined wavelengths and yet were confined to a limited region of space: Resonant standing waves on a string fixed at both ends. The wave on the string is repeatedly reflected from the boundaries, the fixed ends, and it constructively interferes with itself. Resonance is achieved only when the length of the string is some integral multiple of the half-wavelengths; the standing-wave pattern therefore fits between the boundaries.

Consider the elementary wave-mechanical problem that is analogous to that of standing waves on a string. We assume that a particle moves freely back and forth along the X axis but that it encounters an infinitely hard wall at $x = 0$ and another at $x = L$; it is, then, confined between these boundaries. The infinitely hard walls correspond to an infinite potential energy V for all values of x less than zero and greater than L. Because the particle is free between zero and L, its potential energy V in this region is constant. For convenience we choose the constant potential energy to be zero. The situation we have described is that of a *particle in a one-dimensional box*, or a particle in an infinitely deep potential well. Because the walls are infinitely hard, the particle imparts none of its kinetic energy to them, its total energy remains constant, and it continues to bounce back and forth between the walls unabated.

From the point of view of wave mechanics we may say that, if the particle is confined within the limits stated, then the probability of finding it outside these limits is zero. Therefore, the wave function ψ, whose square represents this probability, must be zero for $x \leq 0$ and $x \geq L$.

We may summarize mathematically the conditions of our problem as follows:

$$V = \infty \quad \text{for } x < 0, x > L$$

$$V = 0 \quad \text{for } 0 < x < L$$

$$\psi = 0 \quad \text{for } x \leq 0, x \geq L$$

Only those wave functions which satisfy the boundary conditions are allowed. Since the particle is free and the magnitude of its momentum is constant, we know that it is represented by a sinusoidal wave. To satisfy the conditions at the boundaries, only those wavelengths are allowed which will permit an integral number of half-wavelengths to be fitted between $x = 0$ and $x = L$. The condition for the existence of *stationary,* or standing, waves is, then,

$$L = n \frac{\lambda}{2} \qquad (37\text{-}11)$$

where λ is the wavelength and n is the *quantum number* having the possible values 1, 2, 3, etc.

Figure 37-14 shows the the potential well, the wave function ψ, and the probability ψ^2 plotted against x, for the first three possible *stationary states* of the particle in the box. Note that, whereas ψ can be negative as well as positive, ψ^2 is always positive.

The probability distribution is such that ψ^2 is always zero at the boundaries. For the first state, $n = 1$, the most probable location of the particle is the point midway between the two walls, at $x = L/2$; for the second state, $n = 2$, however, the least probable location is this point, where, in fact, $\psi^2 = 0$, which is to say that it is impossible for the particle to be located there!

The imposition of the boundary conditions on ψ, that is, the fitting of the waves between the walls, has restricted the wavelength of the particle to the values given by (37-11). Now, if only certain wavelengths are permitted, the magnitude of the momentum also is restricted to certain values, since $p = h/\lambda$. Therefore,

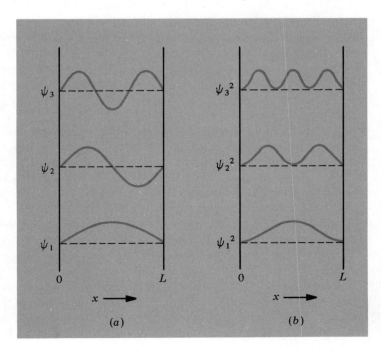

FIG. 37-14. The first three stationary states for a particle in a one-dimensional box with infinitely high sidewalls: *(a)* wave functions and *(b)* probability distributions.

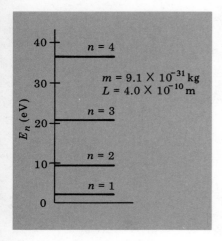

FIG. 37-15. Allowed energies of an electron confined to a one-dimensional box of atomic dimensions.

the permitted momenta are given by

$$p = \frac{h}{\lambda} = \frac{hn}{2L}$$

Finally, the kinetic energy E_k (and, therefore, the total energy E of the particle, since the potential energy is zero) is given by

$$E_k = E = \tfrac{1}{2}mv^2 = \frac{p^2}{2m} = \frac{(hn/2L)^2}{2m}$$

$$E_n = n^2 \frac{h^2}{8mL^2} \qquad (37\text{-}12)$$

where m is the particle's mass (this equation holds only for nonrelativistic speeds). The subscript n signifies that the possible values of the energy depend only on the quantum number n for fixed values of m and L. By this equation we see that the *energy* of the particle in the one-dimensional box is *quantized*. The particle cannot assume just any energy but only those particular energies and speeds which satisfy the boundary conditions placed upon the wave function. The quantization of the energy is analogous to the classical quantization of the frequencies of waves on a string fixed at both ends.

Let us compute the possible values of the energy by assuming that an electron with $m = 9.1 \times 10^{-31}$ kg is constrained to move back and forth within a distance of $L = 4$ Å, or 4×10^{-10} m. Setting these values in (37-12) gives for the energy of the first state, $n = 1$, the value $E_1 = 2.3$ eV. Because $E_n = n^2(h^2/8mL^2) = n^2 E_1$, the next possible energies of the particle are $4E_1$, $9E_1$, $16E_1$, ... The permitted energies of the electron in a 4-Å box are shown in Fig. 37-15, which is called an *energy-level diagram*. It is significant, in relation to atomic structure, that, when an electron is confined to a distance approximately the diameter of an atom, its possible energies are in the range of a few electron volts, comparable to the binding energy of electrons in atoms.

Now consider the allowed energies of a relatively large object confined in a relatively large box; assume that $m = 9.1$ mg $= 9.1 \times 10^{-6}$ kg and that $L = 4$ cm $= 4 \times 10^{-2}$ m. Equation (37-12) shows that for these values $E_1 = 2.3 \times 10^{-41}$ eV, a fantastically small amount of energy! Figure 37-16 is the energy-level diagram for these circumstances, the energy being plotted to the *same* scale as in Fig. 37-15. The spacing between adjacent energies in a diagram for such macroscopic conditions is so very small that energy is effectively continuous. That is why we never see any obvious manifestation of the quantization of the energy of a macroscopic particle; the quantization is there, but it is too fine to be discerned. This result agrees, of course, with the classical requirement that the actually discrete energies of a bound system appear continuous in large-scale phenomena.

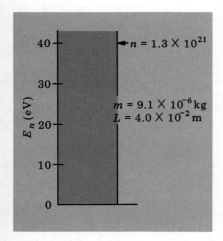

FIG. 37-16. Allowed energies of a 9.1-mg particle confined to a 4-cm one-dimensional box.

The lowest possible energy of a particle in an infinite deep box is not zero, but E_1. This is in accord with the uncertainty principle. If the particle's energy were zero, the particle being at rest somewhere within the box ($\Delta x = L$), both the momentum p and the uncertainty in the momentum Δp_x would be zero. This would violate the uncertainty relation, since the product $\Delta p_x \Delta x$ would be $(0)(L) = 0$, not h. For $x = L$ the uncertainty in momentum is given by $\Delta p_x = h/\Delta x = h/L$. The particle's momentum p_x in one direction must then be at least as great as the uncertainty Δp_x. Moreover, we cannot know whether the particle travels to the left or to the right, so that, all told, $\Delta p_x = 2p_x$. Under these circumstances the particle's energy is $E = p_x^2/2m = (\Delta p_x/2)^2/2m$. With $\Delta p_x = h/L$ we have $E = h^2/8mL$, exactly the energy of the first allowed state, given by (37-12).

For an electron confined within atomic dimensions the energy in the state of lowest energy, or the *ground state*, is a few electron volts. The electron is never at rest but bounces back and forth between the confining walls with its lowest possible energy, the so-called *zero-point energy*. This is, of course, true of any confined particle. Let us compute the minimum speed of the 9.1-mg particle restricted to 4 cm, as in the example. Since $E_1 = 2.3 \times 10^{-41}$ eV $= \frac{1}{2}mv^2$, we find that $v = 9.0 \times 10^{-28}$ m/s, or a mere 10^{-7} Å per millennium. The particle is effectively at rest.

The problem of the particle in the box is somewhat artificial, since there is no such thing as an infinitely great potential energy, and a particle cannot be made completely free from all external influences while thus confined. Nevertheless, the problem is an important one, because it reveals the quantization of the energy. Energy quantization occurs, basically, because only certain discrete values of the wavelength can be fitted between the boundaries.

If a particle has a constant kinetic energy E_k, its wavelength is constant and the wave function is sinusoidal. More generally, a confined particle's kinetic energy is not constant and its wavelength may therefore differ from point to point. Consider a particle of mass m interacting with its surroundings (of infinite mass) through a potential-energy function $V(x)$. Then the particle's total energy E is constant and given by

$$E = E_k + V = \frac{p^2}{2m} + V$$

where p, the particle's momentum along the X direction, is, from the equation above, given by

$$p = \sqrt{2m(E-V)}$$

Since the particle's wavelength λ is in general h/p we may write the wavelength as

$$\lambda = \frac{h}{p} = \frac{h}{\sqrt{(2m)[E-V(x)]}} \qquad (37\text{-}13)$$

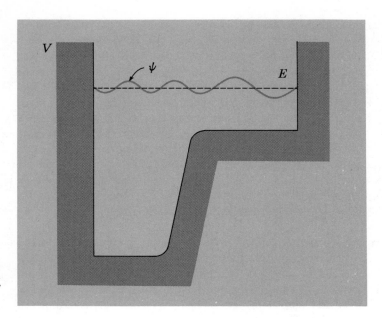

FIG. 37-17. A potential well with infinitely high sidewalls and a two-tier base, together with one of the allowed wave functions.

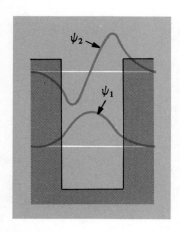

FIG. 37-18. Wave functions for the first two states of a particle in a potential well of finite depth.

Equation (37-13) shows that to the degree that the system's potential energy depends on the particle's location x, so does the particle's wavelength. More specifically, when potential energy V increases, so does the wavelength, and conversely. Figure 37-17 shows one wave function for a potential well with infinitely high sidewalls but with a two-tier base. The wave function is smooth throughout, drops to zero at the walls, and is sinusoidal where the potential V is constant.

Suppose that the potential well confining a particle is not of infinite height. Classically, it is meaningless to speak of a particle being at a location for which the total energy E is less than the potential energy V—that would imply a negative kinetic energy or, from (37-13), a wavelength that is imaginary. But in wave mechanics a particle *may* be found outside the limits defined by classical mechanics. Solution of the Schrödinger equation shows that the wave function decreases exponentially in regions for $V > E$. Wave functions for the first two allowed states for a particle in a potential-energy well of finite depth are shown in Fig. 37-18; the wave functions are sinusoidal inside the well, joined smoothly to exponentially decreasing wave functions at the boundaries.

Since a particle can, wave-mechanically, spill over its classical confines, we have a curious possibility when it encounters a potential wall of finite *width,* as well as height, as in Fig. 37-19. In classical physics a particle whose kinetic energy is less than the height of the potential wall would never get through to the other side simply because of energy conservation. In wave mechanics this *is* a possibility. The wave function decays exponentially when the potential energy exceeds the total energy. Thus, the wave

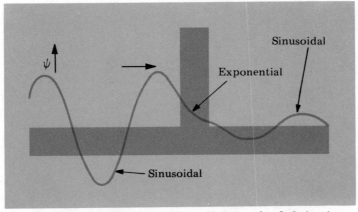

FIG. 37-19. Wave function of a particle incident from the left upon a barrier of finite height and width.

function of a particle approaching the wall from the left is sinusoidal to the left of the wall, exponential through the wall, and sinusoidal, but of much smaller amplitude, to the right of the wall. There is a small but finite probability that the particle that has approached from the left will be on the *right*. In other words, the particle has a high probability of being found on the left, a smaller probability of being found within the wall, and a still smaller probability of being found on the right. The particle—or, more properly, the wave—can penetrate, or tunnel through, the classically insurmountable barrier. The probability of this *tunnel effect* is vanishingly small except at the atomic and nuclear level. The effect is observed, however, in the behavior of certain semiconducting devices (tunnel diodes) and in the emission of α particles from heavy, unstable nuclei.

Two important potential-energy functions, together with the allowed wave functions and energies, are shown in Fig. 37-20: (a) the one-dimensional simple harmonic oscillator in which the particle is bound to the potential-energy function $V(x) = \tfrac{1}{2}kx^2$ and (b) the inverse-square attractive force $V(r) = -ke^2/r$, where r is the distance from the force center, and $V(r)$ the potential-energy function applying to the hydrogen atom.

a. For the simple harmonic oscillator the energy levels are equally spaced, the allowed energies being given by $E_v = (v + \tfrac{1}{2})hf$, where the vibrational quantum number $v = 0, 1, 2, \ldots$ and $f = (1/2\pi)(k/m)^{1/2}$, the classical frequency of oscillation. Note that the wave functions extend beyond the classical confines of the potential well. In progressing from the lowest energy level to higher allowed energies, the number of half-waves fitting between the confining potential boundaries are 1, 2, 3, ...

b. For the inverse-square attractive force only those wave functions are shown which are spherically symmetrical and therefore dependent on the radial distance r (so-called s states). The allowed energies are negative, as they must be for a bound system, and are given by $E_n = -E_I/n^2$, where the principal quantum number $n = 1, 2, 3, \ldots$ and E_I (the ionization energy) is given by $E_I = 2\pi^2 k^2 e^4 m/h^2$. Again the wave

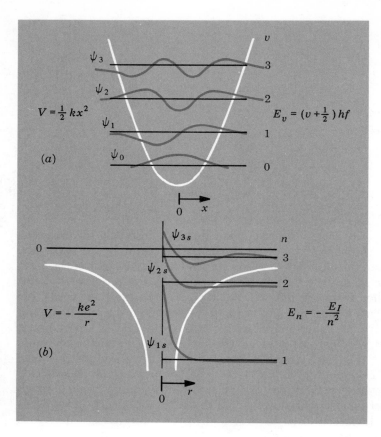

FIG. 37-20. Potential-energy functions, wave functions, and allowed energies for *(a)* a simple harmonic oscillator and *(b)* inverse-square attractive force (hydrogen atom).

function spills outside the classical limits and the number of zeros in the wave function increases by integers as one moves upward in the allowed energies.

Example 37-2

Sketch the wave function of an excited state of a particle moving in the potential shown in Fig. 37-21. This is nothing more than the wave-mechanical version of a standard example in elementary mechanics: a particle sliding on a frictionless inclined plane (here with an infinitely high wall at the bottom of the incline). Solving the problem in complete analytical detail would involve finding solutions to the Schrödinger equation for a potential energy that rises to infinity at $x = 0$ and increases steadily with x, according to $V = ax$, for $x > 0$, but we can show, even without detailed analysis, the general features of the wave-mechanical solution.

Since the potential rises to infinity at $x = 0$, the wave function must be zero there. At the other extreme, where the particle may be found outside the classical upper limit, the wave function must decay to zero. In between the particle has some nonzero kinetic energy, and so the wave function is undulatory there. Further, in the case of a relatively highly excited state the features of the wave-mechanical solution must approach the classical features of a particle sliding on

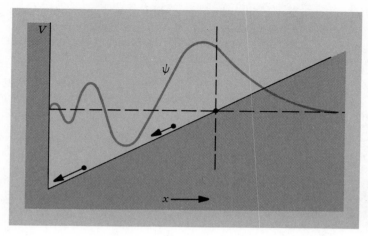

FIG. 37-21. The wave-mechanical inclined plane: potential-energy function and wave function of the fifth allowed state.

an inclined plane and colliding with a hard wall at the bottom. Since the particle moves at high speed at the base of the incline and at less speed at the top, the wavelength of the wave function must be smallest near the base and increasingly large toward the top. Moreover, the classical high speed at the base and low speed at the top imply that the particle is more likely to be found at the top than at the bottom. Given all these features, we know that the wave-mechanical solution of an excited state is as shown in Fig. 37-21.

In the next energy state up we should find one additional half-wavelength fitted between the left and right extremes, and in the next energy state down we should find one less half-wavelength; in the lowest state, we should find a single hump in ψ at the left and right extremes.

SUMMARY

Every particle, whether of finite or zero rest mass, has associated with it a frequency and wavelength given by

$$\nu = \frac{E}{h} \quad \text{and} \quad \lambda = \frac{h}{p} \qquad (37\text{-}1), (37\text{-}2)$$

The wavelengths of 150 eV electrons, x-rays, and the interatomic spacings of a solid are all of the order of 1 Å.

When a wave is incident upon a set of parallel Bragg planes separated by a distance d, constructive interference occurs, according to the Bragg relation, when

$$n\lambda = 2d \sin \theta \qquad (37\text{-}4)$$

X-rays and electrons can be diffracted by crystalline solids, and diffraction effects may be used for measuring x-ray and electron wavelengths and interatomic spacings.

According to Bohr's principle of complementarity, the wave and particle aspects of electromagnetic radiation and of material particles are complementary, but the use of one description precludes the use of the other in a given circumstance.

The wave-mechanical description of material particles parallels that of electromagnetic radiation as follows:

Electromagnetic radiation (photons)   *Material particles*
Wave function: electric field $\mathscr{E}$ The wave function ψ
(or magnetic field)
Probability of observing particle
in interval dx $\mathscr{E}^2\, dx$ $\psi^2\, dx$

The Heisenberg uncertainty principle imposes a limit on the certainty of simultaneous measurements of energy E and time t and of momentum p_x and position x:

$$\Delta E\, \Delta t \geq h \qquad (37\text{-}8)$$

$$\Delta p_x\, \Delta x \geq h \qquad (37\text{-}9)$$

A particle confined to a potential well is restricted to those states in which an integral multiple of the particle's wavelength may be fitted between the boundaries. The system's energy is thereby quantized.

$$\lim_{h \to 0} (\text{quantum physics}) = \text{classical physics}$$

PROBLEMS

37-1. What are the wavelengths of (a) a 1.00-MeV photon, (b) a 1.00-MeV (kinetic energy) electron, and (c) a 1.00-MeV (kinetic energy) proton?

37-2. A proton and electron are both accelerated from rest by an electric potential difference of 50 kV. What is the ratio of the proton's wavelength to the electron's wavelength?

37-3. A *thermal neutron* is one whose average kinetic energy $\tfrac{1}{2}mv^2$ is equal to the average translational kinetic energy per particle $\tfrac{3}{2}kT$ in equilibrium at temperature T, where T is room temperature (300 K). Thermal neutrons in a material are just as likely to gain as to lose energy in collision. Show that the wavelength of a thermal neutron is 1.4 Å. The neutron mass is 1.67×10^{-27} kg.

37-4. (a) At the Stanford Linear Accelerator Center electrons are accelerated to a kinetic energy of 20 GeV; what is their wavelength at that energy? (b) In the high-energy accelerator at Batavia, Ill., protons have energies of 500 GeV; what is their wavelength at that energy?

37-5. Show that the wavelength of a particle of rest energy E_0 and kinetic energy E_k is given (a) by hc/E_k when $E_k \gg E_0$ and (b) $hc/(2E_0 E_k)^{1/2}$ when $E_k \ll E_0$.

37-6. Show that the lattice spacing d between adjacent atoms in the NaCl crystal (see Fig. 37-1) is 2.82 Å. The atomic weight of Na is 23.00, that of Cl is 35.45. The density of NaCl is 2.163 g/cm^3. The number of atoms per g-mol is 6.022×10^{23} (Avogadro's number).

37-7. Monochromatic x-rays consisting of photons with an energy of 5.0 MeV are incident upon a KCl single crystal, whose lattice spacing is 3.14 Å. At what angle with respect to the incident beam would one observe first-order Bragg reflection?

37-8. A narrow beam of thermal neutrons is incident upon a NaCl single crystal whose lattice spacing is 2.82 Å. (a) At what angle with respect to the incident beam must the Bragg planes be oriented so as to produce strong first-order diffraction for those neutrons with a kinetic energy of 0.050 eV? (b) What is then the angle between the incident and the diffracted beams?

37-9. A large-scale model of a crystalline lattice is produced by a collection of small spheres at the corners of densely packed cubes having an edge length of 10 cm. The "crystal" is irradiated with 3.0-cm microwaves. Through what angles must the crystal model be turned in order to produce strong Bragg diffraction? Measure rotation with respect to the configuration in which the incident beam is perpendicular to a face of the cubical elements.

37-10. Two parallel wire-mesh screens are separated by 0.20 m. A monochromatic beam of radio waves is incident upon the screens, and strong first-order Bragg diffraction occurs when the angle between the incident and diffracted beams is 30°. What is the frequency of the radio waves?

37-11. Since the frequency ν and wavelength λ associated with a particle are given by $\nu = E/h$ and $\lambda = h/p$, their product $\nu\lambda$, which gives the *phase speed* v_{ph} of deBroglie waves, is equal to E/p. The speed *of the particle* v, which is the speed with which energy and momentum advance through space, is given by $v = p/m$ where $m = E/c^2$. (a) Show that the phase speed v_{ph} and the particle speed v are related by $v_{\text{ph}} = c^2/v$. (b) Show that for any particle, except one of zero rest mass, the phase speed *exceeds* the speed of light. (c) What is the phase speed for electromagnetic waves through vacuum?

37-12. What intensity of a γ-ray beam consisting of 5.0-MeV photons will correspond to a photon flux of 1.0 photons/mm²-s?

37-13. How do the allowed energies for a particle in a one-dimensional potential well of finite height compare (higher, lower?) with the corresponding energies of a particle in a one-dimensional potential well of infinite height? Compare Fig. 37-14 and 37-18.

37-14. A 1-μg particle moves along the X axis; its speed is uncertain by 6.6×10^{-6} m/s. What are the uncertainties in its position along (a) the X axis and (b) the Y axis? An electron with the same uncertainty in speed moves along the X axis; what are the uncertainties in its position along (c) the X axis and (d) the Y axis?

37-15. A virus is the smallest object that can be "seen" in an electron microscope. Suppose that a small virus, with a size of 10 Å, and a density equal to that of water (1 g/cm³) is localized in a region of space equal to its size. What is the minimum speed of the virus?

37-16. The momentum component along the direction of motion of a 1.02-MeV electron is uncertain by 1 part in 10^2. Over what minimum region of space does the electron extend?

37-17. A camera with a fast shutter takes a photograph during an exposure of 1.0×10^{-5} s. (a) What is the uncertainty in the energy of any one photon passing through the shutter? What is the corresponding fractional uncertainty in the wavelength of (b) of 2-eV photon of visible light and (c) a 2-GeV γ-ray photon?

37-18. What is the minimum kinetic energy of (a) an electron and (b) a proton confined to a region of space the size of a nucleus, about 10^{-14} m? (c) It is known that particles within the atomic nucleus have energies of the order of a few MeV and that the attractive potential energy between a pair of nuclear particles is of the same order; which particles, electrons or protons, might be found within an atomic nucleus?

37-19. The wavelength of a photon is measured to an accuracy of 1 part in 10^8 ($\Delta\lambda/\lambda = 10^{-8}$). What is the uncertainty Δx in the simultaneous localization of (a) a photon of visible light of 6000-Å wavelength, (b) a radio photon of 100-kHz frequency, (c) an x-ray photon of 1.0-Å wavelength, and (d) a γ-ray photon of 12.4-GeV energy?

37-20. A particle is confined to a region having the dimensions of an atomic nucleus (about 2×10^{-15} m). (a) Use the uncertainty relation to compute its approximate momentum. (b) If the particle's kinetic energy is a few MeV, what is its rest energy? (c) What ordinary particle meets the requirements of parts (a) and (b)?

37-21. What is the minimum kinetic energy of a proton confined to an infinitely high one-dimensional potential well of width 1.0 Å?

37-22. In an electron microscope an electron beam replaces a light beam, and electric and magnetic focusing fields replace refracting lenses. The smallest distance that can be resolved by any microscope under optimum conditions (its resolving power) is approximately equal to the wavelength used in the microscope. (a) A typical electron microscope might use 50-keV electrons; compute the minimum distance that can be resolved in such a microscope. (b) By what factor does the actual resolving power of about 20 Å, attained in well-designed electron microscopes, exceed the ultimate resolving power (minimum separation between two point objects distinguishable as two distinct objects), which is limited by the wave properties of electrons?

37-23. The ultimate resolving power (see Prob. 37-22) of any microscope depends solely on the wavelength. Suppose that we wish to study an object having dimensions of 0.50 Å; what are the minimum energy and momentum if we use (a) electrons and (b) photons?

(c) Why is an electron microscope preferable to a γ-ray microscope?

37-24. (a) A particle of mass m is confined to a three-dimensional box of dimensions a, b, and c. Show by imposing boundary conditions on the wave functions in each of the three dimensions that the particle's allowed energies are given by

$$E = \left(\frac{\pi^2\hbar^2}{2m}\right)\left[\left(\frac{n_1}{a}\right)^2 + \left(\frac{n_2}{b}\right)^2 + \left(\frac{n_3}{c}\right)^2\right]$$

where the quantum numbers n_1, n_2, and n_3 can take the integral values 1, 2, 3, ... (b) A billiard ball of mass m is confined to a two-dimensional box (a billiard table) of dimensions L and W. What are the billiard ball's permitted energies? (c) What is a billiard ball's minimum kinetic energy (a ball of mass 0.14 kg on a table 3.7 by 1.9 m)?

37-25. Figure 37-17 shows one of the allowed wave functions for a two-tier one-dimensional potential well. The potential energy at the left is zero; at the right it is V_r. The energy E for the wave function shown is $E = \frac{4}{3}V_r$. (a) Show that the particle's wavelength on the right is twice that on the left. (b) Why must the amplitude of the wave function on the right exceed that on the left?

37-26. A small bead of mass m slides along a straight frictionless rod of length L, whose ends are attached to infinitely massive objects. What are the allowed energies of the bead?

37-27. A small bead of mass m slides freely along a circular ring of radius R. (a) What are the allowed energies of the bead? (b) What are the allowed angular momenta?

37-28. Show that in the classical limit of very large quantum numbers the probability of a particle's being in any small but finite interval Δx in a one-dimensional box is independent of its position within the box. This result is in agreement with the classical expectation that the probability of finding a particle that moves at constant speed within a one-dimensional box is the same at all points.

37-29. (a) Use the fact that the wave function for a particle is undulatory in regions of space for which its total energy exceeds the potential energy but decays for regions in which $(E - V) < 0$ to sketch the first several wave functions for a particle subject to a parabolic potential (a simple harmonic oscillator). (b) Sketch the wave function for a highly excited state, using the fact that the wave-mechanical probability for finding the particle at any point in space must, in the limit of large quantum numbers, approach the corresponding classical probability.

37-30. The permitted energies of a simple harmonic oscillator whose classical frequency of oscillation is f are given by $E_v = (v + \frac{1}{2})hf$, where the vibrational quantum number v can take on the integral values $v = 0, 1, 2, \ldots$. Consequently, the minimum energy of a simple harmonic oscillator is not zero, but $E_0 = \frac{1}{2}hf$, the so-called zero-point vibrational energy. (a) What is the minimum energy of a simple pendulum with a 1.0-kg mass and 1.0 m long? (b) What is the corresponding angular displacement of the string from the equilibrium configuration (the vertical)?

37-31. See Prob. 37-30. Show that when transitions take place between *adjacent* energy levels for the simple harmonic oscillator, the photons emitted have the same frequency as the equivalent classical oscillator.

37-32. The angular momentum of an object such as a dumbbell rotating about an axis perpendicular to the massless rod connecting two masses is given by $L = [J(J+1)]^{1/2}\hbar$, where the rotational quantum number J can take on the integral values $J = 0, 1, 2, \ldots$ and $\hbar$ represents $h/2\pi$. The rotational kinetic energy of such a rotator is given by $E_r = \frac{1}{2}I\omega^2 = (I\omega)^2/2I = L^2/2I$, where ω is angular velocity and I the moment of inertia about the rotation axis. (a) Show that the allowed energies of the rotator are then restricted to the values given by $E_r = J(J+1)\hbar^2/2I$. (b) What is the minimum rotational energy of a rotator consisting of two 10-g masses attached to the ends of a 10-cm rod with rotation about the perpendicular bisector of the rod as axis?

37-33. The interatomic distance in diatomic molecules is typically of the order of a few Å. Using the results of Prob. 37-32, show the spectrum of lines emitted or absorbed by diatomic molecules undergoing quantum transitions in their rotational states (the *pure rotational spectrum*) lies typically in the far infrared or microwave regions of the electromagnetic spectrum. The allowed transitions follow the rule that the rotational quantum number J changes by *one* integer.

37-34. The attractive potential energy between a certain solid substance and electrons entering at its surface is 10 eV. Suppose that 20-eV electrons enter the surface at an angle of incidence (with respect to the normal to the surface) of 30°. (a) What is the angle of refraction of the electrons within the material? (b) What is the electron's wavelength inside?

37-35. According to Prob. 36-14 the wavelength of a photon of energy E is given by $\lambda = (0.0124 \text{ MeV-Å})/E$. Show that the same relation gives the wavelength of a particle of nonzero rest mass when the particle's total energy E greatly exceeds its rest energy E_0.

CHAPTER 38

Atomic Structure

38–1 α-PARTICLE SCATTERING

Our modern concept of the structure of an atom posits these essential features: a *nucleus* occupying a very small region of space, in which all the positive charge and practically all the atom's mass are concentrated, and negatively charged electrons, which surround the nucleus. Let us examine the evidence for this concept of atomic structure, first proposed by Ernest Rutherford in 1911.

At the end of the nineteenth century it was known that the negative electric charge of the atom is carried by electrons, whose mass is but a small fraction of the total mass of the atom. Because atoms as a whole are ordinarily electrically neutral, it follows that, if we were to remove all an atom's electrons, then what would remain would contain all the positive electric charge and essentially all the mass. The question is, then, "How are the mass and positive charge distributed over the volume of the atom?" Atoms are known, from a variety of experiments, to have a "size" (diameter) of the order of 1 Å, and because the positive charge and mass are confined to at least this small a region, it is impossible by any direct measurement to see and observe any details of the atomic structure. An indirect measurement must be resorted to. One of the most powerful methods of studying the distribution of matter or of electric charge is the method of *scattering*, and it was by the α-particle-scattering experiments, suggested by Rutherford, that the existence of atomic nuclei was established.

Consider first a simple example of a scattering experiment. Suppose that we are confronted with a large black box; without looking inside to examine its internal structure, we are asked to determine how the mass is distributed throughout the interior of the box. The box might, for instance, be filled completely with some material of relatively low density, such as wood, or it might be only partly filled with one of high density. To find out whether one of these two possibilities represents the actual distribution of material within the box, we can use a very simple expedient: Shoot bullets into the box and see what happens to them. If we find all bullets emerging in the forward direction with reduced speeds, then we might infer that the box is filled throughout with material which deflects the bullets only slightly as they pass through. On the other hand, if we find a few bullets greatly deflected from their original paths, then we might assume that they had collided with small, hard, and massive objects dispersed throughout the box. Note that it is *not* necessary to aim the bullets; the shots may be fired randomly over the front of the box. This is the essence of the particle-scattering experiments in atomic and nuclear physics.

Rutherford suggested that the mass and positive charge of an atom are essentially a point charge and point mass, called the *nucleus*. He further suggested that his hypothesis could be tested by shooting high-speed, positively charged particles (the bullets) through a thin metallic foil (the black box), and then examining the distribution of the scattered particles. At the time of Rutherford's suggestion the only available and suitable charged particles were α particles, with energies of several million electron volts, from radioactive materials. Rutherford had shown earlier that an α particle is a doubly ionized helium atom; therefore, it has a positive electric charge twice the magnitude of the electron charge and a mass several thousand times greater than the mass of an electron but considerably smaller than the mass of such a heavy atom as gold. To confirm Rutherford's nuclear hypothesis, H. Geiger and E. Marsden in 1913 scattered α particles from thin gold foils.

The essentials of the scattering apparatus are shown in Fig. 38-1. A collimated beam of particles strikes a thin foil of scattering material, and a rotatable detector counts the number of particles scattered at some scattering angle θ from the incident direction.

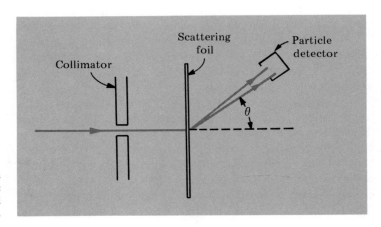

FIG. 38-1. Arrangement for a simple scattering experiment: collimator, scattering foil, and detector.

The experiment consists of measuring the relative number of scattered particles at various scattering angles θ.

Consider the paths of the α particles traversing the interior of the scattering foil. We may dismiss as inconsequential any encounters an α particle may have with electrons within the material, because, since the α-particle mass greatly exceeds that of an electron, the particle is essentially undeflected in such collisions, and a negligible fraction of its energy is transferred to any one electron. Thus the α particles are appreciably deflected and scattered only by close encounters with the nuclei. The mass of the nucleus of a gold atom is considerably greater than (50 times) that of the α particle; therefore, in the collision the nucleus does not recoil appreciably and may be assumed to remain at rest. Since α particles and nuclei are both positively charged, they repel each other. Rutherford assumed that the *only* force acting between a nucleus and an α particle, both regarded as point charges, was the coulomb electrostatic force. As we know, that force varies inversely with the square of the distance between the charges; therefore, although it is never zero (except for at infinite separation between the charges), the α particle is acted upon by a *strong* repulsive force only when it comes quite close to a nucleus.

Fig. 38-2 shows a number of paths of α particles as they move through the interior of a scattering foil. Most particles pass through with only a slight deviation from their original course; the chances of a close encounter with a nucleus, or scattering center, are fairly remote. But those few α particles which barely miss head-on collisions are deflected at sizable angles, and those extremely rare ones which make head-on collisions are deflected through 180°, that is, are brought to rest momentarily and then returned along their paths of incidence. For nuclei as *point* charges, most incident particles are scattered only slightly; a small but significant number are deflected through large angles. (If the positive charge were distributed uniformly throughout the atom, rather than being concentrated in nuclei, virtually no particles would be

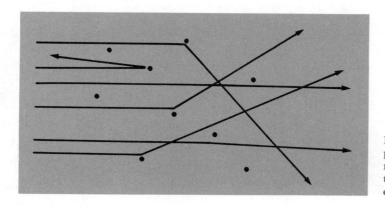

FIG. 38-2. Scattering of α particles by the nuclei of a material (the number scattered through sizable angles is greatly exaggerated).

scattered through large angles.) The nuclear hypothesis of Rutherford was confirmed in the experiments of Geiger and Marsden in that the measured distribution of the scattered α particles was in agreement with the distribution predicted on the assumption of scattering by a Coulomb force by point charges, the number of scattered particles varying with scattering angle θ according to $1/\sin^4(\theta/2)$.† They confirmed the Rutherford theory for a variety of α-particle energies, foil materials, and foil thicknesses.

38–2 THE HYDROGEN SPECTRUM

The simplest atomic system is hydrogen; an atom consists of only two particles—an electron and a much more massive proton as nucleus—interacting by an attractive Coulomb force. Therefore, the theoretical description of atomic structure has concentrated first on hydrogen and on its observed behavior, primarily as exhibited through the electromagnetic radiation it emits and absorbs.

In order to observe the spectrum of isolated hydrogen atoms, it is necessary to use gaseous atomic hydrogen, because in it the atoms are so far apart that each one behaves as an isolated system (molecular hydrogen H_2 and solid hydrogen radiate spectra that reflect some aspects of hydrogen atoms bound together). The visible spectrum emitted by hydrogen may be studied with a prism spectrometer, or a diffraction grating may be used instead of a prism for dispersing the radiation. The hydrogen gas is excited by an electrical discharge or by extreme heating and emits radiation. The dispersed radiation, separated into its various frequency components, falls on a screen or photographic plate, making it possible to measure the frequencies and intensities of the *emission spectrum*.

Any instrument that disperses and measures the various wavelengths of a beam of electromagnetic radiation is called a *spectrometer*. An instrument that disperses the light and photographs the spectrum is known as a *spectrograph;* one that makes the spectrum visible directly to the eye is a *spectroscope*. Spectrometers are designed for the study of each of the several regions of the electromagnetic spectrum, such as radiofrequency, x-rays, and γ rays. That branch of physics which deals with the study of electromagnetic radiation emitted or absorbed by substances is called *spectroscopy*. Spectroscopy is a very powerful method of inquiry into atomic, molecular, and nuclear structure; it is characterized by very great precision (frequencies or wavelengths are

---

†Since from the standpoint of quantum theory a beam of monoenergetic particles is, in effect, a beam of monochromatic waves (Sec. 37-2), the scattering process consists of the diffraction of incident waves by scattering centers. It is a remarkable fact that the wave-mechanical treatment of scattering for an inverse-square force yields precisely the same result as that yielded by a strictly classical analysis. For other types of forces, however, the classical and wave-mechanical results differ.

easily measured to 1 part in 10 million) and very high sensitivity (emission or absorption from samples as small as fractions of a microgram can be observed).

The spectrum emitted by atomic hydrogen consists of a number of sharp, discrete, bright lines on a black background. In fact, the spectra of all chemical elements in monoatomic gaseous form are composed of such bright lines, each spectrum being characteristic of the particular element. The spectrum is known as a *line spectrum.* The *emission spectrum* from atomic hydrogen, then, is a bright-line spectrum characteristic of hydrogen. Since each chemical element has its own characteristic line spectrum, spectroscopy is a particularly sensitive method of identifying the elements.

The line spectrum from atomic hydrogen in the visible region is shown in Fig. 38-3. The lines are labeled H_α, H_β, etc., in the order of increasing frequency and decreasing wavelength. Ordinarily the H_α line is much more intense than the H_β, which is in turn more intense than the H_γ, and so on. The spacings between adjacent lines become smaller as the frequency increases and the discrete lines approach a *series limit,* above which there appears a weak continuous spectrum. This group of hydrogen lines, which appears in the visible region of the electromagnetic spectrum, is known as the *Balmer series,* because in 1885 J. J. Balmer arrived at a simple empirical formula from which all the observed wavelengths in the group could be computed. The Balmer formula, giving the wavelength λ or all spectral lines in this series, may be written

$$\frac{1}{\lambda} = R \left(\frac{1}{2^2} - \frac{1}{n^2} \right) \qquad (38\text{-}1)$$

where $\quad R = 1.0967758 \times 10^7 \text{ m}^{-1} \approx 1.0968 \times 10^{-3} \text{ Å}^{-1}$

and n is an integer having the values 3, 4, 5, ...

Putting $n = 3$ in this formula gives $\lambda = 6564.7$ Å, the H_α line; similarly, putting $n = 4$ gives $\lambda = 4862.7$ Å, the H_β line, and so on up to the series limit ($n = \infty$). The constant R is known as the *Rydberg constant;* its value is chosen by trial to give the best fit for the measured wavelengths. In atomic spectroscopy spectral lines typically are specified by their wavelengths rather than their frequencies, because it is the wavelength that is measured.

Beside the Balmer series in the visible region, hydrogen radiates a series of lines in the ultraviolet region and several series of lines in the infrared. Each series may be represented by a formula similar to the Balmer equation. In fact, one general formula may be written, from which *all* the spectral lines of hydrogen can be computed. It is known as the *Rydberg equation* and is written

$$\frac{1}{\lambda} = R \left(\frac{1}{n_l^2} - \frac{1}{n_n^2} \right) \qquad (38\text{-}2)$$

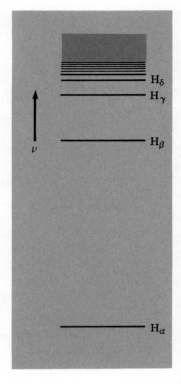

FIG. 38-3. Frequency distribution of radiation from atomic hydrogen in the visible region. This particular group of spectral lines is the Balmer series.

where

$n_l = 1$ and $n_u = 2, 3, 4, \ldots$ *Lyman* series, ultraviolet region

$n_l = 2$ and $n_u = 3, 4, 5, \ldots$ *Balmer* series, visible region

$n_l = 3$ and $n_u = 4, 5, 6, \ldots$ *Paschen* series, infrared region

and so on, to further series lying in the far infrared. The value of the Rydberg constant in this equation is *precisely* the same as its value in (38-1); in fact, (38-2) becomes (38-1) when n_l is set equal to 2. The choice of u ("upper") and l ("lower") as subscripts for the integers in the Rydberg formula will become obvious in Sec. 38-3. The several series of hydrogen lines are named after their discoverers. Although the Rydberg formula is remarkably successful in summarizing the wavelengths radiated by atomic hydrogen, it must be recognized that it is merely an empirical relation, which in itself supplies no information about the structure of hydrogen. On the other hand, a truly successful theory of the hydrogen atom must be capable of predicting the spectral lines; that is, it must be able to yield the Rydberg formula as a result.

We have been concerned with the spectrum given by hydrogen when it is excited by an electrical discharge or by extreme heating. Atomic hydrogen at room temperature does *not*, by itself, emit appreciable electromagnetic radiation, but at room temperature it can selectively absorb electromagnetic radiation, giving an *absorption spectrum*. The absorption spectrum of atomic hydrogen is observed when a beam of white light (all frequencies present) is passed through atomic hydrogen gas and the spectrum of the transmitted light is examined in a spectrometer. What is found is a series of dark lines superimposed on the spectrum of white light; this is known as a *dark-line spectrum*. The gas is transparent to waves of all frequencies except those corresponding to the dark lines, for which it is opaque; that is, the atoms absorb only waves of certain discrete, sharp frequencies from the continuum of waves passing through the gas. The absorbed energy is very quickly radiated by the excited atoms, but in *all directions,* not just in the incident direction. The dark lines in the absorption spectrum of hydrogen occur at precisely the same frequencies as do the bright lines in the emission spectrum, as shown schematically in Fig. 38-4, where the intensity is plotted as a function of frequency. Hydrogen is a radiator of electromagnetic radiation only at the specific frequencies or wavelengths given by the Rydberg formula; it is an absorber of radiation only at the same frequencies.

What we have discussed concerning the emission and absorption spectra of atomic hydrogen holds equally well for the line spectra of all elements. A characteristic set of frequencies is emitted when the atoms radiate energy; the same set of frequencies is absorbed by the atoms when a continuous frequency band of electromagnetic radiation is sent through the gas.

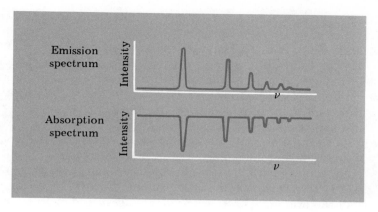

FIG. 38-4. Variation of intensity with frequency in emission and absorption spectra of hydrogen.

The observed spectrum of hydrogen cannot be accounted for by classical mechanics and classical electromagnetism. If the hydrogen atom were merely a miniature solar system with the electron orbiting the nucleus in the fashion of a planet around the sun, then the electrically charged electron, in being accelerated continuously, would radiate continuously (Sec. 30-6) at its orbital frequency. But if the atom loses energy continuously by radiation, the total energy of the atom must decrease continuously — the electron orbiting the nucleus in progressively smaller orbits at progressively higher frequencies. In other words, the atom would collapse rapidly, after having radiated a continuous spectrum.

That classical physics is impotent in accounting for atomic structure is also obvious from considerations of the wave character of electrons. A variety of evidence shows that the "size" of a hydrogen atom is about 1 Å. But as we saw in Sec. 37-7, an electron confined to such atomic dimensions has a minimum kinetic energy of a few electron volts. If one were to localize an electron *within* an atom — that is, confine it to a region of space significantly smaller than 1 Å — its kinetic energy would, through the uncertainty principle, necessarily have to be much greater than several electron volts. Yet, experimental evidence indicates that atomic energies are of the order of only a few electron volts, as must be the kinetic energy of an electron in such an atom as hydrogen. In short, only a quantum treatment of atomic structure, one which invokes the wave nature of particles, can be expected to accord with observation.

38-3 THE BOHR THEORY OF HYDROGEN

The first quantum theory of the hydrogen atom was developed in 1913 by Niels Bohr, a student of Rutherford. The photon nature of electromagnetic radiation had been established, but the wave aspects of material particles were not to be recognized until 1924. The Bohr model of the atom was the first step toward a thoroughgoing wave-mechanical treatment of atomic structure; it should be realized at the start, however, that the Bohr theory has limited applicability. It retains enough classical features for the atomic structure to be readily visualized in terms of a particle model, and it introduces enough quantum features to give a fairly accurate description of the atomic spectrum of hydrogen. Bohr's theory is, therefore, transitional between classical mechanics and the wave mechanics developed during the 1920s.

Like a strictly classical model of hydrogen, the Bohr theory assumes the proton to be at rest and the electron to move in a circular orbit about it. The force maintaining the electron in its orbit is the attractive electrostatic coulomb force between the electron and the proton. For an electron of mass m and charge e, moving at speed v in an orbit of radius r about a nucleus also of electric charge e and effectively infinite mass, we have from Newton's second law

$$F = ma$$

$$\frac{ke^2}{r^2} = \frac{mv^2}{r} \qquad (38\text{-}3)$$

where k is the constant in Coulomb's law.

The atom's total energy E is the sum of the electron's kinetic energy and the system's potential energy:

$$E = \tfrac{1}{2}mv^2 - \frac{ke^2}{r}$$

Using (38-3) in the equation above we may write the total energy as

$$E = \frac{\tfrac{1}{2}ke^2}{r} - \frac{ke^2}{r}$$

$$E = -\frac{ke^2}{2r} \qquad (38\text{-}4)$$

The total energy of the two-particle system is negative when the two particles are bound together. When the electron orbital radius becomes infinite and the atom becomes dissociated into two separate particles, $E = 0$. Since the ionization energy of hydrogen is measured to be 13.6 eV, the total energy of the hydrogen atom in its normal state is $E = -13.6$ eV. Substituting this value in (38-4) yields $r = 0.53$ Å.

One way of imposing quantum conditions on the permitted energies and, therefore, the permitted electron radii, is to use the rule that the orbital angular momentum of the electron as it orbits the nucleus is an integral multiple of Planck's constant h divided by 2π. Since the angular momentum of a particle with linear momentum mv in a circular orbit of radius r is mvr (Sec. 11-3), the rule for quantization of angular momentum may be written

$$mvr = n\frac{h}{2\pi} = n\hbar \qquad (38\text{-}5)$$

where $n = 1, 2, 3, \ldots$ and $\hbar$, read as "aitch bar," represents $h/2\pi$. Equation (38-5) implies that only those electron orbits are permitted for which the angular momentum is an integral multiple of $\hbar$. We must assume, further, that despite classical electromagnetism, an electron in such a stable, allowed, or stationary state does *not* radiate electromagnetic radiation.

Equation 38-5 can be cast into a different form that lends to a simple interpretation of the allowed electron orbits. We have

$$n \frac{h}{mv} = 2\pi r$$

The quantity h/mv is the deBroglie wavelength λ of the electron, so that equation above may be written as

$$n\lambda = 2\pi r \tag{38-6}$$

Equation 38-6 implies that in going the distance $2\pi r$ around the circumference of the circular electron orbit, an integral number $n\lambda$ of electron wavelengths may be fitted; that is, the allowed state is one in which an electron, regarded as a wave wrapped around in a self-completing circle, does not cancel itself out by destructive interference. This latter statement cannot be physically correct, however, because an electron surrounding a nucleus in an atom cannot be regarded as extending only around a sharply defined circular orbit. The electron extends in all three dimensions. Therefore, we must regard (38-6) as a suggestive mnemonic, not as a rigorous application of wave mechanics.

Taking (38-5), or its equivalent (38-6), as the rule for the permitted, stationary states of the hydrogen atom, we can find the allowed orbital radii, and from (38-4), the allowed energies.† Solving for v in (38-5) and substituting in (38-3) we have

$$m \left(\frac{n\hbar}{mr_n} \right)^2 = \frac{ke^2}{r_n} \tag{38-7}$$

where we have added the subscript n to the radius r to indicate that the value of r_n depends on the value of the quantum number $n = 1, 2, 3, \ldots$.

Solving (38-7) for r_n we have

$$r_n = \frac{n^2 \hbar^2}{kme^2} \tag{38-8}$$

---

†In his original theory of hydrogen Bohr used neither the angular-momentum-quantization rule nor the integral-deBroglie-waves-around-the-circumference rule. Instead his argument was based on his *correspondence principle:* Any new, more general theory in physics must yield the same results as the well-established, less general theory when the more general theory is applied under those circumstances for which the less general theory is known to be valid; or, when the circumstances correspond, the results must also correspond. Applied to the quantum theory the correspondence principle implies that the quantum theory must yield the same results as classical physics when the quantum theory is applied to macroscopic phenomena. More specifically, applied to the structure of the hydrogen atom, the correspondence principle requires that the frequency of a photon emitted in a quantum transition for very large values of quantum number n (for which the orbital radii approach macroscopic size) must equal the classical frequency with which the electron orbits the nucleus. See Prob. 38-16.

The smallest allowed radius, the so-called radius of the first Bohr orbit, is given by

$$r_1 = \frac{\hbar^2}{kme^2} = 0.529 \text{ Å} \tag{38-9}$$

in which the values of the known atomic constants have been used. The Bohr model predicts, then, that the size of the hydrogen atom with the smallest stationary orbit is of the order of 1 Å, in good agreement with experimental determinations. We may express all the allowed radii in a simpler form by putting (38-9) in (38-8):

$$r_n = n^2 r_1 \tag{38-10}$$

The radii of the stationary orbits are, therefore, r_1, $4r_1$, $9r_1$, ...

The allowed values of the total energy (excluding the rest energies of the proton and electron) of the hydrogen atom can now be determined easily from (38-4) and (38-10).

$$E_n = -\frac{ke^2}{2r_n} = -\frac{1}{n^2} \frac{ke^2}{2r_1} \tag{38-11}$$

If we represent the quantity $ke^2/2r_1$ by E_I, then this equation becomes

$$E_n = -\frac{E_I}{n^2} \tag{38-12}$$

Thus, the only possible energies of the bound electron-proton system constituting the hydrogen atom are $-E_I$, $-E_I/4$, $-E_I/9$, ... The permitted energies are discrete, and therefore *the energy is quantized.* The lowest energy (that is, the most negative energy) belongs to the state in which the principal quantum number n equals 1, called the *ground state.* In the ground state the energy is $E_1 = -E_I$, and its value computed from (38-11) and (38-9) is

$$E_n = -\frac{k^2 e^4 m}{2n^2 \hbar^2} \tag{38-13}$$

$$E_1 = -E_I = -13.60 \text{ eV} \tag{38-14}$$

Figure 38-5 shows an energy-level diagram for hydrogen. We note from (38-12) that for bound states E is less than zero, and only discrete energies are allowed. As n approaches infinity, the energy difference between adjacent energy levels approaches zero. When n equals infinity, E_n equals zero, and the hydrogen atom may be dissociated into an electron and proton, separated by an infinite distance and both at rest. In this condition the atom is said to be ionized, and the energy that must be added to it when it is in its lowest, or ground, state ($n = 1$) to bring its energy up to $E_n = 0$ is just E_I, the so-called ionization energy. The value predicted by the Bohr theory, $E_I = 13.60$ eV, is in complete agreement with the

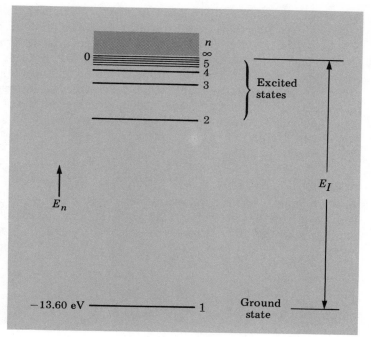

FIG. 38-5. Energy-level diagram of a hydrogen atom.

value obtained by experiment. When the system's total energy is positive, and the electron and proton are unbound, *all* possible energies are allowed, and there is a continuum of energy levels for $E > 0$.

Each of the permitted, or quantized, energies of Fig. 38-5 corresponds to a stationary state in which the atom can exist without radiating. Stationary states above the ground state ($n = 2, 3, 4, \ldots$) are called *excited states;* an atom in one of them tends to make a transition to some lower stationary state. In a downward transition the electron may be imagined to jump suddenly from one orbit to a smaller orbit. When an atom is in some excited state and has an energy E_n, the amount by which this energy exceeds that of the ground state is called the *excitation energy*. The term *binding energy* denotes the energy that must be added to an atom in any state to free the bound particles and thereby make $E_n = 0$.

Consider an atom, initially in an upper, excited, state and having an energy E_u, which makes a transition to a lower state E_l. When the transition occurs, the atom loses an amount of energy $E_u - E_l$. Bohr assumed that in such a transition a single photon having an energy $h\nu$ is created and emitted by the atom. By the conservation of energy,

$$h\nu = E_u - E_l \qquad (38\text{-}15)$$

Whereas the Bohr theory incorporates the particle nature of electromagnetic radiation by assuming that a single photon is created whenever the atom makes a transition to a lower energy, it gives no details of the electron's quantum jump or of the photon's creation. The situation is like that in photon-electron interactions (photoelectric effect, Compton effect, etc.) in that we did not concern ourselves with the details of the interactions but merely applied the conservation laws to the states before and after the interaction.

Let us compute the frequencies and wavelengths of the photons that can, according to the Bohr model, be radiated by a hydrogen atom. Using (38-15) and (38-12), we have

$$\nu = \frac{E_u - E_l}{h} = \left(\frac{-E_I}{n_u^2 h}\right) - \left(\frac{-E_I}{n_l^2 h}\right) = \frac{E_I}{h}\left(\frac{1}{n_l^2} - \frac{1}{n_u^2}\right) \quad (38\text{-}16)$$

where n_u and n_l are the quantum numbers for the upper and lower energy states, respectively. The wavelengths $\lambda = c/\nu$ of emitted photons may then be expressed as

$$\frac{1}{\lambda} = \frac{E_I}{hc}\left(\frac{1}{n_l^2} - \frac{1}{n_u^2}\right)$$

This equation is of precisely the same mathematical form as the empirically derived Rydberg formula:

$$\frac{1}{\lambda} = R\left(\frac{1}{n_i^2} - \frac{1}{n_u^2}\right) \quad (38\text{-}2)$$

By comparing these two equations for $1/\lambda$, we can evaluate the Rydberg constant R from known atomic constants and compare it with the experimentally determined value 1.0968×10^{-3} Å^{-1} for hydrogen.

$$R = \frac{E_I}{hc} = \frac{k^2 e^4 m}{4\pi \hbar^3 c} \quad (38\text{-}17)$$

To arrive at the last term of this equation, we have used (38-9) and (38-11). Setting the known values of the physical constants in this equation, we compute the value of R and find it to be 1.0974×10^{-3} Å^{-1}, in close agreement with the experimental spectroscopic value. Thus, the Rydberg formula, which summarizes the emission and absorption spectrum of hydrogen, follows as a *necessary consequence* of the Bohr atomic model.

The observed spectral lines of hydrogen can now be interpreted in terms of the energy-level diagram Fig. 38-6. The vertical lines represent transitions between stationary states, the lengths of these lines being proportional to the respective photon energies and, therefore, to the frequencies. The lines of the Lyman series correspond to those photons produced when hydrogen atoms in any of the excited states undergo transitions to the ground state ($n_l = 1$). Transitions from the unbound states ($E > 0$) to the ground state account for the observed continuous spectrum lying beyond the series limit. We account in a similar way for the Balmer series,

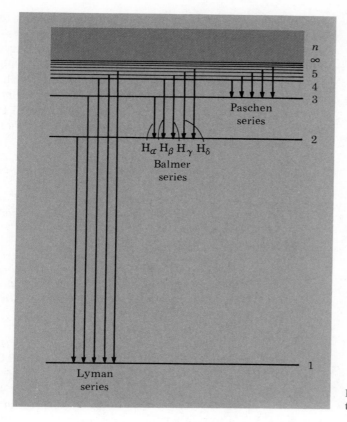

FIG. 38-6. Some possible energy transitions in atomic hydrogen.

which is produced by downward transitions from excited states to the first excited state ($n_l = 2$). Still further emission series involve downward transitions to $n_l = 3$, $n_l = 4$, etc., these series falling progressively toward longer wavelengths.

We have examined the emission from a single hydrogen atom. An atom can exist in only *one* of its quantized energy states at any one time; when it makes a transition from one state to a lower state, it emits a *single* photon. When the entire emission spectrum from an excited hydrogen gas, a collection of a very large number of hydrogen atoms, is observed in a spectroscope, we see the simultaneous emission of many photons produced by downward transitions from each of the excited states. Therefore, to observe the entire emission spectrum, we must have a very large number of hydrogen atoms in each of the excited states, making downward transitions to all lower states.

Now we also have a basis for understanding the characteristics of absorption spectra. When white light passes through a gas, those particular photons having energies equal to the energy difference between stationary states can be removed from the beam. They are annihilated, thereby giving their energy to the internal excitation energy of the atoms. The same set of quantized energy levels participates in both emission and absorption; for this reason the frequencies of the emission and absorption lines are identical. (Because atoms remain in an excited state for only a very short time, the Lyman series is the only one observed in absorption.)

We have used the fundamental postulates of the Bohr theory implicitly in developing a model of the hydrogen atom. It is useful, however, to isolate them, since they are retained in their essential forms in more complete wave-mechanical treatments of atomic structure:

1. *A bound atomic system can exist without radiating only in certain discrete stationary states.*
2. *The stationary states are those in which the orbital angular momentum, mvr, of the atom is an integral multiple of $\hbar$. (This quantization of angular momentum may be related to the wave properties of the electron confined to a semiclassical circular orbit.)*
3. *When an atom undergoes a transition from an upper energy state E_u to a lower energy state E_l, a photon of energy $h\nu$ is emitted, conservation of energy requiring that $h\nu$ be equal to $E_u - E_l$; if a photon is absorbed, the atom will make a transition from the lower to the higher energy state, according to the same relation.*

38–4 THE FOUR QUANTUM NUMBERS FOR ATOMIC STRUCTURE

The Bohr theory of hydrogen, although it incorporates basic quantum features, is only approximately correct. It yields the allowed energies of the hydrogen atom and the hydrogen spectrum fairly satisfactorily, but it does not give the correct quantum relationship for the allowed values of angular momentum, it cannot account for the closely spaced lines (or fine structure) observed in the spectrum, and it cannot successfully account for the structure and spectra of atoms with more than one electron. Worst of all, it takes the electron as a semiclassical particle moving in well-defined classical orbits, not as a wave extending over three-dimensional space.

A thorough-going wave-mechanical analysis of the hydrogen atom involves, as a minimum, finding the allowed wave functions and the corresponding energies by finding the complete solutions of the Schrödinger wave equation for a particle (the electron) subject to an inverse-square attractive Coulomb force from a fixed force center (the nucleus). The analysis is mathematically sophisticated, and we shall not attempt it here. We shall indicate, in an admittedly approximate and qualitative fashion, the principal features which enter, particularly as they are related to the four quantum numbers characterizing atomic structure.

First recall that for the wave-mechanical problem of a particle confined to a one-dimensional potential well (Sec. 37-7) a single quantum number emerged. It specified the allowed wave functions and energies, and it gave, in effect, the number of half-wave segments which could be fitted between the boundaries of the potential well. In this sense, the problem was altogether analogous to that of a wave on a string attached at both ends: The permitted wave patterns, or allowed standing waves, are those for which an integral number of half-wavelengths can be fitted between the reflecting boundaries (Sec. 15-5).

For a wave confined to two dimensions—for example, water waves on the surface of a swimming pool—there are two characteristic integers, or quantum numbers, describing the allowed wave patterns. The allowed wave patterns are, of course, those patterns of standing waves which are consistent with the boundary conditions at the edge of the region over which the waves may extend. For a rectangular swimming pool there are two quantum numbers which give, respectively, the number of half-waves that can be fitted along the length and width of the pool. For a circular pool, two quantum numbers again arise, one giving the number of half-waves that can be fitted going radially outward from the center and the other relating to zeros in the wave function that occur going around in a circle.

For a three-dimensional potential well—for example, sound waves trapped inside a rectangular parallelepiped and reflecting from the three sets of parallel sidewalls—three characteristic numbers enter, each describing in effect the number of half-waves that can be fitted along the three dimensions. In general, the number of characteristic quantum numbers for a wave trapped in a potential well equals the number of dimensions of the well.

And so it is for the electron in a hydrogen atom. Here we have a particle in a three-dimensional potential well. The potential-energy function, $V = -ke^2/r$, depends on the electron's distance r from the nucleus. Although the electron wave function may extend to infinite distances, it must be zero there. Since the potential energy depends upon distance from the force center, so too does the electron wavelength. In fact, as r increases and therefore V increases (becomes less negative), the electron wavelength must also increase, as indicated by (37-13). Thus, the electron wavelength increases as one goes away from the nucleus.

The simplest class of allowed hydrogen wave functions are those which depend only on the radial distance r, and are thereby spherically symmetrical. The first three such s wave functions are illustrated in Fig. 37-20b. In these states one may think of the electron as a rather diffuse spherical ball centered on the nucleus (for $n = 1$) or as a set of concentric spherical shells (for $n > 1$). In classical terms the electron wave may be thought to expand and contract radially, being reflected both at $r = 0$ and $r = \infty$. In general, however, the three-dimensional electron waves need not be spherically symmetrical and such waves then require three quantum numbers for their variation through space.

We give below the symbol, name, allowed values, influence on the atom's energy and angular momentum, and some geometrical characteristics of the associated wave functions for the three quantum numbers associated with the three-dimensional potential-energy function of the hydrogen atom. A crucial fourth quantum number which enters in the relativistic wave-mechanical treatment of four-dimensional space-time, the so-called spin quantum number, has no exact classical counterpart.

Principal Quantum Number n. This number enters in the relation giving the approximate total energy of the atom according to the Bohr formula, $E_n = -E_I/n^2$, (38-12), where E_I is a constant. The allowed values for n are 1, 2, 3, . . . , and as n increases the energy increases, approaching zero for infinite n. The wave functions extend toward progressively larger values of r as n increases, in the same fashion as the size of a classical planetary orbit increases with energy.

Orbital Angular-Momentum Quantum Number l. The magnitude of an electron's angular momentum as it may be thought, classically, to orbit the nucleus is given by

$$L = \sqrt{l(l+1)}\,\hbar \qquad (38\text{-}18)$$

[The rule is different from that appearing in the Bohr theory — (38-5).]

The orbital angular-momentum quantum number l may take on integral values starting with zero and continuing up to $n - 1$; that is, the l values are restricted to

$$l = 0, 1, 2, 3, 4, \ldots n - 1$$
$$s, p, d, f, g, \ldots \qquad (38\text{-}19)$$

The symbols appearing below the numerical values of l may also be used to designate the electron state. For an electron in the state with $n = 1$ (the ground state), the only possible value for l is $l = n - 1 = 1 - 1 = 0$. With $n = 2$, quantum number l may assume the values 0 and 1; the corresponding magnitudes of the orbital angular momentum are, from (38-18), equal to 0 and $\sqrt{2}\,\hbar$.

Any wave function for which $l = 0$ is spherically symmetrical; the electron has no angular momentum relative to the nucleus (and may be thought, classically, as passing in an eccentric ellipse through the force center). When $l \neq 0$, the wave functions are not spherically symmetrical but depend on angle. The simplest of these is the p state with $l = 1$. Then the wave function is such that the probability for finding the electron is high at two "knobs" along a line passing through the nucleus, as shown in Fig. 38-7. For higher values of l the wave function shows a more complicated dependence on angle.

When an atom undergoes a transition from one allowed

FIG. 38-7. Polar plot of the variation of ψ^2 with angle for a p ($l = 1$) state. The pattern is symmetrical with respect to rotation about the symmetry axis.

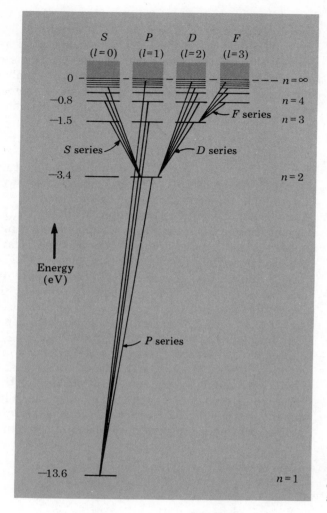

FIG. 38-8. Energy-level diagram of hydrogen showing the S, P, D, and F series.

state to another, and a photon is emitted or absorbed, the state and wave function are changed such that the values of l change by the integer 1, or $\Delta l = \pm 1$ for *allowed transitions*. The permitted energy levels, segregated according to the values of l, together with the allowed transitions, are shown in Fig. 38-8. In hydrogen the several possible l levels for a given n have the *same* total energy; in atoms with more than one electron, however, the $S, P, D, F, \ldots$ levels and the corresponding series of emitted lines differ.

Orbital Magnetic Quantum Number m_l. This quantum number may for a given l assume positive and negative integral values ranging from $+l$ to $-l$; that is,

$$m_l = l, l-1, l-2, \ldots 0, \ldots -l \qquad (38\text{-}20)$$

For example, for a p wave function, with $l = 1$, the allowed values of

The Four Quantum Numbers for Atomic Structure SEC. 38-4

the orbital magnetic quantum number are 1, 0, and −1. For a d state ($l = 2$), we have $m_l = 2, 1, 0, -1$, or -2.

Whereas the orbital angular-momentum number l gives through (38-18) the *magnitude* of the electron's orbital angular momentum, the orbital magnetic quantum number m_l yields the *component* of the electron's orbital angular momentum L_z along some direction Z in space. The angular-momentum component L_z is, in fact,

$$L_z = m_l \hbar \tag{38-21}$$

Thus, for a d state ($l = 2$), the angular-momentum magnitude is $L = \sqrt{l(l+1)}\,\hbar = \sqrt{2(3)}\,\hbar = \sqrt{6}\,\hbar$, and the allowed projections of this angular momentum along the Z axis are $L_z = m_l \hbar = 2\hbar, \hbar, 0, -\hbar$, and $-2\hbar$. To give a more physical interpretation to m_l and the associated angular-momentum component, we may imagine the angular-momentum vector **L** to be oriented relative to the Z in such directions that its components L_z satisfy (38-21). See Fig. 38-9.

What specifies the direction of the Z axis? An external magnetic field can do so. The phenomenon whereby the component of the angular momentum is restricted to certain discrete values, and therefore the vector **L** to certain orientations in space relative to Z, is sometimes referred to as *space quantization*. Since the electron is a negatively charged particle, when an electron's wave function indicates nonzero angular momentum, there must be an associated magnetic moment. Space quantization implies then that not only the angular-momentum vector but also the electron magnetic moment are restricted to certain discrete orientations in space. Further, since the relative orientation of a magnet relative to an

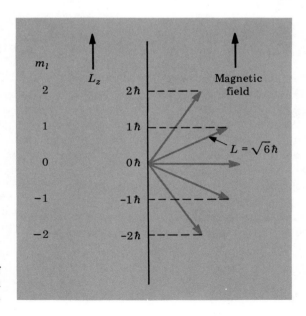

FIG. 38-9. Space-quantization of the orbital angular-momentum vector L for a D state ($l = 2$).

external magnetic field controls the energy of the magnet, the energy of an atom (and the photons emitted in transitions between allowed states) can show a multiplicity which is referred to as the *Zeeman effect*, after its discoverer.

Spin Magnetic Quantum Number m_s. The fourth quantum number specifying the state and wave function of an electron is the spin magnetic quantum number. It can assume just two values (for any values of n, l, or m_l):

$$m_s = +\tfrac{1}{2} \quad \text{and} \quad -\tfrac{1}{2} \tag{38-22}$$

The concept of electron spin first rose in the interpretation of fine structure, closely spaced pairs or triplets of spectral lines where one would otherwise expect to find only a single line. The notion of electron spin was first introduced by S. A. Goudsmit and G. E. Uhlenbeck in 1925 to account for this effect. In essence one imagines the electron with its charge smeared over a region of space to be spinning perpetually about an internal axis of rotation, the spin angular momentum being in addition to and independent of the orbital angular momentum associated with quantum number l. Later P. A. M. Dirac showed that electron spin arises as a necessary consequence of a relativistic treatment of wave mechanics. The magnitude of the spin angular momentum has a *single* value:

$$L_s = \sqrt{s(s+1)}\,\hbar = \sqrt{(\tfrac{1}{2})(\tfrac{3}{2})}\,\hbar = \tfrac{1}{2}\sqrt{3}\,\hbar \tag{38-23}$$

where the spin quantum number for an electron has the *single* value $s = \tfrac{1}{2}$. Quantum number m_s is related to s in a fashion analogous to the relationship between m_l and l: The component of the spin angular momentum along some direction in space $L_{s,z}$ is given by

$$L_{s,z} = m_s \hbar = \tfrac{1}{2}\hbar \quad \text{or} \quad -\tfrac{1}{2}\hbar \tag{38-24}$$

The geometrical interpretation of $L_{s,z}$ is shown in Fig. 38-10.

Since a perpetually spinning charged particle comprises effectively an absolutely permanent magnet, the energy of an atom differs according to the orientation of the spin vector $\mathbf{L}_s$ relative to a magnetic field.

38-5 THE PAULI EXCLUSION PRINCIPLE AND THE PERIODIC TABLE

To specify the state of an electron in an atom is, in the quantum theory, to specify the values of each of the four quantum numbers n, l, m_l, and m_s. By the procedures of the quantum mechanics it is possible to compute an atom's energy, its angular momentum, and other of its measurable characteristics. Indeed, it is possible, at least *in principle*, to predict *all* properties of the chemical elements.

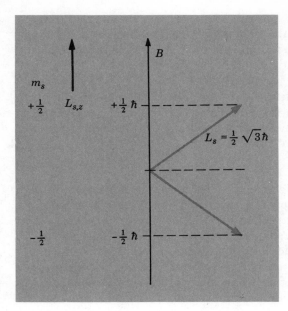

FIG. 38-10. Space quantization of electron-spin angular momentum.

In practice such a program cannot easily be carried out, because formidable mathematical difficulties arise with systems having many component particles. In fact, only the problem of the simplest atom, hydrogen, has been solved completely by relativistic quantum theory. Work on this atom has shown essentially perfect agreement between experiment and theory.

Even though solutions for the other atomic elements are not known exactly, the quantum theory does provide a wealth of information concerning their chemical and physical properties. One of its greatest achievements is a basis for understanding the order of the chemical elements as they appear in the periodic table (the periodic table was first constructed merely by listing elements in the order of their atomic *weights,* and it was found that remarkable periodicities in the properties of the elements were thereby revealed). The key is a principle proposed by W. Pauli in 1924, the Pauli exclusion principle. This principle together with the quantum theory can be used to predict many of the known chemical and physical properties of atoms.

Consider again the energy levels available to the single electron in the hydrogen atom. These energy levels are shown schematically (but *not* to scale) in Fig. 38-11. Each horizontal line corresponds to a particular possible set of values for the quantum numbers n, l, and m_l. For each line there are two possible values of the electron-spin quantum number, $m_s = \pm\frac{1}{2}$. The occupancy of an available state by an electron is indicated by an arrow, whose direction indicates the electron-spin orientation, up for $m_s = +\frac{1}{2}$ and down for $m_s = -\frac{1}{2}$. For brevity only the energy levels with principal quantum numbers 1, 2, and 3 are shown. For a given value of

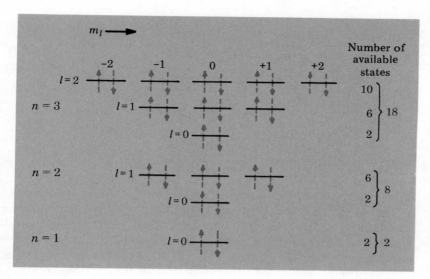

FIG. 38-11. Representation of the states available to the electron in the hydrogen atom (not to scale). There are two states for each horizontal line, corresponding to the two electron-spin orientations.

n the states with $l = 0$ are lowest, states with $l = 1$ next lowest, etc. For a given value of the orbital angular-momentum quantum number l the possible values of the orbital magnetic quantum number m_l are shown horizontally arranged. Every one of the states (two for each dash) is available to the electron in the hydrogen atom. Some are *degenerate,* having the *same* total *energy;* they are, nevertheless, distinguishable when a strong magnetic field or other external influence is applied to the atom.

The rules governing the possible values of the quantum numbers and the number of possible values are these:

For a given n: $l = 0, 1, 2, \ldots, n - 1$ (n possibilities)

For a given l: $m_l = l, l - 1, \ldots, 0, \ldots,$ ($2l + 1$
                              $-(l - 1), -l$ possibilities)

For a given m_l: $m_s = +\frac{1}{2}, -\frac{1}{2}$ (2 possibilities)

(38-25)

When the hydrogen atom is in its lowest, or ground, state, the single electron is in the state in which $n = 1$, $l = 0$, $m_l = 0$, and $m_s = -\frac{1}{2}$. Excitation of the hydrogen atom may promote the electron to any of the higher-lying available states, from which the atom can then decay to the ground state by downward transitions with the emission of one or more photons. Figure 38-12 depicts a hydrogen atom in its ground states and in an excited 3D state ($l = 2$).

Consider next the element lithium, $_3$Li. This atom has three electrons to be placed in the levels shown in Fig. 38-11. If all three electrons were to be placed in the lowest level, that with $n = 0$, $l = 0$, and $m_l = 0$, then two electrons would occupy the state with

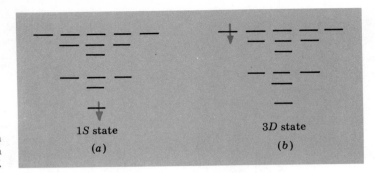

FIG. 38-12. Hydrogen atom *(a)* in its ground state and *(b)* in an excited state.

$m_s = +\frac{1}{2}$ and one would occupy the state with $m_s = -\frac{1}{2}$, or conversely; that is, at least two electrons would have the *same* set of quantum numbers. But spectroscopic and other direct experimental evidence shows that for lithium all three electrons are *never* simultaneously in the state $n = 1$. The lowest-energy configuration, or ground state, for lithium is that in which two electrons, one with $m_s = +\frac{1}{2}$, the other with $m_s = -\frac{1}{2}$, are in the $n = 1$ level, while the third electron occupies a state in the $n = 2$ level. See Fig. 38-13. We can interpret this behavior as follows: Two of the three electrons in a lithium atom cannot have the same set of four quantum numbers; that is, two electrons cannot exist in the same state.

Spectroscopic evidence from all elements is similar; it shows that atoms simply never occur in nature with two electrons occupying the same state. The Pauli exclusion principle formalizes this experimental fact:

No two electrons in an atom can have the same set of quantum numbers n, l, m_l, and m_s; or no two electrons in an atom can exist in the same state.

Exceptions to the exclusion principle, which applies also to systems other than atoms and to particles other than electrons, have never been found. The Pauli principle is analogous, but not equivalent, to the classical assertion that no two particles can be in the same place at the same time (the particles being regarded as impenetrable).

Thus, the two electrons in helium in the normal state occupy the two lowest available states indicated in Fig. 38-11. No more electrons can be added to the $n = 1$ shell; in helium, $_2$He, the $n = 1$ shell is filled, or closed. The electron spins are then oppositely aligned, and the helium atom has no net angular momentum, either orbital or spin. Furthermore, the two electrons are tightly bound to the nucleus, and a considerable amount of energy is required to excite one of them to a higher energy state. It is primarily for these reasons that helium is chemically inactive.

When the values of the quantum numbers of each and every electron in an atom are known, the *electron configuration* of the

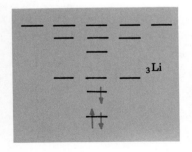

FIG. 38-13. Electron configuration of lithium in the ground state.

Table 38-1

Element	Electron configuration for the ground state					
$_1$H	$1s^1$					
$_2$He	$1s^2$					
$_3$Li	$1s^2$	$2s^1$				
$_4$Be	$1s^2$	$2s^2$				
$_5$B	$1s^2$	$2s^2$	$2p^1$			
$_6$C	$1s^2$	$2s^2$	$2p^2$			
$_7$N	$1s^2$	$2s^2$	$2p^3$			
$_8$O	$1s^2$	$2s^2$	$2p^4$			
$_9$F	$1s^2$	$2s^2$	$2p^5$			
$_{10}$Ne	$1s^2$	$2s^2$	$2p^6$			
$_{11}$Na	$1s^2$	$2s^2$	$2p^6$	$3s^1$		
$_{12}$Mg	$1s^2$	$2s^2$	$2p^6$	$3s^2$		
$_{13}$Al	$1s^2$	$2s^2$	$2p^6$	$3s^2$	$3p^1$	
$_{14}$Si	$1s^2$	$2s^2$	$2p^6$	$3s^2$	$3p^2$	
$_{15}$P	$1s^2$	$2s^2$	$2p^6$	$3s^2$	$3p^3$	
$_{16}$S	$1s^2$	$2s^2$	$2p^6$	$3s^2$	$3p^4$	
$_{17}$Cl	$1s^2$	$2s^2$	$2p^6$	$3s^2$	$3p^5$	
$_{18}$Ar	$1s^2$	$2s^2$	$2p^6$	$3s^2$	$3p^6$	
$_{19}$K	$1s^2$	$2s^2$	$2p^6$	$3s^2$	$3p^6$	$4s^1$
$_{20}$Ca	$1s^2$	$2s^2$	$2p^6$	$3s^2$	$3p^6$	$4s^2$

atom is given. A simple convention is used for specifying an electron configuration, which we illustrate with an example. When a helium atom is in its ground state, the two electrons each have $n = 1$ and $l = 0$, and their configuration is represented by $1s^2$. The leading number specifies the n value, the lowercase letter s designates the orbital quantum number l of *individual* electrons, and the postsuperscript gives the number of electrons having the particular values of n and l.

The element lithium, with three electrons, has the electron configuration $1s^2 2s^1$ — two electrons in a completely filled $n = 1$ shell and the third in the $n = 2$ shell. Proceeding in this fashion — adding one electron as the nuclear charge or atomic number increases by one unit, but always with the restriction that no *two* electrons within the atom can have the same set of quantum numbers — we can confirm the ground-state configurations of the other atoms at the beginning of the periodic table shown in Table 38-1. We see from Fig. 38-11 that two electrons can be accomodated in the s subshell of the $n = 2$ shell and six electrons in the p (or $l = 1$) subshell, after which the $n = 2$ is completely occupied, holding its full quota of eight electrons. With the electron configuration $1s^2 2s^2 2p^6$, corresponding to the rare gas element $_{10}$Ne, the electron wave functions are spherically symmetrical and the atom is chemically inert. In general, a filled subshell for any value of l, with electrons occupying states for all positive, zero, and negative values of m_l and m_s, the atom's net orbital and spin angular momentum is zero and the electron distribution is completely spherical. A closed subshell is effectively a spherical shell of charge.

Chemical properties reflect directly the electron configurations. For example, $_1$H, $_3$Li, $_{11}$Na, and $_{19}$K all have one s electron outside a closed subshell; these elements (the alkali metals) readily relinquish the last s electron to become positive ions, or contribute the electron in chemical combinations and thereby exhibit a valence of +1. On the other hand, the halogen elements $_9$F and $_{17}$Cl, both with electron configurations of p^5, lack one electron for completing a p shell; such elements readily acquire an additional electron to become a negative ion or to form chemical compounds, corresponding to a valence of -1.

SUMMARY

Rutherford's experiments in the scattering of α particles showed that all of the positive charge and essentially all the mass of an atom are confined to a very small region of space called the nucleus.

The Bohr quantum atomic theory assumes (1) stationary states, (2) orbital angular momentum $n\hbar$, and (3) $h\nu = E_u - E_l$ in transitions.

The energies and radii predicted by the Bohr atomic theory for hydrogen are given by

$$E_n = -\frac{E_I}{n^2} \quad \text{where } E_I = \frac{k^2 e^4 m}{2\hbar^2} = 13.60 \text{ eV} \quad (38\text{-}12)$$

$$r_n = n^2 r_1 \quad \text{where } r_1 = \frac{\hbar^2}{kme^2} = 0.529 \text{ Å} \quad (38\text{-}8)$$

In the wave-mechanical treatment of atomic structure the quantum numbers and their possible values are:

The principal quantum number n:

$$n = 1, 2, 3, 4, \ldots$$

For a given n:

$$l = 0, 1, 2, 3, 4, \ldots, n-1 \quad (n \text{ possible values})$$
$$s, p, d, f, g, \ldots$$

For a given l:

$$m_l = l, l-1, \ldots, -(l-1), -l \quad (2l+1 \text{ possible values})$$

For a given m_l:

$$m_s = +\tfrac{1}{2}, -\tfrac{1}{2} \quad (2 \text{ possible values})$$

The Pauli exclusion principle, the basis for understanding the periodic table of chemical elements, specifies that no two electrons in the same atom can have the same set of the four quantum numbers n, l, m_l, m_s.

PROBLEMS

38-1. A beam of 8.0-MeV protons is incident upon a foil of silver atoms (nuclear charge, $+47\,e$). *(a)* What is the potential energy of a proton-silver-nucleus system when an incident proton has been brought to rest in a head-on collision? *(b)* What is the minimum distance separating the centers of the proton and silver nucleus for such a head-on collision?

38-2. What are *(a)* the momentum and *(b)* the wavelength of a 5.0-MeV α particle?

38-3. A beam of 8-MeV protons is incident on a silver foil 10^{-6} m thick and undergoes *coulomb scattering* in accordance with the Rutherford formula. What is the distance of closest approach?

38-4. Twenty thousand small hard spheres, each 2.0 mm in diameter, are dispersed randomly throughout the interior of a cubical box 1.0 m along an edge. *(a)* If 1 million particles, each very small compared with the spheres within the box, are shot randomly over a broad face of the box, how many can be expected to be scattered from the forward direction by collisions with the spheres? *(b)* If 1 million spheres, each 2.0 mm in diameter, are fired at the spheres within the box, how many of them will be scattered?

38-5. An α particle makes a head-on collision with, in turn, *(a)* a gold nucleus, *(b)* an α particle, and *(c)* an electron, each initially at rest. What fraction of the α particle's initial kinetic energy is transferred to the struck particle in each instance?

38-6. An 8.0-MeV α particle makes a head-on collision with an electron initially at rest. What are the kinetic energies after the collision of *(a)* the electron and *(b)* the α particle? *(c)* How many such head-on collisions with electrons initially at rest are required to reduce the α particle's initial kinetic energy by 10 percent?

38-7. Show that a positively charged particle approaches a heavy nucleus more closely in a head-on collision than in a non-head-on one. (*Hint:* Use energy conservation.)

38-8. A hydrogen atom is in an excited state, for which the *binding* energy of the electron to the proton is 3.40 eV. The atom makes a transition to a state for which the *excitation* energy is 12.73 eV. *(a)* What is the energy of the photon associated with this transition? *(b)* Is it emitted or absorbed?

38-9. Assume that 12,000 hydrogen atoms are initially in the $n = 5$ state. The atoms then proceed to make transitions to lower energy states. *(a)* How many distinct spectral lines will be emitted? *(b)* Assuming for simplicity that for any given excited state all possible downward transitions are equally probable, what is the total number of photons emitted?

38-10. *(a)* Compute the electric current for an electron imagined to move in the first Bohr orbit. *(b)* At the site of the proton what is the magnitude of the magnetic field arising from the orbiting electron? *(c)* Is this magnetic field aligned with or against the vector representing the electron's orbital angular momentum?

38-11. *(a)* Show that the speed of the electron in the hydrogen atom, according to the Bohr theory, is for state n equal to $v_n = (ke^2/\hbar)/n$. *(b)* Show that the relative speed v_1/c for the first Bohr orbit is equal in magnitude to approximately $\alpha = \frac{1}{137}$. Indeed, the pure number $\alpha = ke^2/\hbar c$, known as the *fine structure constant,* plays a crucial role in quantum electrodynamics because it gives a relationship involving the fundamental constants of electromagnetism (k and e), of the quantum theory (h), and of relativity (c).

38-12. Show that the quantized energies of the hydrogen atom may be written in the form $E_n = -\frac{1}{2}\alpha^2(mc^2)/n^2$, where α is the fine-structure constant and mc^2 is the electron rest energy. (See Prob. 38-11.)

38-13. Show that the Ritz combination principle is valid. This principle states that the *wave number* of any spectral line radiated or absorbed by an element is equal to the differences or sums of the wave numbers of other *pairs* of lines radiated by the same element. The wave number is the reciprocal of the wavelength.

38-14. The muon is an elementary particle with the

same electric charge as the electron but with a mass 207 times greater. A muon can be captured by a proton (and also by other nuclei), so that a "muonic" atom is formed. *(a)* Calculate the radius of the first Bohr orbit for such an atom. *(b)* What is the ionization energy of a muon-proton atom? *(c)* Compare the speed of the muon in the first orbit with that of the electron in the first orbit.

Because the muonic hydrogen atom is far smaller than an ordinary hydrogen atom, a pair of muonic atoms can come so close together, even at moderate temperatures, that the two nuclei can *attract* one another by the nuclear force between them, and a nuclear reaction can take place. To produce a nuclear reaction between the atoms of a gas of ordinary hydrogen atoms requires temperatures of millions of degrees. Thus, muonic hydrogen atoms can be used in a process known as "cold fusion."

38-15. The total energy of a hydrogen atom may be written as $E = p^2/2m - ke^2/r$, where p is the orbital momentum of the electron and r is its distance from the nucleus. According to the uncertainty principle, if an electron is confined to a distance r its momentum is uncertain by at least an amount $\hbar/r$, so that the orbital momentum of the electron must be at least of the order of $\hbar/r$. Therefore, the total energy of the atom can be written $E = p^2/2m - ke^2p/\hbar$, where p may be regarded as a variable. *(a)* Sketch separately the two terms in the energy expression, together with their sum, as a function of p and also as a function of distance r. *(b)* The total energy is a minimum for some distance r_0. Show that the distance r_0 at which E is a minimum is equal to the first Bohr radius. *(c)* Show that when r_0 is substituted into the general relation for E, the energy is that of the hydrogen atom in the ground state.

38-16. *The correspondence principle.* *(a)* Show that the frequency f of the electron's orbital motion in a classical model of the hydrogen atom is given by $f = (1/2\pi)(ke^2/mr^3)^{1/2}$. *(b)* Show that the frequency ν of the photon emitted by a hydrogen atom in a transition from state n to $n-1$ is, for a large value of n, given by $\nu = 2cR/n^3$, where R is the Rydberg constant. *(c)* According to the correspondence principle of N. Bohr, the results of the quantum theory must correspond to those of classical electromagnetism when the quantum theory is applied to the domain in which classical electromagnetism is known to yield correct results; for the hydrogen atom this implies macroscopic orbits (large r and large n) and the electron's orbital frequency f equal to the frequency ν of the emitted electromagnetic radiation. Show, then, that in the correspondence limit of $n \to \infty$, $f = \nu$. In his original treatment of the hydrogen atom by the quantum theory, Bohr used the correspondence principle, rather than the quantization of orbital angular momentum, to find the allowed quantum states.

38-17. A singly ionized helium atom He$^+$ has a single electron surrounding a nucleus with electric charge $+2e$. *(a)* What are the energies for the ground and first two excited states of He$^+$? *(b)* What is the energy of a photon which can produce a doubly ionized helium atom He^{2+} when absorbed by a He$^+$ atom in the ground state?

38-18. The quantization of the energy of an atomic system such as the hydrogen atom implies, through the mass-energy equivalence of relativity theory, that the mass of the bound system is quantized. Sketch the possible values of the total mass of a hydrogen atom.

38-19. When an atom in an excited state and initially at rest emits a photon, momentum conservation requires that it recoil. Thus, the energy difference between the two stationary states of the atom is, strictly, the energy of the emitted photon plus that of the recoiling atom. *(a)* Show that when the atomic recoil energy is taken into account, the frequency of the emitted photon is reduced by a fraction $h\nu/2Mc^2$, where ν is the photon frequency (computed approximately) and M is the atomic mass. *(b)* What is the fractional correction made to the frequency of Lyman α photons emitted by hydrogen when this effect is taken into account?

38-20. What are the *(a)* energy, *(b)* momentum, and *(c)* wavelength of the photon emitted when a hydrogen atom undergoes the Paschen α transition? *(d)* Assuming the hydrogen atom to be initially at rest, with what energy does it recoil when such a photon is emitted?

38-21. *(a)* What is the frequency of the photon emitted by a hydrogen atom undergoing a transition from the $n = 11$ to the $n = 10$ state? *(b)* What is the frequency of the equivalent classically orbiting electron in the $n = 10$ state? (See Prob. 38-16.)

38-22. An atom remains in a certain excited state for an average time of 10^{-8} s before making a transition to some lower energy state. *(a)* What is the uncertainty in the energy (in electron volts) of the atom for this excited state? *(b)* If the subsequent downward transition results in the emission of a photon of 5000 Å, what is the fractional uncertainty in the frequency, or wavelength, of the emitted radiation? *(c)* Suppose now that a downward transition results in the emission of a photon of frequency 10 MHz, again with a lifetime of 10^{-8} s in the upper state. What is then the fractional uncertainty in the frequency of the emitted radiation?

38-23. The fluorescence phenomenon arises when radiation of one wavelength incident on a material causes it to radiate light of a different wavelength. (For example, in ordinary fluorescent lamps, ultraviolet light causes the emission of visible light.) (a) Show that the fluorescence phenomenon may be interpreted in terms of a high-energy photon exciting atoms to a relatively highly excited state from which the atom makes downward transitions through intermediate excited states. (b) Stokes' rule for fluorescence states that the frequency of light emitted in fluorescence does not exceed that of the exciting light. Show that Stokes' rule follows from the quantum theory as it applies to atoms.

38-24. A particle coasts on a frictionless saucer. (a) If the particle is initially launched from the side of the saucer so that its velocity is not directed toward the saucer's center, show that the particle can never be found at the center of the saucer. (b) Using the classical result of part (a), show that any particle having a finite angular momentum relative to a force center must have a wave function which is zero at the location of the force center. If a particle subject to an attractive central force has zero angular momentum relative to the force center, the corresponding wave function must have a nonzero value at the force center.

38-25. Atoms may be brought to excited states by absorbing photons with the proper energies. That atoms may also make quantum transitions in collisions with particles was first demonstrated in the *Franck-Hertz experiment* of 1914. For example, a hydrogen atom can make a transition from the ground to the first excited state if struck by an electron whose kinetic energy is just equal to the excitation energy, 10.2 eV, the electron being brought to rest in such a collision. In their original experiment Franck and Hertz found that electrons with a kinetic energy of 4.88 eV were effectively brought to rest when striking atoms of mercury. The excitation of the mercury atoms by electron collisions was accompanied by the emission of radiation by the mercury atoms. What was the wavelength of this radiation?

38-26. A fundamental assumption of the kinetic theory of gases is that the molecules make *perfectly elastic* collisions. That such collisions can, in fact, be *perfectly* elastic follows from the quantum nature of atomic and molecular structure: Unless a molecule receives enough energy to permit a quantum transition to an excited state, it must, despite low-energy collisions with other molecules, remain in the ground state. The average translational kinetic energy per molecule for a gas at absolute temperature T is $\tfrac{3}{2}kT$ (See Sec. 18-3); a typical ionization energy for an atom or molecule is a few electron volts. To what approximate temperature must a gas be heated in order to ionize the molecules?

38-27. What is the minimum kinetic energy of electrons for producing (a) emission of the H_α line when they strike hydrogen atoms in the ground state and (b) emission of *all* lines in the hydrogen spectrum?

38-28. What is the minimum kinetic energy of electrons which can make inelastic collisions with doubly ionized lithium atoms ($^7_3$Li) in the ground state?

38-29. (a) Two hydrogen atoms in the ground state make a head-on collision at equal speeds; what is the minimum kinetic energy of either atom (in terms of the hydrogen atom's ionization energy E_I) needed to raise one atom in the first excited state? (b) A hydrogen atom in the ground state collides with a second hydrogen atom also in the ground state but initially at rest; what is the minimum kinetic energy required to raise one of the atoms to the first excited state?

38-30. The phenomena of the aurora borealis (Northern Lights) and the aurora australis (Southern Lights), the luminous displays in the sky near the earth's pole, are produced when charged particles thrown out by the sun collide with oxygen and nitrogen 100 km or more above the earth's atmosphere, thereby exciting and ionizing these atoms. (a) The charged particles are known to travel the 1.5×10^{11} m from the sun to the earth in about 24 h; assuming that they are protons, what is their average kinetic energy? (b) Why do they produce appreciable ionization of oxygen and nitrogen only in the vicinity of the earth's poles?

38-31. Sodium atoms strongly radiate two closely spaced yellow lines with wavelengths of 5,895.944 Å and 5,889.977 Å. They arise from transitions from two closely spaced excited energy levels to the single ground state of sodium. What is the approximate energy difference between the two upper energy levels?

38-32. Spectroscopic observation shows that radiation with a wavelength of 5893 Å is emitted when sodium vapor is bombarded with electrons that have been accelerated from rest through a potential difference of 2.11 V (the so-called sodium D lines thereby emitted arise from transitions between the first excited and the ground state). Compute the value of h/e from these data.

38-33. What is the expected chemical valence of atoms whose only unfilled subshell has the electron configuration p^4?

38-34. The x-ray spectra of atoms are emitted after one of the innermost electrons has been removed, typically by bombardment by electrons of several thousand electron volts; with an inner electron vacancy thereby produced, outer electrons cascade downward in energy and inward toward the nucleus, while the decrease in the atom's energy corresponds to the emission of relatively high-energy, short-wavelength x-ray photons. Assume for simplicity that one of the electrons has been removed from the $n = 1$ shell of a zinc ($_{30}$Zn) atom; then an electron in a $n = 2$ shell "sees" an effective nuclear charge of $Z - 1$, where Z is the number of protons in the nucleus, since the second $n = 1$ electron shields the nuclear charge. What is the approximate wavelength of the photon emitted when an $n = 2$ electron makes a transition to the single vacancy in the $n = 1$ state?

38-35. An atom of $_3$Li is in its ground state. (a) What are the four quantum numbers for each of the three electrons? (b) What are the quantum numbers of the third electron for the two lowest excited states of this atom?

CHAPTER 39

Nuclear Structure

Insofar as atomic structure is concerned, the atomic nucleus may be regarded as a point mass and a point charge. The nucleus contains all the positive charge and nearly all the mass of the atom; it provides, therefore, the center about which electron motion takes place, and it influences atomic structure primarily through its coulomb force of attraction with electrons.

The α-particle scattering experiments of Rutherford established that for distances greater than 10^{-14} m the nucleus interacts with other charged particles by the inverse-square coulomb force. It was found, however, that when the α particles approached the nuclear center closer than 10^{-14} m, the distribution of the scattered particles could not be accounted for simply in terms of Coulomb's law. These experiments showed then that a totally new type of force, the nuclear force, acts at distances smaller than 10^{-14} m.

In this chapter we explore some of the simpler aspects of nuclear structure. We shall see that, apart from the tremendous difference in their relative sizes, 10^{-10} m for atoms but less than 10^{-14} m for nuclei, nuclear structure is different from atomic structure in several significant respects.

39–1 THE NUCLEAR CONSTITUENTS

The particles of which all nuclei are composed are the proton and the neutron. We shall describe here some fundamental properties — charge, mass, and spin — of these particles.

Charge. The proton is the nucleus of the atom $_1^1$H, the light isotope of hydrogen; it carries a single positive charge, equal in magnitude to the charge of the electron.

The neutron is so named because it is electrically neutral. Because it carries no charge, it shows only a feeble interaction with electrons, it produces no direct ionization effects, and it is, therefore, detected and identified only by the effects it produces on charged particles.

Mass. The masses of the proton (the *bare nucleus* of the $^1_1$H atom) and the neutron in unified atomic mass units (see Sec. 35-5), together with the rest energies of these particles in units of MeV are:

Proton rest mass = $1.00727661 \pm 0.00000008$ u
Proton rest energy = 938.259 ± 0.005 MeV
Neutron rest mass = 1.0086652 ± 0.0000001 u
Neutron rest energy = 939.553 ± 0.005 MeV

The proton and neutron have nearly the same mass, the neutron mass exceeding the proton mass by slightly less than 0.1 percent. Both particles have rest energies of about 1 GeV. Because the proton carries an electric charge, its mass can be measured directly with high precision by the methods of mass spectrometry; electric and magnetic fields have virtually no effect on a neutron, and its mass is inferred indirectly from experiments to be described shortly.

Spin. Both the proton and the neutron have intrinsic angular momentum, or the so-called nuclear spin. The nuclear spin can be visualized, as in the case of electron spin, in terms of the spinning of the particle as a whole about some internal rotation axis. The nuclear spin angular momentum L_I corresponds to the nuclear spin quantum number I; its magnitude is given by

$$L_I = \sqrt{I(I+1)}\, \hbar \tag{39-1}$$

which is analogous to (38-23). The nuclear spin quantum numbers of both the proton and the neutron are $I = \tfrac{1}{2}$.

Like electron spin, the nuclear-spin angular momentum is space-quantized by an external magnetic field, the permitted components along the direction of the magnetic field being $+\tfrac{1}{2}\hbar$ and $-\tfrac{1}{2}\hbar$, as in Fig. 38-10.

39-2 THE FORCES BETWEEN NUCLEONS

All nuclei consist of protons and neutrons bound together to form more or less stable systems; therefore, it is important to have some knowledge of the forces that act between these fundamental nuclear constituents. Consider first the force between two protons. The most direct way to examine this force is by means of a proton-proton scattering experiment in which monoenergetic protons strike a target containing mostly hydrogen atoms and, therefore, protons.

From the angular distribution of the scattered protons one can infer the force acting between the incident particles and the target particles—in this instance, both protons. Proton-proton scattering experiments show that the force can be represented approximately by the potential curve shown in Fig. 39-1. At large distances of separation the protons repel one another by the coulomb inverse-square force. At a separation distance of approximately 3×10^{-15} m there is a fairly sharp break in the potential curve. It indicates the onset of the *nuclear force* between a pair of protons. The force is strongly *attractive* at smaller distances (although there is evidence of a repulsive "core" at very small distances). The "size" of the proton can be taken as the *range*, 3×10^{-15} m, of the nuclear proton-proton force. The customary unit for measuring nuclear dimensions is the *fermi*, where 1 fermi = 10^{-15} m.

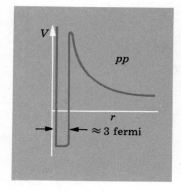

FIG. 39-1. Proton-proton potential.

The force between a neutron and a proton can be investigated by neutron-proton scattering experiments. In these experiments a monoenergetic neutron beam bombards a target containing protons. Again the distribution of the scattered neutrons is analyzed for the force acting between the neutron and the proton. The interaction between a neutron and proton can be represented approximately by the potential curve shown in Fig. 39-2. At large distances of separation there is *no* force between the two particles, but at a distance of about 2 fermi the neutron and proton attract one another by a strong nuclear force having a fairly well-defined range, again with a repulsive inner core. Clearly, this nuclear attraction is in no way dependent on electric charge, inasmuch as the neutron is a neutral particle.

The nuclear force between two neutrons cannot be investigated directly by a neutron-neutron scattering experiment: It is impossible to prepare a target consisting of free neutrons. But a variety of indirect evidence indicates that the force between two neutrons is approximately equal to the neutron-proton and proton-proton force. Because a neutron and a proton are nearly equivalent in their interactions (apart from the coulomb force between protons), it is customary to refer to a neutron *or* a proton as a *nucleon*. The term designates either a proton or a neutron when the distinction between them is of little importance. The independence of the nuclear force from the charge of the particular participating nucleons is known as the *charge independence* of the nuclear force. More sophisticated treatments of the proton-neutron interactions show that it is possible to consider the proton and neutron as two different charge states of the *same* particle.

A variety of scattering experiments, particularly those employing high-speed neutrons, show that the nuclear radius R, defined as the separation distance marking the onset of the nuclear force, is given by

$$R = r_0 A^{1/3} \qquad (39\text{-}2)$$

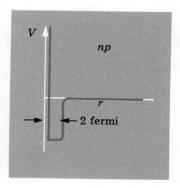

FIG. 39-2. Neutron-proton potential.

where $r_0 = 1.4$ fermi and A is nuclear mass number, or the total number of nucleons. The relationship for nuclear radius leads to an important conclusion concerning the density of nuclear material. Cubing (39-2) and multiplying by $4\pi/3$ we have

$$\frac{4\pi R^3}{3} = \frac{4\pi r_0^3}{3} A$$

The quantity $4\pi R^3/3$ is the volume of the nucleus, assumed to be a sphere, or a near sphere, and we can take $4\pi r_0^3/3$ to be the volume of a single nucleon. Therefore, (39-2) can be written

Volume of nucleus = volume of nucleon × number of nucleons

The total nuclear volume is merely the sum of the volumes of the nucleons composing it. Since all nucleons have nearly the same mass, *all nuclei* have the *same density* of nuclear matter. In this respect nuclear matter, unlike atomic matter, resembles a liquid, in which the molecules can be imagined as touching one another and interacting through a short-range force.

39-3 THE DEUTERON

The simplest nucleus containing more than one particle is the *deuteron,* the nucleus of the deuterium atom. The deuteron consists of a proton and a neutron bound together by the attractive nuclear force to form a stable system. The deuteron has a single positive charge, $+e$. Its mass is approximately twice that of the proton or neutron; more precisely,

Deuteron rest mass = 2.013553 u

It must be emphasized that the deuteron mass given here is that of the *bare* deuterium *nucleus;* the mass of the neutral deuterium atom exceeds that of the deuteron by the mass of an electron, 0.000549 u, and is, therefore, 2.014102 u.

It is interesting to compare the mass of the deuteron, M_d, with the sum of the mass of its constituents, the proton and neutron, M_p and M_n:

$$M_p = 1.007277 \text{ u}$$
$$M_n = 1.008665 \text{ u}$$
$$M_p + M_n = 2.015942 \text{ u}$$
$$M_d = 2.013553 \text{ u}$$

$M_p + M_n - M_d$ (mass difference) = 0.002389 u

The total mass of the proton and neutron when separated *exceeds* the mass of the two particles when they are bound together to form a deuteron. This difference is easily interpreted on the basis of the relativistic conservation of mass-energy (see Sec. 35-4).

When *any* two particles attract one another, the sum of their separate masses exceeds that of the bound system, inasmuch as energy (or mass) must be added to the system to separate it into its component particles. The value of this energy, called the binding energy, can be computed from the mass difference by using the mass-energy conversion factor $1 \text{ u} = 931.5 \text{ MeV}/c^2$. Thus, the binding energy E_b of the neutron-proton forming a deuteron is given by

$$E_b + M_d c^2 = (M_p + M_n)c^2 \tag{39-3}$$

$$E_b = (M_p + M_n - M_d)c^2$$
$$= (0.002389 \text{ u})(931.5 \text{ MeV/u-}c^2)c^2 = 2.225 \text{ MeV}$$

If 2.225 MeV is added to a deuteron, the neutron and proton can be separated from one another, beyond the range of the nuclear force, both particles being left at rest and thus having no kinetic energy.

The binding energy is manifested as a measurable mass difference in nuclear systems because the nuclear force is very strong and the binding energy very great. In fact, nuclear binding energies are roughly a *million* times greater than atomic binding energies.

Suppose that deuterium gas is irradiated with a beam of high-energy monoenergetic γ-ray photons. If the energy of the photons is equal to the deuteron's binding energy, photon absorption will produce a free neutron and a proton. If it exceeds the binding energy, the deuteron will be dissociated into a neutron and proton, each particle having kinetic energy. This *nuclear reaction* may be written as follows:

$$\gamma + d \rightarrow p + n$$

Mass-energy conservation implies that

$$h\nu + M_d c^2 = M_p c^2 + M_n c^2 + K_p + K_n \tag{39-4}$$

where K_p and K_n are the kinetic energies of the freed proton and neutron, respectively. This process, in which the proton and neutron are detached from one another by the absorption of a photon, is a nuclear photoelectric effect, or a nuclear *photodisintegration*. The threshold for the reaction corresponds to $K_p = 0$ and $K_n = 0$; then, from (39-3) and (39-4),

$$h\nu_0 = (M_p + M_n - M_d)c^2 = E_b \tag{39-5}$$

That is, the energy of the photon is equal to the binding energy of the deuteron.† See Fig. 39-3. If the threshold photon energy $h\nu_0$ is measured, the values of M_p and M_d being known, the neutron mass can be computed from (39-5). This is one of several ways in which the neutron mass can be measured by applying energy conservation to nuclear reactions.

---

†Strictly, the photon threshold energy for the photodisintegration of a deuteron initially at rest

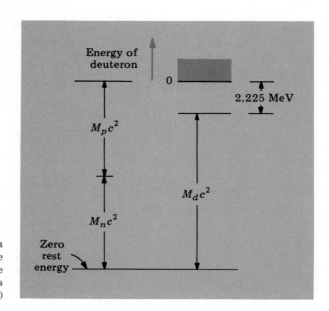

FIG. 39-3. Energy-level diagram of the neutron-proton system, the deuteron. (The magnitude of the deuteron binding energy is exaggerated.)

39-4 STABLE NUCLEI

Consider the stable nuclei containing more than two nucleons. The number of protons in a nucleus is represented by the *atomic number Z*, the total number of nucleons by the *mass number A*, and the number of neutrons by $N = A - Z$. The term *nuclide* designates a particular species of nucleus having the same values of $Z, N,$ or A. Species having the same *proton* number Z are nuclides known as *isotopes*, those having the same *neutron* number N are *isotones*, and those having the same *nuclear number A* are *isobars*. For example, $^{37}_{17}\text{Cl}$, which has 17 protons, 20 neutrons, and 37 nucleons, is an isotope of $^{35}_{17}\text{Cl}$, an isotone of $^{39}_{19}\text{K}$, and an isobar of $^{37}_{18}\text{Ar}$.

The *stable* nuclides found in nature are plotted in Fig. 39-4, where neutron number N is plotted against proton number Z. Each point represents a combination of protons and neutrons forming a stable bound system. If we concentrate on its overall features, we can see a number of interesting and significant regularities in this diagram.

Only those combinations of protons and neutrons which appear as points in the figure are found in nature as stable nuclides in their ground states; all other possible combinations of nucleons are to some degree unstable in that they decay, or disintegrate, into other nuclei. For example, the nuclides $^{16}_{8}\text{O}$, $^{17}_{8}\text{O}$, and $^{18}_{8}\text{O}$, all iso-

---

exceeds the deuteron binding energy because of the requirement of momentum conservation: The total momentum of the photon and neutron coming out of the collision must equal the photon's momentum $h\nu/c$ entering the collision. An analysis shows that the photon threshold energy is $h\nu = E_b/(1 - E_b/M_d c^2)$, which is greater than the deuteron binding energy by about 1 part in 10^3.

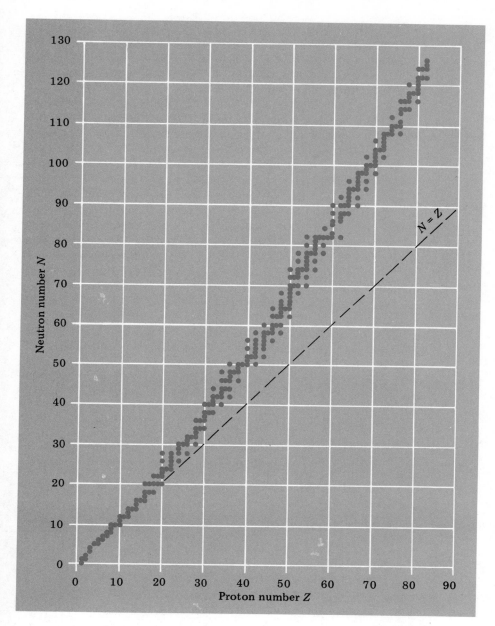

FIG. 39-4. Neutron number vs. proton number for the stable nuclides.

topes of oxygen, exist as stable nuclear systems, but the isotopes $^{15}_{8}O$ and $^{19}_{8}O$ are unstable.

The stable nuclides in Fig. 39-4 constitute a *stability line*, which indicates the general region in which the most stable nuclides fall. We see that at small values of N and Z the stable nuclides lie close to the $N = Z$ line at 45°. For example, $^{16}_{8}O$ has $N = Z = 8$. The most stable light nuclides for a given mass number A

are those whose number of protons is nearly equal to the number of neutrons. Light nuclei prefer to have equal numbers of protons and neutrons because such aggregates are more stable than those in which there is a decided excess of protons or of neutrons. For the heavier nuclides the stability line bends increasingly away from the 45° line; that is, $N > Z$ at large A. For example, the stable nuclide $^{208}_{82}$Pb has $Z = 82$ and $N = 126$. Thus heavy nuclides show a decided preference for neutrons over protons.

The neutron excess can be accounted for through the repulsive coulomb force that acts between protons. If we start with a moderately heavy nucleus and attempt to construct from it a heavier nucleus by adding one nucleon, the binding of the additional neutron will usually be stronger than the binding of an additional proton, because the neutron is only attracted by the nuclear force, whereas the proton is both attracted by the nuclear force and repelled by the coulomb repulsive forces of the protons already in the heavy nucleus. The repulsive coulomb effect competes noticeably with the strong nuclear attractive force only in heavy nuclides.

We can understand the near equality of Z and N at small A and the greater N than Z at large A by considering how the Pauli exclusion principle (Sec. 38-5) operates in the building of stable nuclides. Both the proton and the neutron separately follow this principle: No two identical particles can be placed in the same quantum state in a nucleus. We need not concern ourselves here with details of the quantum theory of nuclear structure, but we shall simply recognize that, if two protons are in a state having the same three spatial quantum numbers, then the protons must differ in their magnetic spin quantum numbers. This implies that two protons can occupy the same quantum state only if their nuclear spins are antialigned. The same rule applies to neutrons: Only two neutrons, one with spin up and one with spin down, can occupy a quantum state that is identical in the three spatial quantum numbers. Apart from the coulomb interaction between protons, the states available to a proton or a neutron are very nearly the same, since the proton and neutron are essentially equivalent in their nuclear interactions.

Since the nuclear force is so strong that the mass of a bound nuclear system is measurably smaller than the sum of the masses of its components, information on the binding energy of any stable nucleus can be arrived at directly from a comparison of masses. An atom of $^{12}_{6}$C has a mass of precisely 12 u by definition. The mass of a nucleus is the atomic mass less Z electron masses (since the number of electrons equals the number of protons). The energy binding the atomic electrons to the nucleus is usually quite small compared with the atom's rest energy, and we can take the neutral atom's mass to be the mass of the nucleus plus the masses of the electrons.

Consider the nucleus of $^{12}_{6}$C, which has 6 protons and 6 neutrons. We wish to calculate the *total* binding energy; that is, the energy required to separate a $^{12}_{6}C$ nucleus into its 12 component nucleons, each nucleon being at rest and effectively out of the range of the forces of the other nucleons. In the following computation we shall use the mass of a *neutral hydrogen atom,* 1.007825 u, and the mass of an electron, 0.000549 u.

$$
\begin{aligned}
6 \text{ protons} &= 6(1.007825 - 0.000549) \text{ u} \\
6 \text{ neutrons} &= 6(1.008665) \text{ u} \\
\text{Total nucleon masses} &= 12.098940 - 6(0.000549) \text{ u} \\
{}^{12}_{6}C \text{ nuclear mass} &= 12.000000 - 6(0.000549) \text{ u} \\
\text{Mass difference} &= 0.098940 \text{ u} \\
\text{Total binding energy} &= 0.098940 \text{ u} \times 931.5 \text{ MeV/u} = 92.16 \text{ MeV}
\end{aligned}
$$

Note that, because the electron masses cancel out, we can use the masses of the *neutral* atoms of hydrogen and carbon 12 rather than the masses of the proton and the carbon 12 nucleus.

We see that 92.16 MeV must be added to a carbon nucleus to separate it completely into its constituent particles; therefore, the 12 nucleons of the carbon atom are bound together with a total binding energy $E_b = 92.16$ MeV to form the nucleus in its lowest energy state. The *average* binding energy per nucleon, E_b/A, is 92.16/12, or 7.68 MeV.

The average binding energy per nucleon, E_b/A, can be computed for any stable nucleus in similar fashion. When the computed values of E_b/A are plotted against the corresponding values of the atomic mass A, we obtain the results shown in Fig. 39-5. Apart from sharp peaks for the especially stable nuclides with

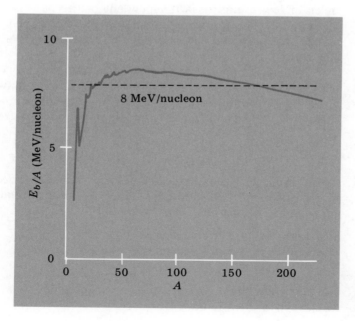

FIG. 39-5. Average binding energy per nucleon as a function of mass number for the stable nuclides.

groups of 2 protons and 2 neutrons ($^4_2$He, $^{12}_6$C, $^{16}_8$O), the curve rises sharply from the lightest stable nuclides to values of $A \approx 20$, then the curve rises slowly, reaches a maximum near the element $^{56}_{26}$Fe, and drops slowly toward the heaviest nuclides. It is approximately horizontal from $A = 20$ onward, E_b/A being roughly constant and equal approximately to 8 MeV/nucleon.

Iron and nuclides close to it represent the most stable configurations of nucleons found in nature; in all elements lighter or heavier than iron the typical nucleon is bound with less energy.

39-5 THE RADIOACTIVE DECAY LAW

Any unstable system, including of course an unstable nucleus, will decay into other parts provided that the fundamental conservation laws — mass-energy, momentum, electric charge, etc. — do not preclude such a decay. We shall consider the principal decay modes for unstable nuclei in some detail in Sec. 39-6. But if unstable nuclei *can* decay, there remains the question as to how rapidly such decays take place. Here we are concerned with the *law of radioactive decay*, which is followed by all unstable nuclei, whatever the differences in the emitted particles.

We call the initial unstable nucleus the parent and the nucleus into which the parent decays the daughter. The death of the parent gives birth to the daughter. The fundamental assumption of the radioactive decay law is that, for an unstable nucleus, one instant of time is like any other, so that the probability that an unstable nucleus will decay spontaneously into one or more particles of lower energy is independent of the parent nucleus' past history, is the same for all nuclei of the same type, and is very nearly independent of external influences (temperature, pressure, etc.).

There is no way of predicting the time that any one unstable nucleus will decay, its survival being subject to the laws of chance, but during an infinitesimally small time interval dt the probability that it will decay is directly proportional to this time interval. Thus,

$$\text{Probability that nucleus } decays \text{ in time } dt = \lambda \, dt$$

where the proportionality constant λ is called the *decay constant* or the *disintegration constant*. Since the total probability that a nucleus will either survive or decay in the time dt is 1 (100 percent),

Probability that nucleus *survives* time $dt = 1 - \lambda \, dt$
Probability that nucleus *survives* time $2 \, dt$
$$= (1 - \lambda \, dt)(1 - \lambda \, dt) = (1 - \lambda \, dt)^2$$

Consider now the probability of the nucleus' surviving n time intervals, each of duration dt:

Probability that nucleus survives time $n\,dt = (1 - \lambda\,dt)^n$

Then putting $n\,dt = t$, the total time elapsed, and noting that, as $dt \to 0$, $n = t/dt \to \infty$, the equation above becomes

Probability that nucleus survives time $t = \lim\limits_{n \to \infty} \left(1 - \dfrac{\lambda t}{n}\right)^n$ (39-6)

Now, the definition of e^{-x}, where e is the base of the natural logarithms, is

$$e^{-x} \equiv \lim_{n \to \infty} \left(1 - \frac{x}{n}\right)^n \qquad (39\text{-}7)$$

Comparing (39-6) and (39-7) we see that

Probability that nucleus survives time $t = e^{-\lambda t}$ (39-8)

This equation is the necessary consequence of the assumption that the decay of nuclei in unstable states is independent of the nucleus' present condition and past history. Although we cannot say precisely when a *single* nucleus will decay, we can predict the statistical decay of a *large* number of identical unstable nuclei. If there are initially N_0 unstable nuclei undergoing a decay process characterized by the decay constant λ, then the number N surviving a period of time t is merely N_0 times the probability that any one nucleus will have survived. Therefore, from (39-8) we have

$$N = N_0 e^{-\lambda t} \qquad (39\text{-}9)$$

The number of unstable nuclei decreases exponentially with time at a rate controlled by the magnitude of λ.

It is customary to measure the rapidity of decay in terms of the *half-life*, $T_{1/2}$, defined as the time in which one-half of the original unstable nuclei have decayed and one-half still survive. Therefore $t = T_{1/2}$ when $N = \tfrac{1}{2}N_0$, and (39-9) gives

$$\tfrac{1}{2}N_0 = N_0 e^{-\lambda T_{1/2}}$$

$$T_{1/2} = \frac{\ln_e 2}{\lambda} = \frac{0.693}{\lambda} \qquad (39\text{-}10)$$

Thus, if a radioactive material decays with a half-life of 3 s, after 3 s one-half of the initial nuclei remain, after 6 s one-quarter of the initial nuclei remain, and after 9 s one-eighth of the initial nuclei remain. The decay constant λ has the units of reciprocal time (for example, s^{-1}), which follows from its definition as the probability per unit time for decay.†

The decay of the parent nuclei and the concomitant growth in the number of daughter nuclei with time are shown in Fig. 39-6.

---

†The half-life is *not* the same as the *average lifetime*, or *mean life* T_{ave}, of an unstable nucleus. A straightforward calculation (see Prob. 39-9) shows that $T_{ave} = 1/\lambda = T_{1/2}/\ln_e 2 = T_{1/2}/0.693$.

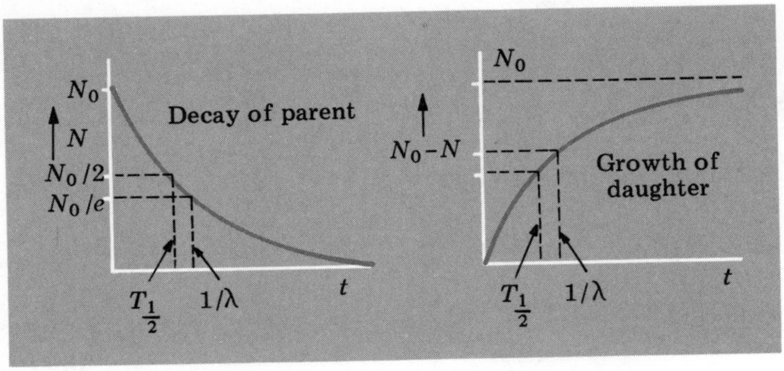

FIG. 39-6. (a) Decay in the number of radioactive parent nuclei and (b) growth in the number of daughter nuclei, as functions of time.

The number of daughter nuclei produced after a time t is $N_0 - N = N_0(1 - e^{-\lambda t})$, where N_0 is again the initial number of parent nuclei.

A useful quantity describing radioactive decay is the *activity*, defined as the number of decays per second. It follows from (39-9) that

$$\frac{dN}{dt} = -\lambda N_0 e^{-\lambda t} = -\lambda N$$

$$\text{Activity} = -\frac{dN}{dt} = \lambda N = (\lambda N_0)e^{-\lambda t} \qquad (39\text{-}11)$$

The minus sign appearing in this equation indicates that the number of unstable nuclei *decreases* with time. The activity λN, originally λN_0, falls off as $e^{-\lambda t}$. The activity of unstable nuclei that radiate particles or photons, the *radioactivity,* can be measured with a nuclear-radiation counter over periods of time that are short compared with the half-life of the decay. This provides, then, a simple and direct method of measuring λ or $T_{1/2}$. The common unit for measuring activity is the *curie,* which is defined as exactly 3.7×10^{10} disintegrations per second.

39–6 α, β, AND γ DECAY

The terms α, β, and γ decay for the three principal decay modes by unstable nuclides were first assigned early in the study of the naturally radioactive materials, which consist mostly of heavy nuclides near the end of the periodic table with half-lives comparable to the age of the universe ($\approx 10^{10}$ yr).

An unstable system, including an unstable nuclide, will undergo decay only if the process is not precluded by the fundamental conservation laws. In addition to the conservation laws of linear momentum, angular momentum, mass-energy, and electric charge, all nuclear decays also satisfy the *conservation of nucleons* (a special case of the more general conservation law of baryons, Sec. 40-4). Thus, in any decay process the total number of nucleons

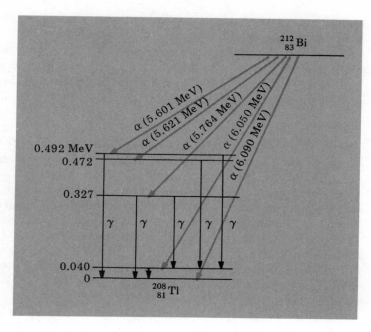

FIG. 39-7. Nuclear energy-level diagram showing the excited states and γ transitions in $^{208}_{81}$Tl, together with α emissions from $^{212}_{83}$Bi.

is unchanged. This conservation law rules out, for example, the decay of a proton into a positron and photon, which would otherwise be permitted. Actually, it is the observation that certain processes, such as the decay of proton, never occur that leads to the formulation of the nucleon conservation law.

γ Decay. Like atoms, stable nuclides are ordinarily in their lowest, or ground, states. If such nuclei are excited and gain energy by photon or particle bombardment, they may exist in any one of a number of excited, quantized energy states. Indeed, all radioactive nuclides are initially in energy states from which they may decay with the emission of photons or particles. The decay of a nucleus from an excited state by the emission of a photon is called γ decay.

The nuclear energy levels of the radioactive element thallium, $^{208}_{81}$Tl, are shown in Fig. 39-7 (we later consider the α transitions displayed in this figure), together with the transitions giving rise to γ rays. The spacings of nuclear energy levels are typically in the range of tens of keV to a few MeV, in contrast to the much smaller energies for atomic transitions. The energy levels may, of course, be inferred from the emitted γ-ray spectrum.

For an excited γ-emitting nucleus initially at rest, momentum conservation requires that the photon and daughter nucleus move in opposite directions so that the total momentum of the isolated system remains zero. Since the photon mass $h\nu/c^2$ is much smaller than the nuclear mass, essentially all of the energy released in the decay appears as the energy $h\nu$ of the photon, the kinetic energy of the recoiling daughter nucleus being negligible by comparison.

In γ decay the particle is created as the nucleus in the excited state, $^AZ^*$, decays to a lower state, say, the ground state ^AZ. We use here the conventional notation, in which a nucleus in an excited state is labeled with an asterisk. We can write symbolically

$$^AZ^* \rightarrow {}^AZ + \gamma$$

The decay is consistent with the conservation of charge (Ze before = Ze after), and the conservation of nucleons (A before = A after).

It is useful to have a criterion for judging the relative rapidity with which nuclear decays take place. For this purpose we may speak of a *nuclear time* t_n as the time required for a typical nucleon, having a kinetic energy of several MeV and therefore traveling at a speed $\approx 0.1c$, to travel a nuclear distance, ≈ 3 fermi. It follows that $t_n = (3 \times 10^{-15} \text{ m})/(3 \times 10^7 \text{ m/s}) \approx 10^{-22}$ s. It is expected, then, that any rapid nuclear decay will have a half-life that is not more than a few orders of magnitude larger than 10^{-22} s, an immeasurably short time.

The half-life in a typical γ decay is predicted to be of the order of 10^{-14} s; such a fast decay cannot be followed in time, but some γ decays are so strongly forbidden that the half-life is greater than 10^{-6} s, which can readily be measured. Such nuclides, having a measurably long half-life in γ decay, are called *isomers*. An isomer is not chemically distinguishable from the lower-energy nucleus into which it slowly decays. An extreme example of isomerism is that of niobium, $^{91}_{41}$Nb, which undergoes γ decay with a half-life of 60 days.

α Decay. Certain radioactive nuclei, those for which $Z > 82$, spontaneously decay into a daughter nucleus and a helium nucleus. Since the α particle has a very stable configuration of nucleons, it is perhaps not surprising that such a group of particles might exist within the parent nucleus prior to α decay. For example, bismuth 212 decays by α emission to thallium 208:

$$^{212}_{83}\text{Bi} \rightarrow {}^{208}_{81}\text{Tl} + {}^4_2\alpha$$

α decay is energetically possible only if the total rest mass of the parent nucleus exceeds the rest mass of daughter nucleus and α particle. The excess mass (multiplied by c^2) is the energy released in the decay and shared by the two outgoing particles which, by momentum conservation, move in opposite directions to ensure a total zero momentum for the system. Most of the released kinetic energy is carried by the light particle, the α particle, and because the released energy is fixed for any one species, all α particles emitted in a particular decay mode have a single energy, as shown in Fig. 39-8. The monoenergetic emitted particles are, in fact, a characteristic feature of any decay into *two* particles.

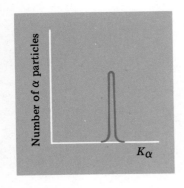

FIG. 39-8. Energy spectrum of α particles from a radioactive substance.

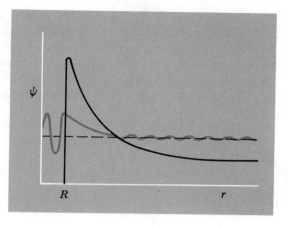

FIG. 39-9. Wave function corresponding to the penetration of an α particle through a nuclear potential barrier.

Figure 39-7 shows the decay by α emission to several of the discrete energy levels of a daughter nucleus.

The aggregate of two protons and two neutrons constituting the emitted α particle may be thought to exist as a discrete entity within the parent nucleus *before* emission. Such a particle confronts a nuclear potential well and would, classically, never escape from the interior of the nucleus if it were not for the quantum-mechanical tunnel effect (Sec. 37-7). Because the α particle's wave function extends through the potential barrier to the exterior of the nucleus, as shown in Fig. 39-9, the α particle can escape, although with very low probability.

About 160 α emitters have been identified. The emitted α particles have discrete energies ranging from about 4 to 10 MeV, a factor of 2, but half-lives ranging from 10^{-6} s to 10^{10} yr, a factor of 10^{23}. Short-lived α emitters have the highest energies, and conversely.

β Decay. β decay may be defined as that radioactive decay process in which the charge of a nucleus is changed without a change in the number of nucleons. As an example of β instability consider the three nuclides boron 12, carbon 12, and nitrogen 12, whose proton and neutron states are shown schematically in Fig. 39-10. All

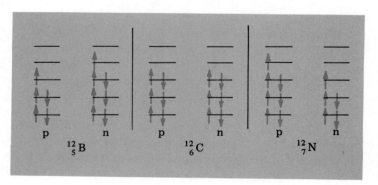

FIG. 39-10. Proton and neutron occupation states for boron 12, carbon 12, and nitrogen 12.

α, β, and γ Decay   SEC. 39-6

have 12 nucleons but they differ in proton and neutron numbers. Only the carbon nucleus, with 6 protons and 6 neutrons, is stable. The boron nucleus has too many neutrons, and the nitrogen nucleus too many protons, to be stable. The boron nucleus can decay to a lower energy state by changing one of its nucleons from a neutron into a proton, the last neutron jumping, as it were, to the lowest available proton level. In this process the $^{12}_{5}B$ nucleus has been transformed into the stable $^{12}_{6}C$ nucleus, and to conserve electric charge one unit of negative charge must be created. An electron cannot exist *within* the nucleus; therefore, the created electron, or β particle, must be emitted from the decaying nucleus, according to the transformation

$$^{12}_{5}B \rightarrow \,^{12}_{6}C + \beta^-$$

where the minus sign indicates the negative charge.

The decay of nitrogen 12 is analogous. It has too many protons and too few neutrons to be stable. Therefore, it decays to a lower energy state by converting one of its nucleons from a proton into a neutron, the last proton jumping to the lowest available neutron level. In this decay the unstable $^{12}_{7}N$ nucleus is transformed into the stable $^{12}_{6}C$ nucleus, and charge is conserved by the creation of a positive beta particle, the positron. Because a positron cannot exist within a nucleus, it must be emitted. The decay may be shown as

$$^{12}_{7}N \rightarrow \,^{12}_{6}C + \beta^+$$

Another type of β decay is *electron capture*. In electron capture an atomic orbital electron combines with a proton of the nucleus to change it into a neutron. Again the number of nucleons is unchanged, but a proton is converted into a neutron, as in β^+ decay. The electrons of the atom have a finite probability of being at the nucleus (see Fig. 37-20b), and one of the innermost, or K, electrons has the highest probability of being captured within the nucleus. β decay resulting from the nuclear capture of a K-shell electron is often referred to as *K capture*.

No charged particle is emitted in the decay by electron capture. The absorption and annihilation of a particle is equivalent to the creation and emission of its antiparticle; in K capture an electron is absorbed, but in β^+ decay an electron antiparticle, or positron, is emitted. Both processes change a proton into a neutron. An example of electron capture is the decay of unstable beryllium 7 to lithium 7:

$$e_K^- + \,^{7}_{4}Be \rightarrow \,^{7}_{3}Li$$

Electron capture cannot, of course, be identified by an emitted charged particle. It may be inferred from the change in the chemical identity of the element undergoing the decay, or it may be detected by observing the *x-ray photons* emitted when the decay takes place. When an electron is absorbed into the nucleus there is a hole, or vacancy, in the innermost electron shell; this vacancy is filled as electrons in outer shells make quantum jumps to inner vacancies, thereby emitting characteristic x-ray spectra. Because the x-ray emission must take place *after* the electron vacancy is created, that is, after the nuclear decay has occurred, the x-rays are characteristic of the *daughter* element, not the parent.

Many hundreds of nuclides are known to decay by emitting an electron or a positron or by capturing an orbital electron. In fact, nearly all unstable nuclides with Z less than 82 decay by at least one of the three processes. β decay differs from α and γ decay in several respects:

The parent and daughter have the same number of nucleons.
Unlike α emission, the electron or positron is created at the time it is emitted.
Whereas γ-ray photons and α particles are emitted with a discrete spectrum of energies, β particles have a continuous energy spectrum.
The half-lives in β decay are never less than about 10^{-2} s, in contrast to γ decay (as small as 10^{-17} s) and α decay (as small as 10^{-7} s).
A neutrino or antineutrino is emitted in every β decay process.

What is the evidence for the appearance of still another particle in β decay? Consider β^- decay, illustrated by the decay of a $^{12}_{5}\text{B}$ nucleus into a $^{12}_{6}\text{C}$ nucleus and electron. If this decay were like α decay in that an unstable nucleus is transformed into just two particles (now an electron and daughter nucleus), then the conservation laws of momentum and energy would impose the following requirements: By momentum conservation the two emitted particles would travel outward from the decay site in opposite directions with the *same* momentum magnitude. By energy conservation, the total energy released in the decay process—the rest energy of the parent nucleus less the sum of the rest energies of the daughter nucleus and electron—would be shared between the two particles in a fashion consistent with the momentum-conservation requirement. Since the electron mass is far less than that of the daughter nucleus, essentially all of the energy released would be carried by the electron. In short, for the assumed decay into *two* particles, electrons from unstable parent nuclei of a particular species would appear with a *single energy,* very nearly equal in magnitude to the total energy released in the decay process.

What does experiment show? The emitted electrons are *not* monoenergetic. See Fig. 39-11. Instead, there is a distribution of electron kinetic energies from zero up to the maximum $K_{\max} = Q$, where Q represents the *total* energy released in the decay. The very few electrons having this maximum energy, and only those electrons, carry the kinetic energy expected on the basis of a two-particle decay. All other electrons—and this means almost all of the emitted electrons—seem to have too little kinetic energy! In

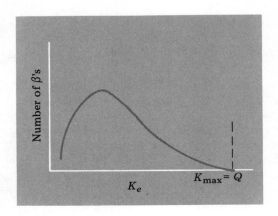

FIG. 39-11. Distribution in energy of emitted β^- particles in boron 12 decay.

short, there is an apparent violation of energy conservation! Furthermore, observations of individual β^- decays show that the electron and daughter nucleus do *not* necessarily leave the site of the decay in opposite directions. Apparently a violation of momentum conservation! In addition, comparison of the angular momenta of the parent nucleus, daughter, and electron indicates a violation of the angular-momentum conservation law!

We hasten to assure the reader that the fundamental laws of energy, linear momentum, and angular momentum are, in fact, not violated in β^- decay. This is so because of the *neutrino* ("little neutral one"), also emitted in β^- decay. Its existence was first suggested by W. Pauli in 1930 as an alternative to abandoning the conservation principles; its existence was directly confirmed by experiment in 1956. It is now known that a radioactive nucleus decays in β^- emission to *three* particles: the daughter nucleus, the electron, and the neutrino. All the difficulties we have mentioned above disappear by virtue of the neutrino's participation in β^- decay.

The neutrino has electric charge 0, rest mass 0, linear momentum p, with *total* relativistic energy $E = pc$, and intrinsic angular momentum $\frac{1}{2}\hbar$.

The neutrino has *zero* electric charge; charge is conserved in β^- decay *without* the neutrino.† The neutrino cannot interact with matter by producing ionization. It interacts very, very weakly with nuclei, and is virtually undetectable.

As we have seen, energy *is* conserved for those very few electrons in β^- decay which are emitted with the maximum kinetic energy $K_{max} = Q$. Therefore, the neutrino mass must be very small compared with the electron mass, and there are good theoretical reasons for taking it to be *exactly zero*. Since the neutrino has a zero rest mass and rest energy, it must, like a photon, always travel

---

†Strictly, the massless particle emitted in β^- decay is the antineutrino, whereas the massless particle emitted in β^+ decay is the neutrino.

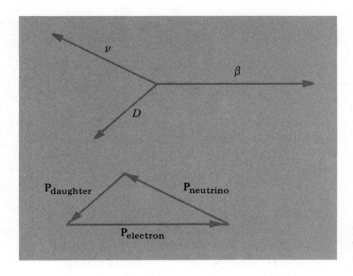

FIG. 39-12. Linear-momentum vectors of the daughter nucleus, electron, and neutrino in β^- decay.

at the speed of light. Therefore, a neutrino's total relativistic energy E is related to its relativistic momentum p by $E = pc$ [see Eq. (35-16)].

Consider again the conservation of energy and of linear momentum in β^- decay, assuming now that a neutrino, as well as an electron, is created in the decay and carries away energy and momentum. Energy conservation requires that

$$Q = K_D + K_e + K_\nu \approx K_e + K_\nu \qquad (39\text{-}12)$$

Note that the kinetic energy of a neutrino, K_ν, is also its total energy: $E_\nu = K_\nu$. Linear momentum conservation requires that the total *vector* momentum of the three particles add up to zero, as shown in Fig. 39-12. With three particles emitted, it is no longer necessary that they leave the site of the decay along the *same* straight line; now there are a variety of ways in which the separate momentum vectors can be arranged to add up to zero, satisfying (39-12) in every instance. The electron and daughter nucleus will usually *not* move along the same straight line in opposite directions, but if they do, the neutrino momentum and energy can be zero and from (39-12) we will have $K_e = K_{\max} = Q$, in agreement with observation for the most energetic electrons. In all other decays the virtually unabsorbable neutrino will carry energy and momentum, and the electron will necessarily have a kinetic energy less than $K_{\max}$. In two-particle decay the emerging particles are monoenergetic; in three-particle decay they are polyenergetic.

Consider finally the angular momentum, or spin, of the neutrino. It is $\frac{1}{2}$ in units of $\hbar$. In the β^- decay of $^{12}_{5}\text{B}$ to $^{12}_{6}\text{C}$, the parent and daughter nuclei both have integral nuclear spins; the electron has a spin of $\frac{1}{2}$. Therefore, when the neutrino's angular momentum is included, total angular momentum is conserved in β^- decay.

The three β decay processes given earlier can then be written more completely as

$$\beta^-: \quad {}^{12}_{5}B \to {}^{12}_{6}C + \beta^- + \nu$$
$$\beta^+: \quad {}^{12}_{7}N \to {}^{12}_{6}C + \beta^+ + \nu$$
$$\text{Electron capture:} \quad e_K^- + {}^{7}_{4}Be \to {}^{7}_{3}Li + \nu$$

The arguments for the appearance of the neutrino also in the β^+ and electron-capture decay processes are these: In positron decay there is also a distribution of positron kinetic energies, and the neutrino plays an analogous role in preserving the conservation laws as in electron emission. In electron capture, a decay into two particles, most of the released energy is carried by monoenergetic neutrinos, while the daughter nuclei recoil with the remaining energy, typically a few electron volts, but again monoenergetic. Without the accompanying neutrino in electron capture this decay process would be completely inexplicable.

The four basic reactions associated with β decay are the following:

$$\begin{aligned} \beta^- \text{ decay:} &\quad n \to p + e + \bar{\nu} \\ \beta^+ \text{ decay:} &\quad p \to n + \bar{e} + \nu \\ \text{Electron capture:} &\quad e + p \to n + \nu \\ \text{Neutrino absorption:} &\quad \bar{\nu} + p \to n + \bar{e} \end{aligned} \quad (39\text{-}13)$$

The symbol e represents the electron (charge, -1), $\bar{e}$ represents the positron (charge, $+1$), the electron's antiparticle, ν represents a neutrino, and $\bar{\nu}$ represents an *antineutrino*.

Up to this point we have recognized only one type of neutrino; there are in reality two, one the antiparticle of the other.† This distinction may seem to be completely formal, but it is not. It has been confirmed in subtle experiments that the antineutrino, the neutrino's antiparticle, has the direction of its spin, or intrinsic angular momentum, along the direction of its linear momentum. For an antineutrino the sense of its spin is clockwise, when viewed from behind, making it "right-handed". On the other hand, the neutrino's angular momentum and linear momentum are in opposite directions, making it left-handed, its spin being counterclockwise when viewed from behind. Nature thus distinguishes between the neutrino and antineutrino. This lack of symmetry — the neutrino is *only* left-handed and the antineutrino is *only* right-handed — is a manifestation of the *nonconservation of parity*, predicted by C. N. Yang and T. D. Lee, and experimentally confirmed in 1957 by C. S. Wu et al. The principle that nature does *not*

---

†In fact, things are still more complicated: A distinctive neutrino and antineutrino are associated with β decay via electron and antielectron emission. Other distinctive neutrinos and antineutrinos are associated with the decay of the unstable elementary particles known as muons (see Table 40-3).

distinguish between left and right, the conservation of parity, is violated in β decay. (See Sec. 40-5.)

The basic process of β^- decay, the decay of the neutron into a proton, electron, and antineutrino, given by (39-13), occurs in a *free neutron*, not merely a neutron bound within a nucleus. The decay is energetically allowed because the neutron mass exceeds that of the proton and electron by 0.78 MeV/c^2. The half-life of this decay is found, in experiments of extreme difficulty, to be 10.8 min. Because a free neutron is typically absorbed in less than 10^{-3} s when it passes through materials, the decay of a neutron is usually unimportant in situations involving free neutrons.

The basic β^+ decay, in which a proton is converted into a neutron, positron, and neutrino, is *not* permitted for a free proton by energy conservation. Positron decay is possible only when protons are bound within a nucleus.

Electron capture is closely related to the β^+ decay, of course. Note that in (39-13) the second reaction becomes the third reaction when the antielectron is transferred to the left side, thereby becoming an electron. This follows the general rule that the emission of a particle is equivalent to the absorption of an antiparticle, and conversely. By using this rule together with the permitted reversal of the arrow, it may be seen that all four β reactions are equivalent.

The last reaction in (39-13) is that in which an antineutrino combines with a proton to become a neutron and a positron. Although the relative probability of neutrino capture is extremely small, the capture was observed directly, with the very large neutrino flux from a nuclear reactor, by C. L. Cowan and F. Reines in 1956, thereby directly confirming the existence of the neutrino (strictly, the antineutrino). The antineutrino is identified by observing the neutron and the positron produced simultaneously when the antineutrino is captured by a proton; the neutron is detected by observing the photon emitted from an excited nucleus that has absorbed the neutron, and the positron is detected by observing annihilation photons. The difficulty of this experiment may be appreciated from the fact that a neutrino or antineutrino has only 1 chance in 10^{12} of being captured while traveling completely through the earth. Because neutrinos have such a very small probability of interacting with matter and being absorbed, a large fraction of the energy released in all β-decay processes is effectively lost.

39–7 HIGH-ENERGY PARTICLE ACCELERATORS

Since the constituent particles of the nucleus are bound together with energies of several *million* electron volts, the internal structure and arrangement of the nuclear constituents can be altered and the structure of the nucleus studied only if the nucleus is given energies of the order of millions of electron volts. The most direct means of altering the structure of nuclei is to bombard targets containing atoms (and therefore also nuclei) with particles that have been accelerated to very high energies. Progress in nuclear physics and in the physics of elementary particles has, therefore, depended upon the invention and design of machines that can accelerate charged particles to kinetic energies measured in MeV or even in several hundred GeV.

The principal motivation for constructing such very-high-energy accelerators is that the particles may be used to create unstable particles not easily found in nature, and these may be studied. For example, antiprotons are created by protons of 6 GeV striking protons at rest. Moreover, as a particle's energy and momentum are increased, its wavelength $\lambda = h/p$ is reduced, so that, for example, an electron of 20 GeV has a wavelength of less than 10^{-16} m, whereas a typical nuclear size is about 10^{-14} m. Indeed, high-energy electrons may be used to probe the distribution of electric charge within a nucleus or even within a proton.

All charged-particle accelerators are based on the fact that a charged particle has its energy changed when it is acted on by an *electric* field. A *constant* magnetic field does *no* work on a moving particle and cannot change its energy; on the other hand, a *changing* magnetic field produces an electric field, which in turn can accelerate a charged particle. Therefore, all high-energy accelerators change the energy of charged particles by subjecting them to an electric field derived either directly from charged particles or indirectly from a changing magnetic field.†

The two general classes of charged-particle accelerators are the *linear accelerators* ("linacs"), in which the particles move along a straight line, and *cyclic accelerators,* in which the particles travel in curved paths and are recycled. We have already described the principal features of the Van de Graaff linear accelerator (Sec. 23-6); it accelerates particles by a *static* potential difference. Singly charged particles can acquire kinetic energies up to about 30 MeV, with multiply charged particles acquiring correspondingly higher energies.

To accelerate particles to significantly higher energies in a linear accelerator requires that the particles receive *multiple* accelerations from electric fields along the single line in which they travel.

In the waveguide-type linear accelerator, very high-frequency electromagnetic waves are guided along a conducting tube, and the charged particles may be thought of as riding the waves in such a fashion that the electric field steadily increases their kinetic energy. See Fig. 39-13, which shows the 20 GeV electron linac of

---

†An accelerating machine that utilizes a *changing* magnetic field to produce an electric field is the *betatron,* an accelerator of high-energy electrons. We omit discussion of its features here,

(a)

(b)

(c)

FIG. 39-13. The 20-GeV Stanford electron linear accelerator. *(a)* The target area at the end of the two-mi-long accelerator where electrons are deflected and magnetically analyzed and are directed to such devices as spark and bubble chambers. *(b)* At the target area and 8 GeV magnetic spectrometer in the foreground and a 20-GeV spectrometer in the rear. *(c)* Interior view of the subsurface accelerator housing. (Courtesy Stanford Linear Accelerator Center, Stanford University.)

the Stanford Linear Accelerator Center; the wavelength of electrons from this machine is so small that the distribution of electric charge within the proton can be determined with it.

The class of accelerators known as cyclic accelerators includes the *cyclotron, synchrocyclotron,* and *synchrotron.* In these machines multiple accelerations are given to charged particles that are restricted to motion in circular arcs by a magnetic field.

The basic relation for a particle of relativistic mass m and charge Q moving at right angles to a magnetic field B in a circular arc of radius r is

$$p = mv = QBr \qquad (35\text{-}10)$$

The particle's angular velocity ω is given by

$$\omega = \frac{v}{r} = \frac{Q}{m} B$$

and the frequency $f = \omega/2\pi$ of the motion, the number of revolutions per unit time, is given by

$$f = \frac{Q}{2\pi m} B \qquad (39\text{-}14)$$

This frequency is known as the *cyclotron frequency* and the expression applies to all cyclic accelerators. Note that f depends on the charge-to-mass ratio and the magnitude of the magnetic field, but not on the particle's speed or the radius of its circular path. Thus, all particles of a given type circle the magnetic field lines at the same frequency, quite apart from differences in their speeds or energies. Strictly, the cyclotron frequency is independent of a particle's kinetic energy only if the relativistic mass m does not differ appreciably from the rest mass m_0.

The simplest cyclic machine is the *cyclotron,* invented by E. O. Lawrence and M. S. Livingston in 1932. In this accelerator a charged particle is subjected to a *constant* magnetic field, which bends it into a circular path, while it is accelerated each half-cycle by an electric field.

Positive ions, such as protons, deuterons, and α particles, are injected into the central region, point C of Fig. 39-14, between two flat, D-shaped, hollow metal conductors (called "dees"). An alternating high-frequency voltage is applied to the dees, producing an alternating electric field in the region between them. During the time that the left dee is positive and the right dee is negative the ions are accelerated to the right by the electric field between the dees. Upon entering the interior of the right dee they are electrically shielded from any electric field and therefore move in a semicircle at a constant speed under the influence of the constant

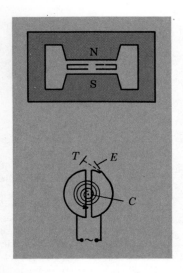

FIG. 39-14. Cyclotron accelerator: (a) top view; (b) side view.

---

inasmuch as the betatron is no longer used primarily as a research tool in nuclear physics, but as a generator of x-rays.

magnetic field. When they emerge from the right dee, they are further accelerated across the gap, if the left dee is now negative. This requires that the frequency of the alternating voltage applied to the dees be equal to the orbital, or cyclotron, frequency of the ions, given by (39-14). During each acceleration the ions gain energy, move at a higher speed, and travel in semicircles of larger radii. As the ions spiral outward in the dees they remain in resonance with the ac source of constant frequency, inasmuch as the time for an ion to move through 180° is independent of its speed or radius, provided only that its mass m in (39-14) remain essentially equal to the rest mass. When the accelerated particles reach the perimeter of the dees, they are deflected by the electric field of an ejector plate E and strike the target T. Their final kinetic energy E_k (for E_k much less than E_0 the rest energy) is

$$E_k = \tfrac{1}{2}mv_{max}^2 = \tfrac{1}{2}m\left(\frac{QBr_{max}}{m}\right)^2 = \frac{Q^2B^2r_{max}^2}{2m}$$

We see that the final kinetic energy of the particle depends on the square of the radius of the dees and on the square of the magnetic field B. To achieve the greatest possible energies, the quantities B and r_{max} are made as large as possible. When the greatest possible magnetic field attainable with iron (about 2 Wb/m²) is used, the frequency f, by means of (39-14), is of the order of megahertz (radio-frequencies). The diameter of the dees, which is also the diameter of the electromagnet's pole faces, may be as large as a few meters; this is enormous (400 tons of iron) and expensive. A typical alternating voltage across the dees is 200 kV.

Massive particles — protons, deuterons, and α particles — can be accelerated in a cyclotron to energies of about 25 MeV. The final kinetic energies of all such ions are much less than their rest energies (a proton's rest energy is about 1 GeV). Therefore, the mass does not increase appreciably, and the particles, if protons, can remain in synchronism with the alternating voltage up to about 12 MeV and, if deuterons, to about 25 MeV. Electrons can be rather easily accelerated to relativistic speeds (their rest energy is only 0.5 MeV); such light particles cannot be synchronized with the applied voltage and therefore cannot be accelerated to high energies by a cyclotron.

An ordinary fixed-frequency cyclotron will work only if the accelerated particle's kinetic energy remains small compared with its rest energy. The cyclotron frequency $f = (Q/2\pi m)B$ of the orbiting particles is constant and in resonance with the alternating electric field between the dees, only if the relativistic mass m is always essentially equal to the rest mass m_0. As a particle's speed and kinetic energy increase, the relativistic mass increases, and its cyclotron frequency in a constant magnetic field decreases. Therefore, as particles spiral outward in a cyclotron with fixed mag-

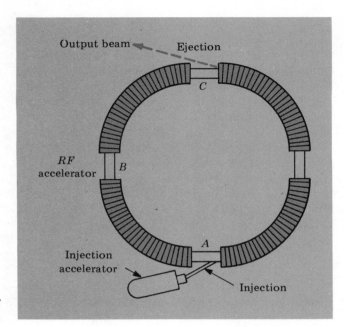

FIG. 39-15. Schematic diagram of synchrotron.

netic field and frequency, they fall increasingly behind the applied frequency and finally arrive at the gap between the dees so late that they are no longer accelerated by the electric field. This limitation is overcome in the *synchrocyclotron*, in which the applied frequency is *decreased* continuously in such a way as to compensate for the decrease in the particles' cyclotron frequency with increased speed. The particles can then remain in synchronism with the alternating electric field.

We finally consider the *synchrotron*. To increase a particle's final kinetic energy in a cyclic accelerating machine, one must increase its final relativistic momentum $p = QBr$. There is a limit on the magnitude of the magnetic field B attainable over moderately large regions of space; because of the properties of magnetic materials, B cannot exceed about 2 Wb/m².† Thus, the only way to increase substantially a particle's momentum and thereby its kinetic energy in a cyclic machine is to make the final orbital radius large. In a synchrocyclotron the particles start at the center of the electromagnet and spiral outward to the final radius; the electromagnet produces a magnetic field over this entire region. So, if a particle's final energy in a synchrocyclotron is to be increased, the radius of the final orbit and, consequently, the radius of the electromagnet must be increased correspondingly. A machine accelerating protons to about 0.7 GeV has a dee diameter of about 5 m; to produce a

---

†Still higher magnetic fields may be achieved with superconductors, materials at low temperatures with *zero* electrical resistance. Superconducting magnets may be employed in future accelerating machines.

FIG. 39-16. Aerial view of main accelerator at the Fermi National Accelerator Laboratory, Batavia, Illinois. The synchrotron has a diameter of 1.24 mi and accelerates protons to more than 500 GeV. (Courtesy Fermi National Accelerator Laboratory.)

large magnetic field over the entire inner region requires an electromagnet of about 4,000 tons. Still larger sizes and energies become economically unfeasible. Therefore, the synchrotron, which utilizes magnetic fields only at *one orbital radius,* has been devised.

Particles, preaccelerated to a fairly high kinetic energy, are injected into the synchrotron and thereafter move in an orbit of *fixed* radius. The basic relations governing the particle motion are

$$p = QBr \qquad (35\text{-}10)$$

$$f = \frac{Q}{2\pi m} B \qquad (39\text{-}14)$$

Since the radius r is fixed, the particle momentum p will increase only if the magnetic field B increases. Equation (39-14), however, shows that if B changes, so too must the frequency f of the accelerating electric field with which the orbiting particles are to remain in synchronism. Thus, in the synchrotron accelerator *both* the magnetic field and the frequency of the accelerating electric field increase with time as the accelerated particle, moving in a fixed circle, gains kinetic energy.

Figure 39-15 shows schematically the principal parts of a proton synchrotron, and Fig. 39-16 is a photograph of one. Protons first are accelerated to an energy of several MeV by a linac serving as an injection accelerator. Then they enter a doughnut-shaped evacuated tube, not more than about 1 m in transverse dimensions, which is contained within an electromagnet producing a deflecting magnetic field at and near the tube but not at interior points. Energy is supplied to the particles once in each revolution by an alternating electric potential difference supplied by a variable radio-frequency source. As the particles acquire speed, momentum, and kinetic energy in successive orbits, the magnitude of the deflecting magnetic field and the frequency of the accelerating electric field both increase with time, so that the particles continue to travel in a path of constant radius and also arrive at the energizing gap at the right time for further acceleration. After the magnetic field has reached its maximum magnitude and the particles their maximum kinetic energy, the particles are deflected and strike an external target. A view of the 500 GeV synchrotron constructed at the Fermi National Accelerator Laboratory, Batavia, Ill., is shown in Fig. 39-16.

39–8 NUCLEAR REACTIONS

Unstable nuclei will decay spontaneously, changing their nuclear structure without external influence. One can, however, induce a change in the identity or characteristics of nuclei by bombarding them with energetic particles. The change is known as a *nuclear reaction*.

Thousands of nuclear reactions have been produced and identified since Rutherford observed the first one in 1919. The bombarding particles were, until the development of charged-particle accelerators in the 1930s, those emitted from radioactive substances. It is now possible to accelerate charged particles to energies up to hundreds of GeV. When particles of such great energy strike nuclei, they severely disrupt them and may create new and strange particles. These so-called high-energy reactions and the particles participating in them will be discussed in Chap. 40.

We shall be concerned here with *low-energy* nuclear reactions, reactions in which the incident particles have energies no greater than, say, 20 MeV, and we shall illustrate several types of nuclear reaction with examples that have been important in the history of nuclear physics.

In the first observed nuclear reaction (1919) Rutherford used α particles of 7.68 MeV from the naturally radioactive element $^{214}_{84}$Po. When the α particles were sent through a nitrogen gas, most of them were either undeflected by the nitrogen nuclei or elastically scattered in close encounters with them. Rutherford found, however, that in a few collisions (about 1 in 50,000) protons were produced, according to the nuclear reaction

$$^{14}_{7}\text{N} + {}^{4}_{2}\text{He} \rightarrow {}^{1}_{1}\text{H} + {}^{17}_{8}\text{O}$$

In this reaction an α particle strikes a nitrogen 14 nucleus, producing a proton and an oxygen 17 nucleus. That the particles emitted in this reaction are protons can be established by measurements of the charge-to-mass ratio with a magnetic field. The reaction represents *induced transmutation* of the element nitrogen into a stable isotope of oxygen; α- or β-radioactive decay represents, of course, *spontaneous transmutation* of one element into another.

The laws of conservation of electric charge and of nucleons are satisfied in all nuclear reactions; therefore, the presubscripts giving the electric charge of the particles and the presuperscripts giving the number of nucleons in each particle each sum to the same amount on both sides of the equation. The reaction may be written in abbreviated form as follows:

$$^{14}_{7}\mathrm{N}(\alpha, p)^{17}_{8}\mathrm{O}$$

where the light particles going into and out of the reaction are written in parentheses between the symbols for the target and product nuclei.

Until 1932 all nuclear reactions were produced by the relatively high-energy α particles or γ rays from naturally radioactive materials. In that year J. D. Cockcroft and E. T. S. Walton, using a 500-keV accelerator, observed the first nuclear reaction produced by artificially accelerated charged particles. They found that α particles were emitted when a lithium target was struck by protons with energies of 500 keV, according to the reaction

$$^{7}_{3}\mathrm{L} + ^{1}_{1}\mathrm{H} \rightarrow ^{4}_{2}\mathrm{He} + ^{4}_{2}\mathrm{He}$$

$$^{7}_{3}\mathrm{Li}(p, \alpha)^{4}_{2}\mathrm{He}$$

The emitted α particles had a total kinetic energy of 17.9 MeV; thus, an energy of 0.5 MeV had been put into the reaction, and 17.8 MeV was released as kinetic energy of the emerging particles. The released energy corresponds exactly to the excess of the rest mass of lithium 7 and a proton over that of two α particles. Here is a striking example of the release of nuclear energy. The total amount of energy released was trifling, of course, since most of the collisions between the incident protons and target nuclei did *not* result in nuclear disintegrations.

In the two reactions described above the product nuclei were stable. The first nuclear reaction leading to an unstable product nucleus was observed by I. Joliot-Curie and F. Joliot in 1934. In the reaction an aluminum target is struck by α particles, leading to

$$^{27}_{13}\mathrm{Al} + ^{4}_{2}\mathrm{He} \rightarrow ^{1}_{0}n + ^{30}_{15}\mathrm{P}$$

$$^{27}_{13}\mathrm{Al}(\alpha, n)^{30}_{15}\mathrm{P}$$

The product nuclide is not stable but decays with a half-life of 2.6 min into a stable isotope of silicon by β^{+} emission:

$$^{30}_{15}P \rightarrow \,^{30}_{14}Si + \beta^+ + \nu$$

where ν is a neutrino. The production of unstable nuclides that spontaneously disintegrate by the law of radioactive decay is a feature of many nuclear reactions. The nuclides are said to exhibit *artificial radioactivity*. Indeed, nuclear reactions are the only means of obtaining artificial radioactive isotopes, or *radioisotopes*. The radioisotopes are chemically identical with the element's stable isotopes. If a small amount of radioisotope is added to stable nuclides of the same element, it can serve, through its radioactivity, as a *tracer* of the element; that is, the presence and concentration of the element can be determined by measuring the radioisotope's activity.

The discovery of the neutron came as a result of a nuclear reaction observed in 1930 by W. Bothe and H. Becker, the bombardment of beryllium by α particles:

$$^{9}_{4}Be + \,^{4}_{2}He \rightarrow \,^{1}_{0}n + \,^{12}_{6}C$$

$$^{9}_{4}Be(\alpha, n)^{12}_{6}C$$

Neutrons are emitted in many nuclear reactions and can themselves be used as bombarding particles. One of the important neutron-induced reactions is that in which a neutron is captured by a target nucleus and a γ-ray photon is emitted. This reaction is known as neutron *radiative capture*. For example,

$$^{27}_{13}Al + \,^{1}_{0}n \rightarrow \gamma + \,^{28}_{13}Al$$

$$^{27}_{13}Al(n, \gamma)^{28}_{13}Al$$

The product nucleus, an unstable isotope of the target nucleus, decays by β^- decay:

$$^{28}_{13}Al \rightarrow \,^{28}_{14}Si + \beta^- + \bar{\nu}$$

where $\bar{\nu}$ is an antineutrino.

Since the neutron has no electric charge, the neutron radiative capture process can occur when a neutron of almost any energy strikes (almost) any nucleus; the heavier isotope thus produced frequently is radioactive, and the absorption of neutrons is, therefore, a common means of producing radioisotopes.

Another important type of reaction resulting from neutron bombardment is that in which a charged particle, such as a proton or α particle, is emitted. Such a reaction offers a method of detecting neutrons, because the emitted charged particles produce detectable ionization. One reaction frequently used in neutron detection is

$$^{10}_{5}B + \,^{1}_{0}n \rightarrow \,^{4}_{2}He + \,^{7}_{3}Li$$

$$^{10}_{5}B(n, \alpha)^{7}_{3}Li$$

Photodisintegration is the nuclear reaction in which the absorption of a γ-ray photon results in the disintegration of the absorbing nucleus. An example is

$$^{25}_{12}\text{Mg} + \gamma \rightarrow {}^{1}_{1}\text{H} + {}^{24}_{11}\text{Na}$$

$$^{25}_{12}\text{Mg}(\gamma, p)^{24}_{11}\text{Na}$$

followed by ${}^{24}_{11}\text{Na} \rightarrow {}^{24}_{12}\text{Mg} + \beta^- + \bar{\nu}$.

A special type of low-energy nuclear reaction is that of *nuclear fission*. In this reaction, a low-energy neutron is captured by a very heavy nucleus, and the resulting aggregate splits into two moderately heavy nuclei along with a few neutrons. Isotope 235 of uranium may undergo fission in capturing a neutron according, for example, to the reactions

$$^{1}_{0}n + {}^{235}_{92}\text{U} \rightarrow {}^{144}_{56}\text{Ba} + {}^{89}_{36}\text{Kr} + 3\,{}^{1}_{0}n$$

or

$$^{1}_{0}n + {}^{235}_{92}\text{U} \rightarrow {}^{140}_{54}\text{Xe} + {}^{94}_{38}\text{Sr} + 2\,{}^{1}_{0}n$$

Each of the moderately heavy fission fragments has a kinetic energy up to 100 MeV; further energy is released as the highly unstable fragments decay, principally by emitting electrons. Note that in addition to the fragments, two or more neutrons are also produced in the fission reaction. These neutrons may initiate still other fission reactions, provided that the surrounding fissionable material has reached the critical mass for a self-sustaining nuclear chain reaction.

Not only does the splitting of the heaviest nuclei into smaller nuclei result in a relatively large energy release, so too does the fusion of the lightest nuclides into more massive species. This follows at once from the fact that the binding energy per nucleon is greatest for nuclides of intermediate mass (see Fig. 39-5). One example of a nuclear fusion process, a thermonuclear reaction releasing energy in the sun and similar stars, is the proton-proton cycle:

$$^{1}\text{H} + {}^{1}\text{H} \rightarrow {}^{2}\text{H} + \beta^+ + \nu$$
$$^{1}\text{H} + {}^{2}\text{H} \rightarrow {}^{3}\text{He} + \gamma$$
$$^{3}\text{He} + {}^{3}\text{He} \rightarrow {}^{4}\text{He} + 2\,{}^{1}\text{H}$$

This cycle, involving three distinct nuclear reactions, fuses four protons into an α particle, two positrons, and two neutrinos. The energy released per nucleon is larger by almost an order of magnitude than that in fission reactions.

Nuclear fusion reactions can take place only if the participating positively charged particles approach one another within the range of the nuclear force despite their mutual repulsion through the Coulomb force. This implies that the particles be in a gas (or, strictly, a *plasma* of negatively and positively charged particles) at temperatures at least of the order of 10^7 K—hence, *thermonuclear* reactions. Although nuclear fission has been used as a practical source of energy for many years, efforts to achieve controlled nuclear fusion as an energy source based on almost inexhaustible available fuel materials have been frustrated.

SUMMARY

Table 39–1. Properties of the nuclear constituents

Property	Proton	Neutron
Mass, u	1.007277	1.008665
Charge, e	1	0
Spin, $\hbar$	$\frac{1}{2}$	$\frac{1}{2}$

Properties of the Nuclear Force

Attractive and much stronger than the coulomb force

Short-range, $\approx$ 3 fermi (3×10^{-15} m)

Charge-independent; all three nucleon interactions, *np, pp,* and *nn,* are approximately equal.

Nomenclature

Nucleon: proton or neutron

Atomic number Z: number of protons

Neutron number N: number of neutrons

Mass number A: total number of nucleons ($Z + N$)

Nuclide: nucleus with a particular Z and a particular N

Isotopes: nuclides with same Z

Isotones: nuclides with same N

Isobars: nuclides with same A

Properties of the Nuclides

Stable nuclides: $N \approx Z$ at small A, and $N > Z$ at large A.

The nuclear radius is given by $R = r_0 A^{1/3}$, where $r_0 = 1.4$ fermi. All nuclei have the same nuclear density.

For $A > 20$ the average binding energy per nucleon is about 8 MeV/nucleon.

In the decay of all unstable nuclei, the laws of conservation of electric charge, nucleons, mass-energy, and momentum are satisfied.

The law of radioactive decay is $N = N_0 e^{-\lambda t}$, where the decay constant λ, the probability per unit time that any one nucleus will decay, is related to the half-life by $T_{1/2} = 0.693/\lambda$.

Table 39–2. *Radioactive decay modes*

	Alpha (helium nucleus)	Beta (electron, positron)	Gamma (photon)
Half-lives	10^{-6} s to 10^{10} yr	$> 10^{-2}$ s	10^{-17} to 10^{5} s (isomer)
Energies	4 to 10 MeV	a few MeV	keV to a few MeV
Decay mode	$^{A}_{Z}P \to \,^{A-4}_{Z-2}D + \,^{4}_{2}\alpha$	β^{-}: $^{A}_{Z}P \to \,^{A}_{Z+1}D + \,^{0}_{-1}e + \bar{\nu}$ β^{+}: $^{A}_{Z}P \to \,^{A}_{Z-1}D + \,^{0}_{+1}e + \nu$ EC: $^{0}_{-1}e + \,^{A}_{Z}P \to \,^{A}_{Z-1}D + \nu$	$^{A}Z^{*} \to \,^{A}Z + \gamma$
Energy distribution of decay products	Monoenergetic	β^{-} and β^{+}: polyenergetic EC: monoenergetic	Monoenergetic

Neutrino Properties

$$\text{Mass: } 0$$
$$\text{Charge: } 0$$
$$\text{Spin: } \tfrac{1}{2}$$
$$\text{Neutrino capture: } \bar{\nu} + p \to n + \beta^{+}$$

The basic relations governing the operation of such cyclic accelerators as the cyclotron and synchrotron is that for the cyclotron frequency

$$f = \frac{Q}{2\pi m} B \qquad (39\text{-}14)$$

and that for the relativistic momentum

$$p = mv = QBr \qquad (35\text{-}10)$$

PROBLEMS

See Appendix IV for values of the atomic masses.

39-1. Show that in a proton-proton or a neutron-proton elastic scattering experiment no particles are scattered from the forward direction by more than 90° (take the neutron and proton masses to be equal, and assume the incident particle kinetic energy to be small compared with the rest energy).

39-2. Show that if a proton collides head-on with a second proton initially at rest, the two particles trade velocities; that is, the proton initially in motion is brought to rest while the other proton acquires the velocity of the proton striking it.

39-3. A 5.0-MeV proton collides head on with a 5.0-MeV neutron and produces a deuteron. (a) What is the energy of the photon emitted? *(b)* With what kinetic energy does the deuteron recoil upon emitting the photon? Take the neutron and proton masses to be equal.

39-4. A proton and neutron are brought together at negligible relative speed until they attract one another by the nuclear force. A photon is emitted as the two particles form a deuteron in its ground state. What is the energy of this photon?

39-5. *(a)* Show that an even-Z nuclide usually has many more stable isotopes than an odd-Z nuclide. *(b)* Between $^{16}_{8}\text{O}$ and $^{32}_{16}\text{S}$ there are one stable isotope for each odd-Z nuclide and three stable isotopes for each even-Z nuclide. Explain this in terms of the filling of neutron and proton shells.

39-6. *(a)* How much energy is required to remove one proton from a nucleus of $^{12}_{6}\text{C}$, leaving the nucleus $^{11}_{5}\text{B}$? *(b)* How much energy is required to remove one neutron from $^{12}_{6}\text{C}$, leaving the nucleus $^{11}_{6}\text{C}$?

39-7. The activity of a certain radioactive material drops by a factor of 10 in a time interval of 1 min. What is the decay constant of this radionuclide?

39-8. Carbon present in the atmosphere is found to be radioactive because of the presence of $^{14}_{6}\text{C}$. Prior to the advent of nuclear weapons, the specific activity of such atmospheric carbon was found to be 15.3 disintegrations/(g)(min). The half-life of $^{14}_{6}\text{C}$ is 5.74×10^3 yr. *(a)* What proportion of atmospheric carbon is $^{14}_{6}\text{C}$? *(b)* The activity of the carbon in a certain biological relic is 2.5 disintegrations/(g)(min). How much time has elapsed since the death of this specimen?

39-9. *(a)* Show that the mean life T_{av} of a radionuclide having a decay constant λ is given by

$$T_{av} = \frac{\int_{N_0}^0 t\, dN}{\int_{N_0}^0 dN} = \frac{1}{\lambda}$$

(b) Of any group of radioactive atoms having the same decay constant, some will survive decay for an essentially infinite period of time. On this basis, convince yourself that the mean life of the species must exceed the half life, and show that the mean life always exceeds the half-life by 44 percent.

39-10. The decay constant, or half-life, for long-lived radioactive materials can be determined, not by noting the decrease with time in the activity, but rather by measuring the activity A from a known number N of unstable nuclei. (Since λ is very small and $T_{1/2}$ very long, neither A nor N changes significantly over moderate periods of time.) Show that the decay constant is then given by $\lambda = A/N$.

39-11. A $^{8}_{4}\text{Be}^*$ nucleus may decay to the ground state with the emission of a 17.6-MeV γ ray. With what kinetic energy does it recoil? (The beryllium 8 nucleus is highly unstable and decays rapidly to two α particles.)

39-12. A nuclide is unstable to α decay and the total energy released in the decay of one nucleus is Q. *(a)* Show that the neutral atomic mass of the parent nucleus must exceed that of the daughter and α by Q/c^2 (assuming the α decay to be to the nuclear ground state of the daughter). *(b)* Show that if a decaying nucleus is initially at rest, the kinetic energy of an emitted α particle is equal to $(A-4)Q/A$, where A is the number of nucleons in the parent nucleus.

39-13. Show that of two neighboring isobars (differing in Z and in N by 1), one must be unstable relative to β decay.

39-14. *(a)* Show that if a nuclide is to undergo positron decay, the neutral atomic mass of the parent nuclide must exceed that of the daughter by two electron masses. *(b)* Show that if a particular nuclide undergoes positron decay then it is always energetically possible for the same nuclide to undergo electron capture.

39-15. The nuclide $^{7}_{4}\text{Be}$ decays to $^{7}_{3}\text{Li}$ by electron capture. *(a)* Show by comparing the masses of the nuclei that the total energy released in the decay is 0.86 MeV. *(b)* What is the approximate energy of the neutrino emitted in the decay? *(c)* What is the neutrino's momentum? *(d)* With what momentum does the $^{7}_{3}\text{Li}$ nucleus recoil? *(e)* With what kinetic energy does the $^{7}_{3}\text{Li}$ nucleus recoil?

39-16. What is the minimum energy an antineutrino may have for capture by a proton so as to produce a neutron and positron?

39-17. What is the maximum possible kinetic energy of the electron emitted in the decay of $^{14}_{6}\text{C}$?

39-18. *(a)* What is the design frequency for deuterons in a cyclotron with dees 1.0 m in diameter and a magnetic field of 1.0 Wb/m²? *(b)* What is the maximum kinetic energy attainable with such a cyclotron?

39-19. A cyclotron is adjusted to accelerate protons. *(a)* If the cyclotron frequency remains fixed, by what factor must the magnetic field be changed to permit the machine to accelerate deuterons? *(b)* What is the ratio of the maximum energy of the deuterons to that of protons (assume the deuteron mass to be twice that of the proton)?

39-20. In the 20-GeV linear electron accelerator at Stanford University electrons accelerate from 30 MeV to 20 GeV along the 2-mi accelerating tube. *(a)* By what percentage is the final electron speed less than c? *(b)* What is the overall length of the accelerating tube as measured by an observer traveling with a 20-GeV electron? *(c)* The electron beam at the target has an electron current of 15 μA and a power of 0.50 MW. What is the average number of electrons striking the target per second?

39-21. Protons from a linear accelerator are injected at 50 MeV into the synchrotron at CERN, the European nuclear research center near Geneva, Switzerland. The protons emerge with a kinetic energy of 30 GeV. By what percentage is a proton's speed increased by the synchrotron?

39-22. By what mode are the unstable products of the following reactions likely to decay: *(a)* (n, γ), *(b)* (p, n), *(c)* (d, p), and *(d)* (α, n)?

39-23. Imagine that a proton synchrotron has a diameter that of the earth, 1.26×10^4 km, and a maximum magnetic field of 1.6 Wb/m²; the field guides protons along a path circling the earth. *(a)* What is the maximum proton kinetic energy for this, the "ultimate" high-energy earth-bound particle accelerator (based on conventional design elements)? *(b)* The 500-GeV proton synchrotron at Batavia, Ill., cost about $\$\frac{1}{4}$ billion; assuming very conservatively for simplicity that the cost of an accelerator is proportional to the kinetic energy of the accelerated particles or, equivalently, to the radius of the orbit, compute the approximate cost of the "ultimate" accelerator in units of the 1970 U.S. Gross National Product (approximately $1 trillion). *(c)* At what approximate date might the maximum energy be achieved, according to extrapolation from the following facts: Starting with a machine for 1-MeV particles in 1932, the development of high-energy accelerators has advanced such that the maximum particle energy attainable has increased by a factor of 20 over a decade?

39-24. Verify by using atomic masses that the energy released in the reaction $^7_3\text{Li}(p, \alpha)^4_2\text{He}$ is 17.9 MeV.

39-25. Before the neutron had been properly identified by Chadwick in 1932, it was thought that the bombardment of ^9_4Be by α particles led to the reaction $^9_4\text{Be}(\alpha, \gamma)^{13}_6\text{C}$. It was found that, when the penetrating radiation from the bombardment struck paraffin, protons were ejected with an energy of 5.7 MeV. *(a)* Show that, if such protons are assumed to have been energized in a Compton collision, the energy of the γ-ray photons must be 55 MeV. *(b)* Chadwick found that, whereas the penetrating radiation produced protons with an energy of 5.7 MeV, the same radiation striking nitrogen atoms imparted a kinetic energy of 1.4 MeV to a nitrogen atom. Show, by applying momentum and energy conservation to a head-on collision between a particle of penetrating radiation (really a neutron) with a proton and a nitrogen atom, with masses in the ratio of 1:14 but kinetic energies in the ratio of 5.7:1.4, that the neutron mass is essentially the same as the proton mass.

39-26. Write at least three nuclear reactions in which targets made of stable nuclides may be used to produce *(a)* nitrogen 13, *(b)* neon 21, and *(c)* iron 57.

39-27. *Definition of cross section.* The decay of unstable nuclei is given by $N = N_0 e^{-\lambda t}$ [Eq. (39-9)]; the number of surviving nuclei N at some one location decreases exponentially with time because, for an unstable nucleus, one instant of time is like any other. An analogous relation applies to the decrease in space along the direction of travel at any instant of time of the number of particles in a beam passing through a material, the incident particles having been removed from the forward direction by scattering or by a nuclear reaction: Within any target material containing a large number of target particles, one location is like any other. The number N of surviving particles in the beam at a distance x from the front surface of the target material is, therefore, given by $N = N_0 e^{-\mu x}$, where μ, known as the *absorption coefficient*, is a constant for a given target. *(a)* At what distance from the front surface of the target material is the number of particles in the forward beam equal to $(1/e)$th of the incident number? *(b)* Since the constant μ must for every material be proportional to the number ρ of target particles per unit volume, we may write $\mu = \rho\sigma$, where σ is a constant for a given material. Show that σ must have the dimensions of area. Indeed, σ is known as the *cross section* of the process removing particles from the incident beam; it represents the effective target area presented by a target particle to an incident particle in the beam. *(c)* Show that for a sufficiently thin target, the fractional number of incident particles removed from the forward beam is equal to $\sigma\rho t$, where t is the thickness of the target.

39-28. Two helium nuclei, each with a kinetic energy of 20 MeV, collide head on and produce the reaction $^4_2\text{He}(\alpha,p)^7_3\text{Li}$. What is the kinetic energy of protons emerging from the reaction?

39-29. Show that approximately 200 MeV is released in nuclear fission reaction by considering the average binding energy per nucleon for the fissionable nucleus and for the fragments (see Fig. 39-5).

39-30. Assume that in nuclear fission the unstable

nucleus splits into two fragments of at least approximately equal mass and that the neutron-proton ratio for the fragments is the same as for the fissionable nucleus. Show that the fragments are highly unstable to β decay and will decay mainly by electron emission. (*Hint:* See Fig. 39-4.)

39-31. A nucleus of $^{240}_{94}$Pu* decays into the fission fragments $^{144}_{56}$Ba and $^{94}_{38}$Sr. Assume that the two fragments are spherical and just touching immediately after their formation. *(a)* What is the coulomb potential energy (in MeV) of this pair of fragments? *(b)* Compare this coulomb energy with the total energy released in the fission process.

39-32. *(a)* Show that the energy released in the proton-proton cycle given in Sec. 39-8 is approximately 25 MeV. *(b)* What is the ratio of the energy released *per nucleon* in this set of reactions to that in a typical fission reaction (200 MeV)?

39-33. All told, there are about 10^{21} kg of water on earth, with one D_2O molecule for every 6,000 H_2O molecules. Assuming that all of the deuterium is used in the fusion reaction $^2_1H(d, p)^3_1H$, what is the total amount of energy that can be extracted?

39-34. Thermonuclear fusion reactions can take place with high probability only if the particles in the plasma collide at such sufficiently high kinetic energies that the particles approach one another to separation distances comparable to the range of the nuclear force. At what temperature T will the average kinetic energy per particle $\frac{3}{2}kT$ equal the electric potential energy ke^2/r between two protons separated by distance r, where r is the range of the nuclear force?

39-35. Approximately 10 percent of the 25 MeV released in a proton-proton cycle is carried by the two neutrinos emitted in each cycle. The intensity of the sun's radiation at the earth's surface is 1.4 kW/m²; the distance from the earth to the sun is 1.5×10^{11} m. *(a)* What is the rate at which neutrinos are produced in the sun's interior by thermonuclear-fusion reactions? *(b)* What is the flux (number/m²-s) of neutrinos at the earth's surface? *(c)* What is the density at the earth's surface (number/m³) of neutrinos originating from the sun?

CHAPTER 40

The Elementary Particles

Man's search for the ultimate building blocks of nature goes back to the Greek notion of four elements—earth, water, air, and fire (and possibly an ethereal fifth element, a "quintessence")—that were supposed to be the basic components of all other materials. Then came the ideas of the chemical elements—molecules and atoms—and, finally, of the particles within atoms, even within the nucleus. Underlying the quest for the elementary particles is the expectation that, if one has identified the truly fundamental particles—hopefully, of only a few distinct types—and learned the rules by which they affect one another, then the remainder of physics will be a straightforward, although possibly very difficult, exercise. At the present time all of chemistry, including the chemical properties and the periodic table, is implicit in a wave-mechanical description of atomic structure.

We have not arrived at the end, though, and possibly never will. The particles that now are thought to be elementary in some sense are many, and they may be grouped in various ways to form coherent patterns, but the grand pattern still eludes physicists. Indeed, the principal motivation for constructing accelerating machines of higher and higher energies is to produce still more particles, and to study the properties of those already identified. Thus, elementary-particle physics is also high-energy physics.

The study of elementary-particle physics is in part a study of the four fundamental forces among particles: the strong, or nuclear, interaction, the electromagnetic interaction, the weak interaction, and the still weaker gravitational interaction. It is also a study of conservation laws, not merely the well-known classical laws of mass-energy, momentum, angular momentum, and electric charge, but also of certain others, somewhat more esoteric. Finally, it is concerned with how the fundamental forces, the conservation laws, the intrinsic properties of the particles, and even the properties of space and time can be fitted together to make some sense.

Table 40-1. Some properties of some elementary particles

Particle	Rest mass,† m_e	Rest energy, MeV	Charge, units of electron charge	Spin, angular momentum, $\times \hbar$	Lifetime, s
Photon γ	0	0	0	1	∞
Electron e^-	1	0.511	-1	$\frac{1}{2}$	∞
Positron e^+	1	0.511	$+1$	$\frac{1}{2}$	∞
Proton p^+	1,836	938.259	$+1$	$\frac{1}{2}$	∞
Antiproton $\bar{p}$	1,836	938.259	-1	$\frac{1}{2}$	∞

†In units of electron mass m_e.

40-1 THE ELECTROMAGNETIC INTERACTION

We consider first those familiar particles that we have already considered to be in some sense elementary, namely the electron, the proton, and photon (the particle of electromagnetic radiation), and the two antiparticles, the positron (designated e^+) and the antiproton (designated p^-). Each has certain intrinsic properties, such as a definite electric charge, a definite rest mass (or rest energy), a definite intrinsic (or spin) angular momentum, and a mean lifetime before decay into other elementary particles. Since all five particles are found to be stable against spontaneous decay, each has an infinite lifetime. See Table 40-1.

As we saw in Chap. 36, we must, in treating the interaction between electromagnetic radiation and a charged particle, regard the radiation as consisting of particlelike photons. The several photon-electron interactions are illustrated in Fig. 36-15.

It is illuminating to represent these interactions on a space-time diagram, a diagram in which time as ordinate is plotted against position as abscissa. For simplicity we show the particle's spatial location in one dimension only; a two-dimensional plot of time vs. a single coordinate reveals all important aspects of the interactions.

The history of a particle is shown in a space-time diagram by a line, known as a *world line*. For constant velocity the line is straight. A vertical line represents a particle whose coordinate x does not change with time; it is a particle at rest. A line inclined with respect to the vertical represents a particle in motion, the angle between it and the vertical increasing with particle speed. Since a photon or any other particle with zero rest mass travels at the maximum possible speed c, the angle of its world line with respect to the vertical is the maximum.

Figure 40-1 shows in schematic fashion space-time diagrams

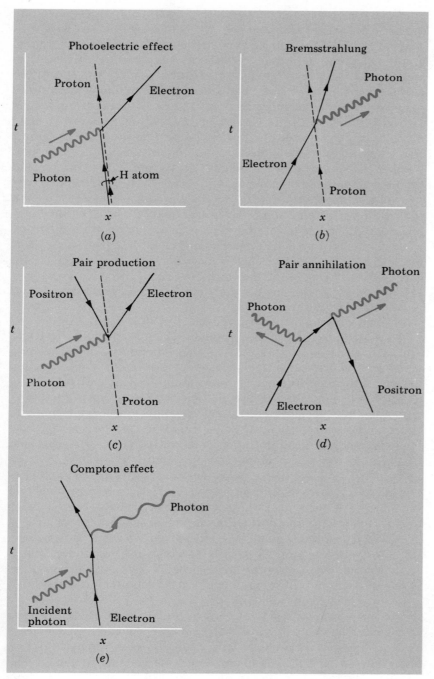

FIG. 40-1. Space-time graphs of the basic photon-electron interactions. An electron is represented by a solid line, a photon by a wiggly line, and a proton by a dashed line. A positron moving forward in time corresponds to an electron moving backward in time. *(a)* Photoelectric effect; *(b)* pair production; *(c)* Compton effect; *(d) bremsstrahlung;* *(e)* pair annihilation.

of the basic electron-photon interactions corresponding to those in Fig. 36-15. Time goes from past to future as the ordinate increases, and we read the events on the graph from bottom to top.

In the *photoelectric effect* (Fig. 40-1a) a hydrogen atom, consisting of an electron and a proton bound together, collides with a photon. After the interaction the photon has been annihilated, and the electron and proton move away as separate particles. The space-time event characterizing the interaction corresponds to the vertex in the figure, where the incoming photon and electron lines coalesce into a single outgoing electron line (the proton's motion is virtually unaffected). Inasmuch as the hydrogen atom consisted initially of an electron and proton bound together, the net result of the interaction is the absorption of one real photon. The total number of electrons (or of protons) going into and coming out of the interaction is unchanged.

In *bremsstrahlung* (Fig. 40-1b) an electron creates a photon in colliding with a proton. Again the interaction is the instantaneous event occurring at the vertex on the space-time diagram, where a photon line joins an electron line; the electron line is bent to indicate that the electron's momentum and energy change. Indeed, we may think of the incoming electron as being annihilated at the vertex while a second outgoing electron of different momentum and energy is simultaneously created.

In *pair production* (Fig. 40-1c) a photon is annihilated and an electron and positron are created. The positron, the electron's antiparticle, is here represented by an electron world line whose arrow is *reversed;* the antiparticle is regarded as an electron moving backward in time. Such a representation—an antiparticle moving forward in time equivalent to a particle moving backward in time— is justified by the considerations of electromagnetic quantum field theory. So is a representation in which the creation of a particle is equivalent to the annihilation of its antiparticle. We see, then, that the electron and photon lines representing the pair-production process are basically the same as those representing the photoelectric effect and *bremsstrahlung:* an inclined electron line joined to a photon line at the vertex. These processes, and still others, differ only in the orientation of the lines on the space-time graph.

In *pair annihilation* (Fig. 40-1d) an electron and positron unite to create a photon. Typically, two or more photons are created in pair annihilation in order to conserve momentum; we may think of the production of first one photon and then another as two distinct, but nearly coincident, events.

We have regarded the *Compton effect* to be that single process in which a photon interacts with a charged particle to produce a scattered photon that is deflected. Actually, as shown in Fig. 40-1e, the Compton effect takes place as two distinct interactions: The incident photon joins the incident electron to produce an intermediate electron; the intermediate electron then produces a photon and an electron. As before, each vertex is the point in space time at which an electron line is joined by a photon line.

Indeed, according to quantum field theory, the basic electromagnetic interaction may be regarded as that instantaneous event in which a charged particle or its antiparticle is created and annihilated, or both are created and annihilated together, and a photon is created or annihilated; see Fig. 40-2. When an electron world line changes direction, we may think of an electron with one energy and momentum as being annihilated and another electron with a different energy and momentum as being created. All photon-electron interactions involve one or more vertices on the same basic graph merely rotated in space-time. We may, of course, draw exactly similar graphs to represent the electromagnetic interactions between protons or antiprotons and photons, the solid world line then representing a proton or antiproton instead of an electron or positron. Still more generally, the same graphs may be used to represent the electromagnetic interaction in schematic fashion between any electrically charged particle and a photon.

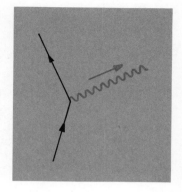

FIG. 40-2. Graph of the basic electromagnetic interaction.

All five particles listed in Table 40-1 are stable against spontaneous decay. In any interaction among them there are a number of fundamental physical properties which remain strictly conserved: linear momentum, mass-energy, angular momentum, and electric charge. Moreover, in every interaction the number of electrons minus the number of positrons is conserved. In graphical terms this means that the world line of an electron does not end: For every electron line into a vertex there is an electron line out of it (an "electron line" going backward in time signifying a positron going forward in time). Similarly, in any interaction the number of protons less the number of antiprotons is constant; graphically, the proton lines are continuous.† Although these additional conservation laws govern the electron-minus-positron and the proton-minus-antiproton numbers, there is no restriction on the number of photons.

We have described the interaction between electromagnetic radiation and a charged particle in terms of the basic space-time graph. What about the interaction between two electrically charged particles, an interaction which, in classical electromagnetic theory, is familiarly described in terms of the electric and magnetic fields produced by the two particles and in terms of an electric and a magnetic force? This, too, is attributable to the creation and annihilation of photons.

Consider Fig. 40-3, which shows a collision between two electrons. One electron creates a photon spontaneously at vertex A (in the fashion of the *bremsstrahlung* process of Fig. 40-1b), and the second electron absorbs the photon at vertex B (in the fashion of the photoelectric effect of Fig. 40-1a). Each of the two interacting electrons has its energy and momentum changed by virtue of the

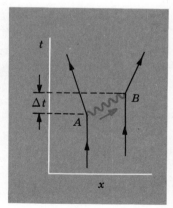

FIG. 40-3. Feynman diagram of the interaction between two electrons.

---

†The conservation laws for electrons and protons are illustrations, respectively, of the more general conservation laws of leptons and baryons. See Sec. 40-4.

exchange of a photon: Each charged particle has been acted upon by an electromagnetic force. The particle whose exchange is responsible for the force between the charged particles is called a *virtual* photon and is itself unobservable. Charged particles exchange energy and momentum by exchanging virtual photons. Graphical representations of interactions in space-time similar to Fig. 40-3 are called *Feynman diagrams,* after R. P. Feynman who introduced such diagrams to represent in simple fashion and also to compute in detail the electrodynamic interactions between quantum charges.

The coulomb force, and electromagnetic forces generally, can be ascribed to the continuous interchange of virtual photons between the charged particles. But virtual photons (which act as the intermediary of the electromagnetic force between charged particles) differ from real photons. Whereas real photons are observable, virtual photons are not.

To see how an unobservable particle is responsible for an interaction, we first recall that if a free electron in empty space were to emit a photon spontaneously, the system's total energy and momentum would not be conserved (that is why a massive particle is required in an actual *bremsstrahlung* collision); likewise, a single free electron cannot, without violating momentum and energy conservation, absorb a photon (that is why the photoelectric effect takes place only when the particle to be freed is initially bound). Although momentum and energy conservation hold in the *overall* electron-electron interaction extending from vertices A to B in Fig. 40-3, both cannot hold simultaneously at each of the separate vertices. It is the unobservability of the virtual photon that allows the nonconservation of energy and momentum during the time interval between photon emission and photon absorption. The violation of these two conservation laws is consistent with the quantum theory as long as the energy ΔE and momentum Δp "borrowed" at the space-time emission event are returned within time and space intervals consistent with Heisenberg's principle of uncertainty (Sec. 37-6), that is, within a time interval $\Delta t \approx \hbar/\Delta E$ and within a space interval $\Delta x \approx \hbar/\Delta p_x$. Thus, the system's total energy may exceed the initial energy of the two incoming electrons by ΔE during the time interval Δt where $\Delta E \approx \hbar/\Delta t$.

The time interval between the emission and the absorption of the virtual photon is Δt. Similarly, the uncertainties in momentum and position are related by $\Delta p_x \approx \hbar/\Delta x$. Virtual photons of all energies, from zero to infinity, may be created. Therefore, the time interval and associated space interval between the emission event and the absorption event can range from very short intervals (associated with interactions separated by small distances and with virtual photons having very high energies) to very long intervals (associated with large distances and photons of very low energies).

In an interaction between two charged particles a virtual photon is created spontaneously by one particle and then absorbed by the other. Can the virtual photon be absorbed by the same charged particle that created it? It *can,* so long as the limits imposed by the uncertainty principle are satisfied. Figure 40-4a shows a single electron (or another electrically charged particle) emitting and then reabsorbing a virtual photon. Moreover, a photon, whether real or virtual, may spontaneously create an electron-positron pair, as shown in Fig. 40-4c, even though its energy is less than the threshold energy for pair production. The process is again possible according to, and limited by, the uncertainty principle. The virtual pair may be annihilated and yield the original photon. Still more complicated processes may be constructed, as shown in Fig. 40-4c; the chain of creation-annihilation processes depicted is, of course, merely a collection of space-time graphs, whose basis is Fig. 40-2. Thus, every electrically charged particle, even if isolated from other particles, may be considered to emit and reabsorb photons, which can become particle-antiparticle pairs. Although virtual particles cannot be observed directly, the validity of the conception is emphatically proved by the success of theoretical field-theory calculations of subtle electromagnetic effects, based on these ideas. The success of the field theory has, in fact, caused it to be the model for understanding fundamental forces besides the electromagnetic interaction and has led to the prediction of particles whose existence was later confirmed in experiment.

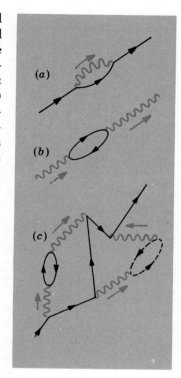

FIG. 40-4. Feynman diagrams of electromagnetic interactions: *(a)* An electron spontaneously creates and then reabsorbs a photon; *(b)* a photon spontaneously creates an electron-positron pair, and the pair is annihilated, and a photon is created; *(c)* a complex chain of annihilation-creation processes.

40-2 OTHER FUNDAMENTAL INTERACTIONS

The electromagnetic interaction between electrically charged particles may be ascribed to the exchange of virtual photons. When energy and momentum need not be "borrowed," as allowed by the uncertainty relation, real photons may, of course, be emitted or absorbed. To allow for the long range (inverse-square force) of the electromagnetic interaction, the associated field particles have *zero* rest mass; then one may have infinitesimally small virtual photon energies to allow for their exchange between charged particles separated by distances up to infinity.

Turning now to the nuclear force, or *strong interaction,* acting between nucleons (as well as between still other particles to be discussed later), we ask whether this force may also be described in terms of the exchange of field particles. As we have seen (Sec. 39-2), the strong interaction goes to zero at distances greater than about 1.4 fermi. What properties must the exchange particles then have? This question was first posed (and answered) by the Japanese physicist H. Yukawa in 1935.

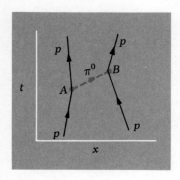

FIG. 40-5. A nucleon interacting with a second nucleon through the exchange of a pion as virtual field particle.

Figure 40-5 shows the interaction between two nucleons arising from the interchange of a virtual field particle here labeled π^0 (a neutral pion). Since the range of the strong interaction is short, field particles cannot be exchanged for vertices A and B spatially separated by more than about $R = 1.4$ fermi. Consequently, there is a *minimum* borrowed energy ΔE corresponding to a *minimum* rest energy $m_\pi c^2 = \Delta E$, where m_π is the rest mass of the exchange particle. Assuming for simplicity that the field particle travels at speed c during the time interval Δt for the exchange, we have

$$\Delta t = \frac{R}{c}$$

and, from the uncertainty principle, the uncertainty in energy, $\Delta E = m_\pi c^2$, is

$$\Delta E \approx \frac{\hbar}{\Delta t}$$

$$m_\pi c^2 = \frac{\hbar}{R/c}$$

or

$$m_\pi = \frac{\hbar}{R} \frac{1}{c} = \frac{10^{-34} \text{ J-s}}{(1.4 \times 10^{-15} \text{ m})(3 \times 10^8 \text{ m/s})} = 2 \times 10^{-28} \text{ kg} \approx 200 \, m_e$$

(Here m_e is the electron mass.) Therefore, on the basis of this argument we expect the field particles associated with the strong interaction to have a *finite* rest mass of the order of 200 electron masses to correspond to the short-range character of the nuclear force.

Mesons, unstable particles with masses intermediate between the electrons (and other light particles in the electron family) and nucleons, have been observed directly. The most familiar are the π mesons (or pions) which appear in three charge states π^+, π^-, and π^0 (in units of the electron charge). Other, more massive mesons are the kaon and eta particles. We shall treat the more detailed properties of these particles later. Suffice it to say here that these particles have the requisite properties to act as agents of the strong interaction, and they may be produced in collisions between high-energy (at least a few hundred MeV) nucleons, such as $p + n \rightarrow p + n + \pi^- + \pi^+$.

The four fundamental interactions which govern the forces between all elementary particles, together with their relative strengths and the field particles which are the agents of the respective forces, are shown in Table 40-2.

The strong interaction, illustrated by the force acting between all nucleons, is the strongest force found in nature; it is short-range and mediated by mesons as field particles. The weaker electromagnetic force acts between all electrically charged particles; it is long-range and mediated by photons, zero-rest-mass particles.

Table 40–2. The Fundamental Interactions

Interaction	Relative strength	Field particle
Strong	1	Meson
Electromagnetic	10^{-2}	Photon
Weak	10^{-13}	W particle (?)
Gravitational	10^{-14}	Graviton (?)

The still weaker force known simply as the *weak interaction* is illustrated by β decay processes, for example, the relatively slow decay of a neutron into a proton, electron, and antineutrino. Its range is no greater than 2×10^{-16} m, which is smaller by an order of magnitude than that of the strong interaction. The corresponding single particle, not yet observed, which acts as the field particle of the weak interaction, would have a rest mass at least three times that of a nucleon. This still undiscovered W (for weak) *particle* is also referred to as the *intermediate vector boson*.

The weakest force in nature now known is the gravitational force. It acts between all particles. Its feeble strength is illustrated, for example, by the fact that an electron and positron attract one another gravitationally by a force that is smaller by a factor of 10^{40} than the Coulomb electric force between them at the same separation distance. Gravitational waves have not yet been observed directly, and their quantization into field particles, a much more subtle effect, has also not yet been observed. But physicists are confident, by analogy with electromagnetism, that field particles, referred to as *gravitons*, must be associated with the gravitational interaction. Because the gravitational force is inverse-square, and therefore of infinite range, the rest mass of the graviton must be zero.

The relative strength of the interactions are indicated by the relative rapidity with which an unstable particle decays into other particles. The stronger the interaction, the more rapid the decay. For example, nuclei in excited states decaying into lower-energy states with the emission of a γ-ray photon typically have half-lives of the order of 10^{-14} s; this is an example of a process taking place through the electromagnetic interaction. On the other hand, unstable nuclei decaying through β-decay processes, that is, through the weak interaction, typically have much longer half-lives, of the order of 10^{-2} s.

40–3 PROPERTIES OF FUNDAMENTAL PARTICLES

Table 40-3 lists those particles which are thought to be fundamental and which have been observed, directly or indirectly, to exist as distinct particles for a time greater than approximately 10^{-21} s (which

Table 40-3. *The Fundamental Particles* ($\tau > 10^{-21}$ s)

Family name	Particle name	Symbol (and electric charge)	Rest energy, MeV (Rest mass × c^2)	Mean-life τ, s
Photon	Photon	γ	0	∞
Lepton	Neutrino	ν_e	0	∞
		ν_μ		
	Electron	e^-	0.51100	∞
	Muon	μ^-	105.660	2.198×10^{-6}
Meson	Pion	π^+	139.58	2.602×10^{-8}
		π^0	134.97	0.8×10^{-16}
	Kaon	K^+	493.8	1.237×10^{-8}
		$K^0 = \frac{1}{2}(K_S^0 + K_L^0)$	497.8	
		(K_S^0)		0.86×10^{-10}
		(K_L^0)		5.17×10^{-8}
	Eta	η^0	548.8	2.5×10^{-19}
Baryon	Proton	p	938.259	∞
	Neutron	n	939.259	0.93×10^3
	Lambda	Λ^0	1,115.6	2.5×10^{-10}
	Sigma	Σ^+	1,189.4	0.80×10^{-10}
		Σ^0	1,192.5	$<10^{-14}$
		Σ^-	1,197.4	1.49×10^{-10}
	Xi	Ξ^0	1,314.7	3.0×10^{-10}
		Ξ^-	1,321.3	1.66×10^{-10}
	Omega	Ω^-	1,672.5	1.3×10^{-10}

is merely 100 times greater than that for light to traverse a nuclear dimension). We shall discuss a few of the many still shorter-lived particles in later sections.

The particles are listed in order of increasing rest energy, or rest mass. They are also grouped according to family. The photon is in a category by itself. The *lepton* family includes the neu-

Spin angular momentum	Strangeness	Isospin	Principal decay modes (fraction if > 5%)
1	0	0 or 1	—
$\frac{1}{2}$	0	—	—
$\frac{1}{2}$	0	—	—
$\frac{1}{2}$	0	—	$\mu^- \to e + \bar{\nu}_e + \nu_\mu$
0	0	1	$\pi^+ \to \bar{\mu} + \nu_\mu$
0	0	1	$\pi^0 \to \gamma + \gamma$ (98.8)
0	+1	$\frac{1}{2}$	$K^+ \to \mu^+ + \bar{\nu}_\mu$ (63.8)
			$K^+ \to \pi^+ + \pi^0$ (20.9)
			$K^+ \to \pi^0 + \pi^+ + \pi^-$ (5.6)
0	+1	$\frac{1}{2}$	—
			$K_S^0 \to \pi^+ + \pi^-$ (68.7)
			$K_S^0 \to \pi^0 + \pi^0$ (31.3)
			$K_L^0 \to \pi^0 + \pi^0 + \pi^0$ (21.4)
			$K_L^0 \to \pi^+ + \pi^- + \pi^0$ (12.6)
			$K_L^0 \to \pi^\pm + \mu^\mp + \nu_\mu$ (26.8)
			$K_L^0 \to \pi^\pm + e^\mp + \nu_e$ (38.9)
0	0	0	$\eta_0 \to \gamma + \gamma$ (38.6)
			$\eta_0 \to \pi^0 + \pi^0 + \pi^0$ (30)
			$\eta_0 \to \pi^+ + \pi^- + \gamma$ (23)
$\frac{1}{2}$	0	$\frac{1}{2}$	—
$\frac{1}{2}$	0	$\frac{1}{2}$	$n \to p + e^- + \bar{\nu}_e$
$\frac{1}{2}$	−1	0	$\Lambda_0 \to p + \pi^-$ (64)
			$\Lambda_0 \to n + \pi^0$ (36)
$\frac{1}{2}$	−1	1	$\Sigma^+ \to p + \pi^0$ (52)
			$\Sigma^+ \to n + \pi^+$ (48)
$\frac{1}{2}$	−1	1	$\Sigma^0 \to \Lambda^0 + \gamma$
$\frac{1}{2}$	−1	1	$\Sigma^- \to n + \pi^-$
$\frac{1}{2}$	−2	$\frac{1}{2}$	$\Xi^0 \to \Lambda^0 + \pi^0$
$\frac{1}{2}$	−2	$\frac{1}{2}$	$\Xi^- \to \Lambda^0 + \pi^-$
$\frac{3}{2}$	−3	0	$\Omega^- \to \Xi^0 + \pi^-$ (?)
			$\Omega^- \to \Xi^- + \pi^0$ (?)
			$\Omega^- \to \Lambda^0 + K^-$ (?)

trinos, electron, and muon. The *meson* family includes pions, kaons, and the eta particle. The *baryon* family consists of the nucleons — proton and neutron — as well as more massive particles called *hyperons* — the lambda, sigma, xi, and omega particles. All baryons and mesons participate in the strong interaction and are referred to collectively as *hadrons*.

All of the more familiar fundamental particles have one of three possible electric-charge values: $+e$, 0, or $-e$, where e is the magnitude of the electron charge. Unless the charge of a particle is implicit in its symbol (the proton p has charge $+e$), the electric charge is indicated in Table 40-3 as a superscript after the symbol.

To every particle listed in the table there corresponds an antiparticle with identical properties save for the sign of its electric charge (and properties, such as magnetic moment, dependent on the sign of the charge). An antiparticle is symbolized by an overbar on the symbol ($\bar{p}$ for antiproton with charge $-e$) or by the opposite sign for electric charge (the antiparticle for the π^+ is the negative pion, π^-). Both the photon γ and the neutral pion π^0 are their own antiparticles.

Every elementary particle is characterized by a distinctive mean-life, which is infinite for such entirely stable particles as the photon, neutrinos, electron, and proton (and their antiparticle counterparts). Strictly, measurements show that the mean-life of the proton, if not infinite, is at least 10^{36} s.

Whereas the *half*-life for some species of unstable particles is the time elapsing until *half* of the initial number have decayed and the other half still survive (Sec. 39-5), the *mean-life* is the *average lifetime* from creation to decay of a large number of identical particles. Since some few particles of any given species may survive decay up to an infinite lifetime under the exponential decay law, the mean-life of a species must exceed the half-life. Specifically, the mean-life τ is related to the half-life $T_{1/2}$ by $\tau = 1.44\ T_{1/2}$ (see Prob. 39-9). The mean-life is indicative of the interaction by which a particle decays. For example, the mean-life for π^+ exceeds that for π^0 by a factor of 10^8; this corresponds to the fact that π^0 decays (primarily) through the electromagnetic interaction whereas π^+ decays much more slowly through the weak interaction.

Every elementary particle has a distinctive intrinsic, or *spin*, *angular momentum*, visualizable classically as the perpetual spinning of the particle about an internal rotation axis. The spin angular momentum is specified by the spin quantum number I, where the magnitude of the spin angular momentum is $\sqrt{I(I+1)}\,\hbar$. Values of I are either integral or half-integral. Such particles as the electron, neutrino, muon, proton, and neutron all have spin $\frac{1}{2}$ (see Secs. 38-4 and 39-1). Particles with half-integral spin values, known as *fermions*, obey the Pauli exclusion principle (Sec. 38-5), which restricts the number of particles in a given quantum state to one and thereby acts as a strong repulsive force between identical fermions. On the other hand, particles with integral spin values, known as *bosons*, are not constrained by the Pauli principle, and may therefore appear without restriction in their number. All field particles mediating the fundamental interactions have integral spin values and are bosons; examples are, the graviton (spin 2), the photon (spin 1), the mesons (spin 0), and the W particle (spin 1).

The meaning of the terms "strangeness" and "isospin" given in Table 40-3 will be treated briefly in Sec. 40-5.

The final column in the table lists the *principal* decay modes. Of course, if a particle is stable and has an infinite meanlife, it does not decay. Still other decay modes, beyond those appearing in Table 40-3, and with great theoretical importance, have been excluded from the listing if the fraction of the decays is less than 5 percent. For every entry in the table there exists a complementary decay in which particles and antiparticles have been interchanged. For example, since a neutron decays into proton, electron, and antineutrino:

$$n \to p + e^- + \bar{\nu}_e$$

then an antineutron decays into an antiproton, positron, and neutrino:

$$\bar{n} \to \bar{p} + e^+ + \nu_e$$

Furthermore, as discussed in Sec. 39-6 for β decay, one may transform one possible process into a second one by reversing the arrow and/or shifting a particle from one side to the other while at the same time changing particle to antiparticle, or conversely. Thus, the decay relation above can be transformed into the allowed reaction:

$$\bar{p} + \nu_e \to \bar{n} + e^-$$

in which an antiproton capturing a sufficiently energetic neutrino yields an antineutron and an electron.

40-4 THE UNIVERSALLY VALID CONSERVATION LAWS

Why do elementary particles appear in only a severely limited number of charge states? Why do they have the specific masses, meanlives, spins, and other basic properties given in Table 40-3, and not other values? What patterns of coherence can be discerned in their properties and interactions? Although many of these and still other fundamental questions remain unanswered, at least one kind of question can now be answered quite straightforwardly: Why do unstable fundamental particles decay in the particular modes shown in Table 40-3 and not other conceivable decay modes? Or, if a particle is stable, why does it not decay?

The answer lies in the conservation laws. First, one must take the attitude that *any* conceivable process may occur unless some basic physical principle precludes it, recognizing, however, that not all conceivable processes will occur with the same probability. Or, conversely, if a conceivable process is never observed, then there must exist some fundamental physical law which prohibits it.

As a simple example consider why a neutron does *not* decay into a proton and photon. Such a process would not violate the conservation laws of mass-energy, linear momentum, and angular momentum, but it would of course be inconsistent with electric-charge conservation. A neutron does not decay into a proton and photon because the net electric charge of an isolated system has (thus far) always been found to remain unchanged. Then how about the possible decay of a proton into a positron and photon? This is consistent with all of the conservation laws thus far named, but it has never been observed. One must conclude that this process (and, as we shall see, other related processes) is ruled out because of the operation of one or more additional conservation laws — laws which are, in fact, formulated simply to summarize the nonoccurence of certain otherwise conceivable processes.

Two additional conservation laws, together with the now classically well-established conservation laws of mass-energy, linear momentum, angular momentum, and electric charge, are now believed to be universally valid and thereby applicable to *all* interactions; these are the conservation laws of *baryon number* and of *lepton number* (the latter being subdivided into separate conservation laws of *electron number* and *muon number*).

A special form of the baryon-number conservation law is the conservation law for nucleons (Sec. 39-6), according to which the total number of nucleons (protons and neutrons) in any nuclear decay or reaction is constant. In the more general baryon-number conservation law one assigns a baryon number as follows: to every particle in the baryon family, +1; to every antiparticle in the baryon family, −1; and to every other particle, 0. Then the conservation law is simply the statement that the *total baryon number* for any process is a *constant*.

For example, energetic (several GeV) protons striking a proton target may create lambda and kaon particles according to:

$$p + p \to p + \pi^+ + \Lambda^0 + K^0$$

Baryon number: $\quad 1 + 1 = 1 + 0 \; + 1 \; + 0$

Another example is that in which a lambda and antilambda particle are created and then subsequently decay when an energetic antiproton interacts with a proton (see Fig. 40-6):

$$\bar{p} + p \to \Lambda^0 + \bar{\Lambda}^0$$

Baryon number: $\quad 1 + (-1) = 1 \; + (-1)$

and then $\quad \Lambda^0 \to p + \pi^- \quad$ and $\quad \bar{\Lambda}^0 \to \bar{p} + \pi^+$

Baryon number: $\quad 1 = 1 + 0 \qquad\qquad -1 = -1 + 0$

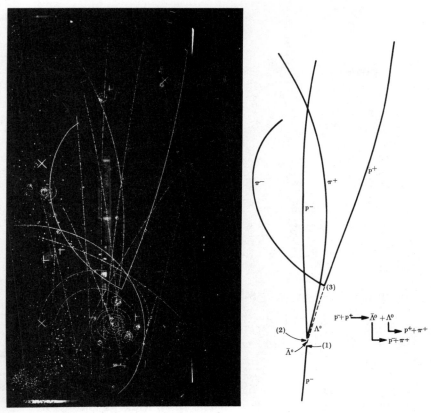

FIG. 40-6. A bubble-chamber photograph of a 3.3-GeV antiproton entering the bubble chamber from the bottom and colliding with a proton at point 1 to create a hyperon (Λ^0). The Λ^0 particle decays into a proton and a π^- pion at point 3; the Λ^0 antiparticle into an antiproton and π^+ pion at point 2. (Courtesy Brookhaven National Laboratory.)

Still other illustrations of the baryon-number conservation are found in the decay schemes given in Table 40-3.

The lepton-number conservation law applies to members of the lepton family, the electron, muon, and *two* varieties of neutrino ν_e and ν_μ, together with their respective antiparticles. (Examples will show why it is, in fact, necessary to recognize, all told, four distinct types of neutrinos: $\nu_e, \bar{\nu}_e, \nu_\mu,$ and $\bar{\nu}_\mu$.) *Electron number* +1 is assigned to the electron e^- and to the electron-type neutrino ν_e, while the positron e^+ and electron-type antineutrino $\bar{\nu}_e$ have electron number -1. All other particles have electron number 0. Similarly, *muon number* +1 is assigned to the muon μ^- and to the muon-type neutrino ν_μ, while the antimuon μ^+ and muon-type antineutrino $\bar{\nu}_\mu$ have muon number -1. All other particles have muon number 0. With these assignments all observations are consistent with the *lepton conservation law: the total electron number is constant,* and the *total muon number is constant.*

For example, in the decay of the three varieties of pion† we have

---

†The most probable decay for the neutral pion is into two photons; decay into an electron and positron is observed in about 1 percent of all decays.

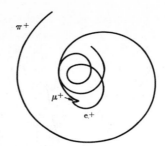
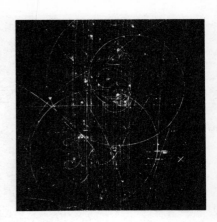

FIG. 40-7. A liquid-hydrogen bubble-chamber photograph of a π^+ decaying into a μ^+, which decays into an e^+. (Courtesy Brookhaven National Laboratory.)

	$\pi^+ \to \mu^+ + \nu_\mu$	$\pi^- \to \mu^- + \bar{\nu}_\mu$	$\pi^0 \to e^- + e^+$
Electron no.:	$0 = 0 + 0$	$0 = 0 + 0$	$0 = 1 + (-1)$
Muon no.:	$0 = (-1) + 1$	$0 = 1 + (-1)$	$0 = 0 + 0$

The muons then decay as follows (see Fig. 40-7):

	$\mu^- \to e^- + \bar{\nu}_e + \nu_\mu$	$\mu^+ \to e^+ + \nu_e + \bar{\nu}_\mu$
Electron no.:	$0 = 1 + (-1) + 0$	$0 = (-1) + 1 + 0$
Muon no.:	$1 = 0 + 0 + 1$	$-1 = 0 + 0 + (-1)$

The decay of the neutron illustrates the conservation of both baryon and electron number, as does the closely related process in which a proton captures an (electron) antineutrino:

	$n \to p + e^- + \bar{\nu}_e$	$\bar{\nu}_e + p \to n + e^+$
Baryon no.:	$1 = 1 + 0 + 0$	$0 + 1 = 1 + 0$
Electron no:	$0 = 0 + 1 + (-1)$	$(-1) + 0 = 0 + (-1)$

When an antineutrino of the muon variety is captured by a proton, the observed process is (see Fig. 40-8)

	$\bar{\nu}_\mu + p \to n + \mu^+$
Baryon no.:	$0 + 1 = 1 + 0$
Muon no.:	$(-1) + 0 = 0 + (-1)$

FIG. 40-8. Two tracks in a spark-chamber photograph used in experiments to establish the existence of the neutrino associated with the muon. (Courtesy Brookhaven National Laboratory.)

Note that when a proton captures an antineutrino of the *muon* variety (one produced in the decay of a pion) an anti*muon* is produced, whereas when a proton captures an antineutrino of the *electron* variety (one produced in β decay), an anti*electron*, or positron, is produced. This observation was crucially important in establishing that the electron varieties of neutrino and antineutrino are quite distinct from the muon neutrino and antineutrino types. In the next section we shall indicate what, in addition to the muon and electron number, distinguishes a neutrino from an antineutrino.

The muon plays a curious role among the fundamental particles. It appears to be identical in all properties to an electron, save for its greater mass (and muon and electron numbers).

In a sense the baryon and lepton conservation laws "explain" why the proton, electron, and neutrino are absolutely stable: There exist no lighter particles into which they can decay without violating these conservation laws. By the same token, there are no conservation laws applying to photon and meson number; such particles may be created in unlimited number, subject only to limitations imposed by other conservation laws.

40-5 ISOSPIN, STRANGENESS, AND PARITY

These three terms are associated with conservation laws which, curiously, apply rigorously to the strong interaction but not universally to other fundamental interactions.

Isospin. Isospin (or isotopic spin) is analogous to the ordinary orbital or spin angular momentum of quantum mechanics. As indicated in Sec. 37-4, for a particle in a state with the orbital angular-momentum quantum number $l = 1$, there are three possible values of the magnetic quantum number m_l: $m_l = +1, 0,$ or -1. The angular-momentum vector may be thought to have three different orientations, or components, in space, corresponding to the three possible values for m_l. In general, for a given l, there are $2l + 1$ possible values for m_l.

We note from Table 40-3 that the mesons and baryons (the strongly interacting particles, or hadrons) are clustered into groups of particles with very nearly equal masses, for example, the two nucleons (p and n), the three pions, the one lambda, the three sigmas. If one attributes the small difference in mass within each group (or multiplet) to the difference in electric charge (and the relatively weaker electromagnetic interaction associated with it), then the particles within each group may be regarded as *different states* of the *same particle*. An isospin quantum number T is assigned to each group such that the multiplicity of particles within it is equal to $2T + 1$. Then for each of the possible orientations, or

Table 40-4

Particle (hadron)	Isospin T	Multiplicity $2T+1$	Particle states	Isospin component T_z
π	1	3	$\begin{cases}\pi^+ \\ \pi^0 \\ \pi^-\end{cases}$	$+1$ 0 -1
K	$\frac{1}{2}$	2	$\begin{cases}K^+ \\ K^0\end{cases}$	$+\frac{1}{2}$ $-\frac{1}{2}$
η	0	1	η^0	0
Nucleon	$\frac{1}{2}$	2	$\begin{cases}p \\ n\end{cases}$	$+\frac{1}{2}$ $-\frac{1}{2}$
Λ	0	1	Λ^0	0
Σ	1	3	$\begin{cases}\Sigma^+ \\ \Sigma^0 \\ \Sigma^-\end{cases}$	$+1$ 0 -1
Ξ	$\frac{1}{2}$	2	$\begin{cases}\Xi^0 \\ \Xi^-\end{cases}$	$\frac{1}{2}$ $-\frac{1}{2}$
Ω	0	1	Ω^-	0

components, T_z of the hypothetical isospin vector **T** in isospace there corresponds a particle. For example, for the two nucleons (p and n), $T = \frac{1}{2}$; then $2T + 1 = 2(\frac{1}{2}) + 1 = 2$, and the proton is identified with $T_z = +\frac{1}{2}$ and the neutron with $T_z = -\frac{1}{2}$. On this basis isospin quantum numbers are assigned as shown in Table 40-3 and again in Table 40-4. Isospin numbers are not assigned to lighter particles, which do not participate in the strong interaction.

The isospin conservation law, which applies *only* to the strong interaction, is this: In magnitude the total isospin out of a reaction is the same as the total isospin into the reaction. In short, *total isospin is conserved.* This simplifies, for example, consideration of the interaction between a π^- and neutron as compared to that between a π^+ and proton:

$\pi^- + n$: $\quad T_z(\pi^-) + T_z(n) = (-1) + (-\frac{1}{2}) = -\frac{3}{2}$

$\pi^+ + p$: $\quad T_z(\pi^+) + T_z(p) = (+1) + (+\frac{1}{2}) = +\frac{3}{2}$

Both systems have the *same magnitude* of total isospin, and their interactions are found to be the same.

Strangeness. Another quantum number, closely related to isospin, which is strictly conserved in all reactions and decays that proceed through the strong interaction, is the *strangeness quantum number.* The assignment of the strangeness quantum number to the several baryon particles and to mesons, and to their respective antiparticles, is indicated in Table 40-5.

The familiar (the *non*-strange) proton-neutron with a "charge center" of $+\frac{1}{2}e$ (midway between $+e$ and 0) is assigned strangeness

Table 40-5

Isospin	Baryon electric charge			Antibaryon electric charge			Meson electric charge			
	$-e$	0	$+e$	$-e$	0	$+e$	$+e$	0		$-e$
$\frac{1}{2}$		n	p	$\bar{p}$	$\bar{n}$					
0		Λ^0			$\bar{\Lambda}^0$					
1	Σ^-	Σ^0	Σ^+	$\bar{\Sigma}^-$	$\bar{\Sigma}^0$	$\bar{\Sigma}^+$	π^+	π^0		π^-
$\frac{1}{2}$	Ξ^-	Ξ^0			$\bar{\Xi}^0$	$\bar{\Xi}^+$	K^+	K^0	$\bar{K}^0$	K^-
0	Ω^-					$\bar{\Omega}^+$		$\eta^0 \overline{\eta^0}$		
Strangeness:	-3 -2 -1 0			0 $+1$ $+2$ $+3$			$+1$	0		-1

0. The neutral lambda Λ^0 lying at charge 0, is assigned strangeness -1, and so on for the other groups of baryons.

The conservation law of strangeness is, then, that the total strangeness into any reaction or decay *governed by the strong interaction* is equal to the total strangeness out of the reaction or decay. For example, when a sufficiently energetic (several GeV) proton collides with a proton in a target, the following reaction can take place:

$$p + p \rightarrow \Lambda^0 + K^0 + p + \pi^+$$

Strangeness: $0 + 0 = (-1) + 1 + 0 + 0$

Two strange particles, the lambda and kaon, with *opposite* strangeness are created. The requirement that at least *two* strange particles appear in a reaction proceeding by the strong interaction is referred to as *associated production*.

Now consider the decay of a lambda particle into a proton and pion:

$$\Lambda^0 \rightarrow p + \pi^-$$

Strangeness: $(-1) \neq 0 + 0$

This process does *not* conserve strangeness, and this nonconservation corresponds to the observation that the decay takes place relatively slowly, as compared to those proceeding by the strong interaction. Indeed, strangeness need not be conserved for processes such as the Λ^0 decay taking place *via* the weak interaction; the strangeness may change by one unit, or none, in such processes.

The several symmetries exhibited through isospin and strangeness have prompted analysis of strong-interacting particles through that branch of mathematics known as group theory. Indeed, the properties of the remarkable particle Ω^- with strangeness -3 were predicted on this basis before the particle was observed. There has, however, thus far not been confirmation of another prediction: That strongly interacting particles are comprised of still more fundamental particles called *quarks*, with electric charges of $\frac{1}{3}e$ and $\frac{2}{3}e$.

Parity. The conservation of parity law in its most primitive form implies that, at least at the microscopic level, nature shows no preference for left or right. Or, if a certain process takes place in nature, then that process which one would see through reflection in a mirror is also a possible process.

Although parity is strictly conserved in the strong and electromagnetic interactions, a variety of experimental evidence shows nonconservation of parity in the weak interaction. The most direct demonstration is in the decay of a pion into antimuon and neutrino. Consider the specific decay

$$\pi^+ \to \bar{\mu} + \nu_\mu$$

We consider the π^+ to be at rest before decay, so that the $\bar{\mu}$ and ν_μ must leave the decay site in opposite directions with equal momentum magnitude to accord with the law of linear-momentum conservation. The π^+ has spin zero, whereas the $\bar{\mu}$ and ν_μ each have spin $\frac{1}{2}$; therefore, the spin senses (or, equivalently, the directions of the angular-momentum vectors) of the outgoing particles must be opposite. There are a priori *two* possible orientations of linear- and angular-momentum vectors, as shown in Fig. 40-9: linear- and angular-momentum vectors aligned (both $\bar{\mu}$ and ν_μ as right-handed particles), or linear- and angular-momentum vectors antialigned (and both $\bar{\mu}$ and ν_μ as left-handed particles). Note that the decay into right-handed particles is merely the mirror image of the decay into left-handed particles; part *(b)* of Fig. 40-9 can be regarded as the image in a vertical mirror of part *(a)*.

Now if parity were strictly conserved in the weak interaction and in the decay of the π^+, one would expect to find from some initial large number of pions equal numbers of right- and left-handed particles. Since neutrinos undergo virtually no absorption through any ordinary absorber, the handedness of the muons must be observed and it is inferred that the handedness of neutrinos is the same. The results of experiment are that *left-handed* particles alone are emitted by decaying pions, never right-handed particles. Pion decay occurs in the left-handed version only, never the right-handed, a violation of the right-left symmetry which is a basic implication of parity conservation.

A related consequence is that *neutrinos* are exclusively *left-handed* whereas *antineutrinos* are exclusively *right-handed*, for both the muon and electron types. Although only left-handed antimuons are emitted in the π^+ decay, it is easy to see that any particle with a *finite* rest mass, such as a muon, cannot have just one intrinsic handedness. Suppose that an antimuon travels up-

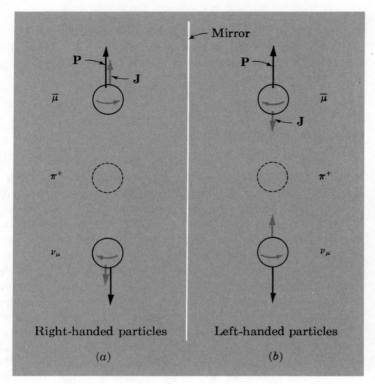

FIG. 40-9. Two a priori possibilities for the decay of π^+ into an antimuon $\bar{\mu}$ and a neutrino ν_μ. (a) Linear momentum P and angular momentum J vectors aligned, corresponding to both particles being *right-handed*; (b) P and J antialigned corresponding to both particles being *left-handed*. Note that (b) is the mirror image of (a), relative to the vertical line dividing them. Experimentally, only decay according to (b), with both particles exclusively left-handed, is observed.

ward with its spin angular momentum also upward, thereby constituting a right-handed particle, as in Fig. 40-9a. If we view the particle from a reference frame traveling upward along the direction of antimuon motion but at a higher speed than the antimuon, then in this reference frame the antimuon moves downward; the particle's linear-momentum vector has been reversed but not its angular-momentum vector, and in the second reference frame it is seen as a left-handed particle. Since antimuons may be both right- and left-handed, the appearance of left-handed particles only in the π^+ decay implies that the neutrino must always be left-handed. This is, moreover, in accord with the fact that, since the neutrino has zero rest mass and always travels at speed c, there is *no* reference frame for which the neutrino's velocity vector can be reversed by observation at a higher speed than that of the particle.

The nonconservation of parity P in the weak interaction has prompted reexamination of two other closely related conservation laws: *Charge conjugation C*, which indicates that interactions and processes are unchanged when every particle is replaced by its antiparticle, and conversely; and *time invariance T* according to which any process can proceed either forward or backward in time without change (or, in more specific terms, if a motion picture portrays a possible process, then that motion picture one sees unfold as the same film is run backward is also a possible process). Suffice it to say here that charge conjugation, time invariance, and parity all appear to be violated in the weak interaction, although the combination *CPT* appears to be preserved as an exact and general conservation principle.

40-6 RESONANCE PARTICLES

In addition to those elementary particles shown in Table 40-3, at least 100 additional fundamental particles with mean-lives shorter than 10^{-21} s,† and known as *resonance particles,* have been observed. Such particles are so very short-lived that they cannot leave a measurably long track in a bubble chamber and their existence and properties must be inferred by less direct means.

To understand one procedure that may be used to identify an extremely short-lived particle, consider first the analogous situation in which electrons of controllable kinetic energy bombard hydrogen atoms initially in the ground state. We know that, since the atom's energy is quantized, the hydrogen atom cannot absorb and convert into internal excitation energy any amount less than 10.2 eV, the excitation energy of the first excited state of hydrogen (Sec. 38-3). Therefore, with electrons of less than 10.2 eV striking hydrogen atoms, the collisions are perfectly elastic until the electron energy reaches 10.2 eV, at which point an electron can effectively be brought to rest in collision while the hydrogen atom it has struck makes a quantum transition to the first excited state, from which it can subsequently decay through the emission of a photon. The phenomenon was first observed in 1914 by J. Franck and G. Hertz in the so-called *Franck-Hertz experiment.*

The excitation to a quantized state by particle bombardment shows a resonance (Sec. 14-5) behavior: The probability‡ for the excitation process is a maximum when the electron kinetic energy equals the excitation energy and falls off for lesser or greater electron energies. Note that when a hydrogen atom exists in, say, the first excited state, it can properly be regarded as being a particle distinct from hydrogen in the ground state. After all, its mass is greater by exactly 10.2 eV/c^2, and the atom's angular momentum may also differ from that of the ground-state atom.

Resonance particles may be produced in analogous fashion

---

†The η^0 particle, with a mean-life of 2.5×10^{-19} s and shown in Table 40-3, is a resonance particle. Unlike most resonance particles, it decays through the electromagnetic, rather than the strong, interaction, and its mean-life is consequently relatively long.

‡The probability for the occurrence of a reaction process is most frequently expressed in terms of the reaction *cross section*, the effective area presented by the target particle to the incident particle (see Prob. 39-27). The resonance peak is in the cross section measured as a function of the energy of the incident particle.

by bombarding nucleons with electrons or pions, but with kinetic energies measured in GeV. As in the atomic case, the existence of the particle is inferred from the energy of the bombarding particle at the resonance peak. For example, 1.2-GeV electrons upon protons produce the resonance particle symbolized by $\Delta(1236)$

$$e^- + p \to \Delta(1236)$$

where the number in parentheses gives the particle's rest energy in MeV. This, the lowest-mass resonance particle in the baryon group, then quickly decays into a proton and electron;

$$\Delta(1236) \to e^- + p$$

The mean-life of a resonance particle can be computed from the width of the resonance peak. If a particle exists for a finite time Δt, then from the uncertainty principle, its rest energy is uncertain by $\Delta E = \hbar/\Delta t$, where ΔE is also the resonance width. For the $\Delta(1236)$, $\Delta E \approx 100$ MeV, and Δt is computed to be 7×10^{-24} s. Other baryon resonance particles are $\Delta(1950)$, $N^*(1520)$, $N^*(1670)$, $N^*(1688)$; for these and still other particles such fundamental properties as the spin angular momentum, isospin, and strangeness, as well as the rest energy, meanlife, and decay schemes, have been determined. Among the meson resonance particles, also decaying chiefly through the strong interaction, are $\rho(765)$, $\omega(784)$, $\phi(1019)$.

PROBLEMS

40-1. Draw graphs on a space-time diagram for the following processes: *(a)* An electron and positron attract one another and finally are annihilated, and photons are produced; *(b)* an antiproton and proton attract one another and finally are annihilated, and pions are produced.

40-2. Figure P40-2 shows four space-time graphs. *(a)* Describe the sequence of events (reading from past to future) for each graph. *(b)* Now imagine each graph to be rotated 90° relative to the space-time axes (the photon lines adjusted to the correct angle with respect to the vertical), and again describe the sequence of events for each graph.

FIG. P40-2

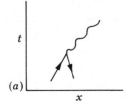

FIG. P40-2, cont.

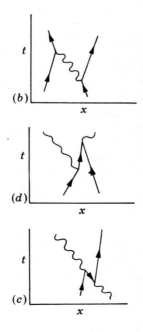

40-3. Consider a proton at rest. If the proton transforms spontaneously into a neutron and π^+, (a) what is the change ΔE in the energy (in MeV) of the system, and (b) over what time interval Δt can this energy be "borrowed"? (c) Show this on a space-time diagram.

40-4. Any unstable particle living for a finite time has an uncertainty in its total energy and, therefore, an uncertainty in its mass. What is the uncertainty in the mass of a particle having a mean life τ?

40-5. The potential energy between two nucleons may be approximated by the relation $V = (a/r)e^{-r/r_0}$, where r is the internucleonic distance and a and r_0 are constants. This is the so-called Yukawa potential. What is the approximate mass of the field particle which mediates the force between nucleons? Give this mass in terms of the parameters which describe the internuclear potential.

40-6. Show that the radius of curvature, in centimeters, of a particle with charge z (in multiples of the electronic charge e) and with a momentum P (in MeV/c), when traveling at right angles to a magnetic field B (in gauss), is given by $r = (3.33 \times 10^3)P/zB$.

40-7. What are the radii of curvature in a magnetic field of 2.0 Wb/m² of (a) 10-keV electrons, (b) 10-GeV electrons, (c) 10-keV protons, and (d) 10-GeV protons?

40-8. Recalling the arguments originally leading to the assumption of the existence of the neutrino in β decay, how could it be verified experimentally that (a) a charged pion decays into a muon and *one* neutrino and (b) a muon decays into an electron and *two* neutrinos?

40-9. A particular 0.50-GeV π^+ exists for a time equal to its mean-life in the reference frame in which the pion is at rest. What length of track would the 0.50-GeV π^+ leave in a bubble chamber, assuming no energy loss along the track?

40-10. What is the radius of curvature in a magnetic field of 2.0 Wb/m² of (a) a 10-GeV proton, (b) a 10-MeV charged pion, and (c) a 1.0-MeV electron, assuming in each instance that the particle moves at right angles to the magnetic field lines.

40-11. If a neutral pion were at rest, what would be energies of the photons into which it decays?

40-12. Consider a neutron at rest. If the neutron transforms spontaneously into a proton and a kaon-minus, (a) what is the change ΔE in energy (in MeV) of the system, and (b) over what time interval Δt can this energy be borrowed? (c) Show this on a space-time diagram.

40-13. One method of measuring the mass difference between the π^- and the π^0 is to observe the capture of low-energy π^- mesons by protons. One possible reaction is $\pi^- + p \to n + \pi^0$, where both incident particles are initially at rest. The neutron leaves the reaction with a kinetic energy 0.60 MeV. (a) Using the known masses of the proton and neutron (Table 40-3) and approximating the pion mass as one-seventh of the nucleon mass, determine the mass difference between the π^- and π^0, and compare it with the data of Table 40-3.

40-14. A 30-MeV π^0 quickly decays into two photons. If the photons move off in directions parallel and antiparallel to the direction of motion of the pion, what will be the energy of each photon, as observed in the laboratory frame of reference?

40-15. A free, negatively charged pion, initially at rest, can decay directly into an electron and neutrino, or it can decay to a muon and neutrino and the muon can then decay to an electron and two neutrinos (see Table 40-3). (a) Calculate the kinetic energy of the electron in the direct decay of the π^-. (b) What is the maximum kinetic energy an electron can have in the two-step decay of the π^-?

40-16. For a K^+ initially at rest, what is the spread in energy of the μ^+ mesons in the decay $K^+ \to \mu^+ + \nu$?

40-17. What experimental evidence shows that the Σ^- is not the antiparticle of the Σ^+, that is, that the Σ^- and $\overline{\Sigma^-}$ are distinct elementary particles?

40-18. If you were given a beam of Σ^0 particles and a second beam of $\overline{\Sigma^0}$ of particles, how could you determine which beam contained the Σ^0's?

40-19. Show that the wavelength of any particle whose kinetic energy is large compared with its rest energy is given by $\lambda = 1.24$ GeV-fermi/E, where E is the particle's total energy in GeV.

40-20. (a) Show that Planck's constant divided by 2π can be expressed as 6.58×10^{-22} MeV-s. (b) What is the approximate linewidth of an excited nuclear state with a lifetime equal to the nuclear time?

40-21. Given a beam of K^0 particles and a beam of $\overline{K^0}$ particles, how could you identify the K^0 beam?

40-22. A beam of K^- mesons strikes a stationary proton target. What is the threshold energy for the production of Ω^- particles?

40-23. Which of the following reactions *cannot* occur? Give the reason(s) for each case.
(a) $5 \text{ GeV} + \Omega^- \to \Sigma^- + \Lambda^0 + \pi^0$
(b) $\pi^0 \to \mu^+ + \mu^- + e^+ + \bar{\nu}_e$
(c) $\Lambda^0 \to p + e^- + \bar{\nu}_e + n$

(d) $\Lambda^0 + \overline{\Lambda^0} \to p^+ + p^- + 3\pi^+ + 3\pi^-$
(e) $\Omega^- \to \Sigma^- + \pi^+ + \pi^- + e^+ + e^-$

40-24. *(a)* Show that, if a particle has electric charge q (in units of the elementary charge e), baryon number B, isotopic spin component T, and strangeness number S, the following relation applies: $S = 2(q - B/2 - T)$.

(b) What is the isotopic spin component T of the Σ^-, Σ^0, and Σ^+?

40-25. Consider an Ω^- particle decaying in a vacuum. *(a)* Write down one possible set of decay equations by which the particle might decay to stable particles. *(b)* Identify the kind and number of stable particles.

APPENDIXES

APPENDIX I

Fundamental Constants

The values listed below are taken from the *Handbook of Chemistry and Physics* (54th ed., 1973–74), The Chemical Rubber Co., Cleveland. The last significant figure may in some instances be indefinite.

Name of quantity	Common symbol	Value
Acceleration due to gravity (standard value)	g	$9.80665 \text{ m/s}^2 = 32.174 \text{ ft/s}^2$
Universal gravitational constant	G	$6.673 \times 10^{-11} \text{ N-m}^2/\text{kg}^2$
Earth's mass	m_E	5.9763×10^{24} kg
Earth's mean radius	r_E	6.370949×10^6 m = 3959 mi
Earth-moon mean distance	r_{EM}	$3.84400 \times 10^8 \text{ m} \approx 60$ earth radii
Earth-sun mean distance (1 astronomical unit)	1 AU	$1.4957 \times 10^{11} \text{ m} = 92.94 \times 10^6$ mi
Sun's mass	m_S	1.991×10^{30} kg
Density of water (at 3.98°C)		1.00000 gm/ml
Standard atmospheric pressure	1 atm	$1.01325 \times 10^5 \text{ N/m}^2 = 76 \text{ cmHg} = 760$ torr
Universal gas constant	R	8.3143 J/g mol–K = 0.0821 liter-atm/g mol–K = 1.986 g cal/g mol–K
Volume of 1 gram mole ideal gas at STP		$2.2414 \times 10^{-2} \text{ m}^3$
Absolute zero of temperature		−273.15°C
Triple point of water		273.16 K
Avogadro's number	N_A	6.02217×10^{23} g-mol^{-1}

Name of quantity	Common symbol	Value
Boltzmann's constant	k	1.38062×10^{-23} J/K
"Mechanical equivalent of heat"	J	4.18400 J/cal
Stefan-Boltzmann constant	σ	5.6696×10^{-8} J/m²-s-K⁴
Coulomb force constant	$k = 1/4\pi\varepsilon_0$	8.987554×10^9 N-m²/C²
Permittivity of free space	ε_0	8.854185×10^{-12} C²/N-m²
Permeability of free space	μ_0	$4\pi \times 10^{-7}$ Wb/A-m
Speed of light (in vacuum)	c	2.99792456×10^8 m/s
Electron charge	e	1.602192×10^{-19} C
Electron volt	1 eV	1.602192×10^{-19} J
Mass-energy conversion factor (atomic mass unit = u)	1 u = $(\frac{1}{12})^{12}_{6}C$	931.481 MeV/c^2 1.6604×10^{-27} kg
Electron mass	m_e	9.10956×10^{-31} kg 0.000548593 u 0.511004 MeV/c^2
Proton mass	m_p	1.67261×10^{-27} kg 1.00727661 u 938.259 MeV/c^2 = 1836.11 m_e
Neutron mass	m_n	1.67492×10^{-27} kg 1.0086652 u 939.553 MeV/c^2
Planck's constant	h	6.626196×10^{-34} J-s
	$\hbar = h/2\pi$	1.054591×10^{-34} J-s

APPENDIX II

Conversion Factors

Length

	m	cm	km
1 meter	1	10^2	10^{-3}
1 centimeter	10^{-2}	1	10^{-5}
1 kilometer	10^3	10^5	1
1 inch	2.540×10^{-2}	2.540	2.540×10^{-5}
1 foot	0.3048	30.48	3.048×10^{-4}
1 mile	1609	1.609×10^5	1.609

	in	ft	mi
1 meter	39.37	3.281	6.214×10^{-4}
1 centimeter	0.3937	3.281×10^{-2}	6.214×10^{-6}
1 kilometer	3.937×10^4	3.281×10^3	0.6214
1 inch	1	8.333×10^{-2}	1.578×10^{-5}
1 foot	12	1	1.894×10^{-4}
1 mile	6.336×10^4	5,280	1

1 foot = $\frac{1200}{3937}$ m
1 light-year = 9.460528×10^{12} km
1 parsec = 3.08572×10^{13} km
1 astronomical unit (mean earth-sun distance) = 1 AU = 1.4957×10^{11} m
1 micron = 1 μm = 10^{-6} m
1 millimicron = 1 mμ = 10^{-9} m
1 angstrom = 1 Å = 10^{-10} m
1 X-unit = 1 XU = 10^{-13} m

Plane Angle
1 rev = 2π rad = $360°$

Mass

	kg	g	slug	u
1 kilogram†	1	10^3	6.852×10^{-2}	6.024×10^{26}
1 gram	10^{-3}	1	6.852×10^{-5}	6.024×10^{23}
1 slug (lb/g)	14.59	1.459×10^4	1	8.789×10^{27}
1 atomic mass unit	1.660×10^{-27}	1.660×10^{-24}	1.137×10^{-28}	1

†1 lb = weight of 0.45359237 kg

Time

	s	min	h
1 second	1	1.667×10^{-2}	2.778×10^{-4}
1 minute	60	1	1.667×10^{-2}
1 hour	3600	60	1
1 day	8.640×10^4	1440	24
1 year (a)	3.156×10^7	5.259×10^5	8.766×10^3

	day	yr
1 second	1.157×10^{-5}	3.169×10^{-8}
1 minute	6.994×10^{-4}	1.901×10^{-6}
1 hour	4.167×10^{-2}	1.141×10^{-4}
1 day	1	2.738×10^{-3}
1 year (a)	365.2	1

Speed

	m/s	cm/s	ft/s	mi/h
1 meter/second	1	10^2	3.281	2.237
1 centimeter/second	10^{-2}	1	3.281×10^{-2}	2.237×10^{-2}
1 foot/second	0.3048	30.48	1	0.6818
1 mile/hour	0.4470	44.70	1.467	1

1 mi/min = 60 mi/h = 88 ft/s

Force

	N	dyn	lb
1 newton	1	10^5	0.2248
1 dyne	10^{-5}	1	2.248×10^{-6}
1 pound	4.448	4.448×10^5	1

Work, Energy, Heat

	J	erg	ft-lb
1 joule	1	10^7	0.7376
1 erg	10^{-7}	1	7.376×10^{-8}
1 ft-lb	1.356	1.356×10^7	1
1 eV	1.602×10^{-19}	1.602×10^{-12}	1.182×10^{-19}
1 g cal	4.186	4.186×10^7	3.087
1 Btu	1.055×10^3	1.055×10^{10}	7.779×10^2
1 kWh	3.600×10^6	3.600×10^{13}	2.655×10^6

	eV	cal	Btu	kWh
1 joule	6.242×10^{18}	0.2389	9.481×10^{-4}	2.778×10^{-7}
1 erg	6.242×10^{11}	2.389×10^{-8}	9.481×10^{-11}	2.778×10^{-14}
1 ft-lb	8.464×10^{18}	0.3239	1.285×10^{-3}	3.766×10^{-7}
1 eV	1	3.827×10^{-20}	1.519×10^{-22}	4.450×10^{-26}
1 g cal	2.613×10^{19}	1	3.968×10^{-3}	1.163×10^{-6}
1 Btu	6.585×10^{21}	2.520×10^2	1	2.930×10^{-4}
1 kWh	2.247×10^{25}	8.601×10^5	3.413×10^2	1

Pressure

	N/m²	dyn/cm²	atm
1 newton/meter²	1	10	9.869×10^{-6}
1 dyne/centimeter²	10^{-1}	1	9.869×10^{-7}
1 atmosphere	1.013×10^5	1.013×10^6	1
1 centimeter mercury†	1.333×10^3	1.333×10^4	1.316×10^{-2}
1 pound/inch²	6.895×10^3	6.895×10^4	6.805×10^{-2}
1 pound/foot²	47.88	4.788×10^2	4.725×10^{-4}

	cmHg	lb/in²	lb/ft²
1 newton/meter²	7.501×10^{-4}	1.450×10^{-4}	2.089×10^{-2}
1 dyne/centimeter²	7.501×10^{-5}	1.450×10^{-5}	2.089×10^{-3}
1 atmosphere	76	14.70	2.116×10^3
1 centimeter mercury†	1	0.1943	27.85
1 pound/inch²	5.171	1	144
1 pound/foot²	3.591×10^{-2}	6.944×10^{-3}	1

†At 0°C and at a location where the acceleration due to gravity has its "standard" value, 9.80665 m/s².

Power
1 hp = 550 ft-lb/s = 0.7457 kW

Conversion Factors

APPENDIX III

Electric and Magnetic Units and Conversion Factors

In addition to the SI system of electromagnetic units (which is also known as the mksa or Giorgi system), the gaussian system of electric and magnetic units is commonly used in scientific work. In the gaussian system all mechanical units are those of the cgs absolute system, for example, length in cm, force in dyn, work in erg. The gaussian system is subdivided into the electrostatic system of units (esu) for strictly electric effects and the electromagnetic system of units (emu) for strictly magnetic effects.

The basic relation defining the unit of electric charge in the esu system is Coulomb's law, written in the form:

$$F = \frac{q_1 q_2}{r^2} \qquad \text{(III-1)}$$

where q_1 and q_2 are in esu (or statcoulomb), r is in cm, and F is in dyn. Thus, by definition, each of two equal point-charges has a charge of 1 esu if they exert an electric force of 1 dyn on one another when separated by a distance of 1 cm. To express electric relations of the SI system in a form appropriate for esu units, one replaces the constant $k = 1/4\pi\varepsilon_0$ by 1. Derived units involve the prefix "stat" followed typically by the SI unit name. Thus, current is expressed in statampere, where 1 statampere = 1 statcoulomb/s; potential difference in statvolt, where 1 statvolt = 1 erg/statcoulomb, etc. Sometimes, however, electric units in the esu system are given simply as esu; thus, 1 esu of charge per second is said to constitute an electric current of 1 esu.

The unit for charge in the emu system is called the abcoulomb, and it is related to the unit for electric current, the abampere, by 1 abampere = 1 abcoulomb/s, where, by definition, the force per unit length between two infinite parallel straight conductors separated by a distance d is given by

$$\frac{F}{L} = \frac{2I^2}{d} \qquad \text{(III-2)}$$

with I in abampere, (F/L) in dyn/cm, and d in cm. Thus, by definition, the equal currents in two parallel straight conductors separated by 1 cm are each 1 abampere when the magnetic force between them is exactly 2 dyn/cm. Other derived emu units typically take the prefix "ab" followed by the SI unit name; for example, 1 abvolt = 1 erg/abcoulomb. Again, however, the units of quantities in the emu system may be given simply as emu; thus, 1 erg per emu charge corresponds to the potential difference which may be expressed merely as 1 emu, as well as 1 abvolt.

The magnitudes and units of electric charge in the esu and emu systems differ. In fact, their numerical ratio is approximately

$$1 \text{ abcoulomb}/1 \text{ statcoulomb} = 3 \times 10^{10}$$

We recognize this numerical factor to be the speed of light in vacuum expressed in cgs units, $c = 3 \times 10^{10}$ cm/s. Relations involving both electric and magnetic quantities, such as Maxwell's equations, therefore may have the speed of light c appearing explicitly in them.

Because Eqs. (III-1) and (III-2), which define the charge units for the esu and emu systems, differ from the corresponding equations in the SI system [Eqs. (20-7) and (27-10), respectively], other equations in electromagnetism may also differ in form. Listed below are some basic relations in the forms appropriate for the SI and gaussian systems. Following this is given a table of conversion factors. Wherever the factor 3 (or its square, 9) appears in conversion factors this number is to be taken as the numerical factor 2.9979256 appearing in the speed of light when expressed in metric units, $c = 2.99792456 \times 10^8$ m/s.

Relation	SI form	Gaussian form
Coulomb's law (20-7)	$F = \dfrac{1}{4\pi\varepsilon_0} \dfrac{q_1 q_2}{r^2}$	$F = \dfrac{q_1 q_2}{r^2}$
Biot-Savart relation (27-4)	$d\mathbf{B} = \dfrac{\mu_0}{4\pi} \dfrac{i\, d\mathbf{l} \times \mathbf{r}}{r^3}$	$d\mathbf{B} = \dfrac{i\, d\mathbf{l} \times \mathbf{r}}{r^3}$
Lorentz force (26-8)	$\mathbf{F} = q(\mathbf{E} + \mathbf{v} \times \mathbf{B})$	$\mathbf{F} = q\left(\mathbf{E} + \dfrac{1}{c}\mathbf{v} \times \mathbf{B}\right)$
Gauss' law (22-3)	$\oint \mathbf{E} \cdot d\mathbf{S} = \dfrac{q}{\varepsilon_0}$	$\oint \mathbf{E} \cdot d\mathbf{S} = 4\pi q$
Faraday's law (28-10)	$\oint \mathbf{E} \cdot d\mathbf{l} = -\dfrac{d\phi_B}{dt}$	$\oint \mathbf{E} \cdot d\mathbf{l} = -\dfrac{1}{c}\dfrac{d\phi_B}{dt}$
Ampère's law (30-6)	$\oint \mathbf{B} \cdot d\mathbf{l} = \mu_0 i + \varepsilon_0 \mu_0 \dfrac{d\phi_E}{dt}$	$\oint \mathbf{H} \cdot d\mathbf{l} = \dfrac{4\pi i}{c} + \dfrac{1}{c}\dfrac{d\phi_E}{dt}$
The field vectors	$\mathbf{D} = \varepsilon_0 \mathbf{E} + \mathbf{P}$	$\mathbf{D} = \mathbf{E} + 4\pi\mathbf{P}$
	$\mathbf{B} = \mu_0(\mathbf{H} + \mathbf{M})$	$\mathbf{B} = \mathbf{H} + 4\pi\mathbf{M}$

Quantity	SI unit	ESU unit	EMU unit
Electric charge	1 coulomb =	3×10^9 statcoulomb =	$\frac{1}{10}$ abcoulomb
Electric field (E)	1 volt/m =	$\frac{1}{3} \times 10^4$ statvolt/cm =	10^6 abvolt/cm
Electric potential	1 volt =	$\frac{1}{300}$ statvolt =	10^8 abvolt
Capacitance	1 farad =	9×10^{11} statfarad =	10^{-9} abfarad
Current	1 ampere =	3×10^9 statampere =	$\frac{1}{10}$ abampere
Magnetic flux	1 weber =	$\frac{1}{300}$ erg/statampere =	10^8 maxwell
Magnetic induction (B)	1 weber/m^2 =	$\frac{1}{3} \times 10^{-6}$ dyn/statampere-cm =	10^4 gauss
Magnetic field intensity (H)	1 ampere-turn/m =	$3(4\pi) \times 10^7$ esu =	$4\pi \times 10^{-3}$ oersted
Resistance	1 ohm =	$\frac{4\pi}{9} \times 10^{-12}$ statohm =	10^9 abohm

APPENDIX IV

The Atomic Masses

Given here are the masses of the neutral atoms of all stable nuclides and a few of the unstable nuclides (designated by an asterisk following the mass number A). Masses are given in unified atomic mass units (u) where M is exactly 12 u for $^{12}_{6}C$, by definition.

These data are derived from J. H. E. Mattauch, W. Thiele, and A. H. Wapstra, *Nuclear Physics, 67,* 1 (1965). The uncertainties are less than 0.000001 u for many nuclides of low mass number and as large as 0.001500 u for some nuclides of high mass number.

Element	A	Atomic mass u	Element	A	Atomic mass u
$_0n$	1*	1.008 665	$_6C$	10*	10.016 810
$_1H$	1	1.007 825		11*	11.011 432
	2	2.014 102		12	12.000 000
	3*	3.016 050		13	13.003 354
				14*	14.003 242
$_2He$	3	3.016 030		15*	15.010 599
	4	4.002 603			
	6*	6.018 893	$_7N$	12*	12.018 641
				13*	13.005 738
$_3Li$	6	6.015 125		14	14.003 074
	7	7.016 004		15	15.000 108
	8*	8.022 487		16*	16.006 103
				17*	17.008 450
$_4Be$	7*	7.016 929			
	9	9.012 186	$_8O$	14*	14.008 597
	10*	10.013 534		15*	15.003 070
$_5B$	8*	8.024 609		16	15.994 915
	10	10.012 939		17	16.999 133
	11	11.009 305		18	17.999 160
	12*	12.014 354		19*	19.003 578

775

Element	A	Atomic mass u	Element	A	Atomic mass u
$_9$F	17*	17.002 095		47	46.951 769
	18*	18.000 937		48	47.947 951
	19	18.998 405		49	48.947 871
	20*	19.999 987		50	49.944 786
	21*	20.999 951	$_{23}$V	48*	47.952 259
$_{10}$Ne	18*	18.005 711		50*	49.947 164
	19*	19.001 881		51	50.943 962
	20	19.992 440	$_{24}$Cr	48*	47.953 760
	21	20.993 849		50	49.946 055
	22	21.991 385		52	51.940 514
	23*	22.994 473		53	52.940 653
$_{11}$Na	22*	21.994 437		54	53.938 882
	23	22.989 771	$_{25}$Mn	54*	53.940 362
$_{12}$Mg	23*	22.994 125		55	54.938 051
	24	23.990 962	$_{26}$Fe	54	53.939 617
	25	24.989 955		56	55.934 937
	26	25.991 740		57	56.935 398
$_{13}$Al	27	26.981 539		58	57.933 282
$_{14}$Si	28	27.976 930	$_{27}$Co	59	58.933 190
	29	28.976 496		60*	59.933 814
	30	29.973 763	$_{28}$Ni	58	57.935 342
$_{15}$P	31	30.973 765		60	59.930 787
$_{16}$S	32	31.972 074		61	60.931 056
	33	32.971 462		62	61.928 342
	34	33.967 865		64	63.927 958
	36	35.967 090	$_{29}$Cu	63	62.929 592
$_{17}$Cl	35	34.968 851		65	64.927 786
	36*	35.968 309	$_{30}$Zn	64	63.929 145
	37	36.965 898		66	65.926 052
$_{18}$Ar	36	35.967 544		67	66.927 145
	38	37.962 728		68	67.924 857
	40	39.962 384		70	69.925 334
$_{19}$K	39	38.963 710	$_{31}$Ga	69	68.925 574
	40*	39.964 000		71	70.924 706
	41	40.961 832	$_{32}$Ge	70	69.924 252
$_{20}$Ca	40	39.962 589		72	71.922 082
	41*	40.962 275		73	72.923 463
	42	41.958 625		74	73.921 181
	43	42.958 780		76	75.921 406
	44	43.955 488	$_{33}$As	75	74.921 597
	46	45.953 689	$_{34}$Se	74	73.922 476
	48	47.952 531		76	75.919 207
$_{21}$Sc	41*	40.969 247		77	76.919 911
	45	44.955 919		78	77.917 314
$_{22}$Ti	46	45.952 632		80	79.916 528

Element	A	Atomic mass u	Element	A	Atomic mass u
	82	81.916 707	$_{48}$Cd	106	105.906 463
$_{35}$Br	79	78.918 330		108	107.904 187
	81	80.916 292		110	109.903 012
$_{36}$Kr	78	77.920 403		111	110.904 189
	80	79.916 380		112	111.902 763
	82	81.913 482		113	112.904 409
	83	82.914 132		114	113.903 361
	84	83.911 504		116	115.904 762
	86	85.910 616	$_{49}$In	113	112.904 089
$_{37}$Rb	85	84.911 800		115*	114.903 871
	87*	86.909 187	$_{50}$Sn	112	111.904 835
$_{38}$Sr	84	83.913 431		114	113.902 773
	86	85.909 285		115	114.903 346
	87	86.908 893		116	115.901 745
	88	87.905 641		117	116.902 959
$_{39}$Y	89	88.905 872		118	117.901 606
$_{40}$Zr	90	89.904 700		119	118.903 314
	91	90.905 642		120	119.902 199
	92	91.905 031		122	121.903 442
	94	93.906 314		124	123.905 272
	96	95.908 286	$_{51}$Sb	121	120.903 817
$_{41}$Nb	93	92.906 382		123	122.904 213
$_{42}$Mo	92	91.906 811	$_{52}$Te	120	119.904 023
	94	93.905 091		122	121.903 066
	95	94.905 839		123	122.904 277
	96	95.904 674		124	123.902 842
	97	96.906 022		125	124.904 418
	98	97.905 409		126	125.903 322
	100	99.907 475		128	127.904 476
$_{44}$Ru	96	95.907 598		130	129.906 238
	98	97.905 289	$_{53}$I	127	126.904 470
	99	98.905 936	$_{54}$Xe	124	123.906 120
	100	99.904 218		126	125.904 288
	101	100.905 577		128	127.903 540
	102	101.904 348		129	128.904 784
	104	103.905 430		130	129.903 509
$_{45}$Rh	103	102.905 511		131	130.905 086
$_{46}$Pd	102	101.905 609		132	131.904 161
	104	103.904 011		134	133.905 398
	105	104.905 064		136	135.907 221
	106	105.903 479	$_{55}$Cs	133	132.905 355
	108	107.903 891	$_{56}$Ba	130	129.906 245
	110	109.905 164		132	131.905 120
$_{47}$Ag	107	106.905 094		134	133.904 612
	109	108.904 756		135	134.905 550
				136	135.904 300

The Atomic Masses

Element	A	Atomic mass u	Element	A	Atomic mass u
	137	136.905 500		167	166.932 060
	138	137.905 000		168	167.932 383
$_{57}$La	138*	137.906 910		170	169.935 560
	139	138.906 140	$_{69}$Tm	169	168.934 245
$_{58}$Ce	136	135.907 100	$_{70}$Yb	168	167.934 160
	138	137.905 830		170	169.935 020
	140	139.905 392		171	170.936 430
	142	141.909 140		172	171.936 060
$_{59}$Pr	141	140.907 596		173	172.938 060
				174	173.938 740
$_{60}$Nd	142	141.907 663		176	175.942 680
	143	142.909 779	$_{71}$Lu	175	174.940 640
	144*	143.910 039		176*	175.942 660
	145	144.912 538	$_{72}$Hf	174	173.940 360
	146	145.913 086		176	175.941 570
	148	147.916 869		177	176.943 400
	150	149.920 915		178	177.943 880
$_{62}$Sm	144	143.911 989		179	178.946 030
	147*	146.914 867		180	179.946 820
	148	147.914 791			
	149	148.917 180	$_{73}$Ta	181	180.948 007
	150	149.917 276	$_{74}$W	180	179.947 000
	152	151.919 756		182	181.948 301
	154	153.922 282		183	182.950 324
$_{63}$Eu	151	150.919 838		184	183.951 025
	153	152.921 242		186	185.954 440
$_{64}$Gd	152	151.919 794	$_{75}$Re	185	184.953 059
	154	153.920 929		187*	186.955 833
	155	154.922 664	$_{76}$Os	184	183.952 750
	156	155.922 175		186	185.953 870
	157	156.924 025		187	186.955 832
	158	157.924 178		188	187.956 081
	160	159.927 115		189	188.958 300
$_{65}$Tb	159	158.925 351		190	189.958 630
$_{66}$Dy	156	155.923 930		192	191.961 450
	158	157.924 449	$_{77}$Ir	191	190.960 640
	160	159.925 202		193	192.963 012
	161	160.926 945	$_{78}$Pt	190*	189.959 950
	162	161.926 803		192	191.961 150
	163	162.928 755		194	193.962 725
	164	163.929 200		195	194.964 813
$_{67}$Ho	165	164.930 421		196	195.964 967
$_{68}$Er	162	161.928 740		198	197.967 895
	164	163.929 287	$_{79}$Au	197	196.966 541
	166	165.930 307	$_{80}$Hg	196	195.965 820

Appendix IV

Element	A	Atomic mass u	Element	A	Atomic mass u
	198	197.966 756		206	205.974 468
	199	198.968 279		207	206.975 903
	200	199.968 327		208	207.976 650
	201	200.970 308	$_{83}$Bi	209	208.981 082
	202	201.970 642			
	204	203.973 495	$_{90}$Th	232*	232.038 124
$_{81}$Tl	203	202.972 353	$_{92}$U	234*	234.040 904
	205	204.974 442		235*	235.043 915
$_{82}$Pb	204	203.973 044		238*	238.050 770

APPENDIX V

Mathematical Relations

Series Expansions

$$(a+b)^n = a^n + \frac{n}{1!}a^{n-1}b + \frac{n(n-1)}{2!}a^{n-2}b^2 + \cdots$$

$$\lim_{n\to\infty}\left(1+\frac{x}{n}\right)^n \equiv e^x = 1 + x + \frac{x^2}{2!} + \frac{x^3}{3!} + \cdots$$

$$\sin x = x - \frac{x^3}{3!} + \frac{x^5}{5!} - \cdots$$

$$\cos x = 1 - \frac{x^2}{2!} + \frac{x^4}{4!} - \cdots$$

$$\tan x = x + \frac{x^3}{3} + \frac{2x^5}{15} + \cdots \qquad \left(\text{for } -\frac{\pi}{2} < x < \frac{\pi}{2}\right)$$

Trigonometric Identities

$\sin^2 a + \cos^2 a = 1$
$\quad\sin(a \pm b) = \sin a \cos b \pm \cos a \sin b$
$\quad\cos(a \pm b) = \cos a \cos b \mp \sin a \sin b$

Numerical Constants

$\pi = 3.14159$
$e = 2.71828$
$\sqrt{2} = 1.414$
$\sqrt{3} = 1.732$

APPENDIX VI

Natural Trigonometric Functions

Angle		Sine	Cosine	Tangent
Degrees	Radians			
0°	0.000	0.000	1.000	0.000
1°	0.018	0.017	1.000	0.018
2°	0.035	0.035	0.999	0.035
3°	0.052	0.052	0.999	0.052
4°	0.070	0.070	0.998	0.070
5°	**0.087**	**0.087**	**0.996**	**0.087**
6°	0.105	0.105	0.995	0.105
7°	0.122	0.122	0.993	0.123
8°	0.140	0.139	0.990	0.141
9°	0.157	0.156	0.988	0.158
10°	**0.175**	**0.174**	**0.985**	**0.176**
11°	0.192	0.191	0.982	0.194
12°	0.209	0.208	0.978	0.213
13°	0.227	0.225	0.974	0.231
14°	0.244	0.242	0.970	0.249
15°	**0.262**	**0.259**	**0.966**	**0.268**
16°	0.279	0.276	0.961	0.287
17°	0.297	0.292	0.956	0.306
18°	0.314	0.309	0.951	0.325
19°	0.332	0.326	0.946	0.344

Angle		Sine	Cosine	Tangent
Degrees	Radians			
20°	**0.349**	**0.342**	**0.940**	**0.364**
21°	0.367	0.358	0.934	0.384
22°	0.384	0.375	0.927	0.404
23°	0.401	0.391	0.921	0.425
24°	0.419	0.407	0.914	0.445
25°	**0.436**	**0.423**	**0.906**	**0.466**
26°	0.454	0.438	0.899	0.488
27°	0.471	0.454	0.891	0.510
28°	0.489	0.469	0.883	0.532
29°	0.506	0.485	0.875	0.554
30°	**0.524**	**0.500**	**0.866**	**0.577**
31°	0.541	0.515	0.857	0.601
32°	0.559	0.530	0.848	0.625
33°	0.576	0.545	0.839	0.649
34°	0.593	0.559	0.829	0.675
35°	**0.611**	**0.574**	**0.819**	**0.700**
36°	0.628	0.588	0.809	0.727
37°	0.646	0.602	0.799	0.754
38°	0.663	0.616	0.788	0.781
39°	0.681	0.629	0.777	0.810

1 rad = 57.3°
For small θ (in radians), $\sin \theta \approx \theta$ and $\tan \theta \approx \theta$

Angle				
Degrees	Radians	Sine	Cosine	Tangent
40°	**0.698**	**0.643**	**0.766**	**0.839**
41°	0.716	0.656	0.755	0.869
42°	0.733	0.669	0.743	0.900
43°	0.751	0.682	0.731	0.933
44°	0.768	0.695	0.719	0.966
45°	**0.785**	**0.707**	**0.707**	**1.000**
46°	0.803	0.719	0.695	1.036
47°	0.820	0.731	0.682	1.072
48°	0.838	0.743	0.669	1.111
49°	0.855	0.755	0.656	1.150
50°	**0.873**	**0.766**	**0.643**	**1.192**
51°	0.890	0.777	0.629	1.235
52°	0.908	0.788	0.616	1.280
53°	0.925	0.799	0.602	1.327
54°	0.942	0.809	0.588	1.376
55°	**0.960**	**0.819**	**0.574**	**1.428**
56°	0.977	0.829	0.559	1.483
57°	0.995	0.839	0.545	1.540
58°	1.012	0.848	0.530	1.600
59°	1.030	0.857	0.515	1.664
60°	**1.047**	**0.866**	**0.500**	**1.732**
61°	1.065	0.875	0.485	1.804
62°	1.082	0.883	0.469	1.881
63°	1.100	0.891	0.454	1.963
64°	1.117	0.899	0.438	2.050

Angle				
Degrees	Radians	Sine	Cosine	Tangent
65°	**1.134**	**0.906**	**0.423**	**2.145**
66°	1.152	0.914	0.407	2.246
67°	1.169	0.921	0.391	2.356
68°	1.187	0.927	0.375	2.475
69°	1.204	0.934	0.358	2.605
70°	**1.222**	**0.940**	**0.342**	**2.748**
71°	1.239	0.946	0.326	2.904
72°	1.257	0.951	0.309	3.078
73°	1.274	0.956	0.292	3.271
74°	1.292	0.961	0.276	3.487
75°	**1.309**	**0.966**	**0.259**	**3.732**
76°	1.326	0.970	0.242	4.011
77°	1.344	0.974	0.225	4.331
78°	1.361	0.978	0.208	4.705
79°	1.379	0.982	0.191	5.145
80°	**1.396**	**0.985**	**0.174**	**5.671**
81°	1.414	0.988	0.156	6.314
82°	1.431	0.990	0.139	7.115
83°	1.449	0.993	0.122	8.144
84°	1.466	0.995	0.105	9.514
85°	**1.484**	**0.996**	**0.087**	**11.43**
86°	1.501	0.998	0.070	14.30
87°	1.518	0.999	0.052	19.08
88°	1.536	0.999	0.035	28.64
89°	1.553	1.000	0.017	57.29
90°	**1.571**	**1.000**	**0.000**	∞

APPENDIX VII

The Greek Alphabet

Alpha	α	A
Beta	β	B
Gamma	γ	Γ
Delta	δ	Δ
Epsilon	ε	E
Zeta	ζ	Z
Eta	η	H
Theta	θ, ϑ	Θ
Iota	ι	I
Kappa	κ	K
Lambda	λ	Λ
Mu	μ	M
Nu	ν	N
Xi	ξ	Ξ
Omicron	o	O
Pi	π	Π
Rho	ρ	P
Sigma	σ, ς	Σ
Tau	τ	T
Upsilon	υ	Υ
Phi	ϕ, φ	Φ
Chi	χ	X
Psi	ψ	Ψ
Omega	ω	Ω

APPENDIX VIII

Moment-of-Inertia Calculations

For a continuous distribution of mass in a solid, the moment of inertia may be written

$$I = \int r^2 \, dm = \int \rho r^2 \, dv \qquad \text{(VIII-1)}$$

where ρ, the density of the solid, is related to the mass element dm and the volume element dv by $dm = \rho \, dv$. If ρ is constant throughout the solid, it can be taken outside of the integral sign in (VIII-1).

The three examples which follow illustrate the usefulness of Eqs. (VIII-1) and (11-14). Formulas for the moment of inertia of several simple shapes are given in Fig. VIII-1.

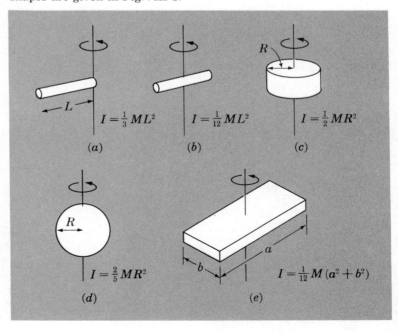

Fig. VIII-1. Moments of inertia for several simple shapes, all of constant density: *(a)* a thin rod about an axis through one end and perpendicular to the rod axis, *(b)* a thin rod about an axis through its center and perpendicular to the rod axis, *(c)* a right circular cylinder through the symmetry axis, *(d)* a sphere through a diametrical axis, and *(e)* a rectangular plate with sides *a* and *b* about an axis perpendicular to the plate and passing through its center.

Example VIII-1

Masses of 3.00 and 1.00 kg are attached to a thin massless rod at $x = 1.00$ m and $x = 1.80$ m, respectively. See Fig. VIII-2. *(a)* What is the moment of inertia of the two masses relative to the Y axis? *(b)* At what distance from the axis would all of the mass have to be placed in order to yield the same moment of inertia? *(c)* What is the location of the center of mass relative to this axis? *(d)* What is the moment of inertia relative to an axis passing through the center of mass and perpendicular to the rod?

(a) From Eq. (12-11),

$$I = \Sigma m_i r_i^2 = (3.00 \text{ kg})(1.00 \text{ m})^2 + (1.00 \text{ kg})(1.80 \text{ m})^2 = 6.24 \text{ kg-m}^2$$

(b) If all of the mass were at a distance k, called the *radius of gyration*, from the rotation axis, the moment of inertia would be $I = Mk^2$. Therefore,

$$k = \sqrt{\frac{I}{M}} = \sqrt{\frac{6.24 \text{ kg-m}^2}{4.00 \text{ kg}}} = 1.25 \text{ m}$$

(c) The location of the center of mass is given by

$$\overline{X} = \frac{\Sigma m_i x_i}{\Sigma m_i} = \frac{(3.00 \text{ kg})(1.00 \text{ m}) + (1.00 \text{ kg})(1.80 \text{ m})}{4.00 \text{ kg}} = 1.20 \text{ m} \quad (6\text{-}4)$$

Note that the center of mass and the radius of gyration are *not* at the same location (see Fig. VIII-2*b*); that is, the point at which all of the mass may be imagined to be concentrated for translational motion is not the same as the point at which all of the mass may be imagined to be concentrated for rotational motion. Furthermore, the center of mass is fixed relative to the rigid body, whereas the radius of gyration depends on the choice of axis.

(d) The moment of inertia relative to the center of mass could be computed by choosing a new origin at the center of mass and then applying Eq. (12-11), but a simpler procedure is to use the parallel-axis theorem, Eq. (12-12),

$$I_{CM} = I - Md^2 = 6.24 \text{ kg-m}^2 - (4.00 \text{ kg})(1.20 \text{ m})^2 = 0.48 \text{ kg-m}^2$$

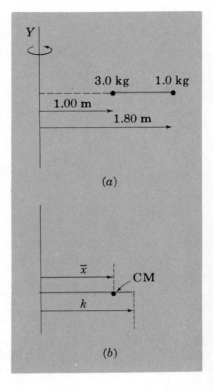

FIG. VIII-2. *(a)* A massless rod with particles on both ends. *(b)* The coordinates of the center of mass $\bar{x}$ and the radius of gyration k.

Example VIII-2

What is the moment of inertia of a thin uniform rod of length L and mass M with respect to an axis of rotation that is perpendicular to the rod and passes through *(a)* the center of the rod and *(b)* one end of the rod?

(a) We consider the contribution to the moment of inertia of a thin slice of thickness dx at a distance x from the axis of rotation. See Fig. VIII-3. For a rod of uniform density, the mass of the thin slice is $dm = M(dx/L)$. Then, with $r = x$, the moment of inertia is written

$$I_{CM} = \int r^2 \, dm = \frac{M}{L} \int_{-L/2}^{+L/2} x^2 \, dx = \frac{M}{L} \frac{L^3}{12} = \tfrac{1}{12} ML^2$$

where the integration has been extended from left to right end of the rod, that is, from $x = -L/2$ to $x = +L/2$.

Moment-of-Inertia Calculations

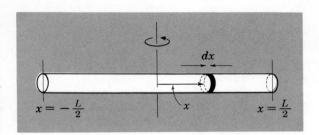

FIG. VIII-3. Computation of the moment of inertia of a thin rod of length L about a perpendicular axis through its center. The thin slice of thickness dx is located at x.

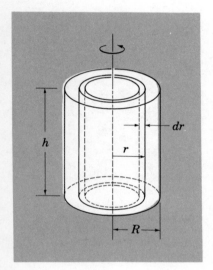

FIG. VIII-4. Computation of the moment of inertia of a right circular cylinder of radius R, and height h, about an axis of rotation through the cylinder's symmetry axis. A thin cylindrical shell of radius r and thickness dr is shown.

(b) We can find the moment of inertia about one end of the rod simply by choosing different limits for the definite integral given above, the rod now extending from $x = 0$ to $x = L$:

$$I_{\text{end}} = \frac{M}{L} \int_0^L x^2 \, dx = \tfrac{1}{3}ML^2$$

This result is consistent with the parallel-axis theorem:

$$I_{\text{end}} = I_{\text{CM}} + Md^2 \qquad (12\text{-}12)$$

$$I_{\text{end}} = \tfrac{1}{12}ML^2 + M\left(\frac{L}{2}\right)^2 = \tfrac{1}{3}ML^2$$

Example VIII-3

What is the moment of inertia of a uniform right circular cylinder of mass M and radius R about the axis of symmetry?

We choose as a volume element a cylindrical shell of thickness dr. See Fig. VIII-4. All mass within such a shell is at the same distance r from the axis of rotation. The volume dv of the shell is the product of its height h, circumference $2\pi r$, and thickness dr; or, $dv = 2\pi h r \, dr$. Then, from Eq. (VIII-1),

$$I = \rho \int r^2 \, dv = \frac{M}{\pi R^2 h} \int_0^R 2\pi h r^3 \, dr = \tfrac{1}{2}MR^2$$

where the density ρ has been replaced by the total mass M divided by the total volume $\pi R^2 h$ and the limits of the integral have been taken from $r = 0$ to $r = R$.

Answers to Odd-numbered Problems

CHAPTER 2

- **2-1.** (a) 12 m south; (b) 6.0 m north; (c) 4.0 m south
- **2-3.** (a) −52.0 m, −30.0 m; (b) 17.3 m, 10.0 m; (c) 156 m, 90.0 m
- **2-5.** (a) 10.4 m; (b) 6.0 m
- **2-7.** 3.4 m at 4.4° north of east
- **2-9.** 9.0 m at 162°
- **2-11.** 12 mi at 17° north of east
- **2-13.** **A** and **B** are perpendicular
- **2-15.** (a) 40; (b) 9.24 units
- **2-21.** (a) −43; (b) 25 downward
- **2-23.** 10

CHAPTER 3

- **3-1.** (a) 6.93 m/s; (b) 15.5 mi/h; (c) 3 min 52.0 s
- **3-3.** (a) 0; (b) 83 km/h; (c) 160 km; (d) 192 km/h
- **3-5.** (a) 26 m/s²; (b) 9.6 s
- **3-7.** (a) 8.0×10^{15} m/s²; (b) 1.0×10^{-8} s
- **3-9.** (a) 12 s; (b) 360 m
- **3-11.** (a) $v = v_0 - \frac{1}{3}At^3, x = v_0 t - \frac{1}{12}At^4, A: LT^{-4}$ (b) $v = v_0 + (B/C) \sin Ct$, $x = v_0 t - (B/C^2) \cos Ct$, $B: LT^{-2}$, $C: T^{-1}$ (c) $v = v_0 + Dt + \frac{1}{4}Et^4$, $x = v_0 t + \frac{1}{2}Dt^2 + \frac{1}{20}Et^5$, $D: LT^{-2}$, $E: LT^{-5}$
- **3-15.** (a) greater than $30\frac{5}{8}$ m ≈ 30.9 m
- **3-17.** (a) 40.6 m; (b) 50.6 m
- **3-21.** (a) 19.6 m/s; (b) 19.6 m
- **3-23.** (a) 1.53 s; (b) 31.4 m
- **3-25.** (a) 0.78 s

CHAPTER 4

- **4-1.** (a) (1 m/s, −2 m/s); (b) (2 m/s, 5 m/s); (c) (−2 m/s, 3 m/s); (d) (0.33 m/s, 2 m/s); (e) (0, 4 m/s)
- **4-3.** (a) reversed directions; (b) unchanged; (c) reversed directions
- **4-5.** 31 m
- **4-7.** (a) $(v_0 \sin \theta)^2/2g$; (b) $(2v_0/g) \sin \theta$
- **4-9.** (a) 6.37 s; (b) 287 m/s directed 49° below the horizontal
- **4-11.** (a), (b), and (c) all larger by factor of 6.1
- **4-13.** 2.0×10^{13} m/s²
- **4-17.** 1.8 h
- **4-21.** (a) 0; (b) $8\pi r/T^2$
- **4-23.** (a) 9.1×10^{22} m/s²; (b) 6.6×10^{15}
- **4-25.** (a) 7.27×10^{-5} rad/s; (b) 4.63×10^2 m/s; (c) 2.32×10^2 m/s
- **4-27.** (c) 3.7×10^2 days
- **4-31.** 670 rev/min

CHAPTER 5

- **5-1.** (a) (100/3) km/h in the direction in which the limousine was moving before impact
- **5-3.** 0.83 m/s
- **5-5.** 125 kg-m/s to the north
- **5-7.** 3.0 m/s
- **5-9.** (b) 6×10^{13} kg
- **5-11.** 46 m/s directed 27° north of west
- **5-13.** 1.85 kg-m/s, down

5-15. 2.4×10^4 m/s
5-17. (b) opposite
5-13. (a) $\frac{1}{6}$ m/s; (b) 0; (c) $\frac{6}{31}$ m

CHAPTER 6

6-1. 545 m/s in direction of electron travel
6-3. (a) 90.1 m/s at 33.7° relative to $+X$ direction; (b) same as part (a)
6-5. (a) $(-1.2$ m, -0.4 m); (b) $(0, 1.2$ m/s)
6-7. $\bar{x} = 1.67$, $\bar{y} = 2.83$ with origin at lower left-hand corner
6-9. 755 km/h 23.4° east of north
6-11. 34.6 m/s moving "into" the rain of hailstones
6-13. (a) 90° heading (motion at 36.9°); (b) 138.6° heading (motion at 48.6°); (c) 180° (doesn't cross)
6-15. 6.5 m/s 19° north of east
6-17. (a) base of mast; (b) 17.1 m/s straight down; (c) 17.9 m/s (directed 73.7° below horizontal); (d) 8.75 m
6-19. Moves left, stops; moves right to initial position, stops. This motion is repeated
6-23. (a) 0, $2v$; (b) 0°; (c) 180°
6-25. (a) 200 kg-m/s to right; (b) 6.7 m/s; (c) 8.3 m/s to right, 16.7 m/s to left; (d) 16.7 kg-m/s right, 16.7 kg-m/s left

CHAPTER 7

7-1. (a) 0; (b) 3 N; (c) 0; (d) -1 N; (e) -2 N; (f) 0
7-3. (a) 6.0×10^3 N; (b) 6.0 mm
7-5. $mv/\Delta t$
7-7. (a) 120 mi/h; (b) 440 lb
7-9. 1.0×10^3 N
7-11. (a) 8.4 N; (b) 0.42 m/s
7-13. (a) 9.9×10^3 N up; (b) 9.8 N down; (c) 0
7-15. (a) 19.8 N-s down; (b) 17.2 kg-m/s at 59° below the horizontal
7-17. (a) 1.99 m/s²; (b) 10.5 s
7-19. (a) $2 mv$; (b) $2 mv \cos \theta$
7-21. 1.3×10^4 kg/s

CHAPTER 8

8-1. (a) $T_{max}/(g + a)$; (b) $T_{max}/(g - a)$
8-3. Aircraft is accelerating to the right at $a = 0.25 g$
8-5. a/b
8-7. (a) 65.3 m from origin at 13.3° west of north; (b) 42.7 m/s directed 6.7° west of north
8-9. 51.5 N
8-11. $a_x = -1.20$ m/s², $a_y = 0.80$ m/s²
8-13. (a) 30 N; (b) 10 N; (c) 1.34 m/s²
8-15. 1.0 m/s²
8-17. 0.75 g
8-19. 0.64
8-21. $(\mu_s g/r)^{1/2}$
8-23. $\tan^{-1}(v^2/rg)$
8-25. (a) 4.2×10^3 N, 1.7×10^4 N, 9.8×10^3 N; (b) leave roadway at point a
8-27. (a) $3 mg_0$ directed radially inward; (b) $-\frac{3}{4}mg$ directed radially outward where g_0 is the acceleration due to the earth's gravity at a distance r_0 from the earth's center

CHAPTER 9

9-1. $p_A/p_B = (m_A/m_B)^{1/2}$
9-3. (a) -270 J; (b) $+270$ J; (c) 0
9-5. (a) $+4.0$ J; (b) 0; (c) -4.0 J; (d) 0
9-7. $\frac{1}{2}mv^2$
9-9. (a) 0; (b) 312 J; (c) 0; (d) 0; (e) -312 J
9-11. (a) $9 D$; (b) $3 \Delta t$
9-13. (a) $mgl(1 - \cos \theta)$; (b) $-mgl(1 - \cos \theta)$; (c) 0
9-15. (a) 5.0 J; (b) 5.0 J; (c) 12.8 J
9-17. $kq_1q_2(1/r_1 - 1/r_2)$
9-19. 4.0 kg
9-21. 5.4 W
9-23. -4.05×10^{-2} J
9-25. (a) $(k_1 + k_2)x^2/2$; (b) $-(k_1 + k_2)x$
9-27. (a) -576 W; (b) -288 W; (c) 0
9-29. (a) 1.5 J by gravity, 0 by tension; (b) 1.5 J, 3.1 m/s; (c) $2 mg = 5.9$ N

CHAPTER 10

10-1. 0.60 m
10-3. 2
10-5. (a) 1.63×10^{14} kg; (b) 5.46 km
10-7. 153 m
10-9. 2
10-11. (a) mv; (b) $\frac{1}{2}mv^2$; (c) $\frac{1}{2}v$; (d) 0; (e) $\frac{1}{4}mv^2$; (f) $v(m/2k)^{1/2}$
10-13. (a) r; (b) $\frac{3}{2}r$
10-15. (a) 7.6×10^2 kg; (b) 8.4×10^5 kg; (c) 2.0×10^5 kg
10-17. 2.4 m/s
10-19. (a) 0.46; (b) 4.0 m/s
10-21. 34 J
10-23. 29 W

10-25. (a) $\frac{1}{2}mv^2$; (b) $\frac{1}{4}mv^2$; (c) $\frac{1}{2}$; (d) $v/2$
10-27. kq_1q_2/r^2
10-31. (a) 12 m/s; (b) explosive
10-33. (a) 4.78 MeV; (b) 86.2 keV

CHAPTER 11

11-1.   2.9 J-s directed to pitcher's left (toward first base)
11-3. $x = -1.5$ m
11-5.   12 J-s perpendicular to the horizontal plane
11-9.   (a) $L = (mv_0^3/2g)\sin^2\theta\cos\theta$, $\tau = mv_0^2\sin\theta$; (b) $L = (2mv_0^3/g)\sin^2\theta\cos\theta$, $\tau = 2mv_0^2\sin\theta\cos\theta$, all along horizontal
11-11. 3.2 J-s
11-13. (a) 15/8 kg-m² ≈ 1.88 kg-m²; (b) 1.67×10^2 kg; (c) 16 rad/s = 2.55 rev/s
11-17. (a) 7.1×10^{33} J-s; (b) 2.7×10^{40} J-s
11-19. 6 parts in 10^5
11-21. (a) 20 rev/min; (b) 658 J; (c) no

CHAPTER 12

12-1.   (a) 15.0π rad/s² = 47.1 rad/s²; (b) 660 rev
12-3.   (a) 0.733 cm/s; (b) 7.68×10^{-2} cm/s²; (c) 0.105 rad/s; (d) 0
12-5.   (a) 2.6×10^6 m/s² ≈ 2.6×10^5 g; (b) -1.7×10^{-4} rad/s²; (c) 1.0×10^4 hour
12-13. (a) 2.6×10^{29} J; (b) 2.7×10^{33} J
12-17. (a) 4.43 rad/s; (b) 4.12 rad/s
12-19. $2FL/\omega^2$
12-25. (a) sphere; (b) cylinder; (c) loop
12-27. (a) 7.5 kg-m²; (b) 3.0 m-N; (c) 0.40 rad/s²; (d) 2.0 rad/s; (e) 6.0 J; (f) 6.0 J
12-31. (a) 945 lb; (b) 824 lb; (c) down 11.5° relative to the vertical
12-33. (a) 1.59×10^3 lb directed 10.9° above the horizontal and to the right; (b) 15 ft
12-35. (a) $w[(2h/R) - (h/R)^2]^{1/2}$ at an angle above the horizontal of $\sin^{-1}[(2h/R) - (h/R)^2]^{1/2}$; (b) wh

CHAPTER 13

13-1. $\sqrt{3}\, Gm^2/d^2$
13-3.   (a) 3.8×10^{12} kg; (b) 1.5×10^{29} N-m²/kg²
13-5. 4.3×10^7 m
13-7.   (a) 1.4×10^5 g/cm³; (b) 5.9×10^5 m/s²
13-9.   (a) 6.8×10^{-3}; (b) 3.4×10^{-3}
13-11. $0.634\, d$
13-15. $-mgr_e$
13-17. (a) 8.94×10^3 km; (b) horizontal
13-19. $(GM/d)^{1/2}$
13-27. 90.0%
13-29. (a) out of the paper; (b) downward in the plane of the paper

CHAPTER 14

14-1.   (a) $\pi/2$; (b) 2.53×10^3 N/m; (c) 0.503 m/s; (d) +1.90 cm; (e) −12.0 m/s²; (f) 15.5 cm/s
14-3.   (a) 142 N/m; (b) 1.90 Hz
14-5.   (a) 6×10^3 m/s; (b) 4×10^{17} m/s²
14-9.   (a) 2.2 N/m; (b) 10.5 cm/s, 11.0 cm/s²; (c) ±8.38 cm/s, 6.58 cm/s²
14-11. 2.48×10^{-7} m
14-13. $2k$
14-17. $\omega^2 r$
14-23. (a) 2.01 s; (b) 1.79 s; (c) infinite
14-25. 7.90 km/s
14-27. $8.4\, g$

CHAPTER 15

15-1.   45 m/s
15-3.   (a) expanded along string by factor $(2)^{1/2}$; (b) unchanged; (c) increased by factor $(2)^{1/2}$
15-7.   (a) $(2.0\text{ cm})(\rho_1/\rho_2)^{1/2}$; (b) 100 Hz; (c) 2.0 m/s; (d) $(2.0\text{ m/s})(\rho_1/\rho_2)^{1/2}$
15-9.   (a) 2.0 mm, 0.63 m, 32 Hz; (b) 20 m/s; (c) 0.28 m/s
15-11. (a) 3.5×10^{-5} kg-m/s; (b) 9.1×10^{-4} N
15-13. Kinetic energy of undisplaced string particles
15-15. $T_D/T_A = 25/16$
15-17. 0.72 m
15-19. 12.6 N
15-21. (a) 500 m/s; (b) 2,000 Hz

CHAPTER 16

16-1.   3.0 km/s
16-3.   71 ms
16-5.   30°C
16-7.   27.6 Hz (at 0°C)
16-9.   (b) +0.606 (m/s)/C°
16-11. $c/400$ s⁻¹ (which is 82.8 cm at 0°C)
16-13. (a) 200 Hz; (b) 360 m/s; (c) 50°C = 323°K

16-15.	334 m/s
16-17.	(a) 60 db; (b) 120 db
16-21.	2.0×10^{-9} J

CHAPTER 17

17-1.	(a) $-40°$; (b) $(2.7 \times 10^5)°$
17-3.	(a) 1 H° = 0.660 C°; (b) $t_H = 1.52\ t_C = 59.0$
17-5.	5×10^{18} kg
17-9.	(a) nR/p; (b) 3.6610×10^{-3}/K°
17-11.	72.0 g/g-mol
17-13.	4.41 kg
17-15.	(a) 909°K = 636°C; (b) 48.5 atm
17-17.	299 m/s

CHAPTER 18

18-1.	$\approx 3 \times 10^{-10}$ m
18-3.	$(2\ mv)N$
18-7.	162 m/s
18-9.	2.7 km/s
18-13.	3.74×10^3 J
18-15.	(a) 2.0 J; (b) rise
18-17.	1.7×10^{-27} kg
18-19.	(a) 10^{-3}; (b) 10^{-24}; (c) $\frac{1}{2}$; (d) $(\frac{1}{2})^{10^8}$; (e) $(10^{-24})(\frac{1}{2})^{10^8}$

CHAPTER 19

19-1.	3.2×10^{-2} C°
19-3.	0.16 cal/g-C°
19-5.	48 grams
19-7.	$\frac{1}{4}\Delta t$
19-9.	3.3 m
19-11.	849 W
19-17.	(a) 11.3 kW; (b) 6.3 kW
19-19.	(a) 53.2 mm²; (b) 2.85×10^3 K
19-21.	4.09×10^3 K

CHAPTER 20

20-3.	5.0 cm from the 0.30 μC charge, 15 cm from the $-2.7\ \mu$C charge
20-5.	2.5×10^9 m
20-7.	2.7×10^{-11} C
20-9.	$+8\ \mu$C and $-4\ \mu$C
20-11.	At center
20-13.	4.8×10^{-15} m
20-15.	$F_A = 12.5$ N, $F_B = 16.1$ N, $F_C = 13.1$ N
20-17.	$Q = 3.4\ \mu$C
20-19.	$(q/\pi d)\sqrt{k/md}$, where d is distance between adjacent charges
20-21.	(a) 8.2×10^{-8} N; (b) 2.2×10^6 m/s
20-25.	Heavy nucleus

CHAPTER 21

21-1.	(a) Along line of centers 30 cm from q_1 and 20 cm from q_2; (b) along line of centers between charges: 6.0 cm from q_1 and 4.0 cm from q_2
21-3.	1.2×10^8 N/C at 77° below $+X$ axis
21-9.	1.3×10^{-13} C/m
21-11.	43
21-15.	1.0 cm for each; 3.4×10^{-7} s for electron, 1.4×10^{-5} s for proton
21-21.	(a) 1.0×10^4 N/C; (b) $\pm 0.18\ \mu$C/m²
21-23.	(a) Upward; (b) 1.6×10^{-15} kg
21-25.	0.10 mm/s upward

CHAPTER 22

22-7.	$1.13 \times 10^{15}\ Q$ lines
22-9.	$f = (1/2\pi)(\frac{4}{3}\pi\rho kq/m)^{1/2}$
22-11.	(a) 0; (b) 5.0×10^6 N/C; (c) 1.25×10^6 N/C
22-13.	(a) $E = \rho x/\varepsilon_0$ for $-\frac{1}{2}t \leq x \leq \frac{1}{2}t$; (b) $f = (1/2\pi)\sqrt{q\rho/m\varepsilon_0}$
22-15.	(a) Zero; (b) 7.2×10^5 N/C radially outward; (c) 2.6×10^5 N/C radially inward
22-17.	1.0×10^{-8} C/m²
22-21.	(a) Zero; (c) largest on spherical surfaces nearest conducting plates
22-23.	(b) charge/(length)$^{n+2} = Q/L^{n+2}$

CHAPTER 23

23-1.	(a) 3.5×10^4 m/s; (b) 6.5 eV
23-3.	(a) 0; (b) $2\ kQ^2/L$; (c) $3\ kQ^2/L$; (d) $3\ kQ^2/L$
23-5.	6.0 V
23-7.	(a) -44 eV; (b) $+10$ eV; (c) 0; (d) $+44$ eV
23-9.	(a) 1/1; (b) 1/42.8; (c) 42.8/1
23-11.	(c) Unstable
23-13.	5.9×10^6 m/s
23-15.	(a) 14 eV; (b) 14 MeV
23-17.	(a) 2×10^7 N/C
23-19.	(a) 0; (b) 28 J; (c) 3.2×10^{-16} kg
23-25.	(b) 7.1×10^{-4} F

23-27. $[(2\kappa_1\kappa_2)/(\kappa_1 + \kappa_2)]\varepsilon_0(A/d)$
23-29. (a) Four in series; (b) 3.0 pF
23-31. (a) Four capacitors; two groups in parallel, each group composed of two capacitors in series; (b) ten capacitors, two groups in parallel, each group composed of five capacitors in series
23-33. (a) Three in series; (b) 33.8×10^{-12} J

CHAPTER 24

24-1. 2.3 μA in electron direction
24-3. (a) 1.6 μA, 3.2 W; (b) 2.0×10^3 years
24-5. (a) 1.2 mA; (b) less than 1.4×10^{-16} s
24-7. 0.10 A, 5.0 V
24-9. 24 hours
24-11. (a) 0.80 A; (b) 80 W; (c) 0.80 A; (d) 160 W
24-13. (a) 72.0×10^{-3} V/m; (b) 0.900 W/m
24-15. 33.4 m
24-17. (a) 49.8 kg; (b) copper costs 5.0 times as much as aluminum
24-19. (a) 100; (b) $\frac{1}{100}$; (c) $\frac{1}{100}$
24-21. B
24-23. 1.2×10^5 m/s
24-25. (b) 1.0×10^6 A/m², 1.7×10^{-2} V/m
24-27. (a) 0.119 kg; (b) 0.123 kg

CHAPTER 25

25-1. 6.0 V
25-3. (a) 8.13 W; (b) 8.80 V; (c) 7.74 W; (d) 0.39 W
25-5. 130 V
25-7. (a) 10.0 Ω; (b) $I_1 = 1.8$ A, $I_2 = 0.60$ A, $I_3 = 0.72$ A, $I_4 = 0.48$ A, $I_5 = 1.20$ A (c) 32.4 W
25-9. (a) $\frac{5}{6}\Omega$; (b) $\frac{1}{3}I$ through each resistor connected to input or output, and $\frac{1}{6}I$ through the other six resistors
25-13. (a) 2.0 A to left through 5-V battery, 1.0 A up through 2-V battery, and 1.0 A to left through 1-Ω resistor; (b) −1.0 V
25-15. (a) 6.7 Ω; (b) 11 Ω
25-17. (a) Zero; (b) 9×10^3 Ω; (c) 9×10^4 Ω
25-19. (a) 0.01 Ω; (b) 46 cm
25-21. (a) $R_{sh} = 3.39$ Ω, $R_{se} = \frac{2}{3} \times 10^3$ Ω; (b) $\frac{1}{3} \times 10^3$ Ω

CHAPTER 26

26-1. kg/C-s
26-3. (a) Positron; (b) 1.8×10^7 m/s; (c) 8.8×10^2 eV
26-5. (a) A circular arc whose plane is perpendicular to the earth's magnetic field; (b) 5.8 m
26-7. (a) 3.8×10^{-8} T; (b) 1.4×10^{-8} T
26-11. 1.41 R
26-13. (a) Horizontal along east-west line; (b) electron current directed west; (c) 14.3×10^3 A
26-15. 10 eV
26-19. 2.8×10^{10} V/m directed east
26-21. $(q/m)B^2R$
26-25. 3.5 A
26-27. 4.7×10^{-5} m-N

CHAPTER 27

27-1. (a) 2.0×10^{-10} T; (b) 2.9×10^{-1} N/C
27-3. (a) 2.3×10^{-14} N; (b) 2.6×10^{-19} N
27-7. (a) 3.46×10^{-7} i/L; (b) 2.0×10^{-7} i/L
27-9. $(\mu_0 iL/2\pi) \ln(1 + L/R)$
27-11. 8.1×10^{22} A-m²
27-13. $\frac{1}{2}\mu_0 ni$
27-15. 1.1×10^{-11} C/m
27-17. 10 cm
27-19. $B = 0$ inside; $B = (\mu_0/4\pi)(2Ni/r)$ outside
27-23. (b) 0.352 a
27-25. $(\mu_0 ni/2)$, where n is the number of wires per unit width; the magnetic field is parallel to the conducting surface and perpendicular to the current
27-27. (a) 10 G; (b) 5.0 G

CHAPTER 28

28-1. (a) 0.22 V; (b) 1.9×10^{-20}
28-5. South pole
28-7. (a) 3.2×10^{-5} V
28-9. 1.3×10^2
28-11. 20 m/s
28-13. The rod emerges at velocity v/2. Upon entering the magnetic field, three-fourths of the original kinetic energy is transformed into heat and electric potential energy due to the charge separation. Upon emerging, the electric potential energy is transformed into heat.
28-15. (a) $(BL)^2 v/R$
28-17. (a) 2.0 H; (b) 5.0 Ω
28-21. (a) 14 mH; (b) 2.8×10^{-2} J
28-23. $\mu_0(Ni)^2/8r^2$

CHAPTER 29

- **29-1.** $1.8\ \mu\text{F}$
- **29-3.** 5
- **29-5.** (a) 5.0 A; (c) 1.2×10^{-2} J
- **29-7.** $2.2 \times 10^{7}\ \Omega$
- **29-9.** 1.2×10^{6} s
- **29-11.** (b) 1.0
- **29-13.** (a) 30 mH; (b) 10 Ω
- **29-15.** 6.0×10^{-2} J
- **29-17.** (a) 4.6×10^{-2} H; (b) 0.40 ms

CHAPTER 30

- **30-1.** (a) 1.0×10^{-11} T; (b) 2.4×10^{-8} W/m² south
- **30-3.** (a) 5.0 A; (b) 5.7×10^{11} V-m/s; (c) 1.8×10^{15} N/C-s; (d) 1.8×10^{12} V/s; (e) 5.0 A
- **30-5.** (a) 4.9×10^{2} V/m; (b) 1.6×10^{-6} T
- **30-7.** 2.0×10^{29} J
- **30-9.** 10 m
- **30-11.** (a) 9.9×10^{3} N/m²; (b) 2.7×10^{-5} N-s/m³
- **30-13.** (a) 9.4×10^{15} W; (b) 3.0×10^{23} W/m²; (c) 1.0×10^{15} J/m³; (d) 1.1×10^{13} V/m, 3.5×10^{4} T
- **30-15.** 16 cm
- **30-17.** 3.9°
- **30-19.** (b) 2.8×10^{-17} m
- **30-21.** 3.4×10^{-11} s

CHAPTER 31

- **31-3.** 3.0 ft
- **31-5.** 20°
- **31-7.** 1.8 m
- **31-9.** (a) 1.1 percent; (b) 5.1 percent
- **31-13.** 0.75 mm
- **31-15.** 1.1 m
- **31-19.** (a) Real image at +60 cm; (b) real image at +80 cm; (c) real image at $+\infty$; (d) virtual image at -40 cm
- **31-25.** 33 cm
- **31-27.** 124 cm from object on opposite side of lenses
- **31-33.** (a) Greater; (b) green

CHAPTER 32

- **32-3.** 1.5 m
- **32-5.** 30° and 60°
- **32-9.** 60° east or west of south
- **32-11.** 2.5 cm
- **32-13.** 1.0 km
- **32-15.** 1.5 mm
- **32-17.** 1.8×10^{-2} mm
- **32-19.** 1.1×10^{-4} mm
- **32-21.** (a) 82°; (b) 38°
- **32-23.** 0.60 mm
- **32-25.** Pattern reduced in size by a factor n for small angles
- **32-27.** 7.6×10^{-2} mm
- **32-29.** 1.3×10^{-2} mm

CHAPTER 33

- **33-1.** 22.5° above X axis
- **33-3.** 45°
- **33-5.** Vertical
- **33-7.** East-west

CHAPTER 34

- **34-3.** (a) 2.23 km; (b) 4.46 km
- **34-5.** (a) 5.3×10^{5} m/s; (b) larger by factor 67
- **34-7.** (a) 17.2×10^{9} yr; (b) 6,173 Å (red)
- **34-9.** 1.2×10^{-8} s
- **34-11.** $\theta_2 = \tan^{-1}\{\tan \theta_1/[1 - (v/c)^2]^{1/2}\}$
- **34-17.** (a) 68 s; (b) 14 s
- **34-19.** (a) 17.6 m; (b) 3.5 m
- **34-21.** (a) 6.67×10^{-5} s; (b) 16.0 km; (c) 9.6 km
- **34-23.** (a) $0.673c$; (b) 2.97×10^{-6} s; (c) 444 m
- **34-25.** $(160/164)c$
- **34-27.** $\lambda = \lambda_0[1 - (v/c)^2]^{-1/2}$
- **34-29.** (a) $0.3c$ along positive X axis; (b) $x = 0$, $t = 0$; $x = 1.14$ km, $t = 2.09 \times 10^{-6}$ s

CHAPTER 35

- **35-3.** (a) 2.45; (b) $0.981c$
- **35-5.** (a) 33.3 m; (b) 0.238 m; (c) 34.1 m
- **35-7.** (a) $0.99999824c$; (b) 8 MW
- **35-9.** (a) 1.00; (b) 9.0×10^{-3}; (c) 26.5
- **35-11.** (a) 5.0 keV; (b) 1 MeV/c

35-13. (a) $\frac{16}{9}$; (b) $\frac{8}{3}$
35-15. (a) 1.2 MeV; (b) 47 MeV/c
35-17. (a) 38 keV; (b) 0.37 c
35-19. (a) 1.0 GeV/c; (b) 5.0 GeV/c
35-21. (a) 24 μA; (b) 1.6×10^{-3} N
35-23. (a) 0.99995c; (b) 5.1 MeV; (c) 93 GeV
35-25. (a) 1.19×10^{22} J; (b) 1.32×10^5 kg; (c) 1.32×10^7 kg ≈ 220 × payload mass
35-29. (a) 2.2×10^{-14} kg decrease; (b) 4.7×10^{-12} kg increase; (c) 3.7×10^{-12} kg increase

CHAPTER 36

36-1. (a) 3.23 eV; (b) 2.97 eV
36-3. 9.0 eV
36-5. (a) 2.0×10^{15} s^{-1}; (b) 3.0×10^{24} s^{-1}; (c) 5.1×10^{11} s^{-1}; (d) 6.3×10^6 s^{-1}
36-7. (a) 7×10^{-30} J/s; (b) 3×10^3 yr (experimentally, ≈ 10^{-9} s)
36-13. ∝ $1/r^2$
36-15. Electron 1
36-17. (a) 7.9×10^5; (b) 1.3×10^{-21} kg-m/s
36-19. (a) 6,200 Å; (b) 8.67 Å
36-23. 0.0016 mm
36-25. (a) 0.08626 Å; (b) 0.1105 Å; (c) 56.2 keV; (d) 8.78 keV
36-27. 1.67×10^{-27} kg (a proton or neutron)
36-29. 55 MeV
36-31. 3.91×10^{-3} Å
36-33. 195

CHAPTER 37

37-1. (a) 1.24×10^{-12} m; (b) 7.09×10^{-13} m; (c) 2.86×10^{-14} m
37-7. 3.95×10^{-4} rad
37-9. 32.8° for diffraction from Bragg planes along 45° diagonal
37-11. (c) c
37-13. Energies for potential well of *finite* height are lower.
37-15. ≈ 1 m/s
37-17. (a) 4.1×10^{-11} eV; (b) 2.1×10^{-11}; (c) 2.1×10^{-20}
37-19. (a) 60 m; (b) 3×10^{11} m ≈ dia. of earth's orbit; (c) 1 cm; (d) 10^{-8} m
37-21. 2.1×10^{-2} eV
37-23. (a) 25 keV/c, 0.61 keV; (b) 25 keV/c, 25 keV; (c) particles have smaller energy for the same wavelength (momentum)
37-25. (b) The particle's speed is low on the right; therefore, the probability for locating the particle there and the amplitude of the wave function must be larger there.
37-27. (a) $n^2\hbar^2/2mR^2$; (b) $n\hbar$

CHAPTER 38

38-1. (a) 8.0 MeV; (b) 8.5×10^{-15} m
38-3. 8.5×10^{-15} m
38-5. (a) 7.8 percent; (b) 100 percent; (c) 0.055 percent
38-9. (a) 10; (b) 25,000
38-17. (a) −54.3 eV and −13.6 eV; (b) 54.3 eV
38-19. (b) 5.4×10^{-9}
38-21. (a) 5.7×10^{12} Hz; (b) 6.6×10^{12} Hz
38-25. 2,536 Å
38-27. (a) 12.1 eV; (b) 13.6 eV
38-29. (a) $\frac{3}{8}E_I$; (b) $\frac{3}{8}E_I$
38-31. 2.2×10^{-3} eV
38-33. −2
38-35. Quantum numbers (n, l, m_l, m_s): (a) $(1, 0, 0, -\frac{1}{2})$, $(1, 0, 0, \frac{1}{2})$ and $(2, 0, 0, -\frac{1}{2})$; (b) $(2, 0, 0, \frac{1}{2})$ and $(2, 1, 0, -\frac{1}{2})$

CHAPTER 39

39-3. (a) 12.2 MeV; (b) 79 keV
39-7. 3.8×10^{-2} s^{-1}
39-11. 20.8 keV
39-15. (b) 0.86 MeV; (c) 0.86 MeV/c; (d) 0.86 MeV/c; (e) 57 eV
39-17. 157 keV
39-19. (a) 2; (b) 2
39-21. ~200 percent
39-23. (a) 3×10^{15} eV; (b) 4 GNP (1970); (c) ≈ A.D. 2,000
39-31. (a) 224 MeV; (b) the same
39-33. ~10^{30} J = 10^{24} kW-h
39-35. (a) 2×10^{38} neutrinos/s; (b) 8×10^{14} neutrinos/m²-s; (c) 3×10^6 neutrinos/m³ ≈ 3 neutrinos/cm³

CHAPTER 40

40-3. (a) 141 MeV; (b) 0.44×10^{-23} s
40-5. $\hbar/r_0 c$
40-7. (a) 0.17 mm; (b) 16.7 m; (c) 7.22 mm; (d) 18.2 m
40-9. 26 m
40-11. 67.49 MeV
40-13. $11 m_e$
40-15. (a) 69 MeV; (b) 52 MeV
40-17. Σ^- and $\overline{\Sigma^-}$ have *different* decay products and strangeness numbers
40-21. A beam of $\overline{K^0}$ mesons striking a target may produce the reaction $\overline{K^0} + p \rightarrow \Lambda^0 + \pi^+$, whereas K^0 mesons could not because it would violate conservation of strangeness.
40-23. Violation of the conservation laws of (a) strangeness; (b) charge; (c) energy, strangeness; (d) energy, strangeness; (e) strangeness
40-25. (b) $2\gamma + 2e^- + 2\bar{\nu}_e + 2\nu_\mu + 2\bar{\nu}_\mu + p$

Index

A

a (*see* Acceleration)
A (*see* Ampere)
Å (Angstrom unit), 281
A (*see* Nuclear mass number), 706, 708
Aberration of starlight, 78, 502
Absolute temperature, 273
   relation to kinetic energy, 285
Absolute zero, 273
Absorption coefficient, 737
Absorption of radiation, 308
Absorption spectrum, 680
Accelerating charges and electromagnetic waves, 495
   polar diagram of intensity, 497
   radiated electric field, 497
   static electric field, 496
Acceleration:
   angular, 177
   average, 22, 36
   center-of-mass, 104
   centripetal, 43
   components, 44
   constant, 24, 28, 37
   as derivative, 23
   dimensions, 23
   of electric charges as radiation source, 308
   due to gravity (earth), 28, 86
   due to gravity (moon), 47
   instantaneous, 22, 36
   linear, 22
   parallel to path, 44
   perpendicular to path, 44
   radial, 43, 45, 179
   and reference frames, 73
   resultant, 179
   rotational (*see* angular)
   as slope of v-t curve, 22
   tangential, 44, 179
   total, 179
   vector character, 36
Accelerators, 723
   cyclotron, 726
   electron synchrotron, 390
   linear, 724
   synchrotron, 728
   Van de Graaff, 367, 376
AC current (Alternating current), 383
Acoustic lens, 527
Acoustics, 259
   spectrum, 259, 263, 264
Action-reaction forces, 85
Activity, 714
Adiabatic process, 276
   sound waves, 256
Air wedge, interference, 546, 559
Alkali metals, 698
Allowed modes of oscillation, 248
Allowed transition, 691
Alpha particle, 65, 676
   tunnel effect, 669
Alternating current (*ac*), 383
Ammeter, 383, 405, 410
Ampère (unit), 383
   definition, 438
Ampère, A.M., 383, 438
Ampère's law, 440, 480, 484
   compared with Faraday's law, 482
   generalized form, 482
   original form, 441
Amplification of light, 542
Amplitude, 220
Analog computer, 404
Analogies:
   electrical-mechanical, 473
   rotation-translation, 187
Analytical method for vectors, 9
Analyzer, 566
Anderson, C.D., 633
Angstrom (unit), 281
Angular acceleration, 177
Angular displacement, 45, 154
   sign conventions, 155
Angular frequency of free oscillations, 470
Angular magnification, 532
Angular momentum:
   angular velocity, 156
   collection of particles, 166
   conservation law, 156, 168
   constant, 163
   dimensions, 161
   direction (sense), 157
   for electromagnetic waves, 495, 570
   Kepler's second law, 207
   magnitude, 160
   as moment of momentum, 167
   nuclear spin, 704
   orbital, 174
   particle, 156, 163
   particles (multiple), 166
   photon, 638
   qualitative features, 155
   quantization, 682
   quantum theory, 174
   rectangular components, 173
   rigid body, 166, 181
   sense (direction), 157
   spin, 174
   and torque, 158
   units, 161
Angular speed, 45
Angular velocity vector, 154
   direction, 155, 157
   magnitude, 155
   sense of rotation, 155
Annihilation, 65, 322, 633

Anode, 391
Antineutrino, 720
Antinode, 257
Antiparticle, 321, 631, 750
Antiproton, 322, 640
Aphelion, 205, 209
Apogee, 205
Apollo 15 mission, 197
Apparent depth, 530
Arm-bones, 195
Arrow of time, 295
Ashkin, A., 491n
Associated production, 757
Associative law, vector addition, 8
Astronomical data:
   (figure), 209
   (tables), 205, 208
Astronomical telescope, 528, 532
   advantage of large diameter, 557
Astronomical unit, 261, 769
atm (unit), 270
Atmospheric pressure, 270
   standard value, 270
Atomic clock, 4
Atomic masses (table), 775
Atomic mass unit, 54, 609
Atomic model, Bohr, 681
Atomic nucleus (*see* Nucleus)
Atomic number, 323, 708
Atomic structure, 314, 675
Atomic weight, 367
AU (*see* Astronomical unit)
Average acceleration, 22
Average velocity, 17, 33
Avogadro's law, 296
Avogadro's number, 280, 285, 323, 384

B

B (*see* Magnetic induction field)
B (*see* Bulk modulus)
Back emf, 461, 468
Balance:
   current, 439
   gravitational, 53, 87
   inertial, 52
   torsion, 201, 316, 495, 530
Ballistic pendulum, 149
Balmer, J.J., 679
Balmer series, 679, 680
Barkla, C.G., 570
Barometer:
   aneroid, 269
   mercury, 271
Barrier penetration, 669
Baryon, 749
   conservation laws, 752
Batavia, Illinois, 729, 737
Battery:
   aging, 395
   cold-cranking power, 409
   discharged, 395

   emf, 393
   internal resistance, 394
   potential difference, 395
   power, 394
   storage (lead-acid), 409
Beam balance (*see* Gravitational balance)
Beams, J.W., 193
Bees' navigation system, 569
Becker, H., 732
Bessel function, 556n
BeV, 608
Bevatron, 640
Biceps (muscle), 195
Binary star system, 216
Binding energy, 359, 605
   nuclear, 711
Binomial theorem, 94
Biot, J.B., 436
Biot-Savart law, 432
Blackbody, 309, 312
   radiation, 614
Blue sky, 568
Bohr, N., 651, 681, 683n, 700
Bohr model of hydrogen atom, 48, 174, 215, 323, 358
   equivalent electron current, 449
   ionization energy, 359
   magnetic field at nucleus, 449
   radius, 684
Bohr theory, 681
   postulates, 688
Boldface symbol, 7
Boltzmann's constant, 285
Bones of the arm, 195
Born, M., 654
Boson, 750
Bothe, W., 732
Boundary conditions, 246
Bound systems, 604
Boyle's law, 275
Bradley, J., 78, 502
Bragg's law, 644, 646
Bragg:
   plane, 645
   reflection, 645
Bragg, W.H., 645
Bragg, W.L., 645
Bremsstrahlung, 622, 742
Bright-line spectrum, 679
British thermal unit, 302
Brown, S.C., 419n
Btu (*see* British thermal unit)
Bubble chamber, 63, 420, 422, 632
Bucherer, A.H., 599
Bulk modulus, 256

C

c (centi), 5
c (*see* Speed of light in vacuum)
c (*see* Specific heat)

C (*see* Capacitance)
C (*see* Celsius)
C (*see* Coulomb)
cal (unit), 302, 771
Calculus, invention, 94
Calculus in kinematics, 26n
Caloric, 265
Calorie, 289, 302
Calorimeter, 311
Calorimetry, 304, 311
Capacitance, 368
   coaxial cable, 380
   coaxial cylinders, 371, 380
   concentric spheres, 380
   earth, 380
   isolated sphere, 380
   parallel circuit, 374
   parallel-plate capacitor, 370, 380
   series circuit, 372
Capacitor, 369
   energy, 375
Capacitors in parallel, 374
Capacitors in series, 372
Carnot cycle, 305
Cascade shower, 637
Cathode, 391
Cathode-ray oscilloscope, 337, 413, 428
Cathode-ray tube, 47
Cavendish, H., 196, 201, 316, 348
Cavendish experiment, 201, 316, 324
Celsius degree vs. degree Celsius, 268n
Celsius temperature, 267
Center of curvature, 234
Center of gravity, 188
Center of mass:
   acceleration of, 104, 180
   and center of gravity, 188
   continuous mass distribution, 69
   and external forces, 105
   location, 68
   and reference frames, 75
   and rotational motion, 186
   and translational motion, 180, 186
   velocity, 66, 105
Centigrade (*see* Celsius)
Central force, 164, 181, 207
   Coulomb, 316
   gravitational, 198, 207
Centrifugal force, 109
Centrifuge, 49
   ultracentrifuge, 192
Centripetal acceleration, 43
Centripetal force, 108
CERN, 737
Cesium-133 time standard, 4, 504
Chadwick, J., 737
Chamberlain, O., 640
Change of state:
   of a gas, 276
   in general, 303

Charge:
  electric, 313
  gravitational, 196
  induced, 349
  source, 326
  test, 326
Charge carriers, 383
Charge conjugation, 759
Charge conservation, 321
  "charging" an object, 315
  Kirchhoff's junction theorem, 402
Charged particles in uniform B and E fields, 419
Charge/mass ratio experiment, 422
Charge quantization, 322, 335
"Charging" an object, 315
Charles, J.A.C., 275
Charles' and Gay-Lussac's law, 275
Chemical force, 394
Chemical potential energy, 361, 394
Circuit:
  LC, 471
  LR, 475, 478
  multiple loop, 402
  RC, 474
  RLC, 474
  single loop, 396
Circularly polarized electromagnetic waves, 562, 570, 638
  left or right, 562
Circular motion, uniform:
  dynamics, 106
  kinematics, 42
  and simple harmonic motion, 221
Classical mechanics, 1, 50
Classical model of hydrogen atom (see Bohr model)
Classical physics, 1, 50
  vs. quantum physics, 663
Clock, atomic, 4
Cloud chamber, 148, 420
cm (unit), 769
CM (see Center of mass)
cm Hg (unit), 771
Coated lens, 560
Cockcroft, J.D., 731
Coefficient of friction, 102
Coherent source, 540
  laser, 541
Collisions:
  and center-of-mass velocity, 66
  completely inelastic, 149
  elastic, 147
  explosive, 147
  hard or soft, 57
  head-on, 55
  inelastic, 147
  perfectly elastic, 147
  in two and three dimensions, 60
  in zero-momentum frame, 75
Commutative law:
  addition of vectors, 8
  scalar product, 14
  vector product, 15
Complementarity principle, 649
Compliance, 232
Component, vector, 9
Compound microscope, 527
Compressional wave, 253-64
Compton, A.H., 627, 629
Compton effect, 625, 742
Compton wavelength, 629
Computations in relativistic mechanics, 607
Concert A, 252
Concert pitch, 252
Condensation, 254
Condenser, 372
Conductance, electrical, 388
Conducting sheet, magnetic field, 448
Conduction electron (see Free electron)
Conductivity, electrical, 388
Conductivity, thermal, 307
Conductor, electrical, 314
  charge distribution, 348, 366
  electric potential, 366
  as perfect shield, 349
Conductor, thermal, 306
  relation to electrical conductors, 306
Configuration, electron, 696
Conjugate points, 525
Conservation of energy and the uncertainty principle, 658
Conservation laws:
  angular momentum, 168
  in classical physics (list), 322
  electric charge, 321
  energy, 131, 144, 356
  first law of thermodynamics, 291
  isospin, 756
  Kirchhoff's rules, 402
  lepton number, 752
  linear momentum, 57, 79, 592
  local conservation in relativity, 589
  mass, 54
  mass-energy, 146, 605
  muon number, 753
  nucleon number, 714
  parity, 758
  strangeness, 757
  and uncertainty principle, 658, 659
  universal validity, 751
Conservative force, 140, 144, 147
  electric, 317, 356
  elastic, 218
  gravitational, 209
Conservative system, 144
Constants, fundamental (table), 767
Constructive interference, 238, 534
Contraction, space, 579
Contrails, 148n
Convection, thermal, 308
Conventional current, 382, 393
Conversion factors (table), 769
  for electric and magnetic units, 772
Coordinate axes, right-hand set, 12
Coordinate transformation relations:
  galilean, 71
  lorentz, 581
  for rotation, 16
Copernicus, N., 78
Copernican cosmology, 78
Corner reflector, 514
Corona discharge, 367
Correspondence principle, 683, 700
Cosmic rays, 416, 578, 635
Coulomb (unit), 319
  definition, 439
Coulomb, C.A., 316
Coulomb force, 316, 318, 678, 744
  in atomic physics, 111
  in nuclear physics, 317, 324
Coulomb force constant, 318
  relation to electric permittivity, 320
Coulomb's law, 316
  experimental test, 348
  range of validity, 316
Coulomb scattering, 316, 378
Couple, 193
Covariance in relativity, 590
Cowan, C.L., 723
Critical angle, 518
Critical launch speed, 213
Crossed E and B fields, 420
Cross product (see Vector cross product), 14
Cross section, 737
Crystal, 644
Curie (unit), 714
Current balance, 439
Current density vector, 391
Current, electric, 382
  conventional, 382
  density, 391
  displacement, 481, 483, 506
  induced, 452
  units, 383
Current-measuring devices:
  ammeter, 383, 405, 410
  electrolytic deposition of mass, 383
  galvanometer, 405, 426
Curvature, 234
Cyclic accelerator, 724
Cyclotron, 726
  frequency, 418, 726
  resonance, 419
Cylindrical wavefronts, 261
  dependence of intensity on distance, 261

D

d (deci), 5
Dalton's law, 297
Damped oscillations, 229
Daniell cell, 391
Dark-line spectrum, 680
Daughter nucleus, 712
Davisson, C., 643, 649
Day:
   sidereal, 193
   solar, 4, 193
DC circuits:
   multiloop, 402
   sign conventions for V and i, 396
   single loop, 396
DC current, 383
DC instruments, 404
   ammeter, 383, 405, 410
   galvanometer, 405, 426
   ohmmeter, 410
   potentiometer, 407, 411
   voltmeter, 405, 410
   Wheatstone bridge, 407
de Broglie, L., 643
de Broglie waves, 642
   probability interpretation, 652
   wavelength, 683
Decay:
   alpha, 716
   beta, 717
   gamma, 715
Decay constant, 712
Decibel, 263
Degenerate states, 695
Delta symbol (Δ), 18
Density:
   current, 391
   electric field energy, 376
   electromagnetic field energy, 489
   magnetic field energy, 464
   mass, 69
Destructive interference, 238, 534
Deuterium, 152
Deuteron, 149, 706
Diamond, 519
Diatomic molecule, 174
   and energy, 299
Diatonic scale, 264
Dichroism, 566
Dielectric constant, 371, (table) 372, 381
Dielectric materials, 371
   bound electrons, 315
   and capacitors, 370
   properties, 315
Diesel engine, 278
Difference in path length, 537, 548
Diffraction, 509
   circular aperture, 556
   double slit, 552
   electron, 647

Fraunhofer, 551
Fresnel, 551
   grating, 552, 644
   intensity variations, 551, 554
   pattern, 549, 550, 552
   quantum theory, 655
   and resolution, 556
   row of point-sources, 548
   single slit, 550, 552
   straight edge, 554
   wavelength dependence, 509
   X-ray, 647
Diffraction grating for sound waves, 552, 560
Diffuse reflection, 512
Diffusion pump, 297
Dilation of time, 574, 587
Dimensional analysis, 224
   necessary test, 27
Dimensions of a physical quantity, 5
Dip angle, 428
Dipole moment:
   electric, 338, 380
   magnetic, 426n, 447
Dirac, P.A.M., 633, 693
Direct current (see dc current)
Disintegration constant, 712
Disordered motion and thermal energy, 287
   and entropy, 305
   of a gas, 288, 298
   of a liquid, 300
   of a solid, 299
Displacement as area, 24
Displacement current, 481, 506
   Hertz' experimental test, 483
Displacement mode, 257
Displacement vector, 6, 33
   distinct from distance, 18
   as prototype vector, 6
Dissipative force, 142
Distributive law:
   scalar product, 14
   vector product, 15
Distribution, maxwellian, 293
Doppler effect, 264, 504
Dot product, 13
Double-slit diffraction, 655
   Young's experiment, 542
Drift speed, 383
   in copper, 384
   in silver, 391
   in ohmic materials, 387
Duality, wave-particle, 649
Dynamic mass, 232
Dynamics, 50
   relativistic, 590
   rotational, 177
dyn (see Dyne)
Dyne (unit), 82, 770

E

e (Magnitude of electron charge), 320
E (see Electric field)
E (Energy), 131
Ear, human:
   audible frequencies, 256, 260
   audible intensities, 259
Earth:
   as approximate inertial frame, 52
   atmosphere, 278
   average mass density, 202
   capacitance, 380
   equatorial bulge, 203, 216
   gravitational torque, 217
   magnetic field, 416, 428, 447
   mass (computation), 202
   mean density, 202
   orbital angular momentum, 175, 193
   precession of spin axis, 217
   rotational kinetic energy, 193
   shape, 203
   spin angular momentum, 175, 193
Earth-moon distance, measurement, 507
Earthquake tremor, compressional, 262
Earth satellite, 44, 107, 111, 174
   dimensional analysis, 232
   energetics, 212, 217
   (graph), 209
   orbital period, 44, 174
   orbital speed, 44, 174, 213
   reentry, 216
   synchronous, 215
   (table), 208
East-west effect of cosmic rays, 416
Einstein, A., 152, 571, 578, 601, 618
Elastic collision, 147
   quantum theory, 701
Elastic wave (see Compressional wave)
Electric charge:
   compared with gravitational charge, 317
   comparison of two charges, 317
   conservation, 321
   Coulomb's halving technique, 318
   induced, 349
   positive and negative, 314
   quantization, 322, 335
   relation to atomic number, 323
   total in universe, 325
Electric charge density:
   linear, 331, 344
   surface, 331, 337, 344
   volume, 347
Electric charge, measuring devices:
   electroscope, 321, 349
   electrometer, 321, 376, 381

Electric-charge quantization (*see* Charge quantization)
Electric circuit (*see* Circuit)
Electric dipole, 338
 dipole moment, 338
 field, 338
 potential, 380
Electric-dipole oscillator, 498, 507
 continuously driven, 563
 polarization properties of emitted waves, 562, 563
 properties, 498
 radiation pattern, 497, 498
Electric field:
 of accelerating charge, 496
 and Ampère's law, 440, 480, 482, 484
 due to changing magnetic flux, 458
 concept, 327
 and conducting spherical shell, 353
 and conductors, 333
 conservative, 458
 definition, 326
 defined by Lorentz force equation, 483
 as derivative of potential, 365
 detached, 328
 dielectric cylinder, 352
 dielectric sphere, 346, 352
 dielectric surface, 334, 352
 as electric flux density, 345
 and electromagnetic induction, 458
 energy density, 376
 and Faraday's law, 458
 flux, 339
 Hall effect, 423
 infinite line of charge, 331, 344
 infinite surface of charge, 331, 344
 lines, 329, 345
 nonconductor, 334
 nonconservative, 459
 parallel plates, 332
 of point-charge, 327, 331
 radiated, 496
 relation to electric potential, 362
 relation to magnetic field (in waves), 488
 speed of propagation, 328
 spherical shell of charge, 346, 352
 surface of charge, 331
 superposition, 327
 transverse, 497
 units, 327
Electric field and electron-shell structure, 352
Electric-field-line concept and Gauss' law, 345

Electric field lines, 329, 345
 changing magnetic flux as source, 459
 density, relation to E, 329
 and electric charges, 458
Electric flux, 339
Electric force, 316
 conservative, 317, 356
 and nonconservative E, 458
Electric force-gravitational force ratio, 317, 321
Electric free oscillator, 468
 angular frequency, 470
 differential equation, 469
Electric generator, 395, 466
Electric intensity, 327
Electric interaction between two charges compared with magnetic interaction, 434
Electric lines of force (*see* Electric field lines)
Electric and magnetic units, 772
Electric monopole (*see* Electric point-charge), 338
Electric oscillator, free, 468
Electric permittivity:
 of dielectric, 372
 of free space, 320
 free space, experimental value, 488
 relation to coulomb force constant, 320
Electric point-charge, 316
Electric potential, 359
 choice of zero, 360
 of conductors, 366
 equipotential surfaces, 364
 gradient, 365
 models, 365
 point-charges, 361
 relation to electric field, 362
 of spherical dielectric shell, 365
 of uniformly charged wire, 363
 units, 360
Electric potential energy, 354
 choice of zero, 355
 of system of point-charges, 357
 of two point-charges, 355
Electric potential, measuring devices:
 electrometer, 321, 376, 381
 potentiometer, 407, 411
 voltmeter, 405, 410
Electrical breakdown, corona discharge, 367
Electrical conductance (*see* Conductance)
Electrical conductivity (*see* Conductivity)
Electrical-mechanical analogs, 473 (tables), 473, 474
Electrical power:
 dissipated by resistor, 387, 397

installed capacity of U.S., 506
 supplied by source, 385, 394
Electrical resistance (*see* Resistance)
Electrochemical cell, 361, 394
 Daniell cell, 391
Electrodynamics, quantum, 635
Electrolysis, 383
Electromagnetic force, 740
Electromagnetic induction, 454
 examples, 455
 Faraday's law, 456
 Lenz' law, 456
 motional emf, 450
Electromagnetic interaction, 740
Electromagnetic radiation, classical theory:
 pulsed, 484
 sinusoidal, 498
Electromagnetic radiation, quantum theory, 613
 particle aspects, 613
Electromagnetic spectrum, 500, 622
Electromagnetic waves, 483, 498
 accelerating charge as source, 495
 double scattering, 569
 energy density, 489
 Hertz' experimental test, 471, 483
 intensity, 489, 534
 linear momentum, 494
 Maxwell's equations, 484
 phase change on reflection, 539
 Poynting vector, 490
 properties, 328
 radiation force, 491
 radiation pressure, 491, 494
 ratio of E to B, 488
 sinusoidal, 498
 spatial relations of c, E, and B, 500
 speed of propagation, 488
 spin angular momentum, 495, 570
 source, 495
 superposition, 533
 wavelength, 499
 wavelength in refractive medium, 545
Electromagnetic waves, scattering:
 classical, 625
 quantum, 627
Electromagnetism, 313
Electro-mechanical analogue analyzer, 473
Electrometer, 321, 376, 381
Electromotive force (*see* Emf)
Electron:
 charge, 320
 conduction, 314
 free, 306, 314, 315, 333

Electron, *cont.*
  intrinsic angular momentum, 693
  magnetic quantum numbers, 691, 693
  microscope, 673
  spin, 693
Electron accelerator:
  synchrotron, 390
  Van de Graaff, 367, 379
Electron-beam oscilloscope (*see* Cathode-ray oscilloscope)
Electron capture, 718
Electron charge, measurement by Millikan, 320
Electron configuration, 696
Electron diffraction, 647
Electron gun, 337
Electron microscope, 673
Electron number, 753
Electron-positron pair:
  annihilation, 322, 633, 741, 742
  charge conservation, 322, 631
  production, 321, 631, 742
Electron synchrotron, 390
Electron volt (unit), 357, 361
Electroscope, 321, 349
Electrostatic force (*see* Coulomb force)
Electrostatic generator, Van de Graaff, 367, 379
Electrostatic lens, 527
Electrostatic potential energy (*see* Electric potential energy)
Electrostatic precipitator, 379
Electrostatics, 314
Elementary particle, 322, 739
  table of properties, 748, 749
Elliptical polarization, 562
e/m experiment, 422
Emf, 394
  back, 461, 468
  battery, 393
  definition, 458
  electric generator, 395
  electromagnetic induction, 458
  induced, 458
  motional, 450
  photovoltaic cell, 395
  seat, 395
  self-induction, 461
  thermoelectric, 395
Emission of radiation, thermal, 308
Emission spectrum, 678, 679
Emu (unit), 772
Energetics:
  projectiles, 136
  simple harmonic motion, 222, 226
Energy:
  atomic, 146, 685
  binding, 359, 605, 685, 711
  chemical, 146
  conservation law, 144

consumption rate in U.S., 151
conversion factors, 771
of electric field, 375
electromagnetic, 146
excitation, 685
external, 685
kinetic, 112, 600
internal, 300
ionization, 684
of magnetic field, 464
maxwellian distribution, 292
mechanical or nonmechanical, 146
nuclear, 146, 711
ordered or disordered, 299
potential, 131, 139
quantization, 666, 684
rest-mass, 601
thermal, 146, 287
total, classical, 131
total, relativistic, 601
zero-point, 667
Energy conservation, 131, 144, 356
  Kirchhoff's loop equation, 403
  Lenz's law, 457
Energy conservation and the uncertainty principle, 658
Energy density (*see* Density)
Energy flux (*see* Poynting vector), 490
Energy-level diagram, 666
  hydrogen, 686
Energy, relativistic, 601
Energy transfer in a gas, 288
English system of units, 3
Entropy, 305
Erg (unit), 113n
Equilibrium, 95, 189, 190, 194
  dynamic, 95, 190
  rigid body, 190
  rotational, 190
  stable and unstable, 189
  static, 95, 190
  thermal, 266
  translational, 95, 190
Equinoxes, precession, 217
Equipotential surface, 364
  relation to electric field lines, 364
Equivalence, principle, 47, 53
Equivalent capacitance:
  parallel circuit, 374
  series circuit, 373
Equivalent resistance:
  parallel circuit, 400
  series circuit, 399
Escape velocity, 212
  from earth, 216
  from the moon, 216
Esu (unit), 772
Eta particle, 749
Ether, 572, 586
Ether-drift experiment, 573, 586
eV (*see* Electron volt)

Event, 2
Excited state, 541, 685
Exclusion principle, 693
Experiment in physics, 1
External agent, 114
External force, 104
External torque, 181
Eye, 525
  corrective lenses, 532
  diffraction, 557
  interpretation of images, 514
  near point, 560
  resolution, 560
  spectral sensitivity, 501
Eyepiece, 528

F

f (*see* Focal length)
f (*see* Frequency)
°F (degrees Fahrenheit)
F° (Fahrenheit degrees)
F (*see* Farad)
F (*see* Faraday constant)
F (*see* Focal point)
F (*see* Force)
Fading in radio reception, 559
Fahrenheit temperature, 267
Farad (unit), 369, 371
Faraday constant, 323
Faraday ice-pail experiment, 348
Faraday's law, 458
  comparison with generalized Ampère's law, 482
Faraday, M., 348, 369
Fermat, P., 531
Fermat's principle, 531
  law of reflection, 531
  Snell's law, 531
Fermi (unit), 705
Fermi National Accelerator Laboratory, 729, 730
Fermion, 750
Feynman, R.P., 744
Feynman diagram, 744
Field:
  electric, 326, 483
  gravitational, 188, 203
  magnetic, 412, 483
Field particle, 745
Film coefficient, 311
Fine-structure constant, 699
First law of thermodynamics, 287, 300
  energy conservation, 291
First-order spectrum, 553
Fission fragment, 733
Fission, nuclear, 733
Fixed points, 267
Fizeau, A.H.L., 503
Fluorescence, 701
Flux:
  electric, 339

energy (*see* Intensity vector or Poynting vector)
   in hydrodynamics, 351
   magnetic induction, 415
Focal length, 522
   equivalent, 532
Focal plane, 522
Focal point (*see* Focus)
Focus, 520
Foot (unit), 3, 769
Foot-pound (unit), 113n, 771
Force:
   action and reaction, 85
   average, 82
   basic kinds, 81
   central, 164, 181, 207
   centrifugal, 109
   centripetal, 108
   conservative, 140, 144, 147
   conversion factors (table), 770
   coulomb, 318
   definition, 80
   diagram, 96
   electric, 81
   electromagnetic, 81, 313, 740, 747
   energy-dissipative, 142
   energy-generative, 142
   external, 104, 180
   friction, 101
   fundamental kinds, 81, 313, 747
   fundamental origins, 80
   gravitational, 81, 86, 313, 740, 747
   impulsive, 89
   inertial, 109
   instantaneous, 82
   intermolecular, 298
   internal, 104, 180
   Lorentz, 419, 483
   magnetic, 81, 413
   nonconservative, 142
   normal, 88
   nuclear, 81, 705
   pairs, 84
   radial, 106
   radiation, 491
   relation to potential energy, 143
   restoring, 121
   of spring, 83
   strong nuclear, 81, 313, 705
   superposition principle, 83
   tangential, 107
   weak interaction, 81, 313, 758
   weight, 86
Force components:
   radial, 95
   tangential, 95
Force constant (*see* Spring constant)
Forced oscillations, 248
Forearm, 195
Forty Eridani B, 312
Foucault, J.B.L., 517

Fourier theorem, 226
Frame of reference, 71
   accelerated (*see* inertial)
   center-of-mass, 75
   galilean transformations, 71, 79
   inertial, 52, 73, 107
   laboratory, 77
   noninertial, 107
   zero-momentum, 75
Franck, J., 760
Franck-Hertz experiment, 760
Franklin, Benjamin, 348
Fraunhofer, J., 552
Fraunhofer diffraction, 551
Free-body diagram, 96
Free election:
   electrical conduction, 306, 314
   role in electrostatic equilibrium, 333
   semiconductors, 315
   thermal conduction, 306
Free fall, 28
   experiments, 197
Free-space wavelength, 545
Frequency:
   of acoustic spectrum, 259, 263
   allowed, 246, 258
   angular, free oscillations, 470
   of electromagnetic spectrum, 500
   fundamental, 246
   harmonic, 246, 258
   natural, 249
   overtone, 248, 259
   resonance, 248
   threshold, 617
Fresnel diffraction, 551
Friction, 101
   coefficients, 102
   kinetic, 184
   negative work, 121
   static, 184
   torque, 171
ft (unit), 3, 769
ft-lb (unit), 113n, 771
Fundamental constants (table), 767
Fundamental forces, 81, 313, 705, 740, 745, 747, 758
Fundamental magnetic effect, 438
Fusion (*see* Nuclear fusion)

G

g (*see* Acceleration due to gravity)
g (*see* Gram)
G (*see* Gauss)
G (giga), 5
G (*see* Gravitational constant)
Galilean transformation relations, 71, 79, 581
   used to derive speed for transverse waves, 235
   for velocity, 584

Galileo, G., 1, 28n, 32, 47, 50, 78, 95
Galvanometer, 405, 426
Gamma rays, 501
   gamma decay, 715
Gas-law constant, 274, 285
   wavespeed of sound, 257
Gas law(s):
   Boyle's, 274
   Charles' and Gay-Lussac's, 275
   general, 273, 285
Gas pressure, 269
Gas thermometer, constant-volume, 271
Gauge pressure, 270, 278
Gauss (unit), 414
Gaussian cgs system of units, 320, 414, 772
Gaussian surface, 339
   and symmetry of charge distribution, 344
Gauss' law for electricity, 341
   conductors, 347
   crucial assumptions, 343
   dielectrics, 347, 352
   and electric field, 341
   point-charge, 341
   relation to electric field lines, 345
   spherically symmetric charge distributions, 346
   uniformly charged infinite plane sheet, 344
   uniformly charged infinite wire, 344
   as volume integral, 352
Gauss' law for magnetism, 439
   magnetic field lines form closed loops, 439
   no magnetic monopoles, 440
Gay-Lussac, J.L., 275
Geiger counter, 379
Geiger, H., 676, 678
General-gas law, 273, 285
General theory of relativity, 47, 447, 591
Generator:
   electric, 395, 466
   electrostatic, 367, 379
Geomagnetic field, 416, 428, 447
   configuration, 416, 419
   dip angle, 428
   horizontal component, 428
   magnetic dipole moment, 447
   positions of geomagnetic poles, 447
   Van Allen belts, 419
Geometrical optics (*see* Ray optics), 508, 661
Geos, 205
Germer, L.H., 643, 649
GeV (unit), 608
Goudsmit, S.A., 693
Gradient, 365

Gradient, *cont.*
  electric field, 365
  temperature, 306
Gram (unit), 770
Gram calorie, 274, 289, 302
Grand tour, solar system, 79
Gravitation, 86, 196
Gravitational:
  balance, 53, 87
  constant, 198
  effects of spherical shell, 217
  field, 188, 203, 204
  field lines, 203
  force, 198
  interaction, 81, 313, 747
  mass, 196, 314
  potential energy, 133, 209
  potential energy of spherical shell, 217
  waves, 204, 747
Gravitational charge (*see* Gravitational mass)
Gravitational force, 198
  in atomic physics, 111
  central, 198, 207
  conservative, 141, 209
  Kepler's laws, 207
Gravitational mass, 196, 314
  comparison of two masses, 318
  relation to inertial mass, 197
Gravitational potential energy, 133, 209
  choice of zero, 210
  definition, 210
  spherical shell, 217
  system of particles, 211
  two particles, 210
Graviton, 747
Gravity:
  acceleration due to, 28, 86
  center of, 188
  effect of earth's shape and spin, 216
  inverse-square dependence, 203
  standard value, 54, 87, 203
  variations, 28, 203, 216
Greek alphabet, 783
Grimaldi, F.M., 554
Ground state, 684

H

h (*see* Hour)
h (*see* Planck's constant)
$\hbar$ ($h/2\pi$), 682
H (*see* Henry)
H (*see* Magnetic field intensity)
Hadron, 749
Half-life, 713
Hall effect, 423
Halley's comet, 207, 217
  and Kepler's third law, 208

Halogen elements, 698
Harmonic frequencies, 246
Heat, 265
  capacity, 303
  defined as thermal-energy transfer, 289
  "direction," 295
  engine, 305
  latent, 303
  pipe, 306
  specific, 302, 303
  rate, 307
  units, 289
  and work compared, 288, 289
Heat-transfer processes:
  conduction, 305
  convection, 308
  radiation, 308
Heavy hydrogen, 152
Heavy-water moderator, 149
Heisenberg, W., 654, 656
Helical path of charge in magnetic field, 417, 418
  pitch of helix, 429
Heliocentric cosmology, Copernican, 78
Helios, 205
Helmholtz coils, 449
Henry, J., 462
Henry (unit), 462, 467
Herapathite, 565
Hertz (unit), 220
Hertz, G., 760
Hertz, H., 471, 483, 500, 616
Hertz' test of Maxwell's theory, 471, 483, 500
High-energy accelerator, 723
High-energy physics, 739
Hooke's law, 121*n*, 218*n*, 253, 255, 426
Hooke, R., 121*n*, 218*n*
Hour (unit), conversion factors, 770
hp (unit), 771
Humerus (bone), 195
Huygens, C., 262
Huygen's:
  construction, 550
  principle, 262, 513
  wavelets, 513
Hydrogen atom:
  Bohr theory, 681
  spectrum, 678
Hyperon, 749
Hz (*see* Hertz)

I

i (*see* Current, electric)
I (*see* Moment of inertia)
I (*see* Nuclear spin quantum number)

I (*see* Wave intensity *and* Poynting vector)
Ice point, 267
Ideal gas, 273
  changes in state, 276, 278, 279
  general-gas law, 274
  macroscopic properties, 273
  other gas laws, 275
  speed of compressional waves, 278
  and work, 279
Ideal heat engine, 305
Image:
  real, 523, 525, 528, 532
  virtual, 514, 525, 528, 532
Image distance, 524
Impulse, 88
Impulse-momentum theorem, 88
Impulsive force, 89
in (unit), 769
Incandescent lamp filament, 390
Inclined plane:
  rough, 103
  smooth, 98
  wave-mechanical treatment, 670
Incoherent sources, 540
  thermal excitation, 541
Index of refraction, 512, 514
  frequency dependence, 516
  relative, 515, 517
  various materials (table), 515
Induced charges, 349
Induced current, 452
Induced emf, 458
Inductance:
  concentric cylindrical conductors, 467
  definition by Ampère's law, 462
  definition by Faraday's law, 462
  inductors in parallel, 467
  inductors in series, 467
  self-inductance, 461
  solenoid, 463
  toroidal solenoid, 462
  units, 462
Inductors, 461
  energy, 464
  energy density, 465
Inelastic collision, 147
  rotational, 172
Inertia, 52
  law of, 51
  moment of, 166, 180, 784
Inertial balance, 52
Inertial force, 109
Inertial frame of reference, 52, 73, 95, 107, 590, 591
Inertial mass, 52, 196
  relation to gravitational mass, 197
Infrared radiation, 501
Instantaneous:
  acceleration, 22, 36
  speed, 20, 34
  velocity, 19, 34

Insulators:
    electrical (*see* Dielectric materials)
    role of free electrons, 306
    thermal, 307, 311
Integral calculus, inventor, 199
Intensity:
    electric, 327
    of electromagnetic waves, 489, 490
    magnetic, 773, 774
    radiated by accelerating charge, 497
    variation with angle (*see* Radiation pattern)
Intensity of *any* wave, 260
    dependence on distance, 261
Intensity pattern (*see* Radiation pattern)
Intensity of solar radiation, 312, 506
Intensity vector (*see* Poynting vector)
Interacting system of particles, 104
Interactions:
    electromagnetic force, 81, 313, 740, 747
    gravitational force, 81, 313, 747
    strong nuclear force, 81, 313, 705, 745, 747
    weak nuclear force, 81, 313, 747, 758
Interference, 533
    air wedge, 546
    constructive and destructive, 238, 244, 257, 534
    diffraction grating, 553
    double-slit, 542, 552
    electromagnetic waves, 533
    fringes, 543
    intensity equations, peaks and zeros, 543, 544, 546, 549, 553
    mechanical waves, 236
    Newton's rings, 546
    row of point-sources, 548
    two point-sources, 535
    radio waves, 560
    thin films, 544
Interferometer:
    in general, 544
    measurement capabilities, 544
    Michelson, 546, 573
    relation to SI length standard, 547
Intermediate vector boson, 747
Intermolecular:
    collisions, 298
    force, 298
    potential energy, 299
    restoring force, 255
Internal energy, 300
Internal energy of a system, 286
    changes and the first law of thermodynamics, 287
Internal force, 169
Internal resistance:
    battery, 394
    coil, 479
    inductor, 476
Internal torque, 170
International pitch, 252
International System of units (SI), 3, 319
    conversion factors, 769, 773, 774
    electric and magnetic units, 772, 773
Intrinsic phase difference, 537, 548
Invariance in relativity, 590
Inverse-square forces:
    between point-charges, 437
    between point-masses, 199, 203
Ionization, 684
Ionization energy, 359
    hydrogen atom, 359
Ionosphere, 559
Irreversible process, 295
Isobar, 708
Isobaric process, 276
Isochronous pendulum, 224
Isomer, 716
Isospin, 755
Isothermal process, 276
Isotone, 708
Isotope, 708
Isotopic spin, 755
Isovolumetric process, 276

J

j (*see* Current density vector)
J (*see* Joule)
J (*see* Mechanical equivalent of heat)
J (*see* Rotational quantum number)
Jerkiness of motion, measure of, 23
Joliot, F., 731
Joliot-Curie, I., 731
Joule (unit), 113
Joule's experiments, 301
Joule heating, 387
Joule, J.P., 301, 387
Junction theorem, Kirchhoff's, 402
Jupiter's moons and Kepler's third law:
    (graph), 209
    (table), 208

K

k (*see* Boltzmann's constant)
k (*see* Coulomb force constant)
k (kilo), 5
k (radius of gyration), 785
k (*see* Wave number)
K (*see* Kelvin temperature)
K (*see* Kinetic energy)
K (*see* Thermal conductivity)
Kaon, 749
K capture, 718
Kelvin's absolute electrometer, 381
Kelvin temperature, 273
Kepler, J., 205
Kepler's laws, 196, 204
keV (unit), 608
kg (*see* Kilogram)
Kilogram (unit), 5, 54, 770
Kilogram calorie (unit), 302
Kinematics, 50
    constant acceleration, 24, 28
    relativistic, 571
    rotational, 177
    straight-line, 17
    in two dimensions, 33
Kinetic energy:
    definition, 112
    and linear momentum, 113
    of a photon, 619
    relativistic, 600
    of rotation, 186
    spin, 168
    total, 186
    units and dimensions, 113
    and work, 114
Kinetic-energy selector, 420, 429
Kinetic friction, 102
    coefficient, 102
Kinetic theory, 281-300
Kirchhoff's rules, 402
Kirchhoff's rules, applications:
    LC circuit, 469
    LR circuit, 476
    multiloop circuits, 403, 410
    RC circuit, 474
km (unit), 769
Krypton-86 spectral line, 3, 504, 547
kWh (unit), 771

L

l (*see* Orbital angular-momentum quantum number)
L (angular-momentum vector), 157
L_s (electron-spin angular-momentum magnitude), 693
L (*see* Inductance)
L (length dimension), 5
L_I (nuclear angular-momentum magnitude), 704
L (orbital angular-momentum magnitude), 690
L_0 (*see* Proper length)
Lambda particle, 749
Laser, 541, 564
    coherent source, 540
    light amplification, 542
    modulation, 559
    Neodymium-glass, 506

nuclear-fusion experiments, 506
optical pumping, 542
properties of emitted light, 542
radiation-pressure measurements, 491n
speed-of-light measurements, 504
stimulated emission, 542
Latent heat, 303
  of fusion, 304, 311
Lateral magnification, 524
Latitude effect for cosmic radiation, 416
Lattice spacing, 644
Law(s):
  Ampère's, 440, 480, 484
  of areas, 165, 207
  Avogadro's, 296
  of Biot and Savart, 432
  Bragg's, 644, 646
  of Charles and Gay-Lussac, 275
  conservation (see Conservation laws)
  Coulomb's, 316
  Dalton's (of partial pressures), 297
  Faraday's, 458, 482
  Gauss', for electricity, 341
  Gauss', for magnetism, 439
  general-gas, 273, 285
  Hooke's, 121n, 218n, 253, 255, 426
  of inertia, 51, 66
  and invariance, 590
  Kepler's, 196, 204
  Lenz', 456
  Malus', 566
  of Maxwell and Ampere, 482
  Newton's, of motion, 84, 94
  Newton's, of universal gravitation, 94
  Ohm's, 386, 391
  in physics, 1, 591
  of Stefan and Boltzmann, 309, 312
  of thermodynamics, 266, 287, 291
  of vector algebra, 7, 9, 10, 12, 13
Law of electromagnetic induction (see Faraday's law)
Lawrence, E.O., 726
Lawrence Livermore Laboratory, 507
Lawton, 348
lb (see Pound)
LC oscillator (see Electric free oscillator)
LCR oscillator, resonance frequency, 474
Least time, principle of, 531
Lebedev, P.N., 491
Lee, T.D., 722
Length:
  as a dimension, 5

conversion factors (tables), 769
  proper, 580
  standard of, 3
Lens equation:
  gaussian, 524n
  newtonian, 524n, 532
  sign conventions, 529
Lenses:
  acoustic, 527
  coated, 560
  combinations, 527, 532
  converging, 522, 527
  diverging, 526
  electrostatic, 527
  eyepiece (ocular), 528
  magnetic, 527
  objective, 528
  ray-construction method, 523
Lenz, H.F.E., 456
Lenz' law, 456
  and energy conservation, 457
  and minus sign in Faraday's law, 458
Lepton, 748
  -number conservation, 752
Lever arm, 181
Lever, class III, 195
Light:
  amplification, 542
  bending of, 47
  infrared, 501
  pipe, 519
  pressure, 63, 491, 494
  spectrum, 501
  speed (see Speed of light)
  ultraviolet, 501
  visible, 501
  weight of, 47
Light-year (unit), 769
Linac, 724
Linear accelerator (see Linac)
Linear charge density, 331
  as measure of source strength, 437
Linear momentum (see Momentum)
Line integral:
  Ampère's law, 441
  and conservative forces, 141
  and emf, 458
  Faraday's law, 458
  in Maxwell's equations (table), 484
  Maxwell's form of Ampère's law, 482
  and nonconservative forces, 142
  and potential energy, 140
  and work, 116
Line spectrum, 679
Liquid, thermal properties, 298
  calorie defined, 302
  (table), 303, 307
Liquid-in-glass thermometer, 267
Livingston, M.S., 726

Lloyd, H., 538
Lloyd's mirror, 538
Load (of electric circuit), 385
Longitudinal wave (see Compressional wave)
Loop (see Antinode)
Loop equation, Kirchhoff's, 403
Lorentz, H.A., 419, 681
Lorentz force, 419, 483
Lorentz transformations:
  coordinate, 581
  velocity, 583
Loudness, 259, 263
Lp record, recorded wavelength, 263
LR circuit, 475, 478
L/R time constant, 476
Luminosity (see Stellar luminosity)
Lyman series, 680

M

m_E (earth's mass), 767
m_e (electron rest mass), 768
m (see Meter)
m (milli), 5
m_n (neutron rest mass), 768
m_p (proton rest mass), 768
m_0 (see Rest mass)
m_l (orbital magnetic quantum number), 691
m_s (spin magnetic quantum number), 693
m_S (sun's mass), 767
M (mass dimension), 5
M (magnetization vector), 773
M (mega), 5
Macroscopic properties, ideal gas, 273
Magnetic bottle, 419
Magnetic compass, 428
Magnetic dipole moment, 426n, 447
Magnetic/electric field ratio, in waves, 488
Magnetic energy density, 464, 467
Magnetic field (see Magnetic induction field)
Magnetic field intensity, 773, 774
Magnetic field lines, 415
  closed loops, 439
  of permanent magnets, 439
  spatial configuration, 439
Magnetic flux, 415
  changing, 454
  definition, 415
  density, 416
  and electric field, 458
  and Gauss' law, 439
Magnetic force:
  between two charges in relative motion, 434
  between two current-carrying conductors, 423, 438

fundamental effect, 438
and magnetic field lines, 415
motional force, 451
noncentral character, 446
no work, 417
on current-carrying conductor, 423
on moving charge, 413
special properties, 413
and torque on current loop, 425
velocity-dependence, 413, 417
Magnetic induction field:
  Biot-Savart law, 432
  of circular coil, 446
  of circular current loop, 435
  of coaxial conductor, 442
  comparison with electric field, 414
  of cylindrical conductor, 447
  current element, 433
  definition, 412
  defined by Lorentz force, 483
  direction, 413
  of earth, 416, 428
  energy density, 464
  flux, 415
  flux density, 416
  of Helmholtz coils, 448
  intensity, 773, 774
  lines, 415
  of moving charge, 431, 433
  of permanent magnet, 440
  of solenoid, 444, 447
  source, 431
  of square coil, 448
  of straight conductor, 436, 448
  of toroid, 448
  units, 414
Magnetic intensity, 773, 774
Magnetic interaction (*see* Magnetic force)
Magnetic interaction between two point-charges, 434
  comparison with electric interaction, 434
Magnetic lens, 430, 527
Magnetic moment (*see* Magnetic dipole moment)
Magnetic monopole, 440
Magnetic quantum number:
  orbital, 691
  spin, 693
Magnetic permeability, 432
Magnetic reflection of charged particles, 419
Magnetic rigidity, 420
Magnetic tape, recorded wavelength, 263
Magnetic torque:
  on a current loop, 425
  on a magnetic dipole, 426
Magnet, permanent (*see* Permanent magnet)

Magnification:
  angular, 532
  lateral, 524
Magnifier, simple, 525, 532, 560
Magnifying glass (*see* Magnifier, simple)
Malus, E.L., 566
Malus' law, 566
Manometer, closed-tube, 271
Marsden, E., 676, 678
Mass:
  atomic unit, 54, 609
  -conservation law, 54
  conversion factors, 770
  of earth, 202
  gravitational, 196, 197, 314
  inertial, 52, 196, 197
  relativistic, 593, 597
  rest, 597
  scalar additivity of, 53
  standard of, 5, 54
  and weight, 87
  units, 54, 770
Mass-energy conservation, 601
Mass-energy equivalence, 604
Mass-equivalent of a photon $(h\nu/c^2)$, 55, 604, 619
Masses, atomic (table), 775
Mass number, 708
Mass ratio:
  of planets and their satellites, 215
  of sun to earth, 200
  of two objects orbiting each other, 206
Mass selector, 421, 429
Mass spectrometer, 429
Mathematics as the language of physics, 1
Mathematical relations, 780
Maxwell, J.C., 293, 481
Maxwell's displacement current (*see* Displacement current)
Maxwell's equations, 483
  prediction of electromagnetic waves, 484-89
  relative intensities, 512
  and speed of light, 488
Maxwell's form of Ampère's law, 482
Maxwellian distribution, 292
Mean life, 713, 736
Measurements in physics, steps, 4
Mechanical-electrical analogs, 473, 474
Mechanical equivalent of heat, 768
Mechanical waves, 233-64
  compressional, 253-64
  definition, 233
  elastic, 253-64
  longitudinal (*see* compressional)
  transverse, on a string, 233-52
Mechanics:
  classical, 1, 50

relativistic, 590-612
statistical, 292, 293n
Medium, elastically deformable, 254
Meson, 746
Meter (unit), 3, 769
Metric system, 3
  conversion factors, 679, 774
  electric and magnetic, 772
  standards, 3, 4, 5, 54
MeV (unit), 362, 608
mho (unit), 388
mi (unit), 769
Michelson, A.A., 507, 546, 573
Michelson interferometer, 546, 573
  and length standard, 547
  in Michelson-Morley experiment, 548, 573
Michelson-Morley experiment, 548, 573, 586
Micron (unit), 769
Microscope:
  compound, 527
  electron, 673
Microscopic state, 293
Microwaves, 501
  interference patterns, 559
Microwave oscillator, 472
Millikan, R.A., 334
Millikan oil-drop experiment, 320, 334, 337, 422
min (unit), 770
Mirror image (*see* Virtual source)
Mirrors, 513, 530
  in lasers, 542
Mksa system of units, 319
Mnemonic for relative velocities, 72
Moderator, 148
Modern physics (defined), 2
Modes, allowed, 248
Modified wave, 629
Modulus, bulk, 256
Mole (unit), 274
Molar specific heat, 303
Molecular chaos and internal energy, 287
Molecular mass (*see* Molecular weight)
Molecular theory of gases, 265
Molecules:
  diameter, 281
  energy distribution, 292
  mass, 274
  size, 280
  temperature and kinetic energy, 285
  "weight," 257, 274
Moment:
  arm, 181
  electric dipole, 338
  of force, 167
  magnetic dipole, 426n, 447
  of momentum (*see* Angular momentum)

Moment of inertia, 166, 180
  calculations, 281, 784
Momentum:
  angular (*see* Angular momentum)
  in center-of-mass frame, 75
  classical form in terms of rest energy, 608
  in classical mechanics, 56
  conservation of, 55, 75, 79, 105, 591
  dimensions, 57
  of electromagnetic waves, 494
  and galilean transformation, 79
  and impulse, 88
  linear, 56
  of light, 63
  of photon, 626
  and reference frames, 79
  relativistic, 58, 593, 597, 608
  of system, 66
  units, 57
  in zero-momentum frame, 73
Momentum conservation and the uncertainty principle, 659
Momentum selector, 420, 429
Moon (earth's):
  diameter (table), 205
  distance from earth, 767
  distance from earth, measurement, 507
  as an earth satellite, 49, 199, 209
  and Kepler's third law, 209
  mass (table), 205
  orbital period (table), 208
Moon, satellites of, 48
  orbital period, 216
Moons of Jupiter, 208
  and Kepler's third law (graph), 209
Morley, E.W., 573
Motional emf, 450
Motion, rotational (*see* Rotational motion)
Motion, translational:
  dynamics, 96-111
  energetics, 129-53
  kinematics in one dimension, 17-32
  kinematics in two dimensions, 33-49
  momentum (*see* Momentum)
  Newton's laws, 94
  periodic, 226
  simple harmonic, 218-32
  work and kinetic energy, 112-28
Mount Palomar telescope, 560
Mount San Antonio, 507
Mount Wilson, 507
Multiples of ten prefixes (table), 5
Multiplication of vectors:
  by a scalar, 9
  scalar product, 13
  vector product, 14

Muon (table), 748
  decay, 578
  number, 753
Muscles:
  biceps and triceps, 195
Musical scales, diatonic and well-tempered, 264

N

n (*see* Index of refraction)
n (nano), 5
n (*see* Principal quantum number)
N_A (*see* Avogadro's number)
N (*see* Neutron number)
N (*see* Newton)
N (*see* Normal force)
Natural frequency (of an oscillator), 249
Naturally radioactive nuclides, 714
Navigation by bees, 569
Near point, 560
Negatively charged object, 315
Negative of a vector, 8
Neutrino, 720, 723
  handedness, 758
  muon and electron types, 755
Neutron, 704
  decay, 723
  discovery, 737
  fast, 148
  moderated, 148
  slow, 47
  star, 216
  thermal, 497, 672
  zero electric charge, 314
Neutron-neutron force, 705
Neutron-proton force, 705
Newton (unit), 82
Newton, I., 1, 44, 48, 50, 94, 196, 199, 201, 205
Newtonian mechanics, 1, 50
  kinetic theory, 281-300
  molecular theory of gases, 265
  applied to rotational motion, 177-95
  applied to translational motion, 94-111
Newton's laws of motion:
  applications, 94-111, 177-95
  first, 94
  in noninertial frames, 107, 109
  relativistic form of second law, 598
  in rotation, 158, 179
  second, 84, 94, 158, 179
  strategy for applications, 96
  third, 84, 94, 169
Newton's law of universal gravitation, 198
Newton's rings, 94, 560
Nodal lines, 537, 539, 559
Node (nodal point), 244, 531

Nonconservative force, 142, 144
Nonconservative system, 145
Noninertial reference frames, 107, 109
Nonlinear (nonohmic) resistor, 386
Normal force, 99
North star, 217
Nuclear:
  constituents, 703
  density, 706
  masses (table), 775
  mass number, 708
  photodisintegration, 707
  radius, 705
  reactions, 730
  reactor, 148, 152
  size, 4
  spin, 704
  structure, 703
  subscripts and superscripts, 352
  time, 4, 716
Nuclear fission, 152, 297, 324, 733
  energy release, 379
  role of coulomb repulsion, 317, 324, 379
Nuclear force, 81, 313, 705, 745, 747
  charge-independence, 705
  range, 199, 705
Nuclear fusion:
  laser-driven, 506
  reactor, 152
  thermonuclear, 733
Nuclei:
  naturally radioactive (unstable), 714
  stable, 708
  unstable, 324, 733
Nucleon, 705
  conservation of, 714
Nucleus, 314, 675
Nuclide, 708
Null instruments:
  inertial balance, 52
  gravitational balance, 53, 87
  potentiometer, 407, 411
  Wheatstone bridge, 407

O

Objective lens, 528, 532, 557
Ocular lens, 528
Oersted effect, 431
Oersted, H.C., 431
Ohm (unit), 386
Ohm, G.S., 386
Ohm's law, 386
  point form, 391
  range of validity, 387
  vector form, 391
Ohmmeter, 410
Oil-drop experiment (*see* Millikan oil-drop experiment)

Omega particle, 749
Open circuit, 395
Optical flatness test, 546
Optical instruments, 525
   astronomical telescope, 528
   camera, 525
   compound microscope, 527
   eye, 525
   simple magnifier, 525
Optical path length, 545
Optical pumping, 541
Optical reversibility (see Reciprocity principle)
Optic axis, 525
Optics:
   ray, 508-32, 661
   transition between ray and wave optics, 555
   wave, 533-60, 661
Orbital angular momentum:
   definition, 174
   of the earth, 175
   quantum number, 690, 691
Orbital magnetic quantum number, 691
Ordered energy and motion, 287, 288
Order of reflection, 646
Orthogonal projection, 9
Oscillations:
   of atoms in a solid, 231
   damped, 229
   driven, 229
   electric, free, 468
   electric-dipole, 497, 498, 507, 562, 563
   energetics, 231
   of LC circuit, 468
   resonance, 225, 229
   small, 225
Oscillator, electric:
   driven, 474
   free, 468
   microwave, 472
Oscillator, electric-dipole, 497, 498, 507, 562, 563
Oscillator, mechanical, 218-32
Oscilloscope (see Cathode-ray oscilloscope)
Overtone, 248

P

p (see Electric-dipole moment)
P (see Momentum)
p (pico), 5
p (see Pressure)
P (total momentum of a system of particles), 52
P (see Power)
Packet, wave, 656, 658
Pair annihilation, 321, 633, 742
Pair production, 321, 631, 742

Parabolic reflector, 511
Parabolic trajectory, 25, 40, 44
Parallel-axis theorem, 180
Paraxial rays, 522
Parent nucleus, 712
Parity, nonconservation, 722
parsec (unit), 769
Partial pressures, Dalton's law of, 297
Particle:
   in box (quantum), 664, 674
   concept, 17, 649
   of electromagnetic radiation, 613
   elementary, 739
   equivalent for system, 66
   field, 745
   resonance, 760
   wave aspect, 842
   zero rest mass, 602
Particle-antiparticle pairs, 321, 322
Paschen series, 680
Path-length difference, 537, 548
Pauli exclusion principle, 693
   for nuclei, 710
Pauli, W., 694, 720
Pendulum:
   ballistic, 149
   (nearly) isochronous, 224
   simple, 223
   torsion (see Torsion balance)
Perfect absorber (see Blackbody)
Perfect gas (see Ideal gas)
Perfect radiator (see Blackbody)
Perigee, 205
Perihelion, 205, 209
Period, 44, 220
Periodic motion, 226
Periodic Table, 693
Permanent magnet, 412, 440
   the earth as, 416, 428, 447
Permeability (see Magnetic permeability)
Permittivity (see Electric permittivity)
Perpendicular-axis theorem, 193
Perpetual-motion machine, 145
Phase:
   adjustment of stereo speakers, 559
   change on reflection, 257, 539
   constant, 220, 243
   lag, 241
   shift, 245
   speed, 242, 673
Phase difference:
   at observation point, 536, 548
   intrinsic, 537, 548
Phonograph record, recorded wavelength, 263
Phosphorescent screen, 413
Photodisintegration, 733
Photoelectric effect, 615, 742
   quantum interpretation, 618

Photoelectron, 615
Photon, 64, 619
   angular momentum, 638
   -electron interactions, 634, 740
   flux, 639, 653
   momentum, 626
   rocket, 494, 507, 639
   virtual, 744
Photosphere (of sun), 297
Photovoltaic cell, 395
Physical law, 1
Physical optics (see Wave optics)
Physics:
   classical, 1
   experimental, 1
   law of, 1
   modern, 2
   theoretical, 2
Pion, 746
Pitch:
   of helical path, 429
   musical, 252, 259
Planck, M., 615
Planck's constant, 174, 619, 663
Plane of polarization, 561
Planetary motion (see Kepler's laws)
Planets:
   average density, 216
   as solar satellites, 49
   (tables), 205, 208, 209
Plane wavefronts, 261
   variation of intensity with distance, 261
   variation of intensity with angle (see Radiation pattern)
Plasma, 733
Plimpton, 348
Point-charge, 316
Polar diagram, 536, 538, 550
Polarization:
   analyzer, 566
   circular, 562, 570
   direction, 564, 565
   direction of easy absorption, 563
   double scattering, 569, 570
   electric dipole, 562
   elliptical, 562
   extinction, 566
   linear, 561
   plane, 561
   polarizer, 566
   reflection, 568
   scattering, 568
   superposition, 561, 567, 570
   unpolarized light, 564
   of visible light, 563
Polaris, 217
Polarizer, 566
Polaroid, 565
   polarization direction, 565
   sun glasses, 570
Positively charged object, 315

Positron, 65, 321, 631
  annihilation, 718
Positronium, 634, 640
Potential (*see* Electric potential)
Potential energy:
  chemical, 361
  definition, 139
  elastic, 132
  electrostatic, 152, 317
  of hill, 213, 217
  gravitational, 133, 153, 209
  intermolecular, 299
  properties, 141
  zero, choice of, 141
Potential well, 143
  in quantum theory, 664, 669
Potentiometer:
  null instrument, 407, 411
  voltage divider, 408
Pound (unit), 54, 82, 87
Power, 124
  average, 124
  electrical (*see* Electrical power)
  of electromagnetic wave, 489, 490
  instantaneous, 124
  resolving, 673
  in rotational motion, 185
  in translational motion, 124
Poynting, J.H., 490
Poynting vector, 490
Precession, 176
  of the equinoxes, 217
Prefixes for multiples of ten (table), 5
Pressure:
  absolute, 270
  atmospheric, 270
  conversion factors, 771
  definition, 270
  of a gas, 282
  gauge, 270
  measuring devices, 269
  radiation, 63, 494
  standard atmospheric, 270
  waves, 256
Pressure loop, 257
Principal focus, 520
Principal quantum number, 690
Principle:
  of complementarity, 649
  of equivalence, 47, 53
  of determinancy, 656
  of Fermat's least time, 531
  of reciprocity, 510, 521
  of superposition (*see* Superposition principle)
Prism, 517, 519, 531
  angle of deviation, 517
  dispersion, 517, 554
  inverting, 531
  minimum angle of deviation, 531
Prism spectrometer, 517

Probability and de Broglie waves, 652
Products involving vectors:
  scalar (dot), 13
  vector (cross), 14
  of vector by a scalar, 9
Projectile motion:
  energetics, 136
  and forces, 119
  kinematics, 39, 47
Proper length, 580
Proper time, 577
Proton, 313, 323, 325
  properties, 703
Proton-neutron force, 705
Proton-proton:
  cycle (in fusion), 733
  force, 705
Pyrometer, radiation, 312

Q

q (electric point-charge), 317, 318
Q (total electric charge on any object), 322
Q (total energy released in beta decay), 719
Quality of sound, 259
Quantization, 613
  of angular momentum, 174, 674, 682, 683, 688, 690-93
  of charge, 322, 335
  of energy, 666, 684
  of space, 692
Quantum description of confined particle:
  in a box, 664
  in one-dimensional potential well, 664
  wavelength in potential well, 667
Quantum effects, 613
  wave aspects of particles, 642, 643, 667
Quantum electrodynamics, 635
Quantum number(s), 665
  and atomic structure, 688
  electron-spin angular-momentum, 693
  isospin, 755
  nuclear spin, 704
  principal, 690
  orbital angular-momentum, 690
  orbital magnetic, 691
  rotational angular-momentum, 674
  spin magnetic (of electron), 693
  strangeness, 756
Quantum physics versus classical, 663
Quantum of radiation (photon), 619

R

r_{EM} (earth-moon mean distance), 767
r (displacement vector), 33
r (*see* Radius vector)
R (*see* Gas-law constant)
R (*see* Radius of curvature)
R (*see* Resistance)
Radial acceleration, 43, 45, 179
Radian, 45
Radiation:
  cosmic, 416, 578, 635
  electromagnetic, 309, 484, 498, 500, 613, 622
  from radioactive substances, 715, 716, 717
  thermal, 310
Radiation force, 491
  measurement, 491, 494
Radiation pattern, 511
  of accelerating charge, 497
  of electric-dipole oscillator, 499
  of row of point-sources, 550
  of two point-sources, 535, 556, 559
  of single point-source, 499
Radiation pressure, 63, 494
Radiation pyrometer, 312
Radiative capture, 732
Radioactive decay, 712
  alpha, 716
  beta, 717
  gamma, 715
  law of, 712
Radioactivity, 153, 714
  artificial, 732
Radiofrequency oscillator, evolution from LC circuit, 471
Radioisotope, 732
Radio receiver:
  fading, 559
  tuning, 478
Radio:
  telescope, 560
  spectrum, 501
Radium, 153
Radius:
  (bone), 195
  of curvature, 234
  vector, 42
Radon, 153
Range, horizontal, 41, 47
Rankine, W., 277
Rankine temperature, 277
Rarefaction, 254
Rationalized units, 320, 433
Ray, 260
Rayleigh's criterion, 557
Ray optics, 508-32, 661
  Fermat's least-time principle, 531
  no information of relative intensities, 512

range of wavelengths, 509
ray-construction, 523
transition from wave optics, 555
RC circuit, 474
RC time constant, 474
Reaction force, 85
Reaction, thermonuclear, 733
Reactor:
  nuclear fission, 148, 152
  nuclear fusion, 152
Real image, 525
Reciprocity principle, 510, 521
Recorded wavelength:
  long-playing record, 263
  magnetic tape, 263
Rectangular components of a vector, 9, 16
Rectangular coordinate axes:
  rotation transformations, 16
  three dimensions, 12
  two dimensions, 9
Rectilinear motion, 17
  at constant acceleration, 24
Red sunsets, 569
Reference frame(s):
  and acceleration, 73, 74
  inertial, 52, 73, 95, 590, 591
  noninertial, 107, 591
Reflection of electromagnetic waves, 493, 511
  from conducting surface, 539
  diffuse, 512
  phase change, 539
  specular, 512
  total internal, 518
  rules, 511
Reflection of mechanical waves:
  longitudinal waves, 257
  phase change, 257, 539
  transverse waves, 238
Refraction of electromagnetic waves, 511, 516
  atomic viewpoint, 518
  index, 512, 514
  rules, 511
  at spherical surface, 520
Refractive index (see Index of refraction)
Reines, F., 442n, 723
Relative index of refraction, 515
Relative motion, 71
  mnemonic notation, 72
Relativistic mechanics:
  computations, 607
  dynamics, 590-612
  energy, 600
  kinematics, 571
  mass, 58n, 593, 597
  momentum, 58n, 593, 597
  relativity of simultaneous events, 582
  units, 607
  velocity relations, 583

Relativity:
  general theory, 47, 447, 591
  special theory, 58n, 571-612
Resistance, 386
  of cylindrical conductor, 387
  equation, 387
  internal, 394
  measurement, 406, 411
  of parallel resistors, 399
  of series resistors, 399
  shunt, 405, 410
  of superconductors, 386
Resistivity, 387
  (table), 388
Resistor:
  nonlinear, 386
  ohmic, 386
  power dissipation, 387
  vacuum-deposited, 390
  wire-wound, 390
Resolution, 556
  of the eye, 560
  Rayleigh's criterion, 557
  resolving power, 573
Resonance, 230
  for waves, 248, 263
  frequency, 248
  particles, 760
Rest energy, 601
Rest mass, 597
  zero rest mass, 602
Resultant vector, 7
Retarding potential, 421
rev (unit), 769
Reversible process, 291, 295
Right-hand coordinate set, 12
Right-hand rule:
  angular velocity, 155
  vector product, 14, 15
Rigid body:
  angular momentum, 166
  center of gravity, 188
  center of mass, 66, 68, 69, 75, 104, 105, 180, 186, 188
  dynamics, 179-94
  equilibrium, 190
  kinematics, 177-78
  kinetic energy, 167
  moment of inertia, 166, 179, 180, 784
Ripple tank, 516, 538
Ritz combination principal, 699
RLC oscillator, resonance frequency, 474
Rms (see Root-mean-square)
Rocket, 65, 86
  booster, 44
  critical launch speed, 213
  energetics, 212
  equation, 65
  exhaust velocity, 65, 93
  photon, 494, 507, 639
  Saturn 5, 93
  space pistol, 93

thrust, 93
Tsiolkovsky, 65
Roemer, O., 502
Roentgen, W., 624
Root-mean-square:
  current, 392
  speed, 284
Rosser, W.G.V., 419
Rotational motion:
  dynamics, 179, 185
  kinematics, 154, 177
  kinetic energy, 185
  Newton's laws, 169, 171, 179, 190
  power, 185
  of top, 176
  work, 185
Rotational spectrum, 674
Rotational quantum number, 674
Rotation and space contraction, 587
Rotation-translation, table of analogous quantities, 187
Rule-of-thumb (see Right-hand rule)
Rutherford, E., 675, 681, 730
Rutherford scattering, 65, 316
Rydberg:
  constant, 679, 686
  equation, 679
  formula, 686

S

s (see Electron-spin angular-momentum quantum number)
s' (image distance), 524
s (object distance), 523
s (see Second)
Sadeh, D., 640
SAE, 409
Satellite(s), 48, 107, 212, 215
  of earth, 44
  of earth, synchronous, 215
  of Jupiter, 208, 209
  Kepler's laws, 204
  of the moon, 48
  of the sun, 49, 208, 209
Saturation of nuclear force, 325
Saturn 5 rocket, 93
Savart, F., 436
Scalar, 6
Scalar additivity of mass, 53
Scalar product, 13
  commutative and distributive laws, 14
  flux, 340, 351, 415, 490
  power, 125
  work, 114
Scattering, alpha particles, 675
Scattering, electromagnetic waves:
  classical, 568, 625
  quantum, 627

Schrödinger, E., 654
Schrödinger equation, 654, 668, 688
Second (unit), 4
Second law of thermodynamics, 291
   directionality of time, 295
Segrè, E., 640
Self-inductance, 461
Semiconductors, 315, 386, 388
   properties, 386, 388
   V-i curve, 386
Semimajor axis, 207
Series:
   Balmer, 680
   Lyman, 680
   Paschen, 680
Series limit, 679
Shunt resistance, 406
SI (*see* International System of Units)
Sidereal day, 193
Sigma particle, 749
Sign conventions, thin lenses, 529
Simple harmonic motion, 218-32
   angular frequency, 221
   dynamics and kinematics, 218
   energetics, 222, 225
   Fourier's theorem, 226
   particle in earth tunnel, 227, 232
   phase constant, 220
   small oscillations, 225
   sinusoidal-wave generator, 240
   superposition, 226, 232, 561
   relation to uniform circular motion, 221, 232
Simple magnifier, 525, 532, 560
Simple pendulum, 223, 232
   dimensional analysis, 224
   energetics, 226
   (nearly) isochronous, 224
Simultaneity, relativity of, 582
Sinusoidal electromagnetic waves, 498
   electric-dipole oscillator source, 498
   superposition, 488, 533, 561
Sinusoidal waves, mechanical
   longitudinal, 253
   superposition, 247, 257
   transverse, 239
Single-slit diffraction, 655
   quantum theory, 659
Sinusoidal:
   electric free oscillations, 268
   electromagnetic waves, 498
   motion, 220
Sirius, 216
Slingshot effect, 78
Slug (unit), 87
Small oscillations, 225
Smoke-filled room, bluish haze, 568

Snell, W., 512
Snell's law, 512, 516
   particle model, 530
Sodium D lines, 701
Solar:
   atmosphere, 297
   constant, 506
   day, 4, 193
   radiation, 152, 264, 312, 506
Solenoid, 444
   inductance, 462, 463
Solids, thermal properties, 298
   conductivity, 307
   energy, 299
   radiation, 308
   specific heats, 302, 303 (table)
Sonar, 262
Sound waves, 256, 263
   intensity, 260
   intensity variation with distance, 260
   wavespeed, 256, 297, 560
Sound waves in pipes:
   closed at one end, 258, 263
   open at both ends, 258, 263
   pipe organ, 263
Sources, coherent and incoherent, 540
Space:
   pistol, 93
   probe, 78
   quantization, 692
   rocket (*see* Rocket)
   in relativity, 571
Space contraction, 579, 587
   and rotation, 587
Space-time:
   continuum, 582
   diagram, 740
Special theory of relativity, 2, 58n, 591
Specific heat, 302
   for various materials (table), 303
Spectrograph, 678
Spectrometer, 678
   diffraction grating, 552, 644
   prism, 517
Spectroscope, 678
Spectroscopy, 678
Spectrum:
   absorption, 680
   acoustic, 259, 263, 264
   audible, 259, 263
   bright-line, 679
   dark-line, 680
   electromagnetic, 500, 501, 622
   emission, 678, 679
   gamma ray, 501, 715
   hydrogen, 678
   line, 679
   order of, 553
   rotational, 674
   X-ray, 501, 625, 702, 719
   zero-order, 553

Specular reflection, 512
Spectacle lenses, 532
Speed:
   angular 45
   at aphelion, 207
   average, 17, 30
   conversion factors, 770
   definition, 20
   instantaneous, 20
   at perihelion, 207
   phase, 242, 673
   rms, 284
   of sound, 256, 259, 263, 297
   tangential, 45, 178
Speed of light:
   constancy of, 571
   current accepted value, 768
   in vacuum, 488, 504, 768
   in water, 517
Speed of light measurements:
   Bradley (stellar aberration), 502
   Fizeau, 503
   Foucault, 517
   Michelson, 507
   National Bureau of Standards (laser resonance), 504
   Roemer, 502
   in water, 517
Speed of sound, 256
   measurement, 263, 560
Spherical wavefront, 261
   intensity-variation with angle (*see* Radiation pattern)
   intensity-variation with distance, 261
Spin angular momentum (*see* Spin angular-momentum quantum number)
Spin angular-momentum quantum number:
   of electron, 693
   of fundamental particles (table), 748, 749
   nuclear, 704
   of photon, 638
Spin magnetic quantum number (of electron), 693
Spring:
   constant, 121, 128
   potential energy, 129, 132, 143
   restoring force, 121
   work, 132
Stability line, 709
Standard:
   acceleration due to gravity, 54, 87, 203
   atmospheric pressure, 270
   cell, 407
   of length, 3
   of mass, 5, 54
   pitch, 252
   of time, 2
Standing waves:
   longitudinal, 257
   transverse, 244

Stanford Linear Accelerator Center, 725
Stars:
    aberration of starlight, 78
    binary system, 216
    luminosity of, 152, 312, 506
    neutron star, 216
    Sirius, 216
    supernova, 216
    thermonuclear fusion, 733
    white dwarf, 216, 312
State:
    of a gas, changes in, 276
    of matter, changes in, 303
Statics, 95
    of rigid bodies, 190
Static friction, 102
Stationary state, 665, 685
Stationary waves (*see* Standing waves)
Statistical mechanics, 292
Steam point, 267
Stefan-Boltzmann law, 309, 312
Stellar aberration, 78, 502
Stellar luminosity, 312, 506
    of the sun, 152, 312, 506
Stereo speakers, phase adjustment, 559
Stevinus, S., 110
Stiffness constant (*see* Spring constant)
Stimulating radiation, 542
Stokes' rule, 701
Stopping potential, 379
STP, 274
Straight line, definition, 17
Straight-line kinematics, 17-30
Strangeness, 756
    quantum number, 756, 757
Strategy:
    for applying Newton's laws, 96
    for solving problems, 27
Stratosphere, 263
Strength, electric field, 327
Strong interaction, 81, 313, 705, 745, 747
    charge independence, 705
    saturation, 325
Strong nuclear force (*see* Strong interaction)
Sun:
    atmosphere, 297
    as blackbody, 312
    luminosity, 152, 312, 506
    photosphere, 297
    radiated energy intercepted by earth, 264
    radiation force on earth, 506
    radiation intensity at earth-distance (*see* solar constant)
    solar constant, 506
Sun/earth mass ratio, 200
Superconductors, 389, 728
Supernova, 216
    radiation force of, 494
Superposition principle:
    electric fields, 343
    electric forces, 319, 327
    electromagnetic waves, 488, 533
    forces, 83, 95, 237
    mechanical waves, 236, 247, 257
    polarized waves, 561
    simple harmonic motions, 226, 232, 561
    waves in general, 534
Surface charge density, 331
Synchrocyclotron, 726, 728
Synchronous earth satellite, 215
Synchrotron:
    electron, 390
    proton, 728, 729, 737
Systems of interacting particles, 104

T

t (time), 18
t_C (*see* Celsius *or* Centigrade temperature)
t_F (*see* Fahrenheit temperature)
t_R (*see* Rankine temperature)
T (*see* Absolute temperature)
T (isospin quantum number), 755
T (*see* Period)
T_0 (*see* Proper time)
T (*see* Tesla)
T (tera), 5
T (*see* Tension)
T (time dimension), 5
Tachyon, 582n
Tandem Van de Graaff, 368, 379
Tangential:
    acceleration, 178
    speed, 178
Telescope:
    astronomical, 528, 532
    Mount Palomar, 560
    radio, 560
Television picture tube, 31
Temperature, 265
    and energy distribution, 292
    gradient, 306
    and kinetic energy, 265
    thermal equilibrium, 266
    and zeroth law of thermodynamics, 266
Temperature scales:
    absolute (Kelvin), 273
    Celsius (centigrade), 267
    Fahrenheit, 267
    Rankine, 277
Tension, 96
    and transverse waves, 234
Terminal velocity, 110, 335, 337
Tesla (unit), 414, 428
Test charge, 326
Theory:
    of general relativity, 47, 447, 591
    of special relativity, 2, 58n, 571-612
Thermal conduction, 305, 311
    air film, 311
    relation to electrical conduction, 306
    through glass, 311
Thermal conductivity, 307
    (table), 307
Thermal energy, 298
    and disorder, 299
    of gases, 298
    in resistor, 387
    of liquids, 298
    of solids, 298
Thermal-energy:
    capacity (*see* Heat capacity)
    transfer (*see* Heat transfer)
Thermal equilibrium, 266
Thermal expansion:
    of gases, 276, 278, 279
    of mercury, 267
Thermal neutron, 47, 672
    rms speed, 297
Thermal radiation, 310
Thermocouple, 395
Thermodynamic processes, 276
Thermodynamics:
    first law, 287
    fundamental concepts, 285
    second law, 291
    zeroth law, 266
Thermometer, 266
    constant-volume, 271
    liquid-in-glass, 267
Thermometric substances, 268
    temperature-dependent properties (table), 269
Thermometry, 267
    calibration, 267, 270
Thin-lens equation (*see* Lens equation)
Thin lenses, 519
    biconcave, 526, 527
    concave-convex, 522
    conjugate points, 525
    convex, 520
    converging, 523
    crude version, 520
    diverging, 526
    focal length, 522
    focal plane, 522
    image distance, 524
    lens axis, 525
    object distance, 523
    optic axis, 525
    planoconvex, 522
    principal focal points, 522
    ray construction, 523
    real image, 525
    real object, 526

Thin lenses, *cont.*
   sign conventions, 526, 529
   spherical surface, 520
   virtual image, 526
Thomson, J.J., 422
Thomson's e/m experiment, 422
Threshold:
   frequency, 617
   of hearing, 259
   of vision, 638
Thrust of a rocket, 93
Time, 3
   arrow of, 295
   conversion factors, 770
   dilation, 574, 587
   of flight, 42, 47
   invariance, 759
   nuclear, 716
   proper, 577
   in relativity, 571
   reversal, 510, 521
Time constant:
   L/R, 476
   RC, 475
Top, motion of, 176
Toroid, 448, 462
   inductance, 463
Toroidal solenoid (*see* Toroid)
Torque, 159
   and angular momentum, 158
   constant, 426
   and equilibrium, 190
   external, 170, 181
   gravitational, 201
   internal, 181
   and kinetic energy, 186
   magnetic, 425, 426
   as moment of force, 167
   and Newton's second law, 181
   and power, 185
   restoring, 201, 426
   and work, 185
   zero, 163
Torr (unit), 271
Torsion balance:
   in Cavendish experiment, 201, 316
   in Coulomb experiment, 316
   mirror, 530
   radiation-force measurement, 494
Torsion pendulum (*see* Torsion balance)
Total energy:
   in classical mechanics, 131
   relativistic, 601
Total internal reflection, 518
   critical angle, 518
Tourmaline, 566
Tower of Pisa, Galileo's experiment, 197
Tracer, 732
Transformation equations:
   galilean, 71, 79, 581

   lorentz, 581, 583
   for rotation of axes, 16
Transformer effect, fundamental, 454
Translational equilibrium, 95, 190
Transmission grating, 552
Transmutation, 731
Transverse wave(s), 233-52
   basic behavior, 233
   interference, 186, 244
   linear momentum, 494
   reflection, 238
   standing waves, 244
   superposition, 136, 247
Triceps (muscle), 195
Trigonometric functions (table), 781
Triple point, 273
Tritium, 152
Tropopause, 263
Tsiolkovsky, 65
Tunnel diode, 669
Tunnel effect, 669, 717
Twin paradox, 578

U

u (*see* Atomic mass unit)
U (internal energy of a gas), 287
U (*see* Potential energy)
Uhlenbeck, G.E., 693
Ulna (bone), 195
Ultracentrifuge, 192
Ultraviolet light, 501
   and resonance of atomic electrons, 568
Uncertainty principle, 656, 744
Underscore, wavy, 7
Uniform circular motion:
   dynamics, 106
   charged particle in magnetic field, 417
   kinematics, 42
   relation to simple harmonic motion, 221
Units:
   conversion tables, 769, 774
   electric and magnetic, 772
   rationalized, 320, 433
   relativistic mechanics, 607
   SI (*see* International System of Units)
   U.S. Customary (English), 3, 4
Universal gas-law constant (*see* Gas-law constant)
Universal gravitational constant (*see* Gravitational constant)
Universal-gravitation law, 94
Universe, 4
Unmodified wave, 629
Uranium fission reactor, 148, 152
   fission reactions, 723

U.S. Customary System of Units, 3, 4

V

v (instantaeous speed), 20
v (instantaneous velocity), 34
V (*see* Electric potential)
V (volume), 274
Van Allen belts, 419, 429
Van de Graaff generator, 367, 379
   tandem, 368, 379
Vector(s), 6-16
   addition and subtraction, 7
   algebra, 6-15
   associative law of addition, 8
   commutative laws, 8, 14, 15
   components, 9
   direction, 7, 10
   distributive laws, 14, 15
   equality of, 8
   equation, 12, 38
   magnitude, 7
   multiplication by scalar, 9
   negative, 8
   product, 14
   properties, 6
   prototype, 6
   radius, 42
   resultant, 7
   sum, 7
Vector current density, 391
Vector field:
   electric, 326
   gravitational, 188, 203
   magnetic induction, 412
Vector intensity for electromagnetic waves (*see* Poynting vector)
Vector product, 14
   angular momentum, 159
   magnetic force, 414, 416, 419, 424, 434
   relation to parallelogram area, 16
   torque, 159
Velocity:
   angular, 154
   as area, 24
   average, 17, 33
   of center of mass, 66, 105
   constant, 19
   as derivative, 34
   dimensions, 19
   escape, 212, 216
   instantaneous, 19, 34
   relative, 71
   rms, 284
   and slope, 19, 24
   terminal, 110, 335, 337
   transverse, 236
   uniform rotational, 190
   uniform translational, 19

units, 19, 770
  as a vector, 33
Velocity relations:
  galilean, 72
  relativistic, 583
Velocity selector, 420, 429
Violin strings, 249, 252
Virtual image, 514, 526
Virtual source, 538
Visible light, 501
Volt (unit), 360
Volta, A., 360
Voltage-current characteristics, 386
Voltage divider "potentiometer," 408
Voltmeter, 405, 410
von Laue, M., 644

W

w (see Weight)
W (see Watt)
W (see Work)
Walton, E.T.S., 731
Watt (unit), 125
Wave(s):
  amplitude, 220
  angular momentum, 570
  aspects of a particle, 642
  compressional, 253-64
  concept, 649
  de Broglie, 642
  diffraction, 509, 552
  elastic, 253-64
  electromagnetic, 471, 483, 498
  energy density, 489
  frequency and period, 220
  function, 654
  in gas, 256, 259
  gravitational, 747
  intensity, 260, 490
  interference, 236
  longitudinal, 253-64
  mechanics, 661
  mechanical, 233-64
  modified, 629
  momentum, 491
  nodes and antinodes, 244, 257, 531
  number, 699
  optical reversibility, 510, 521
  optics, 533, 555, 661
  -packet, 656, 658
  of particles, 642
  -particle duality, 649
  phase, 241
  phase constant, 241
  phase shift, 245
  power, 492
  -pulse, 484
  rays, 260
  reflection, 238, 244, 257, 493, 511
  refraction, 511, 516
  sinusoidal, 239, 251
  sound, 256, 260, 263, 297, 560
  speed, 235, 242
  standing (stationary), 244
  on a string, 233-52
  superposition, 236, 247, 257
  transverse, 233-52, 471, 483, 498
  transverse, velocity, 236
  traveling, 241
  unmodified, 629
  wavefront, 260, 262
  wavelength, 241, 643
  wave number, 241
Wave intensity vector, electromagnetic (see Poynting vector)
Wavelength of particle, 643
Wave mechanics vs. classical mechanics, 661
Wave optics, 533-60
  diffraction, 548, 554, 556
  diffraction grating, 552
  interference, 533, 535
  Michelson interferometer, 546
  phase changes on reflection, 538
  reflection, 538
  resolution, 556
  single slit, 549
  superposition, 533, 535
  thin films, 544
  Young's two-slit experiment, 542
Wavy underscore, 7
Wb (see Weber)
Weak interaction, 81, 313, 747, 758
Weber (unit), 414, 416
Weight, 86
  of a gas, 93, 296
  of light, 47
Weightlessness, 107
Well-tempered scale, 264
Wheatstone bridge, 407
White dwarf star, 216, 312
Wiegend, 640
Work:
  as area, 122
  of conservative force, 140
  definition, 113
  of electric force, 127
  by external agent, 114
  and first law of thermodynamics, 289
  of friction force, 121
  by a gas, 258, 278, 279
  and heat compared, 288, 289
  of magnetic force, 417
  of nonconservative force, 142
  and power, 124
  in rotational motion, 185
  as scalar product, 114
  by a spring, 121
Work function, 619
Work—kinetic-energy theorem, 114-20
  for rotation, 186
World line, 740
W particle, 747
Wu, C.S., 722

X

Xi particle, 749
X-rays, 624
  diffraction, 647
  production, 622
  spectra, 625, 702, 719
XU (see X-unit)
X-unit, 769

Y

Yang, C.N., 722
Yard (unit), 3
yr (unit), 770
Young's double-slit experiment, 542, 553
  diffraction effects, 543
  equation for intensity peaks, 544
  interference pattern, 543
Young, T., 543
Ypsilantis, 640
Yukawa, H., 745
Yukawa potential, 762

Z

Z (atomic number), 708
Zeeman effect, 693
Zero, absolute, 273
Zero-order spectrum, 553
Zero-point energy, 667
  vibration, 674
Zero-rest-mass particle, 602
Zeroth law of thermodynamics, 266

we need your opinion

Dear Reader:

As publisher of this text, we are interested in continuously improving the quality of our texts, and therefore solicit the opinions of you, the user, either professor or student. After completing a course using this book, we urge that you write your comments on the form below and mail to Allyn and Bacon.

We look forward to hearing from you soon.

Arthur B. Conant

Arthur B. Conant
College Division
Allyn and Bacon
470 Atlantic Avenue
Boston, Mass. 02210

Dear Mr. Conant:

Below are my comments regarding your text, *Elementary Physics: Classical and Modern,* by Weidner and Sells.

1. The sections I found most informative, clear, and useful were: _____

2. Sections I found most difficult or unclear were (include criticisms): _____

3. Changes I would like to see in future editions are: _____

4. I found the use of two colors (helpful / not helpful / no opinion): _____

5. Other comments: _____

Name and address _____

School course no. and title _____

Mr. Arthur B. Conant
College Division
Allyn and Bacon, Inc.
470 Atlantic Avenue
Boston, Mass. 02210

Fundamental Constants

Name of quantity	Common symbol	Value
Acceleration due to gravity (standard value)	g	9.80665 m/s² = 32.174 ft/s²
Universal gravitational constant	G	6.673×10^{-11} N-m²/kg²
Earth's mass	m_E	5.9763×10^{24} kg
Earth's mean radius	r_E	6.370949×10^6 m = 3959 mi
Earth-moon mean distance	r_{EM}	3.84400×10^8 m ≈ 60 earth radii
Earth-sun mean distance (1 astronomical unit)	1 AU	1.4957×10^{11} m = 92.94×10^6 mi
Sun's mass	m_S	1.991×10^{30} kg
Density of water (at 3.98°C)		1.00000 gm/ml
Standard atmospheric pressure	1 atm	1.01325×10^5 N/m² = 76 cmHg = 760 torr
Universal gas constant	R	8.3143 J/g mol–K = 0.0821 liter-atm/g mol–K = 1.986 g cal/g mol–K
Volume of 1 gram mole ideal gas at STP		2.2414×10^{-2} m³
Absolute zero of temperature		−273.15°C
Triple point of water		273.16 K
Avogadro's number	N_A	6.02217×10^{23} g-mol^{-1}
Boltzmann's constant	k	1.38062×10^{-23} J/K